MOLECULES TO LIVING CELLS

Readings from

SCIENTIFIC AMERICAN

MOLECULES TO LIVING CELLS

With Introductions by

Philip C. Hanawalt

Stanford University

W. H. Freeman and Company

San Francisco

Most of the SCIENTIFIC AMERICAN articles in *Molecules to Living Cells* are available as separate offprints. For a complete list of articles now available as Offprints, write to W. H. Freeman and Company, 660 Market Street, San Francisco, California 94104.

Library of Congress Cataloging in Publication Data

Main entry under title:

Molecules to living cells.

Bibliography: p.
Includes index.
1. Molecular biology—Addresses, essays, lectures. 2. Genetic regulation—Addresses, essays, lectures. I. Hanawalt, Philip C., 1931– II. Scientific American.
QH506.M666 574.8′8 80–10814
ISBN 0–7167–1208–3
ISBN 0–7167–1209–1 pbk.

Printed in the United States of America

9 8 7 6 5 4 3 2 1

PREFACE

One of our most fascinating intellectual endeavors has been to study the molecular workings of the very cells of which we ourselves are composed. This selection of readings from *Scientific American* unfolds the miracle of life by developing the structural hierarchy of living cells and showing how they are assembled from relatively simple small molecules. Beginning with a successful origin-of-life "event" in the prebiotic milieu 3 billion years ago, the course of evolution on earth has led to the existence of self-reproducing multicellular beings that include a unique species called Human. That highly developed species has now learned enough about the molecular architecture of life that some of its members have begun to manipulate it at very basic levels. The implications of this recently acquired and developing technology are awesome, perhaps terrifying to some. Furthermore, we humans are constantly "enriching" our environment with new chemical combinations that are incompatible with most living forms. Biology no longer should be the specialized domain of a few professional biologists, since we are all called upon to register opinions and make decisions that may affect the future of life on earth. It is incumbent upon all of us to be "biologists" to the extent that we try to understand the possible consequences of our actions on the future of the biosphere. We need to learn how the exercise of our intelligence may enable us to survive even though many other life forms have become extinct. As a start we should all gain some appreciation of the way in which living systems are put together and how they proliferate.

The articles in this collection span the broad field of molecular biology between two other recent *Scientific American* readers on specialized topics—*Life: Origin and Evolution* and *Recombinant DNA*. The emphasis here is upon molecular structures, their interactions in forming supramolecular structures, and the genetic control of their synthesis in the growing cells that constitute life. The story develops from George Wald's thoughtful speculations on the probability of life's origin as a chance occurrence and it leads to Stanley Cohen's discussion of the new technology of gene manipulation in which the element of chance has been minimized if not eliminated.

This is not a text, yet it is definitely intended for use in the classroom. It does not include an essay on each important topic in the field of molecular biology, even though most of those topics have in fact been documented through the years in *Scientific American* articles. To include most of them would have resulted in an unwieldy and costly volume. The articles that are included were selected to highlight material that I consider to be important but which is not normally emphasized in textbooks. I have selected many articles that are much more current and detailed than any textbook in print. Others have been selected for their historical value and because they enable the student to appreciate the perceptiveness of some of the early researchers in the field of molecular biology. Students will learn how the classic experiments were done from the scientists who performed them. My choice of articles has been guided by the experience that my colleagues and I have had in the teaching of molecular and cellular biology at Stanford University and by our use of *Scientific American* readers for many years. Many of my colleagues have been of help, and I particularly wish to acknowledge the counsel of Robert Simoni and Charles Yanofsky in assembling this collection. The collection includes ten articles usually assigned to our students from an earlier reader, *The Chemical Basis of Life*, which I assembled in collaboration with Robert Haynes six years ago. The present volume includes the classic article on the gene by George Beadle, written five years before the proposal of the DNA structure by Watson and Crick. The other selections are all more recent, and they include 12 articles from the past six-year period, of which seven are from the past three years. The earlier articles have been updated and placed in current perspective through my commentary in the Introductions. Some of the authors contributed to the updating task, and I particularly want to thank Harold Morowitz, Daniel Mazia, David C. Phillips, William B. Wood, Masayasu Nomura, Charles Yanofsky, and Arthur Kornberg for their help. A list of references cited and suggested sources for additional reading (including *Scientific American* Offprints) has been provided at the end of each Introduction.

The 26 articles in this volume have been assembled in five sections. Section I considers the origin of life on the barren earth and the present-day range in complexity of cellular types from the simplest free-living cells to the component cells of mammals. Section II describes the structure and function of proteins—the principal building blocks of living organisms and the catalysts that regulate most biological activity. The assembly of subcellular organelles and those ultimate parasites, the viruses, from proteins and nucleic acids is also described. Section III deals with the boundary of the cell—that is, the membrane—and its role in compartmentalization as well as in the organization of certain essential cellular functions, such as the energy transducing systems. In Section IV, an historical flavor is intended through descriptions of classic experiments that elucidate the way in which information is transferred from the primary genetic blueprint to the synthesis of proteins and its expression in the visible properties of organisms. Finally, Section V deals with the nature of the primary genetic material, the DNA, and its transactions. The DNA is replicated, repaired, and manipulated by living cells. These same processes are now carried out in sterile glassware in the laboratories of scientists.

If the five Introductions are read first, that will be helpful in providing an overview as well as an understanding of the rationale for the arrangement of the articles selected for this volume. In the Introductions, I have tried to strike a balance between simplifying background information for the articles and commenting on current developments not discussed otherwise. I suggest that each Introduction be read as though it were the menu in a fine international restaurant, for the articles that follow are indeed delicacies from some of the great scientific "chefs" of the world.

I wish to thank my secretary, Maria Kent, for typing endless revisions and for her enlightened proofreading skill, which included worthwhile editorial suggestions. I thank Dr. Evelyn Parker for preparing the index. My father, J. Donald Hanawalt, has been a continuing source of encouragement, and my wife, Graciela, a source of inspiration.

I wish to dedicate this volume in loving memory of my mother, Lenore Smith Hanawalt, 1902–1978. She taught me to look for the beauty beyond the facts.

February 1980

Philip C. Hanawalt

CONTENTS

I SIMPLE INORGANIC MOLECULES TO COMPLEX FREE-LIVING CELLS

II PROTEIN STRUCTURE AND FUNCTION: ASSEMBLY OF VIRUSES AND RIBOSOMES

III MEMBRANE STRUCTURE AND CELLULAR ENERGETICS

IV NUCLEIC ACID STRUCTURE: EXPRESSION OF THE GENETIC MESSAGE

V REPLICATION, PRESERVATION, AND MANIPULATION OF DNA

Note on cross-references to SCIENTIFIC AMERICAN *articles:* Articles included in this book are referred to by title and page number; articles not included in this book but available as Offprints are referred to by title and offprint number; articles not included in this book and not available as Offprints are referred to by title and date of publication.

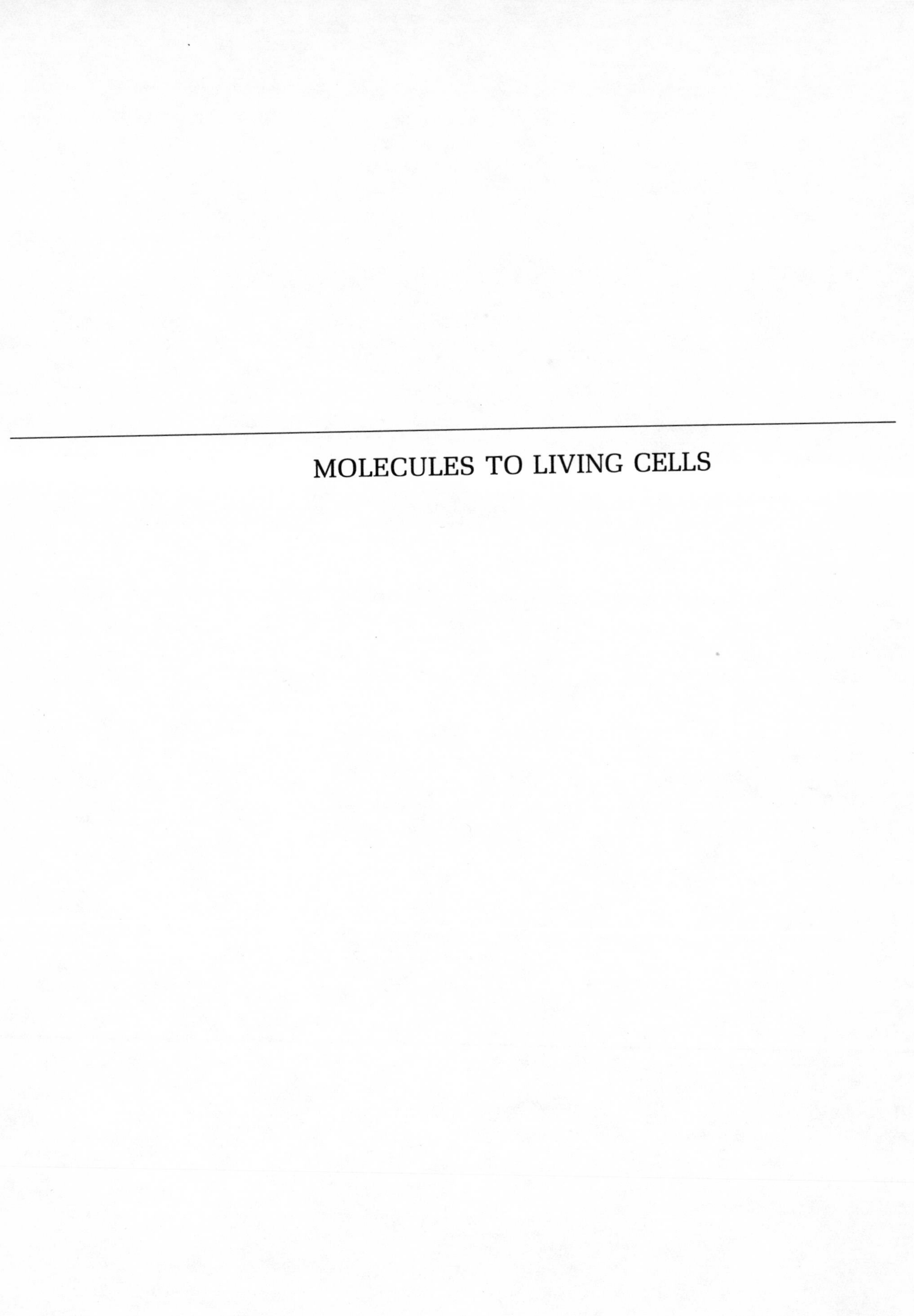

MOLECULES TO LIVING CELLS

I

SIMPLE INORGANIC MOLECULES TO COMPLEX FREE-LIVING CELLS

SIMPLE INORGANIC MOLECULES TO COMPLEX FREE-LIVING CELLS

I

INTRODUCTION

The phenomenon of "life" may be traced to the coordinated behavior of molecules within cells. The cell is the self-contained and self-reproducing fundamental unit of life. Cells may exist independently as free-living organisms like bacteria or as constituent functional units in a multicellular organism. (The human body, for example, consists of 100 trillion cells!) In the construction of cells, a molecular hierarchy exists in which a few types of small molecules are utilized as building blocks for a large variety of macromolecules of several major classes. These macromolecules in turn are assembled into complexes and organelles, which are still further organized into the self-perpetuating units that we call living cells. The molecules in cells are guided by the same basic laws of chemistry and physics that apply throughout the universe. They attract each other, repel each other, and form aggregates with each other. They combine to form stable new compounds that differ in size, shape, and chemical properties. There is nothing mysterious about any of the individual chemical reactions that occur in living systems. The mystery lies rather in the programmed coordination of the myriad of chemical reactions necessary for the metabolic activities of cells. In growing and carrying out its unique mission, each cell regulates its own internal environment and draws selectively from the external environment whatever raw materials it needs. The programmed accomplishments of these tiny chemical factories go far beyond the capabilities of the scientist in his laboratory. Although we now understand many of the molecular interactions and essential chemical processes required for the creation of a living cell, we simply cannot generate "life" in the test tube.

A given cell—indeed, any given organism—has but a fleeting existence in the cosmos. It comes into being and formulates its own identity as it develops from some pre-existing life entity; it carries out specialized functions and prepares for its procreation; then it disappears forever to be replaced by nearly identical descendants, repeating the cycle. Our wonderment and fascination with cellular growth are further stimulated by the realization that the living cell adapts its biochemical habits to a wide range of environments without losing its unique species identity for generation after generation. The progeny from a single bacterial cell are still identifiable with the characteristics of that parent cell after many millions of generations. Likewise, the cells from a carrot or from the liver of a mouse consistently retain their respective tissue and organism identities after countless cycles of reproduction. In the following discussion, we will explore the molecular components and the internal workings of cells. We will most certainly develop an increased appreciation for the complexity of the living cell, but we will also obtain some feeling for how our present understanding of the molecular basis of life may be utilized to benefit mankind and, it is hoped, the entire biosphere.

Even before dissecting a living cell, we can list some essential features that we are certain to find among its components. First, there must be a very detailed master blueprint, or "memory tape," that is transmitted intact from generation to generation. Second, there must be an intricate "read-out" system for translating the instructions from the blueprint into the synthesis of cellular structures. This system must include many contingency plans and the possibility of making use of alternative building materials as available in a particular environment. Third, there must be an energy-generating system and some means for storing chemical energy for allocation as needed in metabolic activity. Fourth, there must be a semipermeable boundary to separate inside from outside and to control which molecules enter and leave the cell. It sounds very simple in principle, doesn't it?

Consider, though, that the normal growth of even the simplest living cell requires that tens of thousands of chemical reactions occur in coordinated fashion. Each of these reactions could be carried out individually in a test tube if the appropriate reactants were present. In the test tube we can control the rates of reactions by altering the concentrations of reactants, changing the temperature, varying the pH (acidity-alkalinity), or in general by adjusting the "environment." But how, within one tiny cell, can 20,000 reactions all be controlled at once? At least part of the answer lies in the different types of environment that exist in different parts of the cell. Although most of the cell contents diffuse freely in an aqueous medium, many reactions occur at designated sites on the surfaces of membranes rather than in the spaces enveloped by them. In fact, many reactions are localized at the sites where highly specific catalysts, called enzymes, are fixed. Enzymes are largely responsible for the control of the molecular processing required for growth of living cells. Most of the biochemical reactions in the cell would take place at immeasurably slow rates were it not for these remarkable biological catalysts.

The essential activities in living cells are generally mediated by large macromolecules of four major classes: proteins, polysaccharides, lipids, and nucleic acids. Their general chemical structures are given in George Wald's article. The detailed structures and functions of these essential molecules are discussed in Sections II, III, and IV. Proteins serve catalytic functions as enzymes, and they are found in structural association with lipids in membranes. The lipids are responsible for cellular compartmentalization; they also serve as a form of stored energy. The polysaccharides are polymers composed of simple sugars like glucose; they form structural materials like cellulose and chitin and function as energy stores in starch or glycogen. The nucleic acids are deoxyribonucleic acid (DNA) and ribonucleic acid (RNA); their primary roles are in the storage and translation of genetic information, respectively. A typical bacterium, *Escherichia coli* (which thrives as a symbiote in the intestine of mammals), contains about a million protein molecules, which make up about 15 percent of the mass of the cell. However, the most abundant molecule in the cell, if not the simplest, is water, which constitutes over 70 percent of the total mass. The lipids and polysaccharides account for only 5 percent, and the nucleic acids make up another 7 percent of the cell mass. Finally, low-molecular-weight precursors, inorganic ions, and other small molecules account for another 1 percent of the total cell mass. This gross tabulation tells us almost nothing about how a cell works. Yet it is the kind of information one obtains from a straightforward chemical analysis of cell composition, and it is an essential early step in learning how molecules organize in cells. A more detailed analysis tells us that those million proteins are of about 3,000 different kinds, that there are several different classes of RNA, but that the DNA exists as one long thread, a chainlike molecule nearly one thousand times the length of the cell in which it is packed. At this point we can stop wondering about how to unravel life's mysteries and worry instead about how to unravel this one long molecule of DNA! It is the master blueprint for the cell; there is only one precious copy, and it is packed into the cell like a convoluted mass of spaghetti rather than carefully wound on a reel! Nevertheless, this apparent organizational disorder poses no evident problem for the growing cell. The information is carefully preserved, used as needed, and the entire thread of DNA is duplicated in time for the parent bacterium to divide into two daughter cells. Furthermore, the rest of the cell substance is also duplicated in that same period of time. We still have much to learn about how the organization of molecules in cells promotes efficient growth.

We have briefly introduced the internal chemical composition of the bacterial cell; how about the external environment? *Escherichia coli* thrives in an aqueous soup containing surprisingly few essential ingredients. Dissolved in the water required for the growth of this bacterium are certain salts and ions; those available in greatest concentration are sodium chloride, potassium chloride, magnesium sulfate, ammonium, and phosphate. In such a salt solution at 37°C and at pH 7 (neutral), an *Escherichia coli* cell will duplicate its substance and divide into two essentially identical cells in a mere 40 minutes—if we additionally provide oxygen, a source of carbon, and a source of energy. The simple sugar glucose is adequate both as a source of carbon and as a source of the chemical energy needed to metabolize it. Most of the cellular energy obtained from the systematic degradation of glucose will go into the synthesis of small building blocks and the assembly of these "monomer" precursors into the large polymers called macromolecules. Incidentally, what we have listed above is not the bare minimum of ingredients required in the growth medium. For example, had we left out the oxygen, the bacteria would have adjusted to that condition, but they would have been able to utilize much less of the available chemical energy in each glucose molecule and consequently would have grown more slowly. Alternatively, we might have supplemented the growth medium with many of the known precursors for cellular structures. The bacteria would then have adapted to growth in this rich broth, turning off some of their now superfluous synthetic machinery, and would consequently have been able to reproduce more efficiently, with a generation time as short as 20 minutes. The weight of one bacterial cell is only 0.00000000000001 gram (customarily written as 10^{-14} gram). Within 14 hours in the enriched growth medium, the mass of progeny from one cell would be greater than 1 gram, and within 36 hours the entire earth could be layered with a frosting of *E. coli* 6 centimeters deep! This exemplifies the fact that optimal growth is an exponential process, following the simple relationship $N = 2^n$, where N is the number of cells present after n generations, or division periods, starting with one cell. Fortunately, exponential growth for any living system cannot continue indefinitely. A theoretical end point would be that time when essentially all the carbon in the universe is tied up in molecules that constitute living cells. No such end point is ever approached, however, as the elements are continually reshuffled and recycled in endless rounds of chemical synthesis and degradation in the present-day biosphere.

How did it all begin? This is an unanswered question that continues to intrigue us and to challenge the ingenuity of organic chemists who would attempt the recreation of that incredible happening. In all likelihood, life arose as a series of spontaneous chemical events in which simple molecules combined to form more complex molecules, which in turn formed supramolecular complexes that eventually appeared in aggregates that were ultimately compartmentalized into cells. From that point on and continuing today, the process of biological evolution has directed the course of life on earth.

Evidence to date singles out the earth as the only planet in our solar system that supports life. Although other solar systems in our galaxy and in the 100 million other galaxies undoubtedly include planets that closely resemble our earth in composition and environment, such planets are generally at distances more than human lifetimes away at present rates of space travel. Thus, although the universe contains 10 million million earthlike planets, our speculations and our experiments are necessarily confined to exploration of the life forms on our own planet. On the geological time scale, life began "shortly" (about a billion years) after the earth was formed, and it has been continuously evolving for 3 billion years since then. What raw materials were available and what conditions prevailed 4 billion years ago? Although the physical laws that govern behavior of molecules were the same then as they are today, the chemical environment as well as the physical conditions (e.g., temperature, radiation levels) were quite different. The prebiotic earth contained water, carbon dioxide, carbon monoxide, nitrogen, and hydrogen, but no free oxygen. The absence of oxygen was essential for the accumulation of cellular precursors, since that highly reactive element, if present, would have readily combined with any organic material to destroy it almost as soon as it was formed. In fact it should be appreciated that the development of photosynthesis in primitive cells was responsible for the eventual appearance of free oxygen in our atmosphere.

We come now to the problem of the initial production on the barren earth of the essential low-molecular-weight precursors for biological macromolecules. In the classic "spark gap experiment" of S. L. Miller and Harold Urey (described in George Wald's article), the simple substances methane, ammonia, hydrogen, and water vapor were placed in a sterile vessel and subjected to a maintained electrical discharge. This simulation of one possible primitive earth environment generated a number of organic substances, including amino acids, the monomer units of which proteins are composed. More recently it has been shown that dilute solutions of hydrogen cyanide (HCN) under another simulated prebiotic environment will form purines and pyrimidines, the precursor building blocks of nucleic acids. The conditions on the primitive earth included energy sources other than electrical discharges—heat pockets from volcanic discharge, a broad spectrum of solar radiations (with no ozone layer in the upper atmosphere to attenuate the ultraviolet rays), and probably some unique combinations of ionizing radiations from the radioactive decay of unstable elements. Each of these energy sources may have been important for particular events in the scenario for prebiotic synthesis of essential organic compounds. The first step in the assembly of a cell—the synthesis of low-molecular-weight precursors—was achieved in the prebiotic milieu, and that part of the puzzle at least may eventually be solved as we experiment with various possible mixtures of simple compounds and different energy sources.

The synthesis of the small-molecule building blocks is complex in itself, but it is child's play in comparison to what must have followed in order to generate the first living cell, George Wald (see p. 13) points out the improbability of the chance occurrence of all of the appropriate precursor molecules being in the right place at the right time to form the first cell, but he argues that such random events might in fact generate a living combination given enough time. (This rationale has been elaborated by Jacques Monod (1971).) One can wonder whether a mere billion years is really enough time for such an improbable series of events! However, as we learn more about the construction of cells, we find that many of the essential assembly events are not improbable at all. Spontaneous assembly of biological structures is quite common in present-day biology—for example, the self-assembly of the protein and nucleic acid components of virus particles and ribosomes, as described in Section II. Even the assembly of lipids to form a micelle, a simple sort of membrane-bounded enclosure, is a natural occurrence. This is not to say that the formation of the first living cell could have been an inevitable event—just that it was aided by some basic chemical principles to be elaborated upon below.

Molecules tend to associate with each other on the basis of different levels of interaction or "recognition." Molecules that carry a net charge and molecules that have electrically charged regions seek interaction with each other and with small polar molecules like water. These are appropriately called hydrophilic (water-loving) interactions. In contrast, nonpolar regions of molecules tend to associate with each other and to exclude water. These are called hydrophobic (water-abhorring) interactions. The aggregation of the hydrophobic lipids into clusters is thus not random at all, but rather an event driven thermodynamically to achieve the lowest potential energy in the system. Yet it is an essential process for the assembly of a cell—even the very first one! In addition to the generalized interactions between molecules due to the distribution of electric charge, there are highly specific recognition features that characterize certain types of macromolecules. These will be discussed in more detail in subsequent sections. Suffice it to say at this point that it is possible to design a protein molecule that will recognize (i.e., bind specifically to) essentially any other molecule, including another protein. Furthermore, it is possible to construct a nucleic acid strand that will bind specifically to any other designated nucleic acid strand to form a highly stable, two-stranded structure.

Since it is appropriate to begin a story at the beginning, we have initiated this selection of readings with Nobel Laureate George Wald's article, "The Origin of Life," which treats that problem with excellent perspective and at a fairly elementary level. Wald elaborates upon the fact that biochemical reactions are reversible and that spontaneous dissolution normally proceeds more readily than spontaneous synthesis of most large biological molecules. He also stresses the formation of complexes in stabilizing macromolecules and the eventual need for a continuous expenditure of energy to maintain biological order. Lastly, he discusses the important role that evolving life forms played in bringing free oxygen to the earth. Even today the concentrations of oxygen and carbon dioxide in our atmosphere are dependent upon the totality of life forms that inhabit our planet.

In "The Evolution of the Earliest Cells," J. William Schopf elaborates further on the indispensible role that oxygen played in the rise of the eukaryotic cell. This article includes an outline showing how a glucose molecule is metabolized in the absence of oxygen (anaerobic fermentation) and in its presence (aerobic respiration). *Escherichia coli* can operate by either the anaerobic or the aerobic mode, as we have discussed. Schopf distinguishes clearly between prokaryotic cells, such as bacteria, and eukaryotic cells in which the DNA is organized in chromosomes in a defined nucleus and in which organelles such as chloroplasts and mitochondria are parts of the more highly developed subcellular organization. It is likely that both cellular types evolved from a much simpler life form, a progenote, in which the relationship between the primary genetic message and the expression of that message in the cell had not yet been fully worked out (Woese and Fox, 1977). The first photosynthesis probably evolved in a group of prokaryotes called the cyanobacteria, and the eukaryotes may have begun to develop only after oxygen was abundantly present. It may seem peculiar that the earliest life forms evolved in the presence of poisonous gases, such as carbon monoxide (CO), hydrogen cyanide (HCN), and hydrogen sulfide (H_2S), but those gases are in fact poisons only for aerobic forms of life. Schopf describes the analysis of single-celled fossils in fascinating detail, placing the first eukaryotic cells at about 1.5 billion years ago. Finally, he emphasizes that "in the Precambrian the influence of life on the environment was at least as important as the influence of the environment on life." Is that not also true today?

What is the lower limit of complexity required for a free-living cell? Harold Morowitz originally suggested that fundamental biological processes may be more accessible to study in very small cells in which the total number of molecules might limit the degree of complexity. In "The Smallest Living Cells," Harold J. Morowitz and Mark E. Tourtellotte describe the mycoplasma. An example of this group, *M. gallisepticum*, is roughly one-tenth the size of an *E. coli,* and it contains fewer than 10 percent as many molecules. Its tiny 0.2 micron sphere is at the very limit of resolution of the optical microscope. Further research since the article was written has revealed no smaller free-living life forms. However, more recent determinations yield a value of about 5×10^8 (500 million) daltons for the molecular weight of the DNA in the "unit genome" of the mycoplasma, which is still only one quarter the size of the *E. coli* genome. (The genome is that amount of DNA required to specify the organism.) Wallace and Morowitz (1973) have suggested an evolutionary scheme in which the mycoplasma are descendants of the "protokaryotes" (or progenotes), which arose before the divergence of prokaryotes and eukaryotes. These simplest of cells require a very complex growth medium supplemented with all of the amino acids and many other low-molecular-weight precursors for cellular structures. They lack the cell wall structure that gives rod-shape form to an *E. coli;* their interiors are separated from the environment by only one delicate membrane. The reader should refer to Section III for our current view of membrane structure. Maniloff and Morowitz (1972) point out that if living cells appreciably smaller than the mycoplasma are ever discovered, "the finding will challenge our ideas of what constitute the necessary biochemical processes for life."

Let us turn now to the more highly structured eukaryotic cells, which are at least 100-fold larger than the mycoplasma. Daniel Mazia, in "The Cell Cycle," explores the period of frenzied biochemical preparation between one cellular fission and the next. Historically, the phases of the cell cycle were identified through microscopic analysis of chromosome configuration in cells at different growth stages. An outline of these stages is given by Mazia, and actual drawings from observations made in the 1880s are reproduced in Mirsky's article in Section V. An abundance of these fascinating sketches from the early microscopic examination of living cells is provided by E. B. Wilson (1896) in a classic text that places our "pre-modern-biology" understanding of life in excellent perspective. The "nucleinic acids," as the nucleic acids were then called, were already identified and localized in chromosomes. It was suspected that they were involved in the "control" of life processes and, in particular, that they were intimately associated with the process of cell division. Mazia stresses the essential role of DNA replication in the cellular commitment to ultimate division. However, many other specific processes must occur in addition to molecular syntheses to prepare for this event. The experimental technique of fusing two cells to form a hybrid cell is described as a powerful approach to an understanding of cell-cycle regulation and of other processes as well.

Up to this point we have given the impression that a living cell controls its own destiny as regards growth and division. That may sometimes be the case for a free-living cell, but it is never true for the component cells of a multicellular organism. The problem of the coordination of cell growth and the differentiation of cells during development to form the specialized organs and structures in multicellular organisms is beyond the scope of this volume and nearly beyond comprehension. Mazia does cite, however, the important role of the cell surface in sensing the presence of other structures and in regulating growth. When normal human cells are cultured *in vitro,* their growth is restricted by "contact inhibition" as a monolayer of cells develops on the surface of the growth vessel. In contrast, cancer cells are not responsive to such external control on proliferation, and they tend to grow into multilayered piles of cells. The cancerous cell thus behaves more like an independent, free-living organism—to the detriment of the "host" organism. It is a cruel irony that the malignant cell contains essentially the same genome and most of the same regulatory machinery as do the normal cells of the victimized host.

The essential concepts of life are very simple and logical. In detail the processes of life are incredibly complex. It is not surprising that we should experience difficulty in comprehending them. As Nobel Laureate Max Delbrück pointed out thirty years ago, " . . . any living cell carries with it the experiences of a billion years of experimentation by its ancestors. You cannot expect to explain so wise an old bird in a few simple words."

REFERENCES CITED AND SUGGESTED FURTHER READING

Balls, M., and F. S. Billett (editors). 1973. *The Cell Cycle in Development and Differentiation.* Cambridge Univ. Press., New York.

Dayhoff, M. O. 1972. *Atlas of Protein Sequence and Structure.* Nat. Biomedical Research Foundation, Silver Spring, MD. (Also consult annual supplements.)

Delbrück, M. 1949. "A Physicist looks at biology." *Trans. Conn. Acad. Arts Sci.* **38:**9–22.

Dickerson, R. E. 1980. "Cytochrome *c* and the evolution of energy metabolism." *Scientific American,* March. (Offprint No. 1464.)

Dirksen, E. R., D. M. Prescott, and C. F. Fox (editors). 1979. *Cell Reproduction.* Academic Press, New York.

Folsome, C. E. 1979. *The Origin of Life: A Warm Little Pond.* W. H. Freeman and Company, San Francisco.

Folsome, C. E. (editor). 1979. *Life: Origin and Evolution. Readings from Scientific American.* W. H. Freeman and Company, San Francisco.

Kenyon, D. H., and G. Steinman. 1969. *Biochemical Predestination.* McGraw-Hill, New York.

Maniloff, J., and H. J. Morowitz. 1972. "Cell biology of the mycoplasmas." *Bacteriol. Rev.* **36:**263–290.

Margulis, L. 1970. *Origin of Eukaryotic Cells.* Yale Univ. Press, New Haven, CT.

Monod, J. 1971. *Chance and Necessity.* Knopf, New York.

Orgel, L. E. 1973. *The Origins of Life.* Wiley, New York.

Prescott, D. M. 1976. *The Reproduction of Eukaryotic Cells.* Academic Press, New York.

Puck, T. T. 1972. *The Mammalian Cell as a Microorganism: Genetic and Biochemical Studies* in Vitro. Holden-Day, Inc., San Francisco.

Razin, S. 1978. "The mycoplasmas." *Microbial Rev.* **42:**414–470.

Schopf, J. W. 1975. "Precambrian paleobiology: Problems and perspectives." In F. A. Donath, F. G. Stehli, and G. A. Wetherill (editors), *Annual Review of Earth and Planetary Sciences* (Vol. 3). Annual Reviews, Inc., Palo Alto, CA.

Schopf, J. W. 1977. "Biostratigraphic usefulness of stromatolitic Precambrian microbiotas: A preliminary analysis." *Precambrian Research* **5**(2):143–173.

Schrödinger, E. 1944. *What is Life? The Physical Aspect of the Living Cell.* Cambridge Univ. Press, New York.

Wallace, D. C., and H. J. Morowitz. 1973. "Genome size and evolution." *Chromosoma* **40:**121–126.

Wilson, A. C., S. S. Carlson, and T. J. White. 1977. "Biochemical evolution." *Ann. Rev. Biochem.* **46:**573–639.

Wilson, E. B. 1896. *The Cell in Development and Inheritance.* Macmillan, New York.

Woese, C. R., and G. E. Fox. 1977. "The concept of cellular evolution." *J. Molec. Evolution* **10:**1–6.

BIBLICAL ACCOUNT of the origin of life is part of the Creation, here illustrated in a 16th-century Bible printed in Lyons. On the first day (*die primo*) God created heaven and the earth. On the second day (*die secundo*) He separated the firmament and the waters. On the third day (*die tertio*) He made the dry land and plants. On the fourth day (*die quarto*) He made the sun, the moon and the stars. On the fifth day (*die quinto*) He made the birds and the fishes. On the sixth day (*die sexto*) He made the land animals and man. In this account there is no theological conflict with spontaneous generation. According to *Genesis* God, rather than creating the animals and plants directly, bade the earth and waters bring them forth. One theological view is that they retain this capacity.

The Origin of Life

1

by George Wald
August 1954

How did living matter first arise on the earth? As natural scientists learn more about nature they are returning to a hypothesis their predecessors gave up almost a century ago: spontaneous generation

About a century ago the question, How did life begin?, which has interested men throughout their history, reached an impasse. Up to that time two answers had been offered: one that life had been created supernaturally, the other that it arises continually from the nonliving. The first explanation lay outside science; the second was now shown to be untenable. For a time scientists felt some discomfort in having no answer at all. Then they stopped asking the question.

Recently ways have been found again to consider the origin of life as a scientific problem—as an event within the order of nature. In part this is the result of new information. But a theory never rises of itself, however rich and secure the facts. It is an act of creation. Our present ideas in this realm were first brought together in a clear and defensible argument by the Russian biochemist A. I. Oparin in a book called *The Origin of Life*, published in 1936. Much can be added now to Oparin's discussion, yet it provides the foundation upon which all of us who are interested in this subject have built.

The attempt to understand how life originated raises a wide variety of scientific questions, which lead in many and diverse directions and should end by casting light into many obscure corners. At the center of the enterprise lies the hope not only of explaining a great past event—important as that should be—but of showing that the explanation is workable. If we can indeed come to understand how a living organism arises from the nonliving, we should be able to construct one—only of the simplest description, to be sure, but still recognizably alive. This is so remote a possibility now that one scarcely dares to acknowledge it; but it is there nevertheless.

One answer to the problem of how life originated is that it was created. This is an understandable confusion of nature with technology. Men are used to making things; it is a ready thought that those things not made by men were made by a superhuman being. Most of the cultures we know contain mythical accounts of a supernatural creation of life. Our own tradition provides such an account in the opening chapters of *Genesis*. There we are told that beginning on the third day of the Creation, God brought forth living creatures—first plants, then fishes and birds, then land animals and finally man.

Spontaneous Generation

The more rational elements of society, however, tended to take a more naturalistic view of the matter. One had only to accept the evidence of one's senses to know that life arises regularly from the nonliving: worms from mud, maggots from decaying meat, mice from refuse of various kinds. This is the view that came to be called spontaneous generation. Few scientists doubted it. Aristotle, Newton, William Harvey, Descartes, van Helmont, all accepted spontaneous generation without serious question. Indeed, even the theologians—witness the English Jesuit John Turberville Needham—could subscribe to this view, for *Genesis* tells us, not that God created plants and most animals directly, but that He bade the earth and waters to bring them forth; since this directive was never rescinded, there is nothing heretical in believing that the process has continued.

But step by step, in a great controversy that spread over two centuries, this belief was whittled away until nothing remained of it. First the Italian Francesco Redi showed in the 17th century that meat placed under a screen, so that flies cannot lay their eggs on it, never develops maggots. Then in the following century the Italian abbé Lazzaro Spallanzani showed that a nutritive broth, sealed off from the air while boiling, never develops microorganisms, and hence never rots. Needham objected that by too much boiling Spallanzani had rendered the broth, and still more the air above it, incompatible with life. Spallanzani could defend his broth; when he broke the seal of his flasks, allowing new air to rush in, the broth promptly began to rot. He could find no way, however, to show that the air in the sealed flask had not been vitiated. This problem finally was solved by Louis Pasteur in 1860, with a simple modification of Spallanzani's experiment. Pasteur too used a flask containing boiling broth, but instead of sealing off the neck he drew it out in a long, S-shaped curve with its end open to the air. While molecules of air could pass back and forth freely, the heavier particles of dust, bacteria and molds in the atmosphere were trapped on the walls of the curved neck and only rarely reached the broth. In such a flask the broth seldom was contaminated; usually it remained clear and sterile indefinitely.

This was only one of Pasteur's experiments. It is no easy matter to deal with so deeply ingrained and common-sense a belief as that in spontaneous generation. One can ask for nothing better in such a pass than a noisy and stubborn opponent, and this Pasteur had in the naturalist Félix Pouchet, whose arguments before the French Academy of Sciences drove Pasteur to more and more rigorous experiments. When he had finished, nothing remained of the belief in spontaneous generation.

We tell this story to beginning students of biology as though it represents a triumph of reason over mysticism. In fact it is very nearly the opposite. The reasonable view was to believe in spontaneous generation; the only alternative, to believe in a single, primary act of supernatural creation. There is no third position. For this reason many scientists a century ago chose to regard the belief in spontaneous generation as a "philosophical necessity." It is a symptom of the philosophical poverty of our time that this necessity is no longer appreciated. Most modern biologists, having reviewed with satisfaction the downfall of the spontaneous generation hypothesis, yet unwilling to accept the alternative belief in special creation, are left with nothing.

I think a scientist has no choice but to approach the origin of life through a hypothesis of spontaneous generation. What the controversy reviewed above showed to be untenable is only the belief that living organisms arise spontaneously under present conditions. We have now to face a somewhat different problem: how organisms may have arisen spontaneously under different conditions in some former period, granted that they do so no longer.

The Task

To make an organism demands the right substances in the right proportions and in the right arrangement. We do not think that anything more is needed—but that is problem enough.

The substances are water, certain salts—as it happens, those found in the ocean—and carbon compounds. The latter are called *organic* compounds because they scarcely occur except as products of living organisms.

Organic compounds consist for the most part of four types of atoms: carbon, oxygen, nitrogen and hydrogen. These four atoms together constitute about 99 per cent of living material, for hydrogen and oxygen also form water. The organic compounds found in organisms fall mainly into four great classes: carbohydrates, fats, proteins and nucleic acids. The illustrations on this and the next three pages give some notion of their composition and degrees of complexity. The fats are simplest, each consisting of three fatty acids joined to glycerol. The starches and glycogens are made of sugar units strung together to form long straight and branched chains. In general only one type of sugar appears in a single starch or glycogen; these molecules are large, but still relatively simple. The principal function of carbohydrates and fats in the organism is to serve as fuel—as a source of energy.

The nucleic acids introduce a further level of complexity. They are very large structures, composed of aggregates of at least four types of unit—the nucleotides—brought together in a great variety of proportions and sequences. An almost endless variety of different nucleic acids is possible, and specific differences among them are believed to be of the highest importance. Indeed, these structures are thought by many to be the main constituents of the genes, the bearers of hereditary constitution.

Variety and specificity, however, are most characteristic of the proteins, which include the largest and most complex molecules known. The units of which their structure is built are about 25 different amino acids. These are strung together in chains hundreds to thousands of units long, in different proportions, in all types of sequence, and with the greatest variety of branching and folding. A virtually infinite number of different proteins is possible. Organisms seem to exploit this potentiality, for no two species of living organism, animal or plant, possess the same proteins.

Organic molecules therefore form a large and formidable array, endless in variety and of the most bewildering complexity. One cannot think of having organisms without them. This is precisely the trouble, for to understand how organisms originated we must first of all explain how such complicated molecules could come into being. And that is only the beginning. To make an organism requires not only a tremendous variety of these substances, in adequate amounts and proper proportions, but also just the right arrangement of them. Structure here is as important as composition—and what a complication of structure! The most complex machine man has devised—say an electronic brain—is child's play compared with the simplest of living organisms. The especially trying thing is that complexity here involves such small dimensions. It is on the molecular level; it consists of a detailed fitting of molecule to molecule such as no chemist can attempt.

The Possible and Impossible

One has only to contemplate the magnitude of this task to concede that the spontaneous generation of a living organism is impossible. Yet here we are—as a result, I believe, of spontaneous generation. It will help to digress for a mo-

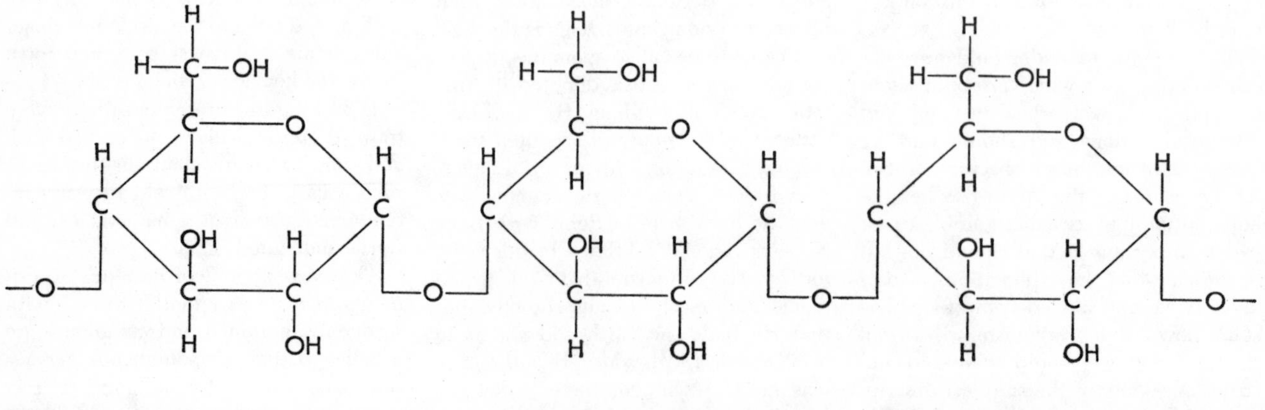

H
H—C—OH
C—O
H H H H H H
C C C C C C
OH H OH H OH H
—O— C—C —O— C—C —O— C—C —O—
H OH H OH H OH

CARBOHYDRATES comprise one of the four principal kinds of carbon compound found in living matter. This structural formula represents part of a characteristic carbohydrate. It is a polysaccharide consisting of six-carbon sugar units, three of which are shown.

ment to ask what one means by "impossible."

With every event one can associate a probability—the chance that it will occur. This is always a fraction, the proportion of times the event occurs in a large number of trials. Sometimes the probability is apparent even without trial. A coin has two faces; the probability of tossing a head is therefore 1/2. A die has six faces; the probability of throwing a deuce is 1/6. When one has no means of estimating the probability beforehand, it must be determined by counting the fraction of successes in a large number of trials.

Our everyday concept of what is impossible, possible or certain derives from our experience: the number of trials that may be encompassed within the space of a human lifetime, or at most within recorded human history. In this colloquial, practical sense I concede the spontaneous origin of life to be "impossible." It is impossible as we judge events in the scale of human experience.

We shall see that this is not a very meaningful concession. For one thing, the time with which our problem is concerned is geological time, and the whole extent of human history is trivial in the balance. We shall have more to say of this later.

But even within the bounds of our own time there is a serious flaw in our judgment of what is possible. It sounds impressive to say that an event has never been observed in the whole of human history. We should tend to regard such an event as at least "practically" impossible, whatever probability is assigned to it on abstract grounds. When we look a little further into such a statement, however, it proves to be almost meaningless. For men are apt to reject reports of very improbable occurrences. Persons of good judgment think it safer to distrust the alleged observer of such an event than to believe him. The result is that events which are merely very extraordinary acquire the reputation of never having occurred at all. Thus the highly improbable is made to appear impossible.

To give an example: Every physicist knows that there is a very small probability, which is easily computed, that the table upon which I am writing will suddenly and spontaneously rise into the air. The event requires no more than that the molecules of which the table is composed, ordinarily in random motion in all directions, should happen by chance to move in the same direction. Every physicist concedes this possibility; but try telling one that you have seen it happen. Recently I asked a friend, a Nobel laureate in physics, what he would say if I told him that. He laughed and said that he would regard it as more probable that I was mistaken than that the event had actually occurred.

We see therefore that it does not mean much to say that a very improbable event has never been observed. There is a conspiracy to suppress such observations, not among scientists alone, but among all judicious persons, who have learned to be skeptical even of what they see, let alone of what they are told. If one group is more skeptical than others, it is perhaps lawyers, who have the harshest experience of the unreliability of human evidence. Least skeptical of all are the scientists, who, cautious as they are, know very well what strange things are possible.

A final aspect of our problem is very important. When we consider the spontaneous origin of a living organism, this is not an event that need happen again and again. It is perhaps enough for it to happen once. The probability with which we are concerned is of a special kind; it is the probability that an event occur *at least once*. To this type of probability a fundamentally important thing happens as one increases the number of trials. However improbable the event in a single trial, it becomes increasingly probable as the trials are multiplied. Eventually the event becomes virtually inevitable. For instance, the chance that a coin will not fall head up in a single toss is 1/2. The chance that no head will appear in a series of tosses is $1/2 \times 1/2 \times 1/2$ . . . as many times over as the number of tosses. In 10 tosses the chance that no head will appear is therefore 1/2 multiplied by itself 10 times, or 1/1,000. Consequently the chance that a head will appear at least once in 10 tosses is 999/1,000. Ten trials have converted what started as a modest probability to a near certainty.

The same effect can be achieved with any probability, however small, by multiplying sufficiently the number of trials. Consider a reasonably improbable event, the chance of which is 1/1,000. The chance that this will not occur in one trial is 999/1,000. The chance that it won't occur in 1,000 trials is 999/1,000 multiplied together 1,000 times. This fraction comes out to be 37/100. The chance that it will happen at least once in 1,000 trials is therefore one minus this number—63/100—a little better than three chances out of five. One thousand trials have transformed this from a highly improbable to a highly probable event. In 10,000 trials the chance that this event will occur at least once comes out to be 19,999/20,000. It is now almost inevitable.

It makes no important change in the argument if we assess the probability that an event occur at least two, three, four or some other small number of

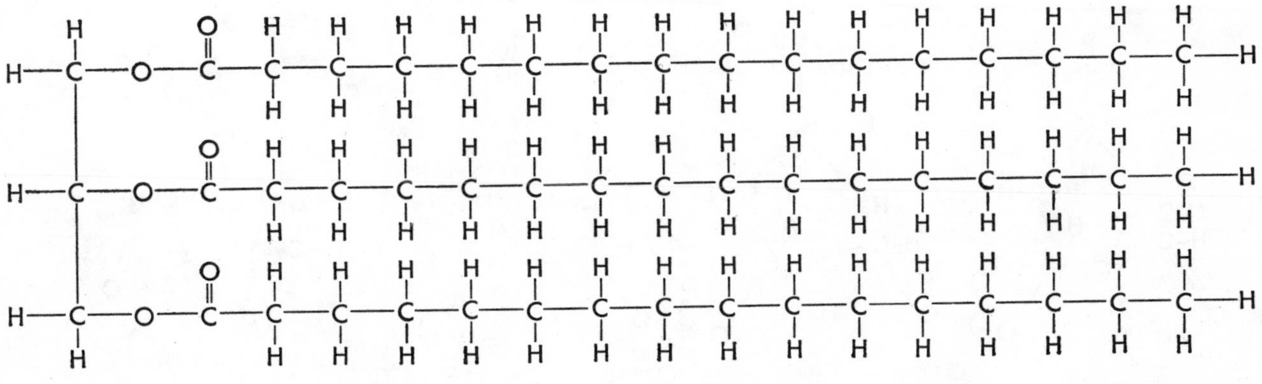

FATS are a second kind of carbon compound found in living matter. This formula represents the whole molecule of palmitin, one of the commonest fats. The molecule consists of glycerol (*11 atoms at the far left*) and fatty acids (*hydrocarbon chains at the right*).

times rather than at least once. It simply means that more trials are needed to achieve any degree of certainty we wish. Otherwise everything is the same.

In such a problem as the spontaneous origin of life we have no way of assessing probabilities beforehand, or even of deciding what we mean by a trial. The origin of a living organism is undoubtedly a stepwise phenomenon, each step with its own probability and its own conditions of trial. Of one thing we can be sure, however: whatever constitutes a trial, more such trials occur the longer the interval of time.

The important point is that since the origin of life belongs in the category of at-least-once phenomena, time is on its side. However improbable we regard this event, or any of the steps which it involves, given enough time it will almost certainly happen at least once. And for life as we know it, with its capacity for growth and reproduction, once may be enough.

Time is in fact the hero of the plot. The time with which we have to deal is of the order of two billion years. What we regard as impossible on the basis of human experience is meaningless here. Given so much time, the "impossible" becomes possible, the possible probable, and the probable virtually certain. One has only to wait: time itself performs the miracles.

Organic Molecules

This brings the argument back to its first stage: the origin of organic compounds. Until a century and a quarter ago the only known source of these substances was the stuff of living organisms. Students of chemistry are usually told that when, in 1828, Friedrich Wöhler synthesized the first organic compound, urea, he proved that organic compounds do not require living organisms to make them. Of course it showed nothing of the kind. Organic chemists are alive; Wöhler merely showed that they can make organic compounds externally as well as internally. It is still true that with almost negligible exceptions all the organic matter we know is the product of living organisms.

The almost negligible exceptions, however, are very important for our argument. It is now recognized that a constant, slow production of organic molecules occurs without the agency of living things. Certain geological phenomena yield simple organic compounds. So, for example, volcanic eruptions bring metal carbides to the surface of the earth, where they react with water vapor to yield simple compounds of carbon and hydrogen. The familiar type of such a reaction is the process used in old-style bicycle lamps in which acetylene is made by mixing iron carbide with water.

Recently Harold Urey, Nobel laureate in chemistry, has become interested in the degree to which electrical discharges in the upper atmosphere may promote the formation of organic compounds. One of his students, S. L. Miller, performed the simple experiment of circulating a mixture of water vapor, methane (CH_4), ammonia (NH_3) and hydrogen—all gases believed to have been present in the early atmosphere of the earth—continuously for a week over an electric spark. The circulation was maintained by boiling the water in one limb of the apparatus and condensing it in the other. At the end of the week the water was analyzed by the delicate method of paper chromatography. It was found to have acquired a mixture of amino acids! Glycine and alanine, the simplest amino acids and the most prevalent in proteins, were definitely identified in the solution, and there were indications it contained aspartic acid and two others. The yield was surprisingly high. This amazing result changes at a stroke our ideas of the probability of the spontaneous formation of amino acids.

A final consideration, however, seems to me more important than all the special processes to which one might appeal for organic syntheses in inanimate nature.

It has already been said that to have organic molecules one ordinarily needs organisms. The synthesis of organic substances, like almost everything else that happens in organisms, is governed by the special class of proteins called enzymes—the organic catalysts which greatly accelerate chemical reactions in the body. Since an enzyme is not used up but is returned at the end of the process, a small amount of enzyme can promote an enormous transformation of material.

Enzymes play such a dominant role in the chemistry of life that it is exceedingly difficult to imagine the synthesis of living material without their help. This poses a dilemma, for enzymes themselves are proteins, and hence among the most complex organic components of the cell. One is asking, in effect, for an apparatus which is the unique property of cells in order to form the first cell.

This is not, however, an insuperable difficulty. An enzyme, after all, is only a catalyst; it can do no more than change the *rate* of a chemical reaction. It cannot make anything happen that would not have happened, though more slowly, in its absence. Every process that is catalyzed by an enzyme, and every product of such a process, would occur without the enzyme. The only difference is one of rate.

Once again the essence of the argument is time. What takes only a few moments in the presence of an enzyme or other catalyst may take days, months or years in its absence; but given time, the end result is the same.

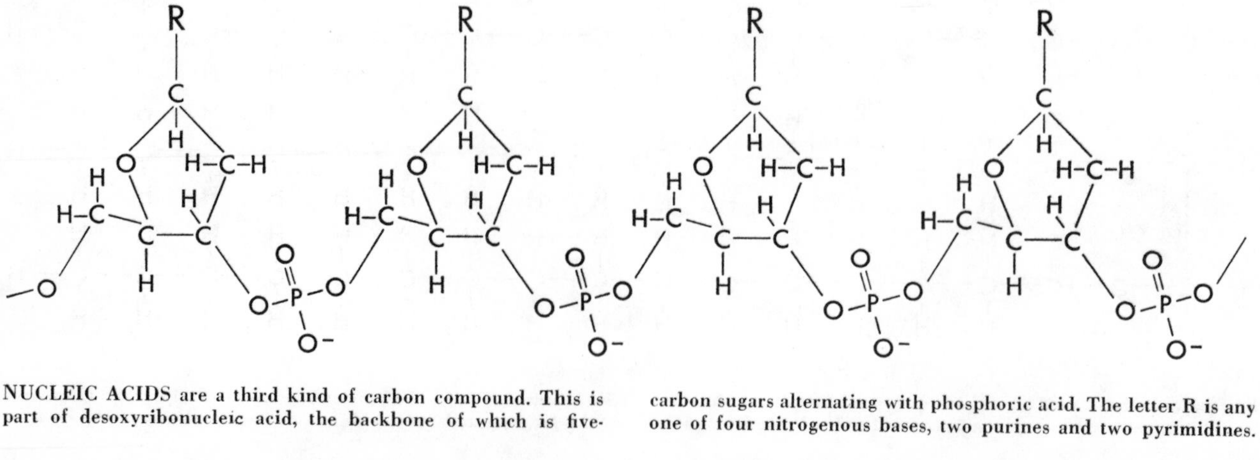

NUCLEIC ACIDS are a third kind of carbon compound. This is part of desoxyribonucleic acid, the backbone of which is five-carbon sugars alternating with phosphoric acid. The letter R is any one of four nitrogenous bases, two purines and two pyrimidines.

Indeed, this great difficulty in conceiving of the spontaneous generation of organic compounds has its positive side. In a sense, organisms demonstrate to us what organic reactions and products are *possible*. We can be certain that, given time, all these things must occur. Every substance that has ever been found in an organism displays thereby the finite probability of its occurrence. Hence, given time, it should arise spontaneously. One has only to wait.

It will be objected at once that this is just what one cannot do. Everyone knows that these substances are highly perishable. Granted that, within long spaces of time, now a sugar molecule, now a fat, now even a protein might form spontaneously, each of these molecules should have only a transitory existence. How are they ever to accumulate; and, unless they do so, how form an organism?

We must turn the question around. What, in our experience, is known to destroy organic compounds? Primarily two agencies: decay and the attack of oxygen. But decay is the work of living organisms, and we are talking of a time before life existed. As for oxygen, this introduces a further and fundamental section of our argument.

It is generally conceded at present that the early atmosphere of our planet contained virtually no free oxygen. Almost all the earth's oxygen was bound in the form of water and metal oxides. If this were not so, it would be very difficult to imagine how organic matter could accumulate over the long stretches of time that alone might make possible the spontaneous origin of life. This is a crucial point, therefore, and the statement that the early atmosphere of the planet was virtually oxygen-free comes forward so opportunely as to raise a suspicion of special pleading. I have for this reason taken care to consult a number of geologists and astronomers on this point, and am relieved to find that it is well defended. I gather that there is a widespread though not universal consensus that this condition did exist. Apparently something similar was true also for another common component of our atmosphere—carbon dioxide. It is believed that most of the carbon on the earth during its early geological history existed as the element or in metal carbides and hydrocarbons; very little was combined with oxygen.

This situation is not without its irony. We tend usually to think that the environment plays the tune to which the organism must dance. The environment is given; the organism's problem is to adapt to it or die. It has become apparent lately, however, that some of the most important features of the physical environment are themselves the work of living organisms. Two such features have just been named. The atmosphere of our planet seems to have contained no oxygen until organisms placed it there by the process of plant photosynthesis. It is estimated that at present all the oxygen of our atmosphere is renewed by photosynthesis once in every 2,000 years, and that all the carbon dioxide passes through the process of photosynthesis once in every 300 years. In the scale of geological time, these intervals are very small indeed. We are left with the realization that all the oxygen and carbon dioxide of our planet are the products of living organisms, and have passed through living organisms over and over again.

Forces of Dissolution

In the early history of our planet, when there were no organisms or any free oxygen, organic compounds should have been stable over very long periods. This is the crucial difference between the period before life existed and our own. If one were to specify a single reason why the spontaneous generation of living organisms was possible once and is so no longer, this is the reason.

We must still reckon, however, with another destructive force which is disposed of less easily. This can be called spontaneous dissolution—the counterpart of spontaneous generation. We have noted that any process catalyzed by an enzyme can occur in time without the enzyme. The trouble is that the processes which synthesize an organic substance are reversible: any chemical reaction which an enzyme may catalyze will go backward as well as forward. We have spoken as though one has only to wait to achieve syntheses of all kinds; it is truer to say that what one achieves by waiting is *equilibria* of all kinds—equilibria in which the synthesis and dissolution of substances come into balance.

In the vast majority of the processes in which we are interested the point of equilibrium lies far over toward the side of dissolution. That is to say, spontaneous dissolution is much more probable, and hence proceeds much more rapidly, than spontaneous synthesis. For example, the spontaneous union, step by step, of amino acid units to form a protein has a certain small probability, and hence might occur over a long stretch of time. But the dissolution of the protein or of an intermediate product into its component amino acids is much more probable, and hence will go ever so much more rapidly. The situation we must face is that of patient Penelope waiting for Odysseus, yet much worse: each night she undid the weaving of the preceding day, but here a night could readily undo the work of a year or a century.

How do present-day organisms manage to synthesize organic compounds against the forces of dissolution? They do so by a continuous expenditure of

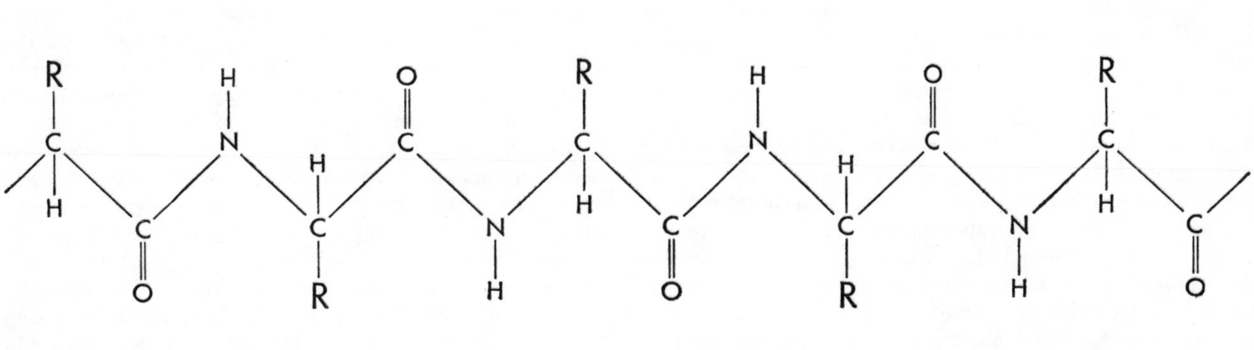

PROTEINS are a fourth kind of carbon compound found in living matter. This formula represents part of a polypeptide chain, the backbone of a protein molecule. The chain is made up of amino acids. Here the letter R represents the side chains of these acids.

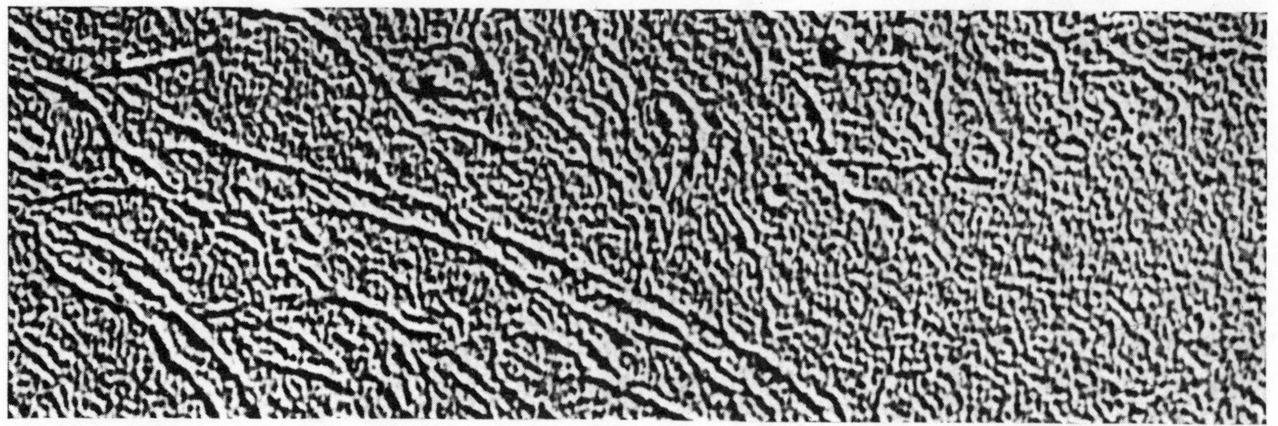

FILAMENTS OF COLLAGEN, a protein which is usually found in long fibrils, were dispersed by placing them in dilute acetic acid. This electron micrograph, which enlarges the filaments 75,000 times, was made by Jerome Gross of the Harvard Medical School.

energy. Indeed, living organisms commonly do better than oppose the forces of dissolution; they grow in spite of them. They do so, however, only at enormous expense to their surroundings. They need a constant supply of material and energy merely to maintain themselves, and much more of both to grow and reproduce. A living organism is an intricate machine for performing exactly this function. When, for want of fuel or through some internal failure in its mechanism, an organism stops actively synthesizing itself in opposition to the processes which continuously decompose it, it dies and rapidly disintegrates.

What we ask here is to synthesize organic molecules without such a machine. I believe this to be the most stubborn problem that confronts us—the weakest link at present in our argument. I do not think it by any means disastrous, but it calls for phenomena and forces some of which are as yet only partly understood and some probably still to be discovered.

Forces of Integration

At present we can make only a beginning with this problem. We know that it is possible on occasion to protect molecules from dissolution by precipitation or by attachment to other molecules. A wide variety of such precipitation and "trapping" reactions is used in modern chemistry and biochemistry to promote syntheses. Some molecules appear to acquire a degree of resistance to disintegration simply through their size. So, for example, the larger molecules composed of amino acids—polypeptides and proteins—seem to display much less tendency to disintegrate into their units than do smaller compounds of two or three amino acids.

Again, many organic molecules display still another type of integrating force—a spontaneous impulse toward structure formation. Certain types of fatty molecules—lecithins and cephalins—spin themselves out in water to form highly oriented and well-shaped structures—the so-called myelin figures. Proteins sometimes orient even in solution, and also may aggregate in the solid state in highly organized formations. Such spontaneous architectonic tendencies are still largely unexplored, particularly as they may occur in complex mixtures of substances, and they involve forces the strength of which has not yet been estimated.

What we are saying is that possibilities exist for opposing *intra*molecular dissolution by *inter*molecular aggregations of various kinds. The equilibrium between union and disunion of the amino acids that make up a protein is all to the advantage of disunion, but the aggregation of the protein with itself or other molecules might swing the equilibrium in the opposite direction: perhaps by removing the protein from access to the water which would be required to disintegrate it or by providing some particularly stable type of molecular association.

In such a scheme the protein appears only as a transient intermediate, an unstable way-station, which can either fall back to a mixture of its constituent amino acids or enter into the formation of a complex structural aggregate: amino acids $\leftrightarrows$ protein $\rightarrow$ aggregate.

Such molecular aggregates, of various degrees of material and architectural complexity, are indispensable intermediates between molecules and organisms. We have no need to try to imagine the spontaneous formation of an organism by one grand collision of its component molecules. The whole process must be gradual. The molecules form aggregates, small and large. The aggregates add further molecules, thus growing in size and complexity. Aggregates of various kinds interact with one another to form still larger and more complex structures. In this way we imagine the ascent, not by jumps or master strokes, but gradually, piecemeal, to the first living organisms.

First Organisms

Where may this have happened? It is easiest to suppose that life first arose in the sea. Here were the necessary salts and the water. The latter is not only the principal component of organisms, but prior to their formation provided a medium which could dissolve molecules of the widest variety and ceaselessly mix and circulate them. It is this constant mixture and collision of organic molecules of every sort that constituted in large part the "trials" of our earlier discussion of probabilities.

The sea in fact gradually turned into a dilute broth, sterile and oxygen-free. In this broth molecules came together in increasing number and variety, sometimes merely to collide and separate, sometimes to react with one another to produce new combinations, sometimes to aggregate into multimolecular formations of increasing size and complexity.

What brought order into such complexes? For order is as essential here as composition. To form an organism, molecules must enter into intricate designs and connections; they must eventually form a self-repairing, self-constructing dynamic machine. For a time this problem of molecular arrangement seemed to present an almost insuperable obstacle in the way of imagining a spontaneous origin of life, or indeed the laboratory

FIBRILS OF COLLAGEN formed spontaneously out of filaments such as those shown on the opposite page when 1 per cent of sodium chloride was added to the dilute acetic acid. These long fibrils are identical in appearance with those of collagen before dispersion.

synthesis of a living organism. It is still a large and mysterious problem, but it no longer seems insuperable. The change in view has come about because we now realize that it is not altogether necessary to *bring* order into this situation; a great deal of order is implicit in the molecules themselves.

The epitome of molecular order is a crystal. In a perfect crystal the molecules display complete regularity of position and orientation in all planes of space. At the other extreme are fluids—liquids or gases—in which the molecules are in ceaseless motion and in wholly random orientations and positions.

Lately it has become clear that very little of a living cell is truly fluid. Most of it consists of molecules which have taken up various degrees of orientation with regard to one another. That is, most of the cell represents various degrees of approach to crystallinity—often, however, with very important differences from the crystals most familiar to us. Much of the cell's crystallinity involves molecules which are still in solution—so-called liquid crystals—and much of the dynamic, plastic quality of cellular structure, the capacity for constant change of shape and interchange of material, derives from this condition. Our familiar crystals, furthermore, involve only one or a very few types of molecule, while in the cell a great variety of different molecules come together in some degree of regular spacing and orientation—*i.e.*, some degree of crystallinity. We are dealing in the cell with highly mixed crystals and near-crystals, solid and liquid. The laboratory study of this type of formation has scarcely begun. Its further exploration is of the highest importance for our problem.

In a fluid such as water the molecules are in very rapid motion. Any molecules dissolved in such a medium are under a constant barrage of collisions with water molecules. This keeps small and moderately sized molecules in a constant turmoil; they are knocked about at random, colliding again and again, never holding any position or orientation for more than an instant. The larger a molecule is relative to water, the less it is disturbed by such collisions. Many protein and nucleic acid molecules are so large that even in solution their motions are very sluggish, and since they carry large numbers of electric charges distributed about their surfaces, they tend even in solution to align with respect to one another. It is so that they tend to form liquid crystals.

We have spoken above of architectonic tendencies even among some of the relatively small molecules: the lecithins and cephalins. Such molecules are insoluble in water yet possess special groups which have a high affinity for water. As a result they tend to form surface layers, in which their water-seeking groups project into the water phase, while their water-repelling portions project into the air, or into an oil phase, or unite to form an oil phase. The result is that quite spontaneously such molecules, when exposed to water, take up highly oriented positions to form surface membranes, myelin figures and other quasi-crystalline structures.

Recently several particularly striking examples have been reported of the spontaneous production of familiar types of biological structure by protein molecules. Cartilage and muscle offer some of the most intricate and regular patterns of structure to be found in organisms. A fiber from either type of tissue presents under the electron microscope a beautiful pattern of cross striations of various widths and densities, very regularly spaced. The proteins that form these structures can be coaxed into free solution and stirred into completely random orientation. Yet on precipitating, under proper conditions, the molecules realign with regard to one another to regenerate with extraordinary fidelity the original patterns of the tissues [*see illustration above*].

We have therefore a genuine basis for the view that the molecules of our oceanic broth will not only come together spontaneously to form aggregates but in doing so will spontaneously achieve various types and degrees of order. This greatly simplifies our problem. What it means is that, given the right molecules, one does not have to do everything for them; they do a great deal for themselves.

Oparin has made the ingenious suggestion that natural selection, which Darwin proposed to be the driving force of organic evolution, begins to operate at this level. He suggests that as the molecules come together to form colloidal aggregates, the latter begin to compete with one another for material. Some aggregates, by virtue of especially favorable composition or internal arrangement, acquire new molecules more rapidly than others. They eventually emerge as the dominant types. Oparin suggests further that considerations of optimal size enter at this level. A growing colloidal particle may reach a point at which it becomes unstable and breaks down into smaller particles, each of which grows and redivides. All these phenomena lie within the bounds of known processes in nonliving systems.

The Sources of Energy

We suppose that all these forces and factors, and others perhaps yet to be revealed, together give us eventually the first living organism. That achieved, how does the organism continue to live?

We have already noted that a living organism is a dynamic structure. It is the site of a continuous influx and outflow of matter and energy. This is the very sign of life, its cessation the best evidence of death. What is the primal organism to use as food, and how derive the energy it needs to maintain itself and grow?

For the primal organism, generated under the conditions we have described, only one answer is possible. Having arisen in an oceanic broth of organic molecules, its only recourse is to live upon them. There is only one way of doing that in the absence of oxygen. It is called fermentation: the process by which organisms derive energy by breaking organic molecules and rearranging their parts. The most familiar example of such a process is the fermentation of sugar by yeast, which yields alcohol as one of the products. Animal cells also ferment sugar, not to alcohol but to lactic acid. These are two examples from a host of known fermentations.

The yeast fermentation has the following over-all equation: $C_6H_{12}O_6 \rightarrow 2\ CO_2 + 2\ C_2H_5OH$ + energy. The result of fragmenting 180 grams of sugar into 88 grams of carbon dioxide and 92 grams of alcohol is to make available about 20,000 calories of energy for the use of the cell. The energy is all that the cell derives by this transaction; the carbon dioxide and alcohol are waste products which must be got rid of somehow if the cell is to survive.

The cell, having arisen in a broth of organic compounds accumulated over the ages, must consume these molecules by fermentation in order to acquire the energy it needs to live, grow and reproduce. In doing so, it and its descendants are living on borrowed time. They are consuming their heritage, just as we in our time have nearly consumed our heritage of coal and oil. Eventually such a process must come to an end, and with that life also should have ended. It would have been necessary to start the entire development again.

Fortunately, however, the waste product carbon dioxide saved this situation. This gas entered the ocean and the atmosphere in ever-increasing quantity. Some time before the cell exhausted the supply of organic molecules, it succeeded in inventing the process of photosynthesis. This enabled it, with the energy of sunlight, to make its own organic molecules: first sugar from carbon dioxide and water, then, with ammonia and nitrates as sources of nitrogen, the entire array of organic compounds which it requires. The sugar synthesis equation is: $6\ CO_2 + 6\ H_2O$ + sunlight $\rightarrow C_6H_{12}O_6 + 6\ O_2$. Here 264 grams of carbon dioxide plus 108 grams of water plus about 700,000 calories of sunlight yield 180 grams of sugar and 192 grams of oxygen.

This is an enormous step forward. Living organisms no longer needed to depend upon the accumulation of organic matter from past ages; they could make their own. With the energy of sunlight they could accomplish the fundamental organic syntheses that provide their substance, and by fermentation they could produce what energy they needed.

Fermentation, however, is an extraordinarily inefficient source of energy. It leaves most of the energy potential of organic compounds unexploited; consequently huge amounts of organic material must be fermented to provide a modicum of energy. It produces also various poisonous waste products—alcohol, lactic acid, acetic acid, formic acid and so on. In the sea such products are readily washed away, but if organisms were ever to penetrate to the air and land, these products must prove a serious embarrassment.

One of the by-products of photosynthesis, however, is oxygen. Once this was available, organisms could invent a new way to acquire energy, many times as efficient as fermentation. This is the

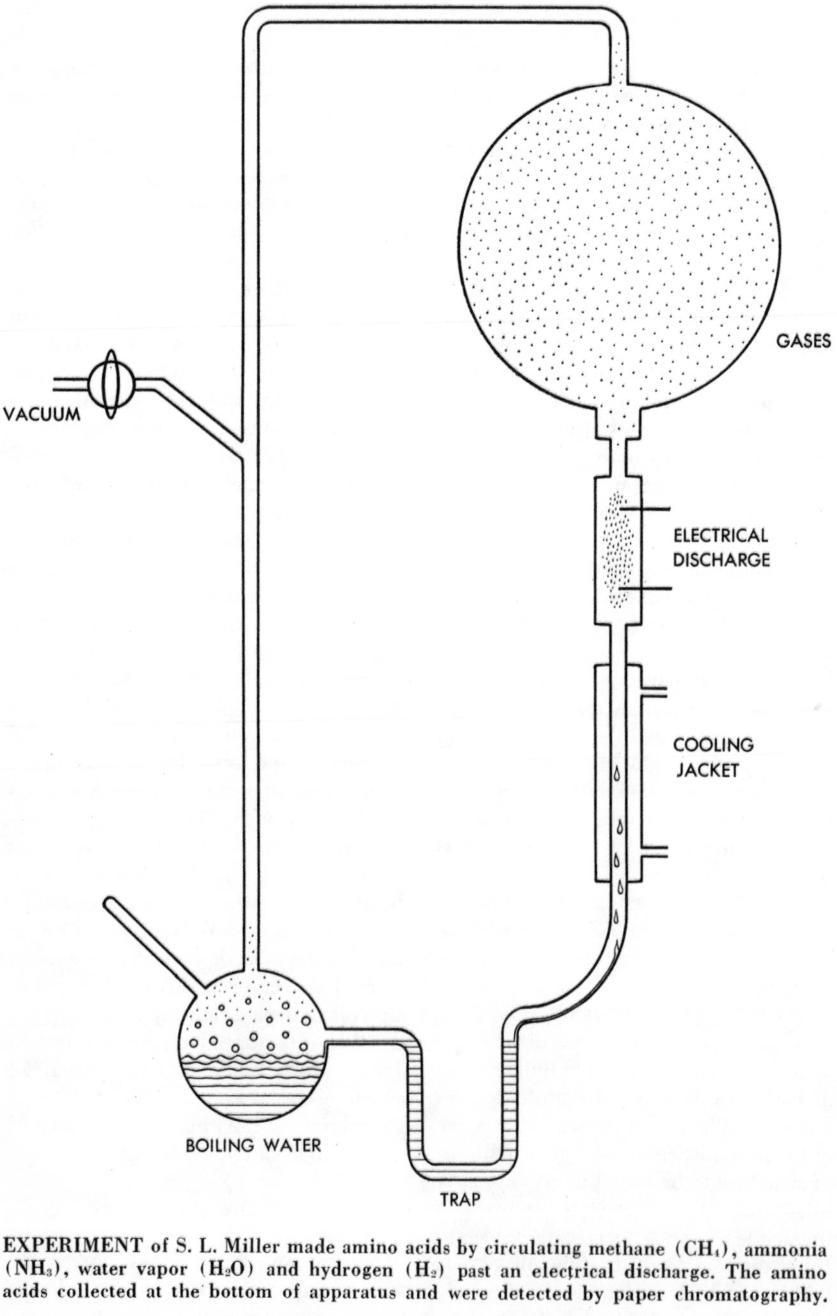

EXPERIMENT of S. L. Miller made amino acids by circulating methane (CH_4), ammonia (NH_3), water vapor (H_2O) and hydrogen (H_2) past an electrical discharge. The amino acids collected at the bottom of apparatus and were detected by paper chromatography.

process of cold combustion called respiration: $C_6H_{12}O_6 + 6\ O_2 \rightarrow 6\ CO_2 + 6\ H_2O$ + energy. The burning of 180 grams of sugar in cellular respiration yields about 700,000 calories, as compared with the approximately 20,000 calories produced by fermentation of the same quantity of sugar. This process of combustion extracts all the energy that can possibly be derived from the molecules which it consumes. With this process at its disposal, the cell can meet its energy requirements with a minimum expenditure of substance. It is a further advantage that the products of respiration—water and carbon dioxide—are innocuous and easily disposed of in any environment.

Life's Capital

It is difficult to overestimate the degree to which the invention of cellular respiration released the forces of living organisms. No organism that relies wholly upon fermentation has ever amounted to much. Even after the advent of photosynthesis, organisms could have led only a marginal existence. They could indeed produce their own organic materials, but only in quantities sufficient to survive. Fermentation is so profligate a way of life that photosynthesis could do little more than keep up with it. Respiration used the material of organisms with such enormously greater efficiency as for the first time to leave something over. Coupled with fermentation, photosynthesis made organisms self-sustaining; coupled with respiration, it provided a surplus. To use an economic analogy, photosynthesis brought organisms to the subsistence level; respiration provided them with capital. It is mainly this capital that they invested in the great enterprise of organic evolution.

The entry of oxygen into the atmosphere also liberated organisms in another sense. The sun's radiation contains ultraviolet components which no living cell can tolerate. We are sometimes told that if this radiation were to reach the earth's surface, life must cease. That is not quite true. Water absorbs ultraviolet radiation very effectively, and one must conclude that as long as these rays penetrated in quantity to the surface of the earth, life had to remain under water. With the appearance of oxygen, however, a layer of ozone formed high in the atmosphere and absorbed this radiation. Now organisms could for the first time emerge from the water and begin to populate the earth and air. Oxygen provided not only the means of obtaining adequate energy for evolution but the protective blanket of ozone which alone made possible terrestrial life.

This is really the end of our story. Yet not quite the end. Our entire concern in this argument has been to bring the origin of life within the compass of natural phenomena. It is of the essence of such phenomena to be repetitive, and hence, given time, to be inevitable.

This is by far our most significant conclusion—that life, as an orderly natural event on such a planet as ours, was inevitable. The same can be said of the whole of organic evolution. All of it lies within the order of nature, and apart from details all of it was inevitable.

Astronomers have reason to believe that a planet such as ours—of about the earth's size and temperature, and about as well-lighted—is a rare event in the universe. Indeed, filled as our story is with improbable phenomena, one of the least probable is to have had such a body as the earth to begin with. Yet though this probability is small, the universe is so large that it is conservatively estimated at least 100,000 planets like the earth exist in our galaxy alone. Some 100 million galaxies lie within the range of our most powerful telescopes, so that throughout observable space we can count apparently on the existence of at least 10 million million planets like our own.

What it means to bring the origin of life within the realm of natural phenomena is to imply that in all these places life probably exists—life as we know it. Indeed, I am convinced that there can be no way of composing and constructing living organisms which is fundamentally different from the one we know—though this is another argument, and must await another occasion. Wherever life is possible, given time, it should arise. It should then ramify into a wide array of forms, differing in detail from those we now observe (as did earlier organisms on the earth) yet including many which should look familiar to us—perhaps even men.

We are not alone in the universe, and do not bear alone the whole burden of life and what comes of it. Life is a cosmic event—so far as we know the most complex state of organization that matter has achieved in our cosmos. It has come many times, in many places—places closed off from us by impenetrable distances, probably never to be crossed even with a signal. As men we can attempt to understand it, and even somewhat to control and guide its local manifestations. On this planet that is our home, we have every reason to wish it well. Yet should we fail, all is not lost. Our kind will try again elsewhere.

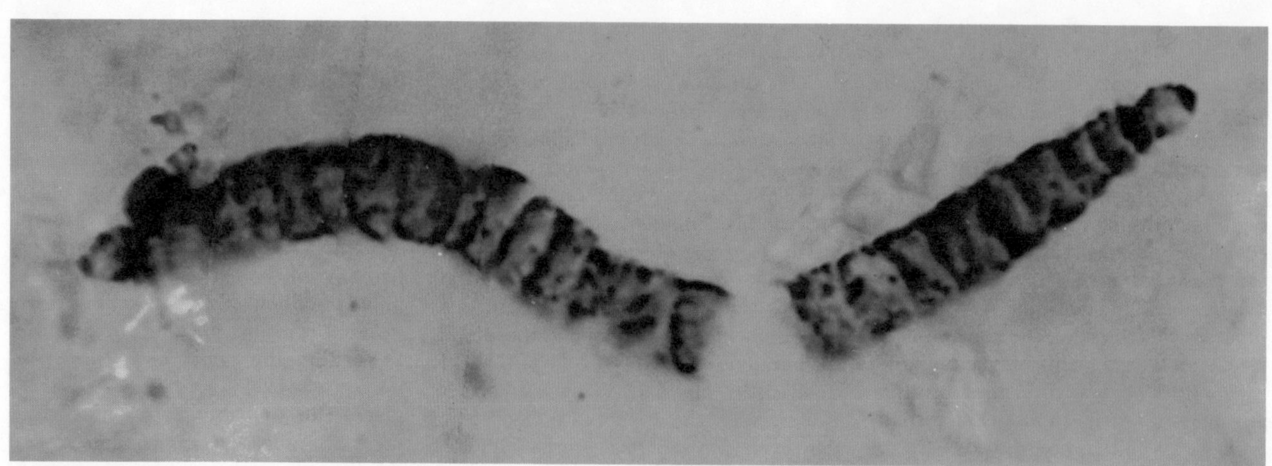
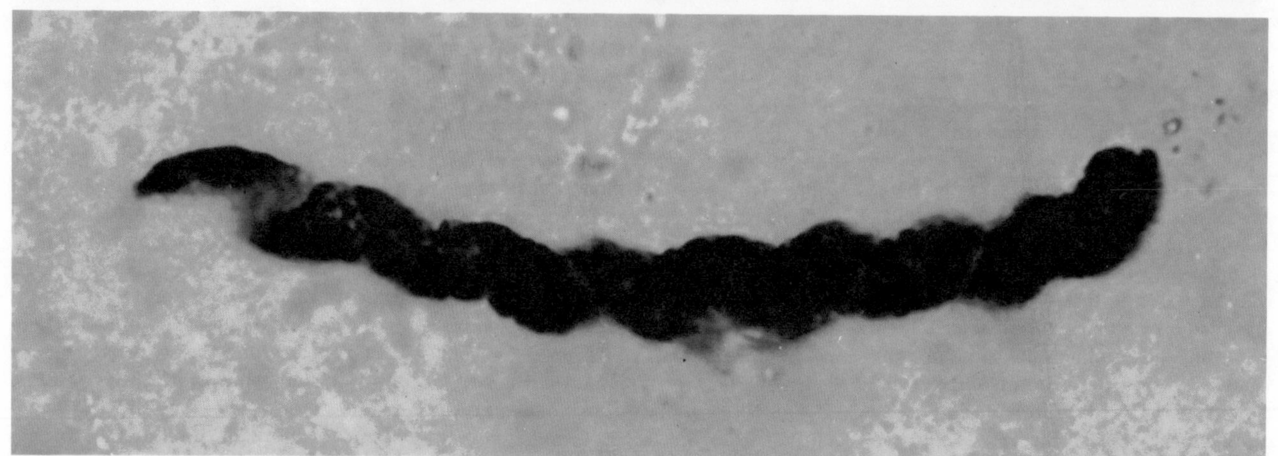
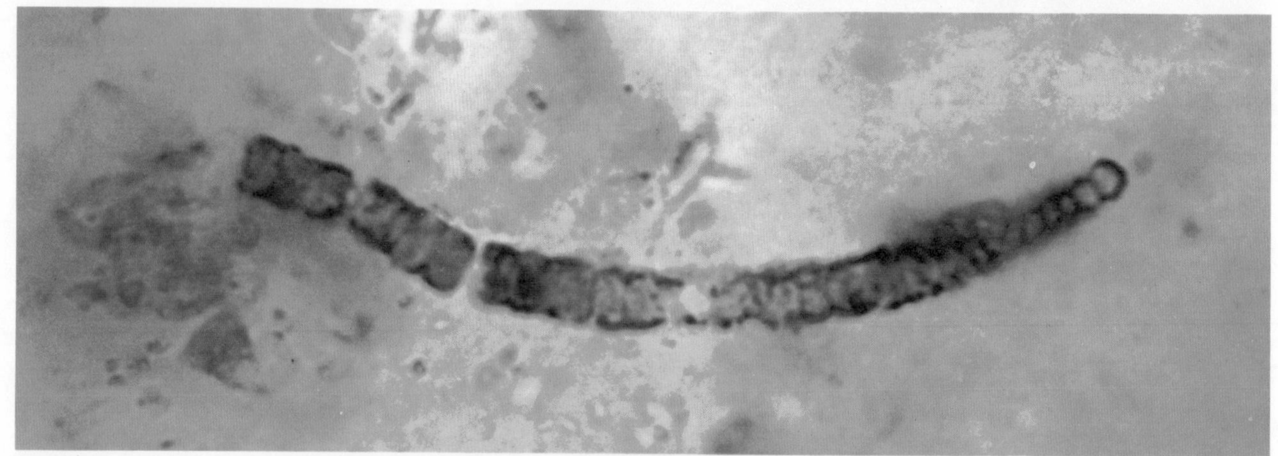
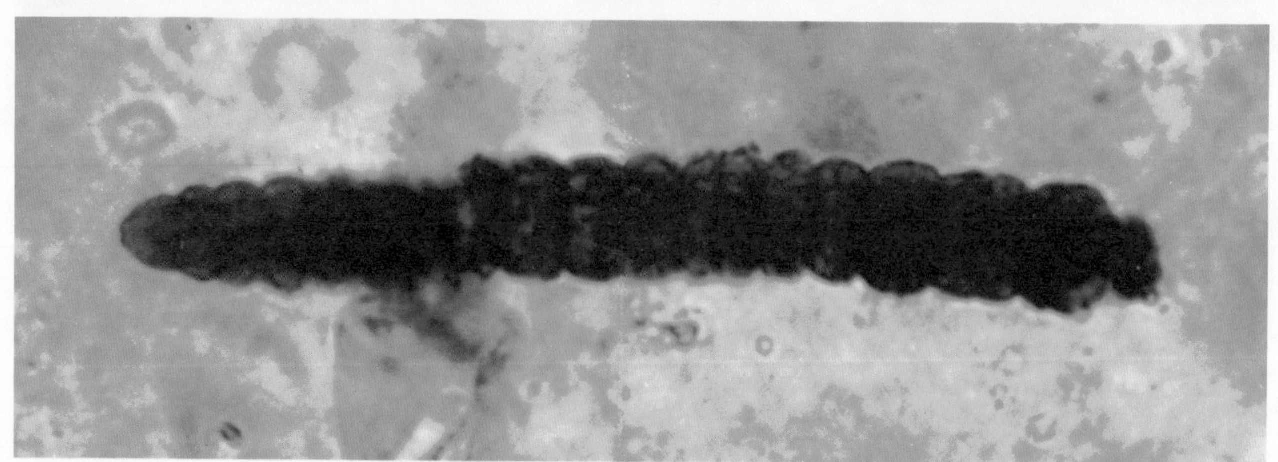

The Evolution of the Earliest Cells

2

by J. William Schopf
September 1978

For some three billion years the only living things were primitive microorganisms. These early cells gave rise to biochemical systems and the oxygen-enriched atmosphere on which modern life depends

When *On the Origin of Species* appeared in 1859, the history of life could be traced back to the beginning of the Cambrian period of geologic time, to the earliest recognized fossils, forms that are now known to have lived more than 500 million years ago. A far longer prehistory of life has since been discovered: it extends back through geologic time almost three billion years more. During most of that long Precambrian interval the only inhabitants of the earth were simple microscopic organisms, many of them comparable in size and complexity to modern bacteria. The conditions under which these organisms lived differed greatly from those prevailing today, but the mechanisms of evolution were the same. Genetic variations made some individuals better fitted than others to survive and to reproduce in a given environment, and so the heritable traits of the better-adapted organisms were more often represented in succeeding generations. The emergence of new forms of life through this principle of natural selection worked great changes in turn on the physical environment, thereby altering the conditions of evolution.

One momentous event in Precambrian evolution was the development of the biochemical apparatus of oxygen-generating photosynthesis. Oxygen released as a by-product of photosynthesis accumulated in the atmosphere and effected a new cycle of biological adaptation. The first organisms to evolve in response to this environmental change could merely tolerate oxygen; later cells could actively employ oxygen in metabolism and were thereby enabled to extract more energy from foodstuff.

A second important episode in Precambrian history led to the emergence of a new kind of cell, in which the genetic material is aggregated in a distinct nucleus and is bounded by a membrane. Such nucleated cells are more highly organized than those without nuclei. What is most important, only nucleated cells are capable of advanced sexual reproduction, the process whereby the genetic variations of the parents can be passed on to the offspring in new combinations. Because sexual reproduction allows novel adaptations to spread quickly through a population its development accelerated the pace of evolutionary change. The large, complex, multicellular forms of life that have appeared and quickly diversified since the beginning of the Cambrian period are without exception made up of nucleated cells.

The history of life in its later phases, since the start of the Cambrian period, has been reconstructed mainly from the study of fossils preserved in sedimentary rocks. In the 18th and 19th centuries it gradually became apparent that the fossil record has appreciable chronological and geographical continuity. The fossil deposits form recognizable layers, which can be identified in widely separated geological formations. Boundaries between such layers, where one characteristic suite of fossils gives way to another, provide the basis for dividing geologic time into eras, periods and epochs.

MICROSCOPIC FOSSILS on the opposite page are the remains of organisms that were once the dominant form of life on the earth. The fossils are from silica-rich rocks in the Bitter Springs formation of central Australia, deposited about 850 million years ago, or late in the Precambrian era. The rocks have the layered structure of stromatolites, sedimentary deposits that were formed by matlike communities of microorganisms. Among Precambrian fossils these specimens are exceptionally well preserved; their petrified cell walls are composed of organic matter and have retained their three-dimensional form. In size, structure and ecological setting they resemble living cyanobacteria, or blue-green algae. Like their modern counterparts, the fossil forms were presumably capable of photosynthesis, and similar cyanobacteria some billion years earlier were evidently responsible for the first rapid release of oxygen into the earth's atmosphere. Organisms in these photomicrographs are about 60 micrometers long.

One of the most dramatic boundaries in the rock record is the one that separates the Cambrian period from all that came before. The 11 periods of geologic time since the start of the Cambrian are referred to collectively as the Phanerozoic era, which might be translated from the Greek as the era of manifest life. The preceding era is called simply the Precambrian.

By itself the geologic time scale cannot provide dates for fossil deposits; it only lists their sequence. Ages can be calculated, however, from the constant rate of decay of radioactive isotopes in the earth's crust. By determining how much of an isotope has decayed since the minerals in a rock unit crystallized, a date can be assigned to that unit and to nearby strata containing fossils. Radioactive-isotope studies of this kind, carried out on rocks from many parts of the world, have established a rather well-defined date for the start of the Phanerozoic era: it began about 570 million years ago. The same method indicates that the earth itself and the rest of the solar system are 4.6 billion years old. Thus the Precambrian era encompasses some seven-eighths of the earth's entire history.

The boundary between the Precambrian era and the Cambrian period has traditionally been viewed as a sharp discontinuity. In Cambrian strata there are abundant fossils of marine plants and animals: seaweeds, worms, sponges, mollusks, lampshells and, what are perhaps most characteristic of the period, the early arthropods called trilobites. It was thought for many years that fossils were entirely absent in the underlying Precambrian strata. The Cambrian fauna seemed to come into existence abruptly and without known predecessors.

Life could not have begun with organisms as complex as trilobites. In *On the Origin of Species* Darwin wrote: "To the question why we do not find rich fossiliferous deposits belonging to... periods prior to the Cambrian system, I can give no satisfactory answer.... The case at present must remain inexplicable; and may be truly urged as a valid argument against the views here entertained." The argument is no longer valid, but it is only in the past 20 years or so that a definitive answer to it has been found.

One part of the answer lies in the discovery of primitive fossil animals in rocks below the earliest Cambrian strata. The fossils include the remains of jellyfishes, various kinds of worms and possibly sponges, and they make up a fauna quite distinct from that of the predominantly shelled animals of the Cambrian period. These discoveries, however, extend the fossil record by only about 100 million years, less than four percent of the Precambrian era. It can still be asked: What came before?

Since the 1950's a far-reaching explanation has emerged. It has come to be recognized that not only are many Precambrian rocks fossil-bearing but also Precambrian fossils can be found even in some of the most ancient sedimentary deposits known. These fossils had escaped notice earlier largely because they are the remains only of microscopic forms of life.

An important clue in the search for Precambrian life was discovered in the early years of the 20th century, but its significance was not fully appreciated until much later. The clue came in the form of masses of thinly layered limestone rock discovered by Charles Doolittle Walcott in Precambrian strata from western North America. Walcott found numerous moundlike or pillarlike structures made up of many draped horizontal layers, like tall stacks of pancakes. These structures are now called stromatolites, from the Greek *stroma,* meaning bed or coverlet, and *lithos,* meaning stone.

Walcott interpreted the stromatolites as being fossilized reefs that had probably been formed by various types of algae. Other workers were skeptical, and for many years the stromatolites were widely attributed to some nonbiological origin. The first convincing evidence substantiating Walcott's hypothesis came in 1954, when Stanley A. Tyler of the University of Wisconsin and Elso S. Barghoorn of Harvard University reported the discovery of fossil microscopic plants in an outcropping of Precambrian rocks called the Gunflint Iron formation near Lake Superior in Ontario. Most of the Gunflint fossils, which form the layers of dome-shaped and pillarlike stromatolites, resemble modern blue-green algae and bacteria. More recently, living stromatolites have been identified in several coastal habitats, most notably in a lagoon at Shark Bay on the western coast of Australia. They are indeed built up by communities of blue-green algae and bacteria, and they are strikingly similar in form to the fossilized Precambrian structures.

Today microfossils have been identified in some 45 stromatolitic deposits. (All but three of these fossilized communities have been found in the past 10 years.) The fossils are often well preserved, the cell walls being petrified in three-dimensional form, and they have become a prime source of documentation for the early history of life. In recent years the search for Precambrian microfossils in other kinds of sediments, such as shales deposited in offshore environments, has also been rewarded. These fossils are generally not as well preserved as the ones in stromatolites, most of them having been flattened by pressure; on the other hand, they supply information about Precambrian life in a habitat quite different from that of the shallow-water stromatolites.

A surprising amount of information can be derived from the fossil remains of a microorganism. Size, shape and degree of morphological complexity are among the most easily recognized features, but under favorable circumstances even details of the internal structure of cells can be discerned. In retracing the course of Precambrian evolution, however, there is no need to rely exclusively on the fossil record. An entirely independent archive has been preserved in the metabolism and the biochemical pathways of modern, living cells. No living organism is biochemically identical with its Precambri-

FOSSIL STROMATOLITES typically exhibit the appearance of mounds or pillars made up of many thin layers piled one on top of another. The stromatolites were formed by communities of cyanobacteria and other prokaryotes (cells without a nucleus) in shallow water; each layer represents a stage in the growth of the community. Stromatolites formed throughout much of the Precambrian era. They are an important source of Precambrian fossils. These specimens are in limestone about 1,300 million years old in Glacier National Park.

LIVING STROMATOLITES were photographed at Shark Bay in Australia. Elsewhere stromatolites are rare because of grazing by invertebrates. Here the invertebrates are excluded because the water is too salty for them; in the Precambrian era they had not yet evolved. In size and form the modern stromatolites are much like the fossil structures, and they are produced by the growth of cyanobacteria and other prokaryotes in matlike communities. The discovery of such living stromatolites has confirmed the biological origin of the fossil ones.

an antecedents, but vestiges of earlier biochemistries have been retained. By studying their distribution in modern forms of life it is sometimes possible to deduce when certain biochemical capabilities first appeared in the evolutionary sequence.

Still another independent source of information about the early evolutionary progression is based neither on living nor on fossil organisms but on the inorganic geological record. The nature of the minerals found there reflects physical conditions at the time the minerals were deposited, conditions that may have been influenced by biological innovations. In order to understand the introduction of oxygen into the early atmosphere, for example, all three fields of study must be called on to testify. The mineral record tells when the change took place, the fossil record reveals the organisms responsible and the distribution of biochemical capabilities among modern organisms puts the development in its proper evolutionary context.

Since the 1960's it has become apparent that the greatest division among living organisms is not between plants and animals but between organisms whose cells have nuclei and those that lack nuclei. In terms of biochemistry, metabolism, genetics and intracellular organization, plants and animals are very similar; all such higher organisms, however, are quite different in these features from bacteria and blue-green algae, the principal types of non-nucleated life. Recognition of this discontinuity has been important for understanding the early stages of biological history.

Organisms whose cells have nuclei are called eukaryotes, from the Greek roots *eu-*, meaning well or true, and *karyon,* meaning kernel or nut. Cells without nuclei are prokaryotes, the prefix *pro-* meaning before. All green plants and all animals are eukaryotes. So are the fungi, including the molds and the yeasts, and protists such as *Paramecium* and *Euglena.* The prokaryotes include only two groups of organisms, the bacteria and the blue-green algae. The latter produce oxygen through photosynthesis like other algae and higher plants, but they have much stronger affinities with the bacteria than they do with eukaryotic forms of life. I shall therefore refer to blue-green algae by an alternative and more descriptive name, the cyanobacteria.

Several important traits distinguish eukaryotes from prokaryotes. In the nucleus of a eukaryotic cell the DNA is organized in chromosomes and is enclosed by an intracellular membrane; many prokaryotes have only a single loop of DNA, which is loose in the cytoplasm of the cell. Prokaryotes reproduce asexually by the comparatively simple process of binary fission. In contrast, asexual reproduction in eukaryot-

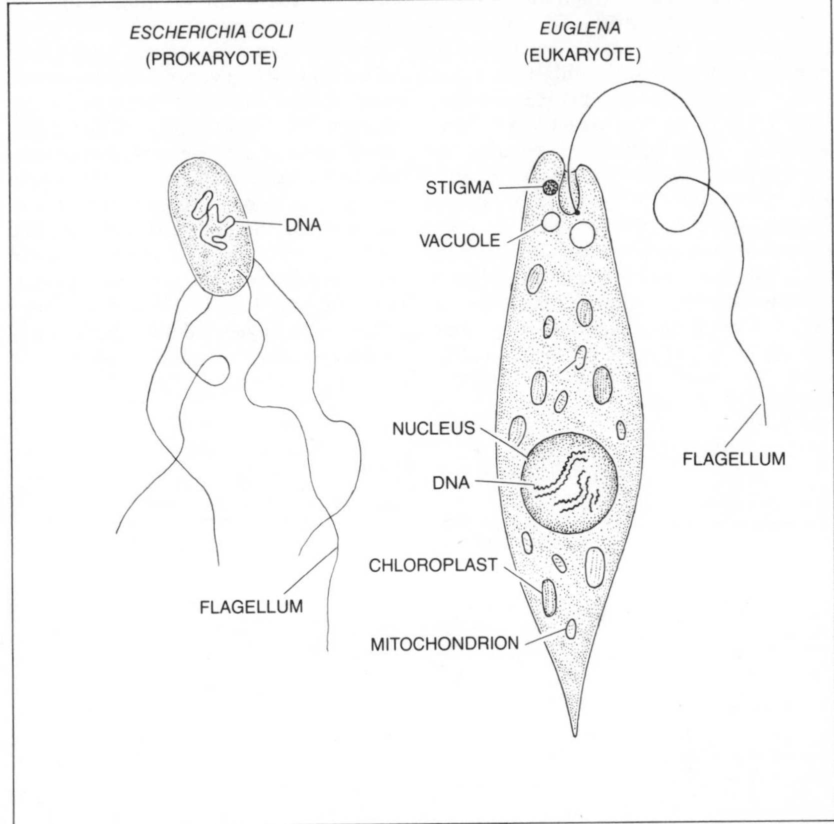

	PROKARYOTES	EUKARYOTES
ORGANISMS REPRESENTED	BACTERIA AND CYANOBACTERIA	PROTISTS, FUNGI, PLANTS AND ANIMALS
CELL SIZE	SMALL, GENERALLY 1 TO 10 MICROMETERS	LARGE, GENERALLY 10 TO 100 MICROMETERS
METABOLISM AND PHOTOSYNTHESIS	ANAEROBIC OR AEROBIC	AEROBIC
MOTILITY	NONMOTILE OR WITH FLAGELLA MADE OF THE PROTEIN FLAGELLIN	USUALLY MOTILE, CILIA OR FLAGELLA CONSTRUCTED OF MICROTUBULES
CELL WALLS	OF CHARACTERISTIC SUGARS AND PEPTIDES	OF CELLULOSE OR CHITIN, BUT LACKING IN ANIMALS
ORGANELLES	NO MEMBRANE-BOUNDED ORGANELLES	MITOCHONDRIA AND CHLOROPLASTS
GENETIC ORGANIZATION	LOOP OF DNA IN CYTOPLASM	DNA ORGANIZED IN CHROMOSOMES AND BOUNDED BY NUCLEAR MEMBRANE
REPRODUCTION	BY BINARY FISSION	BY MITOSIS OR MEIOSIS
CELLULAR ORGANIZATION	MAINLY UNICELLULAR	MAINLY MULTICELLULAR, WITH DIFFERENTIATION OF CELLS

GREATEST DIVISION among organisms is the one separating cells with nuclei (eukaryotes) from those without nuclei (prokaryotes). The only prokaryotes are bacteria and cyanobacteria, and here they are represented by the bacterium *Escherichia coli* (*top left*). All other organisms are eukaryotes, including higher plants and animals, fungi and protists such as *Euglena* (*top right*). Eukaryotic cells are by far the more complex ones, and some of the organelles they contain, such as mitochondria and chloroplasts, may be derived from prokaryotes that established a symbiotic relationship with the host cell. Prokaryotes vary widely in their tolerance of or requirement for free oxygen, and they are thought to have evolved during a period of fluctuating oxygen. All eukaryotes require oxygen for metabolism and for the synthesis of various substances, and they must have emerged after an atmosphere rich in oxygen became established.

ic cells takes place through the complicated process of mitosis, and most eukaryotes can also reproduce sexually through meiosis and the subsequent fusion of sex cells. (The "parasexual" reproduction of some prokaryotes differs markedly from advanced eukaryotic sexuality.) Eukaryotic cells are generally larger than prokaryotic ones, although the range of sizes overlaps, and almost all prokaryotes are unicellular organisms whereas the majority of eukaryotes are large, complex and many-celled. A mammalian animal, for example, can be made up of billions of cells, which are highly differentiated in both structure and function.

An intriguing feature of eukaryotic cells is that they have within them smaller membrane-bounded subunits, or organelles, the most notable being mitochondria and chloroplasts. Mitochondria are present in all eukaryotes, where they play a central role in the energy economy of the cell. Chloroplasts are present in some protists and in all green plants and are responsible for the photosynthetic activities of those organisms. It has been suggested that both mitochondria and chloroplasts may be evolutionary derivatives of what were once free-living microorganisms, an idea discussed in particular by Lynn Margulis of Boston University. The modern chloroplast, for example, may be derived from a cyanobacterium that was engulfed by another cell and that later established a symbiotic relationship with it. In support of this hypothesis it has been noted that both mitochondria and chloroplasts contain a small fragment of DNA whose organization is somewhat like that of prokaryotic DNA. In the past several years the testing of this hypothesis has generated a large body of

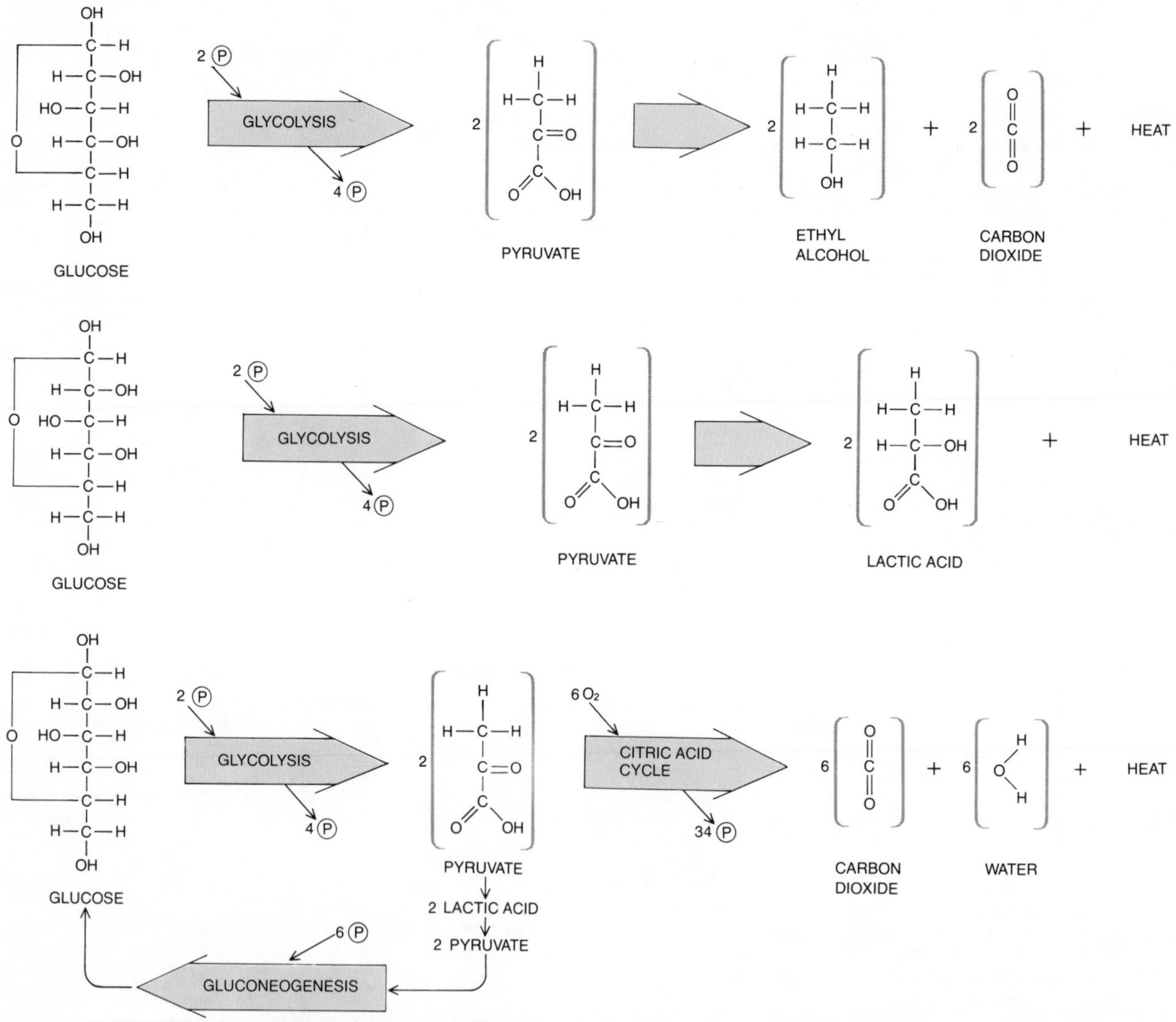

METABOLIC PATHWAYS by which cells extract energy from foodstuff apparently evolved in response to an increase in free oxygen. In all organisms the only usable energy derived from the breakdown of carbohydrates such as glucose is the fraction stored in high-energy phosphate bonds, denoted ~P; the rest is lost as heat. In anaerobic organisms (those that live without oxygen) glucose is broken down through fermentation: each molecule of glucose is split into two molecules of pyruvate, the process called glycolysis, with a net gain of two phosphate bonds. In bacterial fermentation the pyruvate is converted, in a step that provides no usable energy, into products such as lactic acid or ethyl alcohol and carbon dioxide, which are excreted as wastes. The metabolic system of aerobic organisms (those that require oxygen) is respiration. It begins with glycolysis, but the pyruvate is treated not as a waste but as a substrate for a further series of reactions that make up the citric acid cycle. In these reactions pyruvate is decomposed one carbon atom at a time and combined with oxygen, the ultimate products being carbon dioxide and water. Respiration releases far more energy than fermentation, and the proportion of the energy recovered in useful form is also greater; as a result 36 phosphate bonds are formed instead of two. Respiratory metab-

data on the comparative biochemistry of modern microorganisms, data that also provide clues to the evolution of life in the Precambrian.

One further difference between prokaryotes and eukaryotes is of particular importance in the study of their evolution: the extent to which the two types of organisms tolerate oxygen. Among the prokaryotes oxygen requirements are quite variable. Some bacteria cannot grow or reproduce in the presence of oxygen; they are classified as obligate anaerobes. Others can tolerate oxygen but can also survive in its absence; they are facultative anaerobes. There are also prokaryotes that grow best in the presence of oxygen but only at low concentrations, far below that of the present atmosphere. Finally, there are fully aerobic prokaryotes, forms that cannot survive without oxygen.

In contrast to this variety of adaptations the eukaryotes present a pattern of great consistency: with very few exceptions they have an absolute requirement for oxygen, and even the exceptions seem to be evolutionary derivatives of oxygen-dependent organisms. This observation leads to a simple hypothesis: the prokaryotes evolved during a period when environmental oxygen concentrations were changing, but by the time the eukaryotes arose the oxygen content was stable and relatively high.

One indication that eukaryotic cells have always been aerobic is provided by mitotic cell division, a process that can be considered a definitive characteristic of the group. Many eukaryotic cells can survive temporary deprivation of oxygen and can even carry on some metabolic functions; it appears that no cell, however, can undergo mitosis unless oxygen is available at least in low concentration.

The pathways of metabolism itself—the biochemical mechanisms by which an organism extracts energy from foodstuff—provide more detailed evidence. In eukaryotes the central metabolic process is respiration, which in overall terms can be described as the burning of the sugar glucose with oxygen to yield carbon dioxide, water and energy. Some prokaryotes (the aerobic or facultative ones) are also capable of respiration, but many derive their energy solely from the simpler process of fermentation. In bacterial fermentation glucose is not combined with oxygen (or with any other substance from outside the cell) but is simply broken down into smaller molecules. In both respiration and fermentation part of the energy released through the decomposition of glucose is captured in the form of high-energy phosphate bonds, usually in molecules of adenosine triphosphate (ATP). The rest of the energy is lost from the cell as heat.

Respiratory metabolism has two main components: a short series of chemical reactions, collectively called glycolysis, and a longer series called the citric acid cycle. In glycolysis a glucose molecule, with six carbon atoms, is broken down into two molecules of pyruvate, each having three carbon atoms. No oxygen is required for glycolysis, but on the other hand it releases only a little energy with a net gain of only two molecules of ATP.

The fuel for the citric acid cycle is the pyruvate formed by glycolysis. Through a series of enzyme-controlled reactions the carbon atoms of the pyruvate are oxidized and the oxidations are coupled to other reactions that result in the synthesis of ATP. For each two molecules of pyruvate (and hence for each molecule of glucose entering the sequence) 34 additional molecules of ATP are formed. The complete respiratory pathway is thus far more effective than glycolysis alone. In respiration the proportion of energy released that can be recovered in useful form (as ATP) is higher than it is in fermentation, about 38 percent instead of only some 30 percent, and in respiration the net energy yield to the cell is some 18 times greater. By breaking down the glucose to simple inorganic molecules (carbon dioxide and water) respiration liberates virtually all the biologically usable energy stored in the chemical bonds of the sugar.

The metabolism of the prokaryotes immediately suggests an evolutionary relationship between them and the eukaryotes: up to a point fermentation is indistinguishable from glycolysis. In bacterial fermentation a molecule of glucose is split into two molecules of pyruvate, with a net yield of two molecules of ATP. As in glycolysis, no oxygen is required for the process. In anaerobic prokaryotes, however, the metabolic pathway essentially ends at pyruvate. The only further reactions transform the pyruvate into such compounds as lactic acid, ethyl alcohol or carbon dioxide, which are excreted by the cell as wastes.

The similarity of fermentation in prokaryotes to glycolysis in eukaryotes seems too close to be a coincidence, and the assumption of an evolutionary relationship between the two groups provides a ready explanation. It seems likely that anaerobic fermentation became established as an energy-yielding process early in the history of life. When atmospheric oxygen became available for metabolism, it offered the potential for extracting 18 times as much useful energy from carbohydrate: a net yield of 36 molecules of ATP instead of only two molecules. The oxygen-dependent reactions did not, however, simply replace the anaerobic ones; they were appended to the existing anaerobic pathway.

Further evidence for this proposed evolutionary sequence can be found in the behavior of some eukaryotic cells under conditions of oxygen deprivation. In mammalian muscle cells, for example, prolonged exertion can demand more oxygen than the lungs and the blood can supply. The citric acid cycle is then disabled, but the cells continue to function, albeit at reduced efficiency, through glycolysis alone. Under such conditions of oxygen debt pyruvate is not consumed in the cell, but in the liver it can be converted back into glucose (at a cost in energy of six ATP molecules). Significantly the pyruvate itself is not transported to the liver but instead is

GAIN IN Ⓟ	TOTAL ENERGY RELEASED Ⓟ PLUS HEAT (KILOCALORIES)	AVAILABLE ENERGY Ⓟ (KILOCALORIES)	CALCULATED EFFICIENCY (PERCENT)
2	57	14.6	26
2	47	14.6	31
36	686	262.8	38

olism could have evolved, however, only when free oxygen became readily available; it appears to have developed simply by appending the citric acid cycle to the glycolytic pathway. When aerobic cells are deprived of oxygen, many revert to fermentative metabolism, converting pyruvate into lactic acid. In vertebrates the lactic acid from muscle cells is exported to the liver, where it is returned to the form of pyruvate and converted into glucose.

converted into lactic acid, which in the liver must then be returned to the form of pyruvate. This use of lactic acid may represent a vestige of an earlier, bacterial pathway that under aerobic conditions has been suppressed. Indeed, the oxygen-starved muscle cell seems to revert to a more primitive, entirely anaerobic form of metabolism.

The development of an oxygen-dependent biochemistry can also be traced through a consideration of reaction sequences in the synthesis of various biological molecules. Once again stages in the synthetic pathway that emerged early in the Precambrian can be expected to proceed in the absence of oxygen. Reaction steps nearer the end product of the pathway, which were presumably added at a later age, might with

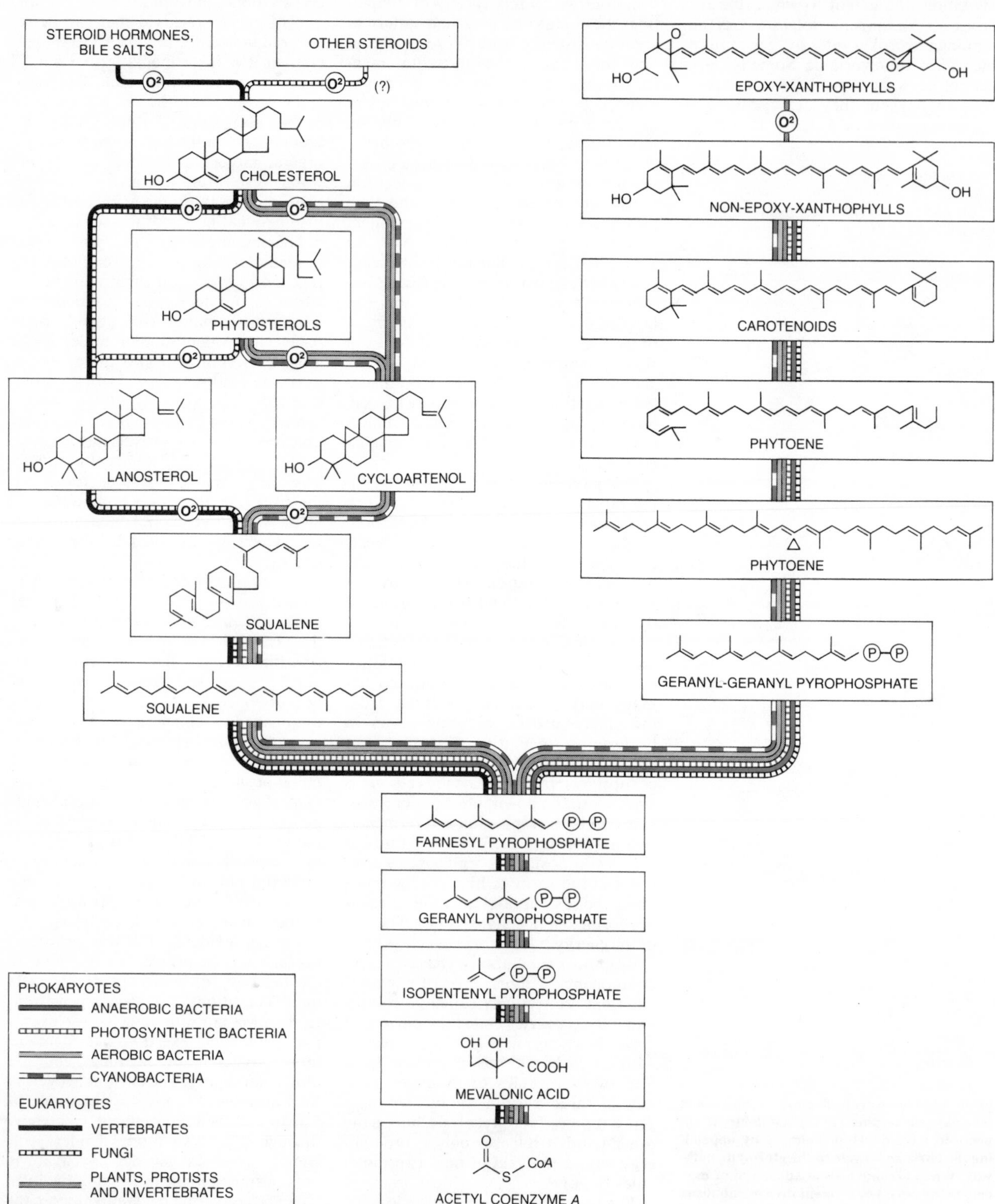

SYNTHESIS OF STEROLS and of related compounds such as the carotenoid pigments of plants requires molecular oxygen (O_2) only for steps near the end of the reaction sequence. Oxygen-dependent steps can be carried out only by aerobic organisms that evolved comparatively late in history of Precambrian life. Organic molecules are in schematic form with most carbon and hydrogen atoms omitted.

increasing frequency require oxygen. The distribution of the oxygen-demanding steps among various kinds of organisms could also have evolutionary significance. If only one pathway has evolved for the synthesis of a class of biochemical substances, then primitive forms of life might be expected to exhibit only the initial, anaerobic steps. Organisms that arose later might exhibit progressively longer, oxygen-dependent synthetic sequences.

In aerobic organisms it might seem at first that virtually all biochemical syntheses require oxygen; eukaryotic cells exhibit relatively little synthetic activity under anoxic conditions. For the most part, however, the oxygen requirement of such syntheses is simply for metabolism: the construction of biological molecules demands energy in the form

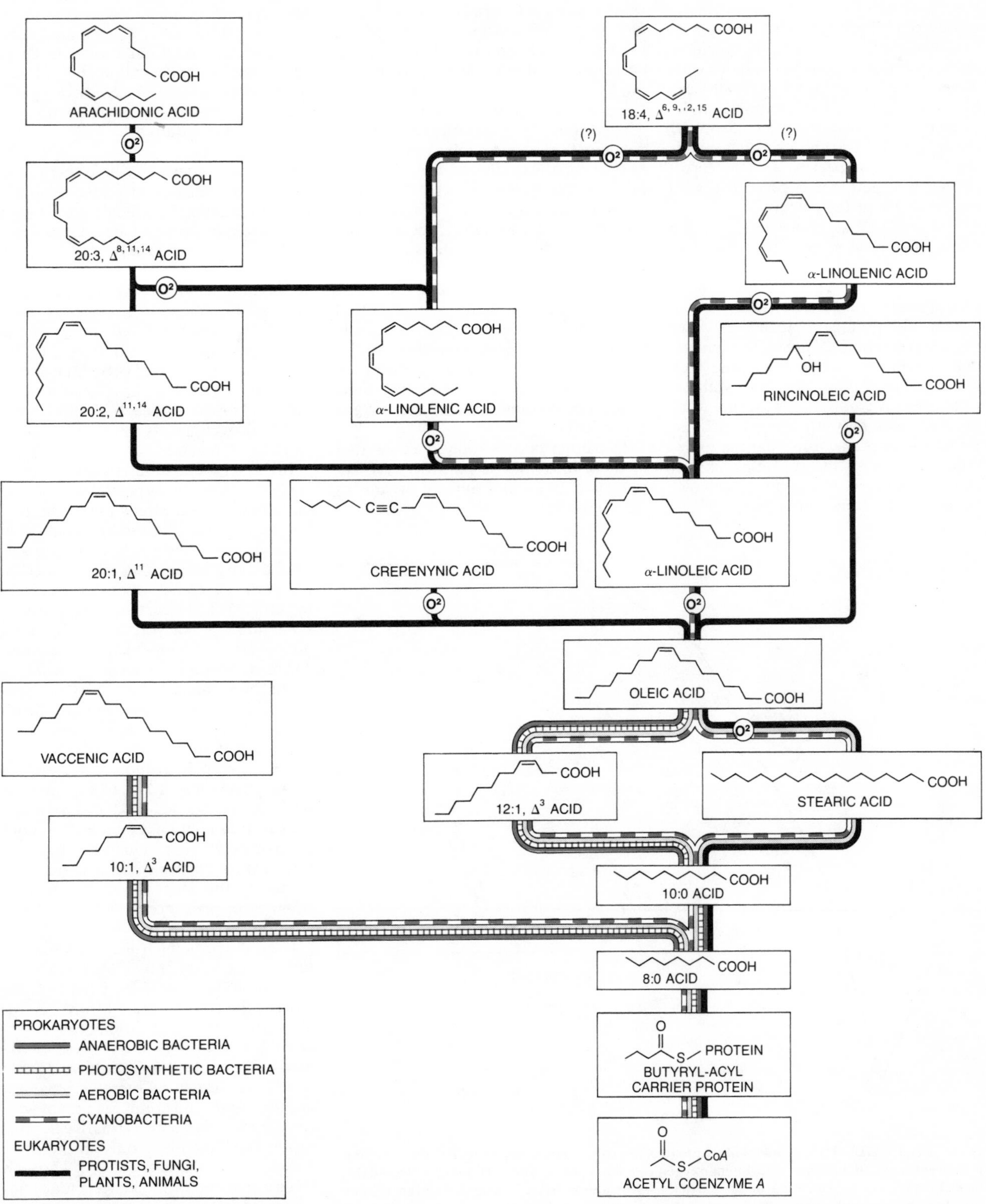

FATTY-ACID SYNTHESIS also follows a pattern suggesting the late addition of oxygen-dependent steps. Most prokaryotes can make mono-unsaturated fatty acids (those having one double bond) by inserting double bond during elongation of molecule. Eukaryotes and some prokaryotes first make a fully saturated molecule, stearic acid, then introduce double bonds by the process of oxidative desaturation.

of ATP, and most of the ATP is supplied through the oxygen-dependent citric acid cycle. If ATP is made available from some other source, many synthetic pathways can proceed unimpaired.

Some syntheses, however, have an intrinsic requirement for oxygen, quite apart from metabolic demands. Molecular oxygen is needed, for example, in the synthesis of bile pigments in vertebrates, of chlorophyll *a* in higher plants and of the amino acids hydroxyproline and, in animals, tyrosine. The oxygen dependence of two synthetic pathways in particular has been determined in detail. One of these pathways controls the manufacture of a class of compounds that includes the sterols and the carotenoids and the other is concerned with the synthesis of fatty acids.

Sterols, such as cholesterol and the steroid hormones, are flat, platelike molecules derived from the compound squalene, which has 30 carbon atoms. Carotenoids are derived from the 40-carbon compound phytoene; they are pigments, such as carotene, the orange-yellow compound in carrots, and they are found in virtually all photosynthetic organisms. A common starting point for the synthesis of both groups of compounds is isoprene, a five-carbon molecule that is also the repeating unit in synthetic rubber. In the biological synthesis two isoprene subunits are joined head to tail; then a third isoprene is added to form a 15-carbon polymer, farnesyl pyrophosphate. At this point there is a fork in the pathway. In one continuation of the synthesis two farnesyl chains are joined to form squalene, the 30-carbon precursor of the sterols. In the other continuation a fourth isoprene subunit is added, and only then are two of the chains joined. The product in this case is phytoene, the 40-carbon precursor of the carotenoids and of other pigments derived from them, such as the xanthophylls.

Up to this step in the synthetic pathway none of the reactions requires the participation of molecular oxygen. The next step in the synthesis of sterols, however, is the conversion of the linear squalene molecule to a 30-carbon ring, and this transformation does require oxygen; so do most of the subsequent steps in sterol synthesis. On the other branch of the pathway there are a few more anaerobic reactions, and indeed carotenoids can be made from phytoene without oxygen. Several further modifications of the carotenoids, however, such as the production of the pigments called epoxy-xanthophylls, are oxygen dependent.

Two observations about the evolution of these biosynthetic pathways are appropriate. Even in groups of organisms that have long been aerobic the first steps in the synthesis are independent of the oxygen supply: molecular oxygen enters the reaction sequence only at later stages. In a similar way the most primitive living organisms, the anaerobic bacteria, are capable only of the first segments of the pathway, the anaerobic segments. The more complex aerobic bacteria and the photosynthetic cyanobacteria have longer synthetic pathways, including some steps that require oxygen. Advanced eukaryotes, such as vertebrate animals and higher plants, have long, branched synthetic pathways, with many steps in which molecular oxygen is required.

A similar pattern can be discerned in the synthesis of fatty acids and their derivatives. The fatty acids are straight carbon-chain compounds that have a carboxyl group (COOH) at one end. A fatty acid is said to be saturated if there are no double bonds between carbon atoms in the chain; it is saturated with hydrogen, which fills all the available bonding positions. An unsaturated fatty acid has a double bond between two carbon atoms or it may have several such double bonds; for each double bond two hydrogen atoms must be removed from the molecule.

In the synthesis of fatty acids the molecule grows by the repeated addition of units two carbon atoms long. The first few steps in the synthesis are identical in all organisms, and they yield fully saturated fatty acids. The first branch in the pathway comes when the developing chain is eight carbons long. At that point many prokaryotes can introduce a double bond, which eukaryotes cannot. There is a second branch at the next step when the saturated chain is 10 carbons long; a double bond can similarly be introduced at that point by many prokaryotes but not by eukaryotes. No matter which branch is followed, elongation of the chain ends at 18 carbons. At that point the fatty acids produced by many prokaryotes contain a double bond, but in eukaryotes the product is always the fully saturated molecule, stearic acid. None of the steps in this sequence, whether in prokaryotes or eukaryotes, requires molecular oxygen.

If no subsequent transformations of fatty acids were possible, eukaryotic cells would be incapable of synthesizing any but the fully saturated forms. Actually extensive modifications can be accomplished through the process of oxidative desaturation, in which double bonds are formed by removing two hydrogen atoms and combining them with oxygen to form water. Oxidative desaturation can take place only in the presence of molecular oxygen (O_2). Through this mechanism cyanobacteria make unsaturated fatty acids with two, three and four double bonds, and eukaryotes form

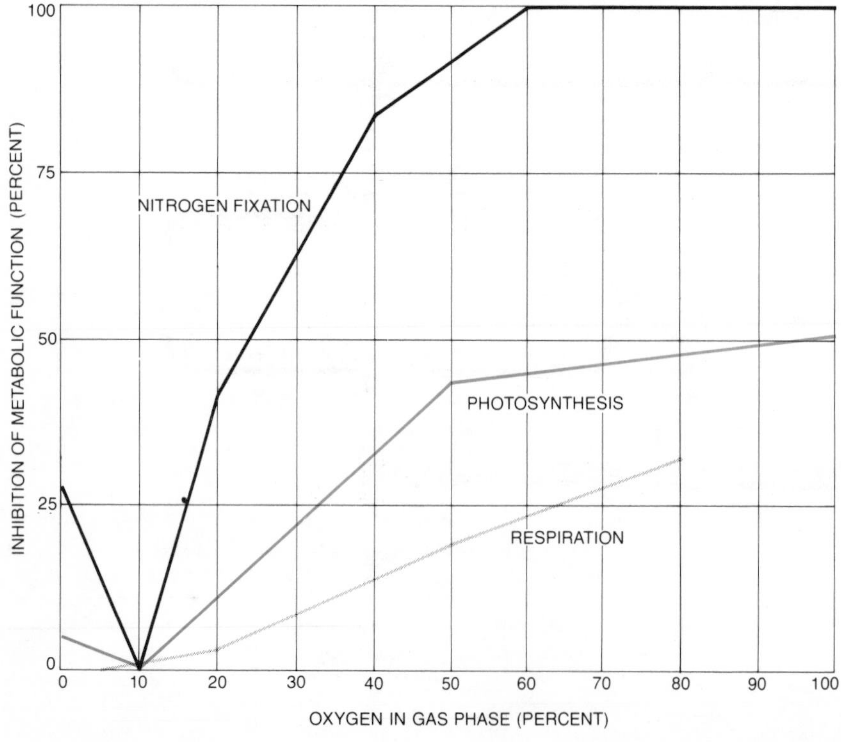

OXYGEN INHIBITION of metabolic functions in cyanobacteria suggests that these aerobic prokaryotes are adapted to an optimum oxygen concentration of about 10 percent, or roughly half the oxygen concentration of the earth's present atmosphere. Nitrogen fixation is completely halted by high oxygen levels, but even respiration, which requires oxygen, can be partially inhibited. Data are for the heterocyst-forming cyanobacterium *Anabaena flos-aquae.*

polyunsaturated fatty acids (with multiple double bonds).

As in the sterol-carotenoid synthesis, an analysis of the fatty-acid pathway argues for a pattern of biochemical evolution in which the increasing availability of atmospheric oxygen played a central role. The first steps in the synthetic sequence are common to all organisms capable of making fatty acids, and in the most primitive organisms those are the only steps. Hence the reactions that come first in the biochemical sequence apparently also developed early in the history of life; these first steps are all anaerobic. Organisms that presumably emerged somewhat later (such as aerobic bacteria and cyanobacteria) have longer pathways, including a few steps of oxidative desaturation. In advanced eukaryotes a substantial proportion of the steps are oxygen dependent.

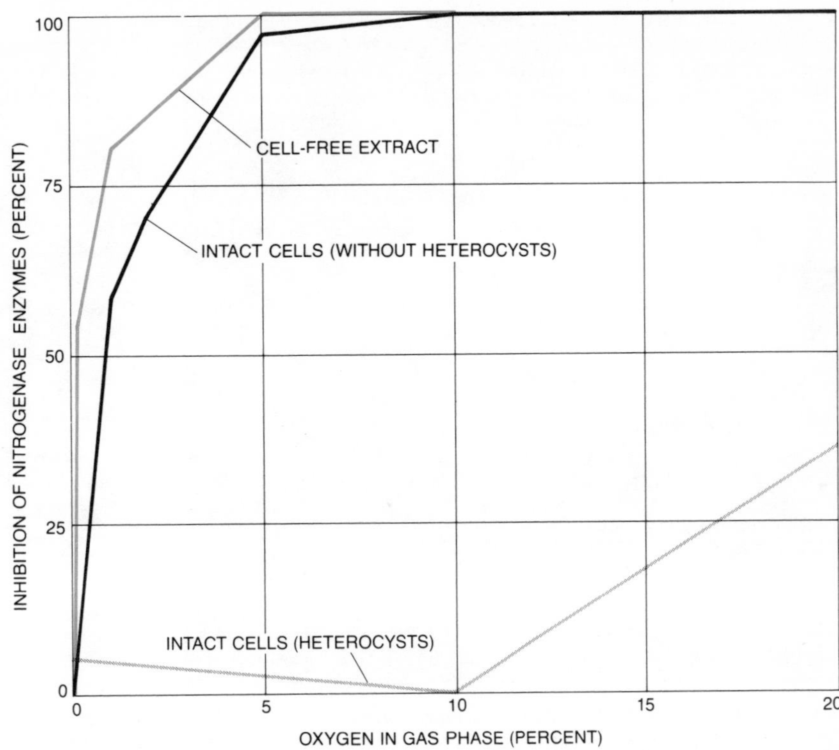

INHIBITION OF NITROGEN FIXATION in the presence of oxygen is caused by the deactivation of the nitrogenase enzymes. In cell-free extracts of the cyanobacterium *Plectonema boryanum* the nitrogenase enzymes are inhibited by even minute quantities of oxygen, and the intact cells of this species, which does not form heterocysts, offer little protection against the inhibition; such organisms can fix nitrogen only in an anoxic habitat. The thick cell walls and other special features of heterocyst cells, such as those formed by *Nostoc muscorum*, allow fixation to continue in a fully aerobic environment. Data suggest that the capability for nitrogen fixation evolved before significant quantities of oxygen had accumulated in the atmosphere.

Comparisons of the metabolism and biochemistry of prokaryotes and eukaryotes thus provide strong evidence that the latter group arose only after a substantial quantity of oxygen had accumulated in the atmosphere. Hence it is of interest to ask when eukaryotic cells first appeared. It seems apparent that an oxygen-rich atmosphere cannot have developed later than this signal evolutionary event.

The primary means of assigning a date to the origin of the eukaryotes is through the fossil record. Because this field of study is so new, however, the available information is scanty and often difficult to interpret. It is rarely a straightforward task to identify a microscopic, single-cell organism as being eukaryotic merely from an examination of its fossilized remains. And even when a fossil has been identified as unequivocally eukaryotic the available radioactive-isotope methods of dating can rarely assign it a precise age. At best such methods have an accuracy of only about plus or minus 5 percent. What is more, the age determinations are generally carried out on rocks that were once molten, such as volcanic lavas, whereas the fossils are found in sedimentary deposits. Consequently the stratum of the fossil itself usually cannot be dated; it is merely assigned an age somewhere between the ages of the nearest underlying and overlying datable rock units.

In spite of these difficulties there is now substantial evidence for the existence of eukaryotic fossils in rocks hundreds of millions of years older than the earliest Phanerozoic strata. The evidence is of two kinds: microfossils that display a morphological or organizational complexity judged to be of eukaryotic character, and the presence of fossil cells whose size is typical only of eukaryotes.

The evidence from relatively complex microscopic fossils includes the following: (1) branched filaments, made up of cells with distinct cross walls and resembling modern fungi or green algae, from the Olkhin formation of Siberia, a deposit thought to be about 725 million years old (but with a known age of between 680 and 800 million years); (2) complex, flask-shaped microfossils from the Kwagunt formation in the eastern Grand Canyon, thought to be about 800 (or 650 to 1,150) million years old; (3) fossils of unicellular algae containing intracellular membranes and small, dense bodies that may represent preserved organelles, from the Bitter Springs formation of central Australia, dated at approximately 850 (or 740 to 950) million years; (4) a group of four sporelike cells in a tetrahedral configuration that may have been produced by mitosis or possibly meiosis, also from Bitter Springs rocks; (5) spiny cells or algal cysts several hundred micrometers in diameter and with unquestionable affinities to eukaryotic organisms, from Siberian shales that are reportedly 950 (or 750 to 1,050) million years old; (6) highly branched filaments of large diameter and with rare cross walls, similar in some respects to certain green or golden-green eukaryotic algae, from the Beck Spring dolomite of southeastern California (1,300, or 1,200 to 1,400 million years old) and from the Skillogalee dolomite of South Australia (850, or 740 to 867 million years old); (7) spheroidal microfossils described as exhibiting two-layered walls and having "medial splits" on their surface, and which may represent an encystment stage of a eukaryotic alga, from shales 1,400 (or 1,280 to 1,450) million years old in the McMinn formation of northern Australia; (8) a tetrahedral group of four small cells, resembling spores produced by mitotic cell division of some green algae, from the Amelia dolomite of northern Australia, approaching 1,500 (or 1,390 to 1,575) million years in age; (9) unicellular fossils that appear to be exceptionally well preserved and that are reported to contain small membrane-bounded structures that could be remnants of organelles, from the Bungle-Bungle dolomite in the same region as the Amelia dolomite and of approximately the same age.

Thus the earliest of these eukaryote-like fossils are probably somewhat less than 1,500 million years old. Numerous types of microfossils have been discovered in older sediments, but none of them seems to be a strong candidate for

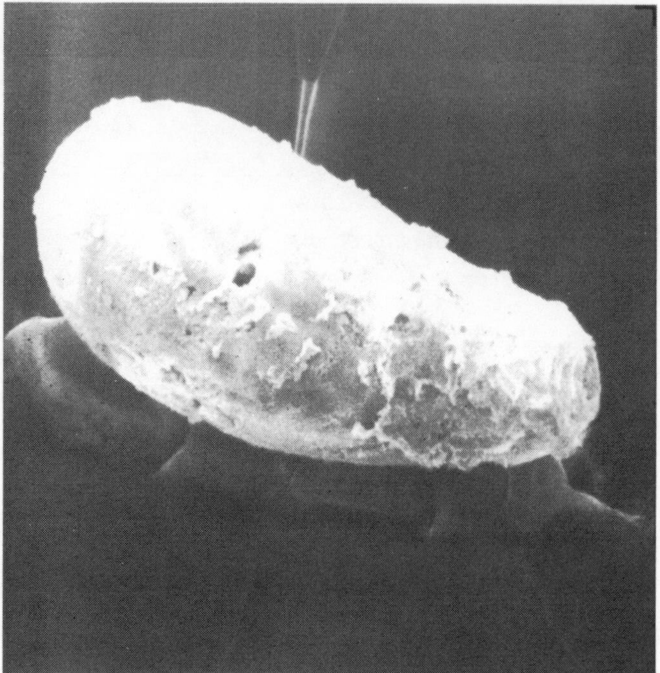

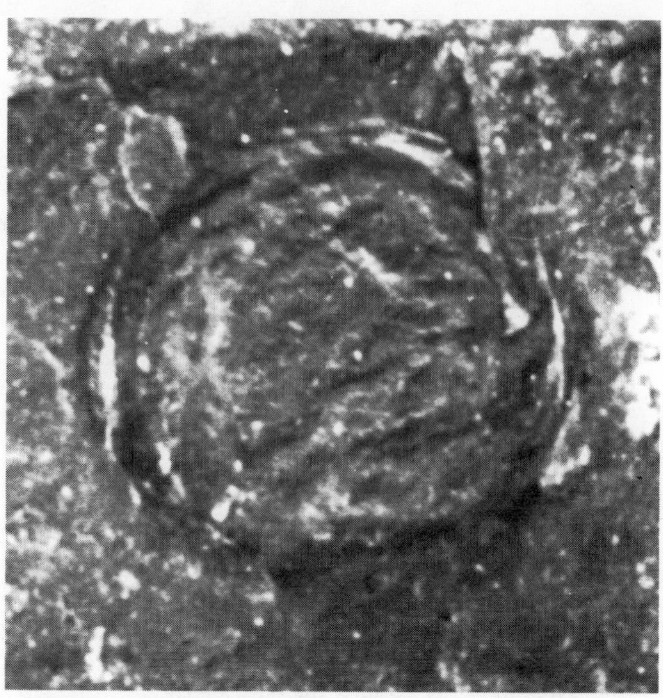

EARLY EUKARYOTIC CELLS may be represented among Precambrian microfossils. The gourd-shaped cell at the left, from shales in the Grand Canyon thought to be 800 million years old, is morphologically more complex than any known prokaryote; it is also larger, about 100 micrometers long. The second cell from the left is some two millimeters in diameter and hence is more than 30 times the size of

identification as eukaryotic. For example, the well-studied Canadian fossils of the Gunflint and Belcher Island iron formations, which are about two billion years old, have been interpreted as exclusively prokaryotic.

The testimony of these as yet rare and unusual specimens can be checked through statistical studies of the sizes of known Precambrian microfossils. The size ranges of prokaryotes and eukaryotes overlap, so that a particular fossil cannot always be classified unambiguously on the basis of size alone; by cataloguing the measured sizes in a large sample of fossils, however, it may be possible to determine whether or not eukaryotic cells are present. Among modern species of spheroidal cyanobacteria about 60 percent are very small, less than five micrometers in diameter; of the remaining species only a few are larger than 20 micrometers and none is larger than 60 micrometers. Unicellular eukaryotes, such as green or red algae, can be much larger. Typically they fall in the range between five and 60 micrometers, but several percent of living species are larger than 60 micrometers and a few are larger than 1,000 micrometers (one millimeter).

Systematic size measurements have been made on some 8,000 fossil cells from 18 widely dispersed Precambrian deposits. On the basis of those data certain tentative conclusions can be drawn. Cells larger than 100 micrometers, and hence of distinctly eukaryotic dimensions, are unknown in rocks older than about 1,450 million years. Virtually all the unicellular fossils from rocks of that age, whether they grew in shallow-water stromatolites or were deposited in offshore shales, are of prokaryotic size.

Cells larger than modern prokaryotes (greater than 60 micrometers in diameter) first become abundant in rocks about 1,400 million years old. Algae of this type were apparently free-floating rather than mat-forming species, and they are therefore particularly common in shales, sediments deposited in deeper water. Such eukaryote-size fossils have been known for several years from shales of this age in China and in the U.S.S.R. Recently cells more than 100 micrometers in diameter have also been discovered in the Newland limestone of Montana, and cells more than 600 micrometers in size (10 times the size of the largest spheroidal prokaryote) have been found in the McMinn formation of Australia; the age of both of these fossil-bearing deposits is about 1,400 million years.

In somewhat younger Precambrian sediments there are still larger cells, fossils greater than one millimeter in diameter (with some as large as eight millimeters). They were first described in 1899 by Walcott, who discovered them in rocks from the Grand Canyon. They have since been found in nearly a dozen other rock units throughout the world. The oldest seem to be those from Utah and from Siberia, each about 950 million years old, and those from northern India, which could be even older (from 910 to 1,150 million years old).

Studies of both the morphology and the size of unicellular fossils therefore suggest that there is a break in the fossil record between 1,400 and 1,500 million years ago. Below this horizon cells with eukaryotelike traits are rare or absent; above it they become increasingly common. Moreover, the data suggest that the diversification of the eukaryotes began shortly after the cell type first appeared, apparently within the next few hundred million years. By a billion years ago there had been substantial increases in cell size, in morphological complexity and in the diversity of species. All these indicators also suggest, of course, that oxygen-dependent metabolism, which is highly developed even in the most primitive eukaryotes, had already become established by about 1.5 billion years ago.

The prokaryotes that must have held exclusive sway over the earth before the development of eukaryotic cells were less diverse in form, but they were probably more varied in metabolism and biochemistry than their eukaryotic descendants. Like modern prokaryotes, the ancient species presumably varied over a broad range in their tolerance of oxygen, all the way from complete intolerance to absolute need. In this regard one group of prokaryotes, the cyanobacteria, are of particular interest in that they were largely responsible for the development of an oxygen-rich atmosphere.

Like higher plants, cyanobacteria carry out aerobic photosynthesis, a process

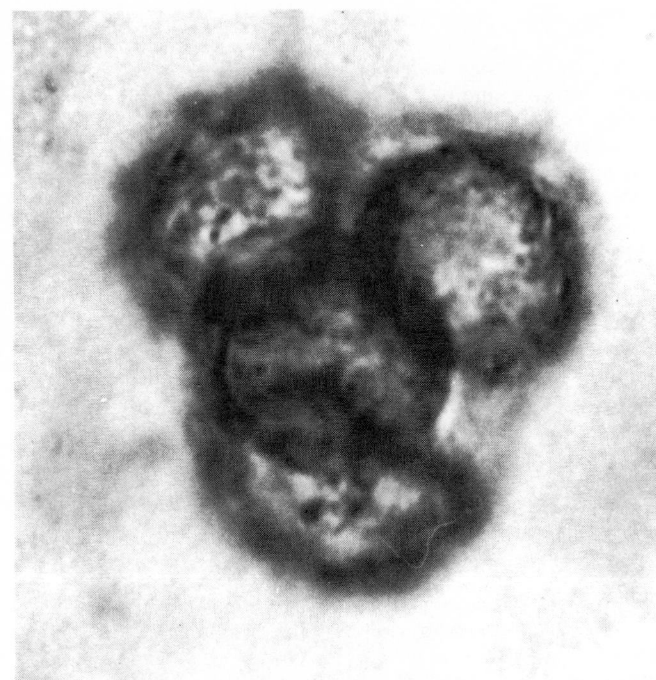

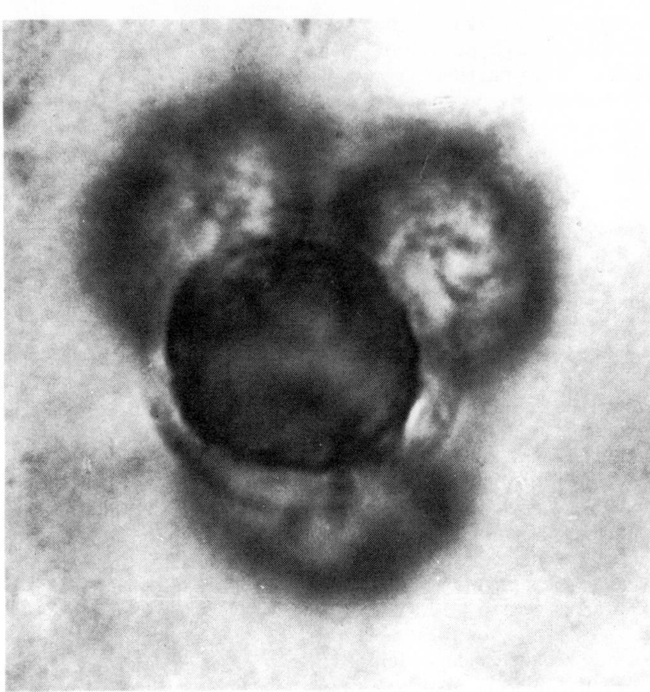

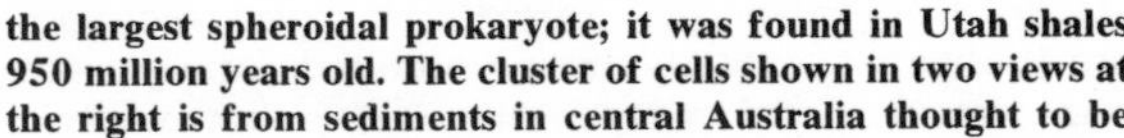

the largest spheroidal prokaryote; it was found in Utah shales 950 million years old. The cluster of cells shown in two views at the right is from sediments in central Australia thought to be 850 million years old. The cells are only 10 micrometers across, but their tetrahedral arrangement suggests they formed as a result of mitosis or possibly meiosis, mechanisms of cell division known only in eukaryotes.

that in overall effect (although not in mechanism) is the reverse of respiration. The energy of sunlight is employed to make carbohydrates from water and carbon dioxide, and molecular oxygen is released as a by-product. The cyanobacteria can tolerate the oxygen they produce and can make use of it both metabolically (in aerobic respiration) and in synthetic pathways that seem to be oxygen dependent (as in the synthesis of chlorophyll *a*). Nevertheless, the biochemistry of the cyanobacteria differs from that of green, eukaryotic plants and suggests that the group originated during a time of fluctuating oxygen concentration. For example, although many cyanobacteria can make unsaturated fatty acids by oxidative desaturation, some of them can also employ the anaerobic mechanism of adding a double bond during the elongation of the chain. In a similar manner oxygen-dependent syntheses of certain sterols can be carried out by some cyanobacteria, but the amounts of the sterols made in this way are minuscule compared with the amounts typical of eukaryotes. In other cyanobacteria those sterols are not found at all, the biosynthetic pathway being terminated after the last anaerobic step: the formation of squalene. Hence in their biochemistry the cyanobacteria seem to occupy a middle ground between the anaerobes and the eukaryotes.

In metabolism too the cyanobacteria occupy an intermediate position. They flourish today in fully oxygenated environments, but physiological experiments indicate that for many species optimum growth is obtained at an oxygen concentration of about 10 percent, which is only half that of the present atmosphere. Both photosynthesis and respiration are increasingly inhibited when the oxygen concentration exceeds that optimum level. It has recently been discovered that some cyanobacteria can switch the cellular machinery of aerobic metabolism on and off according to the availability of oxygen. Under anoxic conditions these species not only halt respiration but also adopt an anaerobic mode of photosynthesis, employing hydrogen sulfide (H_2S) instead of water and releasing sulfur instead of oxygen. This capability for anaerobic metabolism is probably a relic of an earlier stage in the evolutionary development of the group.

Another activity of some cyanobacteria that seems to reflect an earlier adaptation to anoxic conditions is nitrogen fixation. Nitrogen is an essential element of life, but it is biologically useful only in "fixed" form, for example combined with hydrogen in ammonia (NH_3). Only prokaryotes are capable of fixing nitrogen (although they often do so in symbiotic relationships with higher plants). The crucial complex of enzymes for fixation, the nitrogenases, is highly sensitive to oxygen. In cell-free extracts nitrogenases are partially inhibited by as little as .1 percent of free oxygen, and they are irreversibly inactivated in minutes by exposure to oxygen concentrations of only about 5 percent.

Such a complex of enzymes could have originated only under anoxic conditions, and it can operate today only if it is protected from exposure to the atmosphere. Many nitrogen-fixing bacteria provide that protection simply by adopting an anaerobic habitat, but among the cyanobacteria a different strategy has developed: the nitrogenase enzymes are protected in specialized cells, called heterocysts, whose internal milieu is anoxic. The heterocysts lack certain pigments essential for photosynthesis, and so they generate no oxygen of their own. They have thick cell walls and are surrounded by a mucilaginous envelope that retards the diffusion of oxygen into the cell. Finally, they are equipped with respiratory enzymes that quickly consume any uncombined oxygen that may leak in.

Because of the thick cell walls heterocysts should be comparatively easy to recognize in fossil material. Indeed, possible heterocysts have been reported from several Precambrian rock units, the oldest being about 2.2 billion years in age. If these cells are indeed heterocysts, they may be taken as a sign that free oxygen was present by then, at least in small concentrations.

Nitrogen fixation has a high cost in energy, and the capability for it would therefore seem to confer a selective advantage only when fixed nitrogen is a scarce resource. Today the main sources of fixed nitrogen are biological and industrial, but biologically usable nitrate (NO_3^-) is formed by the reaction of atmospheric nitrogen and oxygen. In the anoxic atmosphere of the early Precambrian the latter mechanism would obviously have been impossible. The lack of atmospheric oxygen would also have indirectly reduced the concentration of ammonia to very low levels. Ammonia is dissociated into nitrogen and hydrogen by ultraviolet radiation, most of which is filtered out today by a layer of ozone (O_3) high in the atmosphere; without free oxygen there would have been little ozone, and without this protective shield atmospheric ammonia would have been quickly destroyed.

It is likely that the capability for nitrogen fixation developed early in the Precambrian among primitive prokaryotic organisms and in an environment where fixed nitrogen was in short supply. The vulnerability of the nitrogenase enzymes to oxidation was of no consequence then, since the atmosphere had little oxygen. Later, as the photosynthetic activities of the cyanobacteria led to an increase in atmospheric oxygen, some nitrogen fixers adopted an anaerobic habitat and others developed heterocysts. By the time eukaryotes appeared, apparently more than half a billion years later, oxygen was abundant and fixed nitrogen (both NH_3 and NO_3^-) was probably less scarce, and so the eukaryotes never developed the enzymes needed for nitrogen fixation.

At present oxygen-releasing photosynthesis by green plants, cyanobacteria and some protists is responsible for the synthesis of most of the world's organic matter. It is not, however, the only mechanism of photosynthesis. The alternative systems are confined to a few groups of bacteria that on a global scale seem to be of minor importance today but that may have been far more significant in the geological past.

The several groups of photosynthetic bacteria differ from one another in their pigmentation, but they are alike in one important respect: unlike the photosynthesis of cyanobacteria and eukaryotes, all bacterial photosynthesis is a totally anaerobic process. Oxygen is not given off as a by-product of the reaction, and the photosynthesis cannot proceed in the presence of oxygen. Whereas oxygen appears to be required in green plants for the synthesis of chlorophyll *a*, oxygen inhibits the synthesis of bacteriochlorophylls.

The anaerobic nature of bacterial photosynthesis seems to present a paradox: photosynthetic organisms thrive where light is abundant, but such environments are also generally ones having high concentrations of oxygen, which poisons bacterial photosynthesis. These contradictory needs can be explained if it is assumed that anaerobic photosynthesis evolved among primitive bacteria early in the Precambrian, when the atmosphere was essentially anoxic. The photosynthesizers could thus have lived in matlike communities in shallow water and in full sunlight.

Somewhat later such bacteria gave rise to the first organisms capable of aerobic photosynthesis, the precursors of modern cyanobacteria. For the anaerobic photosynthetic bacteria the molecular oxygen released by this mutant strain was a toxin, and as a result the aerobic photosynthesizers were able to supplant the anaerobic ones in the upper portions of the mat communities. The anaerobic species became adapted to the lower parts of the mat, where there is less light but also a lower concentration of oxygen. Many photosynthetic bacteria occupy such habitats today.

Photosynthetic bacteria were surely not the first living organisms, but the history of life in the period that preceded their appearance is still obscure. What little information can be inferred about that early period, however, is consistent with the idea that the environment was then largely anoxic. One tentative line of evidence rests on the assumption that among organisms living today those that are simplest in structure and in biochemistry are probably the most closely related to the earliest forms of life. Those simplest organisms are bacteria of the clostridial and methanogenic types, and they are all obligate anaerobes.

There is even a basis for arguing that anoxic conditions must have prevailed during the time when life first emerged on the earth. The argument is based on the many laboratory experiments that have demonstrated the synthesis of organic compounds under conditions simulating those of the primitive planet. These syntheses are inhibited by even small concentrations of molecular oxygen. Hence it appears that life probably would not have developed at all if the early atmosphere had been oxygen-rich. It is also significant that the starting materials for such experiments often include hydrogen sulfide and carbon monoxide (CO), and that an intermediate in many of the reactions is hydrogen cyanide (HCN). All three compounds are poisonous gases, and it seems paradoxical that they should be forerunners of the earliest biochemistry. They are poisonous, however, only for aerobic forms of life; indeed, for many anaerobes hydrogen sulfide not only is harmless but also is an important metabolite.

It was argued above that oxygen must have been freely available by the time the first eukaryotic cells appeared, probably 1,400 to 1,500 million years ago. Hence the proliferation of cyanobacteria that released the oxygen must have taken place earlier in the Precambrian. How much earlier remains in question. The best available evidence bearing on the issue comes from the study of sedimentary minerals, some of which may have been influenced by the concentration of free oxygen at the time they were deposited. In recent years a number of workers have investigated this possibility, most notably Preston E. Cloud, Jr., of the University of California at Santa Barbara and the U.S. Geological Survey.

One mineral of significance in this argument is uraninite (UO_2), which is found in several deposits that were laid down in Precambrian streambeds. In the presence of oxygen, grains of uraninite are readily oxidized (to U_3O_8) and are thereby dissolved. David E. Grandstaff of Temple University has shown that streambed deposits of the mineral probably could not have accumulated if the concentration of atmospheric oxygen was greater than about 1 percent. Uraninite-bearing deposits of this type are found in sediments older than about two billion years but not in younger strata, suggesting that the transition in oxygen concentration may have come at about that time.

Another kind of mineral deposit, the iron-rich formations called red beds, exhibits the opposite temporal pattern: red beds are known in sedimentary sequences younger than about two billion years but not in older ones. The red beds are composed of particles coated with iron oxides (mostly the mineral hematite, Fe_2O_3), and many are thought to have formed by exposure to oxygen in the atmosphere rather than under water. It has been proposed that the oxygen may have been biologically generated. This hypothesis is consistent with several lines of evidence, but objections to it have also been raised. For example, most red beds are continental deposits rather than marine ones and are therefore susceptible to erosion; it is thus conceivable that red beds were formed earlier than two billion years ago as well as later but that the earlier beds have been destroyed. It is also possible that the oxygen in the red beds had a nonbiological origin; it may have come from the splitting of water by ultraviolet radiation. This has apparently happened on Mars to create a vast red bed across the surface of that planet, where there are only traces of free oxygen and there is no evidence of life.

Perhaps the most intriguing mineral evidence for the date of the oxygen transition comes from another kind of iron-rich deposit: the banded iron formation. These deposits include some tens of billions of tons of iron in the form of oxides embedded in a silica-rich matrix; they are the world's chief economic reserves of iron. A major fraction of them was deposited within a comparatively brief period of a few hundred million years beginning somewhat earlier than two billion years ago.

A transition in oxygen concentration could explain this major episode of iron sedimentation through the following hypothetical sequence of events. In a primitive, anoxic ocean, iron existed in the ferrous state (that is, with a valence of +2) and in that form was soluble in seawater. With the development of aerobic photosynthesis small concentrations of oxygen began diffusing into the upper portions of the ocean, where it reacted with the dissolved iron. The iron was thereby converted to the ferric form (with a valence of +3), and as a result hydrous ferric oxides were precipitated and accumulated with silica to form rusty layers on the ocean floor. As the process continued virtually all the dissolved iron in the ocean basins was precipitated: in a matter of a few hundred million years the world's oceans rusted.

As in the deposition of red beds, an inorganic origin could also be proposed for the oxygen in the banded iron formations; the oxygen in some formations laid down in the very early Precambrian might well have come from such a source. For the extensive iron formations of about two billion years ago, however, inorganic processes such as the photochemical splitting of water do not appear to be adequate; they probably could not have produced the necessary quantity of oxygen quickly enough to account for the enormous volume of iron ores deposited at about that time. Indeed, only one mechanism is known that could release oxygen at the required rate: aerobic photosynthesis, followed by the sedimentation and burial of the organic matter thereby produced. (The burial is a necessary condition, since aerobic decomposition of the organic remains would use up as much oxygen as had been generated.)

In relation to this hypothesis it is notable that fossil stromatolites first become abundant in sediments deposited about 2,300 million years ago, shortly before the major episode of iron-ore deposition. It is therefore possible that the first widespread appearance of stromatolites might mark the origin and the earliest diversification of oxygen-producing cyanobacteria. Even at that early date the cyanobacteria would probably have released oxygen at a high rate, but for several hundred million years the iron dissolved in the oceans would have served as a buffer for the oxygen concentration of the atmosphere, reacting with the gas and precipitating it as ferric oxides almost as quickly as it was generated. Only when the oceans had been swept free of unoxidized iron and similar materials would the concentration of oxygen in the atmosphere have begun to rise toward modern levels.

Although much remains uncertain, evidence from the fossil record, from modern biochemistry and from geology and mineralogy make possible a tentative outline for the history of Precam-

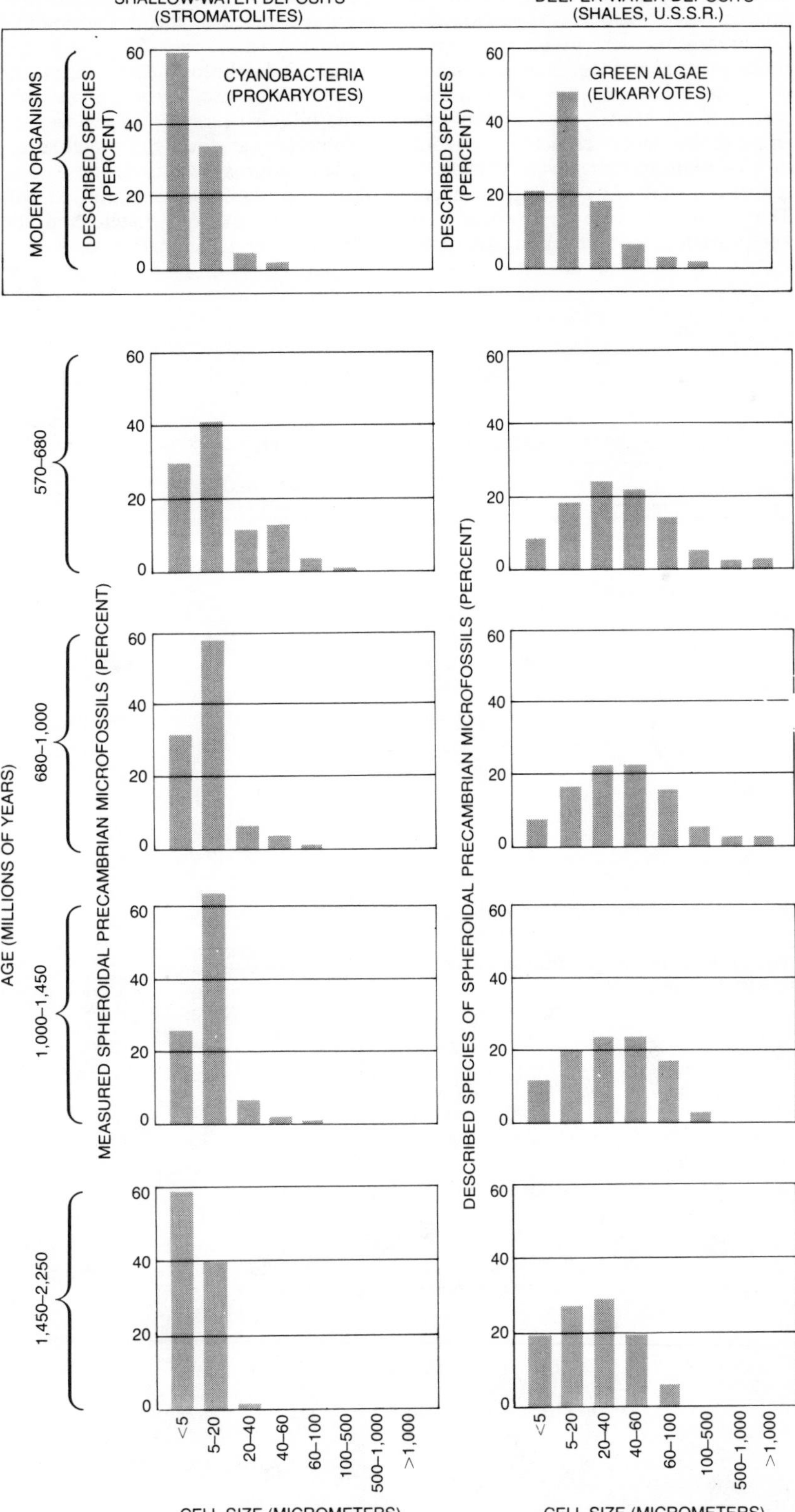

SIZE OF FOSSIL CELLS provides evidence on the origin of the eukaryotes. Spheroidal microfossils of various ages were measured and assigned to eight size categories; a similar procedure was followed with spheroidal members of two groups of modern microorganisms, the prokaryotic cyanobacteria and the eukaryotic green algae. The range of sizes for the modern species overlaps, but the largest cells are observed only among the eukaryotes. The oldest fossils examined have a distribution of sizes similar to that of prokaryotes, but Precambrian rock units younger than 1,450 million years include larger cells that are probably eukaryotic, and the proportion of larger cells increases in later periods. The larger cells tend to be more abundant in shales than in stromatolitic sediments. Because shales are deposited offshore that fact would be explained if early eukaryotes were predominantly free-floating rather than mat-forming.

brian life. The most primitive forms of life with recognizable affinities to modern organisms were presumably spheroidal prokaryotes, perhaps comparable to modern bacteria of the clostridial type. Initially at least they probably derived their energy from the fermentation of materials that were organic in nature but were of nonbiological origin. These materials were synthesized in the anoxic early atmosphere and were of the type that during the age of chemical evolution had led to the development of the first cells.

The first photosynthetic organisms apparently arose earlier than about three billion years ago. They were anaerobic prokaryotes, the precursors of modern photosynthetic bacteria. Most of them probably lived in matlike communities in shallow water, and they may have been responsible for building the earliest fossil stromatolites known, which are estimated to be about three billion years old.

The rise of aerobic photosynthesis in the mid-Precambrian introduced a change in the global environment that was to influence all subsequent evolution. The resulting increase in oxygen concentration probably led to the extinction of many anaerobic organisms, and others were forced to adopt mar-

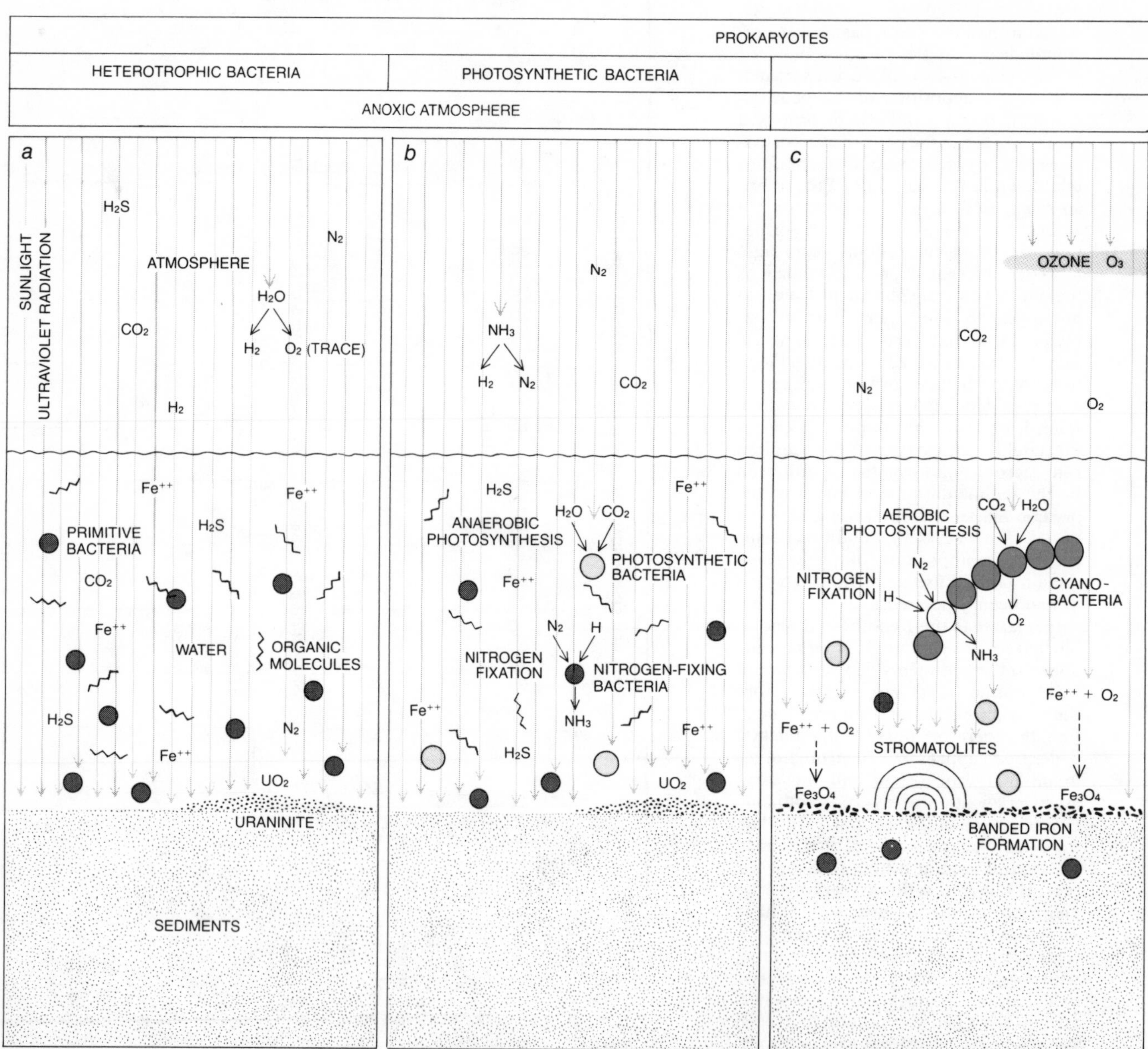

ORGANISM AND ENVIRONMENT evolved in counterpoint during the Precambrian. The first living cells (*a*) were presumably small, spheroidal anaerobes. Only traces of oxygen were present. They survived by fermenting organic molecules formed nonbiologically in the anoxic environment. The role of such ready-made nutrients was diminished, however, when the first photosynthetic organisms evolved (*b*). This earliest mode of photosynthesis was entirely anaerobic. Another early development was nitrogen fixation, required in part because ultraviolet radiation that could then freely penetrate the atmosphere would have quickly destroyed any ammonia (NH_3) present. A little more than two billion years ago (*c*) aerobic photosynthesis began in the precursors of modern cyanobacteria. Oxygen was generated by these stromatolite-building microorganisms, but for some 100 million years little of it accumulated in the atmosphere; instead it reacted with iron dissolved in the oceans, which was then precipitated to create massive banded iron formations. Only when the oceans had

ginal habitats, such as the lower reaches of bacterial mat communities. Nitrogen-fixing organisms also retreated to anaerobic habitats or developed heterocyst cells. With little competition for those regions having optimum light the cyanobacteria were able to spread rapidly and came to dominate virtually all accessible habitats. With the development of the citric acid cycle and its more efficient extraction of energy from foodstuff, the dominance of the biological community by aerobic organisms was confirmed. When the major episode of deposition of banded iron formations ended some 1,800 million years ago, the trend toward increasing oxygen concentration became irreversible.

By the time eukaryotic cells arose 1,500 to 1,400 million years ago a stable, oxygen-rich atmosphere had long prevailed. Adaptive strategies needed by earlier organisms to cope with fluctuations in the oxygen level were unnecessary for eukaryotes, which were from the start fully aerobic. The diversity of eukaryote cell types present by about a billion years ago suggests that some form of sexual reproduction may have evolved by then. Within the next 400 million years the rapid diversification of eukaryotic organisms had led to the emergence of multicellular forms of

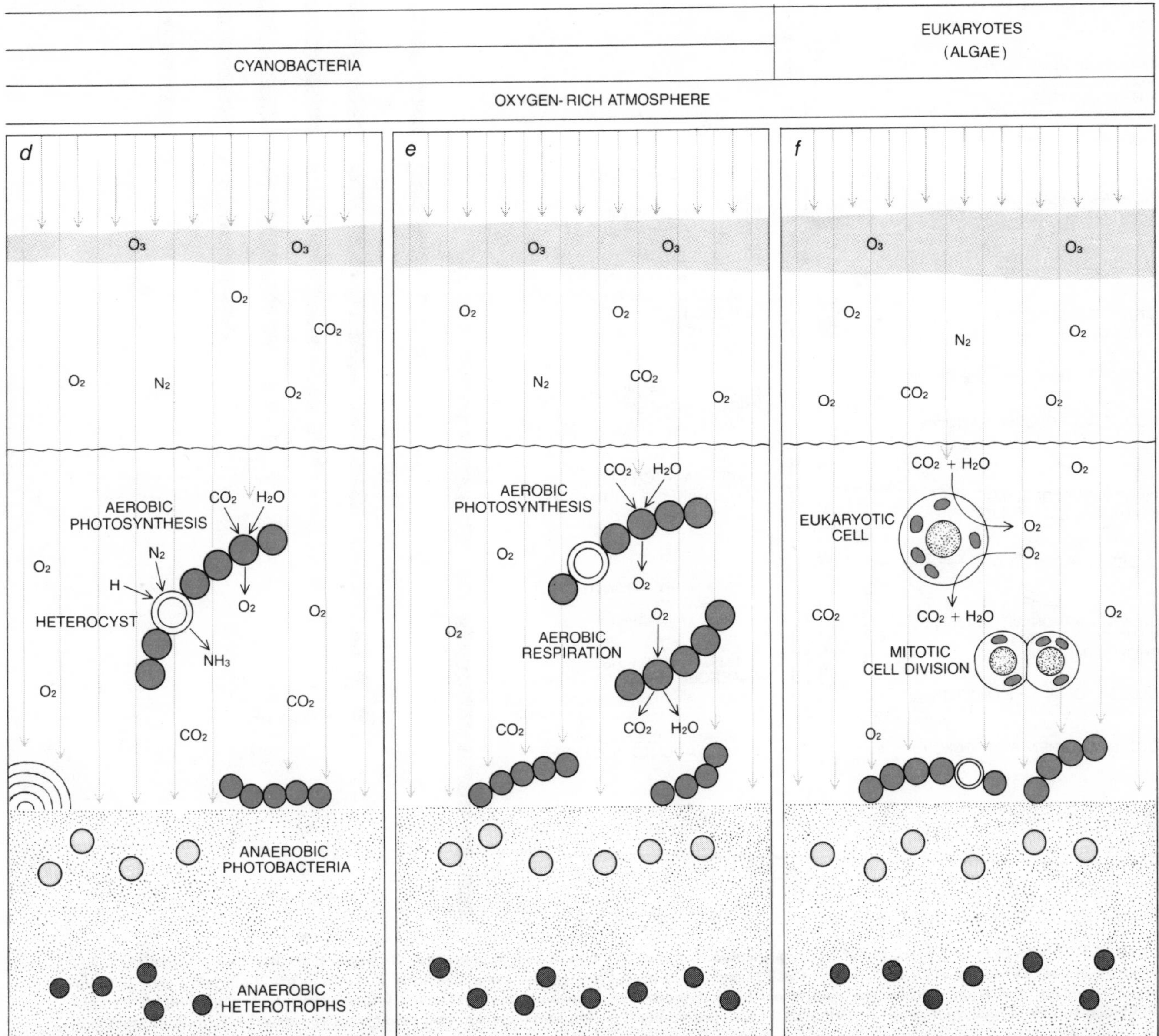

been swept free of iron and similar materials (*d*) did the concentration of free oxygen begin to rise toward modern levels. This biologically induced change in the environment had several effects on biological development. Anaerobic organisms were forced to retreat to anoxic habitats, leaving the best spaces for photosynthesis to the cyanobacteria. In a similar manner nitrogen-fixing organisms had to adopt an anaerobic way of life or develop protective heterocyst cells. Atmospheric oxygen also created a layer of ozone (O_3) that filtered out most ultraviolet radiation. Once the oxygen-rich atmosphere was fully established (*e*) cells evolved that not only could tolerate oxygen but also could employ it in respiration. The result was a great improvement in metabolic efficiency. Finally, about 1,450 million years ago, the first eukaryotic cells emerged (*f*). From the start they were adapted to a fully aerobic environment. The new modes of reproduction possible in eukaryotes, in particular the advanced sexual reproduction that evolved later, led to rapid diversification of the group.

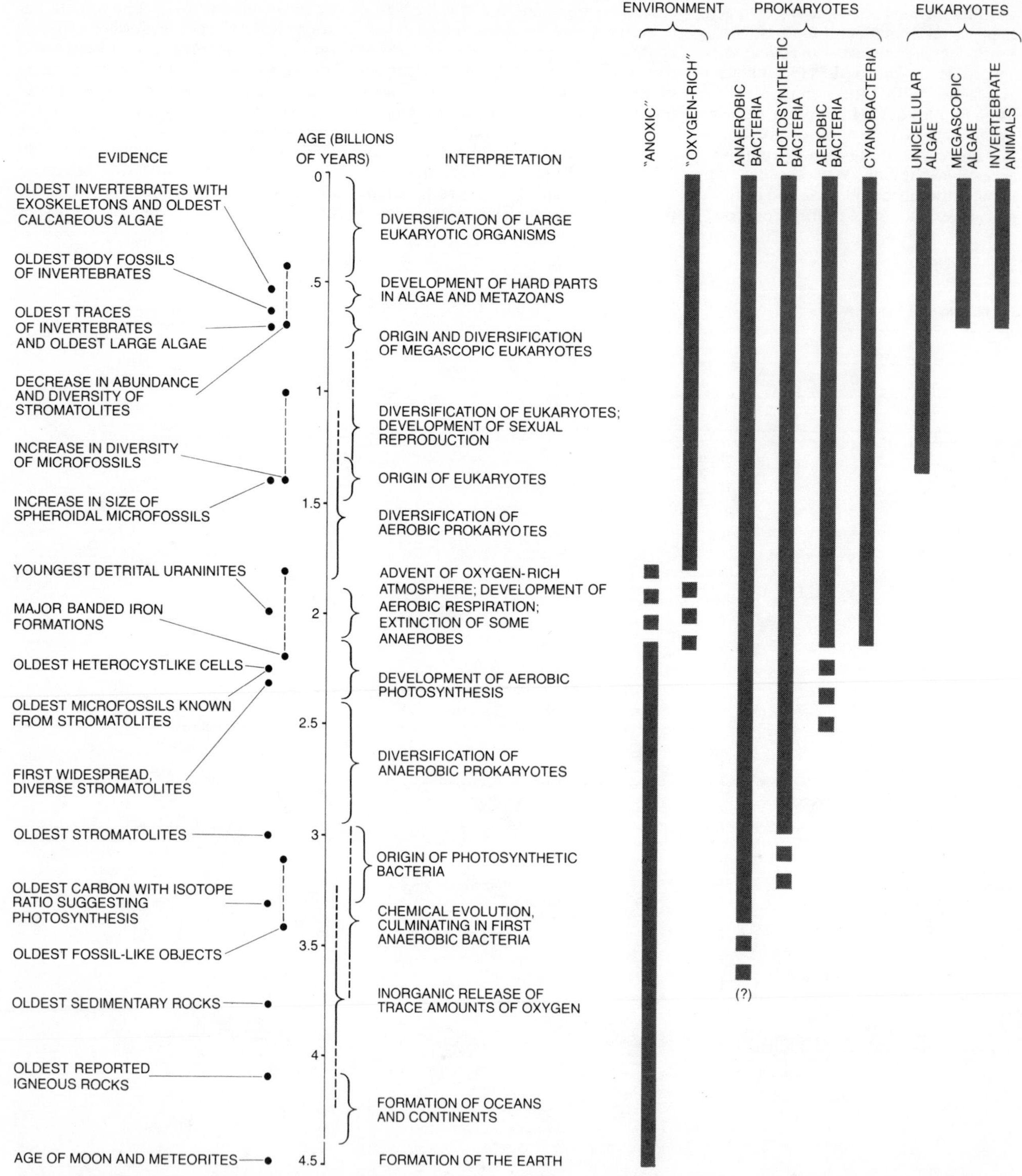

MAJOR EVENTS in Precambrian evolution are presented in chronological sequence based on evidence from the fossil record, from inorganic geology and from comparative studies of the metabolism and biochemistry of modern organisms. Although the conclusions are tentative, it appears that life began more than three billion years ago (when the earth was little more than a billion years old), that the transition to an oxygen-rich atmosphere took place roughly two billion years ago and that eukaryotes appeared by 1.5 billion years ago.

life, some of them recognizable antecedents of modern plants and animals.

In style and in tempo evolution in the Precambrian was distinctly different from that in the later, Phanerozoic era. The Precambrian was an age in which the dominant organisms were microscopic and prokaryotic, and until near the end of the era the rate of evolutionary change was limited by the absence of advanced sexual reproduction. It was an age in which the major benchmarks in the history of life were the result of biochemical and metabolic innovations rather than of morphological changes. Above all, in the Precambrian the influence of life on the environment was at least as important as the influence of the environment on life. Indeed, the metabolism of all the plants and animals that subsequently evolved was made possible by the photosynthetic activities of primitive cyanobacteria some two billion years ago.

The Smallest Living Cells

3

by Harold J. Morowitz and Mark E. Tourtellotte
March 1962

A microbe known as the pleuropneumonia-like organism gives rise to free-living cells smaller than some viruses. They suggest the question: What are the smallest dimensions compatible with life?

What is the smallest free-living organism? The most likely candidate for this niche in the order of nature was discovered by Louis Pasteur when he recognized that bovine pleuropneumonia, a highly contagious disease of cattle, must be caused by a microbial agent. But Pasteur was unable to isolate the microbe: he could not grow it in nutrient broth nor could he see it under the microscope. Apparently it was too small to be seen.

Then, in 1892, the Russian investigator D. Iwanowsky succeeded in demonstrating that certain infectious agents were so small that they could pass easily through the porcelain filters used to trap bacteria. The size of the microbes postulated by Pasteur was comparable to that of Iwanowsky's organisms, which were subsequently named viruses. All viruses, however, are parasites of the living cell. The pleuropneumonia agent, on the other hand, is not. In 1898 Pasteur's successors E. I. E. Nocard and P. P. E. Roux were able to grow the pleuropneumonia agent in a complex, but cell-free, medium. In this respect the agent seemed more like a bacterium than a virus. In 1931 W. J. Elford of the National Institute for Medical Research in London, who developed the first filters in which pore size could be precisely determined, showed that cultures of the pleuropneumonia agent contained viable particles only .125 to .150 micron (.0000125 to .000015 centimeter) in diameter. Thus the particles were smaller than many viruses. Yet, as subsequent investigations have shown, the particles fully satisfy the definition "free-living": they have the ability to take molecules out of a nonliving medium and to give rise to two or more replicas of themselves.

More than 30 strains of this tiny organism have now been isolated from soil and sewage, as contaminants from tissue cultures and from a number of animals,

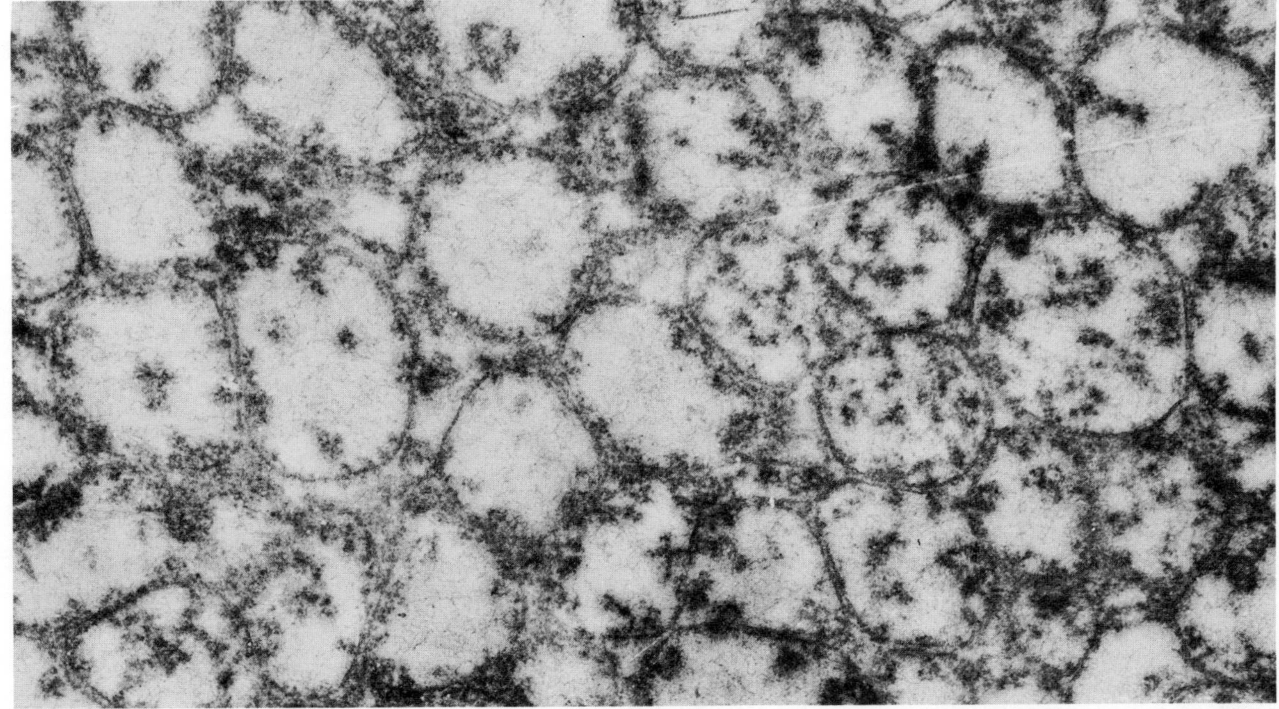

CELLS OF PLEUROPNEUMONIA-LIKE ORGANISM, abbreviated PPLO, are seen in cross section in this electron micrograph made by Woutera van Iterson of the University of Amsterdam. The cells, which are enlarged 72,000 diameters, are not the smallest PPLO's that have been observed. Nevertheless, they are only about 50 per cent larger than the vaccinia virus. Unlike the virus, however, these cells and smaller PPLO's meet a biologist's criterion for life: they are able to grow and reproduce in a medium free of other cells.

including man. In veterinary medicine one or another of them has been identified as the cause of a respiratory disease in poultry, of a type of arthritis in swine and of an udder infection in sheep. Although a pleuropneumonia organism was implicated in cases of human urethritis (inflammation of the urethra), it was not until January of this year that one of them was positively identified as an agent of disease in man. Robert M. Chanock and Michael F. Barile of the National Institutes of Health and Leonard Hayflick of the Wistar Institute of Anatomy and Biology then published their finding that an organism called the Eaton agent, first isolated in 1944, is actually a member of the pleuropneumonia group and is the cause of a common type of pneumonia. Because these organisms pass through filters (like viruses) and grow in nonliving media (like bacteria) they are considered by some workers to be a bridge between these two large classes of organism, and because they show obvious differences from both bacteria and viruses they have been accorded the status of a separate and distinct order: *Mycoplasmatales*. Because of their similarity to the original pleuropneumonia

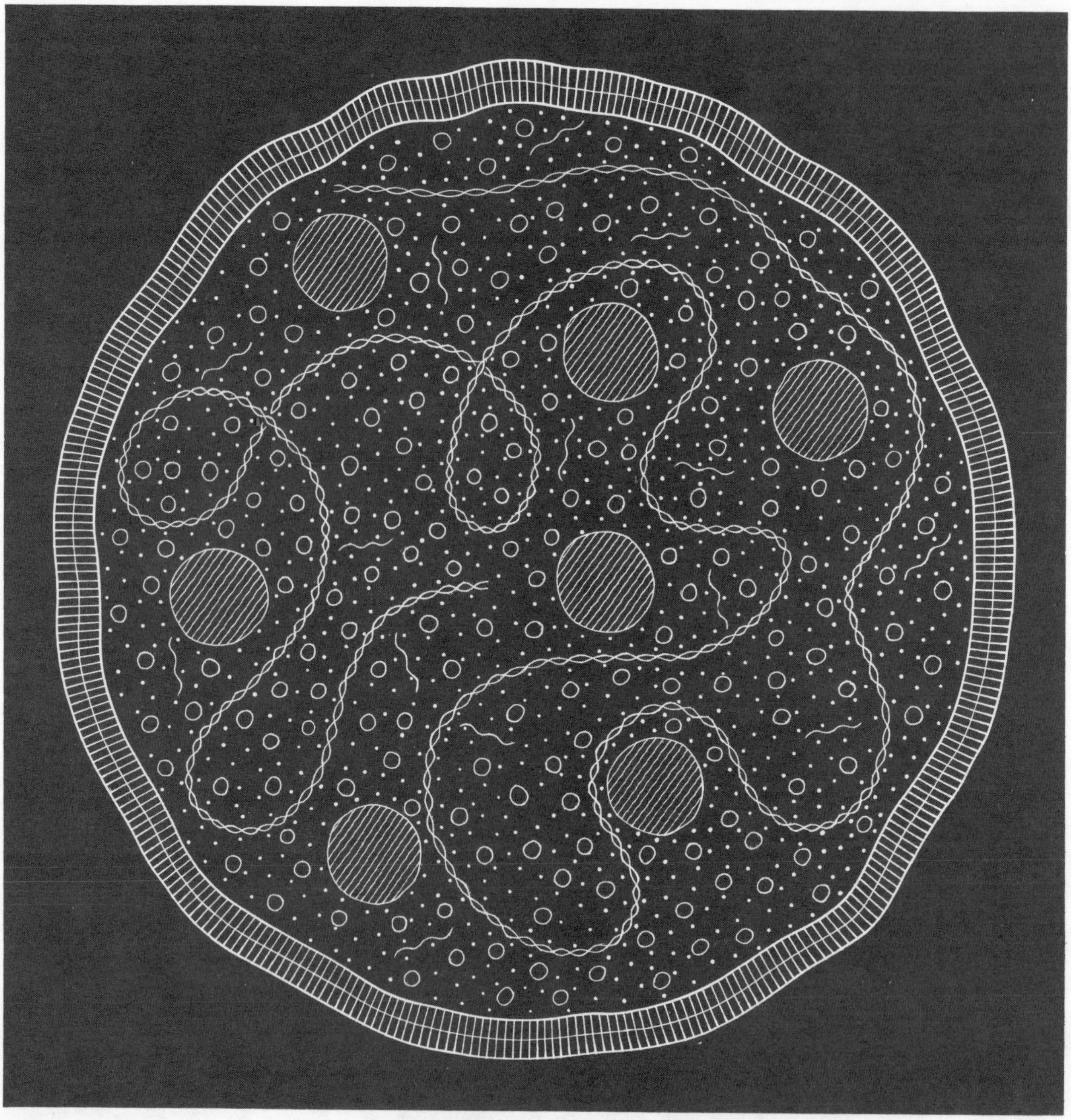

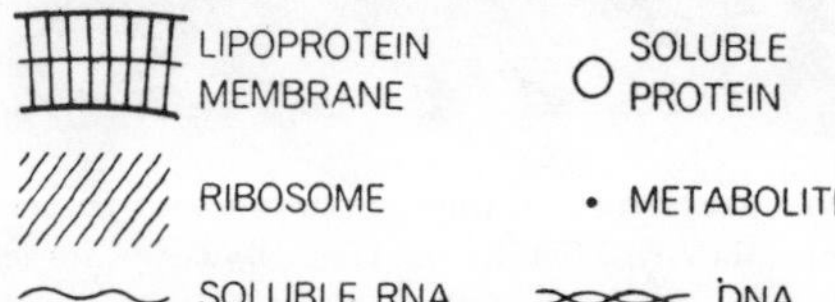

SCHEMATIC REPRESENTATION of a single cell of a PPLO is based on the authors' chemical analysis of *Mycoplasma gallisepticum*, which causes a respiratory disease in poultry. Deoxyribonucleic acid (DNA) and ribonucleic acid (RNA), found both in the ribosomes and in soluble particles, constitute 12 per cent of the total weight of the cell. The soluble proteins are similar to those in larger cells. The delicate cell membrane is composed of successive layers of protein, lipid, lipid and protein.

organism they are usually referred to as pleuropneumonia-like organisms, abbreviated to PPLO.

Although some very small bacteria are smaller than the larger PPLO, none is as small as the smaller PPLO: .1 micron (.00001 centimeter) in diameter. This is a tenth the size of the average bacterium; it is only a hundredth the size of a mammalian tissue cell and a thousandth the size of a protozoon such as an amoeba. But as the British mathematical biologist D'Arcy Wentworth Thompson observed some years ago, a major factor in any comparison of living things is mass, and mass varies as the cube of linear dimension. By such reckoning a protozoon is a billion times heavier than a PPLO. This vast gap in size gains vividness in the mind's eye from the reckoning that a laboratory rat is about a billion times heavier than a protozoon. A protozoon weighs .0000005 gram; a PPLO weighs a billionth as much: 5×10^{-16} gram.

In terms of linear dimensions the smallest PPLO is as close in size to an atom as it is to a 100-micron protozoon. A hydrogen atom measures one angstrom unit (.0001 micron) in diameter; a PPLO cell .1 micron in diameter is only 1,000 times larger. The existence of such a small cell raises intimate questions about the relationship of molecular physics to biology. Does a living system only a few orders of magnitude larger than atomic dimensions possess sufficient molecular equipment to carry on the full range of biochemical activity found in the life processes of larger cells? Or does the minuscule amount of molecular information it can carry compel it to operate in a simpler way? What biological or physical factors place a lower limit on the size of living cells?

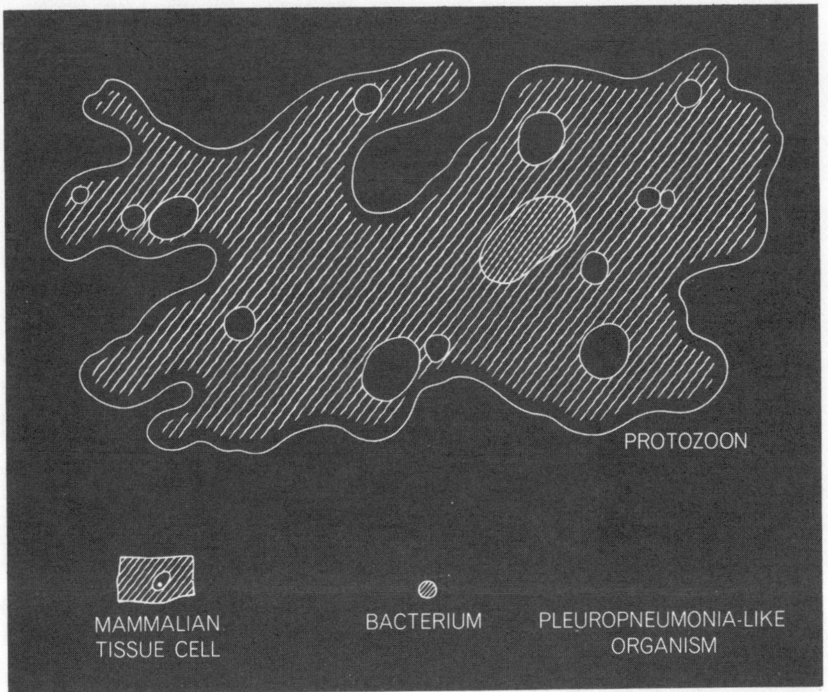

SIZES OF VARIOUS CELLS are compared. A protozoon, with a diameter of .01 centimeter, is 10 times bigger than a tissue cell, 100 times bigger than a bacterium and 1,000 times bigger than the smallest PPLO, with a diameter of .1 micron, or .00001 centimeter.

In our laboratory at Yale University we have cultured 10 distinct strains of PPLO, clearly distinguished from one another by their metabolic behavior and by the antibody responses they produce in rabbits. Our work so far has been concentrated primarily on two of these strains: *Mycoplasma laidlawii,* a strain that is normally free-living in nature, and *Mycoplasma gallisepticum,* which causes chronic respiratory disease in poultry. In the first, which contains the smallest cells we have thus far studied, we have been able to follow the life cycle. In the second we have been able to determine details of chemical composition and structure.

At many stages in its life cycle the individual PPLO cell is too small to be seen in the light microscope. In the electron microscope, however, we have been able to examine at least four different types of cell in *M. laidlawii.* One, called an elementary body, is a small sphere between .1 and .2 micron in diameter. A second is somewhat larger than this. A third is still larger: up to a full micron in diameter, about the size of a bacterium. A fourth type, which is of similar size, contains inclusions that are about the size of elementary bodies. In addition to observing the cell sizes directly, we measured them by forcing the cultures through filters with pores of various sizes and then examining in the electron microscope the material that had gone through the filters [*see illustration on page 44*]. To determine the size of the smallest PPLO cells we calibrated our filters by performing filtrations on two viruses of known size: the influenza virus, which is .08 to .1 micron in diameter, and the vaccinia virus, which is .22 by .26 micron in size. The smallest PPLO cells lie between these two; they are larger than the influenza virus but smaller than the vaccinia virus.

To separate the smallest cells of the strain from the others we had to employ the method of density-gradient centrifugation [*see bottom illustration on next two pages*]. This technique derives its effectiveness from the fact that cells as small as the PPLO vary in density as well as in size as they go through their life cycle. The density at each phase depends on the changing chemical composition of the cell and closely approximates the mean of the densities of its constituents. Salt solutions of different concentration are layered in a centrifuge tube, and the cell culture is added at the top. When the tube is inserted in the centrifuge and spun at high speed, cells of various sizes settle in the layer of salt solution that has a density equal to their own. Centrifuging a 72-hour culture of *M. laidlawii* in solutions that varied in density from 1.2 to 1.4 (the density of water is 1.0) revealed three bands. Examination of these bands in the electron microscope showed the bottom band contained large cells; the top band, elementary bodies; and the middle band, cells of intermediate size and large cells with inclusions.

Starting with elementary bodies thus isolated from a culture, we have been able to follow a culture of *M. laidlawii* through its life cycle. Our method is to sample the culture at periodic intervals and inspect the samples in the electron microscope. Young cultures—about 24 hours old—are primarily composed of large cells. Cultures about six days old, on the other hand, are predominantly elementary bodies. Samples

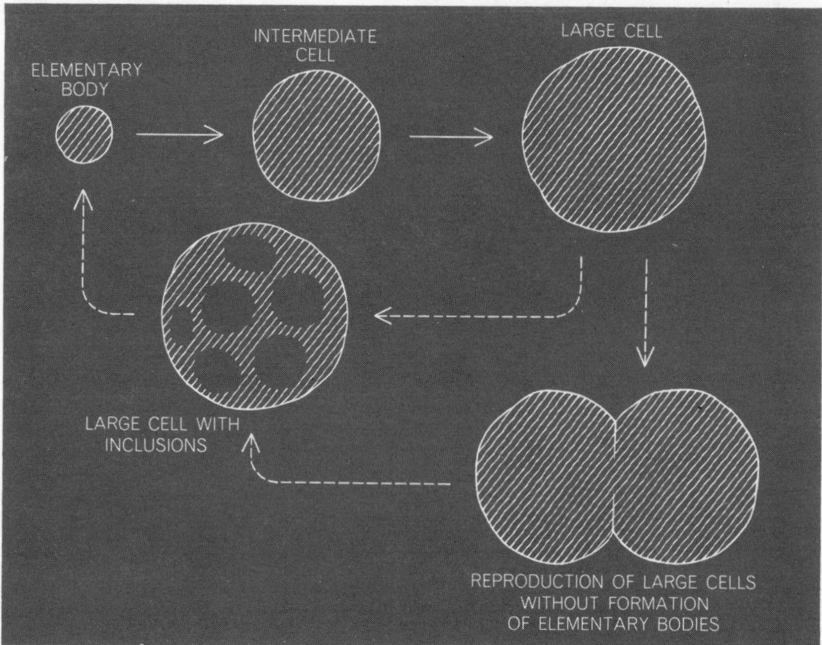

LIFE CYCLE of the PPLO *Mycoplasma laidlawii* is outlined. Elementary bodies grow to intermediate cells and then large ones. The large cells may divide, some developing inclusions released as elementary bodies, or may develop and release inclusions directly.

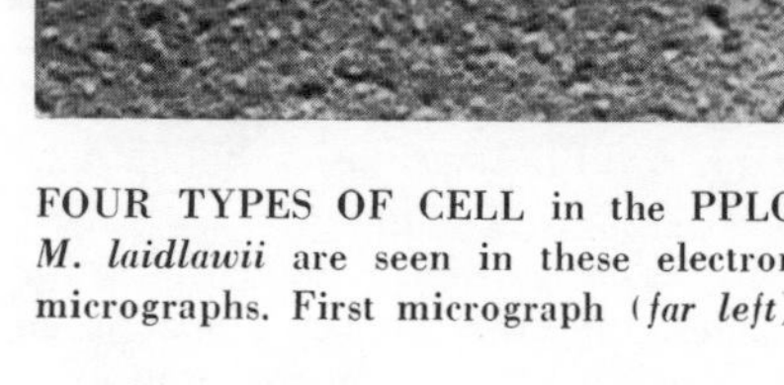

FOUR TYPES OF CELL in the PPLO *M. laidlawii* are seen in these electron micrographs. First micrograph (*far left*)

taken over the course of the five-day interval suggest that this strain has two methods of reproduction. In both cases the organism goes through a cycle in which elementary bodies are transformed first into intermediate cells, then into large cells and then back into elementary bodies again. Differences in composition between young and old cultures show, however, that the organism can probably adopt one of two courses once it has reached the large cell stage. In one cycle the large cells develop inclusions, which are apparently released as elementary bodies. In the second the large cells seem to reproduce by binary fission. Thereafter it appears that some of them form inclusions from which new elementary bodies are liberated. In either case the new elementary bodies begin the life cycle all over again.

We have not so far been able to establish the mode of reproduction in *M. gallisepticum*. None of our cultures has

SEPARATION OF PPLO CELLS BY SIZE AND TYPE is achieved by density-gradient centrifugation. This method can be used because small cells have different densities at different times in their life cycle.

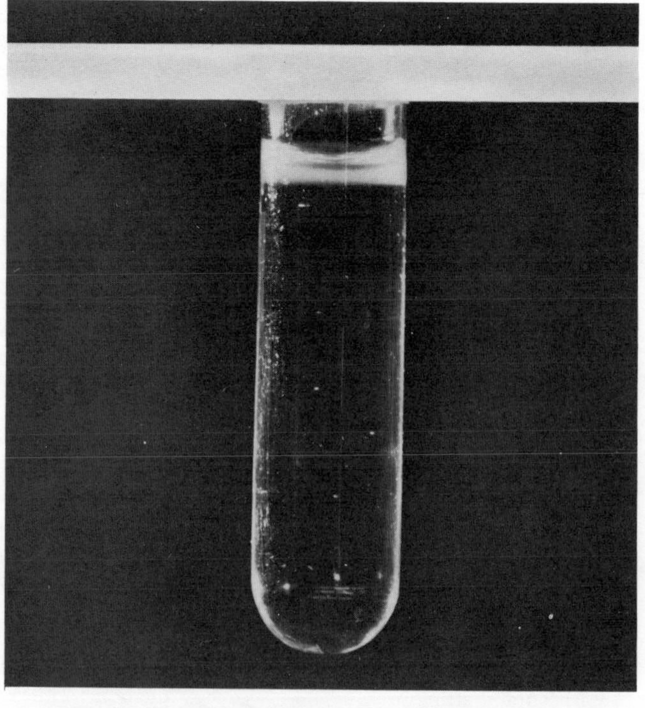

In photograph at far left two densities of salt solution are layered in a centrifuge tube. In the second photograph a PPLO culture has been added at the top of the tube. In the third photograph the

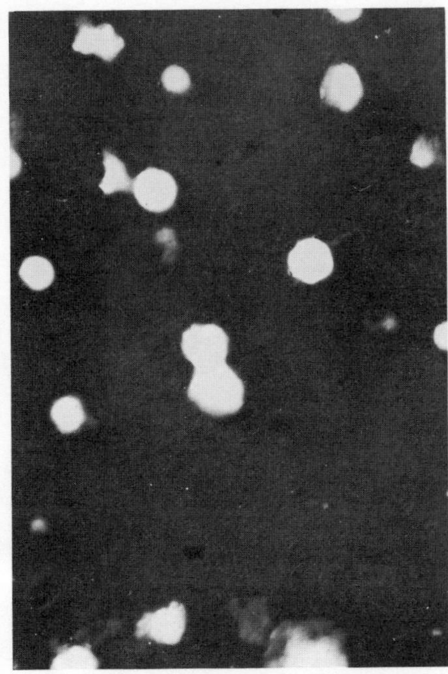

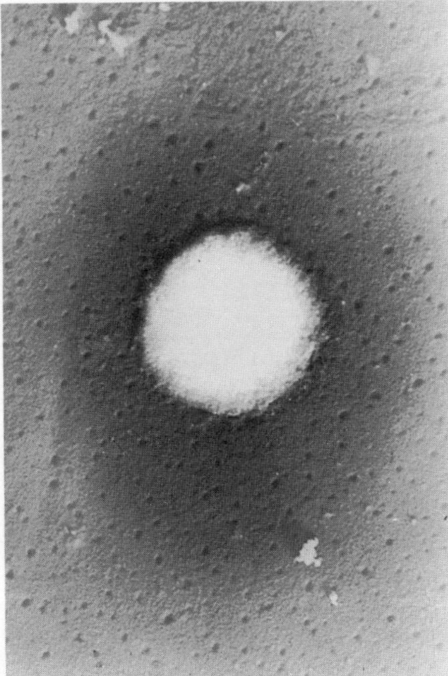

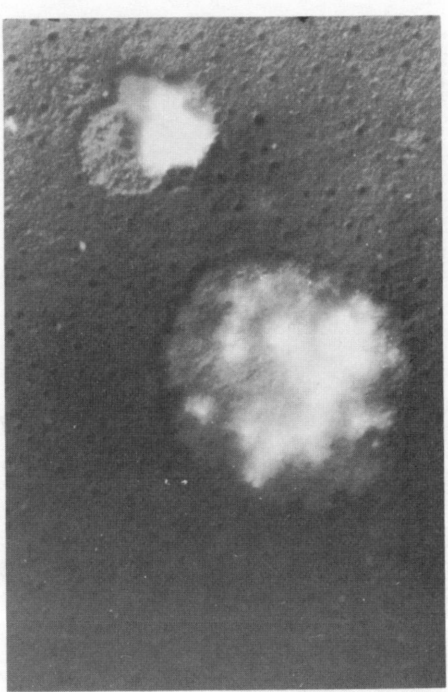

shows elementary bodies about .1 micron in diameter. The second shows intermediate cells. The third shows large cells about 1 micron in diameter, and fourth shows large cells that have developed inclusions. The inclusions may be released as elementary bodies to begin the life cycle again. All the micrographs, which enlarge the cells 17,750 diameters, were made by the authors.

revealed either elementary bodies or large cells. All the cells we have seen appear uniformly spherical and all appear to be about .25 micron in diameter [*see illustration on page 38*]. Our work with *M. gallisepticum* has helped, however, to settle the question of whether or not these tiny organisms conduct the same biochemical processes as larger cells.

Chemical analysis shows that the *M. gallisepticum* cell has the full complement of molecular machinery. In the first place, the nonaqueous substance of the cell contains 4 per cent deoxyribonucleic acid (DNA) and 8 per cent ribonucleic acid (RNA). These large molecules have been identified in larger cells as the bearers of the genetic information that governs the synthesis of the other components of a cell. Moreover, we find that in the tiny cells the composition of these molecules, the so-called base ratios, falls within the normal range. The DNA ap-

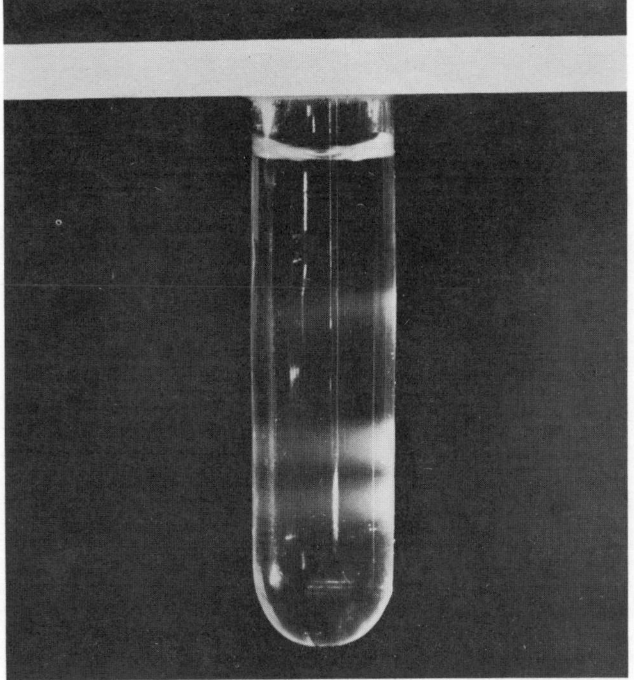

tube is ready to be placed in a container (*left*) and fastened to a rotor, part of which is seen at right. The rotor is then placed in a centrifuge. The last photograph shows the tube after centrifugation. The PPLO's have settled in three bands. The bottom band contains large cells; the middle band, intermediate cells and large cells with inclusions; the top band, elementary bodies.

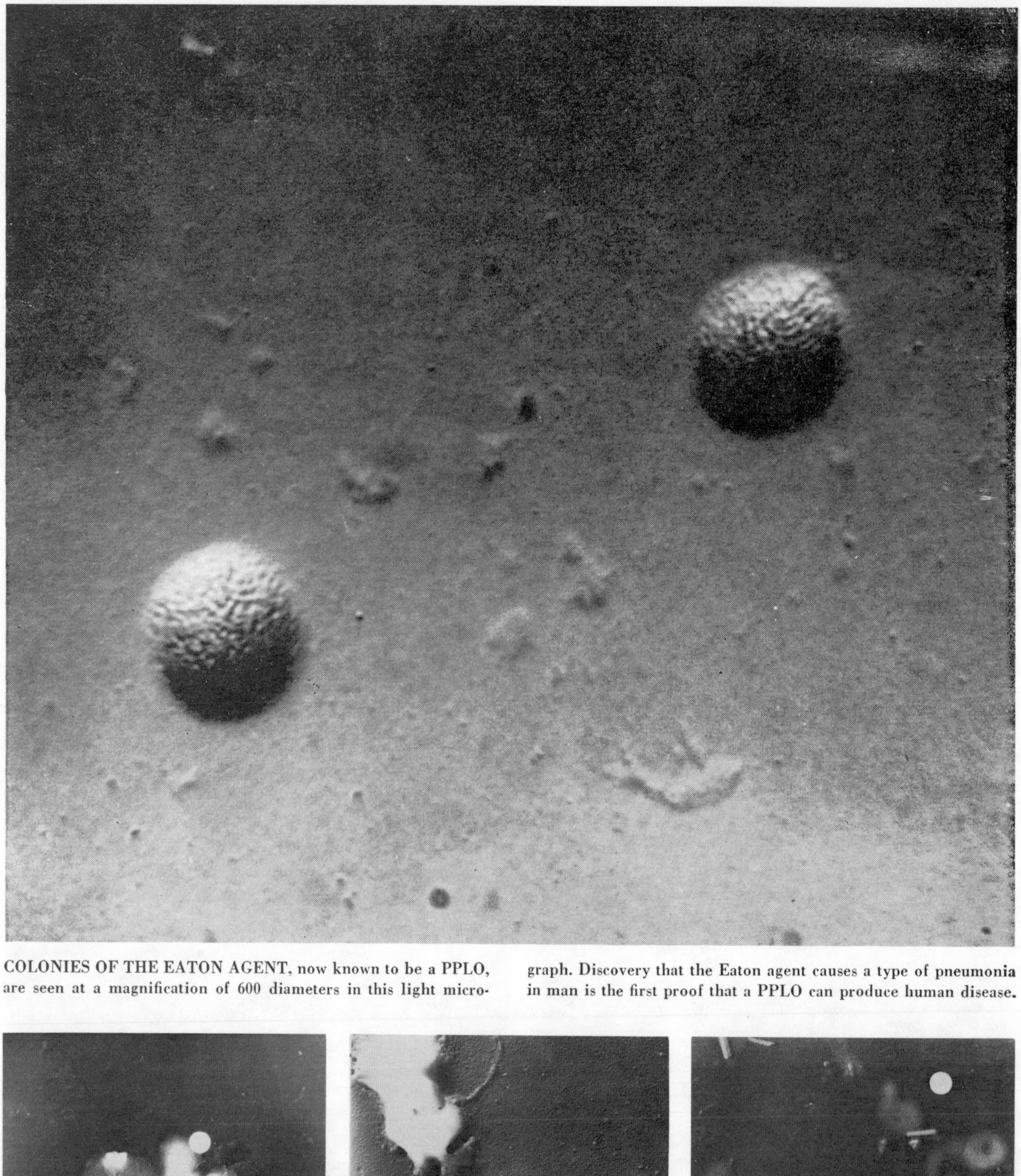

COLONIES OF THE EATON AGENT, now known to be a PPLO, are seen at a magnification of 600 diameters in this light micrograph. Discovery that the Eaton agent causes a type of pneumonia in man is the first proof that a PPLO can produce human disease.

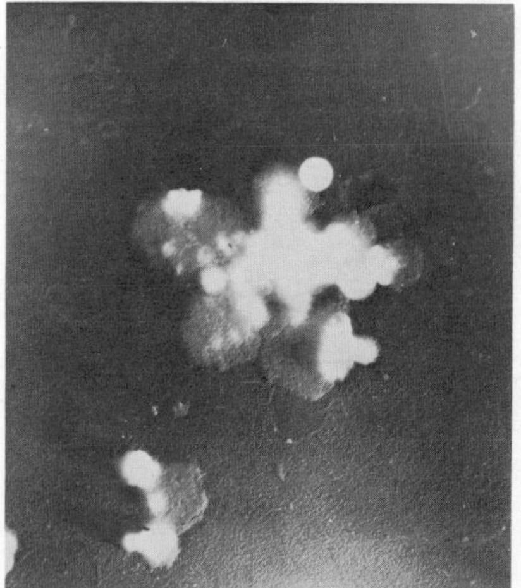

PPLO CELLS from a rat strain are contrasted in size with a .26-micron sphere. The large cell seen at the center contains inclusions.

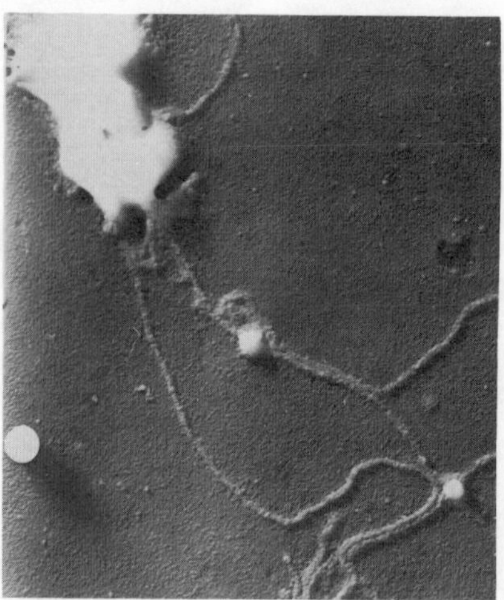

FILAMENTS are observed in many strains of PPLO. The cells shown here are the same strain as those in micrograph at left.

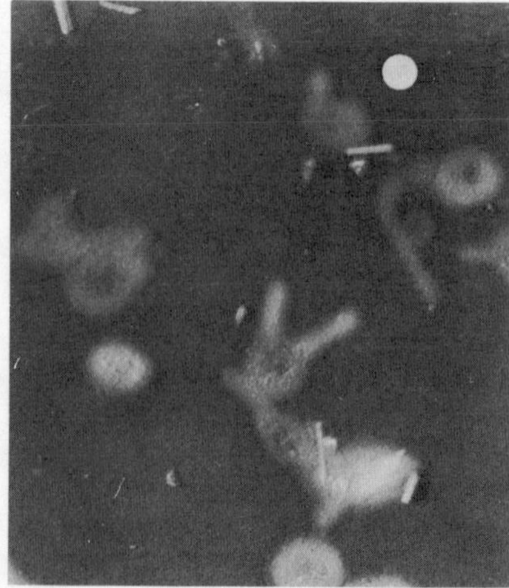

HUMAN PPLO is seen in this electron micrograph. All three micrographs, magnified 16,000 diameters, were made by the authors.

pears as the familiar double-stranded helix found in the chromosomes of larger cells, and most of the RNA appears to be in the form of particles resembling ribosomes, the organelles that are believed to conduct protein synthesis in larger cells. The soluble proteins in the cell seem to have the usual range of size and variety, and the amino acid units of which they are composed occur in the expected ratios.

In several respects this PPLO cell appears to resemble animal cells more than it does plant cells or bacteria. The composition of its fatty substances, including cholesterol and cholesterol esters as an essential element, is characteristic of animal cells. More important, it has no rigid cell wall but has instead a flexible membrane that, in other strains of PPLO, permits the cells to assume a great variety of shapes. In spite of its delicacy the PPLO membrane is able to fulfill the functions of a cell membrane. It effectively distinguishes the cell from its environment and it is firm enough to contain the cell's internal structures in a coherent way. Indirect measurement of its electrical properties shows that they fall within the normal range. At a sufficiently high magnification the membrane can be seen in the electron microscope. It measures about 100 angstrom units (.01 micron) in thickness, which is typical of many animal cells.

Thus far we have been able to demonstrate more than 40 different enzymatic functions in *M. gallisepticum*. These include the entire system of enzymes necessary for the metabolism of glucose to pyruvic acid, one of the processes by which cells extract energy from their nutrients. Therefore the evidence points to a considerable biochemical complexity in these organisms. In spite of their size they seem to compare in structure and function with other known cells.

The demonstration that these tiny cells are indeed free-living compels a further question: Can there be other cells, even smaller than the PPLO and as yet undiscovered, that possess the capabilities for growth and reproduction in a cell-free medium? The mere detection of such cells presents a challenge to the ingenuity and technical resources of the biologist. If the cells happen to be pathogens, they might be discovered by the diseases they produced. If they are harmless to other forms of life, they might put in a visible appearance by causing turbidity in a culture medium through mass growth, or they might

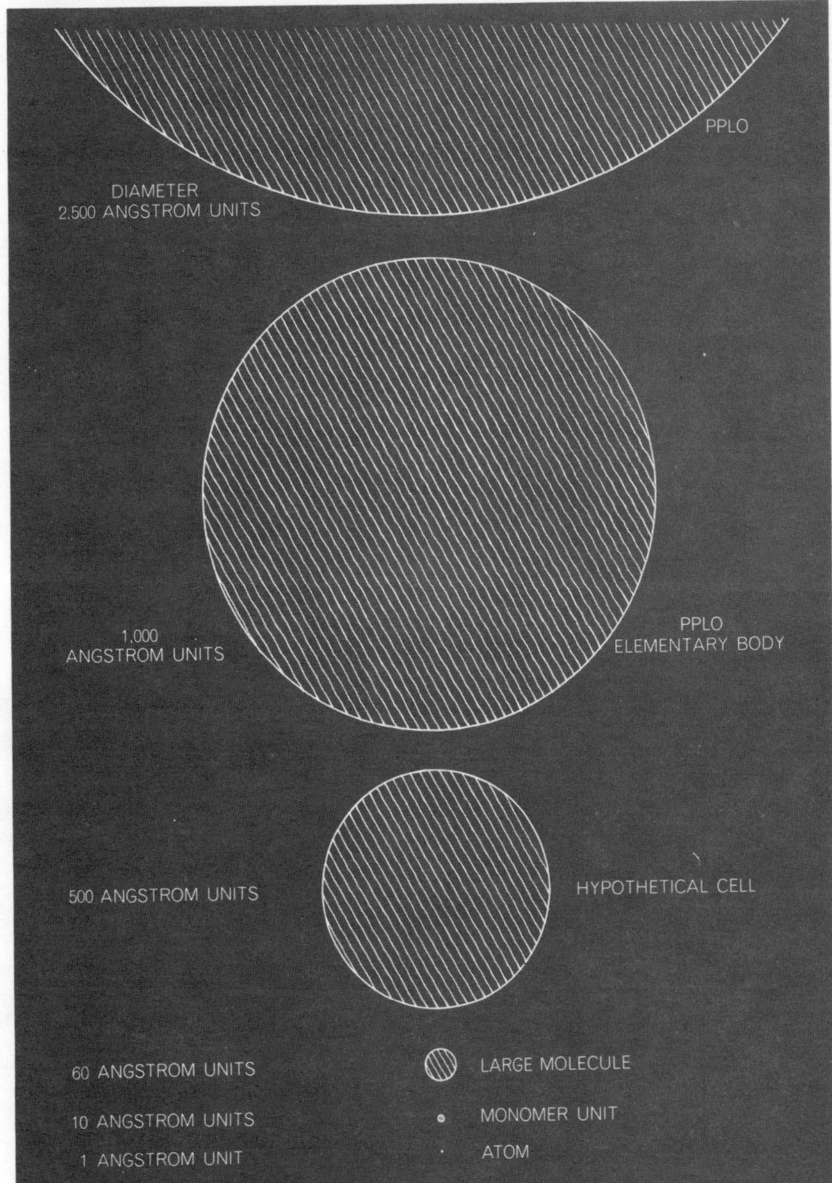

	NUMBER OF ATOMS IN DRY PORTION OF CELL	MOLECULAR WEIGHT OF DNA (DNA = 4 PER CENT OF CELL CONTENT)	NUMBER OF MONOMER UNITS (AMINO ACIDS AND NUCLEOTIDES)	NUMBER OF LARGE MOLECULES
PLEUROPNEUMONIA-LIKE ORGANISM (DIAMETER: 2,500 ANGSTROM UNITS)	187,500,000	45,000,000	9,375,000	18,750
PLEUROPNEUMONIA-LIKE ORGANISM ELEMENTARY BODY (DIAMETER: 1,000 ANGSTROM UNITS)	12,000,000	2,880,000	600,000	1,200
HYPOTHETICAL CELL (DIAMETER: 500 ANGSTROM UNITS)	1,500,000	360,000	75,000	150

PPLO CELLS AND ATOMS can be shown on the same scale; a PPLO elementary body is only 1,000 times larger than a hydrogen atom. Table shows number of atoms, molecular weight of DNA, number of monomer units (the repeating units of a large molecule) and number of large molecules anticipated in PPLO cells and in smallest theoretical cell.

show up in electron microscope preparations. A very small cell, however, might escape detection by any of these methods. If the population it formed grew to a concentration of only 100,000 cells per cubic centimeter of the growth medium, the chances of finding it would be slight. There is no assurance that the proper growth medium could be found to culture such cells. It may well be, therefore, that the PPLO is not the smallest living cell.

Yet there are lower limits, in theory at least, to the size of a living organism. Biological considerations suggest one such limit. A cell must have a membrane, if only to provide coherence for its structure. Since all cell membranes so far studied appear to be on the order of 100 angstrom units (.01 micron) in thickness, it would seem that no cell could exist that had a diameter less than 200 to 300 angstrom units (.02 to .03 micron), or about a tenth the diameter of the *M. gallisepticum* cell. Biochemistry suggests that the smallest cell would have to be somewhat larger in size. The complexity of function necessary to growth and reproduction indicates that the minimal organism must be equipped to conduct at least 100 enzymatically catalyzed reactions. If each reaction were mediated by a single enzyme molecule, the molecules would require a sphere 400 angstrom units in diameter to encompass them and the raw materials on which they operate.

Biophysics suggests another limit to the smallest size. In his little book *What Is Life?* the physicist Erwin Schrödinger pointed out that a cell has to survive against the ceaseless internal deterioration caused by the random thermal motion of its constituent molecules. In a very small cell even small motions are large in proportion to the entire system, and small motions are statistically more likely to occur than large ones. With only one or at most a few molecules of each essential kind present in the smallest conceivable cell, the most minute dislocation might be enough to disable the cell.

The foregoing reasoning seems to set 500 angstrom units as the minimum diameter of a living cell. A cell of this size would have, in its nonaqueous substance, about 1.5 million atoms. Combined in groups of about 20 each, these atoms would form 75,000 amino acids and nucleotides, the building blocks from which the large molecules of the cell's metabolic and reproductive apparatus would be composed. Since these large molecules each incorporate about

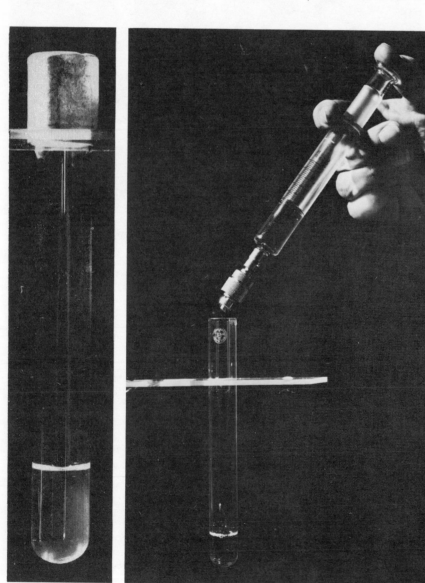

FILTRATION of a PPLO culture requires filters calibrated for viruses. At left is a tube holding an unfiltered culture of PPLO's. In middle, PPLO culture is forced through a filter with .22-micron pores. Tube at right contains filtered PPLO elementary bodies.

500 building blocks, the cell would have a complement of 150 large molecules. This purely theoretical cell would be delicate in the extreme, its ability to reproduce successfully always threatened by the random thermal motion of its constituents.

The smallest living organism actually observed—the .1-micron, or 1,000-angstrom, elementary body of *M. laidlawii*—is only twice this diameter. Its mass, of course, is eight times larger, and calculation from its observed density shows that it may contain 1,200 large molecules. This is a quite finite number, and since the organism grows to considerably larger size in the course of its reproductive cycle it cannot be said that 1,200 large molecules constitute its complete biochemical equipment. In the case of *M. gallisepticum*, however, we know that a diameter of .25 micron—only five times the theoretical lower limit—does encompass an autonomous metabolic and reproductive system. Our chemical analysis shows that the entire system is embodied in something less than 20,000 large molecules [*see chart on page 43*].

This is still an exceedingly small amount of material to sustain the complexity of biochemical function necessary to life. In fact, the portion of it allotted to the genetic function seems inadequate to the task. The 4 per cent of its dry substance that is DNA has a total molecular weight of 45 million. Since, according to current views of genetic coding, it takes an amount of DNA with a molecular weight of one million to encode the information for the synthesis of one enzyme molecule, *M. gallisepticum* would seem to contain enough genetic material to encode only a few enzymes beyond the 40 we have identified so far. That is far short of the 100 enzymes thought to be the minimum for cellular functions. It may be that the enzymes of very small cells are less specific and hence more versatile in their action than the enzymes of larger ones. On the other hand, it may prove necessary to re-examine prevailing ideas about the way information is encoded in the genetic material.

Questions of this kind suggest the principal challenge of very small cells. If they are indeed simpler than other cells, they can tell much about the basic mechanisms of cell function. If, on the other hand, they are functionally as complicated as other cells, they pose the fundamental question of how such functional complexity can be carried in such tiny pieces of genetic material.

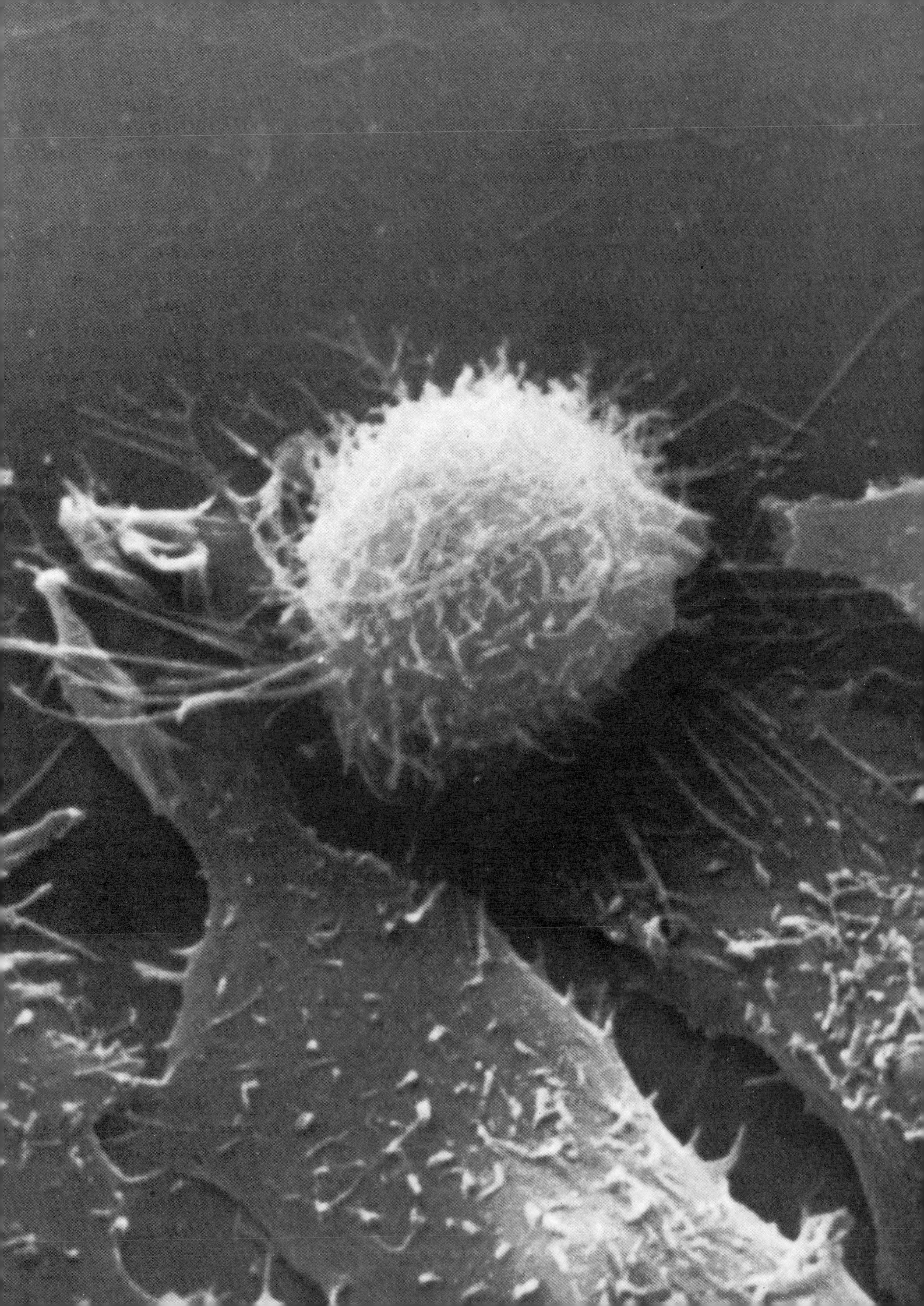

The Cell Cycle

4

by Daniel Mazia
January 1974

What happens in the living cell between the time it is born in the division of another cell and the time it divides again? New methods of investigation reveal four phases in the cycle

Double or nothing. With few exceptions a living cell either reproduces or dies; the principle is so simple that no one has bothered to call it a principle. A cell is born in the division of a parent cell. It then doubles in every respect: in every part, in every kind of molecule, even in the amount of water it contains. Thereafter it divides with such equal justice that each new daughter cell is an identical copy of the parent. This doubling and halving, the cycle of growth and division, is known generally as the cell cycle.

Although these facts have been known for a long time, the concerted study of the cell cycle began only recently. Scarcely 20 years ago, when I published a lecture called "The Life History of the Cell," I thought I was saying something that needed saying when I concluded: "We are no longer satisfied with a good measurement or a good observation or the clever elucidation of some mechanism in the cell unless we can locate it on a time axis, and the time I refer to is not our time but that of the cell itself, as expressed in its life history." The past two decades have seen rapid progress along this very line. Although the first comprehensive book on the cell cycle, J. M. Mitchison's outstanding volume, was not published until 1971, he was able by then to cite 1,000 other cell-cycle studies that he had selected from an even larger literature.

The objective of all these studies has generally been to define as exactly as possible the events within a cell that characterize a given phase of its life history. Going further, the goal has been to discover just which events are the ones that drive the cell through time, moving it from one phase of its history to the next.

Growth and Division

When we say that cells double in all their constituents and then divide in half, we are obviously describing the average case. When we observe individual cells, however, we find deviations from the average. For example, a number of kinds of cells have been weighed at various intervals throughout their period of growth. If the process of growth consisted in a steady production of all the constituents of the cell, including those required for further growth, we would expect a series of weight measurements to yield a smoothly accelerating growth curve, showing a rate of growth that increases in a regular way. In fact, such exponential growth curves have been found, but they are not always found. In some instances the growth rate is constant rather than accelerating. In others the growth rate is high at first and then decelerates. In still others the growth rate increases abruptly at some point. Of course, from the cell's point of view these variations might be matters of indifference as long as the cell has doubled in every way by the end of the growth period.

Is growth really an exact doubling? If it were, we could propose some enticing theories. Observation reveals, however, that the doubling and halving process is only approximate. When Dick Killander and Anders Zetterberg of the Royal Caroline Institute in Stockholm weighed sister cells at the time of division in mouse connective tissue, they found an average 10 percent difference in their weight. David M. Prescott, who was then working at the University of California at Berkeley (and later Killander and Zetterberg as well), followed the growth of unequal sister cells. They did not find the small one precisely doubling its lesser volume and the large one doubling its greater volume. Instead both cells tended to reach what I shall call an "adult" size before dividing, and it simply took the smaller cells longer to do so. Finally, the adult size was not absolutely fixed; Killander and Zetterberg found a variation of about 12 percent in the size of mouse connective-tissue cells that were ready to divide.

These findings showed that a mass of cells doubles and halves only in an average way and not in an exact way. This led us to a conclusion that sounds banal but has in fact provided the basis for more incisive research: evidently cells divide when they are "ready," and they are ready only when they have completed certain preparations for division. A further conclusion was that the preparations for division, whatever they may be, are hidden in the overall growth process.

If a cell fails to divide, it does not continue to grow; something limits cell size and halts the growth process. An old and not quite correct theory called attention to the cell nucleus as the limiting factor. In 1908 the German biologist Richard

DOUBLING CELL (*center on opposite page*) **has changed its surface configuration from flat to round as it enters the final phase of the cell cycle: mitosis, or cell division. The surface of the cell is covered by many long, thin projections; a number of these secure the cell to the substrate. The cells are from the ovary of a hamster. The adjacent flattened cells in this scanning electron micrograph have passed through much of the preceding (G_2) phase of the cell cycle. The micrograph was made by Keith R. Porter, David M. Prescott and Jearl F. Frye of the University of Colorado; the cells are seen magnified 8,300 diameters.**

G_1 (INITIAL GROWTH: ABOUT EIGHT HOURS)

S (CHROMOSOME REPLICATION: ABOUT SIX HOURS)

G_2 (CONTINUED GROWTH: ABOUT FIVE HOURS)

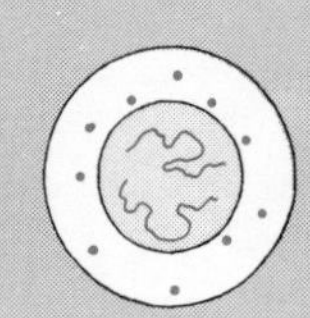
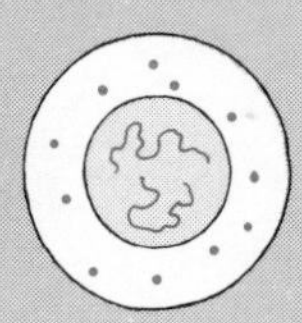
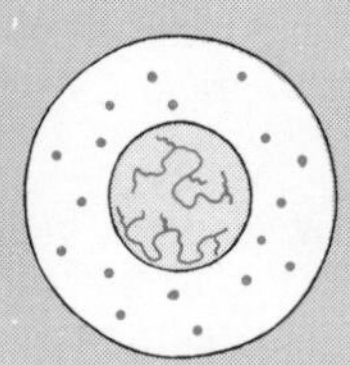
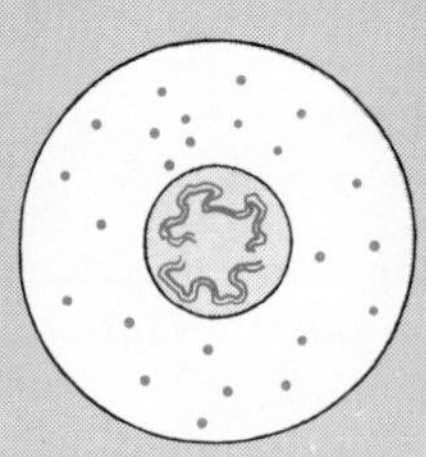
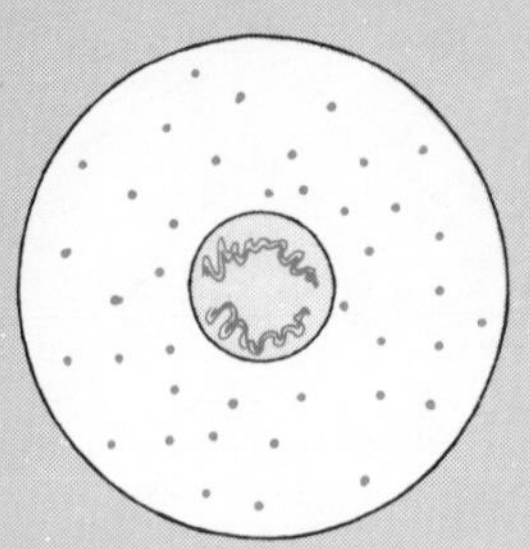

PHASES OF THE CELL CYCLE are shown schematically, beginning at the far left with the first and longest of three "growing" phases: G_1. Space in the diagram is not proportional to time (*see time scale on top*). In a typical mammalian cell the first growing phase takes from six to eight hours, but it can take much longer. The G_1 phase ends and the second, or *S*, phase begins (*left*) when the cell begins to synthesize DNA. During the six hours of the *S* phase the DNA, the genetic material of the chromosomes, replicates. (The diagram shows the replication of only two chromosomes, whereas a human nucleus would have 46 chromosomes.)

von Hertwig proposed what I call the theory of the critical mass. In Hertwig's view a growing cell eventually reaches a size at which the ratio between the masses represented by cytoplasm on the one hand and the cell nucleus on the other becomes limiting. When that limiting ratio is reached, he proposed, some instability sets in and triggers cell division. The nucleus-cytoplasm relation theory, as Hertwig called it, does not hold up very well when mass alone is considered to be the factor that triggers cell division. As we have seen, individual cells may vary substantially in mass at the time of division. Moreover, cells can be made to divide before they have doubled in size. For example, Mitchison and I once performed an experiment that involved starving the cells of a kind of yeast that grows by cell division rather than by budding. We did this by depriving the yeast of nitrogen; the deprived cells could not grow because in the absence of nitrogen they could not make the proteins that growth requires. What they could still do, however, was convert some of the proteins in their possession into the kinds needed for cell division. Although starved, the yeast cells continued the cell cycle and divided into abnormally small daughter cells.

Even if the critical-mass theory is not tenable, the fact remains that the size of a cell appears to be limited by the capacity of a single cell nucleus to support growth. If in the laboratory we manufacture a cell containing two nuclei, or two sets of chromosomes in a single nucleus, the cell can grow to twice its normal adult size. One example of this behavior is the way mammalian cells in laboratory culture give rise to gigantic cells when exposed to an appropriate dose of ionizing radiation. The irradiated cells go through the normal cycle repeatedly but cannot divide; as the amount of genetic material contained in the cell nucleus increases, so does the size of the cell increase.

The Tempo of the Cell Cycle

Like other historians, the student of cell-cycle history finds it useful to recognize epochs. Furthermore, like other historians, he must distinguish between the markers provided by real events (for example a revolution or a decisive battle) and the arbitrary markers he invents for his own convenience (for example the Age of Exploration or the Renaissance). When he divides the cell cycle into two parts, a period of growth between cell divisions known generally as the interphase and a second period when division takes place, the student of the cell cycle is recognizing real events. The growing cell is different from the dividing cell in almost every way.

In terms of real time the interphase period is generally much longer than the period of division. In most plant and animal cells the entire cycle is accomplished in a day or less; a typical cycle takes about 20 hours. Of this time only an hour or so is devoted to cell division; the rest of the time is occupied by interphase growth. Some "hothouse" varieties of animal cells that have evolved under laboratory conditions may have a cycle as short as 10 hours, and some one-celled organisms such as yeasts and certain protozoans have an even faster cycle. (I exclude bacteria and other procaryotic organisms whose cells have no nuclei; their cycle can be completed in as little as 20 minutes.) The protozoan *Tetrahymena,* for example, is a favorite among students of the cell cycle because interphase and division together take no more than 150 to 180 minutes.

There is nothing arbitrary about the tempo of the cell cycle. Given optimum temperature and nutrition, the length of the cycle for a given kind of cell is always the same. Furthermore, the tempo is difficult to alter. Of course, under unfavorable conditions it will be slowed. If we wanted to speed it up and make cells grow faster, however, we would not know how to go about it. We have come to think of the duration of the cell cycle as the time required for the execution of a precise program that has been built into each kind of cell.

The cell cycle is a bicycle. One wheel is the wheel of reproduction; its responsibilities include not only the events that take place at division but also the definite preparations that precede division. The preparations are easily characterized: They are not vital to the survival of the cell at the time they take place, and they do not take place at all in cells that are destined not to divide. Perhaps the best example of such a preparatory activity is the replication of the genetic material in the chromosomes. A cell does not need to replicate in order to stay alive. Indeed, the replication has one meaning only: it expresses the intention of the cell to become two cells instead of one. The second wheel of the cell cycle is overall cell growth, the doubling of all the supplies and equipment that keep

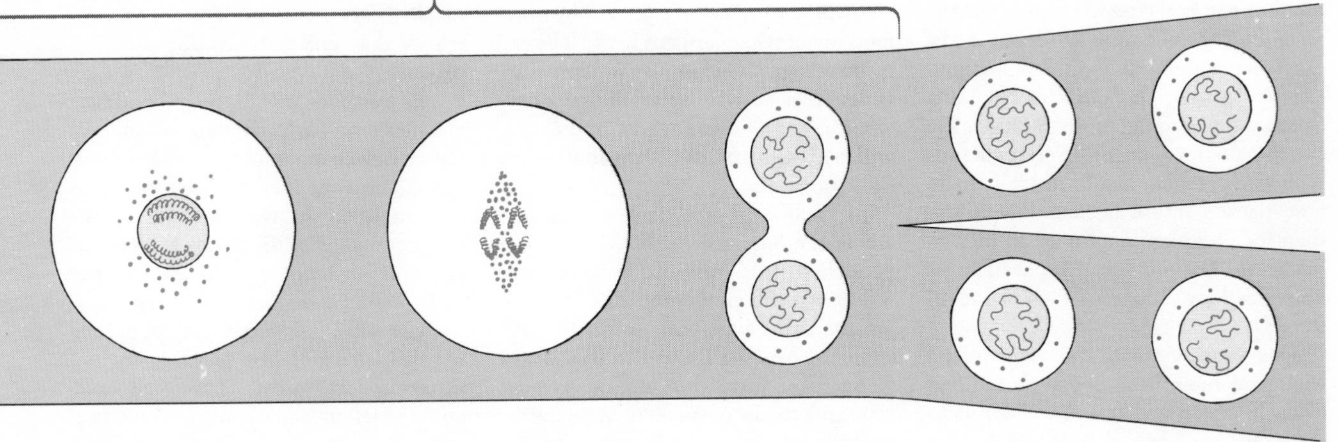

When the replication of DNA is terminated, the cell enters the G_2 phase (*left of center*). In five more hours the cell has become twice its original size and is ready to enter the *M* phase (*right of center*). In this phase, which is the period of mitosis, the chromosomes have condensed into visible threads, the contents of the nucleus have mingled with the cytoplasm, and the molecules in the cytoplasm needed to move the chromosomes (*color*) have assembled into a mitotic apparatus. The chromosomes move apart; the parent cell, having doubled, is now halved by being pinched in two, and identical daughter cells begin the cycle all over again.

the cell viable and that will become the dowries of the daughter cells.

Nature tells us some of the reasons the cell-cycle program requires a certain minimum amount of time. Consider the eggs of animals: special kinds of cells that have been fattened up in advance and are full of the kinds of molecules that are usually produced only by growing cells. When the egg cell begins to divide, it divides again and again into smaller and still smaller cells. Its cell cycle, which requires no growth but only preparations for division and the act of division itself, may be completed in an hour or less. Short time intervals such as this tell us that the reproductive "wheel" of the cell cycle can rotate quite rapidly all by itself, but that the two wheels together travel less swiftly because the growth wheel requires time to produce the raw materials and machinery that the reproductive wheel must have.

Today's analysis of the individual phases that make up the cell cycle uses as its principal signposts the replication and division of deoxyribonucleic acid (DNA), the genetic material in the cell nucleus. This is logical; the mere numerical multiplication of cells can only be called true reproduction because the genetic material of the cells is exactly doubled and equally divided in the course of each cycle. Although the process of cell division had been studied intensively for more than half a century, not until about 1950 was it realized that the chromosomes, which carry the DNA, replicate during interphase and only separate during division. It can be proved that DNA doubles during only one part of interphase. That proof came in the 1950's with the development of refined techniques for autoradiography. Newly available radioactive isotopes made it possible to supply cells with radioactive ingredients out of which DNA would be made. The nuclei of those cells in which the DNA was being replicated, and only those cells, became radioactive and registered their presence on a photographic emulsion. Following cells through their cycles, we found that they began to double their DNA at a certain time in interphase and completed the doubling before the end of interphase.

The phases now used to express the timetable of the cell cycle were first defined in 1953 by Alma Howard and Stephen Pelc, who were working at Hammersmith Hospital in London. They called the period of DNA replication the S phase; this period provides the landmarks by which other phases are defined. Usually some time elapses between the "birth" of the cell at division and the beginning of DNA replication; that interval is called the G_1 phase. By the time the S phase is completed the cell is usually not ready to divide, and the interval between the end of S and the onset of cell division is called the G_2 phase. The completion of G_2 is marked by the beginning of the period of cell division variously known as the *M* phase (for mitosis) or the *D* phase (for division).

The duration of the G_1, S, G_2 and *M* phases is different in different kinds of cells, but the variations between individual cells of the same kind are small. For that matter, the variations between different kinds of cells are not huge. We can even, with due caution, speak of a typical cell cycle. For example, the timetable of a human cancer cell grown in culture is as follows: eight hours in the G_1 phase, six hours in the S phase, four and a half hours in the G_2 phase and one hour in the *M* phase. This does not differ very radically from the timetable of the cells in the root tip of the broad bean: G_1 four hours, S nine hours, G_2 three and a half hours and *M* 114 minutes. Even when there are variations between cell cycles, they show some uniformity. For example, the duration of all the phases may vary, but by far the greatest variation is found in the G_1 phase. When a cell cycle is long, most of the prolongation is in the G_1 phase; when a cell cycle is very short, as it is in egg cells, there is no measurable G_1 phase at all. Conversely, the duration of the S and G_2 phases can be remarkably constant. For example, in most mammalian cells the S phase lasts from six to eight hours and the G_2 phase lasts from three to five hours. The G_1 phase, however, may be as short as a few hours or may take days or even weeks.

The Events of Interphase

In its variability the G_1 phase is trying to tell us something important about the cell cycle, but we do not understand the message very clearly. Hidden somewhere in the cell's overall growth during G_1 must be some key processes that make it possible for the cell to enter the S phase and thereby commit itself to division in the future. In speaking of the events of the S phase it is an understatement to characterize them solely as "DNA synthesis" or "DNA doubling."

This, to be sure, is the phenomenon we measure, but the significance of the S phase is not in the mere doubling of the amount of the chemical DNA but in the exact replication of the chromosomes. Let us consider the DNA of a human cell at the beginning of the S phase. We can think of it as consisting of 46 threads with an aggregate length of six or more feet, packed into a nucleus that is less than four ten-thousandths of an inch in diameter. The process of replication requires making a copy of each of the 46 threads.

This process would be complicated enough if replication began at one end of each thread and traveled along it to the other end, but that is not how it works. Instead each thread replicates in segments according to a definite program. The segments do not replicate in tandem in any one chromosome, nor does any one chromosome complete its replication before another chromosome begins the process. Because the replication program is reasonably constant we are beginning to know which parts of which chromosome replicate first and which last. We also know that when the last segments are replicated, DNA synthesis shuts down and cannot resume until the next cycle.

Turning briefly to the G_2 phase, all we now know is that it embraces the final steps in the cell's preparations for division. If we arrest the synthesis of proteins by the cell during the G_2 phase, the cell will not divide. We know few other facts about what happens in G_2. The cell is nonetheless telling us in blunt language that it will enter division only when it is good and ready, and that it will not even try to divide if it is not fully prepared.

By following the increase in the amounts of enzymes within the cell it is possible to study the life history of the cell at the chemical level. The enzymes are the substances that define the biochemical capabilities of the cell at any given time. Just as we might visualize cell growth as a steadily accelerating process, so we might conceive of the cell's biochemical history as a record of steadily making more of everything. This is not so; different enzymes have different histories. Mitchison has found that enzymes fall into groups with three kinds of history. Some he calls continuous enzymes; they follow the exponential path and increase in a predictable way as the cell grows. Others Mitchison calls step enzymes; these increase abruptly at certain points in the cell cycle, different step enzymes increasing at different points. The third group Mitchison calls peak enzymes; they increase at certain points in the cycle and decrease thereafter, as if they were meant to carry out a particular job at a particular time and were discarded after the job was done. For example, the enzymes involved in making DNA are evidently peak enzymes that appear only during the S phase.

The existence within the cell of different programs for producing enzymes in the cell cycle leads us to broader questions. The very idea of programs turns our attention to the ultimate source of reproducibility in living things: the genes. It is a tenet of molecular genetics that a cell's production of any enzyme is the end result of a "readout" of a gene for that enzyme. The program for enzyme production during the cell cycle ought to tell us something about the program of the gene readout. Going a step further, we shall recall the consistency of the timetable of the phases of the cycle itself and ask whether genes determine the advance of the cell cycle from one phase to the next. Several geneticists who use one-celled organisms in their experiments have answered the question affirmatively by identifying specific genes that control specific steps in the cell cycle. When the genes are made unworkable by means of mutation, the cell simply "gets stuck" at some point in the cell cycle and cannot advance beyond that point. Leland H. Hartwell of the University of Washington has described a number of such gene mutations in yeasts; each of the mutant genes, when

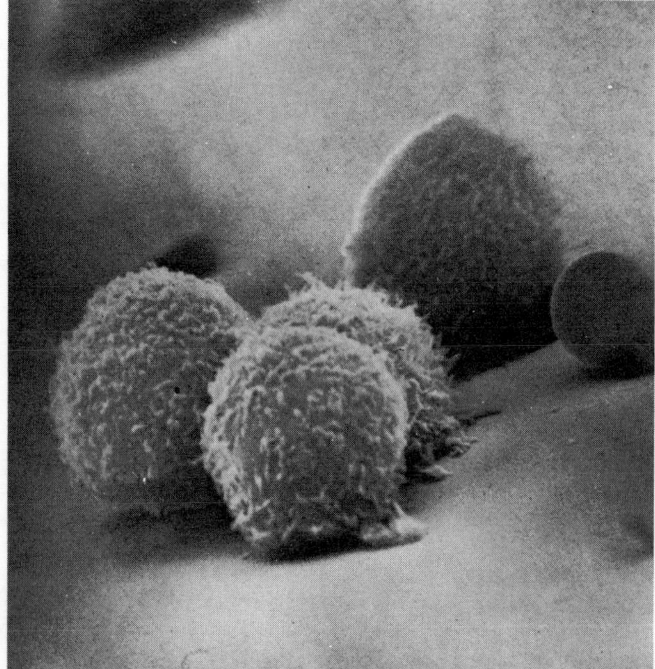

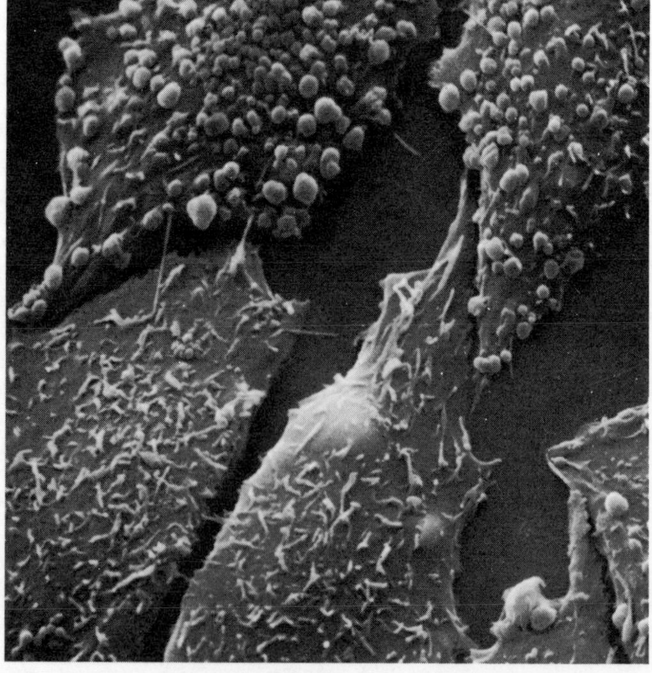

SURFACE CHANGES in shape and texture are characteristic of hamster ovary cells grown in the laboratory. Daughter cells (*left*), just entering the G_1 phase, still have "hairy" surfaces; the projections may be remnants of the longer strands that are abundant in the *M* phase (*see photograph on page 46*). The newborn cells are just beginning to spread out on the substrate; the smaller smooth sphere is a fragment of cell cytoplasm and not a living cell. The flattened cells (*left of center*) have reached a late stage in the G_1 phase; their surfaces are covered with microvilli (small hairs) and blebs (blister-like spheres). Several have established contact with adjacent cells.

normal, had acted at a different point in the yeast cell cycle.

Changes in the cell as it goes through the cell cycle can be seen even in its superficial appearance. Working with the scanning electron microscope, Keith R. Porter, David Prescott and Jearl F. Frye of the University of Colorado have made portraits of one kind of cell in various phases of the cycle. The cells were Chinese hamster ovary cells grown on laboratory cultures. During the *M* phase the cells are spherical and are not strongly attached to the substrate they grow on. As the cells enter the G_1 phase they begin to flatten, and the cell surface exhibits bubblelike blisters and fingerlike projections known as microvilli. As the G_1 phase progresses, the margins of the cells become thin and active and present a ruffled appearance. During the *S* phase the appearance of the cell surface changes again: the cells flatten out and their "complexion" becomes smooth. Going through the G_2 phase the cells thicken once more: their surfaces again show microvilli and ruffles, although the blisters of the G_1 phase are much less abundant. Thus at least one kind of cell from one kind of mammal seems to reflect in its physiognomy the sequence of events in progress within it.

I have characterized the cell cycle as a progress through time; the cell changes radically in the course of the cycle and we are able to recognize distinct phases along the way. The phases, of course, could be of our own invention. The progress of the cell cycle could in fact be perfectly smooth, and the phases we have detected could be no more significant than signposts on a straight highway or numbers on the face of a clock. At the same time it is also possible that the phases we recognize are both real and distinct one from the other and that the progress from one phase to the next requires a definite event; a metaphorical whistle may have to be blown, a switch pulled or a signal given. The highway may be a toll road and the cell may be required to pay its way through a series of gates.

At present much of our evidence favors this second view and suggests that the transition from one phase of the cell cycle to the next calls for a distinct event. We have long been aware of the existence of what I have called "points of no return" in the cell cycle. For example, work in Erik Zeuthen's laboratory in Copenhagen has demonstrated that cell division in the protozoan *Tetrahymena* could be prevented by raising the temperature of the environment before a definite point in the cell cycle. Once that point was passed, however, exposure to high temperatures would not prevent division; evidently the protozoans had passed a point of no return. Moreover, the decisive point was passed some time before any sign of division appeared.

The Two Crucial Signals

Of all the transitions in the cell cycle that we know the two most important are the transition from the G_1 phase to the *S*, when the replication of the chromosomes starts, and the transition from the G_2 phase to the *M*, when the chromosomes condense and mitosis begins. A discovery in the field of animal virology has provided a new and powerful means of studying the two crucial transitions. It appears that certain viruses, sometimes called Sendai viruses, alter the membranes of cells in such a way that two or more adjacent cells can fuse into one cell. The new cell contains the cytoplasm and the nuclei of all the cells within a common membrane. Even more remarkable, very different kinds of cells, and even cells from different species of animals, can be fused by the action of the virus and form hybrids that not only conduct normal processes but also reproduce as hybrids.

By fusing cells that are in different phases of the cell cycle we can obtain direct answers to the question of whether "switches" exist that move a cell from one phase of the cycle to the next. Working at the University of Colorado Medical Center in Denver, Potu N. Rao and

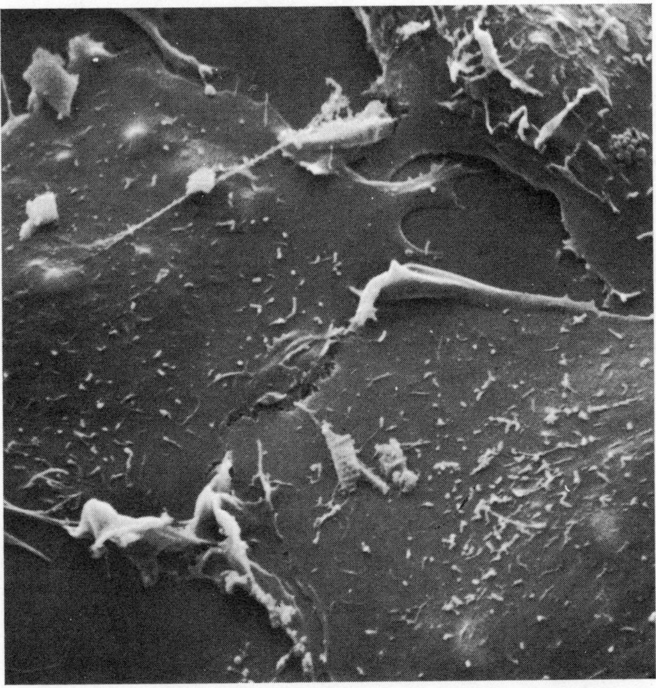

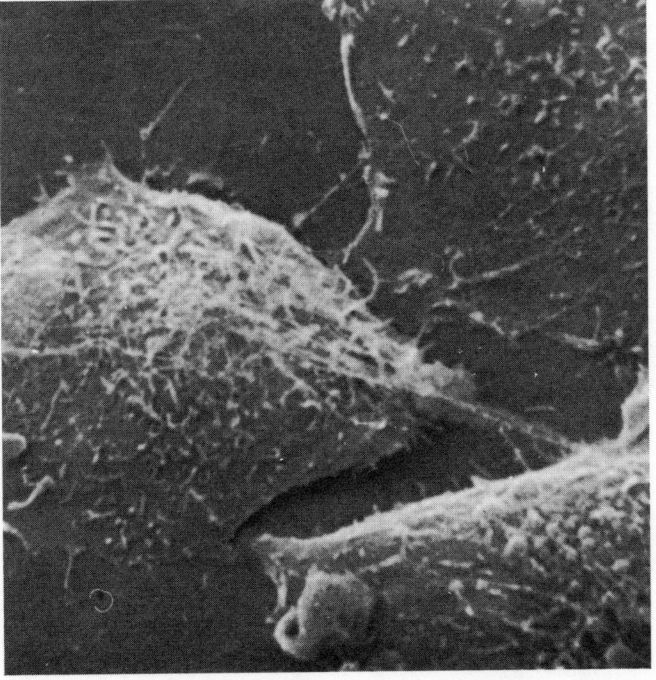

The even flatter cells (*right of center*) have entered the *S* phase. Their surfaces are free of blebs and the microvilli are fewer and less prominent. On some parts of the cells' edges vertical "ruffles" (*light areas*) have appeared. Next (*right*) the cells have entered the G_2 phase. Ruffles are still present; the microvilli are more prominent and one cell (*foreground*) shows an array of blebs. The significance of the surface changes is not yet wholly understood, but their coincidence with the four cell-cycle phases suggests that cell membrane is active in the cycle. Like the micrograph on page 46, these are by Keith R. Porter and his colleagues.

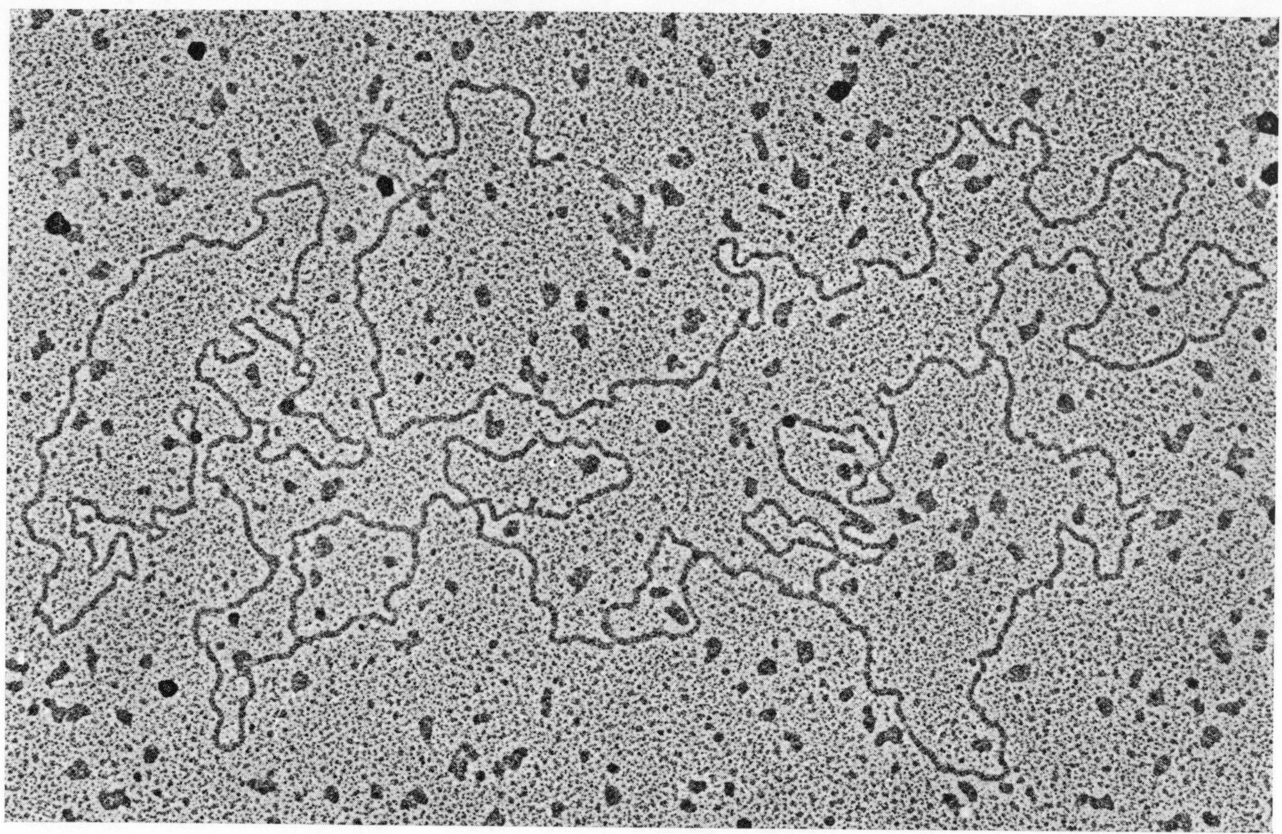

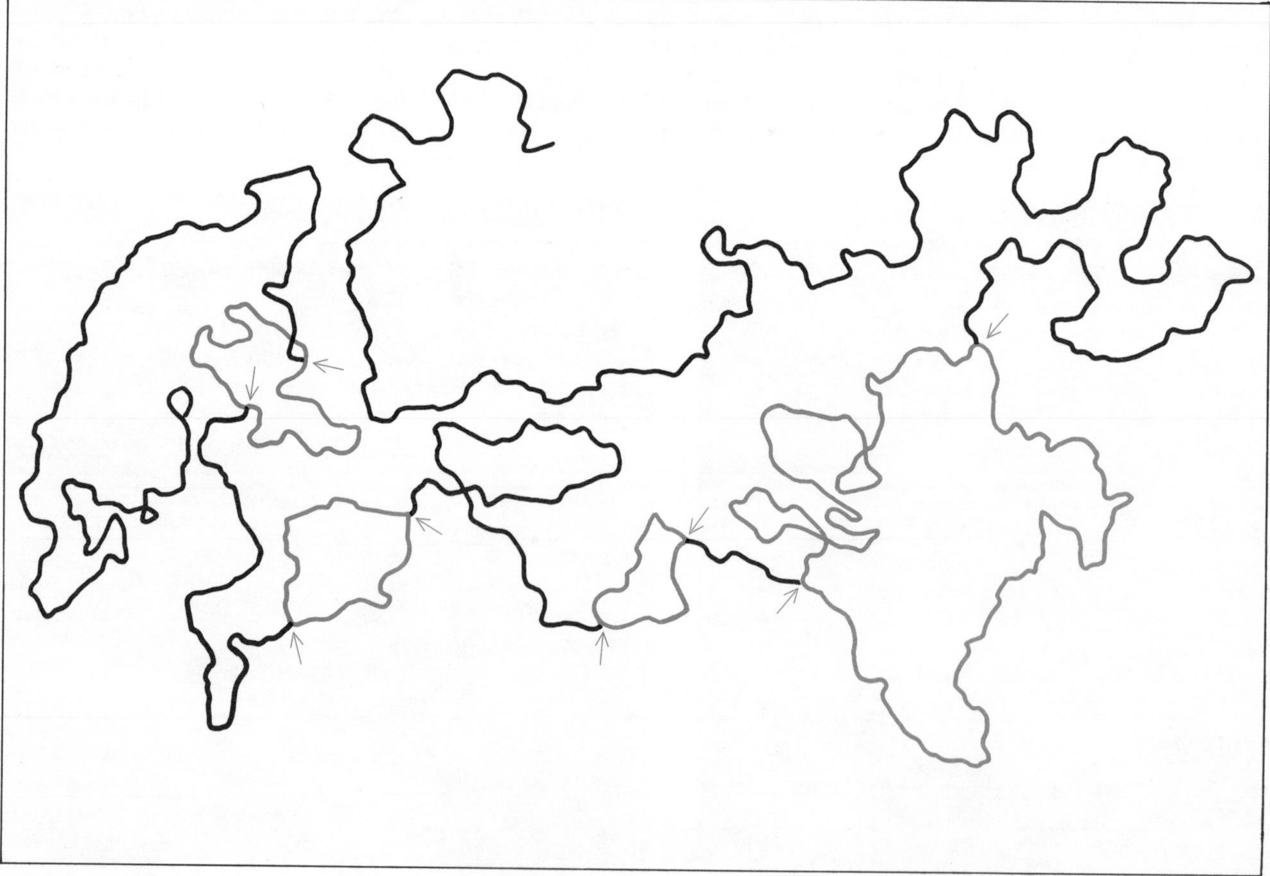

CHROMOSOME REPLICATION, the key event in the *S* phase of the cell cycle, is seen in the micrograph (*top*) as it progresses in a fruit-fly egg. The seemingly tangled strand in the micrograph is the DNA from the replicating chromosomes. The accompanying map of the strand (*bottom*) locates the four areas where segments have begun the process of replication (*color*); the arrows point to the forks in the replicating segments. Both the micrograph and the map are the work of David R. Wolstenholme of the University of Utah.

Robert T. Johnson have merged cells in the S phase with cells in the G_1 phase and found that the nuclei of the G_1 cells began to make DNA long before they would normally have done so.

This result clearly demonstrated that a cell in the S phase contains something that triggers DNA synthesis, whereas this something is absent from a cell that has not completed its G_1 phase. Exactly the opposite result could easily have been predicted. For example, it is logical to propose that a cell remains in the G_1 phase because it contains something that keeps DNA synthesis turned off. Such a hypothetical inhibitor, if placed in contact with the nuclei of cells in the S phase, would presumably halt any further synthesis of DNA by the S-phase nuclei.

Further insight into phase-control switches comes from experiments involving the hybrids made by the fusion of kinds of cells whose normal cell-cycle schedules are different. For example, Jennifer Graves of the University of California at Berkeley grew hybrids of mouse and hamster cells. Hamster cells normally have a shorter G_1 phase than mouse cells, so that if each cell nucleus in the hybrid cells had followed its normal timetable, the hamster nucleus would have begun DNA synthesis some time earlier than the mouse nucleus. Actually both nuclei began synthesis at the same time. Moreover, the shift to the S phase occurred after the shorter G_1 interval of hamster cells. Evidently whatever turned on DNA synthesis in the hamster nuclei also turned it on in the mouse nuclei.

In a further experiment Graves followed the S phase to the end. Once the nuclei had entered the S phase, the nucleus of each species followed its own characteristic replication timetable and its own replication program. Those parts of the chromosomes that replicated first in the parent cell replicated first in the hybrids and those that normally replicated last in the parent were last in the hybrid. Evidently the signal for starting the S phase has no control over what happens once the phase has started. Instead the replication program characteristic of each species must somehow be contained in the nucleus of that species' cell.

These experiments with fused cells provide some answers about the role of switches in the cell cycle. They tell us that there is a something that is responsible for starting the replication of chromosomes. The something, moreover, pervades the cell; it can force any nucleus, ready or not, to enter the S phase. If the something is a special kind of molecule that can switch on replication, then its identification will be an important discovery indeed. It is also possible that the signal for starting chromosome replication is not a special molecule but is instead some change in the internal environment of the cell. This is a watery medium; its variables are the concentrations of ions and small molecules and its acidity or alkalinity. Assuming that the ingredients and enzymes needed for chromosome replication are all present, then a change in such simple factors as these could be responsible for switching on the replication process. "Simple" is not the word. We do know that most of the reactions and functions in cells are extremely sensitive to such variables but we do not know much about how they act or how the cell changes them.

An example of such an environmental signal comes from my own recent work with the unfertilized eggs of sea urchins. Egg cells are loaded with the ingredients and enzymes needed for replication. No replication occurs in the unfertilized egg, but within minutes after fertilization the replication process is turned on and running at high speed. I found that I could turn on the process in unfertilized eggs simply by adding a little ammonia to the seawater that bathed them. The ammonia easily penetrates the cell membrane and has the effect of making the cell interior more alkaline.

As I interpret this result, a change in the cells' internal environment (in this instance the environment of a cell that is otherwise ready to begin chromosome replication) switches on the replication process. The ammonia-treated egg nonetheless remains an unfertilized egg and can even be fertilized later, showing that the ammonia treatment is not a substitute for fertilization. Instead the ammonia treatment is a bypass, mimicking something that normally happens only after fertilization. Many other experiments show that the cell cycle is sensitive to simple factors present in the environment external to the cell, for example ions of hydrogen, potassium and calcium. Learning how factors in the external environment act through the cell membrane to influence the cell's internal environment is vital to an understanding of cell-cycle controls.

Virus-induced cell fusion has also been used to study the other key event in the cell cycle: entrance into the *M* phase. This is the period of cell division, characterized by condensation of the chromosomes. In condensation the chromosomes pack themselves into threads that are visible under the microscope. At Denver, Johnson and Rao have induced cells in the *M* phase to fuse with cells that are at other phases in the cell cycle. If an *M*-phase cell is fused with a cell in the G_2 phase, the G_2 nucleus gives rise to normal-looking condensed chromosomes. The chromosomes are double, as would be expected because they had undergone replication in the preceding S phase. When, however, an *M*-phase cell is fused with a cell in the G_1 phase, thereby forcing the chromosomes of the G_1 nucleus to condense before replication, the chromosomes are seen to be single rather than double.

The third possible combination, the fusion of an *M*-phase cell with a cell in the S phase, produces a result so startling that it has still to be digested. The chromosomes of the S-phase nucleus do condense, but they condense in small fragments. The effect is called "pulverization" [*see illustration on page 55*]. Evidently in the *M* phase there is a something that forces chromosomes to condense even when they are obviously not ready to do so. Whatever the something is, it appears to work on any kind of cell; hybrids have been produced that combine the cells of animals that are only remotely related, for example the cells of men and toads, and the experimental results are the same. If one cell contains condensed chromosomes, it can force premature chromosome condensation in the other cell.

The Cycle in Societies of Cells

The governance of an organism such as a higher animal, which is a society of cells, dictates that some of the cells in the society will reproduce and that others will not. In general the cells of tissues that perform special services for the entire organism, such as cells of the nervous system and the muscular system, do not reproduce at all. In other tissues, such as skin, the blood-forming system and epithelial linings, the rate of production of new cells is nicely modulated to compensate for the continual loss of old cells. In still other instances, notably in immunity reactions and the healing of wounds, cell multiplication is turned on as a bodily response to external provocation. Only in instances of malignant cancer do body cells defy the governance of the organism and go through their cycles anarchically. There is nothing special about the cell cycle in cancer cells compared with other cells that reproduce; the tempo and the phases of the cycle are the same. It is only that cancer cells repeat the cycle without restraint. The private impulses of the individual cells have become a malignancy for the society of cells as a whole.

Our clues to the means by which reproduction in a society of cells is governed come from detailed analyses of the cell cycle. The chief clue is a very simple one: cells that do not divide never enter the S phase. Conversely, cells that do enter the S phase almost always complete that phase and go on to divide; replication of the chromosomes is evidently a commitment to ultimate division. Thus the control of cell division lies in the supervision of one decision: whether or not a cell will enter into replication. This is true but it is too simple. Cells that are not destined to divide are not cells stuck at the transition from the G_1 phase to the S phase, waiting indefinitely for the signal to begin replication. Some students of the cell cycle prefer to think of them as cells that have not entered the cycle at all. In this view two kinds of cells are distinguished: cycling and noncycling. In the body many kinds of specialized cells are noncycling cells whose progress toward division is permanently shut down. We think of the shutdown as the result of the action of the body's government. Many kinds of cells that have left the cycle can be made to reenter the cycle by putting them under culture in the laboratory, thus removing them from the governance of the body. Agents that cause cancer must cause noncycling cells to reenter the cycle.

At the same time the conversion of noncycling cells to cycling cells can be normal and important. An example is the lymphocyte in higher animals. In these small cells growth and division are shut down most of the time. In immunity reactions, however, lymphocytes begin to grow and divide, producing cells that contribute to antibody formation. Noncycling lymphocytes can also be transformed into cycling ones in the laboratory by exposing them to certain plant proteins known as lectins. When noncycling lymphocytes are thus artificially stimulated, nearly a day passes before they enter the S phase and begin chromosome replication. The transformation has been observed in some detail. For example, Lawrence A. Loeb and his collaborators at the Institute for Cancer Research in Philadelphia have learned that the stimulated lymphocytes need to produce the enzymes required for the replication process as a preliminary to entering the S phase. They are far from being stuck in the G_1 phase, ready to go into the S phase when signaled. When stimulated, it appears, a noncycling cell must first do the things that a cycling cell does in the G_1 phase.

The influence a cell society exercises over the cell cycles of its individual members is observable in quite simple social situations. One example is the "contact inhibition" that is studied in laboratory cultures of mammalian cells. The cells normally grow and multiply on a solid substrate, such as the bottom of a dish or flask. Their growth ceases, however, at the point where a certain population density is reached. This is when the substrate has become covered by a single layer of cells. As Harry Rubin of the University of California at Berkeley has shown, contact inhibition can be overcome by infecting the cell culture with a cancer virus. The infection transforms the cells in a number of ways. They may change in appearance. They are no longer confined to a single-layer conformation but will grow on top of one another. If put into a host animal, they will grow in its body as cancer cells.

As in the body, the control of cell reproduction in contact inhibition is achieved by shutting down the cell cycle before the cell enters the S phase and

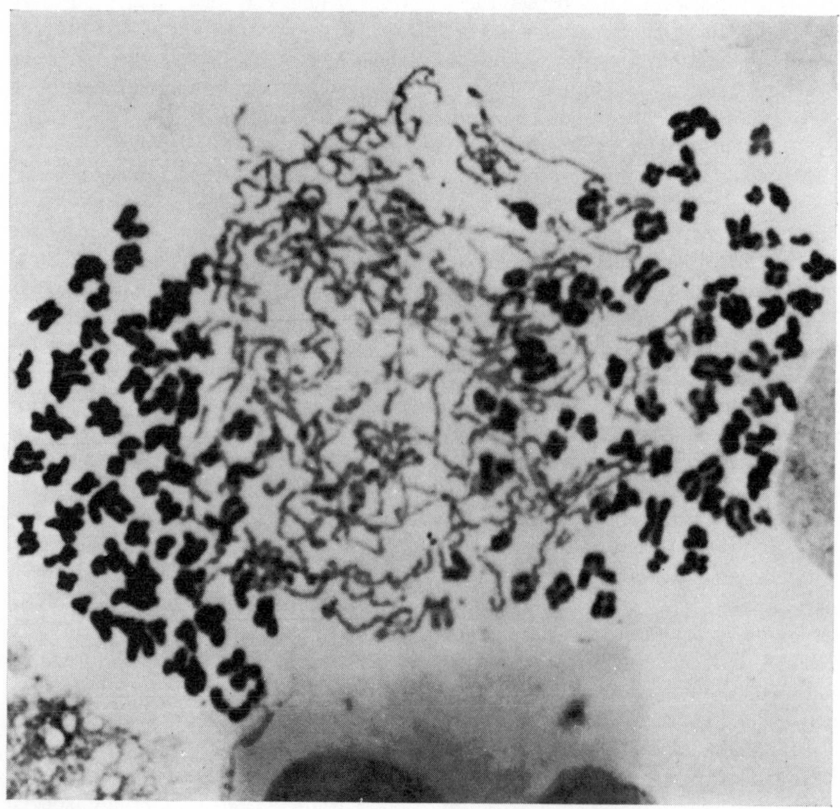

CHROMOSOME CONDENSATION can be forced on cells in other phases of the cycle by inducing fusion with a cell in the *M* phase. Here prematurely condensed chromosomes of a cell in the G_1 phase appear as strands of chromatin (*center*). Not having replicated, the strands are single, unlike the doubled chromosomes of the *M* phase cell (*left and right*).

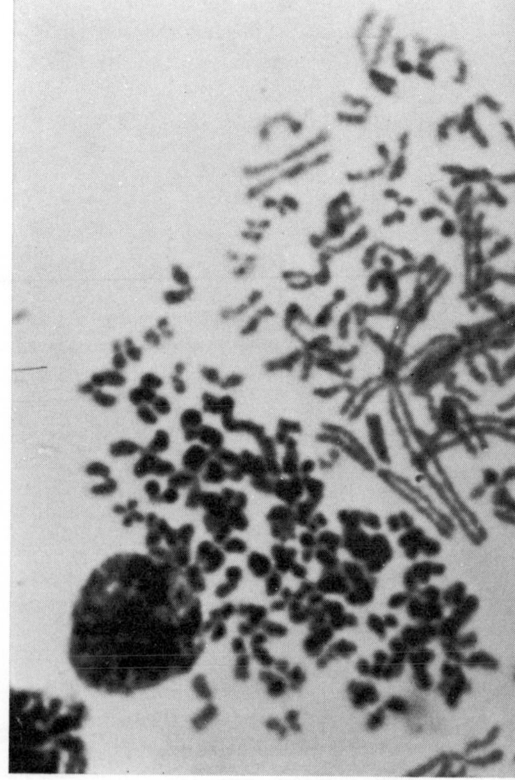

SECOND KIND OF FUSION, between an *M*-phase cell and a cell that has undergone S-phase chromosome replication and entered the G_2 phase, also brings about a premature

chromosome replication begins. Whatever the means may be, it is an effect of the cell society as a whole on its individual members. This can be demonstrated experimentally as follows. If a single-layer sheet of cells is punctured or otherwise wounded, the cell cycle resumes. This happens in all parts of the culture and not only among the cells along the edge of the wound, and reproduction persists until the empty space is filled. The cell cycle is then shut down.

We can hardly imagine that the cycle is shut down merely by the exchange of gentle caresses between adjacent cells. For one thing, different kinds of cells will grow to different degrees of crowding under different conditions before the cycle is halted. For another, contact inhibition can be overcome by making changes in the environment, for example by adding large amounts of blood serum. Moreover, some kinds of cells are not subject to contact inhibition at all.

The study of the cell cycle has led us to some understanding of the governance of cell reproduction as seen from inside the cell. We know that the key event will be found in the supervision of the sequence leading to the switching on of chromosome replication. At the same time studies of the cell cycle in whole organisms, or even in simple societies such as the contact-inhibitable cell cultures, direct us to a further question: How are inside events dictated from the outside?

We can narrow the question in various ways. Does the society give its commands to the cell cycle by circulating hormonelike agents? Examples of such possible agents are the chalones investigated by W. S. Bullough of the University of London and the promine and retine being studied by Albert Szent-Gyorgyi at the Marine Biological Laboratory in Woods Hole, Mass. There is also much evidence that variations in the ordinary ions and small molecules of the cell's external environment can dictate cell cycles.

The Cycle and the Cell Surface

Any influences the outside may have on the inside are monitored by the cell membrane. The part played by the cell membrane in the governance of cell reproduction is now being examined with both interest and excitement. The cell biologists' passionate concern with the cell membrane can be summarized in a statement that sounds sarcastic: Whenever our ordinary explanations break down, we like to say that the final explanation will be found in the cell membrane. This is not really sarcasm. The cell membrane is not a wall or a skin or a sieve. It is an active and responsive part of the cell; it decides what is inside and what is outside and what the outside does to the inside. Cell membranes have "faces" that enable cells to recognize and influence one another. The membranes are also communications systems. Things outside a cell do not necessarily act on the cell interior by passing through the membrane; they may simply change the membrane in some way that causes the membrane, in turn, to make changes in the cell interior.

Today cell surfaces are being observed in a variety of ways; for example, we have seen how scanning electron microscopy is used to examine the outer aspect. Another example is the measurement of membrane potentials: the use of electrodes to determine the difference in voltage between the interior and exterior environments. One group of workers in the Department of Embryology of the Carnegie Institution of Washington has found that the membrane potential changes during the cell cycle, and furthermore that the cell cycle will be halt-

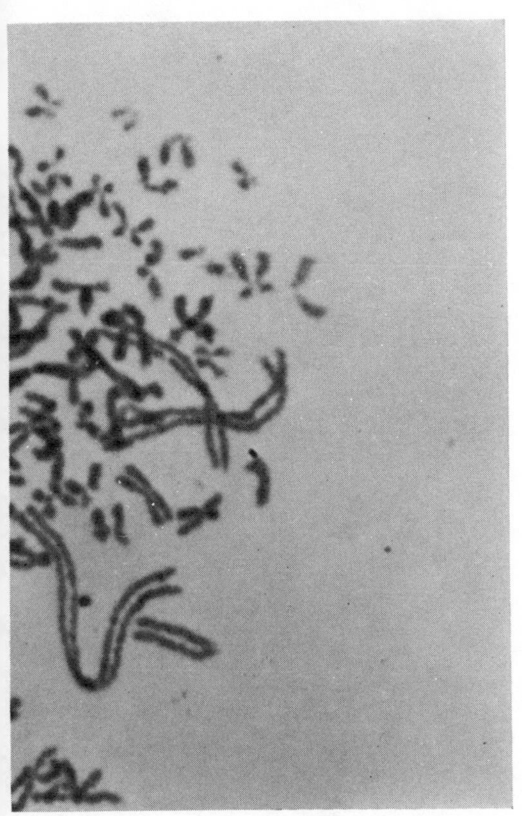

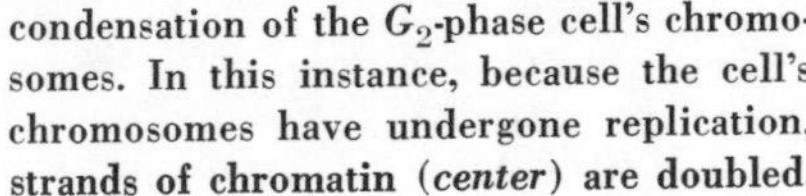

condensation of the G_2-phase cell's chromosomes. In this instance, because the cell's chromosomes have undergone replication, strands of chromatin (*center*) are doubled.

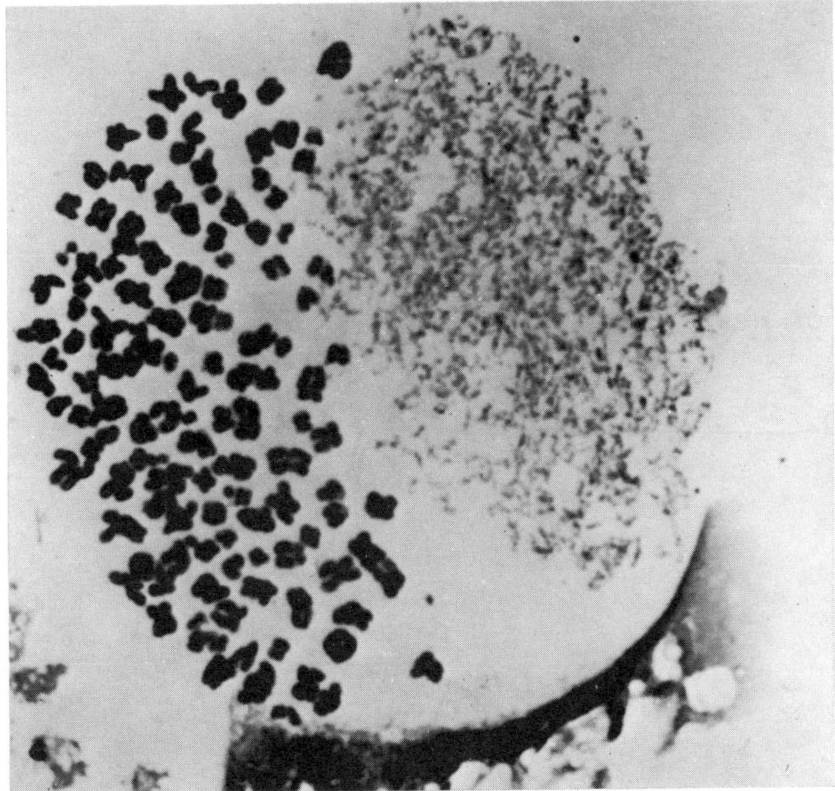

THIRD KIND OF FUSION, between a cell in the *M* phase and a cell in the *S* phase, has a peculiar outcome. The prematurely condensed chromatin does not form strands but instead is fragmented and pulverized (*right*). The three experiments were conducted by Potu N. Rao and Robert T. Johnson at the University of Colorado Medical Center in Denver.

ed by changes in the external environment that induce changes in membrane potential. At the University of Miami School of Medicine, Werner R. Loewenstein's group has made electrode studies of the junctions between cells in contact with each other. The channels that the junctions provide allow ions and small molecules to pass between the cells. They have found evidence suggesting that contact-inhibitable cells make junctions, whereas junctions are not formed between cells that are not.

The study of surface differences between cells that are contact-inhibitable and cells that are not is being hotly pursued because one key to the governance of the cell cycle may be found here. The surfaces of the two kinds of cells are different. Max M. Burger of Princeton University was the first to find that the membrane of contact-inhibitable cells reacts with those plant proteins that induce cycling in noncycling lymphocytes. The reaction is easy to observe; when the protein is present, the cell surfaces become sticky and the cells clump together. Cells that are not contact-inhibitable are unaffected by the presence of the protein. Cell-surface studies such as these are progressing rapidly. If the governance of the cell cycle can be revealed by the cell's surface, then there is every reason to look to the cell membrane for the links between the government and the governed.

Today any student of the cell cycle can expect to be asked just how the cell cycle relates to the cancer problem. If understanding plays some part in the solution of problems, it can be said that the cancer problem *is* the cell-cycle problem. Cells either grow and divide without external restraint or they do not. The many kinds of malignant growth that are called cancer have only one lethal attribute in common: all such cells pursue the cell cycle without restraint.

Studies of the cell cycle as viewed from the interior of the cell tell us what must be restrained: the entrance of the cell into the S phase and chromosome replication. The restraint must come from outside the cell. Cells alone do not switch their cycles on and off, but the governance of the cell cycle from the outside is seen in simple collectives, such as contact-inhibited cultures, as well as in the complex society represented by the whole body. Finally, studies of the cell membrane search for the means whereby outside things control inside events. Inside, outside or surface; any of the three could equally well give us the weapons needed for an attack on uncontrolled growth.

II

PROTEIN STRUCTURE AND FUNCTION: ASSEMBLY OF VIRUSES AND RIBOSOMES

PROTEIN STRUCTURE AND FUNCTION: ASSEMBLY OF VIRUSES AND RIBOSOMES

II

INTRODUCTION

The proteins are undoubtedly the most versatile molecules in the living cell in terms of their structural and functional possibilities. Proteins are macromolecules made up fundamentally of long chains of amino acids linked together in "head-to-tail" fashion by peptide bonds. (These chains are called polypeptides.) Only 20 different amino acids are commonly found in proteins (Fig. 1), but this makes possible an astronomical number of unique polypeptide chains. Since a typical protein may contain as many as 200 amino acids in sequence, 20^{200} unique variants are possible if any one of the 20 distinct monomers can occupy each of the 200 sites. (20^{200} is far larger than 10^{55}, the number of electrons equal in mass to the entire earth!) Furthermore, the lengths of the polypeptide chains can also be varied; proteins range in size from small hormones like insulin, with only 51 amino acids and a molecular weight under 5,000 daltons, to contractile proteins like myosin, with a molecular weight of half a million daltons. The amino acids themselves possess a diversity of "side groups" with different chemical properties; some are hydrophobic and others hydrophilic. By selecting the right proportion of the different amino acids and arranging them in the appropriate sequence, a protein can be constructed that will bind to essentially any other molecule. The particular sequence of amino acids in a protein is thus ultimately responsible for its three-dimensional structure and its biological activity. As enzymes, these large molecules help to "process" small ones—to guide them to assembly sites, to bend them, to break them apart, and to fuse them together in new combinations. For example, the enormous enzyme urease (482,700 daltons), binds the tiny urea molecule (60 daltons) to catalyze its degradation to carbon dioxide and ammonium. Some important insights into the properties of proteins are provided in this section, as we focus upon several examples of catalytic proteins and show how their structures are essential to their regulated activity.

Proteins clearly exemplify a number of important general concepts in the molecular design of living systems. The subunit construction mode, as we have seen, provides for a vast array of diverse structures while requiring a minimum of specialized assembly information: Only 20 different amino acids are needed, and they are connected to one another primarily by the same type of covalent bond. The linear assembly of amino acids in a polypeptide chain constitutes the primary structure of a protein. In three dimensions these chains can assume a variety of secondary configurations in which the structures are stabilized by ionic interactions (hydrogen bonds) between the polar peptide linkages. Hydrogen bonds are weaker connections than the peptide bonds. Thus the linear polypeptide may condense into a helical configuration (the alpha helix) in which hydrogen bonds form between every fourth peptide group along the chain, or it may remain extended (the beta conformation) to form hydrogen-bonded structures with other, adjacent chains. The alpha helix is the major structural component of the α-keratin in wool and in hair, whereas the beta conformation makes up most of the fibroin protein of silk. Collagen, the most abundant protein in the animal kingdom, is made up of three characteristic polypeptide chains wound about each other to form a cable of high tensile strength. Its distinctive fiber properties render it suitable for skin, bone, cartilage, and connective tissue. These cables of collagen aggregate spontaneously to form filaments, which then unite in a systematic way to form long fibrils. Such spontaneous associations are examples of another important construction principle—the principle of self-assembly, in which subunits combine in the correct manner with no energy input and with no information other than that provided by the subunits themselves. The possible and likely secondary and tertiary configurations of proteins are determined solely by their amino acid sequences. The subunit construction mode has an intrinsic quality control feature: Defective units can be rejected at different stages in the assembly process, so that the survival of the cell does not depend upon every amino acid or even every protein being perfect. In summary, the principles of efficient design that have evolved and which apply to biological construction generally are that polymers are built by the enzyme-catalyzed joining of structurally similar monomers followed by the self-assembly of successive substages to produce the finished structure.

Thus far we have mentioned the catalytic roles of proteins as enzymes and have given a few examples of their specialized structural functions. Proteins also provide many other essential services in living systems, of which only a few are documented in this volume. Some proteins are hormones, such as insulin, with its profound regulatory effects on the metabolism of sugar (Thompson, 1955). Others, such as actin and myosin, form the structural basis of various contractile assemblies; they serve as the working components of muscle and

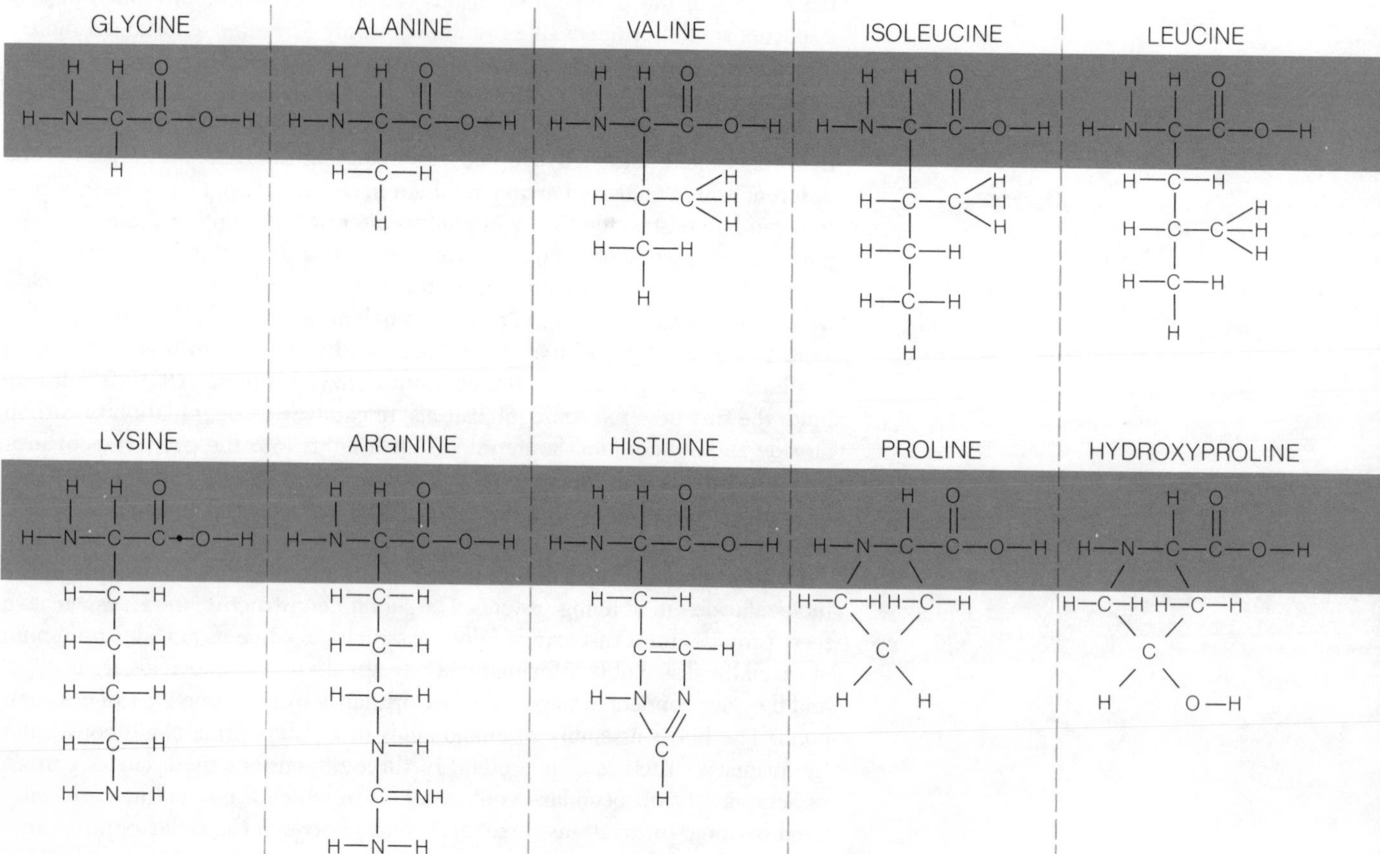

Figure 1. Schematic structures of the 20 amino acids commonly found in proteins. [From "Proteins" by Paul Doty. Copyright © 1957 by Scientific American, Inc. All rights reserved.]

are responsible for the motility of cells and the organized motion of organelles within cells (Murray and Weber, 1974). Oxygen-carrying proteins like myoglobin and hemoglobin are essential to overcome the size limitation otherwise imposed on organisms by the low solubility of oxygen in water. Receptor proteins are involved in neural transmission, and "nerve growth factor" regulates the development of neural networks. Genetic repressors are proteins that operate as regulators of expression of the genetic blueprint at the level of DNA. (See Section IV.) Cytochrome *c* is a protein that first appeared in the biosphere nearly 2 billion years ago to harness the energy produced by oxidation of glucose and other foodstuffs (Dickerson, 1972). (See Section III.) Finally, some specialized proteins serve as antibodies to provide immunity by neutralizing foreign elements that could be deleterious to the organism (Edelman, 1970). The ubiquity of proteins and their essential roles in mediating the processes of life are abundantly clear.

We would like to understand how the structures of different proteins enable them to fulfill their respective roles. A first step is to determine the amino acid composition of a purified protein; using hydrolysis to sever all of the peptide bonds in a protein, we can find out what proportion of the various classes of amino acids are present. The next step is to find out the actual sequences of amino acids in the polypeptide chains. This is a considerably more formidable task and it was a Nobel-prize-winning achievement when the English biochemist Frederick Sanger and co-workers at Cambridge University, after 10 years of intensive effort, reported the amino acid sequences in the two polypeptides that constitute insulin. Innovations in technique and automation now render protein sequencing a straightforward cookbook process. Unfortunately, although the amino acid sequence tells the protein how to fold in three dimensions it doesn't tell *us* how it folds. The direct and only exact approach to that problem is the technique of x-ray diffraction, by which the position of

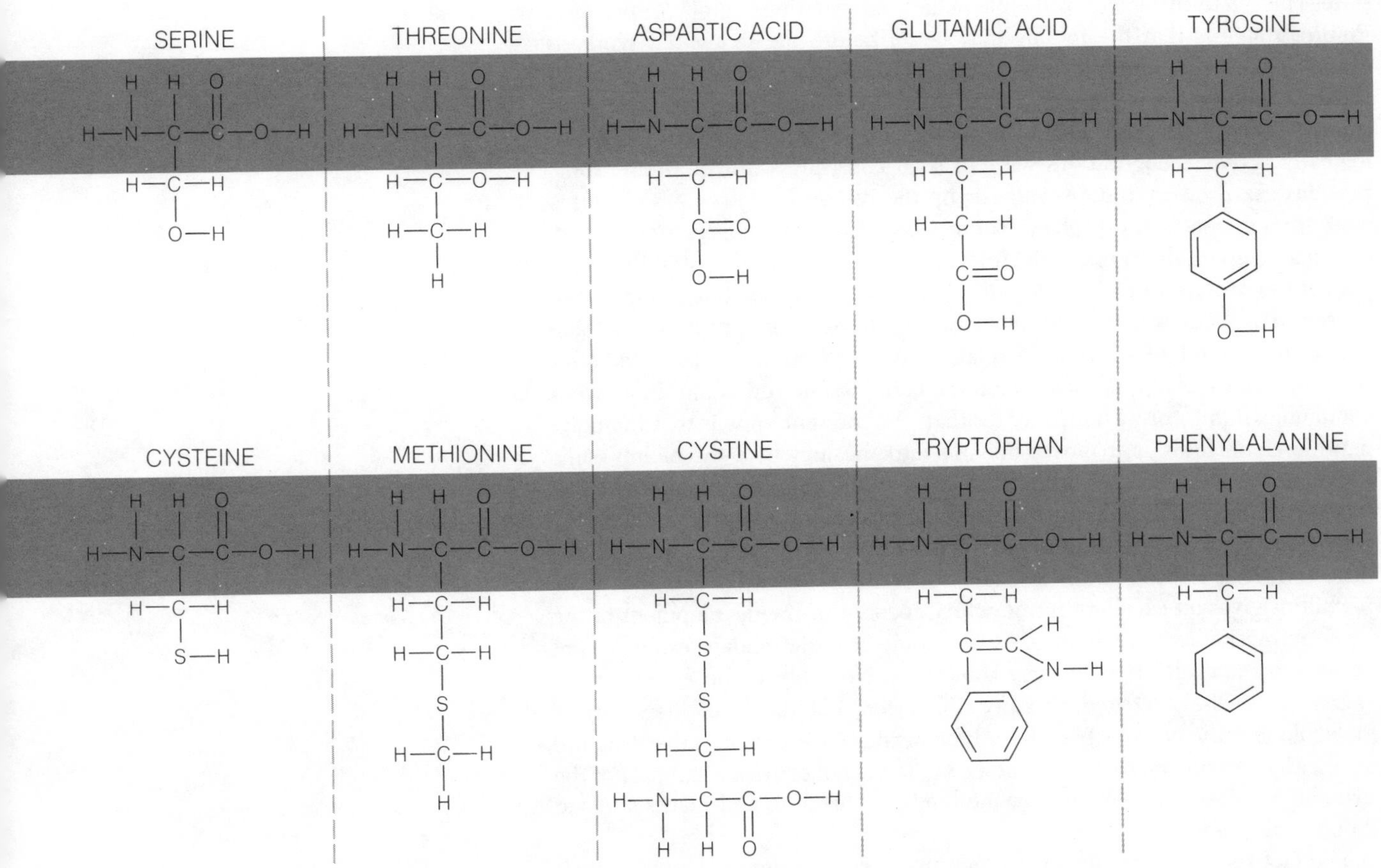

SERINE | THREONINE | ASPARTIC ACID | GLUTAMIC ACID | TYROSINE

CYSTEINE | METHIONINE | CYSTINE | TRYPTOPHAN | PHENYLALANINE

each atom in a crystal can be precisely determined. Unfortunately, proteins in cells are not crystalline. However, when isolated pure proteins are crystallized from solution, enough water is trapped in the structure that the tertiary configuration is normally retained. At the time that Max Perutz, an audacious graduate student at Cambridge, set out to map the positions of the 10,000 atoms in hemoglobin by x-ray crystallography, the most complex organic molecule that had been analyzed contained only 58 atoms. That herculean project took over 20 years and was fully worth the Nobel prize, which Perutz shared with his equally persistent colleague John Kendrew, who solved the structure of myoglobin, a "simpler" protein with only 2,500 atoms! In the first two articles in this section, we can share in the excitement of realizing atomic detail in complex protein molecules. We also will come to appreciate that these giant molecules are not rigid but dynamic in operation. For example, the conformation of hemoglobin is altered slightly as it binds the miniscule oxygen molecule, and then it snaps back as the oxygen is released (Perutz, 1978).

Still another English crystallographer, David C. Phillips, began work on the x-ray analysis of egg white lysozyme in 1960 at the Royal Institution in London, where William Bragg, J. D. Bernal, W. T. Astbury, and others had originally developed this technique. Within two years the general structure of this enzyme molecule was apparent, and within five years a detailed map was constructed that resolved atomic groups at the level of 2 angstroms (about 1/20 the molecular dimensions of the enzyme). The results of his work are summarized in "The Three-Dimensional Structure of an Enzyme Molecule." More recent studies of lysozyme structure by nuclear magnetic resonance confirm that the crystalline form of the enzyme is closely similar to that in solution. The folding pattern of the 129-amino-acid polypeptide chain reveals that hydrophobic groups tend to concentrate in the interior of the molecule, minimizing water contact and thereby attaining the minimum free energy state. This is a fairly general principle in the conformation of globular proteins. Phillips suggests that the folding may begin before the sequential synthesis of the chain is complete. The initial folds at the starting end may form a matrix for the condensation of the growing chain as it is spewed forth from the ribosome. Many polypeptide chains, however, are known to assume their biologically active configurations without such constraints. Furthermore, some proteins are processed after synthesis by the removal of stretches of amino acids from the polypeptide chain. Before these deleted sequences are released, they may help establish the initial folding pattern as well as assist in the transport of the protein to its intended site of action, such as on a membrane. (See Section III.) We now know that lysozyme is synthesized as "pre-lysozyme" and that a short stretch of amino acids is later removed from the starting end after the polypeptide chain has penetrated the endoplasmic reticulum. In a recent communication to me, Phillips writes that "My present view is that the polypeptide chain folds as it moves linearly into the lumen of the endoplasmic reticulum so that the drawings in the article, which suggest folding on the surface of the ribosome, are to some extent misleading." He adds that "the *Scientific American* article has stood the test of time very well . . . there is very little in it that I would want to change if writing it now."

Phillips gives an excellent introductory account of the development of the x-ray diffraction technique, its basis, and some of the attendant experimental hurdles in its application to large molecules. Not only is the active site of lysozyme precisely located by x-ray diffraction, but the fit of the substrate molecule into the active site cleft and the resultant distortion of the structure are clearly demonstrated. Such conformational changes are a usual part of the dynamic process by which an enzyme binds its substrate and places stress at designated bonds.

The fact that enzymes undergo conformational changes as they bind their substrates led D. E. Koshland, Jr., to propose the "induced fit" theory, which holds that the protein is altered to a shape complementary to its substrate, just as "a hand induces a change in the shape of a glove." In his article, "Protein Shape and Biological Control," Koshland explains how effectors interacting with distant sites on an enzyme can alter the active-site conformation to regulate the efficiency with which the catalytic activity is carried out. The principle of "allosteric" control, in which regulatory elements and substrates fit different sites in the protein, was originally conceived by Jacques Monod, Francois Jacob, and Jean-Pierre Changeux at the Pasteur Institute in Paris. Koshland describes allosteric interactions in which a conformational change induced in one subunit of a multisubunit enzyme is transmitted to another subunit. He also explains how the interaction between subunits can cause positive cooperativity in the binding of successive oxygen molecules to hemoglobin and provides a basis for the rationale that a protein with positive cooperativity can be much more sensitive to small fluctuations in the environment than one with a simple binding pattern. The superb illustrations in this article provide an excellent introduction to the mechanism of enzyme action and the ways in which the biochemical pathways in cells are regulated. Finally, the ideas developed with the examples of enzymes and hemoglobin are extended to other specialized proteins.

Before leaving the discussion of hemoglobin, we should mention the classic example of a hereditary molecular disease, sickle-cell anemia, in which one amino acid change in the beta subunit of this protein causes profound clinical symptoms in humans: resistance to malaria if one beta subunit is altered, but chronic hemolytic anemia if both are abnormal. It was shown by V. M. Ingram at the Cavendish Laboratory in Cambridge that the polar amino acid glutamine is replaced by the hydrophobic amino acid valine at position 6 in the beta subunit of the defective hemoglobin. Thus a diseased condition can result from a "mutation" affecting just one particular amino acid in a particular peptide chain in an organism as complex as man. (See section IV.)

We turn next to higher levels of molecular assembly in living cells, citing three examples of multiprotein complexes that additionally include nucleic acids: a bacterial virus, a plant virus, and the ribosome. These examples again illustrate the principles of subassembly and self-assembly, and they further develop the concept that macromolecules may be processed after they are synthesized.

Just below the borderline between the free-living and the pre-living is the virus, a macromolecular assembly that includes a nucleic acid genome, a few proteins, and a semipermeable outer capsid, usually composed of protein. The virus is a kind of regressed parasite, in that it is unable to proliferate without the metabolic machinery of a free-living cell as "host." In particular, the energy-generating equipment and the protein-synthesizing facility must normally be provided by the host. Some simple viruses, such as ϕX174, even rely on the host for replication of their genome, whereas others, such as the T4 bacteriophage, duplicate many biochemical functions that the host could otherwise provide. Different viruses utilize different nucleic acid types and structures as their genomes; these can be either DNA or RNA, either single-stranded or double-stranded, and in either closed circular or linear conformations. A still simpler "parasite," probably the simplest, is the viroid—merely a free single strand of RNA with no protein coat. The known viroids have genomes with molecular weights in the range of 10^5 daltons, and they have been implicated in certain important diseases in cultivated plants. In a sense the one molecule of a viroid becomes a living entity when it is appropriately assisted in the host cell. But in another sense it is no more a living entity than is the resident genome of that host cell.

A typical viral life cycle is outlined by William B. Wood and R. S. Edgar in their article, "Building a Bacterial Virus." The viral DNA is injected into the host cell, the capsid being left behind. The viral genome is replicated, and that genome also provides the essential information for construction of new capsids. The new DNA is packaged in the new capsids, and the cell bursts open to release the progeny viruses. However, for many types of viruses this lytic mode of infection is just one possible outcome. Some viruses, such as ϕX174, continue to proliferate and are exuded into the medium with nearly negligible effect on the host cells. The DNA of certain other viruses, such as phage lambda, may be physically incorporated into the host genome. (See Campbell's article in Section V.) Some cancer-producing RNA viruses undergo a process called "reverse transcription," in which a DNA copy is made from the RNA master blueprint so that this viral DNA transcript may then be integrated into the host genome DNA (Temin, 1972). Viruses exert various levels of control over the growth of the infected cell, with consequences ranging from redirected metabolic goals to "cellucide."

What is required to assemble an infective virus structure from its component parts once those parts have been synthesized in the host cell? Twenty-five years ago the reconstitution of functional tobacco mosaic virus (TMV) was accomplished in the test tube by Heinz Fraenkel-Conrat and Robley Williams at the University of California in Berkeley. The single-strand RNA genome appropriately combined with the sheath proteins to form the infective rod-shaped structure. This is clearly a demonstration of self-assembly. However, TMV is a very simple virus with less than half a dozen genes in its genome and 2,130 *identical* protein subunits to form its capsid. Consider in contrast the size of the genome and the complexity of the capsid for the T4 bacteriophage, as described by Wood and Edgar. It is now known that over 50 different genes are essential for morphogenesis, a number that exceeds the estimate of 40 different proteins in the T4 structure (Wood, 1979). Thus some accessory proteins are evidently required that are not destined to become a part of the final virus particle. Such nonstructural accessory proteins include "scaffolding proteins" (possibly analogous to the support framework that falls away when a rocket is launched); proteins that may stabilize weak associations of subunits until stable connections are made; and true enzymes, particularly "proteases," which cleave portions of specified polypeptide chains at particular stages in the assembly process. Wood and Edgar illustrate the complex morphogenetic pathway that was elucidated through painstaking effort by them and their associates at Caltech. The only substantive correction to the scheme proposed in 1967 is that the gene No. 9 product is now known to operate along with the products of genes No. 11 and No. 12 in tail end-plate assembly rather than at a later stage. T4 morphogenesis serves as a carefully worked out example of the step-by-step intricacy with which living cells construct supramolecular complexes. This genetic dissection of T4 phage assembly was made possible by Edgar and Epstein (1967) and their associates, who first exploited conditionally lethal mutants for the study of essential components in the viral replication cycle.

In "The Assembly of a Virus," P. Jonathan G. Butler and Aaron Klug present a detailed view of the plant virus TMV, which contains only one kind of protein. Although the construction proceeds without the need for accessory proteins, it does require a unique structural feature of the RNA genome—a hairpin configuration in which the RNA strand folds back upon itself. In contrast with the packaging of T4, in which the capsid is assembled before the nucleic acid is added, the packaging of TMV proceeds hand in hand with the construction of the capsid. One can only marvel at the stepwise self-assembly of this simple protein-nucleic acid structure and wonder how many millions of years were needed for its evolution to the present stage of perfection.

The supreme example of self-assembly within living cells is the ribosome, that essential workbench upon which proteins are synthesized. However, the ribosome is much more than a workbench, both in its complexity and in its active participation in decoding the genetic message. Ribosomes are constructed from over 50 different proteins plus three distinct RNA molecules.

These are built into the architecture of two separable subunits that combine to complete the structure after the messenger-RNA instruction tape has attached to one of these subunits to initiate protein synthesis. In *E. coli* the number of ribosomes is adjusted from 5,000 to 30,000 per cell, depending upon the growth conditions. Clearly the number of active ribosomes is a prime determinant for the rate at which new proteins can be made. The *E. coli* ribosome consists of a 30S and a 50S subunit, which unite to form the 70S complete organelle. (The "S" value refers to the sedimentation rate in a centrifuge tube; it is related to the size and shape of the particle.) It was a monumental achievement when Masayasu Nomura and his research group at the University of Wisconsin first reconstituted functional 30S ribosomal subunits by combining the 16S ribosomal RNA molecule with the 19 distinct proteins that constitute the structure. He describes his systematic approach to this task in his article on "Ribosomes." The reconstitution of the 50S subunit was even more difficult, but was accomplished by Nomura with the more thermo-stable structure from *Bacillus stearothermophilus* in 1970. Four years later K. H. Nierhaus and F. Dohme in Berlin succeeded in assembling the *E. coli* 50S subunit in the test tube.

What general features about the assembly of biological structures have we learned from the studies on ribosome reconstitution? Careful analysis of the reconstituted particles has shown that they are not identical to those assembled *in vivo*. The slight differences are probably due to the processing of both the protein and RNA components as the particle is built in cells. The *in-vivo* assembly of the 30S ribosomal subunit uses a precursor RNA molecule somewhat larger than 16S. This is cleaved near each end and also subjected to methylation at certain sites. The 16S RNA for the 30S subunit, the 23S and the 5S RNA for the 50S subunit, a number of transfer RNA molecules, and a few sequences of "spacer" RNA with no evident biological role are all transcribed from the master template DNA as one long, flexible coil of RNA. The precursor RNA is then systematically cut up and processed for the respective structures and roles. We now know that in the construction of ribosomal subunits, the RNA operates in part as a structural element for organizing the assembly. Thus the first half dozen proteins must bind to the 16S RNA before the next group is added. Nevertheless, the overall construction of ribosomes is apparently designed to require only that information provided by the sequences of amino acids in the proteins and the unique ribosomal RNA species.

REFERENCES CITED AND SUGGESTED FURTHER READING

Bunn, H. F., B. G. Forget, and H. M. Ranney. 1977. *Human Hemoglobins*. Saunders, Philadelphia.

Butler, P. J. G., and A. C. H. Durham. 1977. "Tobacco mosaic virus protein aggregation and the virus assembly." *Adv. Protein Chem.* **31**:187–251.

Champness, J. N., A. C. Bloomer, G. Bricogne, P. J. G. Butler, and A. Klug. 1976. "The structure of the protein disk of tobacco mosaic virus to 5 Å resolution." Nature **259**:20–24.

Dickerson, R. E. 1972. "The structure and history of an ancient protein." *Scientific American*, April. (Offprint No. 1245.)

Dickerson, R. E., and I. Geis. 1973. *The Structure and Action of Proteins*. Benjamin, Menlo Park, CA.

Diener, T. O. 1979. "Viroids." *Science* **205**:859–866.

Edelman, G. M. 1970. "The structure and function of antibodies." *Scientific American*, August. (Offprint No. 1185.)

Edgar, R. S., and Epstein, R. H. 1965. "The genetics of a bacterial virus." *Scientific American*, February. (Offprint No. 1004.)

Edgar, R. S., and R. H. Epstein. 1963. "Conditional mutations in bacteriophage." In S. J. Geerts (editor), *Genetics Today*. Pergamon Press, Elmsford, NY.

Koshland, D. E., Jr., and K. E. Neet. 1968. "The catalytic and regulatory properties of enzymes." *Ann. Rev. Biochem.* **37**:359–410.

Kurland, C. G. 1977. "Structure and function of the bacterial ribosome." *Ann Rev. Biochem.* **46**:173–200.

Monod, J., J. Wyman, and J. P. Changeux. 1965. "On the nature of allosteric transition: A plausible model." *J. Mol. Biol.* **12**:88–118.

Murray, J., and Weber, A. 1974. "The cooperative action of muscle proteins." *Scientific American*, February. (Offprint No. 1290.)

Nomura, M. 1973. "Assembly of bacterial ribosomes." *Science* **179**:864–873.

Nomura, M., A. Tissieres, and P. Lengyel (editors). 1975. *Ribosomes.* Cold Spring Harbor Laboratory, Long Island.

Perutz, M. F. 1978. "Hemoglobin structure and respiratory transport." *Scientific American*, December. (Offprint No. 1413.)

Richards, K. E., and R. C. Williams. 1976. "Assembly of tobacco mosaic virus *in vitro.*" In H. Fraenkel-Conrat and R. R. Wagner (editors), *Comprehensive Virology* (Vol. 6). Plenum Press, New York.

Stryer, L. 1975. *Biochemistry.* W. H. Freeman and Company, San Francisco.

Temin, H. 1972. "RNA-directed DNA synthesis." *Scientific American*, January. (Offprint No. 1239.)

Thompson, E. O. P. 1955. "The insulin molecule." *Scientific American*, January. (Offprint No. 469.)

Vogel, H. J. (editor). 1977. *Nucleic Acid-Protein Recognition.* Academic Press, New York.

Wood, W. B. 1974. "Molecular design in living systems" (Unit 2). In P. R. Ehrlich, R. W. Holm, and P. C. Hanawalt (editors), *Biocore.* McGraw-Hill, New York.

Wood, W. B. 1979. "Bacteriophage T4 assembly and the morphogenesis of subcellular structure." The Harvey Lectures, Series 73, pp. 203–223.

Zimmern, D., and P. J. G. Butler. 1977. "The isolation of tobacco mosaic virus RNA fragments containing the origin for viral assembly." *Cell* **11**:455–462.

The Three-Dimensional Structure of an Enzyme Molecule

5

by David C. Phillips
November 1966

The arrangement of atoms in an enzyme molecule has been worked out for the first time. The enzyme is lysozyme, which breaks open cells of bacteria. The study has shown how lysozyme performs its task.

One day in 1922 Alexander Fleming was suffering from a cold. This is not unusual in London, but Fleming was a most unusual man and he took advantage of the cold in a characteristic way. He allowed a few drops of his nasal mucus to fall on a culture of bacteria he was working with and then put the plate to one side to see what would happen. Imagine his excitement when he discovered some time later that the bacteria near the mucus had dissolved away. For a while he thought his ambition of finding a universal antibiotic had been realized. In a burst of activity he quickly established that the antibacterial action of the mucus was due to the presence in it of an enzyme; he called this substance lysozyme because of its capacity to lyse, or dissolve, the bacterial cells. Lysozyme was soon discovered in many tissues and secretions of the human body, in plants and most plentifully of all in the white of egg. Unfortunately Fleming found that it is not effective against the most harmful bacteria. He had to wait seven years before a strangely similar experiment revealed the existence of a genuinely effective antibiotic: penicillin.

Nevertheless, Fleming's lysozyme has proved a more valuable discovery than he can have expected when its properties were first established. With it, for example, bacterial anatomists have been able to study many details of bacterial structure [see "Fleming's Lysozyme," by Robert F. Acker and S. E. Hartsell; SCIENTIFIC AMERICAN, June, 1960]. It has now turned out that lysozyme is the first enzyme whose three-dimensional structure has been determined and whose properties are understood in atomic detail. Among these properties is the way in which the enzyme combines with the substance on which it acts—a complex sugar in the wall of the bacterial cell.

Like all enzymes, lysozyme is a protein. Its chemical makeup has been established by Pierre Jollès and his colleagues at the University of Paris and by Robert E. Canfield of the Columbia University College of Physicians and Surgeons. They have found that each molecule of lysozyme obtained from egg white consists of a single polypeptide chain of 129 amino acid subunits of 20 different kinds. A peptide bond is formed when two amino acids are joined following the removal of a molecule of water. It is customary to call the portion of the amino acid incorporated into a polypeptide chain a residue, and each residue has its own characteristic side chain. The 129-residue lysozyme molecule is cross-linked in four places by disulfide bridges formed by the combination of sulfur-containing side chains in different parts of the molecule [*see illustration on page 69*].

The properties of the molecule cannot be understood from its chemical constitution alone; they depend most critically on what parts of the molecule are brought close together in the folded three-dimensional structure. Some form of microscope is needed to examine the structure of the molecule. Fortunately one is effectively provided by the techniques of X-ray crystal-structure analysis pioneered by Sir Lawrence Bragg and his father Sir William Bragg.

The difficulties of examining molecules in atomic detail arise, of course, from the fact that molecules are very small. Within a molecule each atom is usually separated from its neighbor by about 1.5 angstrom units (1.5×10^{-8} centimeter). The lysozyme molecule, which contains some 1,950 atoms, is about 40 angstroms in its largest dimension. The first problem is to find a microscope in which the atoms can be resolved from one another, or seen separately.

The resolving power of a microscope depends fundamentally on the wavelength of the radiation it employs. In general no two objects can be seen separately if they are closer together than about half this wavelength. The shortest wavelength transmitted by optical microscopes (those working in the ultraviolet end of the spectrum) is about 2,000 times longer than the distance between atoms. In order to "see" atoms one must use radiation with a much shorter wavelength: X rays, which have a wavelength closely comparable to interatomic distances. The employment of X rays, however, creates other difficulties: no satisfactory way has yet been found to make lenses or mirrors that will focus them into an image. The problem, then, is the apparently impossible one of designing an X-ray microscope without lenses or mirrors.

Consideration of the diffraction theory of microscope optics, as developed by Ernst Abbe in the latter part of the 19th century, shows that the problem can be solved. Abbe taught us that the formation of an image in the microscope can be regarded as a two-stage process. First, the object under examination scatters the light or other radiation falling on it in all directions, forming a diffraction pattern. This pattern arises because the light waves scattered from different parts of the object combine so as to produce a wave of large or small amplitude in any direction according to whether the waves are in or out of phase—in or out of step—with one another. (This effect is seen most easily in light waves scattered by a regularly repeating structure, such as a diffraction grating made of lines scribed at regular intervals on a glass plate.) In the second stage of image formation, according to Abbe, the objective lens of the microscope collects the diffracted waves and recombines them to form an image of the object. Most important, the nature of the image depends critically on how much of the diffraction pattern is used in its formation.

X-Ray Structure Analysis

In essence X-ray structure analysis makes use of a microscope in which the two stages of image formation have been separated. Since the X rays cannot be focused to form an image directly, the diffraction pattern is recorded and the image is obtained from it by calculation. Historically the method was not developed on the basis of this reasoning, but this way of regarding it (which was first suggested by Lawrence Bragg) brings out its essential features and also introduces the main difficulty of applying it. In recording the intensities of the diffracted waves, instead of focusing them to form an image, some crucial information is lost, namely the phase relations among the various diffracted waves. Without this information the image cannot be formed, and some means of recovering it has to be found. This is the well-known phase problem of X-ray crystallography. It is on the solution of the problem that the utility of the method depends.

The term "X-ray crystallography" reminds us that in practice the method was developed (and is still applied) in the study of single crystals. Crystals suitable for study may contain some 10^{15} identical molecules in a regular array; in effect the molecules in such a crystal diffract the X radiation as though they were a single giant molecule. The crystal acts as a three-dimensional diffraction grating, so that the waves scattered by them are confined to a number of discrete directions. In order to obtain a three-dimensional image of the structure the intensity of the X rays scattered in these different directions must be measured, the phase problem must be solved somehow and the measurements must be combined by a computer.

The recent successes of this method in the study of protein structures have depended a great deal on the development of electronic computers capable of performing the calculations. They are due most of all, however, to the discovery in 1953, by M. F. Perutz of the Medical Research Council Laboratory of Molecular Biology in Cambridge, that the method of "isomorphous replacement" can be used to solve the phase problem in the study of protein crystals. The method depends on the preparation and study of a series of protein crystals into which additional heavy atoms, such as atoms of uranium, have been introduced without otherwise affecting the crystal structure. The first successes of this method were in the study of sperm-whale myoglobin by John C. Kendrew of the Medical Research Council Laboratory and in Perutz' own study of horse hemoglobin. For their work the two men received the Nobel prize for chemistry in 1962 [see "The Three-dimensional Structure of a Protein Molecule," by John C. Kendrew; SCIENTIFIC AMERICAN Offprint 121, and "The Hemoglobin Molecule," by M. F. Perutz; SCIENTIFIC AMERICAN Offprint 196].

Because the X rays are scattered by the electrons within the molecules, the image calculated from the diffraction pattern reveals the distribution of electrons within the crystal. The electron density is usually calculated at a regular array of points, and the image is made visible by drawing contour lines through points of equal electron density. If these contour maps are drawn on clear plastic sheets, one can obtain a three-dimensional image by assembling the maps one above the other in a stack. The amount of detail that can be seen in such an image depends on the resolving power of the effective microscope, that is, on its "aperture," or the extent of the diffraction pattern that has been included in the formation of the image. If the waves diffracted through sufficiently high angles are included (corresponding to a large aperture), the atoms appear as individual peaks in the image map. At lower resolution groups of unresolved atoms appear with characteristic shapes by which they can be recognized.

The three-dimensional structure of lysozyme crystallized from the white of hen's egg has been determined in atomic detail with the X-ray method by our group at the Royal Institution in London. This is the laboratory in which Humphry Davy and Michael Faraday made their fundamental discoveries during the 19th century, and in which the X-ray method of structure analysis was developed between the two world wars by the brilliant group of workers led by William Bragg, including J. D. Bernal, Kathleen Lonsdale, W. T. Astbury, J. M. Robertson and many others. Our work on lysozyme was begun in 1960 when Roberto J. Poljak, a visiting worker from Argentina, demonstrated that suitable crystals containing heavy atoms could be prepared. Since then C. C. F. Blake, A. C. T. North, V. R. Sarma, Ruth Fenn, D. F. Koenig, Louise N. Johnson and G. A. Mair have played important roles in the work.

ALA	ALANINE	GLY	GLYCINE	PRO	PROLINE
ARG	ARGININE	HIS	HISTIDINE	SER	SERINE
ASN	ASPARAGINE	ILEU	ISOLEUCINE	THR	THREONINE
ASP	ASPARTIC ACID	LEU	LEUCINE	TRY	TRYPTOPHAN
CYS	CYSTEINE	LYS	LYSINE	TYR	TYROSINE
GLN	GLUTAMINE	MET	METHIONINE	VAL	VALINE
GLU	GLUTAMIC ACID	PHE	PHENYLALANINE		

TWO-DIMENSIONAL MODEL of the lysozyme molecule is shown on the opposite page. Lysozyme is a protein containing 129 amino acid subunits, commonly called residues (*see key to abbreviations above*). These residues from a polypeptide chain that is cross-linked at four places by disulfide (—S—S—) bonds. The amino acid sequence of lysozyme was determined independently by Pierre Jollès and his co-workers at the University of Paris and by Robert E. Canfield of the Columbia University College of Physicians and Surgeons. The three-dimensional structure of the lysozyme molecule has now been established with the help of X-ray crystallography by the author and his colleagues at the Royal Institution in London. A painting of the molecule's three-dimensional structure appears on pages 70 and 71. The function of lysozyme is to split a particular long-chain molecule, a complex sugar, found in the outer membrane of many living cells. Molecules that are acted on by enzymes are known as substrates. The substrate of lysozyme fits into a cleft, or pocket, formed by the three-dimensional structure of the lysozyme molecule. In the two-dimensional model on the opposite page the amino acid residues that line the pocket are shown in dark green.

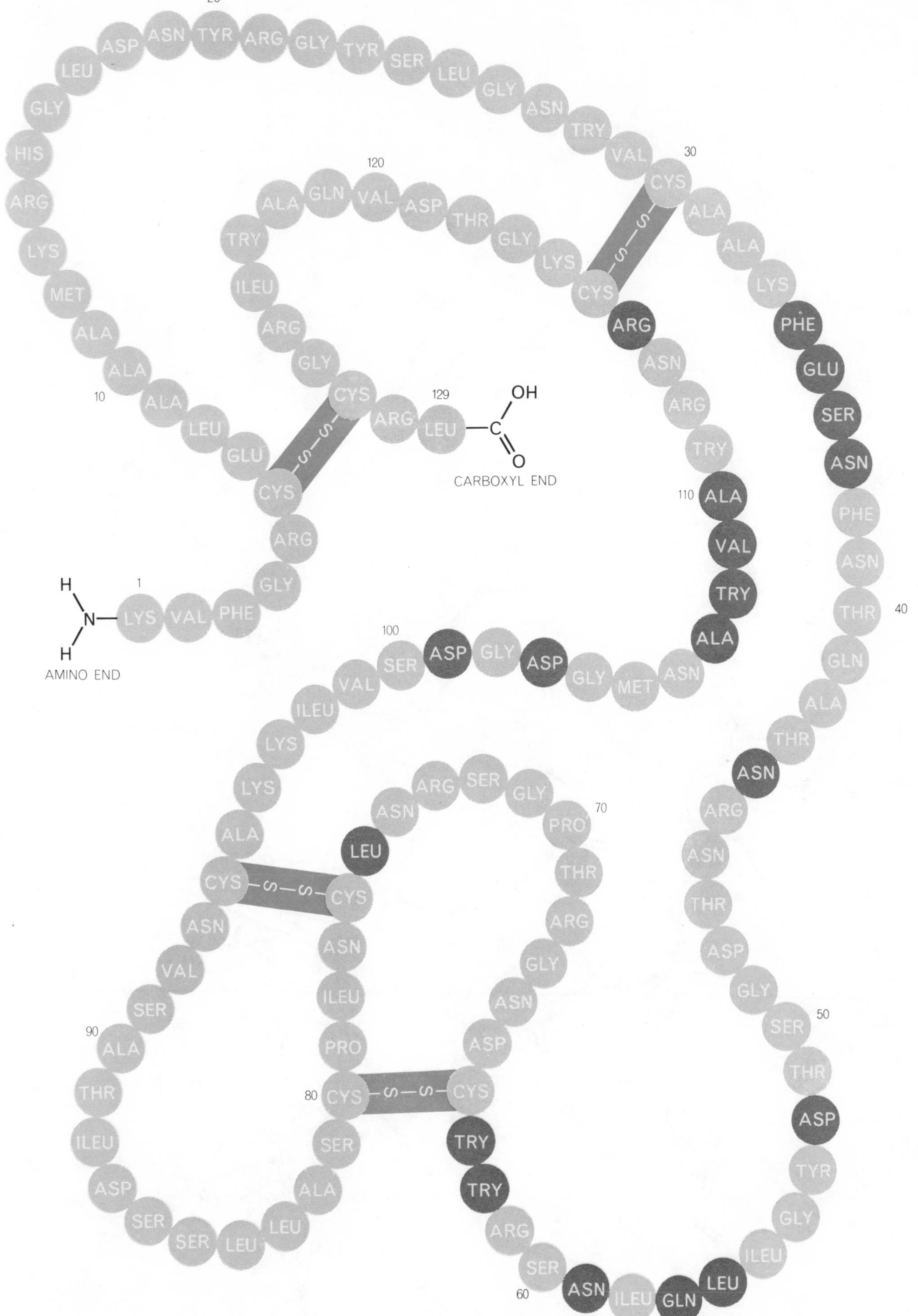
H
N
H
AMINO END
1
LYS
VAL
PHE
GLY
ARG
CYS
GLU
LEU
ALA
10
ALA
ALA
MET
LYS
ARG
HIS
GLY
LEU
ASP
ASN
20
TYR
ARG
GLY
TYR
SER
LEU
GLY
ASN
TRY
VAL
30
CYS
ALA
ALA
LYS
PHE
GLU
SER
ASN
PHE
ASN
40
THR
GLN
ALA
THR
ASN
ARG
ASN
THR
ASP
GLY
50
SER
THR
ASP
TYR
GLY
ILEU
LEU
GLN
ILEU
ASN
60
SER
ARG
TRY
TRY
CYS
ASP
ASN
GLY
ARG
THR
70
PRO
GLY
SER
ARG
ASN
LEU
CYS
ASN
ILEU
PRO
80
CYS
SER
ALA
LEU
LEU
SER
SER
ASP
ILEU
THR
90
ALA
SER
VAL
ASN
CYS
ALA
LYS
LYS
ILEU
VAL
100
SER
ASP
GLY
ASP
GLY
MET
ASN
ALA
TRY
VAL
110
ALA
TRY
ARG
ASN
ARG
CYS
LYS
GLY
THR
ASP
120
VAL
GLN
ALA
TRY
ILEU
ARG
GLY
CYS
ARG
129
LEU
C
OH
O
CARBOXYL END
S
S

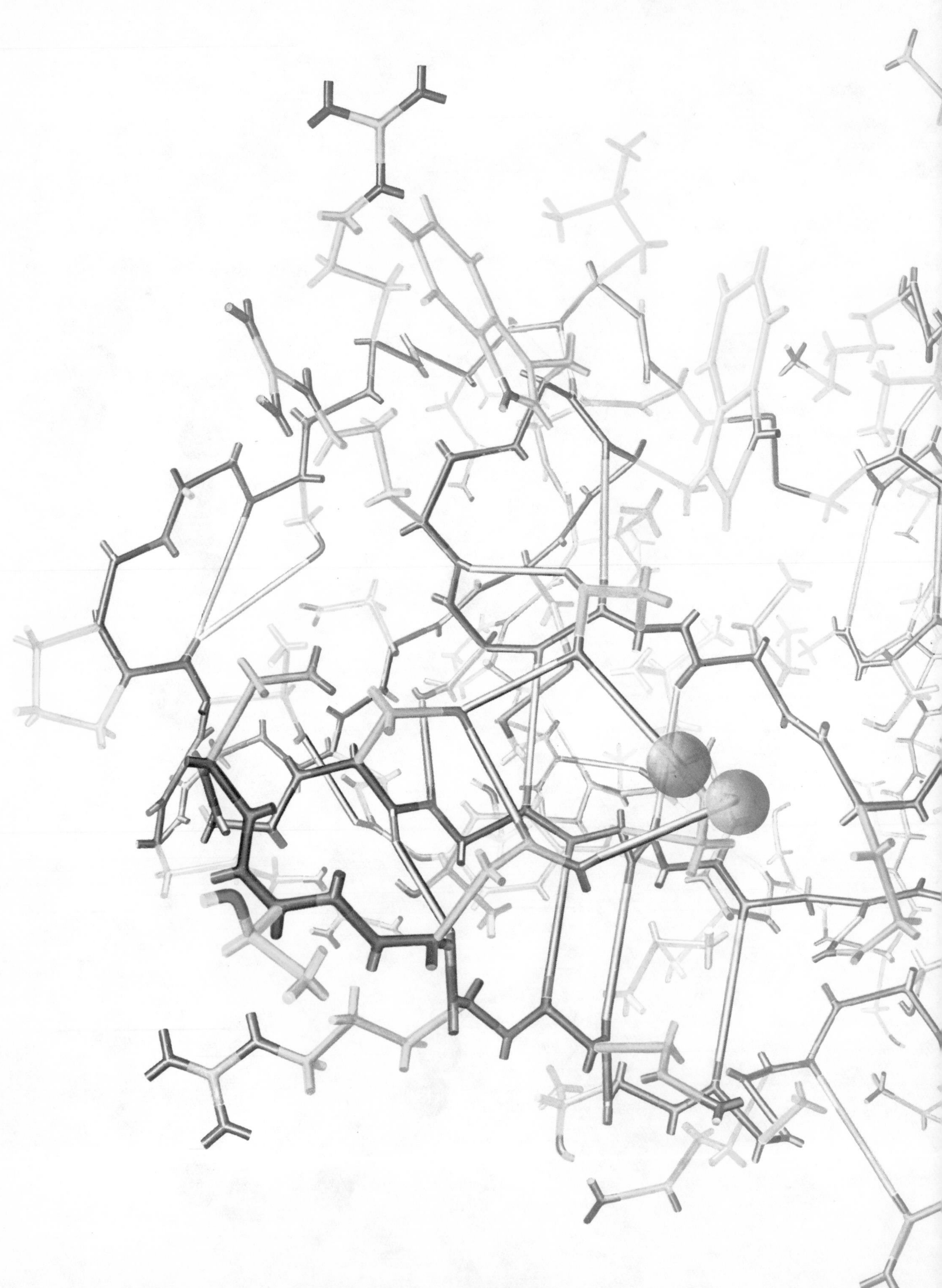

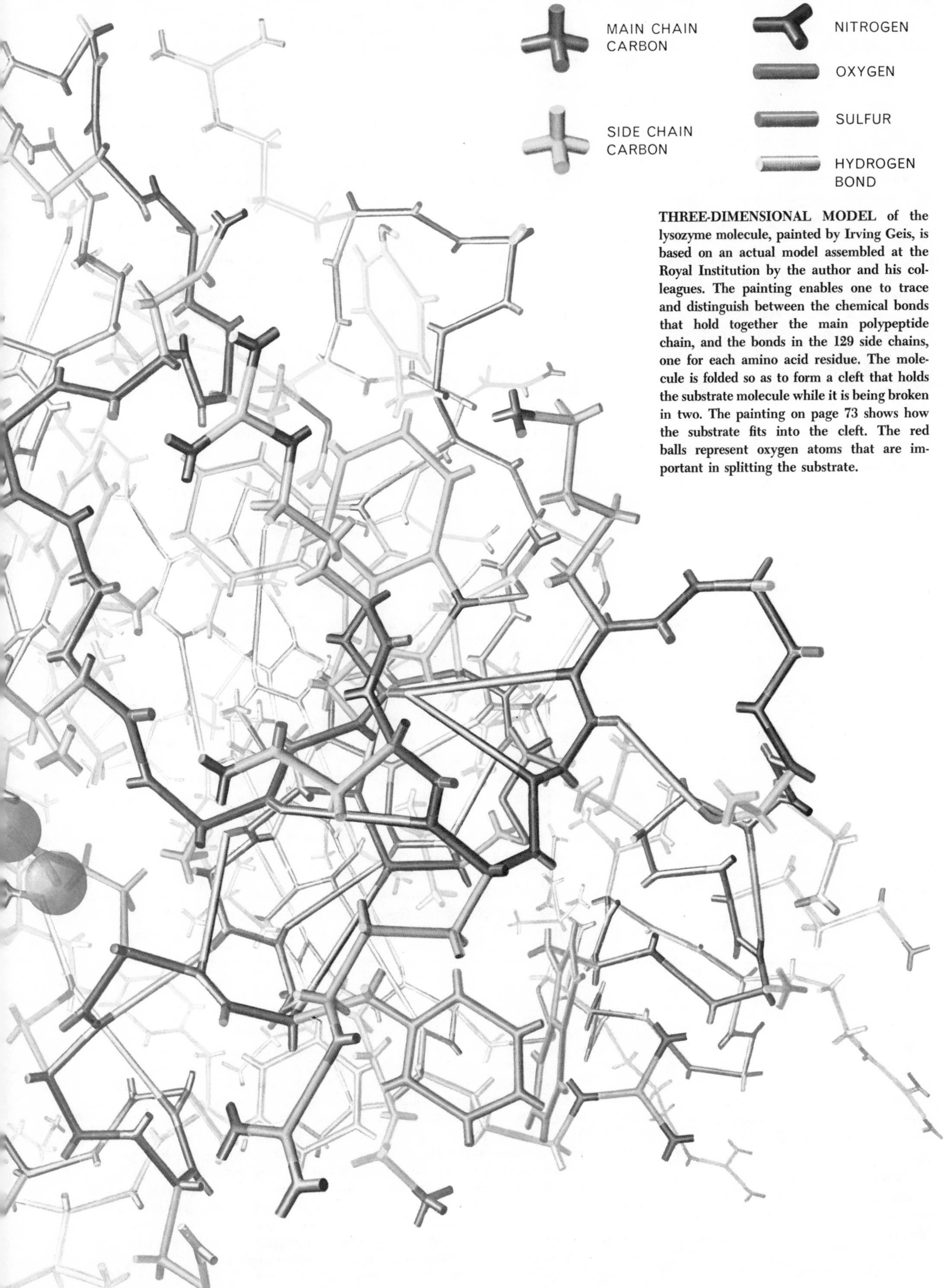

THREE-DIMENSIONAL MODEL of the lysozyme molecule, painted by Irving Geis, is based on an actual model assembled at the Royal Institution by the author and his colleagues. The painting enables one to trace and distinguish between the chemical bonds that hold together the main polypeptide chain, and the bonds in the 129 side chains, one for each amino acid residue. The molecule is folded so as to form a cleft that holds the substrate molecule while it is being broken in two. The painting on page 73 shows how the substrate fits into the cleft. The red balls represent oxygen atoms that are important in splitting the substrate.

In 1962 a low-resolution image of the structure was obtained that revealed the general shape of the molecule and showed that the arrangement of the polypeptide chain is even more complex than it is in myoglobin. This low-resolution image was calculated from the amplitudes of about 400 diffraction maxima measured from native protein crystals and from crystals containing each of three different heavy atoms. In 1965, after the development of more efficient methods of measurement and computation, an image was calculated on the basis of nearly 10,000 diffraction maxima, which resolved features separated by two angstroms. Apart from showing a few well-separated chloride ions, which are present because the lysozyme is crystallized from a solution containing sodium chloride, the two-angstrom image still does not show individual atoms as separate maxima in the electron-density map. The level of resolution is high enough, however, for many of the groups of atoms to be clearly recognizable.

The Lysozyme Molecule

The main polypeptide chain appears as a continuous ribbon of electron density running through the image with regularly spaced promontories on it that are characteristic of the carbonyl groups (CO) that mark each peptide bond. In some regions the chain is folded in ways that are familiar from theoretical studies of polypeptide configurations and from the structure analyses of myoglobin and fibrous proteins such as the keratin of hair. The amino acid residues in lysozyme have now been designated by number; the residues numbered 5 through 15, 24 through 34 and 88 through 96 form three lengths of "alpha helix," the conformation that was proposed by Linus Pauling and Robert B. Corey in 1951 and that was found by Kendrew and his colleagues to be the most common arrangement of the chain in myoglobin. The helixes in lysozyme, however, appear to be somewhat distorted from the "classical" form, in which four atoms (carbon, oxygen, nitrogen and hydrogen) of each peptide group lie in a plane that is parallel to the axis of the alpha helix. In the lysozyme molecule the peptide groups in the helical sections tend to be rotated slightly in such a way that their CO groups point outward from the helix axes and their imino groups (NH) inward.

The amount of rotation varies, being slight in the helix formed by residues 5 through 15 and considerable in the one formed by residues 24 through 34. The effect of the rotation is that each NH group does not point directly at the CO group four residues back along the chain but points instead between the CO groups of the residues three and four back. When the NH group points directly at the CO group four residues back, as it does in the classical alpha helix, it forms with the CO group a hydrogen bond (the weak chemical bond in which a hydrogen atom acts as a bridge). In the lysozyme helixes the hydrogen bond is formed somewhere between two CO groups, giving rise to a structure intermediate between that of an alpha helix and that of a more symmetrical helix with a three-fold symmetry axis that was discussed by Lawrence Bragg, Kendrew and Perutz in 1950. There is a further short length of helix (residues 80 through 85) in which the hydrogen-bonding arrangement is quite close to that in the three-fold helix, and also an isolated turn (residues 119 through 122) of three-fold helix. Furthermore, the peptide at the far end of helix 5 through 15 is in the conformation of the three-fold helix, and the hydrogen bond from its NH group is made to the CO three residues back rather than four.

Partly because of these irregularities in the structure of lysozyme, the proportion of its polypeptide chain in the alpha-helix conformation is difficult to calculate in a meaningful way for comparison with the estimates obtained by other methods, but it is clearly less than half the proportion observed in myoglobin, in which helical regions make up about 75 percent of the chain. The lysozyme molecule does include, however, an example of another regular conformation predicted by Pauling and Corey. This is the "antiparallel pleated sheet," which is believed to be the basic structure of the fibrous protein silk and in which, as the name suggests, two lengths of polypeptide chain run parallel to each other in opposite directions. This structure again is stabilized by hydrogen bonds between the NH and CO groups of the main chain. Residues 41 through 45 and 50 through 54 in the lysozyme molecule form such a structure, with the connecting residues 46 through 49 folded into a hairpin bend between the two lengths of comparatively extended chain. The remainder of the polypeptide chain is folded in irregular ways that have no simple short description.

Even though the level of resolution achieved in our present image was not enough to resolve individual atoms, many of the side chains characteristic of the amino acid residues were readily identifiable from their general shape. The four disulfide bridges, for example, are marked by short rods of high electron density corresponding to the two relatively dense sulfur atoms within them. The six tryptophan residues also were easily recognized by the extended electron density produced by the large double-ring structures in their side chains. Many of the other residues also were easily identifiable, but it was nevertheless most important for the rapid and reliable interpretation of the image that the results of the chemical analysis were already available. With their help more than 95 percent of the atoms in the molecule were readily identified and located within about .25 angstrom.

Further efforts at improving the accuracy with which the atoms have been located is in progress, but an almost complete description of the lysozyme molecule now exists [*see illustration on pages 70 and 71*]. By studying it and the

MODEL OF SUBSTRATE shows how it fits into the cleft in the lysozyme molecule. All the carbon atoms in the substrate are shown in purple. The portion of the substrate in intimate contact with the underlying enzyme is a polysaccharide chain consisting of six ringlike structures, each a residue of an amino-sugar molecule. The substrate in the model is made up of six identical residues of the amino sugar called N-acetylglucosamine (NAG). In the actual substrate every other residue is an amino sugar known as N-acetylmuramic acid (NAM). The illustration is based on X-ray studies of the way the enzyme is bound to a trisaccharide made of three NAG units, which fills the top of the cleft; the arrangement of NAG units in the bottom of the cleft was worked out with the aid of three-dimensional models. The substrate is held to the enzyme by a complex network of hydrogen bonds. In this style of model-making each straight section of chain represents a bond between atoms. The atoms themselves lie at the intersections and elbows of the structure. Except for the four red balls representing oxygen atoms that are active in splitting the polysaccharide substrate, no attempt is made to represent the electron shells of atoms because they would merge into a solid mass.

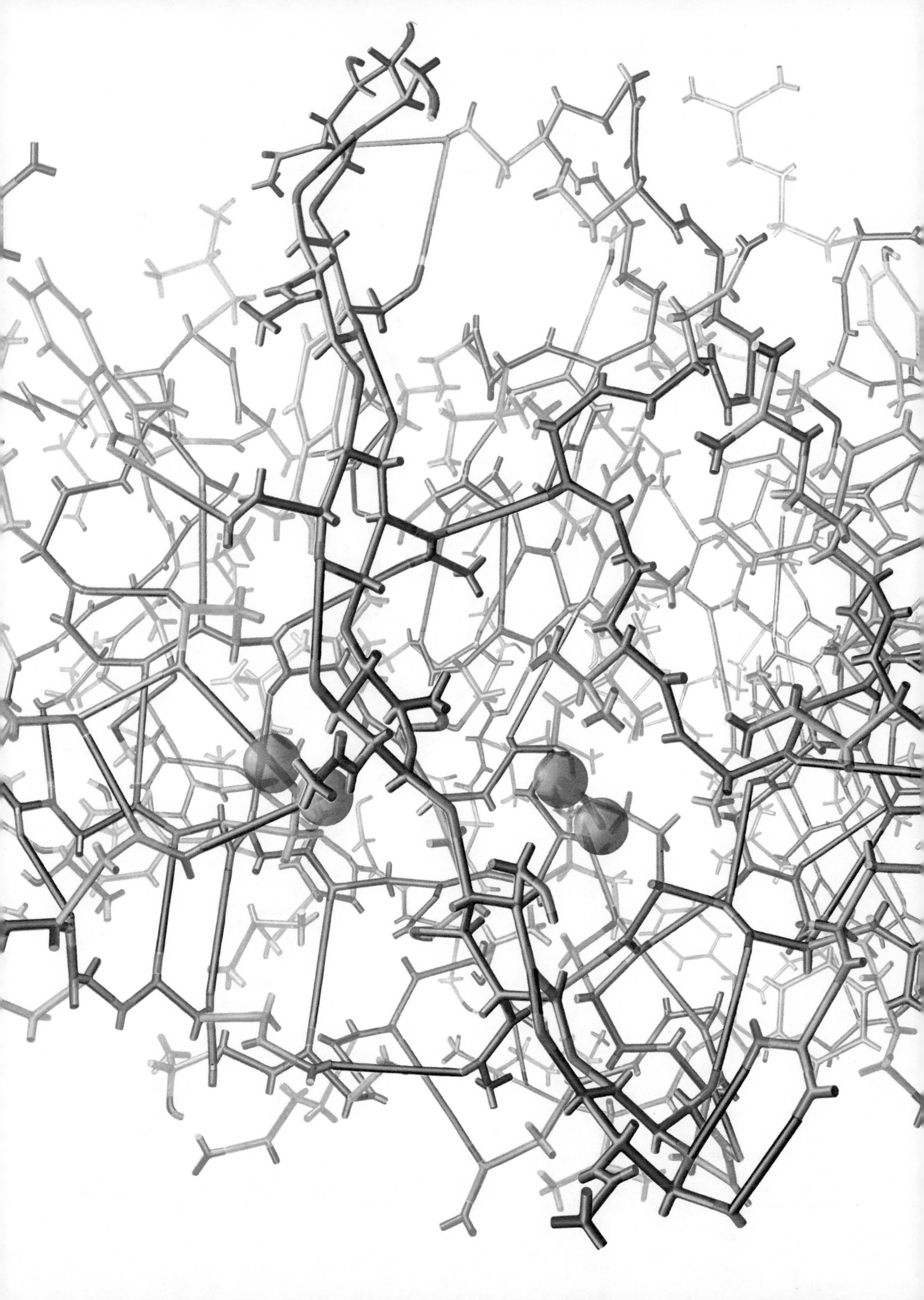

results of some further experiments we can begin to suggest answers to two important questions: How does a molecule such as this one attain its observed conformation? How does it function as an enzyme, or biological catalyst?

Inspection of the lysozyme molecule immediately suggests two generalizations about its conformation that agree well with those arrived at earlier in the study of myoglobin. It is obvious that certain residues with acidic and basic side chains that ionize, or dissociate, on contact with water are all on the surface of the molecule more or less readily accessible to the surrounding liquid. Such "polar" side chains are hydrophilic—attracted to water; they are found in aspartic acid and glutamic acid residues and in lysine, arginine and histidine residues, which have basic side groups. On the other hand, most of the markedly nonpolar and hydrophobic side chains (for example those found in leucine and isoleucine residues) are shielded from the surrounding liquid by more polar parts of the mole-

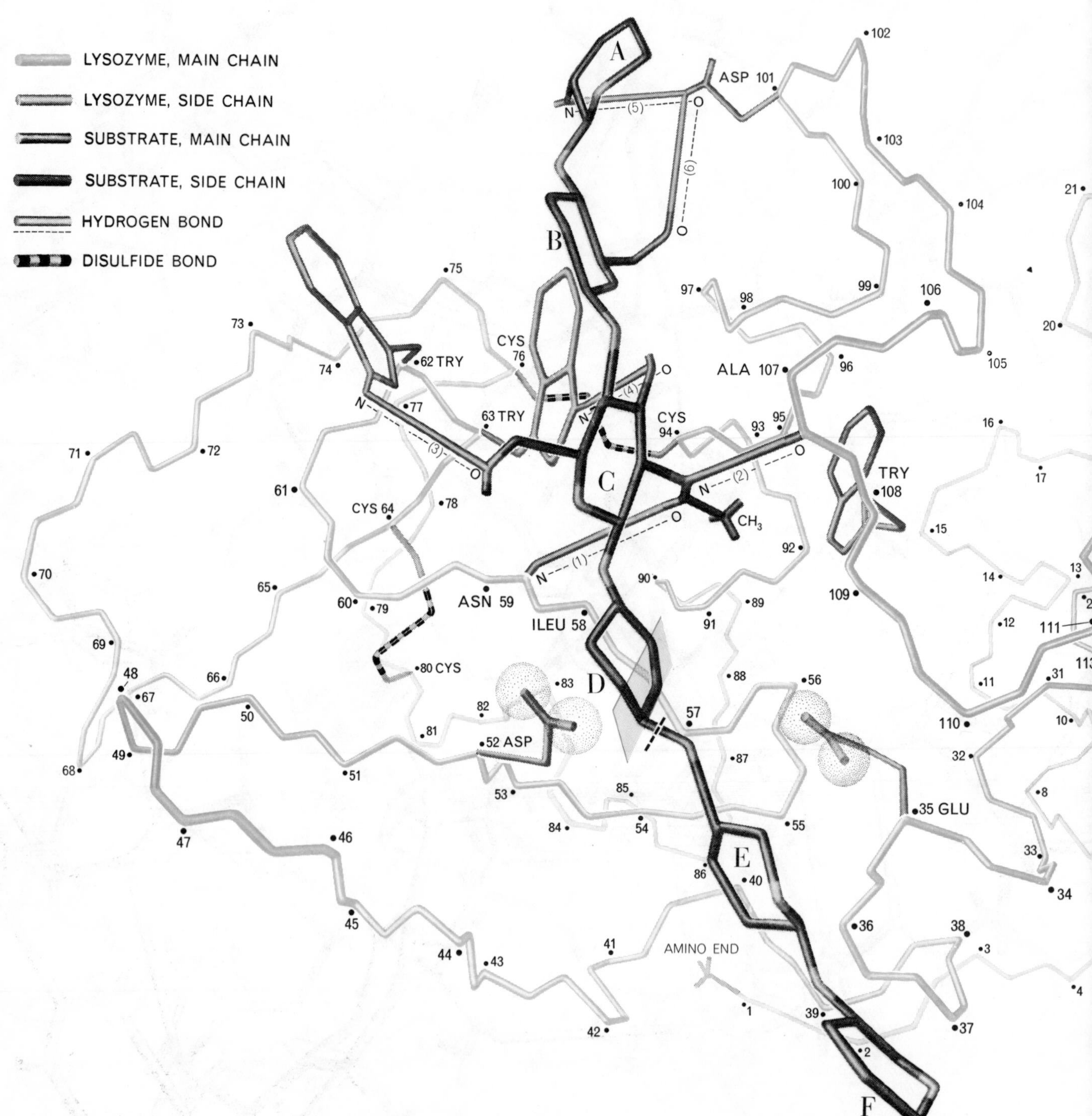

MAP OF LYSOZYME AND SUBSTRATE depicts in color the central chain of each molecule. Side chains have been omitted except for those that produce the four disulfide bonds clipping the lysozyme molecule together and those that supply the terminal connections for hydrogen bonds holding the substrate to the lysozyme. The top three rings of the substrate (*A*, *B*, *C*) are held to the underlying enzyme by six principal hydrogen bonds, which are identified by number to key with the description in the text. The lyso-

cule. In fact, as was predicted by Sir Eric Rideal (who was at one time director of the Royal Institution) and Irving Langmuir, lysozyme, like myoglobin, is quite well described as an oil drop with a polar coat. Here it is important to note that the environment of each molecule in the crystalline state is not significantly different from its natural environment in the living cell. The crystals themselves include a large proportion (some 35 percent by weight) of mostly watery liquid of crystallization. The effect of the surrounding liquid on the protein conformation thus is likely to be much the same in the crystals as it is in solution.

It appears, then, that the observed conformation is preferred because in it

FIRST 56 RESIDUES in lysozyme molecule contain a higher proportion of symmetrically organized regions than does all the rest of the molecule. Residues 5 through 15 and 24 through 34 (*right*) form two regions in which hydrogen bonds (*gray*) hold the residues in a helical configuration close to that of the "classical" alpha helix. Residues 41 through 45 and 50 through 54 (*left*) fold back against each other to form a "pleated sheet," also held together by hydrogen bonds. In addition the hydrogen bond between residues 1 and 40 ties the first 40 residues into a compact structure that may have been folded in this way before the molecule was fully synthesized (*see illustration at the bottom of the next two pages*).

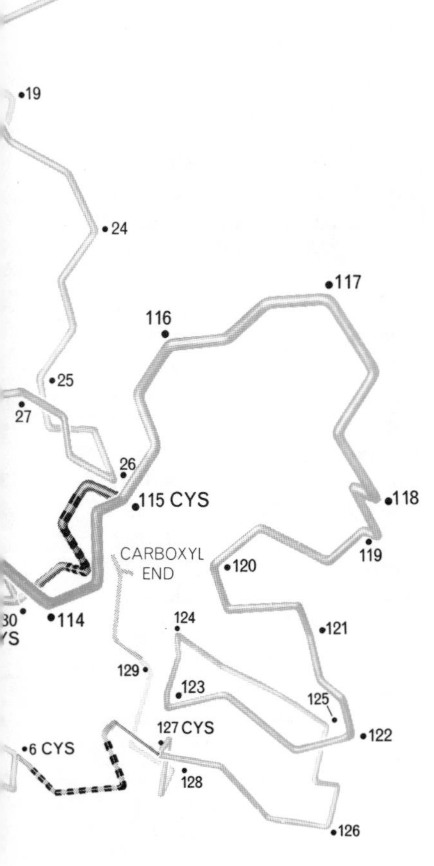

zyme molecule fulfills its function when it cleaves the substrate between the *D* and the *E* ring. Note the distortion of the *D* ring, which pushes four of its atoms into a plane.

the hydrophobic side chains are kept out of contact with the surrounding liquid whereas the polar side chains are generally exposed to it. In this way the system consisting of the protein and the solvent attains a minimum free energy, partly because of the large number of favorable interactions of like groups within the protein molecule and between it and the surrounding liquid, and partly because of the relatively high disorder of the water molecules that are in contact only with other polar groups of atoms.

Guided by these generalizations, many workers are now interested in the possibility of predicting the conformation of a protein molecule from its chemical formula alone [see "Molecular Model-building by Computer," by Cyrus Levinthal; SCIENTIFIC AMERICAN Offprint 1043]. The task of exploring all possible conformations in the search for the one of lowest free energy seems likely, however, to remain beyond the power of any imaginable computer. On a conservative estimate it would be necessary to consider some 10^{129} different conformations for the lysozyme molecule in any general search for the one with minimum free energy. Since this number is far greater than the number of particles in the observable universe, it is clear that simplifying assumptions will have to be made if calculations of this kind are to succeed.

The Folding of Lysozyme

For some time Peter Dunnill and I have been trying to develop a model of protein-folding that promises to make practicable calculations of the minimum energy conformation and that is, at the same time, qualitatively consistent with the observed structure of myoglobin and lysozyme. This model makes use of our present knowledge of the way in which proteins are synthesized in the living cell. For example, it is well known, from experiments by Howard M. Dintzis and by Christian B. Anfinsen and Robert Canfield, that protein molecules are synthesized from the terminal amino end of their polypeptide chain. The nature of the synthetic mechanism, which involves the intracellular particles called ribosomes working in collaboration with two forms of ribonucleic acid ("messenger" RNA and "transfer" RNA), is increasingly well understood in principle, although the detailed environment of the growing protein chain remains unknown. Nevertheless, it seems a reasonable assumption that, as the synthesis proceeds, the amino end of the chain becomes separated by an increasing distance from the point of attachment to the ribosome, and that the folding of the protein chain to its native conformation begins at this end even before the synthesis is complete. According to our present ideas, parts of the polypeptide chain, particularly those near the terminal amino end, may fold into stable conformations that can still be recognized in the finished molecule and that act as "internal templates," or centers, around which the rest of the chain is folded [*see illustration at bottom of these two pages*]. It may therefore be useful to look for the stable conformations of parts of the polypeptide chain and to avoid studying all the possible conformations of the whole molecule.

Inspection of the lysozyme molecule provides qualitative support for these ideas [*see top illustration on page* 75]. The first 40 residues from the terminal amino end form a compact structure (residues 1 and 40 are linked by a hydrogen bond) with a hydrophobic interior and a relatively hydrophilic surface that seems likely to have been folded in this way, or in a simply related way, before the molecule was fully synthesized. It may also be important to observe that this part of the molecule includes more alpha helix than the remainder does.

These first 40 residues include a mixture of hydrophobic and hydrophilic side chains, but the next 14 residues in the sequence are all hydrophilic; it is interesting, and possibly significant, that these are the residues in the antiparallel pleated sheet, which lies out of contact with the globular submolecule formed by the earlier residues. In the light of our model of protein folding the obvious speculation is that there is no incentive to fold these hydrophilic residues in contact with the first part of the chain until the hydrophobic residues 55 (isoleucine) and 56 (leucine) have to be shielded from contact with the surrounding liquid. It seems reasonable to suppose that at this stage residues 41 through 54 fold back on themselves, forming the pleated-sheet structure and burying the hydrophobic side chains in the initial hydrophobic pocket.

Similar considerations appear to govern the folding of the rest of the molecule. In brief, residues 57 through 86 are folded in contact with the pleated-sheet structure so that at this stage of the process—if indeed it follows this course—the folded chain forms a structure with two wings lying at an angle to each other. Residues 86 through 96 form a length of alpha helix, one side of which is predominantly hydrophobic, because of an appropriate alternation of polar and nonpolar residues in that part of the sequence. This helix lies in the gap between the two wings formed by the earlier residues, with its hydrophobic side buried within the molecule. The gap between the two wings is not completely filled by the helix, however; it is transformed into a deep cleft running up one side of the molecule. As we shall see, this cleft forms the active site of the enzyme. The remaining residues are folded around the globular unit formed by the terminal amino end of the polypeptide chain.

This model of protein-folding can be tested in a number of ways, for example by studying the conformation of the first 40 residues in isolation both directly (after removal of the rest of the molecule) and by computation. Ultimately, of course, the model will be regarded as satisfactory only if it helps us to predict how other protein molecules are folded from a knowledge of their chemical structure alone.

The Activity of Lysozyme

In order to understand how lysozyme brings about the dissolution of bacteria we must consider the structure of the bacterial cell wall in some detail. Through the pioneer and independent studies of Karl Meyer and E. B. Chain, followed up by M. R. J. Salton of the University of Manchester and many others, the structures of bacterial cell walls and the effect of lysozyme on them are now quite well known. The important part of the cell wall, as far as lysozyme is concerned, is made up of glucose-like amino-sugar molecules linked together into long polysaccharide chains, which are themselves cross-connected by short lengths of polypeptide chain. This part of each cell wall probably forms one enormous molecule—a "bag-shaped macromolecule," as W. Weidel and H. Pelzer have called it.

The amino-sugar molecules concerned in these polysaccharide structures are of two kinds; each contains an acetamido ($-NH \cdot CO \cdot CH_3$) side group, but one of them contains an additional major group, a lactyl side chain [*see illustration at top of opposite page*]. One of these amino sugars is known as N-acetylglucosamine (NAG) and the other as N-acetylmuramic acid (NAM). They occur alternately in the polysaccharide chains, being connected by bridges that include an oxygen atom (glycosidic linkages) between carbon atoms 1 and 4 of consecutive sugar rings; this is the same linkage that joins glucose residues in

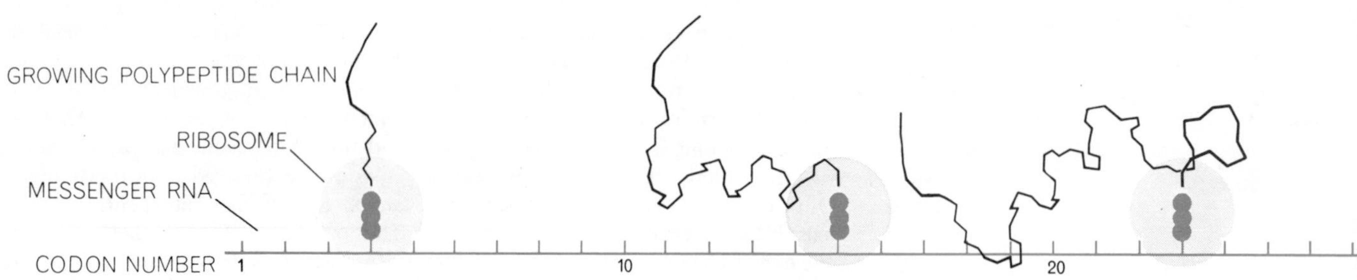

FOLDING OF PROTEIN MOLECULE may take place as the growing polypeptide chain is being synthesized by the intracellular particles called ribosomes. The genetic message specifying the amino acid sequence of each protein is coded in "messenger" ribonucleic acid (RNA). It is believed several ribosomes travel simultaneously along this long-chain molecule, reading the message as they go.

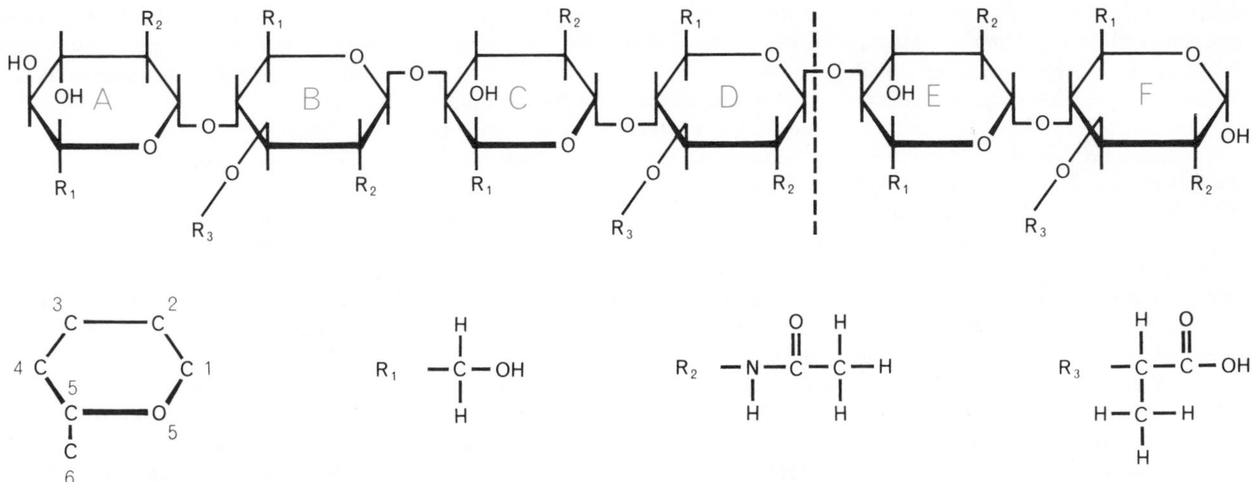

POLYSACCHARIDE MOLECULE found in the walls of certain bacterial cells is the substrate broken by the lysozyme molecule. The polysaccharide consists of alternating residues of two kinds of amino sugar: N-acetylglucosamine (NAG) and N-acetylmuramic acid (NAM). In the length of polysaccharide chain shown here *A*, *C* and *E* are NAG residues; *B*, *D* and *F* are NAM residues. The inset at left shows the numbering scheme for identifying the principal atoms in each sugar ring. Six rings of the polysaccharide fit into the cleft of the lysozyme molecule, which effects a cleavage between rings *D* and *E* (*see illustration on pages 74 and 75*).

cellulose. The polypeptide chains that cross-connect these polysaccharides are attached to the NAM residues through the lactyl side chain attached to carbon atom 3 in each NAM ring.

Lysozyme has been shown to break the linkages in which carbon 1 in NAM is linked to carbon 4 in NAG but not the other linkages. It has also been shown to break down chitin, another common natural polysaccharide that is found in lobster shell and that contains only NAG.

Ever since the work of Svante Arrhenius of Sweden in the late 19th century enzymes have been thought to work by forming intermediate compounds with their substrates: the substances whose chemical reactions they catalyze. A proper theory of the enzyme-substrate complex, which underlies all present thinking about enzyme activity, was clearly propounded by Leonor Michaelis and Maude Menton in a remarkable paper published in 1913. The idea, in its simplest form, is that an enzyme molecule provides a site on its surface to which its substrate molecule can bind in a quite precise way. Reactive groups of atoms in the enzyme then promote the required chemical reaction in the substrate. Our immediate objective, therefore, was to find the structure of a reactive complex between lysozyme and its polysaccharide substrate, in the hope that we would then be able to recognize the active groups of atoms in the enzyme and understand how they function.

Our studies began with the observation by Martin Wenzel and his colleagues at the Free University of Berlin that the enzyme is prevented from functioning by the presence of NAG itself. This small molecule acts as a competitive inhibitor of the enzyme's activity and, since it is a part of the large substrate molecule normally acted on by the enzyme, it seems likely to do this by binding to the enzyme in the way that part of the substrate does. It prevents the enzyme from working by preventing the substrate from binding to the enzyme. Other simple amino-sugar molecules, including the trisaccharide made of three NAG units, behave in the same way. We therefore decided to study the

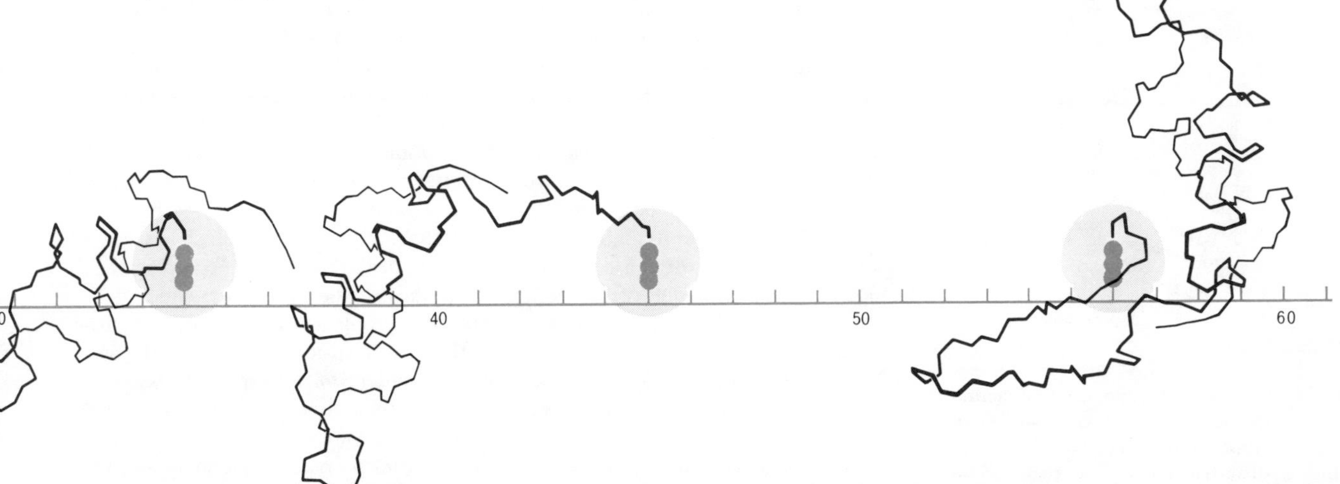

Presumably the messenger RNA for lysozyme contains 129 "codons," one for each amino acid. Amino acids are delivered to the site of synthesis by molecules of "transfer" RNA (*dark color*). The illustration shows how the lysozyme chain would lengthen as a ribosome travels along the messenger RNA molecule. Here, hypothetically, the polypeptide is shown folding directly into its final shape.

binding of these sugar molecules to the lysozyme molecules in our crystals in the hope of learning something about the structure of the enzyme-substrate complex itself.

My colleague Louise Johnson soon found that crystals containing the sugar molecules bound to lysozyme can be prepared very simply by adding the sugar to the solution from which the lysozyme crystals have been grown and in which they are kept suspended. The small molecules diffuse into the protein crystals along the channels filled with water that run through the crystals. Fortunately the resulting change in the crystal structure can be studied quite simply. A useful image of the electron-density changes can be calculated from measurements of the changes in amplitude of the diffracted waves, on the assumption that their phase relations have not changed from those determined for the pure protein crystals. The image shows the difference in electron density between crystals that contain the added sugar molecules and those that do not.

In this way the binding to lysozyme of eight different amino sugars was studied at low resolution (that is, through the measurement of changes in the amplitude of 400 diffracted waves). The results showed that the sugars bind to lysozyme at a number of different places in the cleft of the enzyme. The investigation was hurried on to higher resolution in an attempt to discover the exact nature of the binding. Happily these studies at two-angstrom resolution (which required the measurement of 10,000 diffracted waves) have now shown in detail how the trisaccharide made of three NAG units is bound to the enzyme.

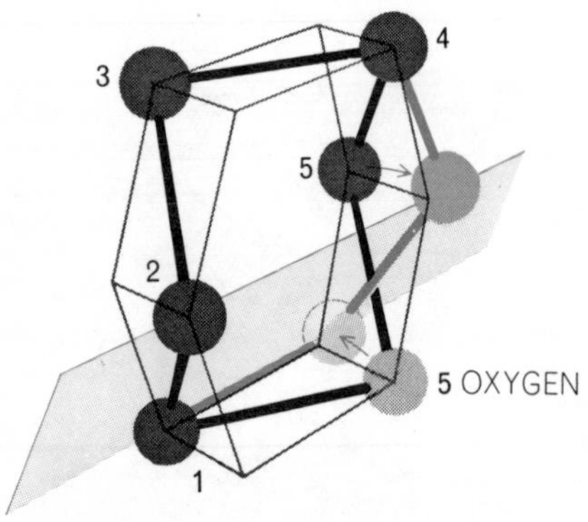

"CHAIR" CONFIGURATION (*gray*) is that normally assumed by the rings of amino sugar in the polysaccharide substrate. When bound against the lysozyme, however, the *D* ring is distorted (*color*) so that carbon atoms *1*, *2* and *5* and oxygen atom *5* lie in a plane. The distortion evidently assists in breaking the substrate below the *D* ring.

The trisaccharide fills the top half of the cleft and is bound to the enzyme by a number of interactions, which can be followed with the help of the illustration on pages 74 and 75. In this illustration six important hydrogen bonds, to be described presently, are identified by number. The most critical of these interactions appear to involve the acetamido group of sugar residue *C* [*third from top*], whose carbon atom 1 is not linked to another sugar residue. There are hydrogen bonds from the CO group of this side chain to the main-chain NH group of amino acid residue 59 in the enzyme molecule [*bond No. 1*] and from its NH group to the main-chain CO group of residue 107 (alanine) in the enzyme molecule [*bond No. 2*]. Its terminal CH_3 group makes contact with the side chain of residue 108 (tryptophan). Hydrogen bonds [*No. 3 and No. 4*] are also formed between two oxygen atoms adjacent to carbon atoms 6 and 3 of sugar residue *C* and the side chains of residues 62 and 63 (both tryptophan) respectively. Another hydrogen bond [*No. 5*] is formed between the acetamido side chain of sugar residue *A* and residue 101 (aspartic acid) in the enzyme molecule. From residue 101 there is a hydrogen bond [*No. 6*] to the oxygen adjacent to carbon atom 6 of sugar residue *B*. These polar interactions are supplemented by a large number of nonpolar interactions that are more difficult to summarize briefly. Among the more important nonpolar interactions, however, are those between sugar residue *B* and the ring system of residue 62; these deserve special mention because they are affected by a small change in the conformation of the enzyme molecule that occurs when the trisaccharide is bound to it. The electron-density map showing the change in electron density when tri-NAG is bound in the protein crystal reveals clearly that parts of the enzyme molecule have moved with respect to one another. These changes in conformation are largely restricted to the part of the enzyme structure to the left of the cleft, which appears to tilt more or less as a whole in such a way as to close the cleft slightly. As a result the side chain of residue 62 moves about .75 angstrom toward the position of sugar residue *B*. Such changes in enzyme conformation have been discussed for some time, notably by Daniel E. Koshland, Jr., of the University of California at Berkeley, whose "induced fit" theory of the enzyme-substrate interaction is supported in some degree by this observation in lysozyme.

The Enzyme-Substrate Complex

At this stage in the investigation excitement grew high. Could we tell how the enzyme works? I believe we can. Unfortunately, however, we cannot see this dynamic process in our X-ray images. We have to work out what must happen from our static pictures. First of all it is clear that the complex formed by tri-NAG and the enzyme is not the enzyme-substrate complex involved in catalysis because it is stable. At low concentrations tri-NAG is known to behave as an inhibitor rather than as a substrate that is broken down; clearly we have been looking at the way in which it binds as an inhibitor. It is noticeable, however, that tri-NAG fills only half of the cleft. The possibility emerges that more sugar residues, filling the remainder of the cleft, are required for the formation of a reactive enzyme-substrate complex. The assumption here is that the observed binding of tri-NAG as an inhibitor involves interactions with the enzyme molecule that also play a part in the formation of the functioning enzyme-substrate complex.

Accordingly we have built a model that shows that another three sugar residues can be added to the tri-NAG in such a way that there are satisfactory interactions of the atoms in the proposed substrate and the enzyme. There is only one difficulty: carbon atom 6 and its adjacent oxygen atom in sugar residue *D* make uncomfortably close contacts with atoms in the enzyme molecule, unless this sugar residue is distorted a little out of its most stable "chair" conformation into a conformation in which carbon atoms 1, 2 and 5 and oxygen atom 5 all lie in a plane [*see illustration at left*]. Otherwise satisfactory interactions immediately suggest themselves, and the model falls into place.

At this point it seemed reasonable to assume that the model shows the structure of the functioning complex between the enzyme and a hexasaccharide. The next problem was to decide which of the five glycosidic linkages would be broken under the influence of the enzyme. Fortunately evidence was at hand to suggest the answer. As we have seen, the cell-wall polysaccharide includes alternate sugar residues of two kinds, NAG and NAM, and the bond broken is between NAM and NAG. It was therefore important to decide which of the six sugar residues in our model could be NAM, which is the same as NAG except for the lactyl side chain appended to carbon atom 3. The answer was clear-cut. Sugar residue *C* cannot be NAM because there is no room for this additional group of atoms. Therefore the bond broken must be between sugar residues *B* and *C* or *D* and *E*. We already knew that the glycosidic linkage between residues *B* and *C* is stable when tri-NAG is bound. The conclusion was inescapable: the linkage that must be broken is the one between sugar residues *D* and *E*.

Now it was possible to search for the origin of the catalytic activity in the neighborhood of this linkage. Our task was made easier by the fact that John A. Rupley of the University of Arizona had shown that the chemical bond broken under the influence of lysozyme is the one between carbon atom 1 and oxygen in the glycosidic link rather than the link between oxygen and carbon atom 4. The most reactive-looking group of atoms in the vicinity of this bond are the side chains of residue 52 (aspartic acid) and residue 35 (glutamic acid). One of the oxygen atoms of residue 52 is about three angstroms from carbon atom 1 of sugar residue *D* as well as from the ring oxygen atom 5 of that residue. Residue 35, on the other hand, is about three angstroms from the oxygen in the glycosidic linkage. Furthermore, these two amino acid residues have markedly different environments. Residue 52 has a number of polar neighbors and appears to be involved in a network of hydrogen bonds linking it with residues 46 and 59 (both asparagine) and, through them, with residue 50 (serine). In this environment residue 52 seems likely to give up a terminal hydrogen atom and thus be negatively charged under most conditions, even when it is in a markedly acid solution, whereas residue 35, situated in a nonpolar environment, is likely to retain its terminal hydrogen atom.

A little reflection suggests that the concerted influence of these two amino acid residues, together with a contribution from the distortion to sugar residue *D* that has already been mentioned, is enough to explain the catalytic activity of lysozyme. The events leading to the rupture of a bacterial cell wall probably take the following course [*see illustration on this page*].

First, a lysozyme molecule attaches itself to the bacterial cell wall by interacting with six exposed amino-sugar residues. In the process sugar residue *D* is somewhat distorted from its usual conformation.

Second, residue 35 transfers its terminal hydrogen atom in the form of a hydrogen ion to the glycosidic oxygen, thus bringing about cleavage of the bond between that oxygen and carbon atom 1 of sugar residue *D*. This creates a positively charged carbonium ion (C^+) where the oxygen has been severed from carbon atom 1.

Third, this carbonium ion is stabilized by its interaction with the negatively charged aspartic acid side chain of residue 52 until it can combine with a hydroxyl ion (OH^-) that happens to diffuse into position from the surrounding water, thereby completing the reaction. The lysozyme molecule then falls away,

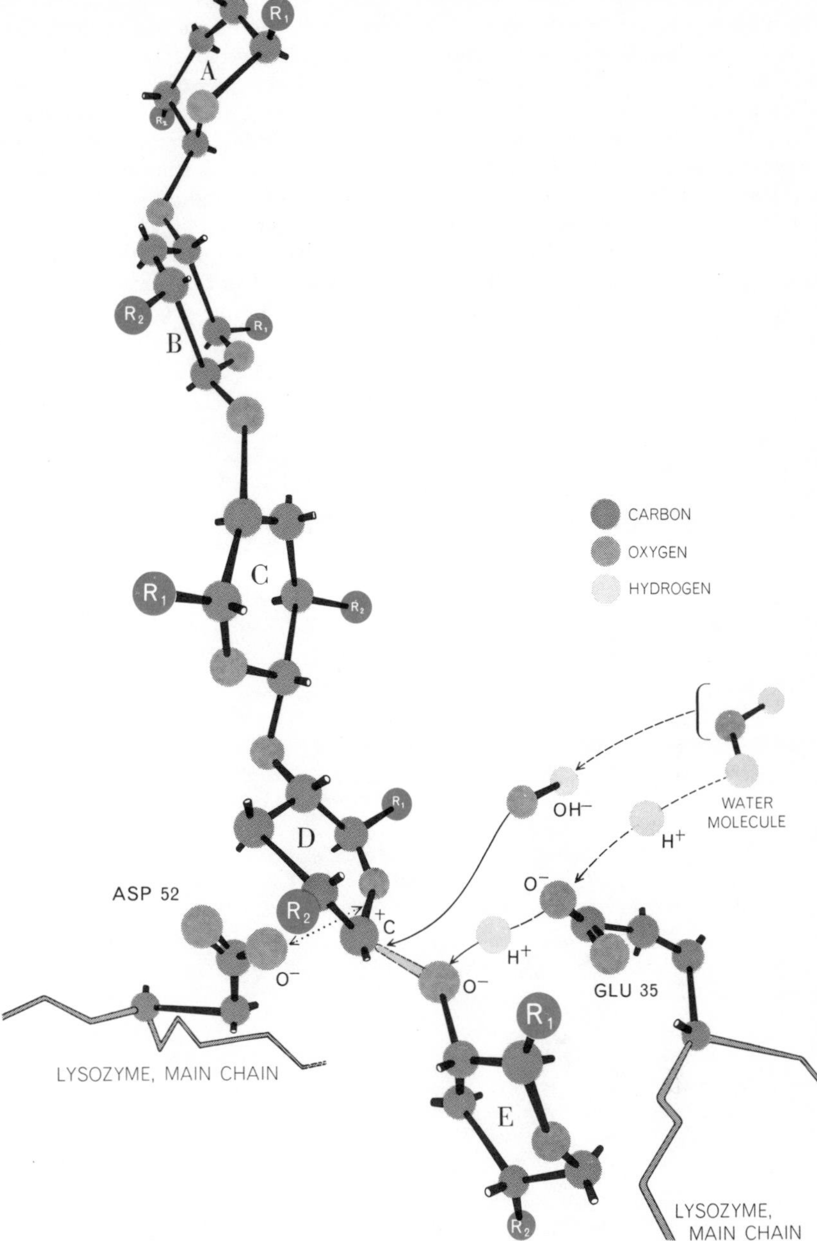

SPLITTING OF SUBSTRATE BY LYSOZYME is believed to involve the proximity and activity of two side chains, residue 35 (glutamic acid) and residue 52 (aspartic acid). It is proposed that a hydrogen ion (H^+) becomes detached from the OH group of residue 35 and attaches itself to the oxygen atom that joins rings *D* and *E*, thus breaking the bond between the two rings. This leaves carbon atom 1 of the *D* ring with a positive charge, in which form it is known as a carbonium ion. It is stabilized in this condition by the negatively charged side chain of residue 52. The surrounding water supplies an OH^- ion to combine with the carbonium ion and an H^+ ion to replace the one lost by residue 35. The two parts of the substrate then fall away, leaving the enzyme free to cleave another polysaccharide chain.

leaving behind a punctured bacterial cell wall.

It is not clear from this description that the distortion of sugar residue *D* plays any part in the reaction, but in fact it probably does so for a very interesting reason. R. H. Lemieux and G. Huber of the National Research Council of Canada showed in 1955 that when a sugar molecule such as NAG incorporates a carbonium ion at the carbon-1 position, it tends to take up the same conformation that is forced on ring *D* by its interaction with the enzyme molecule. This seems to be an example, therefore, of activation of the substrate by distortion, which has long been a favorite idea of enzymologists. The binding of the substrate to the enzyme itself favors the formation of the carbonium ion in ring *D* that seems to play an important part in the reaction.

It will be clear from this account that although lysozyme has not been seen in action, we have succeeded in building up a detailed picture of how it may work. There is already a great deal of chemical evidence in agreement with this picture, and as the result of all the work now in progress we can be sure that the activity of Fleming's lysozyme will soon be fully understood. Best of all, it is clear that methods now exist for uncovering the secrets of enzyme action.

Protein Shape and Biological Control

6

by Daniel E. Koshland, Jr.
October 1973

The processes of life are turned on and off by means of a universal control mechanism that depends on the ability of protein molecules to bend flexibly from one shape to another under external influences

A living system must have both the capacity to act and the capacity to control its actions. We humans, for example, must be able to digest food, but we cannot be eating all the time. Hence we need both a positive control to turn on the process (a desire to eat when food is needed) and a negative control to turn off the process (a desire to stop eating when we have had enough). Similarly, the process of blood clotting must be turned on when we bleed from a wound but must be turned off afterward so that it does not lead to coronary thrombosis. We must also be able to turn muscles on in order to move and to turn them off in order to relax. In short, every biological system has built-in controls to initiate or accelerate a process under some conditions and to terminate or decelerate it under other conditions.

Considering the diversity of processes that must be regulated and the diversity of environmental conditions to which an organism must react, it might be expected that the controls in biological systems are enormously complex. That is true, and yet when one examines the processes more closely, it appears that the fundamental elements of control are remarkably simple and universal.

The fundamental control element in all living systems—from the smallest bacterium to man—is the protein molecule. Enzymes, the biological catalysts that control all the chemical processes of living systems, are proteins. Sensory receptors, which enable us to see, hear, taste and smell, are proteins. Antibodies, which provide immunity against infection, are proteins. Recent experiments have established that it is the ability of these proteins to change shape under external influences that provides the "on-off" controls that are so vital to the living system.

The concept of protein shape as a control mechanism arose from studies of enzymes, a development that is hardly surprising, since enzymes are the easiest to study of all the regulating proteins. It has been estimated that an average living cell contains some 3,000 different enzymes. Each of them catalyzes a distinct chemical reaction in which compounds called substrates are converted into other compounds called products. Fortunately for our understanding of biological systems many enzymes are quite sturdy molecules and can be extracted from a physiological system without destroying their biological properties. Hence they can be studied in the test tube and made to perform the same catalytic role there that they perform in the living organism. Moreover, one can subject them to the same environmental influences in the test tube that they experience in the living cell, thereby getting clues to their role in biological regulation. Finally, enzymes can be obtained in large enough amounts for their physical properties to be studied. However, even though our understanding of the role of shape in protein regulation began with enzymes, the principles of regulation worked out for enzymes appear to be universal and can be applied to other proteins that are more difficult to obtain in bulk.

Shape Changes in Enzyme Catalysis

It has long been known that the basic mechanism by which enzymes catalyze chemical reactions begins with the binding of the substrate (or substrates) to the surface of the enzyme. The enzyme then polarizes the chemical bonds in the substrate, causing a reaction that leads to the formation of products on the surface. The release of these products from the surface regenerates the free enzyme and allows the cycle to be repeated.

Unlike catalysts made in the laboratory, enzymes have the special property called specificity, which means that only one chemical compound or a very few can react with a particular enzyme. This property can be explained by the template, or lock-and-key, hypothesis put forward in 1894 by Emil Fischer, which postulates that the enzyme is designed to allow only special compounds to fit on its surface, just as a key fits a lock or as two pieces of a jigsaw puzzle fit together. The complementary shapes allow one compound to fit and exclude other compounds that lack the correct size, shape or charge distribution [*see illustration on pages 82 and 83*]. Modern X-ray crystallography has revealed in detail precisely such a fit between enzyme and substrate [*see illustration on page 89*].

Although this concept could explain much of the specificity data, some glaring discrepancies were found. For instance, certain oversized and undersized compounds were found to bind to the surface of the enzyme even though they failed to form products. Furthermore, it was difficult on the basis of the rigid-template theory to explain how sugars can compete with water in enzymatic reactions. It was also difficult to explain why substrates bind in a specific order in many enzymatic reactions. These facts and others like them led to the hypothesis that the enzyme does not exist initially in a shape complementary to that of the substrate but rather is induced to take the complementary shape in much the same way that a hand induces a change in the shape of a glove [*see illustration on pages 84 and 85*]. This "induced fit" theory assumes that the substrate plays a role in determining the final shape of the enzyme and that the enzyme is therefore flexible. Proof that proteins do in fact change their shape

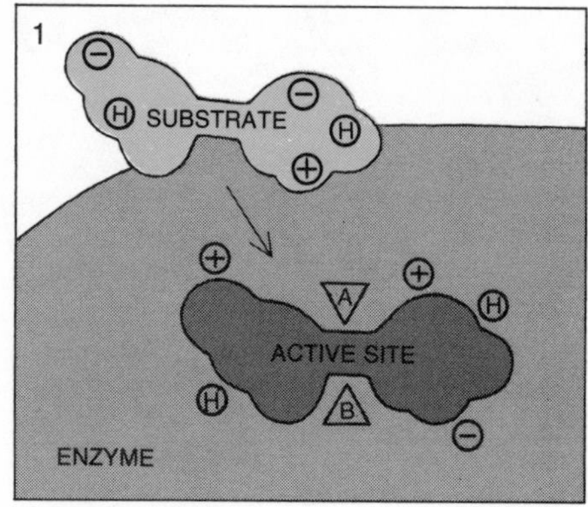

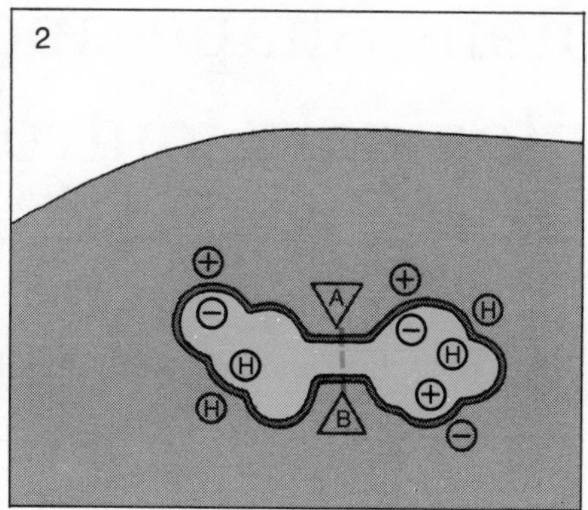

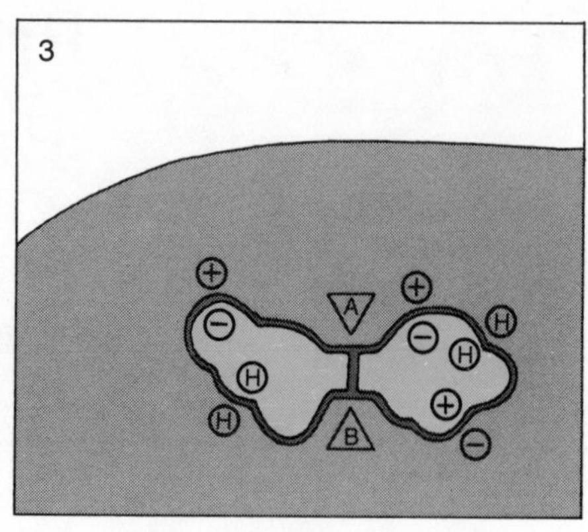

LOCK-AND-KEY MODEL of the mechanism by which enzymes catalyze chemical reactions, put forward in 1894 by Emil Fischer, postulates that the active site of the enzyme is a rigid, templatelike structure that allows only special compounds to fit, just as a key fits a lock. Such a compound, called a substrate, is attracted to the enzyme's active site by mutually attractive groups, such as the electrostatic charges on certain amino acid side chains (+, −), and by the coalescing tendency of adjacent hydrophobic side chains (*H*). In the lock-and-key model the catalytic groups (*A*, *B*) are poised in advance (*1*) to cause the reaction that ruptures the chemical bond

under the influence of small molecules was initially obtained by chemical studies showing differences in the reactivity to protein reagents of the amino acid side chains that are arrayed along the spine of the protein molecule. The hypothesis has since been verified with the aid of advanced physical techniques, most notably X-ray crystallography.

Once the concept of a flexible enzyme was entertained, the puzzling behavior of the special classes of compounds that were found to bind to the enzyme without forming products could be explained. One type of oversized molecule, for example, binds to the surface of the enzyme, but in doing so it distorts the protein into a shape that does not allow the catalytic groups to be properly aligned. Other compounds can bind to the enzyme too, but they do not have either sufficient size or the correct chemical characteristics to induce the proper alignment. Hence even though both types of compound bind, neither reacts with the enzyme. (Some molecules are of course too big to be bound even to a flexible enzyme; the examples cited here are chosen to show the differences between the rigid-template theory and the induced-fit theory.)

Shape Changes in Regulation

This finding of flexibility does not mean that all proteins must be flexible. Some may indeed be quite rigid, and these are explained very well by the lock-and-key hypothesis. Nor does it mean that enzymes that exhibit flexibility must do so with all chemical compounds. The finding does mean, however, that protein flexibility is a key feature of enzyme action. Indeed, the capacity to induce a change in shape has been found to be a widespread and vital feature of most, if not all, enzymes.

The concept of protein flexibility led to the deduction that small molecules not themselves involved in the chemical reaction could help to make a deficient molecule act as a substrate by altering the shape of the enzyme [*see illustration on page 86*]. In the case of a molecule that is too small to induce the proper alignment of catalytic groups, for example, certain molecules that are not consumed in the reaction can be added to produce a stable shape with the right alignment of catalytic groups. One way for this realignment to occur is for a second molecule to bind immediately adjacent to the deficient molecule, thus inducing the proper shape at the active site. This prediction of the flexible-enzyme theory has been confirmed in many cases (for example in the case of the carbohydrate-splitting enzyme hexokinase by Alberto Sols and his co-workers in Spain and in the case of the digestive enzyme trypsin by T. Inagami and T. Murachi in Japan). Molecules that bind far away from the active catalytic site can also induce a proper shape. In that case the induced change is transmitted through the protein like a row of falling dominoes until the active site is altered appropriately.

The reverse process can also occur. Flexible enzymes can be distorted out of the active shape by molecules called inhibitors. These molecules can cause a disruption of either the catalytic function or the binding function of the enzyme, in either case giving rise to an inactive shape [*see illustration on page 87*].

In short, a regulatory molecule that is not itself involved in the chemical reaction can control the activity of an enzyme by changing its shape. It can turn the enzyme on by inducing the correct shape or turn it off by inducing an incorrect shape. In biological systems one of the most important groups of such molecules is the hormones. Although hormones are secreted in small amounts, much too small to be important directly as foodstuffs or sources of energy, they have a tremendous influence on the regulatory processes of the cell. The manner in which hormones exert control is easily explained by the flexible-protein hypothesis. Since these molecules are not consumed, they can be used again and again to activate the enzyme molecules, unlike the substrate molecules, which are consumed. Therefore such regulators need be present only in very small amounts.

Sometimes, as in the case of adrenalin, the initial hormone induces the formation of a second molecule, cyclic AMP, which acts as a regulator for many enzymes by changing their shape [see "Cyclic AMP," by Ira Pastan; SCIENTIFIC AMERICAN Offprint 1256]. In general the shape changes induced by the regulator molecules are similar to the shape changes induced by the substrate.

The Regulation of Pathways

In discussing the shapes of proteins it is difficult, tedious and often unnecessary to show the detailed parts of a large protein molecule. For convenience the different molecular shapes can be symbolized by geometric figures: squares, triangles, circles and so on [*see illustration on page 91*]. It is nonetheless important to remember that even though such shapes can be expressed by simple line drawings, a change from a circle to a square, say, designates subtle and com-

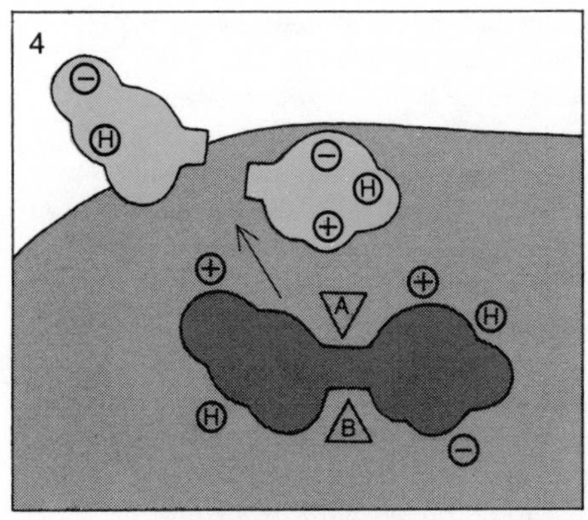

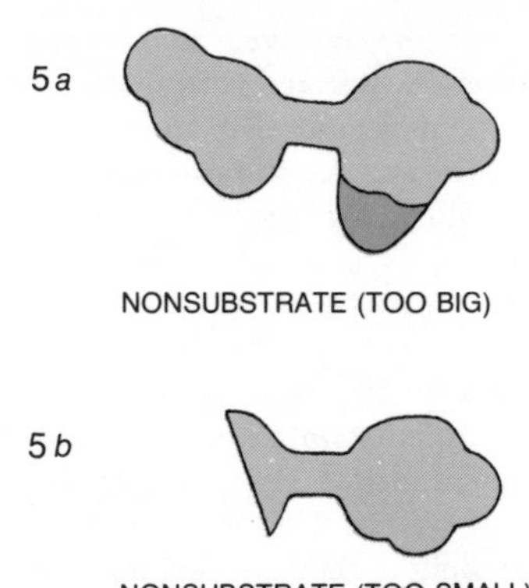

in the neck of the substrate (*2*), converting the substrate into the compounds called products (*3*). The release of these products from the surface (*4*) regenerates the free enzyme and allows the cycle to be repeated. According to this view, the precisely complementary shapes of the substrate and the enzyme that acts on it exclude reactions with compounds that are either too large to fit into the active site (*5a*) or too small to be attracted to the active site (*5b*).

plex changes in the orientation of the many amino acid side chains that constitute a protein. This procedure is analogous to using the symbol *C* for a carbon atom rather than writing out the complete quantum-mechanical description of the electrons, protons and neutrons that constitute the atom. The different shapes of proteins are referred to as conformations, because the protein changes fulfill a chemist's definition of a conformational change, that is, a change in the shape of a molecule caused by rotation around a single chemical bond. The terms "conformation" and "conformational change" are technical synonyms for the terms "shape" and "shape change" employed so far in this article.

The site on the surface of an enzyme at which the catalytic action takes place is called the active site. The binding site for the regulatory molecule is called the regulatory site or the allosteric site to distinguish it from the catalytic site. (The term "allosteric," meaning "the other site," was coined by the French biochemist Jacques Monod, a leader in this area of investigation, and it has gradually come to be used as a general term for regulatory proteins.) Regulatory molecules are also called effectors, modifiers or allosteric effectors.

Let us now consider how these ideas of protein shape help to explain some types of regulatory control. One of the most important decisions a living system must make involves how to process a food substance. We need energy in the form of molecules of adenosine triphosphate, or ATP, for all our bodily processes: to contract our muscles, to see with our eyes, to activate our nerves and to generate our structural materials. This energy comes from the combustion of food. A certain amount of ATP is constantly being used to maintain the system even in a resting state; the heart, for example, continues pumping even when we are asleep.

It is obviously desirable during periods of low energy demand to store energy for future needs. One of the ways of storing energy is in the form of glycogen, a chain of sugar molecules. Thus when a molecule of glucose, say, is ingested by a living system, it can be directed along alternate pathways: either it can be oxidized immediately to form carbon dioxide, giving off large amounts of ATP, or it can be stored in the form of glycogen to be released on future demand [*see illustration on page 88*]. The enzyme phosphofructokinase, which is involved in the first of these pathways, is turned on and off by variations in the level of ATP in the system. Another enzyme, glycogen synthetase, is involved as part of the second pathway in the regulatory control of the synthetic process; it can be turned on and off by the presence of regulatory compounds in the cell. A third enzyme, glycogen phosphorylase, catalyzes the reaction from glycogen to glucose-1-phosphate in the third pathway and is the enzyme used when the need arises for the retrieval of glucose from storage.

The way these enzymes work under the influence of supply and demand is simple and ingenious. When ATP levels are high, the phosphofructokinase enzyme of pathway 1 and the phosphorylase enzyme of pathway 3 are turned off and the glycogen synthetase of pathway 2 is activated. Glucose is therefore diverted to glycogen for storage. When an animal is frightened, however, it secretes adrenalin, which ultimately activates phosphorylase (pathway 3) and phosphofructokinase (pathway 1) and deactivates glycogen synthetase (pathway 2). In that case glycogen is removed from storage and converted into ATP energy to help the animal escape from the danger. In other words, when we are frightened, the hormone demands that our reserves of glycogen be made available as a source of energy. More than one demand can release glycogen from storage. Hunger and certain stimulators of muscular activity (such as calcium ions) activate phosphorylase and thus can also generate energy. In principle such influences act in the same way that adrenalin does.

This particular regulatory system can be used to illustrate two important points. First, enzymes must be available in every living system and yet they must not be equally active at all times. The alteration in shape under the influence of metabolites and hormones therefore provides a mechanism for turning these enzymes on and off under different external conditions (ranging, say, from starvation to satiety).

The second principle is more subtle. If all the enzymes involved in carbohydrate metabolism were to be activated simultaneously, one would simply have a short circuit going around and around pathways 1, 2 and 3, storing and burning glucose to no avail. It has usually been found for alternate pathways of this kind that a molecule that activates one pathway inhibits the other. In the case of glycogen storage, for example, the regulatory molecule inhibits the enzyme in pathway 2 and activates enzymes in pathways 1 and 3; the pathways for the oxidation of glucose and the removal of glucose from storage are accelerated and the pathway for the storage of glucose is blocked. In the absence of this regulatory molecule, equilibrium favors the inactive form of the enzyme in pathways 1 and 3 and the active form of the enzyme in pathway 2; in other words, the system favors glucose storage. Thus a short circuit is avoided by changing the shape of the enzymes reciprocally so that synthesis to glycogen is favored when ATP levels are high and degradation to carbon dioxide and ATP is favored when energy is needed.

Protein Structure

At this point it is worth considering in a little more detail how the design of proteins enables this regulation to proceed. The proteins that act as enzymes range in weight from approximately 10,000 daltons to many millions of daltons. (One dalton is roughly equal to the weight of one hydrogen atom.) The higher figure is deceptive, however, because all such large proteins are made up of peptide subunits, which usually range between 15,000 and 100,000 daltons. For example, the enzyme phosphorylase, which is important in glycogen storage and degradation, is a dimer (a two-peptide polymer) composed of two identical subunits each with a molecular weight of 96,000 daltons. Similarly, aspartyl transcarbamylase, the first enzyme in the pathway leading to the synthesis of cytidine triphosphate, or CTP, is a dodecamer (a 12-peptide polymer) composed of six subunits of one kind (each with a molecular weight of 35,000 daltons) and six subunits of another kind (each with a molecular weight of 17,000 daltons). These subunits are attracted to one another by noncovalent forces: largely electrostatic attractions or hydrophobic bonds (a name used to describe the tendency of oily regions of a structure to be forced together in the same way that oil droplets tend to coalesce in water). Fortunately qualitative features do not change with the size of the protein, so that by studying the simpler proteins it is possible to understand the properties of proteins in general.

A peptide chain with a molecular weight of only 25,000 daltons is still large compared with the molecular weight of most substrates, which are usually compounds in the molecular-weight range of 100 to 1,000 daltons. Occasionally enzymes act on very large molecules such as DNA (deoxyribonucleic acid), cellulose or other proteins, but when they do, they usually bind only a small portion of these large molecules, so that the effective substrate size is still only about 1,000 daltons or less. This difference in relative size means that only a small portion of the enzyme's surface is actually involved in catalysis. The rest of the surface is available for binding the molecules that are involved in regulation and for the association of subunits with one another.

Cooperativity

The concept of protein flexibility provided an explanation for a long-known phenomenon that had been originally discovered by the Danish physiologist Christian Bohr in the 19th century. Bohr noted that the binding of oxygen to hemoglobin could be described by a sigmoid, or S-shaped, curve instead of the normal hyperbolic curve observed for the binding pattern of most enzymes [*see illustration on page 90*]. He correctly deduced that this unusual type of binding curve would result if the first molecule bound made it easier for the next molecule to bind, and so forth; hence he called the process cooperative. Since then hemoglobin has been the subject of intensive study by many prominent investigators, and much of our knowledge of the phenomenon of cooperativity results from examination of this vital protein. Cooperativity is not limited to hemoglobin, however. Sigmoid binding curves are also common and important features of the regulatory proteins.

The appearance of a cooperative binding pattern in the case of a regulatory protein can be explained by means of the flexible-protein hypothesis, using the simplest multisubunit protein, a dimer made up of two identical subunits [*see illustration on page 92*]. A number of such dimer proteins exist in nature; the types of interaction they are involved in are similar to those of the more complex proteins. The binding of the substrate induces conformational changes that depend on the structure of both the substrate and the protein.

Three general types of conformational change are known. In one type the first molecule of substrate alters the subunit to which it is bound but does not alter the interactions between the subunits. The second subunit therefore binds substrate in just the same way and with the same affinity as the first. There is no cooperation between the subunit sites.

The next type of conformational change is quite different. Here the first molecule to be bound induces a conformational change in the first subunit, which induces shape changes in the second subunit. These changes, which are transmitted through the protein structure, change the active site in the second subunit so that it becomes more receptive to the substrate; hence the second molecule of substrate is bound more readily than the first. This phenomenon, called positive cooperativity, explains the sigmoid binding curve discovered by Bohr.

In the third case the first molecule induces a conformational change that makes the binding site in the second subunit less attractive to the substrate because of their incompatible geometries. The binding of the second molecule is therefore discriminated against in favor of the first and proceeds much less readily. This phenomenon is called negative cooperativity (cooperativity because of the subunit interactions, negative because the first molecule has a negative effect on the second).

The induced-fit hypothesis can readily explain the sigmoid curve of multisubunit proteins; a mathematical adaptation of the induced-fit approach was devised by George Némethy, David Filmer and me to do just that. Our solution is not the only possible explanation, but

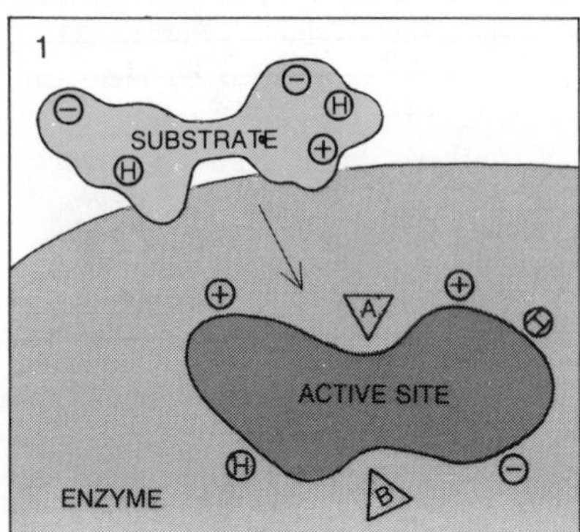

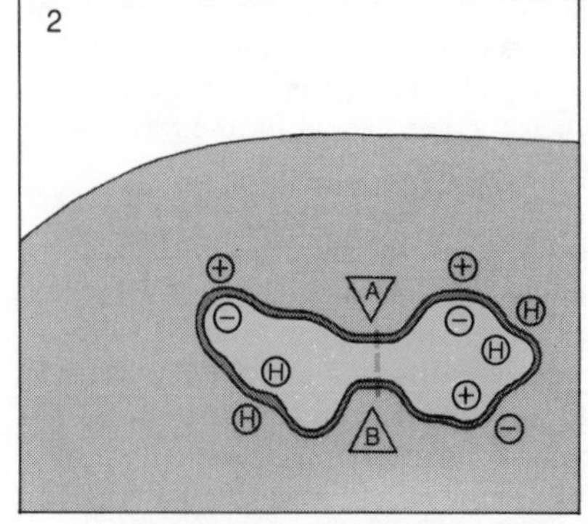

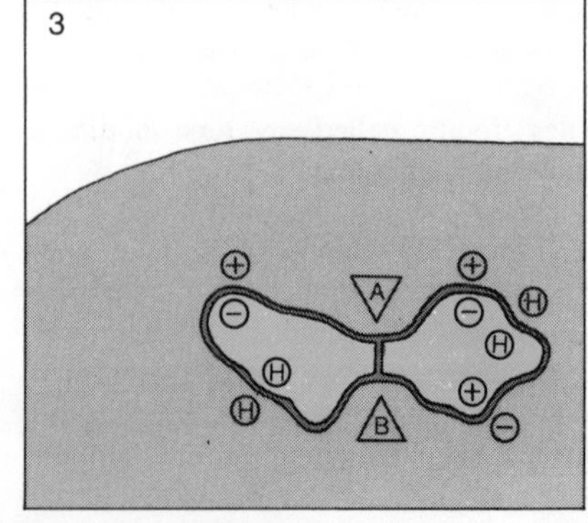

INDUCED-FIT MODEL of enzyme action, developed recently by the author and his colleagues at the University of California at Berkeley, assumes that the enzyme does not exist initially in a shape complementary to that of the substrate (*1*) but rather is induced to take the complementary shape in much the same way that a hand induces a change in the shape of a glove (2). Once the substrate is bound, the catalytic groups of the enzyme are in position to cut the chemical bond in the substrate's neck, forming the reac-

there is substantial evidence that it is valid for a large number of enzymes exhibiting such sigmoid curves. Besides being able to explain a puzzling phenomenon in terms of the protein structure, the induced-fit approach predicted that a different type of interaction—negative cooperativity—should exist (in other words, that the induced conformational changes would make the second molecule bind less readily than the first). Such a phenomenon was not known in nature at the time, but its prediction from theory led to a determined search for it. In 1968 Abby Conway and I discovered an example of this strange phenomenon in an enzyme that participates in carbohydrate metabolism: glyceraldehyde 3-phosphate dehydrogenase. Negative cooperativity has since been found in many other enzymes.

The mathematical analysis of such a negative-cooperativity pattern explains in part why the phenomenon escaped detection. On superficial inspection a negative-cooperativity binding curve looks like a hyperbola. A more careful analysis shows that in reality the curve is not a true hyperbola. Such a curve can be explained only by assuming that the second molecule binds to the protein less readily than the first. Several examples of each type of cooperativity have now been established. Induced conformational changes can be caused by activators and inhibitors as well as by substrates, and therefore all three types of molecule can give rise to cooperative interactions.

All these concepts apply to molecules with more than two subunits in the same way. In positive interactions the first molecule makes it easier for the second to bind, the second makes it easier for the third and so on. In negative interactions each interacting molecule makes it more difficult for the next molecule to interact. The small molecule (a substrate, an inhibitor or an activator) induces the change and the effect is then transmitted to the neighboring subunits. The greater the number of subunits is, the more dramatic the cooperativity can be, but the actual cooperativity pattern observed depends on the details of the individual protein structure.

Why should such cooperative phenomena exist in nature? Would it not be better for the protein to be designed correctly in the first place, so that these induced conformational changes need not alter the interactions between the subunits or the shape of the protein? One of the reasons protein flexibility is so important, of course, has already been mentioned: flexibility makes it possible for an enzyme to be regulated by molecules that are not themselves consumed in the enzymatic reaction.

The concept of cooperativity, however, apparently provides another reason for protein flexibility. If one examines the binding curves for the three types of cooperativity and compares the change in concentration of a compound to the change in the activity, one finds that approximately an 81-fold change in the concentration of the substrate is needed to go from an activity level of 10 percent to one of 90 percent, assuming that the protein follows the hyperbolic binding curve of a normal noninteractive protein. In sharp contrast, only a ninefold change in concentration is needed if one assumes that the protein is designed with rather mild positive cooperativity. (In hemoglobin a threefold change in concentration will do the job, and in CTP synthetase a 1.5-fold change is sufficient.) In other words, a protein with positive cooperativity is much more sensitive to small fluctuations in the environment than a protein with a normal binding pattern. The sensitivity is increased for inhibitors and activators as well as for substrates. Positive cooperativity is thus an amplification device to make a small signal have a much larger regulatory effect. This increased sensitivity is extraordinarily important for the regulatory function of these proteins.

One example of the physiological importance of the phenomenon of cooperativity can be observed in patients who have a mutant hemoglobin that lacks the positive cooperativity of normal hemoglobin. Positive cooperativity enables hemoglobin molecules to absorb large amounts of oxygen in the lungs and to deposit large amounts of oxygen in the tissues, even though the pressure of oxygen does not vary much from one location to the other. This situation follows from the steepness of the sigmoid curve associated with positive cooperativity. The noncooperative protein is far less efficient in transporting oxygen, and patients possessing it are very sick. In a crude sense it might be said that the positively cooperative protein resembles a truck that takes on a full load of dirt at one location and empties all of it at another, whereas a protein without cooperativity cannot take on a full load to begin with and can only get rid of part of the load.

What, then, is the role of negative cooperativity? A look at its binding curve provides the answer to that question too. Here the protein is less sensitive to fluctuations in the environment than a noncooperative protein; in other words, a much greater change in the concentration of the substrate, the inhibitor or the activator is needed to go from an activity level of 10 percent to one of 90 percent. Some proteins should not be subject to fluctuations in the environ-

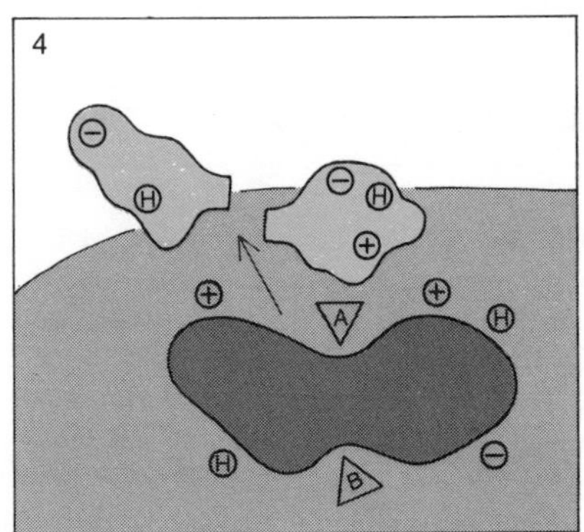

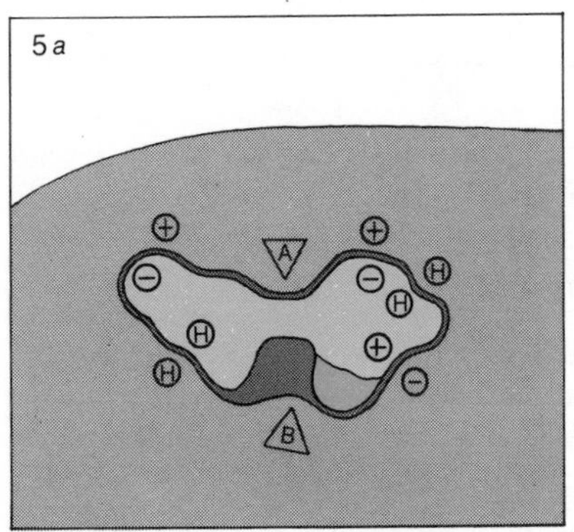

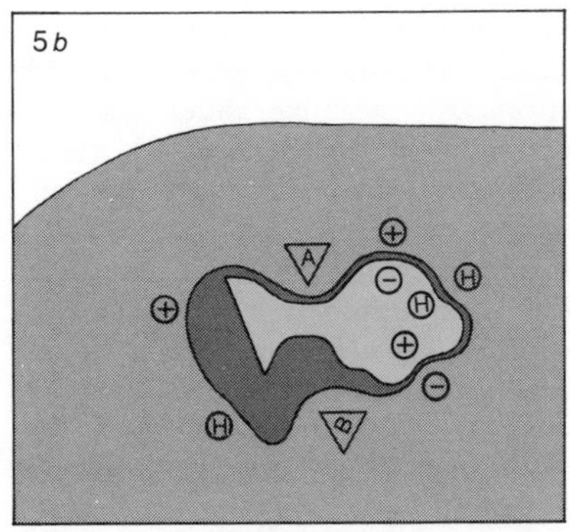

tion products (3), which leave the surface of the enzyme, returning the active site to its original noncomplementary shape (4). This concept of a flexible enzyme made it possible to explain the previously puzzling observation that certain oversized and undersized compounds were able to bind to the surface of the enzyme without forming products (5a, 5b). Even though both types of nonsubstrate compound bind, neither succeeds in inducing the proper alignment of the catalytic groups and hence neither reacts with the enzyme.

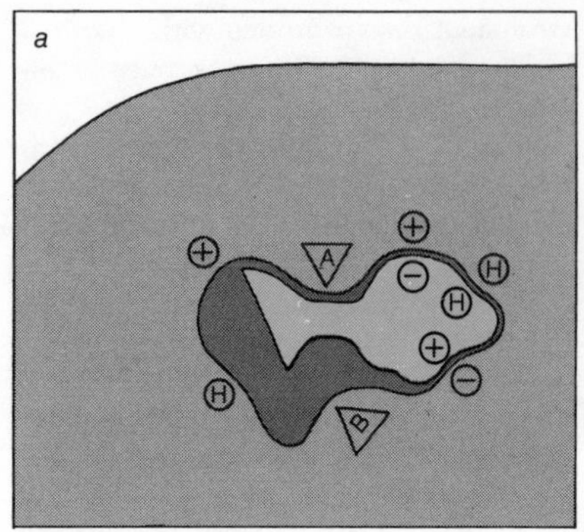

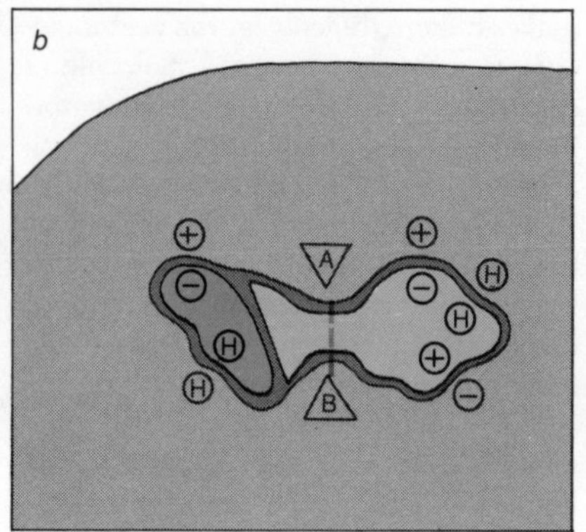

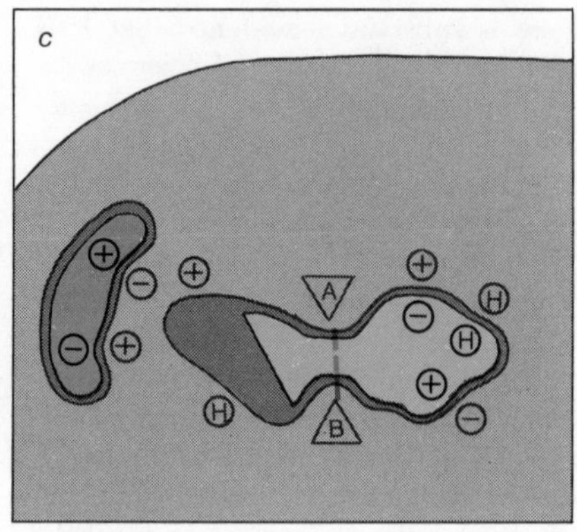

ACTIVATOR MOLECULES can, according to the flexible model of enzyme action, help to make a deficient molecule act as a substrate by altering the shape of the enzyme. For example, in the case of a molecule that is too small to induce the proper alignment of catalytic groups (*a*) a second molecule can bind immediately adjacent to the deficient molecule inside the active site, thereby inducing a stable shape with the proper alignment of catalytic groups (*b*). A molecule that binds at a site outside the active site can also induce the proper shape (*c*). In neither case is the activator molecule itself consumed in the ensuing chemical reaction.

ment, and the proteins that regulate such processes may use negative cooperativity to damp their sensitivity and make them less subject to environmental changes.

The Nature of the Shape Changes

How are conformational changes propagated through the protein molecule and how extensive are they? An understanding of these processes has flowed from two sources: chemical studies of the composition of proteins and X-ray studies of protein structure.

Some of the important structural data were obtained as a result of the pioneering work of M. F. Perutz and his coworkers on hemoglobin [see "The Hemoglobin Molecule," by M. F. Perutz; SCIENTIFIC AMERICAN Offprint 196]. Hemoglobin consists of four similar but not identical subunits arranged in a tetrahedral array. When oxygen is bound to the heme, or iron-containing, binding group of one of these subunits, there is a small shift in the position of the iron atom, which in turn causes a small shift of an adjacent histidine side chain. The histidine shift then compresses several amino acid side chains in a helical portion of the protein and squeezes a tyrosine side chain out of a small pocket. The movement of the tyrosine group dislocates the subunit to which it is attached and breaks the salt linkage between the subunits.

These are not the only regions that are affected by the binding of the oxygen atom, but they are illustrative of the type of change that can occur in a typical protein molecule. The point is that the protein is a tightly structured molecule with intimate close-packing relations between the atoms of the amino acid side chains. A shift of one such group by even a fraction of an angstrom causes realignments of other side chains, generating a "domino" effect that can extend through the entire protein molecule. The changes within the hemoglobin subunit cause a realignment of the subunits with respect to one another.

The importance of these protein shifts can be demonstrated with mutant hemoglobins isolated from sick patients. When the mutant hemoglobins were examined by Perutz and his colleagues, the amino acid side chains involved in the shape changes were found to be altered.

A further understanding of these processes has been obtained from experiments conducted in my laboratory at the University of California at Berkeley by Alexander Levitzki and William Stallcup. They worked with CTP synthetase, a key enzyme that converts uridine triphosphate, or UTP, to CTP in the metabolism of nucleic acids. This regulatory protein usually exists in the form of four identical subunits, but it can be studied as a dimer. Each subunit has binding sites for its three substrates: UTP, ATP and glutamine, and also for its regulatory effector, guanosine triphosphate, or GTP [*see illustration on page 93*]. The glutamine reacts at the glutamic site on the enzyme, forming what is called the glutamyl enzyme. In doing so it liberates ammonia, which binds to the ammonia site on the enzyme. The ammonia in turn reacts with UTP, which reacts with ATP. The close coupling of the chemical steps indicates that the sites are immediately adjacent to one another.

When the glutamyl enzyme is formed, the reaction produces a covalent bond between glutamine and a cysteine side chain on the surface of the protein. At the beginning the two cysteine groups of the two active sites are equally reactive. When glutamine reacts with the same side chain, however, a strange thing happens. The reaction of the substrate with one subunit turns off the cysteine side chain on the other subunit. The effect is only temporary; the glutamyl enzyme reacts further to regenerate free enzyme.

As it happens, this change in the shape of the enzyme can be "frozen" (somewhat like stopping a motion-picture film at a single frame) with the aid of an "affinity label," dioxoazonorleucine. The molecule of dioxoazonorleucine is enough like the molecule of glutamine so that it forms a covalent bond with the cysteine side chain but enough different so that it cannot react further. The result is that only one of the two initially identical subunits reacts with the dioxoazonorleucine. A reaction at one subunit turns off its neighbor, so that only one of the two potential subunits reacts at any one time, giving rise to what we have termed the "half of the sites" phenomenon.

This phenomenon is observed in a number of enzymes and indicates that in many cases the shape change can be transmitted over long molecular distances. On the basis of the size of the enzyme subunits the cysteine side chains in CTP synthetase are probably between 20 and 60 angstroms apart.

The most surprising part of the finding is that the immediately adjacent sites—the ammonia, UTP and ATP sites—are not altered at all when the cysteine group is modified. All their properties remain the same. Hence the formation of a bond apparently transmits a signal that has dramatic consequences as much as 60 angstroms away without perturbing structures within four or five angstroms! When the reactivity of amino acid side chains to protein reagents is examined, many side chains are found to change position and many others to remain unchanged.

A similar pattern is evoked by GTP, the regulatory molecule. GTP activates the enzyme by altering the reactivity of the glutamine site on the same subunit, so that the glutamyl enzyme is formed more rapidly. The same shape change that activates the glutamyl site deactivates the GTP site of the neighboring subunit. Moreover, the shape changes induced by GTP do not alter the properties of the ammonia, ATP or UTP sites. Thus the GTP merely by binding to the surface of the protein can direct the alteration of some side chains and leave others unchanged. The same is true of many other activators, inhibitors and substrates.

The Scale of the Shape Change

Hence it appears that induced conformational changes in proteins are not like the ever widening concentric ripples produced by throwing a pebble into a pond. They are more like a spider web in which the strands are devised to transmit a perturbation occurring in one corner of the web to another corner. The perturbation can be transmitted over long distances and can alter the positions of many strands, but a clever design can ensure that some strands will remain unchanged at the same time that others are shifting appreciably. The protein, like the spider web, is designed to transmit information in a focused manner to some regions and to leave others unchanged.

The schematic drawings used to illustrate the alteration of protein shapes in this article tend to exaggerate the relative movements necessary to achieve this control. Actually we do not know precisely how big the movement of catalytic or binding side chains must be to achieve on-off switching. There are strong suggestions, however, that the movements do not have to be very large. The length of a carbon-oxygen bond, for example, is about 1.3 angstroms. A catalytic group that needs to be positioned close enough to an oxygen atom to pull electrons out of it would therefore be ineffective as a catalyst if it were positioned next to the carbon atom. Hence a movement on the order of an angstrom or two would appear to be sufficient to make the difference between an effective catalyst and an ineffective one. The effective movement is probably somewhat less than that, particularly because most regulators do not completely turn off the enzyme but rather reduce its function in the direction of either catalysis or binding.

Movements of side chains by 10 or 12 angstroms have been observed in carboxypeptidase molecules by William N. Lipscomb, Jr., and his colleagues at Harvard University, using the X-ray-crystallographic approach. It seems likely that the large movements observed in this type of protein are more than are necessary for regulation in general and that many of the atoms move much less or not at all. It has already been found that the movement of certain atoms such as the iron atom in hemoglobin and the sulfur atom in CTP synthetase may be less than an angstrom, and yet such a movement can trigger large conformational changes. Model experiments with simpler compounds suggest that an alteration of less than half an angstrom can also alter reactivity greatly. In short, the schematic drawings shown here should be taken to indicate that the movements are large enough to alter the function of a protein but that they do not represent the actual scale of conformational changes. Overall observations of protein molecules also indicate that no tremendously large shape changes occur in the subunit as a whole even though a few of the atoms move by several angstroms. By focusing on a few groups and using schematic pictures an erroneous impression could be obtained if one does not remember that one is looking at only a very small portion of the spider web. The gross anatomy of the web is unchanged; only a few strands shift with respect to one another, and yet these shifts are highly significant in terms of function.

Advantages of Induced Changes

One of the advantages of having enzymes that can be induced to change their shape in response to the proper stimulus can be seen in the example we have been considering: CTP synthetase. I have mentioned that the substrates ATP and UTP show strong positive cooperativity. In the case of ATP, for example, the first molecule of ATP alters the shape of the enzyme so that the subsequent molecules of ATP bind rapidly. Actually CTP synthetase is one of the most cooperative enzymes known (even more so than hemoglobin). The binding of the first molecule of substrate has a triggering effect that allows the rapid and complete binding of the subsequent molecules of ATP, so that only the free enzyme and a form of the enzyme with four ATP molecules bound are found in appreciable amounts. This positive cooperativity means that the enzyme is

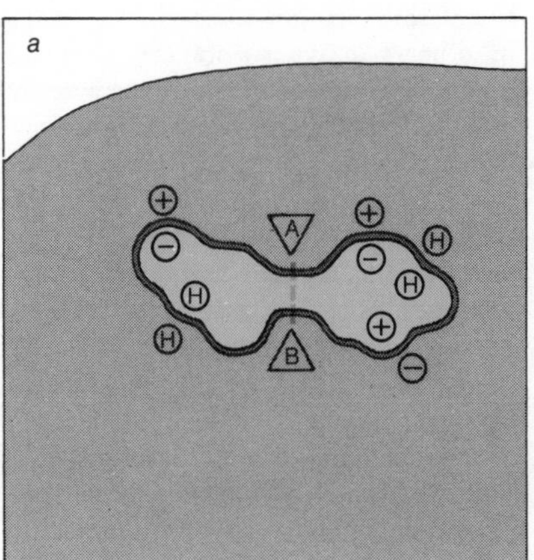

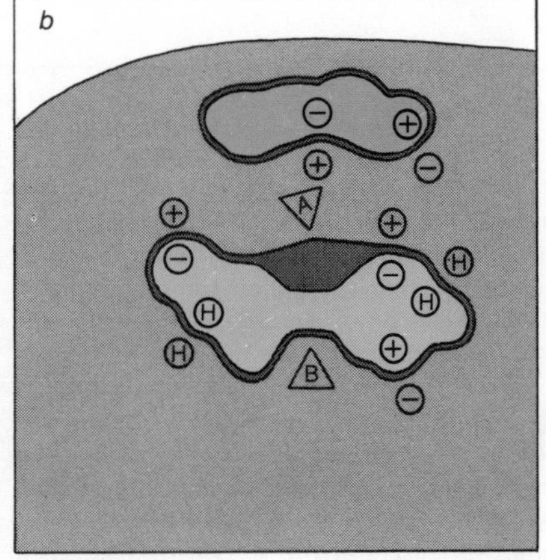

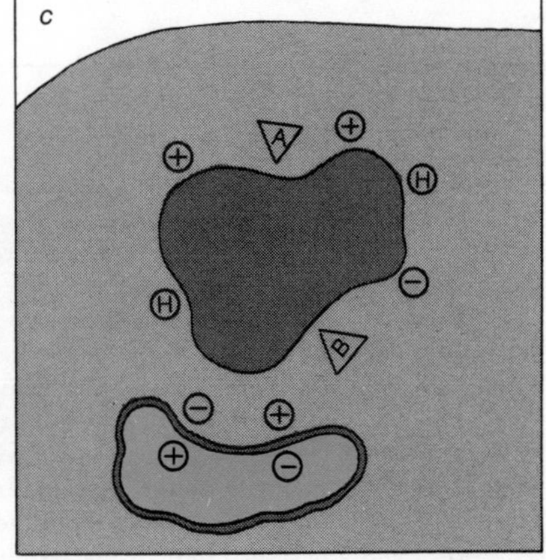

INHIBITOR MOLECULES can, in the reverse process, distort a flexible enzyme out of the active shape by binding at sites outside the active site. In the case of a substrate that would otherwise react with the enzyme (*a*) such inhibitors can cause a disruption of either the catalytic function (*b*) or the binding function (*c*) of the enzyme, in effect "turning off" the enzyme-substrate reaction.

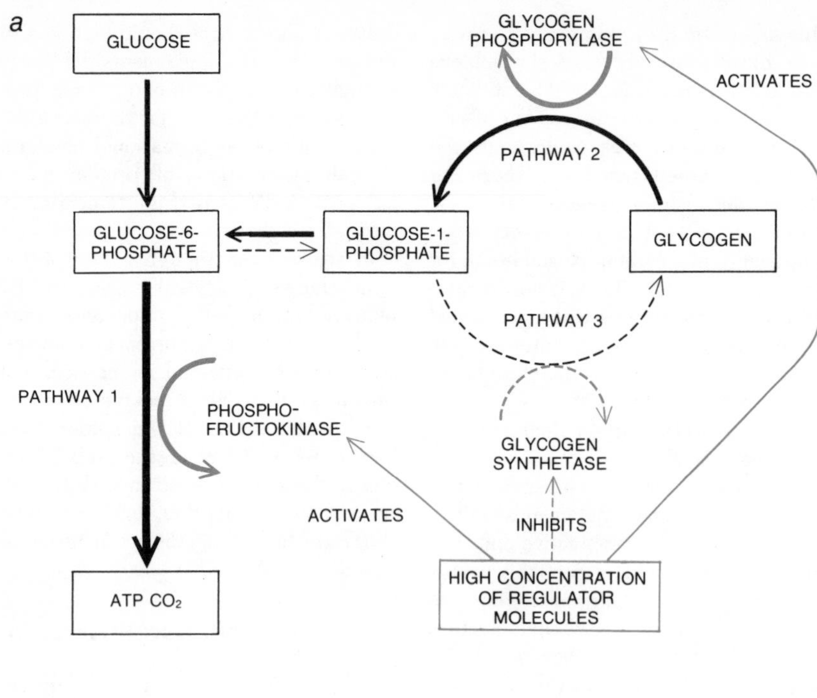

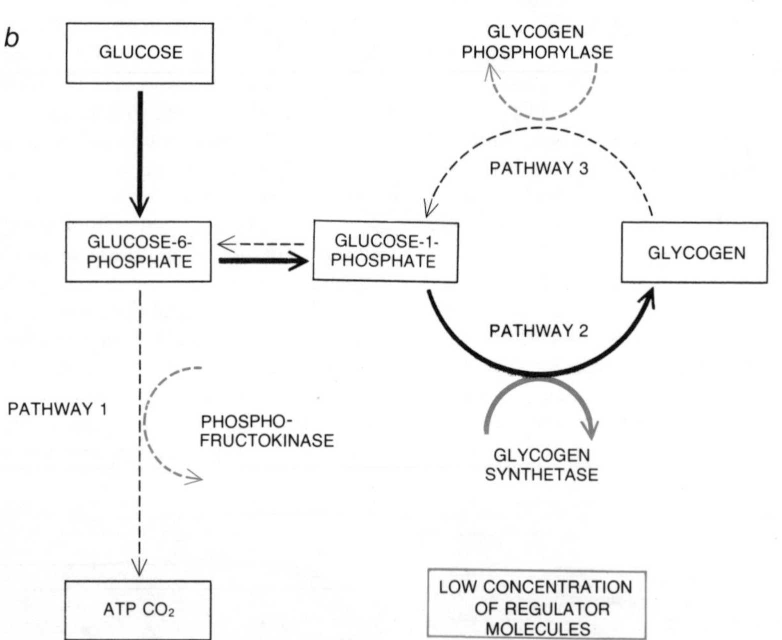

ALTERNATE BIOLOGICAL PATHWAYS are controlled by regulatory molecules through the alteration of enzyme shapes. In the human body, for example, ingested glucose is normally converted to glucose-6-phosphate, which can either be oxidized immediately to provide energy in the form of molecules of ATP (adenosine triphosphate) or stored in the form of glycogen (a chain of sugar molecules) to be released on future demand. Which pathways are chosen depends on the concentration of regulatory molecules. In times of energy demand high concentrations of these molecules convert the enzyme of pathway 1 (phosphofructokinase) and the enzyme of pathway 3 (glycogen phosphorylase) to active forms while at the same time converting the enzyme of pathway 2 (glycogen synthetase) to an inactive form, thereby channeling both the ingested glucose and the stored glycogen into the production of ATP (*top*). Low concentrations of the regulatory molecules enable the enzymes of pathways 1 and 3 to return to their inactive forms while enabling the enzyme of pathway 2 to become active again, thereby diverting the glucose into glycogen for storage (*bottom*). The reciprocal effect of the regulatory molecules on pathways 2 and 3 prevents their both being active at the same time and therefore rules out a futile short circuit in the system.

very susceptible to small fluctuations in the concentration of ATP and hence is highly sensitive to environmental influences on this molecule. In contrast, the regulatory molecule, GTP, induces conformational changes leading to negative cooperativity, which desensitizes the enzyme toward fluctuations of GTP in the environment.

What does all this mean in terms of enzyme action? It means that the enzyme has been programmed by its design through evolutionary time so that it can be responsive to small changes in ATP levels and at the same time be desensitized to rather wide fluctuations of GTP levels. A conceivable reason for this ability is that ATP is highly controlled in the biological system, since it is such a central compound for so many pathways. As a result enzymes such as CTP synthetase must respond readily to even small changes in the levels of ATP when they are regulated by the ATP level itself. On the other hand, GTP may fluctuate greatly from time to time in the organism's life cycle because it plays a role in different but less vital pathways. If a fairly constant level of CTP-synthetase activity is needed throughout these fluctuations, the negative cooperativity of GTP will ensure this desensitization to fluctuations in GTP levels. Of course, complete desensitization would allow the production of CTP in the absence of GTP, which would be a wasteful operation because both are needed for the synthesis of RNA. The protein ensures GTP control by requiring it to serve as the activator but eliminates excessive sensitivity through the device of negative cooperativity.

Another advantage of these induced conformational changes arises from the sequence of steps on an enzyme surface. High-energy intermediates are frequently formed in chemical syntheses, but if they are not isolated from water or other reactive substances, they decompose in side reactions that lower the yield of the reaction. Induced conformational changes can enable one step in the reaction to trigger the next step, which in turn triggers the third step and so on. In this way the high-energy intermediates exist for only brief intervals and are nestled in the protective harbor of the active site during the chemical changes. Wasteful side reactions are prevented, and the characteristically high yields of enzymatic reactions are achieved. In short, the induced-fit conformational change both explains the anomaly of a required order in the binding of substrates and provides a reason for it.

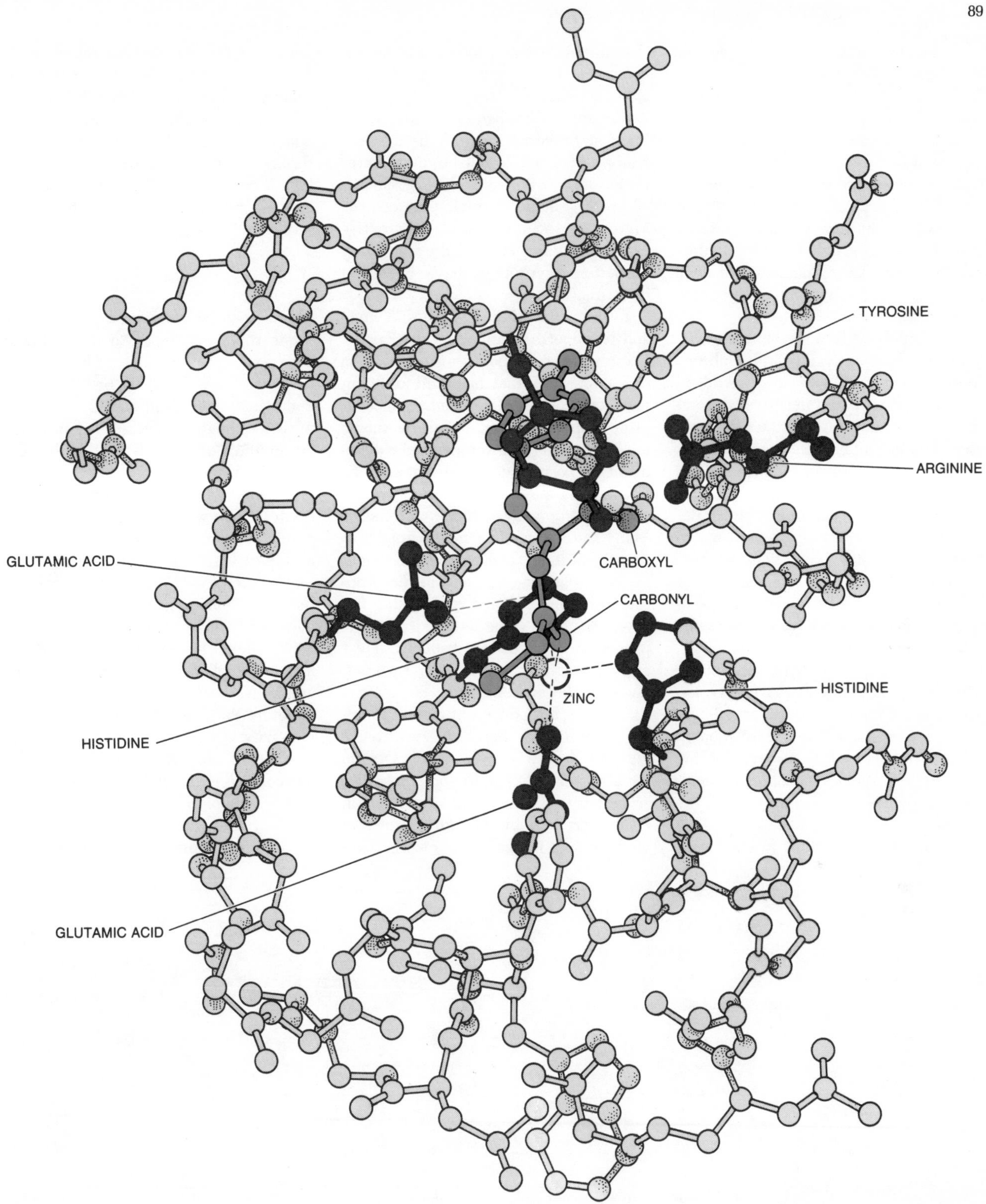

PRECISE FIT between the active site on the surface of a large protein molecule and the specific substrate molecule with which the protein reacts is evident in this simplified three-dimensional drawing, based on X-ray-crystallographic data obtained by William N. Lipscomb, Jr., and his colleagues at Harvard University. The protein, rendered in shades of gray, is carboxypeptidase *A*, a digestive enzyme that (as its name implies) works by cutting the polypeptide chain of the substrate near its carboxyl end. The substrate, rendered in color, is carbobenzoxyalanylalanyl tyrosine. Approximately a fourth of the total number of atoms in the polypeptide chains that comprise the two molecules are represented in this view. The atoms shown are mostly carbon, with a small admixture of nitrogen and oxygen; all hydrogen atoms have been omitted. The six active-site side chains that specifically interact with the substrate are the darker gray. For example, a positively charged arginine side chain is shown attracting the negatively charged carboxyl group of the substrate. In addition certain hydrophobic, or oily, regions on the substrate are attracted to similar regions on the enzyme, strengthening the attraction between the two molecules. A zinc atom (*white*) forms an additional "coordination bond" involving a carbonyl group on the substrate and three other amino acid side chains (glutamic acid and two histidines) extending from inside the enzyme's bowl-shaped active site. The tyrosine side chain and the second glutamic acid side chain of the active site are catalytic groups that polarize the electrons in one of the substrate's chemical bonds, splitting that bond (*broken colored line*) and thereby dividing the substrate into the two parts that are the reaction products of this particular enzyme-substrate combination.

It therefore seems likely that flexible enzymes arose early in evolutionary time because of catalytic needs and certain specificity requirements. One of the greatest of these needs was the exclusion of water from some reactions. The cooperative property probably developed later because of the multisubunit structure of the enzyme; cooperativity survived because it had a useful function of its own: the amplification of some responses and the damping of other responses.

It is important to note that the parallelism between activators and inhibitors (positive effectors and negative effectors) and positive and negative cooperativity should not obscure their different and complementary roles. An activator is designed to turn an enzyme on and an inhibitor is designed to turn it off. Cooperativity is designed to increase or decrease the sensitivity of the enzyme to the environmental fluctuations of these regulators. An activator may show either positive cooperativity, negative cooperativity or noncooperativity, and the same is true of a substrate or an inhibitor. GTP, for example, is an activator of enzyme action but shows negative cooperativity in its binding pattern. In a simple sense one might say that activation or inhibition defines the key role of the regulatory molecules, whereas cooperativity provides the fine tuning of the system.

Finally, it is important to emphasize that although this account of the development of the induced-fit theory of enzyme action concentrates on the particular approach taken in our laboratory, many other workers in both the U.S. and Europe have made outstanding contributions to our understanding of the role of protein shape in biological control. Moreover, several possible alternative explanations for some of the regulatory properties of proteins have been put forward.

In particular Monod, Jeffries Wyman and Jean-Pierre Changeux at the Pasteur Institute have proposed an explanation of cooperativity that has quite different features from the induced-fit mechanism proposed above. It can be shown mathematically that their model can account for positive cooperativity and can also explain many features of activation and inhibition by effector molecules. Many features of their model appear to be present in the cooperative binding of oxygen to hemoglobin and in the properties of some enzymes; moreover, it has been of great value in clarifying the properties of regulatory proteins. Their approach cannot apply to all enzymes, however, because negative cooperativity requires an induced-fit model. In addition, Sidney A. Bernhard and his coworkers at the University of Oregon have postulated that in some cases an asymmetry of the type observed in insulin crystals by Dorothy Crowfoot Hodgkin and her colleagues at the University of Oxford may be important in regulatory proteins. In spite of the usefulness of these alternatives in describing some properties of proteins, it would appear that the main mechanism used throughout nature in the control of biological processes is the induced alteration of protein structure I have described.

Triggering Events

I have so far mentioned two types of event that can trigger conformational changes: (1) the reaction of a substrate to form a new covalent bond with an amino acid side chain and (2) the binding of a regulatory molecule (a substrate, an inhibitor or an activator) to the surface of a protein without the formation of a covalent bond. Both of these events occur in the reaction of substrates and in regulatory processes. In the preceding examples I used only the binding of effectors such as GTP to illustrate regulation, but regulation by the formation of covalent bonds can also occur at regulatory sites. Phosphorylation of the enzyme phosphorylase by ATP, for example, can lock the enzyme into a new structure that has different properties of reactivity and sensitivity to regulatory compounds. Earl W. Sutherland, Jr., and his colleagues at the Vanderbilt University School of Medicine showed that this covalent change is controlled by the hormone adrenalin through its mes-

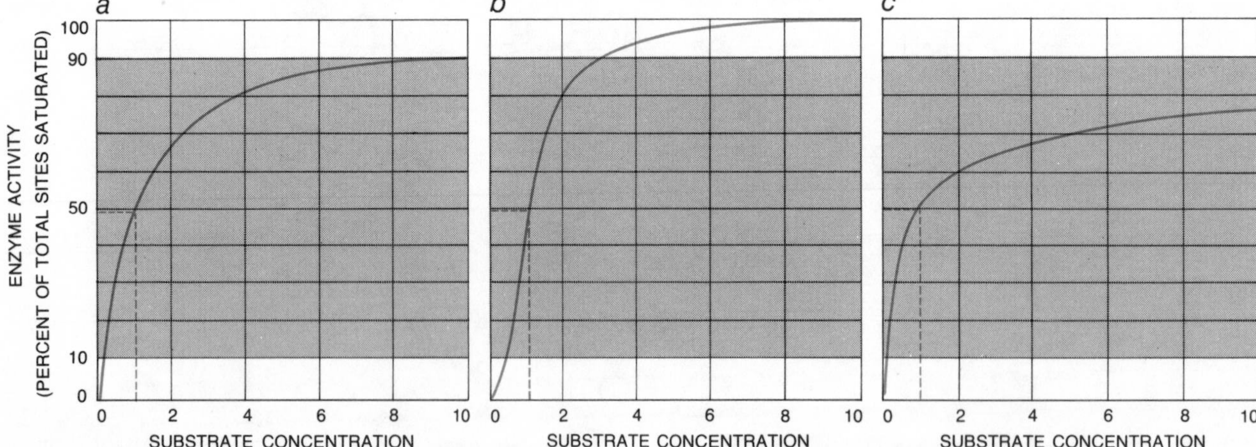

THREE TYPES OF COOPERATIVITY observed in the binding patterns of different enzyme-substrate combinations are represented by these curves, which relate the change in the concentration of a substrate to the change in the level of enzyme activity induced by the substrate. Curve *A* depicts the normal hyperbolic binding pattern exhibited by most enzymes; in this case an 81-fold increase in concentration of the substrate is needed to go from an activity level of 10 percent to one of 90 percent. Curve *B* shows the sigmoid, or *S*-shaped, binding pattern associated with an enzyme that exhibits positive cooperativity (the tendency for the first molecule bound to make it easier for the next molecule to bind, and so forth); in this case only a ninefold change in substrate concentration is needed to go from an activity level of 10 percent to one of 90 percent. Curve *C* shows the binding pattern associated with an enzyme that exhibits negative cooperativity (the tendency for the first molecule bound to make it more difficult for the second molecule to be bound, and so forth); in this case the curve looks somewhat like a hyperbola, but in reality it is not. The curve approaches the final saturation state so slowly that it does not even reach the 90 percent activity level in this graph. Actually an increase in substrate concentration of 6,541-fold would be needed for such an enzyme to go from an activity level of 10 percent to one of 90 percent. To clarify relations between types of cooperativity all three curves are presented so that the situation at which half of the sites are occupied corresponds to a substrate concentration of 1. Similar curves can be obtained for activator and inhibitor molecules.

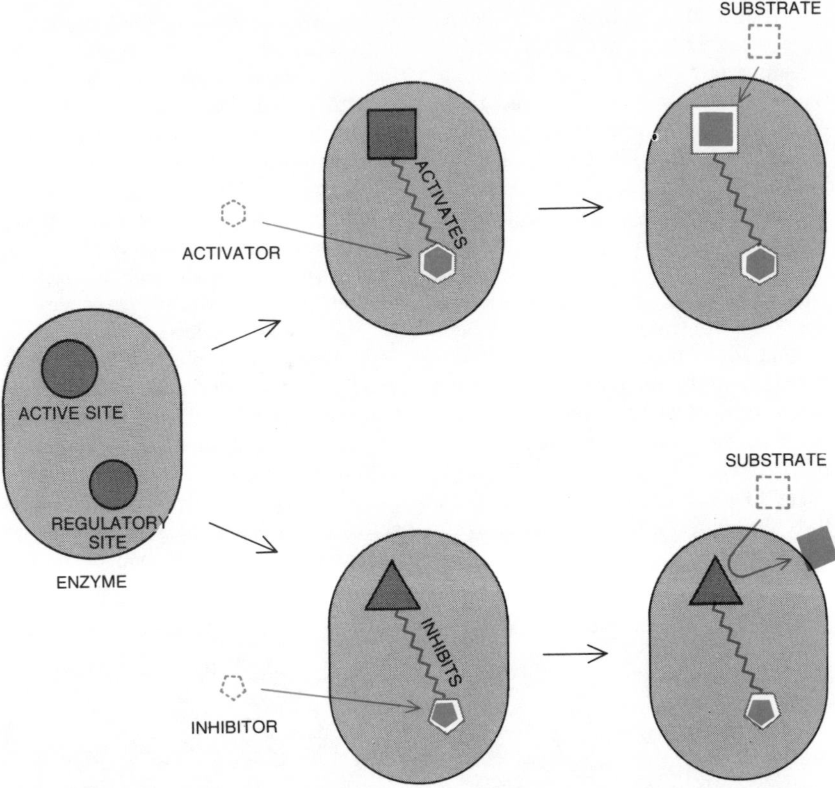

SIMPLIFIED SCHEME is presented here for illustrating the mechanism by which enzyme activity is controlled by the induced conformational changes brought about by small regulatory molecules that are not themselves involved in the primary enzymatic reaction. The complex polypeptide chain that comprises the actual enzyme is represented in this convention by the light gray area; the darker gray circles symbolize the shapes of the enzyme's active site and regulatory site in the absence of any bound molecules. The binding of the activator molecule (*colored hexagon*) induces a conformational change at the regulatory site (*symbolized by the change from a circle to a hexagon*) and also alters the enzyme's structure (*symbolized by the colored zigzag line*), which in turn changes the shape of the active site (*symbolized by the change from a circle to a square*). As a result the substrate (*colored square*) can now bind more easily than it can in the absence of the activator, since no further induced conformational change is needed for binding. In the reverse process the inhibitor molecule (*colored pentagon*) induces a different type of conformational change in the regulatory site (*symbolized by the change from a circle to a pentagon*), which is transmitted through the enzyme's structure, so that the resulting alteration at the active site (*symbolized by the change from a circle to a triangle*) makes the active site repel the substrate.

senger, cyclic AMP. Similar covalent changes are caused by the adenylation of the enzyme glutamine synthetase, as shown by Earl R. Stadtman, Helmut Holzer and their co-workers at the National Heart and Lung Institute. These covalent changes can be used to activate or to inhibit the enzyme. In such instances the reversal to the original enzyme is achieved by a second enzyme, which breaks the covalent bond of the regulatory molecule.

Why should nature use two devices, covalent changes and noncovalent binding, to induce the same kind of shape change for regulation? The answer is not known with certainty, but the suspicion arises that time is one factor that is involved. If an animal is frightened and needs large amounts of energy for a short period of time, it would be desirable to have an override mechanism that would ensure the high activity of some crucial enzymes until the crisis is past. The phosphorylation of phosphorylase induced by adrenalin would seem to provide such a mechanism. The formation of the covalent phosphoryl-enzyme bond converts the enzyme to a more active shape. When the crisis is over, the phosphate group can be removed by a second enzymatic process to regenerate the original, more placid enzymes, but for a time the normal instantaneous controls are eliminated to mobilize glucose for the crisis.

In a similar fashion, conformational changes that occur by the aggregation of subunits may also play an important role in regulation. In many instances it is found that a protein is active as a monomer but inactive as a dimer, or inactive as a dimer and active as a tetramer. The association of the polypeptide chains with one another causes a shape change with a resulting alteration in the activity of the protein. In essence, then, one subunit becomes the regulator of the other. If that is true, compounds that cause shape changes directly should also induce indirect changes in the polymeric structure of proteins. Indeed, some compounds induce changes in shape that cause them to dissociate. Such association-dissociation reactions appear to be important for many enzymes in metabolic pathways. The types of shape change in these associations and dissociations are exactly analogous to those induced by small molecules. It is also quite possible that certain hormones act as polypeptide regulators in much the same way that small molecules or subunits act in other instances.

Shape changes in proteins can occur very rapidly, in some cases in only a billionth of a second. Other shape changes occur in the millisecond range and have been shown to limit the rate of some enzymatic reactions. Still others take minutes to be effected; they are largely involved in association-dissociation reactions. This range of speeds serves the purposes of regulation well because some processes require an instantaneous correction in response to a stimulus, whereas a slower response is required to prevent an overshoot or to "lock in" a response until a crisis is passed. Furthermore, a slow conformational change has recently been suggested as a memory device in bacterial chemotaxis, in which a bacterium moves in response to a gradient in the concentration of a chemical substance. Such a change may have its counterpart in the neural systems of animals. Shape changes, in short, may occur over many time intervals from very fast to very slow, and each may have its usefulness.

Application to Other Systems

Up to this point I have emphasized the regulatory control of enzymes because they are the regulatory proteins that have been studied the most intensively and they are readily available in the laboratory. As biochemistry has progressed, however, other molecules that have key regulatory roles in biological systems have been isolated. Receptor molecules involved in sensory systems have been shown to be similar to enzymes in terms of structure and binding properties. These molecules have the specificity characteristic of enzymes, and it is generally believed, but not absolutely proved, that induced conformational changes are the signals that trigger the sensory impulse. When we smell or taste a compound, the compound induces a change in the shape of the receptor molecule, which triggers a response in our nervous system. Light induces a change in the shape of a protein in our eyes. Sensory phenomena are the highest form of regulation, and the brain is the ultimate regulatory apparatus of the most complex biological system: man.

Similarly, molecules that control protein synthesis have been isolated. Repressor molecules, for example, bind to DNA and prevent the reading of the genetic message unless they are removed by inducers. Again it is generally believed, although not yet absolutely proved, that the inducer causes a conformational change that peels the repressor molecule off the DNA, thereby initiating protein synthesis. Conformational changes are also believed to be crucial in promoter molecules, such as the cyclic-AMP-binding protein, which binds to DNA and aids in the initiation of protein synthesis. Such initiation and control of the reading of selected portions of the DNA allow different proteins to be made according to circumstances. A control of this type has been postulated to be a mechanism of differentiation, the reason nerve cells have one mixture of proteins and muscle cells have another. Differentiation, the key process that allows multicellular organisms to have specialized functions, is thus also dependent on changes in protein shape.

Antibody molecules, which protect us against invasion by foreign substances, induce a series of reactions in the apparatus called the complement-fixation system. That system is activated to destroy harmful cells and proteins by digesting them. Moreover, the presence of a foreign protein, such as a diphtheria toxin, can induce antibody-producing cells to multiply rapidly so that antibody to that specific harmful agent can be generated. This selective inducing of certain types of protective behavior is another regulatory device of living systems and is again believed to be caused by induced conformational changes of antibody molecules or antibody-receptor molecules. The transport of foodstuffs across cell walls is also thought to be triggered by induced changes in the

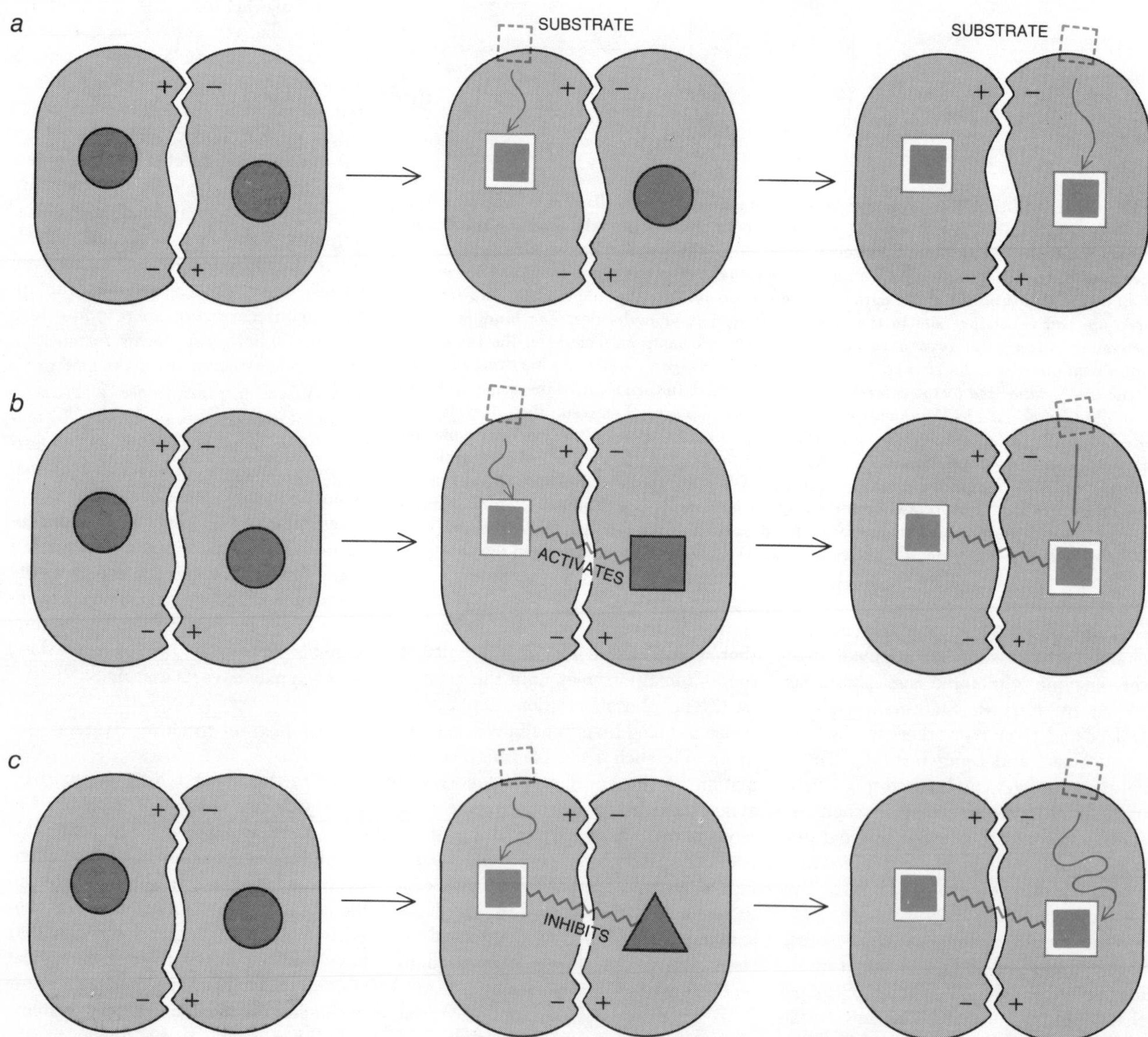

COOPERATIVITY IS EXPLAINED according to the induced-fit model of enzyme action in these diagrams, which use the simplified schematic convention introduced in the illustration on the opposite page. The type of protein represented here is the simplest known multisubunit enzyme: a dimer (that is, a two-peptide polymer) made up of two identical subunits. The three different enzymes shown exhibit respectively noncooperativity (*top row*), positive cooperativity (*middle row*) and negative cooperativity (*bottom row*).

shapes of proteins, as is the contraction of muscle.

There is no single regulatory process in the living system any more than there is a single type of control in a big city. The controls in the city are mediated by traffic lights, judges, payrolls, telephones, foremen, mayors and so on. The controls in biological systems are mediated by enzymes, antibodies, receptors, repressors and so on. Each of these protein molecules is designed differently and therefore each has a special history, giving rise to enormous diversity. Nevertheless, out of all this complexity comes a great simplicity. The flexibility of proteins and the shape changes they can undergo are the key features of regulatory control. Because the protein can exist in more than one shape and because the shape can be altered by external agents, the living system can respond to external stimuli and protect itself against environmental changes.

Nature thus places in the living cell the most powerful catalysts known and tames them to obey commands by subtle shifts in their structure, shifts measured in angstroms. These changes in shape can be imagined in a gross way as the alteration of a glove by the insertion of a hand, but on closer examination they seem much more like the delicate responses of a spider web designed with exquisite balance and interweaving. The delicate web we call a protein can be altered in shape by subtle perturbations, and through these perturbations functions can be turned on and off. As a result we feel like eating when we need food and lose our appetite when we are full. We can control the use of energy and the growth of specialized tissues, protect ourselves against hostile invading substances and develop a brain. All these essential regulations rest on the ability of a protein molecule to bend flexibly from one shape to another.

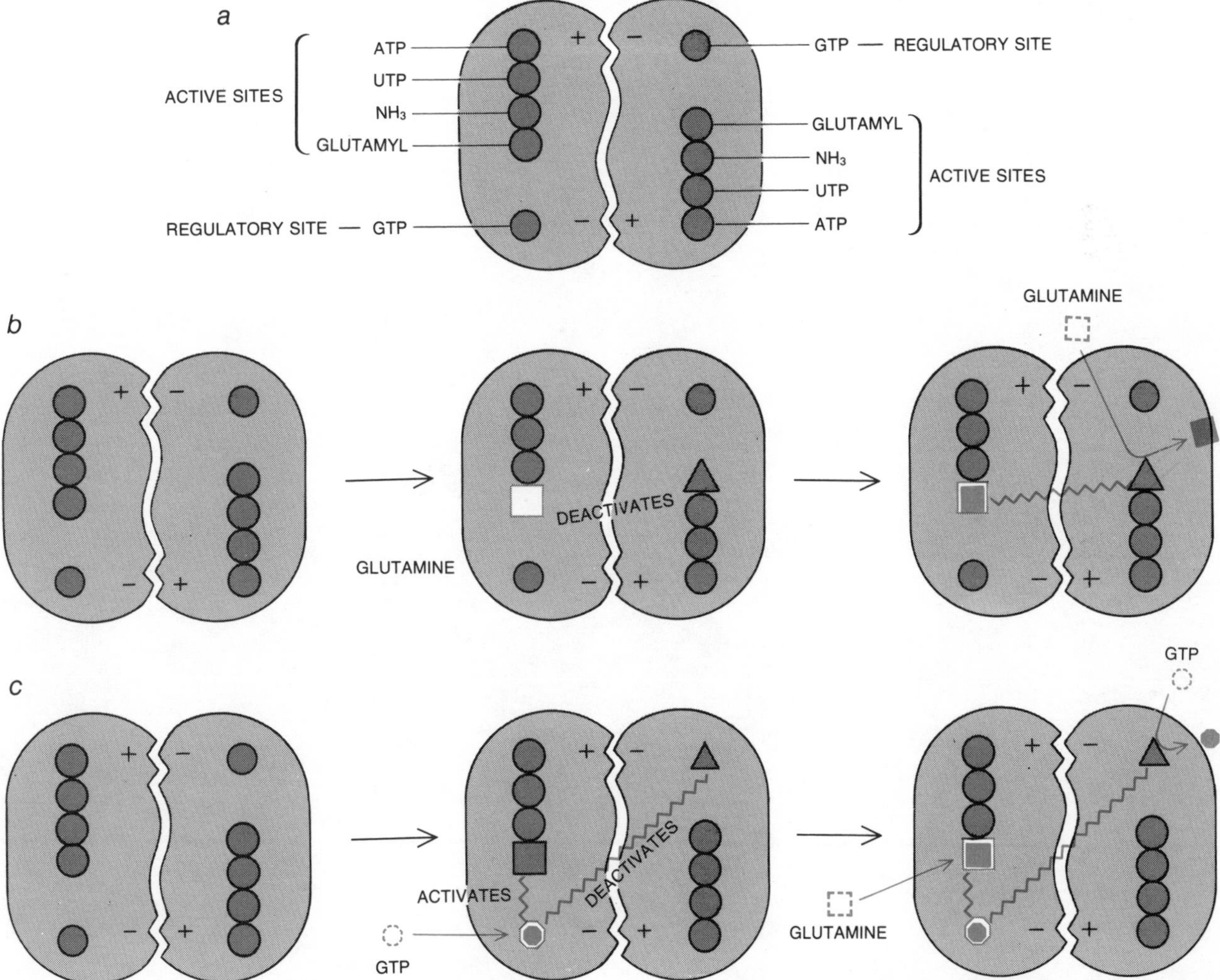

HIGHLY SPECIFIC NATURE of the changes induced in the structure and action of proteins as a result of the binding of selected substrate and regulator molecules is demonstrated in these schematic diagrams, which portray the behavior of the dimer of CTP synthetase, a key enzyme in the metabolism of nucleic acids. Experiments conducted in the author's laboratory have demonstrated the existence of specific binding sites on the surface of this enzyme for four substrate molecules and one regulator molecule (*a*). When the amino acid glutamine binds to the glutamyl site on one of the enzyme's subunits (*b*), it forms a glutamyl-enzyme intermediate. The conformational changes induced in this subunit in turn cause changes in the shape of the glutamyl site on the other subunit, making it less reactive. Since only one subunit of every dimer can engage in this reaction, the process is sometimes referred to as the "half of the sites" phenomenon. In contrast the binding of the regulatory molecule, GTP (guanosine triphosphate), to one of the subunits induces a conformational change that makes the glutamyl site of the subunit react more readily with glutamine at the same time that it causes the GTP site in the other subunit to assume a new shape that has less affinity for the second GTP molecule (*c*). In this case the design of the protein enables GTP to act as an activator of a catalytic reaction and an inhibitor of further GTP binding. The binding of both glutamine and GTP causes negligible changes in the shapes of the ATP, UTP or NH_3 (ammonia) sites.

Building a Bacterial Virus

by William B. Wood and R. S. Edgar
July 1967

T4 viruses with mutations in certain genes produce unassembled viral components. These particles are combined in the test tube in an effort to learn how the genes of a virus specify its shape

Slice an orange in half, squeeze the juice into a pitcher and then drop in the rind. It comes as no surprise that the orange does not reconstitute itself. If, on the other hand, the components of the virus that causes the mosaic disease of tobacco are gently dissociated and then brought together under the proper conditions, they do reassociate, forming complete, infectious virus particles. The tobacco mosaic virus consists of a single strand of ribonucleic acid with several thousand identical protein subunits assembled around it in a tubular casing. The orange, of course, is a large and complex structure composed of a variety of cell types incorporating many different kinds of proteins and other materials. Yet both orange and virus are examples of biological architecture that must arise as a consequence of the action of genes.

Molecular biologists have now provided a fairly complete picture of how

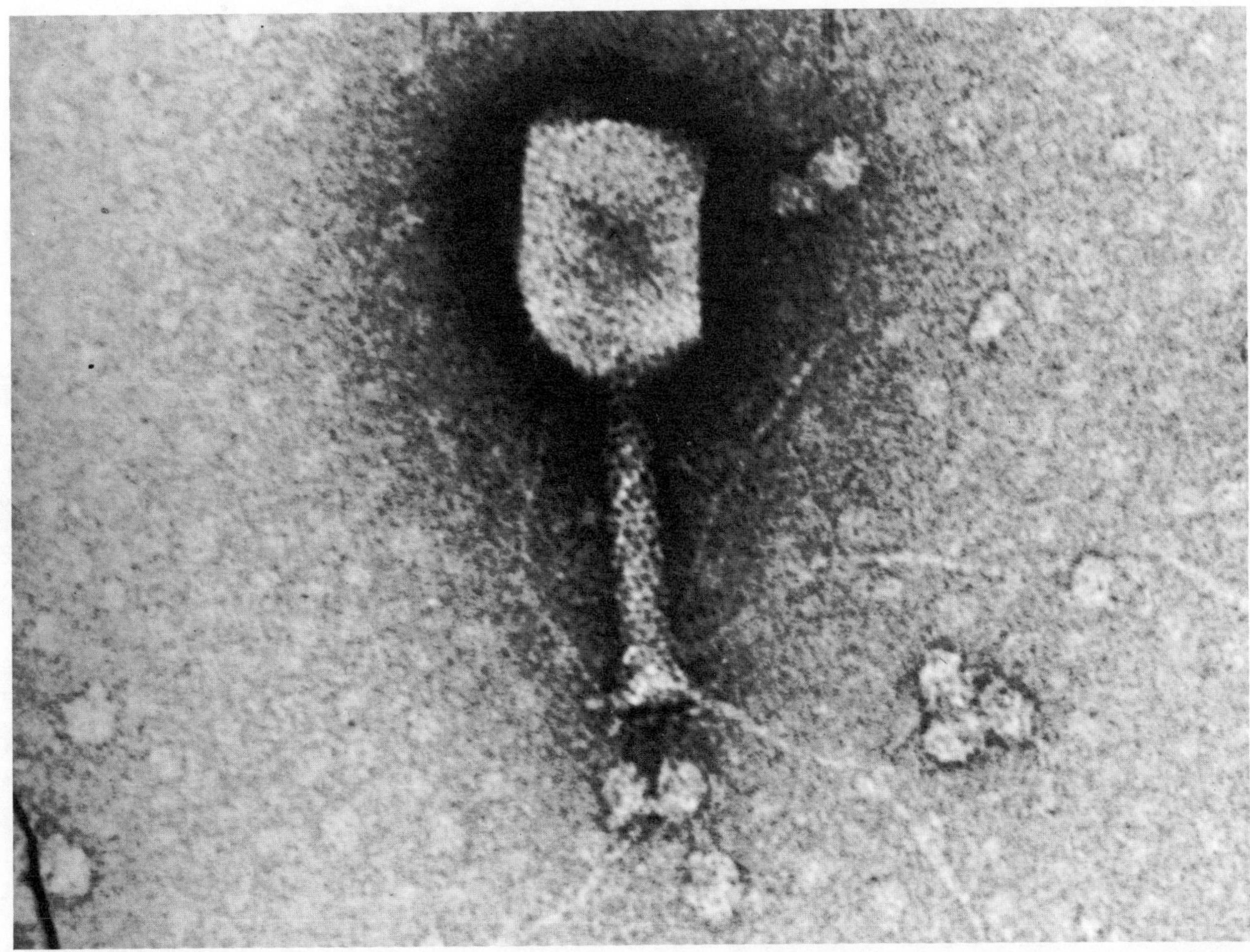

COMPLETE T4 PARTICLE was built by assembling component parts in the test tube. The virus is enlarged about 300,000 diameters in this electron micrograph made, like the ones on the next page, by Jonathan King of the California Institute of Technology.

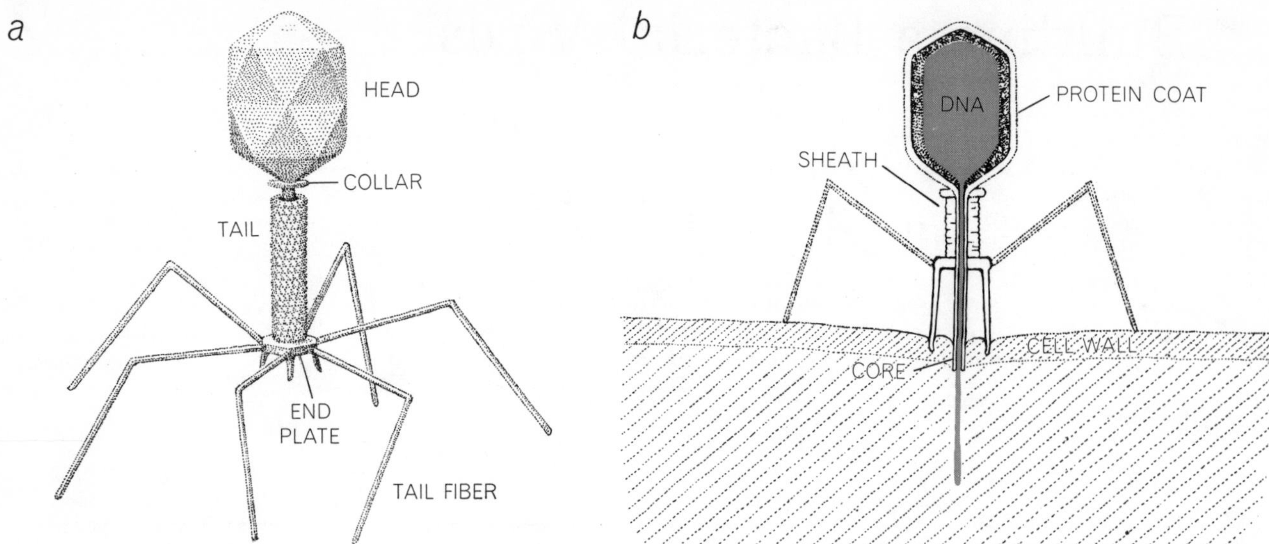

T4 BACTERIAL VIRUS is an assembly of protein components (*a*). The head is a protein membrane, shaped like a kind of prolate icosahedron with 30 facets and filled with deoxyribonucleic acid (DNA). It is attached by a neck to a tail consisting of a hollow core surrounded by a contractile sheath and based on a spiked end plate to which six fibers are attached. The spikes and fibers affix the virus to a bacterial cell wall (*b*). The sheath contracts, driving the core through the wall, and viral DNA enters the cell.

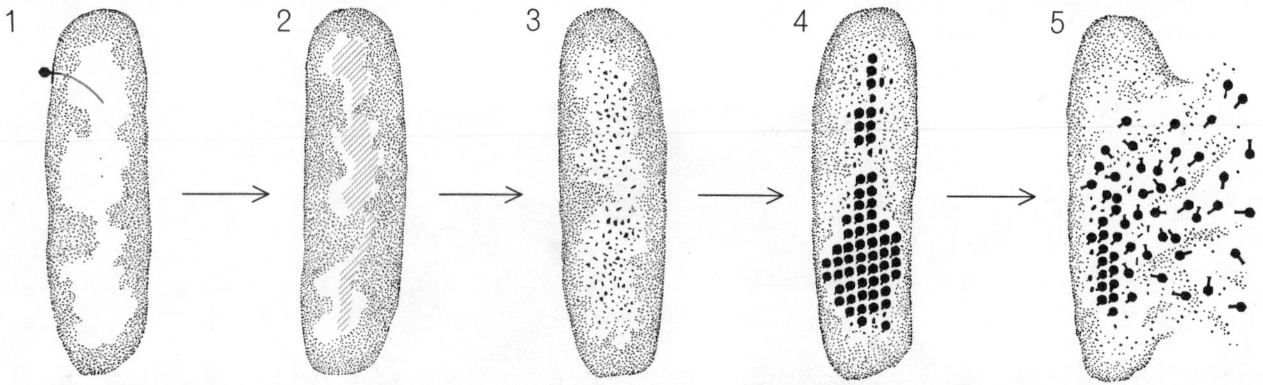

VIRAL INFECTION begins when viral DNA (*color*) enters a bacterium (*1*). Bacterial DNA is disrupted and viral DNA replicated (*2*). Synthesis of viral structural proteins (*3*) and their assembly into virus (*4*) continues until the cell bursts, releasing particles (*5*).

genes carry out their primary function: the specification of protein structure. The segment of nucleic acid (DNA or RNA) that constitutes a single gene specifies the chain of amino acids that comprises a protein molecule. Interactions among the amino acids cause the chain to fold into a unique configuration appropriate to the enzymatic or structural role for which it is destined. In this way the information in one gene determines the three-dimensional structure of a single protein molecule.

Where does the information come from to direct the next step: the assembly of many kinds of protein molecules into more complex structures? To build the relatively simple tobacco mosaic virus no further information is required; the inherent properties of the strand of RNA and the protein subunits cause them to interact in a unique way that results in the formation of virus particles. Clearly such a self-assembly process cannot explain the morphogenesis of an orange. At some intermediate stage on the scale of biological complexity there must be a point at which self-assembly becomes inadequate to the task of directing the building process. Working with a virus that may be just beyond that point, the T4 virus that infects the colon bacillus, we have been trying to learn how genes supply the required additional information.

Although the T4 virus is only a few rungs up the biological ladder from the tobacco mosaic virus, it is considerably more complex. Its DNA, which comprises more than 100 genes (compared with five or six in the tobacco mosaic virus), is coiled tightly inside a protein membrane to form a polyhedral head. Connected to the head by a short neck is a springlike tail consisting of a contractile sheath surrounding a central core and attached to an end plate, or base, from which protrude six short spikes and six long, slender fibers.

The life cycle of the T4 virus begins with its attachment to the surface of a colon bacillus by the tail fibers and spikes on its end plate. The sheath then contracts, driving the tubular core of the tail through the wall of the bacterial cell and providing an entry through which the DNA in the head of the virus can pass into the bacterium. Once inside, the genetic material of the virus quickly takes over the machinery of the cell. The bacterial DNA is broken down, production of bacterial protein stops and within less than a minute the cell has begun to manufacture viral proteins under the control of the injected virus genes. Among the first proteins to be made are the enzymes needed for viral DNA replication, which begins five minutes after infection. Three minutes later a second set of genes starts to direct the synthesis of the structural proteins that will form the head components and the tail components, and the process of viral morphogenesis begins. The first completed virus particle materializes 13 minutes after infection. Synthesis of both the DNA and the protein components continues for 12 more minutes until about 200 virus particles have accumulated within the cell. At this point a viral enzyme, lysozyme, attacks the cell wall from the inside to break open the bacterium and liberate the new viruses for a subsequent round of infection.

Additional insight into this process has come from studying strains of T4 carrying mutations—molecular defects that arise randomly and infrequently in the viral DNA during the course of its replication [see "The Genetics of a Bacterial Virus," by R. S. Edgar and R. H. Epstein; SCIENTIFIC AMERICAN Offprint 1004]. When a mutation is present, the protein specified by the mutant gene is synthesized in an altered form. This new protein is often nonfunctional, in which case the development of the virus stops at the point where the protein is required. Normally such a mutation has little experimental use, since the virus in

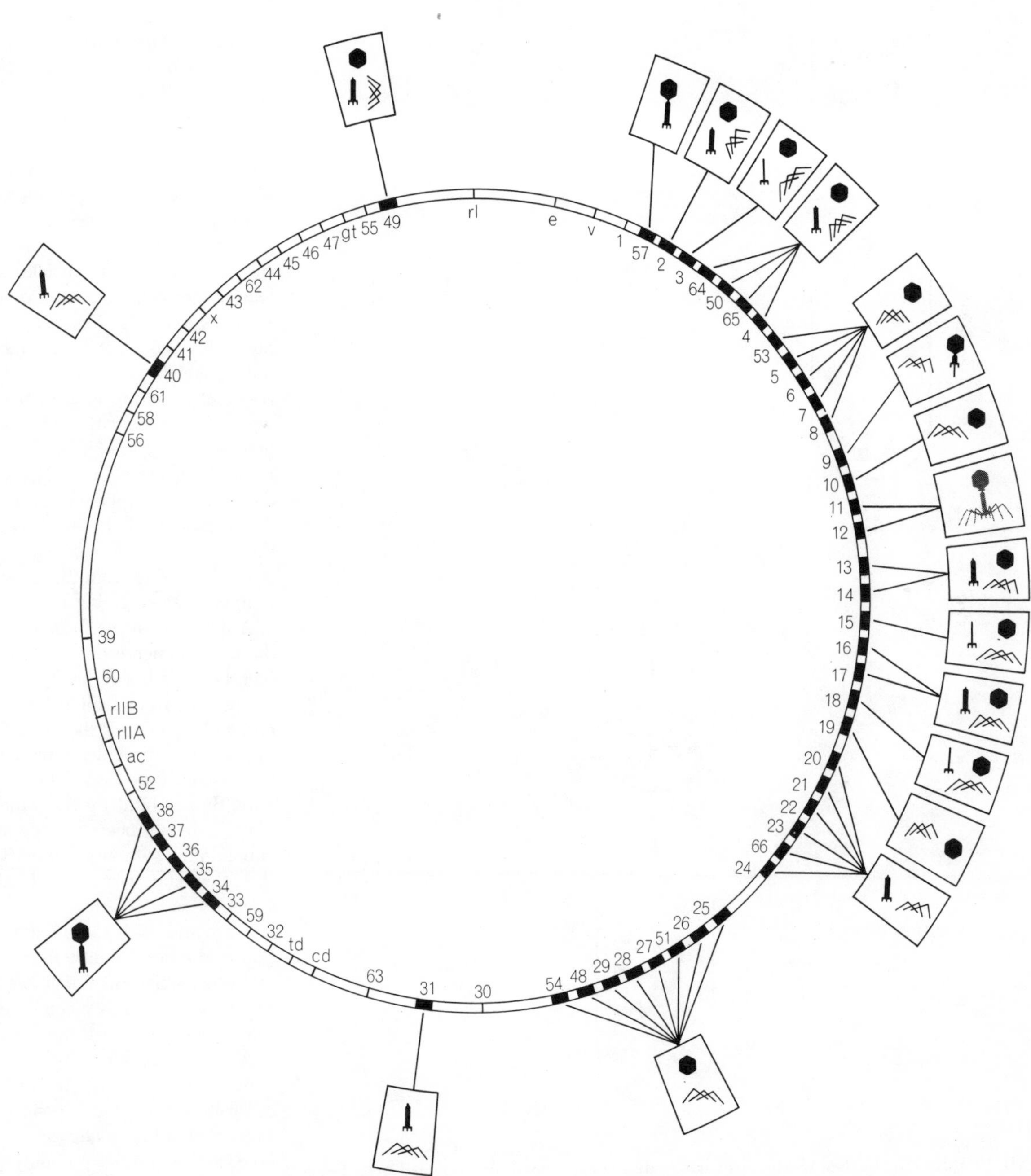

GENETIC MAP of the T4 virus shows the relative positions of more than 75 genes so far identified on the basis of mutations. The solid black segments of the circle indicate genes with morphogenetic functions. The boxed diagrams show which viral components are seen in micrographs of extracts of cells infected by mutants defective in each morphogenetic gene. A defect in gene No. 11 or 12 produces a complete but fragile particle. Heads, all tail parts, sheaths or fibers are the missing components in other extracts.

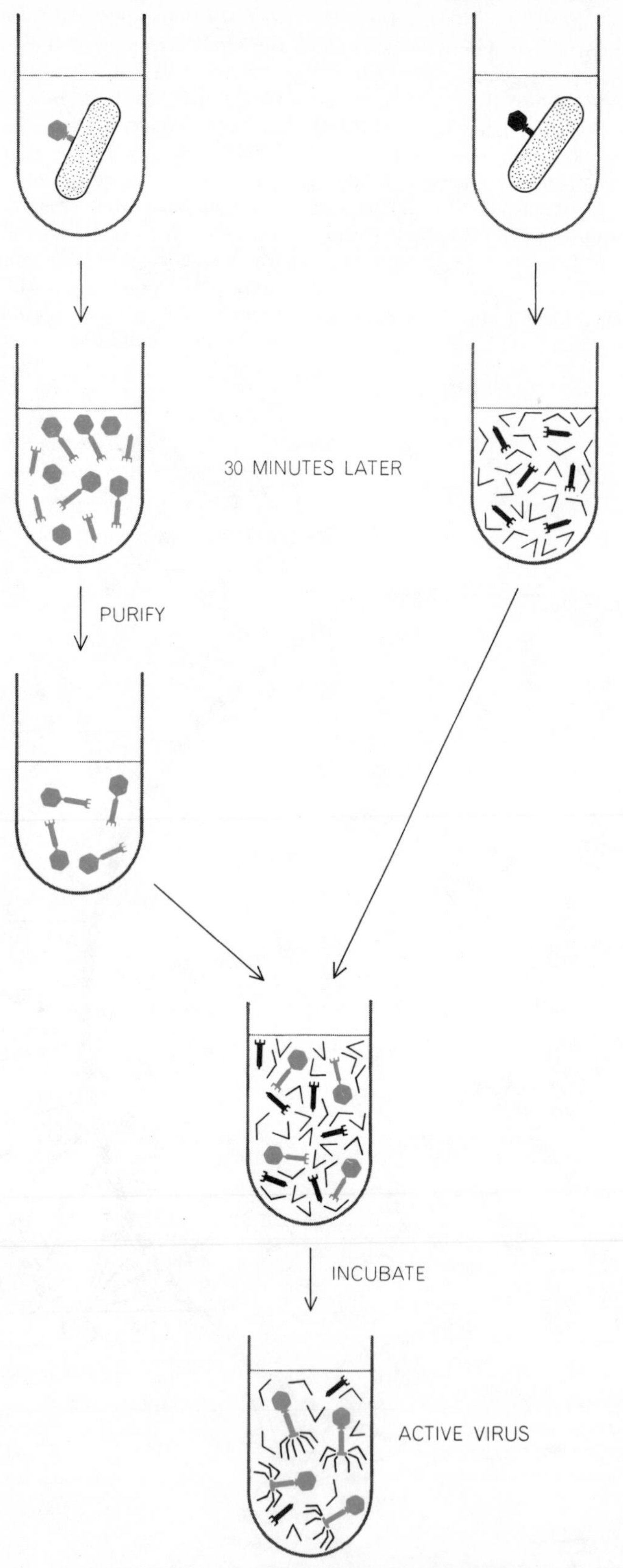

TAIL FIBERS are attached to fiberless particles in the experiment diagrammed here. Cells are infected with a virus (*color*) bearing defective tail-fiber genes. The progeny particles, lacking fibers, are isolated with a centrifuge. A virus with a head-gene mutation (*black*) infects a second bacterial culture, providing an extract containing free tails and fibers. When the two preparations are mixed and incubated at 30 degrees centigrade, the fiberless particles are converted to infectious virus particles by the attachment of the free fibers.

which it arises is dead and hence cannot be recovered for study. Edgar and Epstein, however, found mutations that are only "conditionally lethal": the mutant protein is produced in either a functional or a nonfunctional form, depending on the conditions of growth chosen by the experimenter. Under "permissive" conditions reproduction is normal, so that the mutants can be cultured and crossed for genetic studies. Under "restrictive" conditions, however, viral development comes to a halt at the step where the protein is needed, and by determining the point at which development is blocked the investigator can infer the normal function of the mutated gene. In this way a number of conditionally lethal mutations have been assigned to different genes, have been genetically mapped and have been tested for their effects on viral development under restrictive conditions [*see illustration on page* 97].

In the case of genes that control the later stages of the life cycle, involving the assembly of virus particles, mutations lead to the accumulation of unassembled viral components. These can be identified with the electron microscope. By noting which structures are absent as a result of mutation in a particular gene, we learn about that morphogenetic gene's normal function. For example, genes designated No. 23, No. 27 and No. 34 respectively appear to control steps in the formation of the head, the tail and the tail fibers; these are the structures that are missing from the corresponding mutant-infected cells.

A blockage in the formation of one of these components does not seem to affect the assembly of the other two, which accumulate in the cell as seemingly normal and complete structures. This information alone provides some insight into the assembly process. The virus is apparently not built up the way a sock is knitted—by a process starting at one end and adding subunits sequentially until the other end is reached. Instead, construction seems to follow an assembly-line process, with three major branches that lead independently to the formation of heads, tails and tail fibers. The finished components are combined in subsequent steps to form the virus particle.

A second striking aspect of the genetic map is the large number of genes controlling the morphogenetic process. More than 40 have already been discovered, and a number probably remain to be identified. If all these genes specify proteins that are component parts of the virus, then the virus is considerably more complex than it appears to be. Alternatively, however, some gene products may play directive roles in the assembly process without contributing materially to the virus itself. Studies of seven genes controlling formation of the virus's head support this possibility [see "The Genetic Control of the Shape of a Virus," by Edouard Kellenberger; SCIENTIFIC AMERICAN Offprint 1058].

In order to determine the specific functions of the many gene products involved in morphogenesis, it seemed necessary to seek a way to study individual assembly steps under controlled conditions outside the cell. One of us (Edgar) is a geneticist by training, the other (Wood) a biochemist. The geneticist is inclined to let reproductive processes take their normal course and then, by analyzing the progeny, to deduce the molecular events that must have occurred within the organism. The biochemist is eager to break the organism open and search among the remains for more direct clues to what is going on inside. For our current task a synthesis of these two approaches has proved to be most fruitful. Since it seemed inconceivable that the T4 virus could be built from scratch like the tobacco mosaic virus, starting with nucleic acid and individual protein molecules, we decided to let cells infected with mutants serve as sources of preformed viral components. Then we would break open the cells and, by determining how the free parts could be assembled into complete infectious virus, learn the sequence of steps in assembly, the role of each gene product and perhaps its precise mode of action.

Our first experiment was an attempt to attach tail fibers to the otherwise complete virus particle—a reaction we suspected was the terminal step in morphogenesis. Cells infected with a virus bearing mutations in several tail fiber genes (No. 34, 35, 37 and 38) were broken open, and the resulting particles—complete except for fibers and noninfectious—were isolated by being spun in a high-speed centrifuge. Other cells, infected with a gene No. 23 mutant that was defective in head formation, were similarly disrupted to make an extract containing free fibers and tails but no heads. When a sample of the particles was incubated with the extract, the level of infectious virus in the mixture increased rapidly to 1,000 times its initial value. Electron micrographs of samples taken from the mixture at various times showed that the particles were indeed acquiring tail fibers as the reaction proceeded.

In that first experiment the production of infectious virus required only one kind of assembly reaction—the attachment of completed fibers to completed particles. We went on to test more demanding mixtures of defective cell extracts. For example, with a mutant blocked in head formation and another one blocked in tail formation we prepared two extracts, one containing tails and free tail fibers but no heads and another containing heads and free tail fibers but no tails. When a mixture of these two extracts also gave rise to a large number of infectious viruses, we concluded that at least two reactions must have occurred: the attachment of heads to tails and the attachment of fibers to the resulting particles.

By infecting bacilli with mutants bearing defects in different genes con-

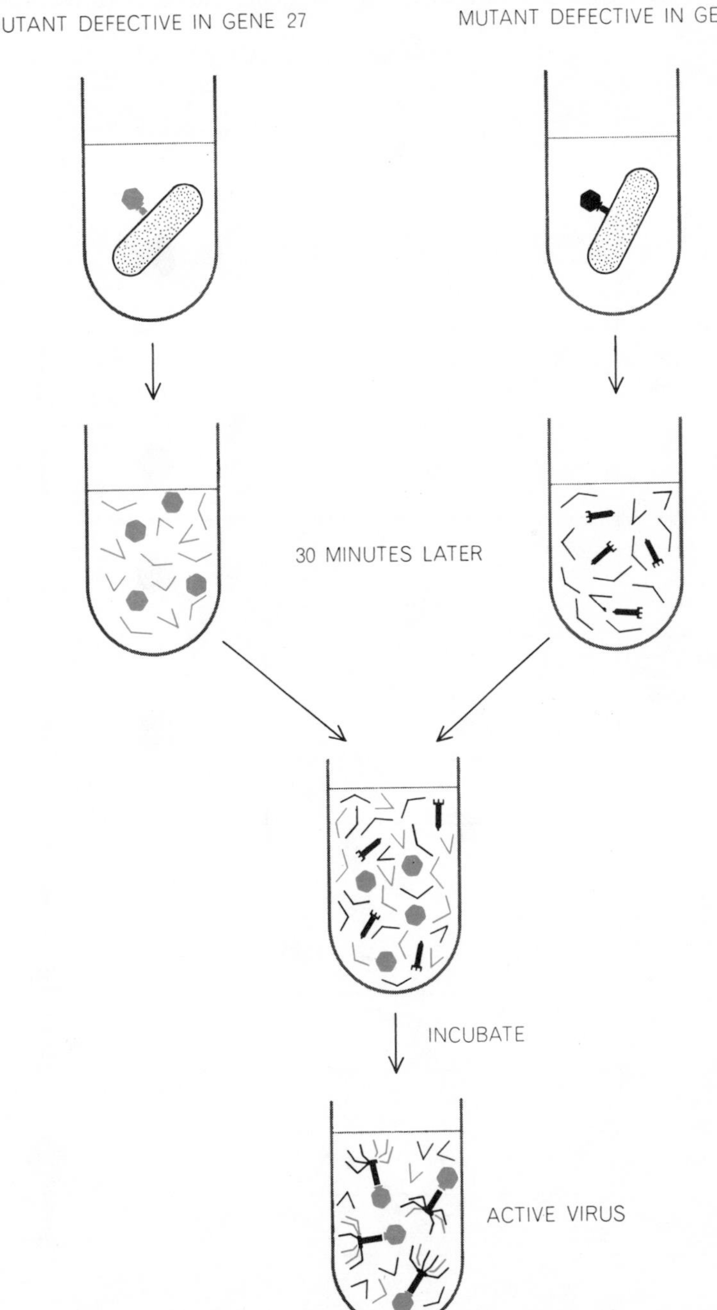

TWO ASSEMBLY REACTIONS occur in this experiment: union of heads and tails and attachment of fibers. One virus (*color*), with a defective tail gene, produces heads and fibers. Another (*black*), with a mutation in a head gene, produces tails and fibers. When the two extracts are mixed and incubated, the parts assemble to produce infectious virus.

cerned with assembly, we prepared 40 different extracts containing viral components but no infectious virus. When we tested the extracts by mixing pairs of them in many of the appropriate combinations, some mixtures produced active virus and others showed no detectable activity. The production of infective virus implied that the two extracts were complementing each other in the test tube, that each was supplying a component that was missing or defective in the other and that could be assembled into complete, active virus under our experimental conditions. Lack of activity, on the other hand, suggested that both extracts were deficient in the same viral component—a component being defined as a subassembly unit that functions in our experimental system. By analyzing the pattern of positive and negative results we could find out how many functional components we were dealing with.

It developed that there are at least 13 such components. That is, analysis of our pair combinations produced 13 complementation groups, the members of which did not complement one another but did complement any member of any other group. Two of these groups were quite large [*see illustration below*]. Since one gene produces one protein and since each extract has a different defective gene product, a mixture of any two extracts should include all the proteins required for building the virus. The fact that members of these large groups do not complement one another must mean that our experimental system is not as efficient as an infected cell; whatever the gene products that are missing in each of these extracts do, they cannot do it in the test tube.

The idea that a complementation group consisted of extracts deficient in

EXTRACT GROUP	MUTANT GENES	COMPONENTS PRESENT	INFERRED DEFECT
I	5, 6, 7, 8, 10, 25, 26, 27, 28, 29, 48, 51, 53		TAIL
II	20, 21, 22, 23, 24, 31	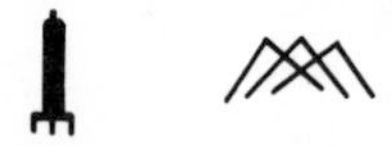	HEAD (FORMATION)
	2, 4, 16, 17, 49, 50, 64, 65		HEAD (COMPLETION)
III	54		TAIL CORE
IV	13, 14		?
V	15		
VI	18		
VII	9		?
VIII	11	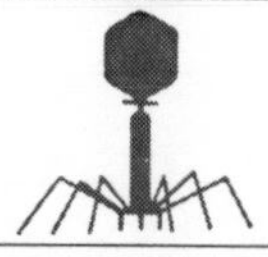	?
IX	12		
X	37, 38		TAIL FIBERS
XI	36		
XII	35		
XIII	34		

COMPLEMENTATION TESTS defined 13 groups of defective extracts, as described in the text. Mixing any two extracts in a single group fails to produce infectious virus in the test tube, but mixing any two members of different groups yields infectious virus. Apparently each group represents the genes concerned with the synthesis of a component that is functional under experimental conditions. The precise nature of the defect in some extracts, and hence the function of the missing gene product, could not be identified on the basis of the structures recognized in electron micrographs and remained to be determined by additional experiments.

the same functional component could be checked against the earlier electron micrograph results. Micrographs of the 12 defective extracts of Group I, for example, all show virus heads and tail fibers but no tails. Each of these extracts must therefore be deficient in a gene product that has to do with a stage of tail formation that cannot be carried out in our extracts. The second large complementation group appeared at first to be anomalous in terms of electron micrography: some extracts contained only tails and tail fibers, whereas others contained heads as well. Tests against extracts known to contain active tails revealed, however, that these heads—although they looked whole—could not combine to produce active virus in the test tube. In other words, heads, like tails, must be nearly completed within an infected cell before they become active for complementation. The early stages of head formation are still inaccessible to study in mixed extracts.

The remaining defective extracts gave rise to active virus in almost all possible pair combinations, segregating into another 11 complementation groups. With a total of 13 groups, there must be at least 12 assembly steps that can occur in mixtures of extracts. The defects recognizable in micrographs suggest what some of these steps must be: the completion and union of heads and tails, the assembly of tail fibers and the attachment of fibers to head-tail particles. These, then, are the steps that can be studied further in our present experimental system. We have in effect a virus-building kit, some of whose more intricate parts have been preassembled at the cellular factory.

Our next experiments were designed to determine the normal sequence of assembly reactions and further characterize those whose nature remained ambiguous. Examples of the latter were the steps controlled by genes No. 13, 14, 15 and 18. Defects in the corresponding gene products resulted in the accumulation of free heads and tails, suggesting that they are somehow involved in head-tail union. It was unclear, however, whether these gene products are required for the attachment process itself or for completion of the head or the tail before attachment. We could distinguish the alternatives by complementation tests using complete heads and tails. These we isolated from the appropriate extracts in the centrifuge, taking advantage of their large size in relation to the other materials present. On the basis of the evidence for the independent assembly of heads and tails, we assumed that

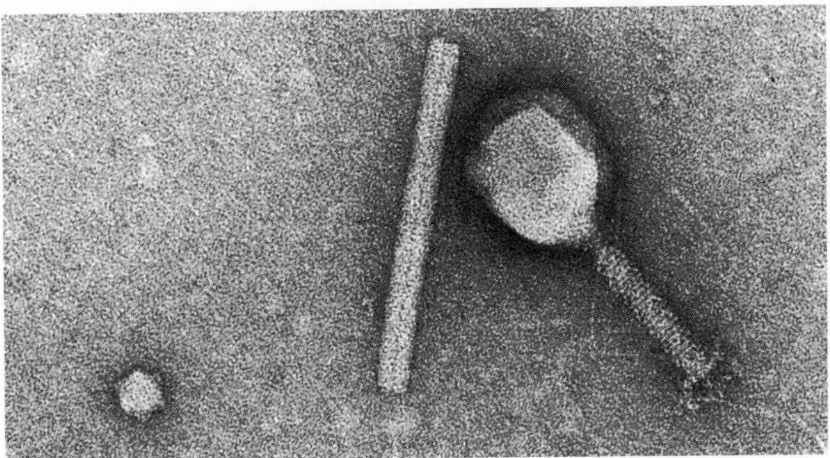

TWO SIMPLER VIRUSES are shown with the T4 in an electron micrograph made by Fred Eiserling of the University of California at Los Angeles. The icosahedral ΦX174 virus (*left*) infects the colon bacillus, as does the T4. The rod-shaped tobacco mosaic virus reassembles itself in the test tube after dissociation. The enlargement is 200,000 diameters.

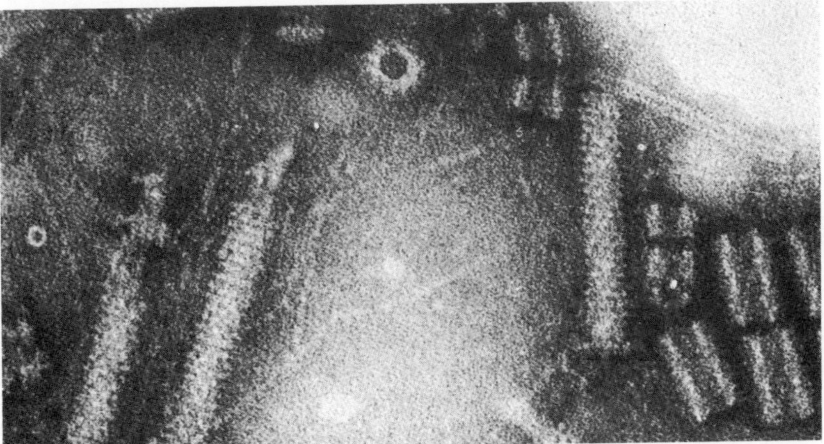

COMPLEX STRUCTURE of the T4 tail is shown in an electron micrograph made by E. Boy de la Tour of the University of Geneva. The parts were obtained by breaking down virus particles, not by synthesis, which is why fibers are attached to tails. The hollow interiors of the free core (*top right*) and pieces of sheath are delineated by dark stain that has flowed into them. There are end-on views of pieces of core (*left*) and sheath (*top center*).

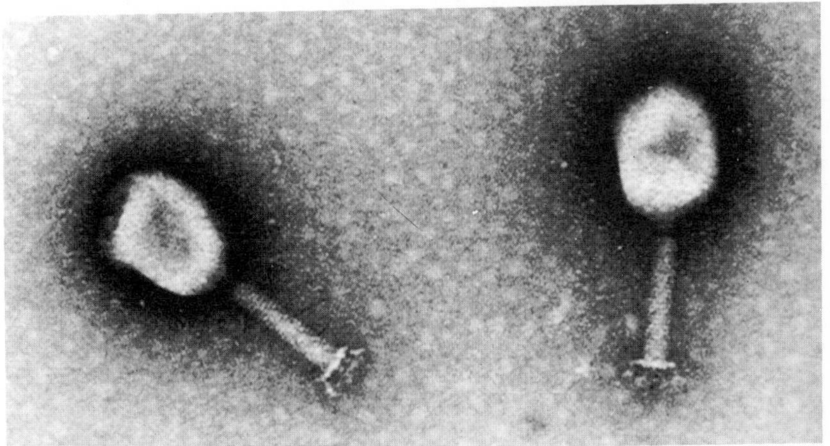

FIBERLESS PARTICLES, otherwise complete, are the products of infection by a mutant defective in one of the fiber-forming genes. Heads, tails and fibers are each formed by a subassembly line (*see illustration on page 103*). The electron micrograph was made by King.

the heads we isolated from a tail-defective extract would be complete, as would tails isolated from a head-defective extract.

The results of the tests were unambiguous. The addition of isolated heads to extracts lacking the products of gene No. 13 or 14 resulted in virus production, whereas the addition of tails did not. We could therefore conclude that the components missing from these extracts normally affect the head structure, and that genes No. 13 and 14 control head completion rather than tail completion or head-tail union. The remaining two of the four extracts gave the opposite result; these were active with added tails but not with added heads, indicating that genes No. 15 and 18 are involved in the completion of the tail. All four of these steps must precede the attachment of heads to tails, since defects in any of the corresponding genes block head-tail union.

By manipulating extracts blocked at other stages we worked out the remaining steps in the assembly process with the help of Jonathan King and Jeffrey Flatgaard. The various reactions were characterized and their sequence determined by many experiments similar to those described above. In addition, more detailed electron micrographs of defective components helped to clarify the nature of some individual steps. For example, knowing that genes No. 15 and 18 were concerned with tail completion, we went on to find just what each one did. Electron micrographs showed that in the absence of the No. 18 product no contractile sheaths were made. If No. 18 was functional but No. 15 was defective, the sheath units were assembled on the core but were unstable and could fall away. The addition of the product of gene No. 15 (and of No. 3 also, as it turned out) supplied a kind of "button" at the upper end of the core and thus apparently stabilized the sheath.

The results to date of this line of investigation can be summarized in the form of a morphogenetic pathway [*see illustration on next page*]. As we had thought, it consists of three principal independent branches that lead respectively to the formation of the head, the tail and the tail fibers.

The earliest stages of head morphogenesis are controlled by six genes. These genes direct the formation of a precursor that is identifiable as a head in electron micrographs but is not yet functional in extract-complementation experiments. Eight more gene products must act on this precursor to produce a

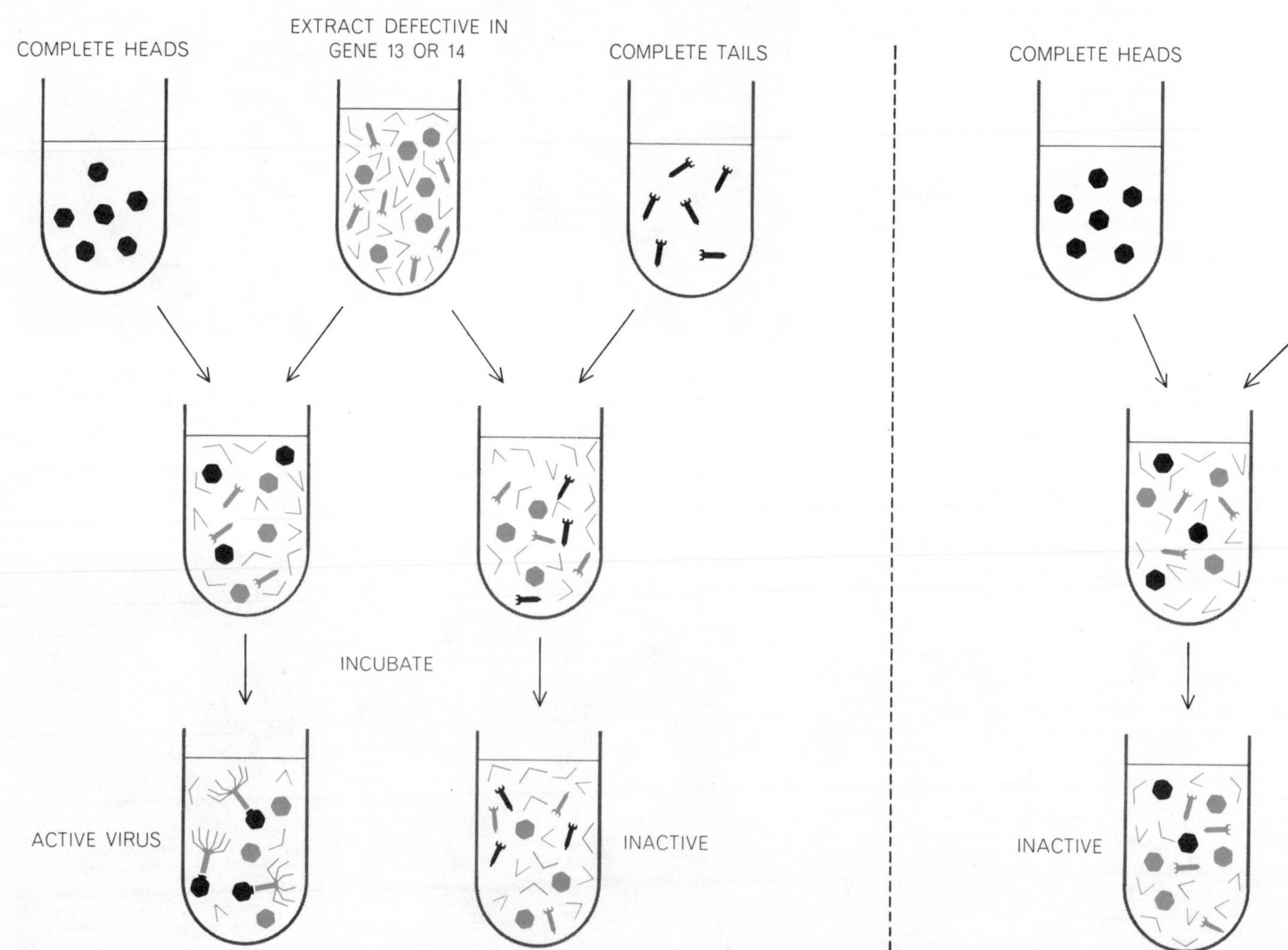

ASSEMBLY DEFECTS of mutants (*color*) that seem to produce complete heads, tails and fibers are identified, using isolated complete heads and tails (*black*) as test reagents. When complete heads are added to some extracts to be tested (*left*), infectious virus is produced, but the addition of complete tails is ineffective. This indicates that the tails made by these mutants must be functional,

head structure that is active in complementation experiments. This active structure undergoes the terminal step in head formation (the only one so far demonstrated in the test tube): conversion to the complete head that is able to unite with the tail. The nature of this conversion, which is controlled by genes No. 13 and 14, remains unclear. A likely possibility would be that these genes control the formation of the upper neck and collar, but evidence on this point is lacking. The attachment of head structures to tails has never been observed in extracts prepared with mutants defective in gene No. 13 or 14, or with any of the preceding class of eight genes. It therefore appears that completion of the head is a prerequisite for the union of heads and tails.

The earliest structure so far identified in the morphogenesis of the tail is the

EXTRACT DEFECTIVE IN GENE 15 OR 18
COMPLETE TAILS
INCUBATE
ACTIVE VIRUS

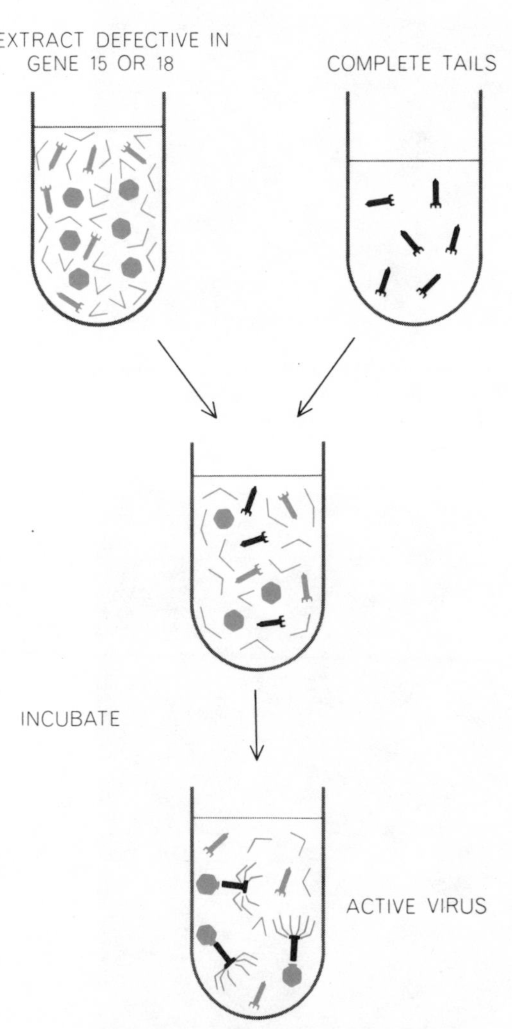

implying that the heads must be defective. In the case of other mutants (*right*), such tests indicate that the tails must be defective.

HEAD
20, 21, 22, 23, 24, 31
TAIL
5, 6, 7, 8, 10, 25, 26, 27, 28, 29, 48, 51, 53
11
12
54
18
2, 4, 16, 17, 49, 50, 64, 65
13, 14
3, 15
(SPONTANEOUS)
TAIL FIBER
37
38
9
36
35
34
"LABILE FACTOR"

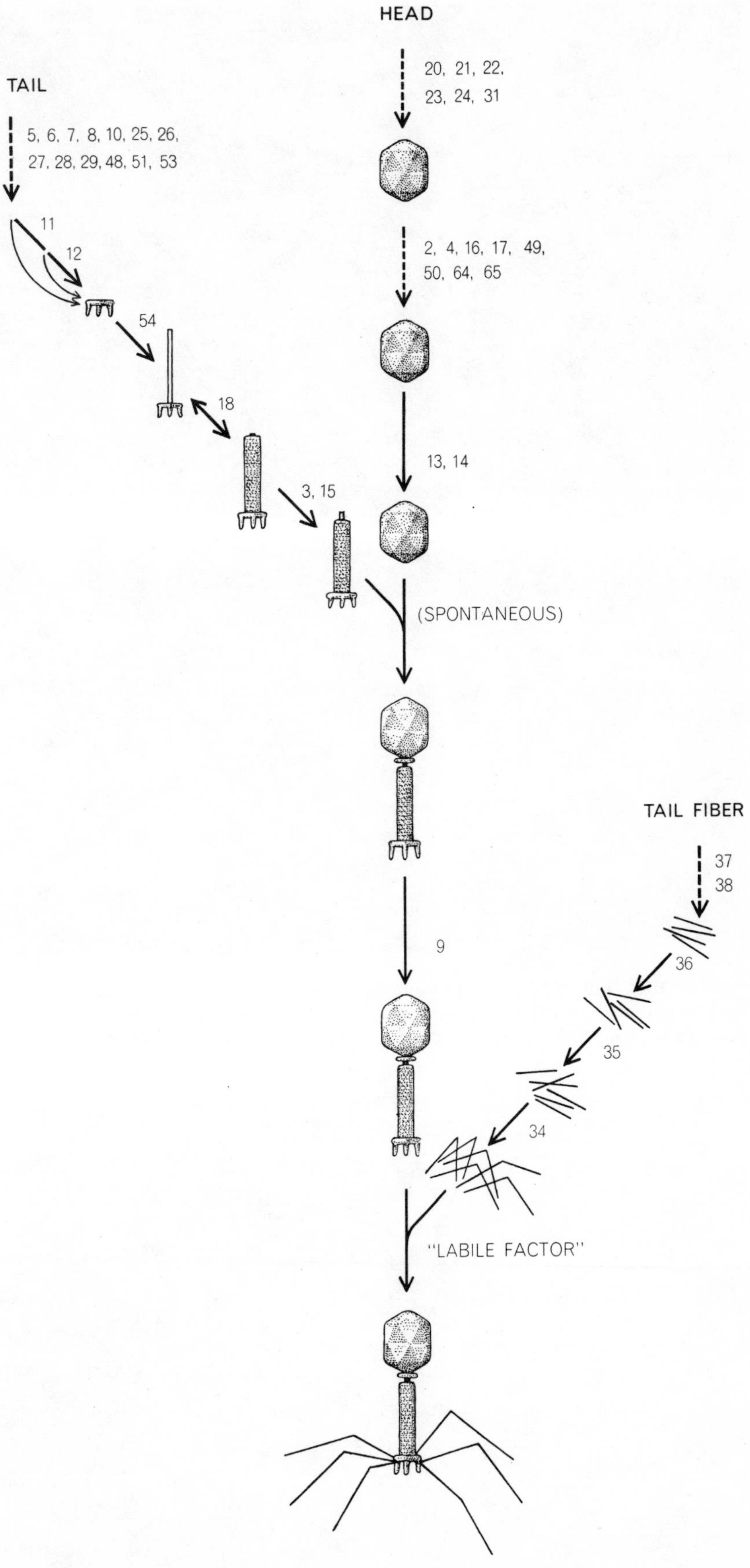

MORPHOGENETIC PATHWAY has three principal branches leading independently to the formation of heads, tails and tail fibers, which then combine to form complete virus particles. The numbers refer to the gene product or products involved at each step. The solid portions of the arrows indicate the steps that have been shown to occur in extracts.

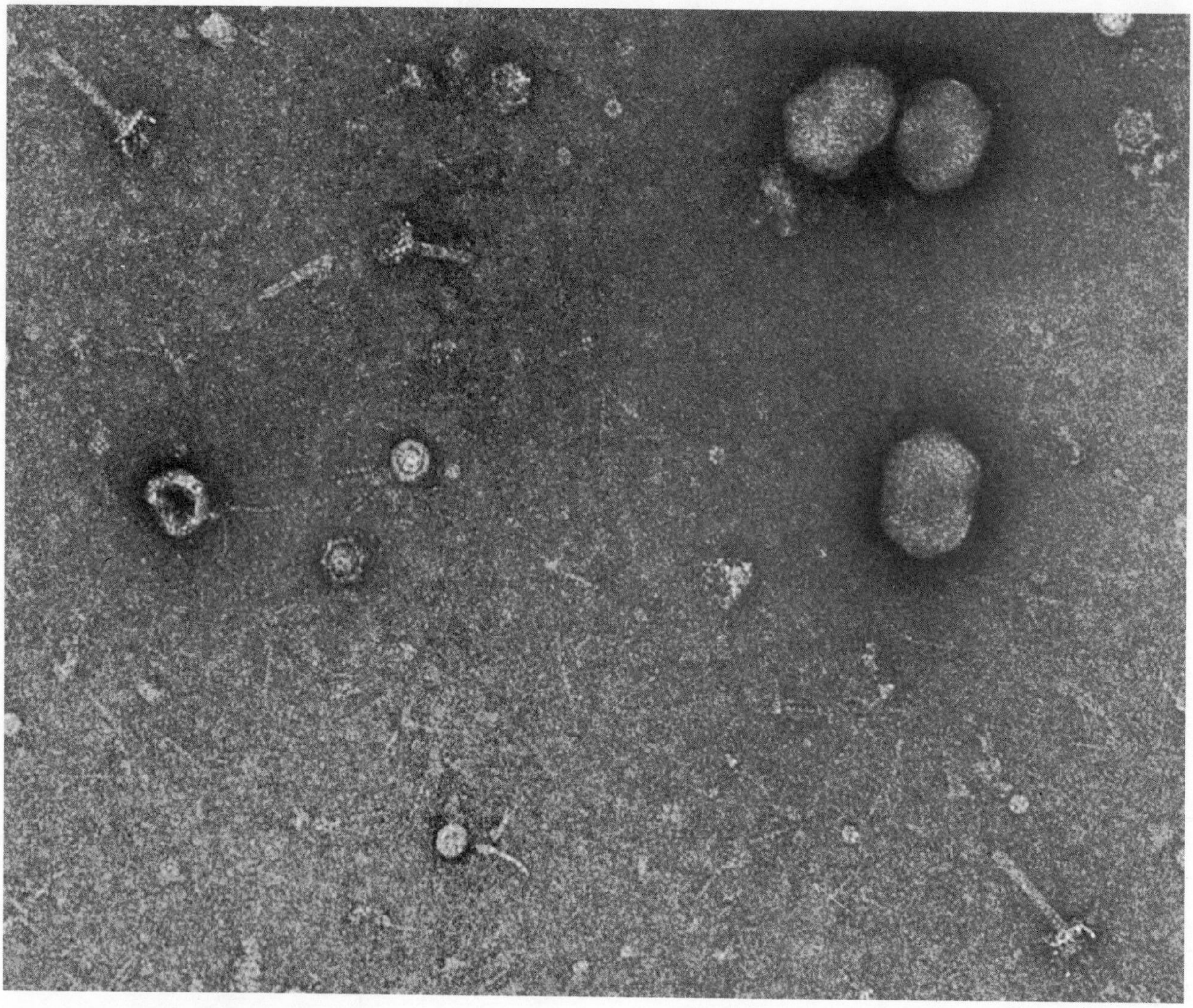

UNASSEMBLED PARTS of the T4 virus are present in this extract. It was prepared by infecting colon bacilli with a mutant virus defective in gene No. 18, which specifies the synthesis of the sheath (*see upper illustration on page 96*). The result is the accumulation of all major components except the sheath: heads, free tail fibers and "naked" tails consisting of cores and end plates.

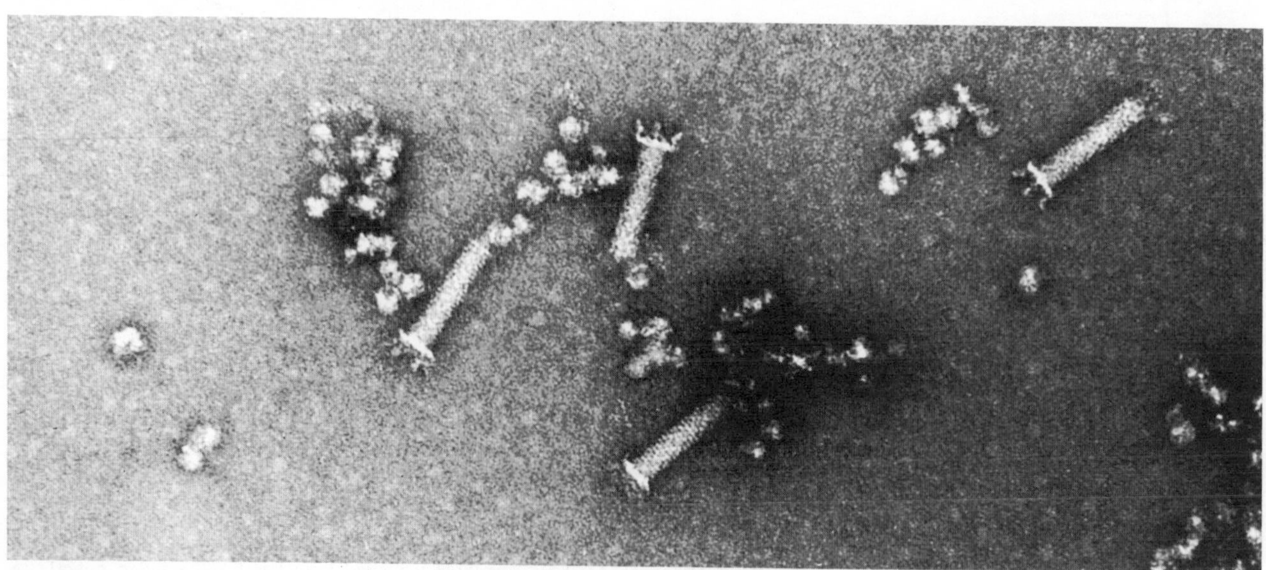

COMPLETE TAILS, enclosed in sheaths, were produced by a different mutant, defective in a gene involved in head formation. The tails were separated from the resulting extract (along with some spherical bacterial ribosomes) by being spun in a centrifuge. If the tails are added to the extract (*top photograph*), they combine with the heads and free fibers in it to form infectious virus.

end plate. It is apparently an intricate bit of machinery, since 15 different gene products participate in its formation. All the subsequent steps in tail formation can be demonstrated in the test tube. The core is assembled on the end plate under the control of the products of gene No. 54 and probably No. 19; the resulting structure appears as a tail without a sheath. The product of gene No. 18 is the principal structural component of the sheath, which is somehow stabilized by the products of genes No. 3 and 15. Tails without sheaths do not attach themselves to head structures, indicating that the tail as well as the head must be completed before head-tail union can occur. Moreover, unattached tail structures are never fitted with fibers, suggesting that these can be added only at a later stage of assembly.

Completed heads and tails unite spontaneously, in the absence of any additional factors, to produce a precursor particle that interacts in a still undetermined manner with the product of gene No. 9, resulting in the complete head-plus-tail particle. It is only at this point that tail fibers can become attached to the end plate.

At least five gene products participate in the formation of the tail fiber. In the first step, which has not yet been demonstrated in extracts, the products of genes No. 37 and 38 combine to form a precursor corresponding in dimensions to one segment of the finished fiber. This precursor then interacts sequentially with the products of genes No. 36, 35 and 34 to produce the complete structure. Again the completion of a major component—in this case the tail fiber—appears to be a prerequisite for its attachment, since we have never seen the short segments linked to particles.

The final step in building the virus is the attachment of completed tail fibers to the otherwise finished particle. We have studied this process in reaction mixtures consisting of purified particles and a defective extract containing complete tail fibers but no heads or tails. When we divided the extract into various fractions, we found that it supplies two components, both of which are necessary for the production of active virus. One of these of course is the tail fiber. The other is a factor whose properties suggest that it might be an enzyme. For one thing, the rate at which fibers are attached depends on the level of this factor present in the reaction mixture, and yet the factor does not appear to be used up in the process. Moreover, the rate of attachment depends on the temperature of incubation—increasing by a factor of about two with every rise in temperature of 10 degrees centigrade. These characteristics suggest that the factor could be catalyzing the formation of bonds between the fibers and the tail end plate. At the moment we can only speculate on its possible mechanism of action, since the chemical nature of these bonds is not yet known; we call it simply a "labile factor," not an enzyme. Although no gene controlling the factor has yet been discovered, we assume that its synthesis must be directed by the virus, since it is not found in extracts of uninfected bacteria.

The T4 assembly steps so far accomplished and studied in the test tube represent only a fraction of the total number. Already, however, it is apparent that there is a high degree of sequential order in the assembly process; restrictions are somehow imposed at each step that prevent its occurrence until the preceding step has been completed. Only two exceptions to this rule have been discovered. The steps controlled by genes No. 11 and 12, which normally occur early in the tail pathway, can be bypassed when these gene products are lacking. In that case the tail is completed, attaches itself to a head and acquires tail fibers, but the result is a fragile, defective particle. The particle can, however, be converted to a normal active virus by exposure to an extract containing the missing gene products. These are the only components whose point of action in the pathway appears to be unimportant.

The problem has now reached a tantalizing stage. A partial sequence of gene-controlled assembly steps can be written, but the manner in which the corresponding gene products contribute to the process remains unclear, and the questions posed at the beginning of this article cannot yet be answered definitively. There is the suggestion that the attachment of tail fibers is catalyzed by a virus-induced enzyme. If this finding is substantiated, it would overthrow the notion that T4 morphogenesis is entirely a self-assembly process. Continued investigation of this reaction and the assembly steps that precede it can be expected to provide further insight into how genes control the building of biological structures.

8 The Assembly of a Virus

by P. Jonathan G. Butler and Aaron Klug
November 1978

The tobacco-mosaic virus is make up of a strand of nucleic acid encased in a rod of one kind of protein. The two components come together spontaneously but in a way that is unexpectedly complex

Viruses are complex particles made up of inert giant molecules: proteins and nucleic acids (DNA or RNA). They are dead in the sense that they lack any internal metabolism, but they come alive on entering the living cell. For this reason they are obligate parasites, able to reproduce only by taking over the enzymatic machinery of the host cell.

Because of their extreme simplicity viruses have proved invaluable to molecular biologists interested in the structure and function of genes. Viruses also provide a simple model of cell development, because their multiplication inside host cells involves the controlled expression of a small number of genes and the assembly of a small number of proteins into a highly ordered structure. The assembly of viruses has therefore become a paradigm for the construction of large molecular structures within living cells.

The work of our group at the Medical Research Council Laboratory of Molecular Biology in Cambridge, England, has focused on the tobacco-mosaic virus, which infects the cells of the tobacco leaf. One of the simplest and most viable viruses known, it consists of a single strand of RNA packaged within a rod of protein with a hollow center. The viral RNA is 6,400 nucleotides long and is intercalated within the turns of a closely coiled helix of 2,130 identical protein subunits, so that about three RNA nucleotides are bound to each protein subunit. The protein surrounds and isolates the RNA, protecting it from damage until the virus has successfully infected the host cell. Once the RNA is inside the cell it is released from the protein and the viral genes set to work generating large numbers of new virus particles.

In 1955 a classic series of experiments conducted by Heinz Fraenkel-Conrat and Robley C. Williams of the University of California at Berkeley demonstrated that tobacco-mosaic virus could be reconstituted in the test tube from its isolated protein and RNA components [see "Rebuilding a Virus," by Heinz Fraenkel-Conrat; SCIENTIFIC AMERICAN Offprint 9]. On simple remixing, infectious virus particles were formed that were structurally indistinguishable from the original virus. Therefore all the information necessary for constructing the virus is inherent in its parts, which "self-assemble" spontaneously in solution.

The self-assembly of a helical structure such as a tobacco-mosaic virus may not seem particularly impressive. One can postulate that the protein subunits have a precise surface geometry so that they can assemble only in a unique way; the subunits make identical contacts with one another that are repeated over and over to yield a regular structure. According to the most obvious assembly scheme, the free RNA interacts with individual protein subunits to get the helix started. Then the subunits simply add themselves, one or a few at a time, to the "step" at the end of the growing helix, much as a crystal grows at a screw dislocation, but in this case trapping the RNA as they go along. Since both the virus particle and the RNA have distinct ends, it is also logical to expect that growth starts at one end of the RNA and proceeds toward the other. However plausible, these simple ideas are now known to be wrong. With the benefit of hindsight one can understand why the virus has adopted what at first seemed to be a puzzlingly more complex strategy.

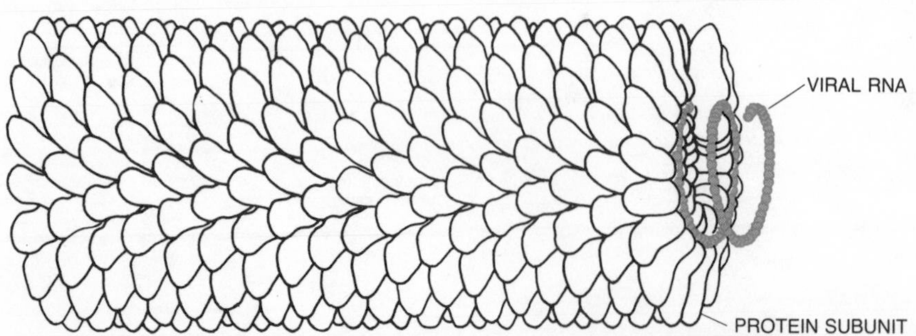

HELICAL STRUCTURE of the tobacco-mosaic virus is apparent in this drawing, which shows about a sixth of the length of the rod-shaped virus particle. The virus consists of a single long strand of RNA (*color*), representing perhaps four genes, packed between the turns of a helical protein coat made up of 2,130 identical subunits. The final length of the rod is determined by the length of the RNA. Until the virus infects its host cell the protein helix protects the RNA from damage; after infection the RNA is released from the protein and the viral genes are expressed by the host's enzymes. Central hole in the rod of the virus particle, once thought to be a trivial consequence of protein packing, plays an essential role in assembly of the virus.

The early reconstitution experiments appeared to have many features in common with the natural assembly process. The artificial reassembly of tobacco-mosaic virus from its protein and RNA components proceeded only at room temperature and at neutral *p*H, conditions similar to those found in the cells of the host plant. Moreover, reassembly was quite specific: it proceeded most readily with RNA from the same strain of the virus or from a closely related strain but poorly or not at all with other natural RNA's or synthetic RNA's. The only puzzling aspect of the reassembly process was its low rate: six hours or more were required to get maximum yields of assembled virus particles. That period of time seemed too long for the normal assembly of the virus, because the viral RNA is protected

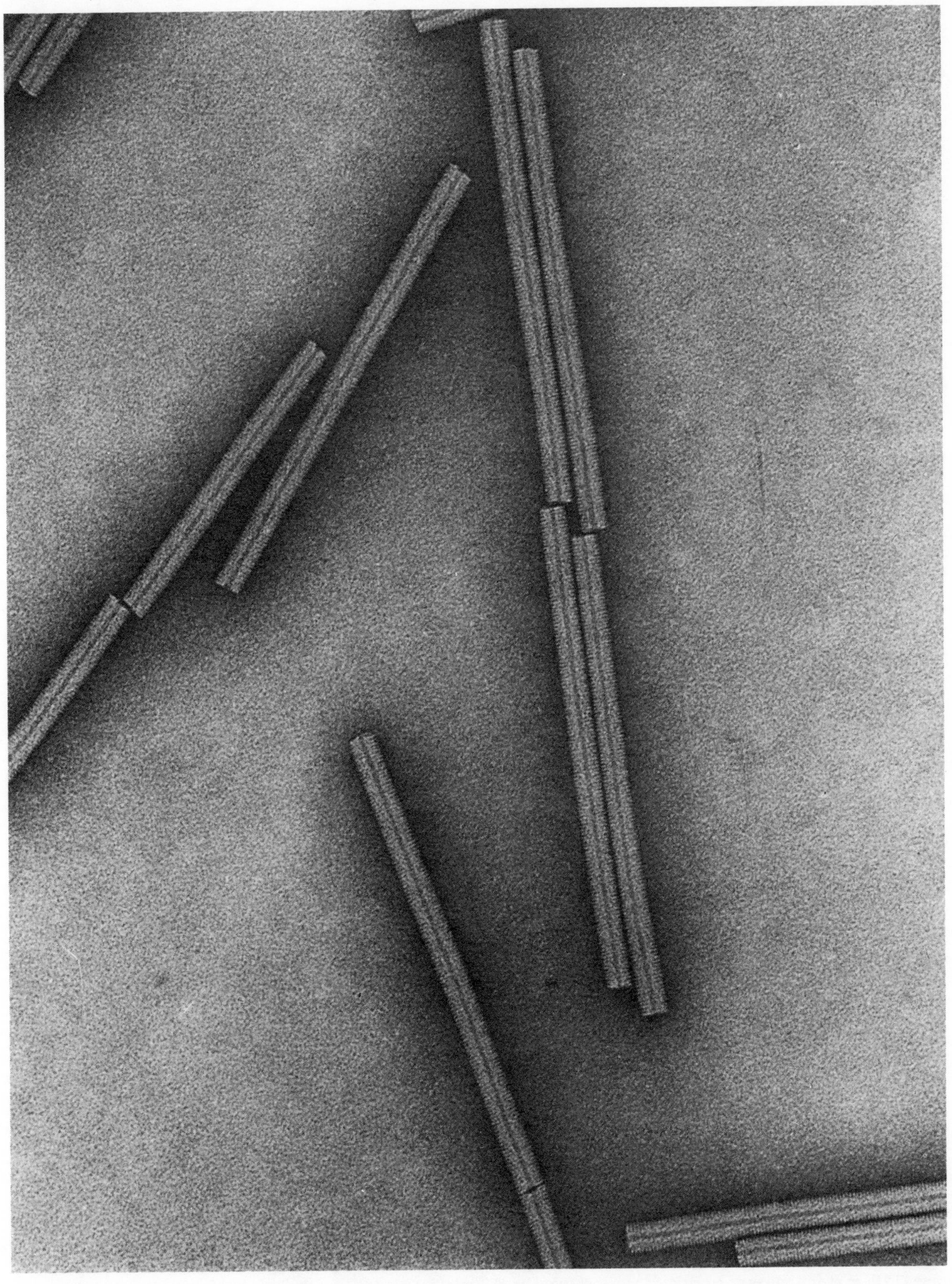

ROD-SHAPED PARTICLES of tobacco-mosaic virus are magnified some 300,000 diameters in this electron micrograph made by John T. Finch of the Medical Research Council Laboratory of Molecular Biology in Cambridge, England. The virus "self-assembles" spontaneously in the test tube from its constituent RNA molecule and protein subunits, giving rise to infective virus particles indistinguishable from those found in nature. The assembly of the virus provides a model for how large structures are built within living cells.

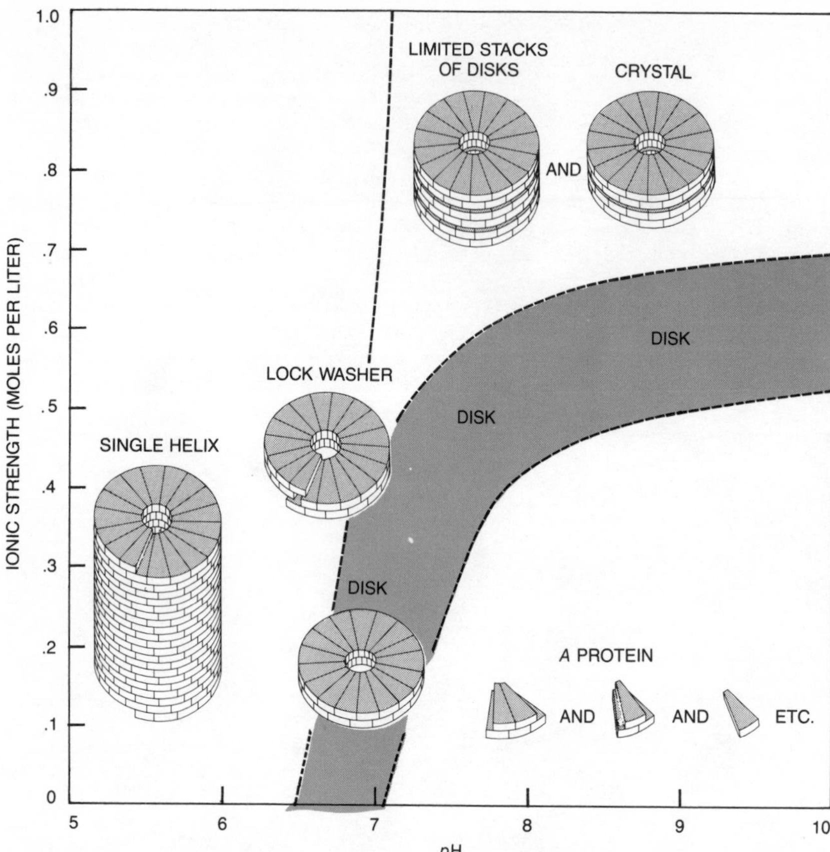

PROTEIN SUBUNITS AGGREGATE in different forms depending on the *p*H (the concentration of protons, or hydrogen ions) and the ionic strength (the concentration of salt) of the surrounding medium. Individual protein subunits are found only under conditions of very low ionic strength and strong alkalinity (*p*H 10). Under less alkaline conditions (*p*H 8) clusters of three or more subunits appear. In a neutral solution (*p*H 7), similar to the condition found in the host cell, the subunits arrange themselves in flat disks consisting of 34 subunits in two rings. If the solution is made acidic (about *p*H 6.5), the protein forms long helixes devoid of RNA. At neutral *p*H, however, the helixes assemble only when the viral RNA is present. The "lock washer" consisting of a single disk is an intermediate form that arises as disks convert into helixes.

from damage only when it is completely surrounded by protein.

The assembly of any large aggregate of identical subunits, such as a crystal, can be considered in two stages: nucleation and growth, or in the case of the tobacco-mosaic virus, initiation and elongation. The rate-limiting step in the assembly of the virus, as it is in most other instances, is initiation. Because of the large number of protein subunits per turn of the helix ($16\frac{1}{3}$) about 18 separate subunits would have to bind to the flexible RNA molecule before the assembling structure could close on itself and become more than a linear aggregate of protein along the RNA. The difficulty could be avoided if some kind of jig were available on which the first few turns of the viral helix could assemble until it grew large enough to be stable.

The solution to the problem was found to lie in an intriguing observation: the coat protein by itself, free of the viral RNA, can aggregate in a number of distinct yet related forms rather than only in a helix. Donald L. D. Caspar of Brandeis University foresaw that some of the forms might provide clues to the way the virus assembles. The various aggregation states were first examined in detail by our group (including Anthony C. H. Durham and John T. Finch), and other workers have since contributed to the picture. Although there is some disagreement about the details, the broad outline is now clear: the coat protein is so designed that it knows not only where it is going (into the viral helix) but also how to get there.

The dominant factor controlling the state of aggregation of the coat protein is the *p*H of the surrounding medium (the negative logarithm of the medium's concentration of free protons, or hydrogen ions). In a slightly alkaline solution (above *p*H 7) the coat protein tends to exist as a mixture of small aggregates of several subunits; this mixture is referred to as *A* protein. Near neutrality (*p*H 7) a different and specific structure appears: disks consisting of two layers of subunits. Each layer is a ring of 17 subunits, which is nearly the same number of subunits as there are in one turn of the viral helix. At conditions like those expected within the host cell as much as 80 percent of the coat protein is incorporated into disks; the rest is *A* protein. Making the solution abruptly more acidic (down to *p*H 5) converts the disks directly into short helical "lock washers" of just over two turns in length; the lock washers then stack in imperfect register and eventually anneal to yield helixes of indefinite length that are structurally very similar to the virus particle except that they are devoid of viral RNA.

The disk aggregate of the coat protein has a number of significant properties. It is the dominant form of the protein under conditions that are known to be optimal for the reassembly of the virus in the test tube and that are plausible for its natural assembly in the host cell. Moreover, the size and structure of the disk suggest that it might be ideal to act as the jig for the initiation of virus assembly. On the hypothesis that the disk might serve as a nucleating center, we looked at its effect on the reassembly reaction. The results were dramatic. Complete virus particles were formed within 10 minutes rather than the six hours that had been required for reconstitution experiments in which the protein was in disaggregated form. We reasoned that if the disks were needed for initiation, much of this time would be spent waiting for the disaggregated subunits to assemble spontaneously into disks before the growth of the virus particles could begin.

The notion that disks are involved in the natural biological process of initiation was strengthened by experiments in which assembly was carried out with RNA's from different sources. We found that the disks interacted much more readily with tobacco-mosaic RNA than with foreign or synthetic RNA's, ensuring that only the viral RNA was picked out for being coated with the viral protein. The structure of the disk allows up to a complete turn of RNA to bind to it during the first step, so that it clearly provides for a much greater discrimination than three nucleotides binding to a single protein subunit could.

Specificity also calls for a unique sequence of nucleotides in the viral RNA that interacts strongly with the protein disk. This sequence must be a significant stretch of the RNA in order to account for the high selectivity observed; about 50 nucleotides can interact with the 17 protein subunits in one turn of the first disk. We therefore set about isolating the initiation region of the RNA by supplying just enough coat protein to allow initiation but not growth and then digesting away the uncoated ends of the

RNA with an enzyme. Together with our colleague David Zimmern we found we could isolate a series of RNA fragments, all of which contained a common core sequence with variable lengths at each end. The shortest of the fragments was about 65 nucleotides long, just over the length necessary to bind to one complete turn of the disk, and we found that it bound to disks tightly and specifically. We concluded that this short fragment contains all the information necessary to specify the normal initiation reaction.

The large size and relatively low yield of the initiation region of the tobacco-mosaic RNA made determining its nucleotide sequence technically difficult. As we were working on the isolation and sequencing of this region Léon Hirth and his colleagues at the University of Strasbourg had begun to determine the sequence of nucleotides in various tobacco-mosaic RNA fragments they had isolated from uncoated viral RNA that had been partly digested by an enzyme. In this way they could obtain relatively good yields of these shorter fragments, and the determination of the nucleotide sequences was not too difficult. By chance one of their fragments overlapped the initiation region, and from our joint results it was possible to identify and complete the sequence.

The sequence of the initiation region suggests that it has a hairpin structure: a stem consisting of a double helix with weakly paired nucleotides and at the top of the stem a loop with unpaired nucleotides. The loop and the adjacent part of the stem consist of an unusual series of nucleotide bases with a repeating triplet motif of guanine (*G*), adenine (*A*) and uracil (*U*), with guanine in every third position: *AGAAGAAGUUGUUGAUGA*. Since there are three nucleotide binding sites per protein subunit, we speculated that such a triplet pattern could lead to the recognition of the exposed RNA loop by the protein disk during the initiation process.

With the initiation region identified

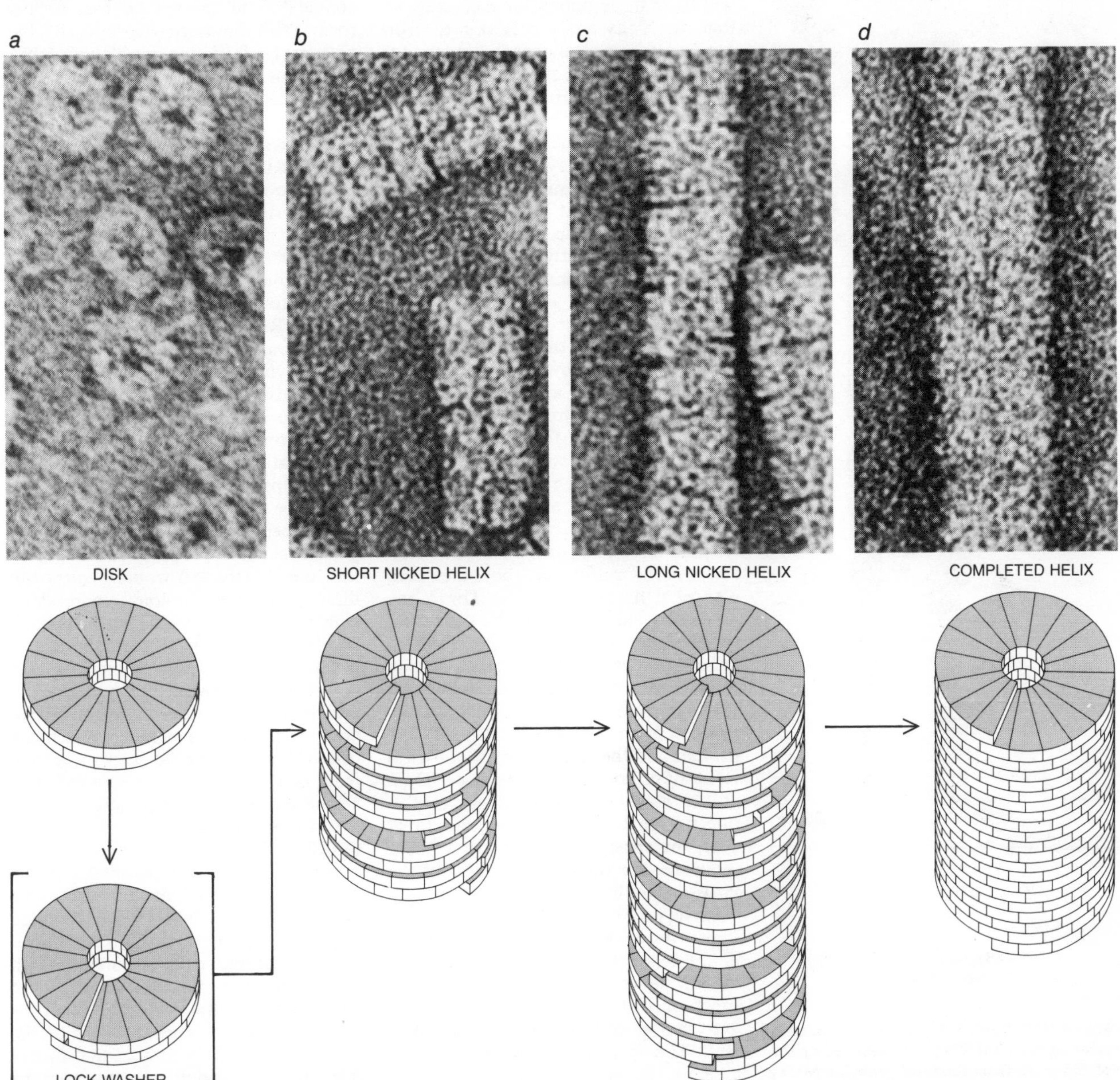

TRANSFORMATION OF DISKS TO A HELIX in the absence of the viral RNA can be brought about by lowering the *p*H of the solution to stabilize the helical aggregate, as is shown in these electron micrographs made by Finch. (In the lower half of the illustration the various structures are shown diagrammatically.) If the *p*H is lowered rapidly enough, the protein disks are converted into short helixes of just over two turns (lock washers) without dissociation into subunits (*a*). Within a few minutes the lock washers stack in random vertical orientation to yield short "nicked" helixes (*b*). These short stacks slowly aggregate over a period of about 15 minutes, forming a long nicked helix (*c*). Finally the imperfections in the rods are annealed out over a further period of hours to yield finished protein helix (*d*).

and the sequence of its nucleotides known, it became possible to locate the region along the RNA strand. Again the obvious expectation—that the initiation sequence would be near one end of the RNA—turned out to be wrong. Zimmern and T. Michael A. Wilson showed that the initiation region is about a sixth of the way along the RNA, so that more than 5,000 nucleotides have to be coated in one direction along the RNA strand and about 1,000 nucleotides have to be coated in the opposite direction.

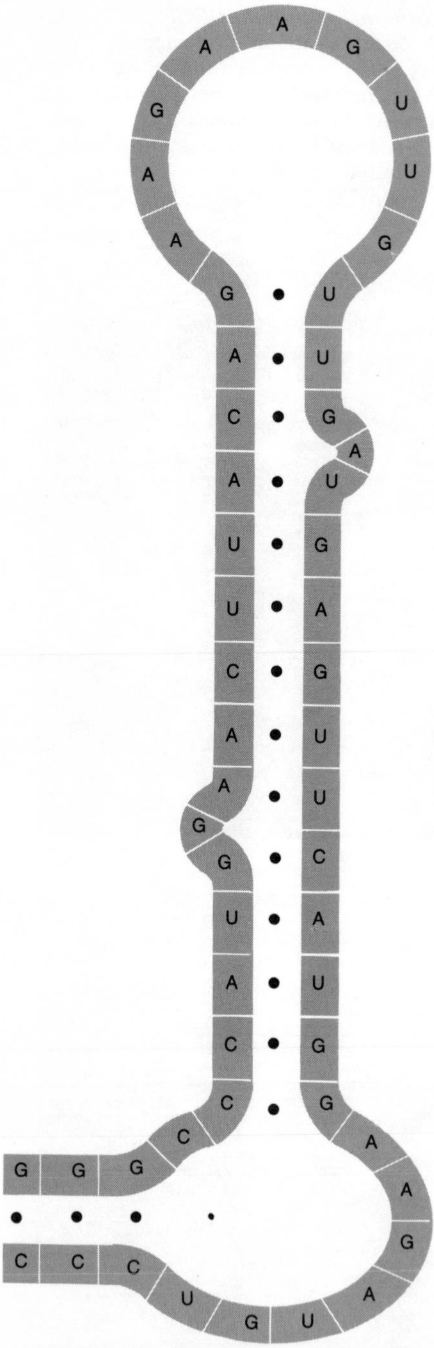

"HAIRPIN" STRUCTURE of the initiation region of the viral RNA was deduced by David Zimmern from the most probable pairing of the nucleotide bases in the RNA strand. This arrangement gives rise to a weakly bonded double-helical stem that has a loop at the top with a special sequence of bases. The loop is thought to bind to the first protein disk to begin the assembly of the virus. It has a distinctive motif of three nucleotide bases, with guanine (*G*) in the middle position and either adenine (*A*) or uracil (*U*) in outer positions.

In parallel with the RNA-sequence studies we and our colleagues investigated the detailed structure of the protein disk with X-ray-diffraction techniques. (Meanwhile Kenneth C. Holmes of the Max Planck Institute for Medical Research in Heidelberg pursued the X-ray studies on the virus itself begun by J. D. Bernal at the University of Cambridge as far back as 1936.) The disks will form three-dimensional crystals, so that the X-ray analysis is akin to ordinary protein crystallography except for the extremely large size of the repeating unit. The disk, which consists of 34 protein subunits and about 50 nucleotides and has a molecular weight of about 600,000 daltons, was the first very large structure to be examined in detail by X-ray analysis, and it has taken a dozen years for our group in Cambridge (including most recently Ann C. Bloomer, Gerard Bricogne and John N. Champness) to carry the analysis through to high resolution. The formidable technical problems were overcome only after the development by other workers in our laboratory of a special apparatus for tackling structures of this size.

The disk has a 17-fold rotational symmetry, which gave rise to redundant information in the X-ray data that could be exploited in the analysis. The structure is now known at a resolution of better than three angstrom units. (One angstrom is 10^{-7} millimeter, about the diameter of a hydrogen atom.) The X-ray analysis yields a map of the density of the electrons in the protein disk, and we have interpreted the electron-density map to get a detailed atomic model for the coat protein. The individual interactions between the amino acids in the protein and the nucleotides in the viral RNA have not yet been deduced. Even at the stage of five-angstrom resolution we reached a few years ago, however, the structure had important consequences for our understanding of the virus-assembly process.

In the protein disk the subunits of the upper ring are almost flat, whereas those of the lower ring are tilted downward toward the center, so that the two rings touch only at the outer part of the disk and open toward the center like a pair of jaws. This geometry is further accentuated by the flexibility of the chain of amino acids in the region of the protein inward from the RNA binding site; in this region the chain is disordered and not tightly packed into a regular structure. When the disk is converted into the viral helix, however, the subunits from both rings tilt upward toward the center until they are all at the same angle, with adjacent turns parallel and in tight contact. The inner regions of the subunits pack together tightly, assuming a definite structure and completely enclosing and protecting the viral RNA. The structural transition therefore closes the jaws of the disk, with the RNA trapped inside.

At this stage the efforts to determine the sequence of nucleotides in the viral RNA and to elucidate the structure of the protein disk by X-ray analysis drew together. We considered the structure of the disk and how the hairpin-shaped initiation region of the viral RNA might interact with it to trigger its conversion to the helical lock-washer form. With Zimmern we proposed a hypothesis for initiation in which the RNA hairpin inserts itself through the central hole of the disk into the jaws between the rings of subunits. (The dimensions of the disk and of the RNA hairpin are right for this process.) The nucleotides in the double-helical stem then unpair, and the stem opens out as more of the RNA is bound within the jaws of the disk. Some still unknown feature of this interaction then causes the disk to transform into the lock-washer form, trapping the RNA within the rings of protein subunits as the jaws close.

Our hypothesis about the mechanism of initiation led to a prediction that could be tested experimentally. If an RNA loop is inserted into the center of the initiating protein disk, both "tails," that is, projecting ends, of the RNA should be associated with the end of the rod-shaped virus particle from which growth began. Moreover, we expected that the shorter tail would project out directly and that the longer one (along which most of the elongation occurs) would be doubled back down the central hole of the virus. These predictions have now been confirmed. Hirth's group in Strasbourg has obtained electron micrographs of virus particles in which the two RNA tails can be seen projecting from one end of many of the rods [*see bottom illustration on opposite page*].

In Cambridge we have utilized high-resolution electron microscopy, with which the two ends of the virus particle can be distinguished, to confirm that the longer tail of the RNA is indeed doubled back through the growing rod. Moreover, we prepared virus particles with the longer RNA tail free by making the *p*H of the solution alkaline, thereby stripping protein from one end of the assembled particles; then the longer RNA tail extended directly out from the particle rather than being doubled back through the central hole. The partly stripped rods grew at less than a tenth the rate of the partly assembled rods, demonstrating the importance of the loop geometry.

Our model of initiation has an important consequence. The special configu-

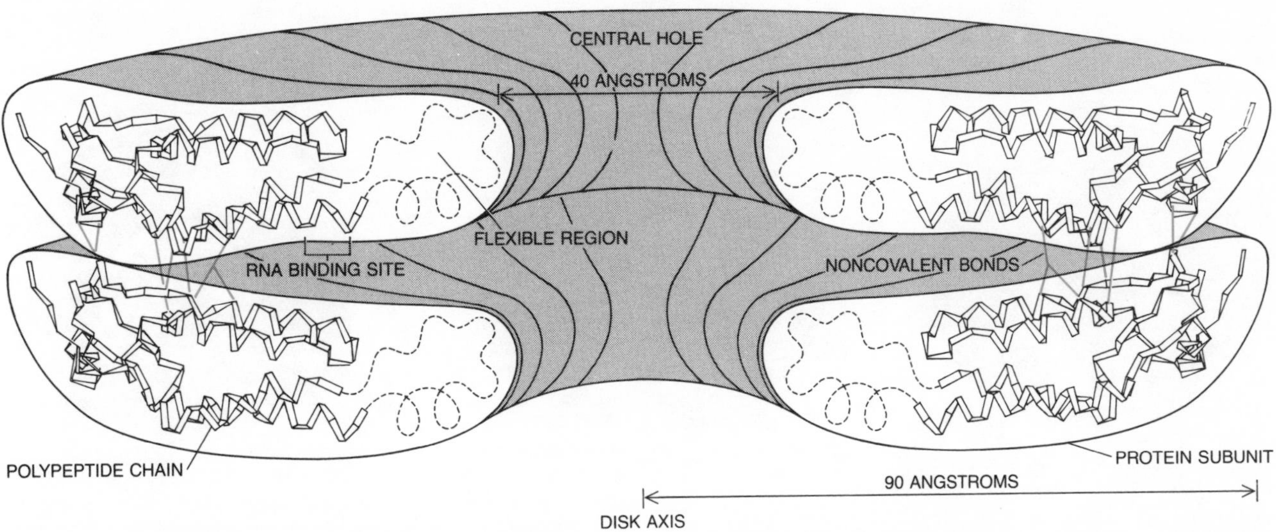

CROSS SECTION THROUGH A DISK was reconstructed from the results of an X-ray-diffraction analysis to a resolution of 2.8 angstrom units. Bent ribbons indicate the polypeptide chains that make up the protein subunits. The two-layered structure of the disk is evident: the subunits of the two stacked rings touch over a small area near the outer rim of the disk but open up toward the center like a pair of jaws. During the assembly of the virus the viral RNA binds within the jaws. The broken lines indicate the flexible portion of the protein chains extending in from the RNA binding site. Because this chain segment is in constant motion its structure cannot be resolved.

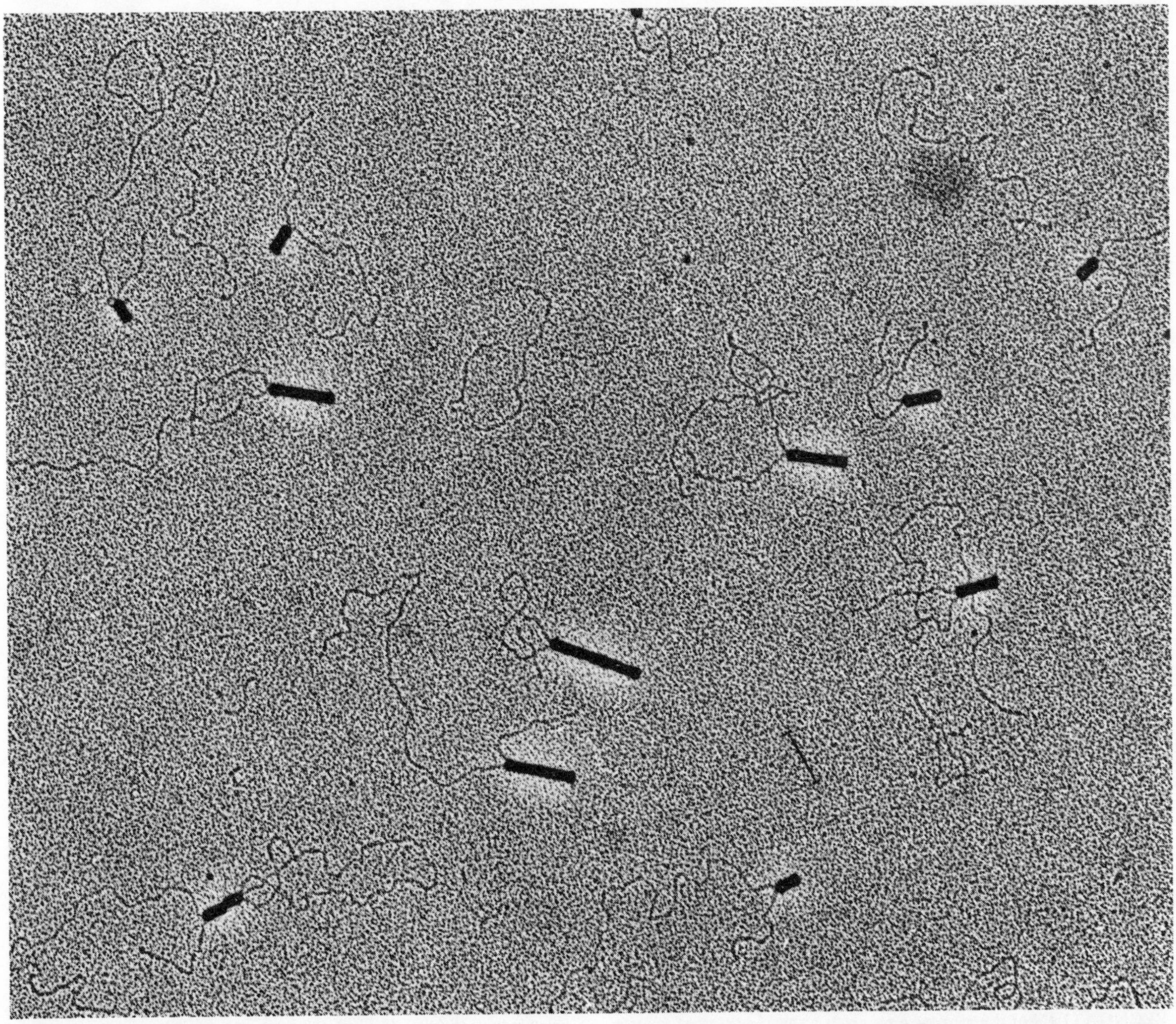

GROWING VIRUS PARTICLES have two RNA "tails" at one end, as is shown in this electron micrograph made by Geneviève Lebeurier of the University of Strasbourg. Other evidence indicates that the shorter RNA tail extends out directly from the particle, whereas the longer tail is doubled back inside the hollow core of the particle. The virus elongates primarily at the end of the particle opposite the tails.

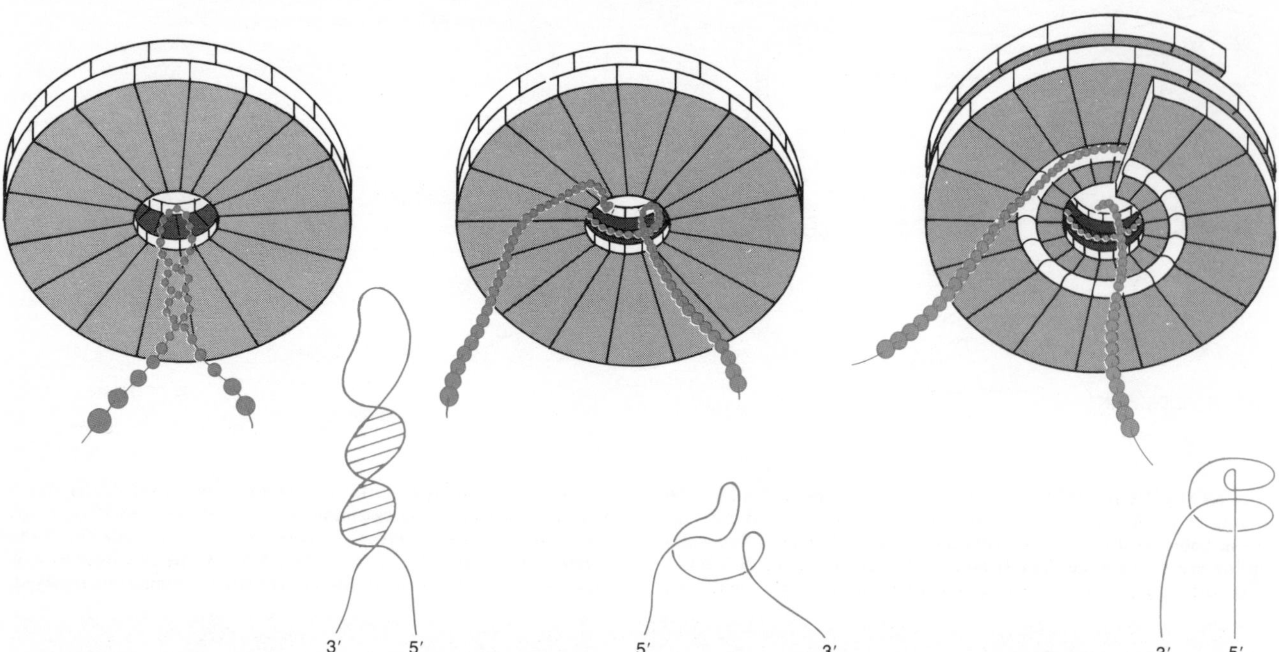

NUCLEATION of the tobacco-mosaic virus begins with the insertion of the hairpin loop formed by the initiation region of the viral RNA into the central hole of the protein disk (*a*). The loop intercalates between the two layers of subunits and binds around the first turn of the disk, opening up the base-paired stem as it does so (*b*). Some feature of the interaction causes the disk to dislocate into the helical lock-washer form (*c*). This structural transformation closes the jaws made by the rings of subunits, trapping the viral RNA inside

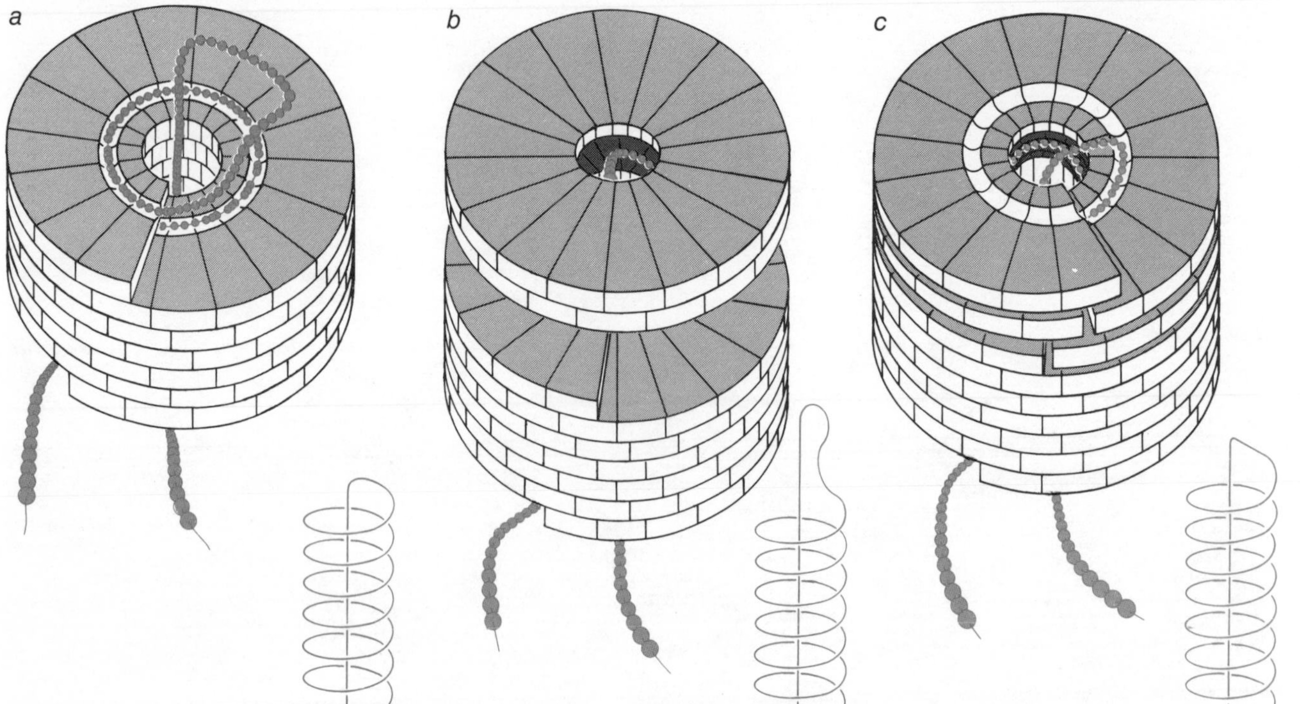

ELONGATION of the virus also proceeds by the addition of protein disks. As a result of the mode of initiation the longer RNA tail is doubled back through the central hole of the growing rod, forming a traveling loop at the growing end of the particle (*a*). The loop inserts itself into the center of an incoming disk and binds within open jaws of the rings (*b*). This interaction converts the new disk into a helical lock

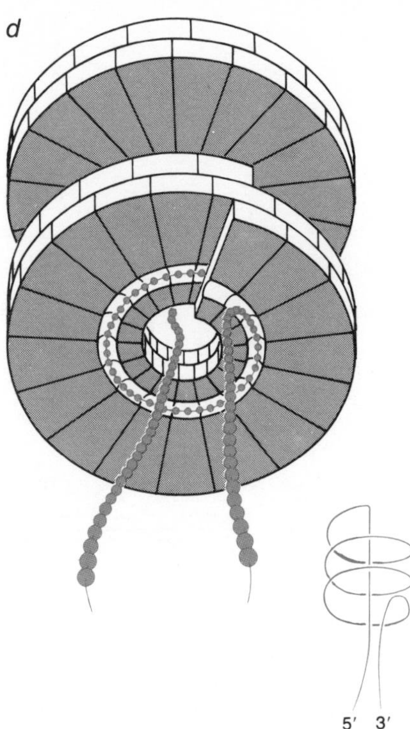

(*d*). The lock-washer-RNA complex provides the start of the helix. Additional disks then add rapidly to the nucleating complex, so that the helix elongates to a minimum stable length.

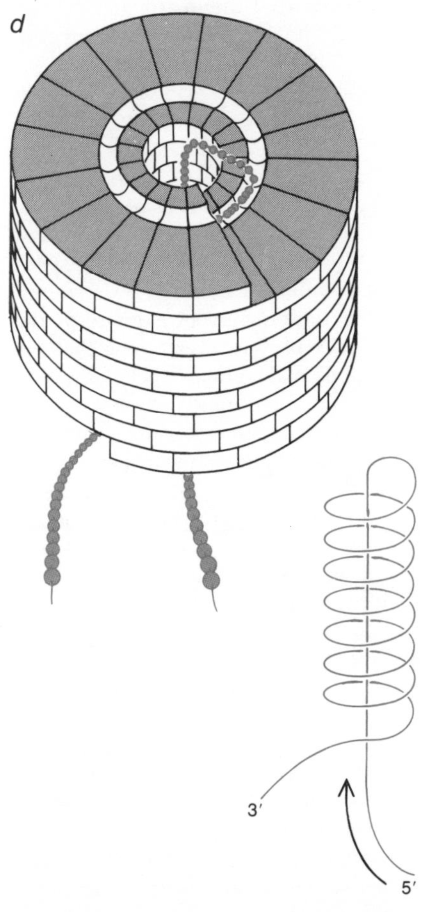

washer (*c*). The transformed disk then stacks onto the rod, providing two turns of helix (*d*). Process repeats until assembly is complete.

ration generated by the insertion of the RNA loop into the central hole of the initiating disk could subsequently be repeated during the addition of further disks on top of the growing helix; the loop could be perpetuated by drawing more of the longer tail of the RNA up through the central hole of the growing virus-particle rod. Hence the particle could elongate by a mechanism similar to initiation, only now instead of the specific initiation loop there would be a traveling loop of RNA at the main growing end of the virus particle. This loop would insert itself into the central hole of the next incoming disk, causing its conversion to the lock-washer form and continuing the growth of the virus particle. The helical geometry of the growing rod would tend to facilitate the conversion, so that elongation would be quite rapid. Apart from the obvious advantage of delivering a package of 34 subunits at a time rather than individual ones, this mode of coating the RNA would be less affected by unfavorable nucleotide sequences than the binding of individual subunits would be.

In the elongation of the virus particle there is no obvious requirement for specific nucleotide sequences, and indeed the protein must be able to coat any sequence of nucleotides that occurs in the RNA. As the longer RNA tail is being rapidly coated with protein, the virus-particle rod must also elongate along the shorter RNA tail at the opposite end of the rod to yield the complete particle. Growth in the direction of the shorter tail of the RNA is known to be slower than that in the direction of the longer tail, but as yet little is known about how it proceeds.

There is now direct evidence that the tobacco-mosaic virus grows mainly by the addition of entire blocks of subunits in the form of disks rather than by the addition of individual subunits. In recent experiments with George P. Lomonossoff we looked at the length of viral RNA in partly assembled rods that was protected against enzymatic digestion. It turns out that the RNA is protected in steps of 50 or 100 nucleotides, corresponding to one or two turns of the RNA in the viral helix. When growth is well established, the step size tends to be 100 nucleotides, the amount of RNA associated with a single protein disk. These findings point firmly to the conclusion that the disk aggregate is involved not only in the virus's initiation but also in its elongation.

What prevents the protein disks from forming a helix devoid of the viral RNA under normal conditions? As we have seen, the aggregation of the protein subunits is controlled by the *p*H of the surrounding solution. This control appears to be achieved through the presence on each subunit of chemical groups that bind protons, such as two closely spaced carboxyl groups (COOH). At the *p*H of the host cell one of these groups in the disk is ionized (negatively); in the helical lock-washer form both are ionized. The disk form is therefore favored because of the electrostatic repulsion between the two negatively charged groups. When the viral RNA is present, however, it binds to the protein disk, providing enough free energy to overcome the electrostatic repulsion in the lock-washer form. In this way the carboxyl groups act as a negative switch to block the formation of the protein helix in the absence of the viral RNA. Moreover, only the viral RNA is coated, because it alone has the specific initiation sequence that can start the process of assembly by converting the first protein disk into the lock washer.

We have seen that the special properties of the protein disk are the key to the mechanism by which the tobacco-mosaic virus is assembled. Indeed, one might say that the protein subunit is designed to form not an endless helix but a closed two-layer variant of it—the disk—that is stable but can be converted into a helical form. The disk therefore represents an intermediate subassembly whereby the thermodynamically difficult task of initiating helical growth is overcome. At the same time the protein disk furnishes a mechanism for the recognition of the viral RNA (and the rejection of foreign RNA's) by providing a long stretch of sites that interact with a specific sequence of RNA nucleotides. In short, as an obligatory intermediate in the assembly of the tobacco-mosaic virus, the protein disk simultaneously fulfills the physical requirement for initiating the growth of the RNA-protein helix and the biological requirement for specifically recognizing the viral RNA.

The development of the disk aggregate of the coat protein and the hairpin configuration of the RNA for initiating the viral helix was apparently too good a system to abandon in favor of subsequent elongation by the addition of single subunits. The protein disks therefore participate in the growth of the helix in the principal direction of elongation. On the other hand, the virus particle must also be disassembled in order to liberate the viral RNA in the infection of the host cell. The disassembly is probably accomplished by the sequential removal of individual subunits from one end of the virus particle. Therefore the construction of the virus is not left to the driving power of an unbalanced biochemical equilibrium, as it would be if assembly and disassembly were simple reversals of each other. Instead an intricate structural mechanism has evolved to give the process an efficiency and certainty whose basis is now understood.

9 Ribosomes

by Masayasu Nomura
October 1969

They are the organelles that conduct the synthesis of proteins in the living cell. Their structure and functioning are studied by taking them apart and seeing how they reassemble themselves

It is one thing to discover the basic principles of a life process and quite another to know in detail the chemical mechanisms that underlie it. In order to genuinely understand a cellular function one must study the machinery that performs it, and in many cases that means studying a highly organized cellular element, or organelle, that provides the machinery. One must first determine the organelle's structure and learn how it operates and then find out how the organelle itself is generated in the cell. In this article I shall relate how the structural and functional description of the ribosome, the organelle that conducts protein synthesis, has been attempted and is even now being achieved.

The story of the ribosome goes back to the discovery some years ago that the capacity of various types of cell to synthesize proteins was correlated with the cells' content of ribonucleic acid (RNA), and that most of the cellular RNA was in the form of small particles (then known as microsomes) in the cytoplasm of the cell. This suggested that the particles must play some role in protein synthesis, but the real importance of ribosomes emerged only after intensive biochemical investigation.

The pioneer work was done by Paul C. Zamecnik and his collaborators at the Massachusetts General Hospital in the 1950's [see "The Microsome," by Paul C. Zamecnik; SCIENTIFIC AMERICAN Offprint 52]. They homogenized rat-liver cells, added amino acids labeled with atoms of the radioactive isotope carbon 14, fortified the homogenate with adenosine triphosphate (ATP) to provide chemical energy and were able to detect the formation of small amounts of protein. By a process of elimination they established that several cellular organelles, including the nucleus and the mitochondrion, were not necessary for protein synthesis but that the microsomes were essential. They were able to identify other cellular components required for protein synthesis, including the small RNA molecules called transfer RNA and enzymes that attach amino acids to transfer-RNA molecules. These early test-tube assembly systems, however, made only very small amounts of protein. Then in 1961 Marshall W. Nirenberg and J. Heinrich Matthaei of the National Institutes of Health found that in order to obtain intensive synthesis of protein in cell-free extracts of the bacterium *Escherichia coli* it was necessary to include a third type of RNA, called messenger RNA, that had been postulated by François Jacob and Jacques Monod of the Pasteur Institute in Paris.

Once a complete cell-free protein-synthesizing system could be assembled it was possible to study the functioning of its several components. One of the most interesting of these was the ribosome. It was now clear that this particle coordinates the translation of the genetic information in the sequence of nucleotide bases in the messenger RNA (transcribed from the DNA molecule, the gene) to the sequence of amino acids in each protein manufactured by the cell [*see illustration on page 116*]. The first systematic studies of ribosomes were initiated about 1957 by several groups, notably one at Harvard University led by Alfred Tissières and James D. Watson and one in the Carnegie Institution of Washington that included Ellis T. Bolton, Roy Britten and Richard B. Roberts. (I should add that my association with Watson's group at that time, although it was brief, had a great influence on my later research on the ribosome.) The initial studies were done mainly on *E. coli* ribosomes, which consist of two subunits of unequal size that are designated 30S and 50S. The size is determined by the rate, measured in Svedberg units (S), at which a particle sediments when it is spun at high speed in an ultracentrifuge. Together these particles constitute the functional unit in protein synthesis: the 70S ribosome. (The reason the two S values are not additive is that the shape of a particle influences its rate of sedimentation.) In each of these subunits proteins represent about a third of the total mass; the rest is RNA. The 50S ribosome subunit contains a 23S RNA molecule and a 5S RNA molecule. The 30S ribosome subunit incorporates one 16S RNA molecule. In 1961 J. P. Waller and J. I. Harris of Harvard observed that the ribosome contains different kinds of protein molecules, indicating that its structure must be quite complex. Subsequent experiments conducted by many workers, including those in my own group at the University of Wisconsin, show that the 30S ribosomal subunit includes either 19 or 20 different protein molecules and that the 50S subunit apparently has more than 30 protein molecules.

As the early work on the structure of the ribosome was proceeding, a general picture of its functional properties had begun to emerge. The existence of specific ribosomal binding sites for transfer and messenger RNA was demonstrated, forcing the conclusion that the ribosome plays an active role in protein synthesis and is not merely an inert workbench on whose surface amino acids are assembled. The observed physical complexity of the ribosome must therefore reflect the complexity of its function. What we needed to establish was the relation between structure and function. Yet as far as the actual roles of the RNA and proteins and their critical interrelations were concerned, the ribosome was still a mysterious "black box." How was one to understand this complicated piece of machinery? One could take it apart and try to reassemble it, but what tools were delicate enough to avoid destroying the machine in the process?

In 1961 Jacob, Sydney Brenner and Matthew S. Meselson, in a paper describing the classic experiments that proved the messenger-RNA theory, noted the presence of two kinds of ribonucleoprotein particle in mixtures that were centrifuged in a solution of the salt cesium chloride, which forms a density gradient in the centrifuge cell. (Density-gradient centrifugation, originally developed by Meselson, Franklin W. Stahl and Jerome R. Vinograd at the California Institute of Technology, separates large molecules on the basis of their different buoyancy in such a solution.) When they centrifuged bacterial extracts containing ribosomes, they observed two bands containing ribosomal particles. The lighter band (the *B* band), corresponding to a density of 1.61, contained messenger RNA as well as proteins being synthesized; the heavier band (the *A* band), corresponding to the density of 1.65, did not.

The presence of the *A* band was not relevant to the main theme of the paper, and no reason was given for its presence. At the time the paper was published I was working at the University of Osaka on both ribosomes and messenger RNA, and I was quite curious about this phenomenon. We knew then that the 30S and 50S subunits and their aggregate, the 70S ribosome, all have the same chemical composition: 65 percent RNA and 35 percent protein. Since the buoyancy of complex molecules in solution usually reflects their chemical composition, why should density-gradient centrifugation reveal two kinds of ribonucleoprotein particle?

In the summer of 1962 I had an opportunity to visit Meselson's laboratory at Harvard to look into the question. When we recovered the particles from the two bands, we found that the *B* band contained undegraded 50S and 30S ribosomal subunits. The denser *A* band, however, consisted of a mixture of smaller 40S and 23S "core" particles that had been created from the usual ribosomal subunits by the splitting off of about 40 percent of the protein during the density-gradient centrifugation; the split proteins could be found in a protein fraction at the top of the gradient. The explanation for the *B* band was apparently that in crude bacterial extracts some of the ribosomes are resistant to this splitting, perhaps because they are stabilized by messenger RNA and growing protein chains.

On returning to Osaka I continued experiments with Robert K. Fujimura to characterize these core particles. Initially we prepared 40S and 23S particles (the latter are not to be confused with 23S RNA molecules) by respectively centrifuging purified 50S and 30S ribo-

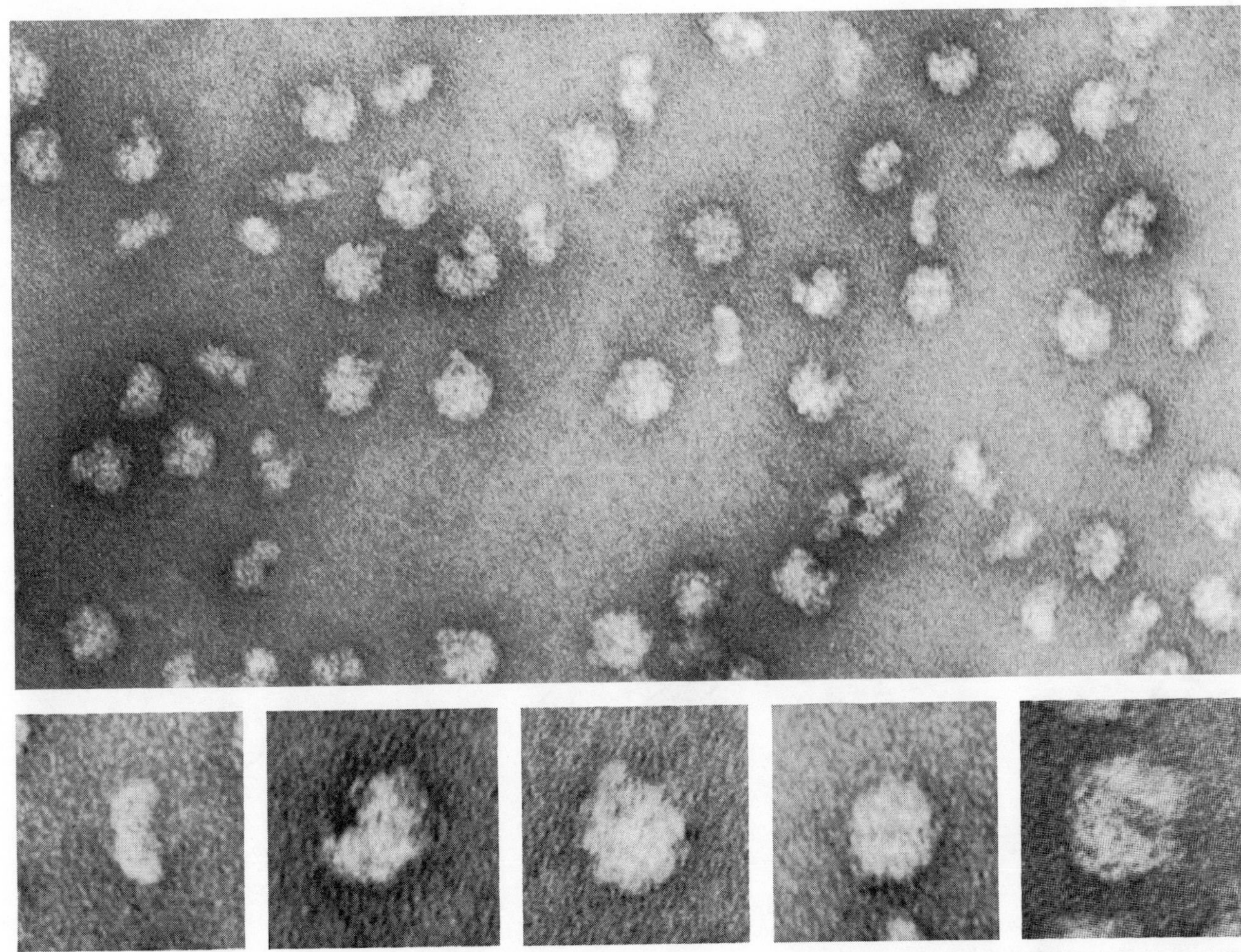

RIBOSOMAL SUBUNITS are enlarged 450,000 diameters in an electron micrograph (*top*) made by Martin Lubin of the Dartmouth Medical School. To prepare them, intact (70S) ribosomes were dissociated in solutions with a low magnesium concentration and the subunits were negatively stained with uranyl formate. Individual particles are enlarged 800,000 diameters (*bottom*). The smaller of the two subunits is the 30S (*left*). The larger is the 50S, seen in three different views characterized by a kidney shape (*second from left*), a "nose" (*third from left*) and a groove (*fourth from left*). The two subunits join to form a 70S ribosome (*right*).

somal particles in cold cesium chloride solution for 36 hours and then recovering the core particles from the band in the middle of the centrifuge cell [*see middle illustration on following page*]. This procedure was troublesome, however, and unsuitable for large-scale preparation of the particles. We therefore tried omitting the centrifuge step. Reasoning that it was surely the particular salt solution and not the physical centrifugation that disrupted the subunits, we simply kept 50S subunits in the solution for 36 hours, expecting that irreversible splitting of the protein would take place, yielding the 40S core particles we wanted to study.

When we removed the cesium chloride and examined the products, however, we found to our surprise that the recovered particles behaved just like the original 50S ribosomal subunits. Why had there been no splitting of the proteins? We immediately realized the important implication of this experimental observation: The splitting of the 50S unit is reversible; the reaction is pushed in the direction of dissociation only by separation in the centrifuge. To test this supposition we prepared core particles and split proteins by the usual centrifugation method. Then we mixed them together and removed the cesium chloride. We found complete conversion of the core particles to intact ribosomal particles. In this way we succeeded in reconstituting the 50S ribosomal subunit from the 40S core and split proteins that had been derived from the 50S, and also in reconstituting the 30S subunit from the 23S core and the homologous split proteins.

In order to prove that the reconstituted ribosomal particles really had the

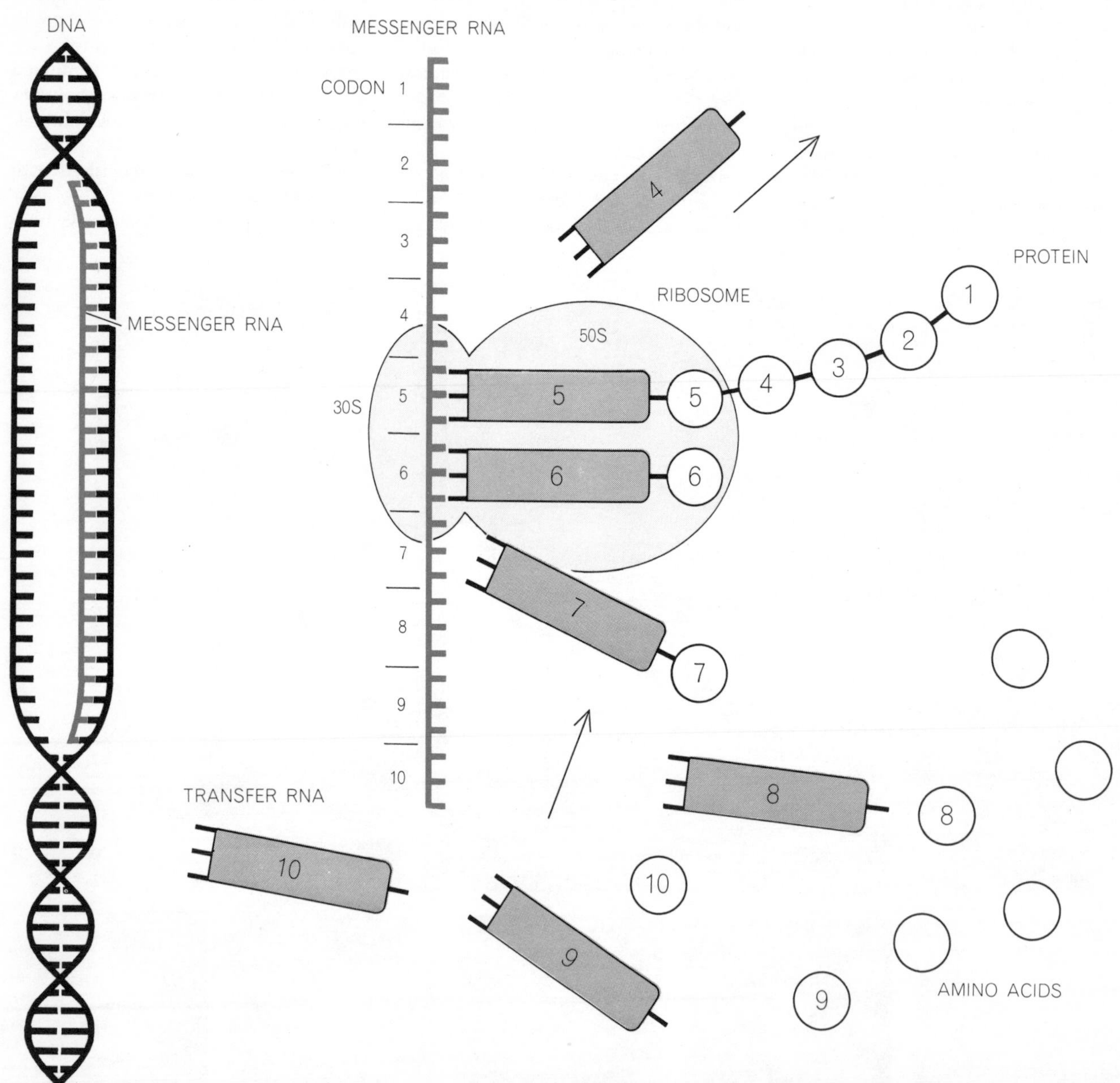

RIBOSOMES conduct protein synthesis. Genetic information is encoded in the sequence of bases (*horizontal elements*) in the double helix of DNA (*left*). This information is transcribed into a complementary sequence of RNA bases to form messenger RNA (*dark color*). Each group of three bases in the messenger RNA constitutes a codon, which specifies a particular amino acid and is recognized by a complementary anticodon on a transfer-RNA molecule (*lighter color*) that has previously been charged with that amino acid. Here amino acid No. 6, specified by the sixth codon, has just been bound to its site on the ribosome by the corresponding transfer RNA. It will bond to amino acid No. 5, thus extending the growing peptide chain. Then the ribosome will move along the messenger RNA the length of one codon and so come into position to bind transfer RNA No. 7 with its amino acid.

same specific structure as the original ones, we had to demonstrate the functional integrity of the reconstituted particles. Before we could succeed in such experiments I left Osaka and moved to the University of Wisconsin. There Keiichi Hosokawa and I were able to show in 1965 that, whereas neither the 23S nor the 40S cores have any activity in a cell-free protein-synthesizing system, reconstituted 30S and 50S particles have activities comparable to the original intact 30S and 50S ribosomal subunits. At the same time Theophil Staehelin and Meselson, who were taking a similar approach at Harvard, independently succeeded in demonstrating the reconstitution of the ribosomal subunits from core particles and split proteins. The functional capabilities of the reconstituted particles can be assayed in various ways. For example, the function of 30S particles is usually assayed by measuring the rate of protein synthesis directed by messenger RNA in the presence of intact 50S subunits and other necessary components. One can also test the subunits' ability to bind several different transfer RNA's in the presence of various messenger RNA's and the ability to bind messenger RNA itself.

The success in reconstitution, although it involved the dissociation and reassociation of only some of the ribosomal proteins (the split proteins), had several important implications. First, the experiment showed that at least part of the ribosome assembly in the test tube is spontaneous; no extraribosomal template or enzyme is required. Second, it provided a system in which the functional roles of individual split proteins could be analyzed. The 30S split-protein fraction consists of seven proteins. By column chromatography we separated the proteins into five pure protein components and one fraction containing two proteins, and we showed that the five proteins differ from one another in amino acid composition. Then we determined the functional need for each of the purified proteins by omitting one of them at a time in reconstitution experiments. We found that three of the five purified proteins are essential for reconstitution, and that the omission of either one of the others has only a partial effect. From this type of experiment we could conclude that all five of the purified proteins are chemically and functionally distinct, and that some of them are absolutely required whereas others are not (although they are required for full activity in protein synthesis).

The partial-reconstitution system was a first step toward the functional analy-

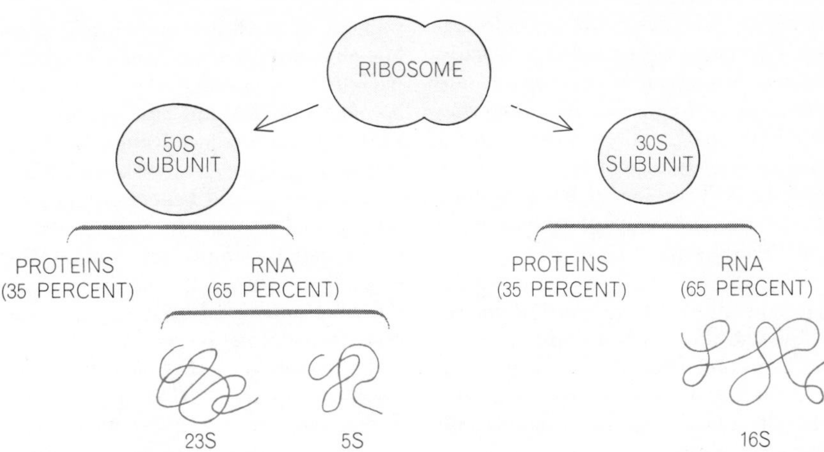

TWO SUBUNITS of a ribosome can be separated by spinning ribosomes in a centrifuge because the subunits are different sizes and move through the centrifuge cell at different rates. Both subunits are about 35 percent protein and 65 percent RNA. The 50*S* subunit contains a 23*S* and a 5*S* RNA molecule and the smaller 30*S* subunit has a 16*S* RNA molecule.

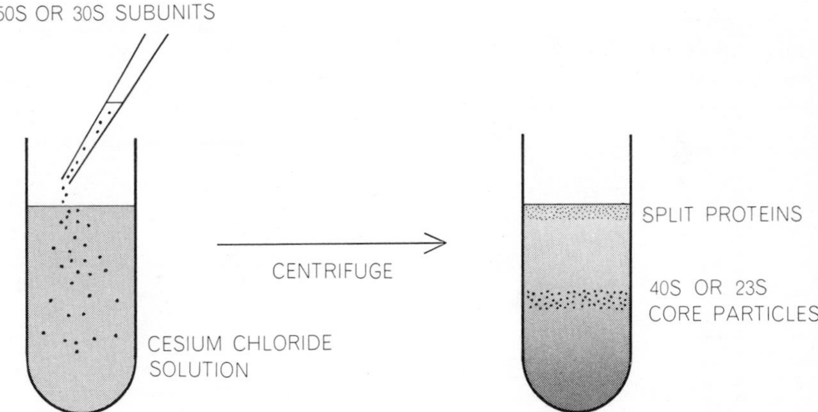

FURTHER CENTRIFUGATION of the subunits breaks them down. The subunits are added to a cesium chloride solution (*left*). Centrifugation establishes a stable density gradient in the solution (*right*), within which the subunit components form layers according to density. Some proteins split off, leaving "core" particles of RNA and other proteins.

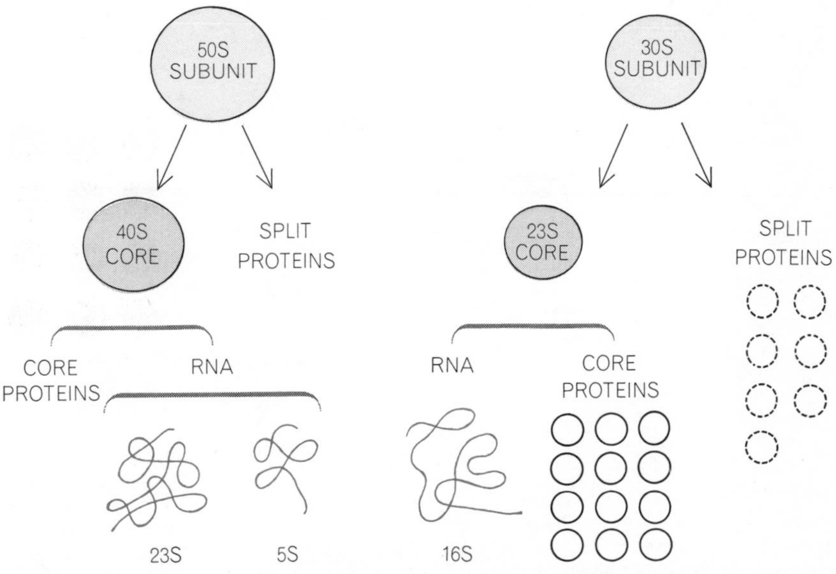

DISSOCIATION of the 50*S* subunit yields a 40*S* core particle and split proteins, dissociation of the 30*S* subunit a 23*S* core particle and split proteins. In the case of the 30*S* subunit there appear to be seven split proteins, with about 12 proteins remaining in the core.

sis of the ribosome, but the information that could be provided by such a system was limited. To accomplish complete analysis we needed a way to reconstitute ribosomes entirely from free RNA and completely separated proteins. In 1967 Peter Traub and I began systematic attempts at complete reconstitution with the 30S subunit.

We assumed that we must do the reconstitution in two steps, first making 23S cores from 16S RNA and proteins and then making complete 30S subunits from the 23S cores and the split proteins. We therefore prepared 16S RNA by treating 23S cores with phenol. We separated proteins from other 23S cores by treating the core particles with urea and a high concentration of lithium chloride. We mixed these core proteins with the 16S RNA under several different conditions, hoping to obtain 23S core particles. Then we added the split proteins to the reaction mixture, recovered the particles by centrifugation and assayed the activity of the recovered particles in a cell-free protein-synthesizing system. As typical enzyme chemists, we felt it was essential to protect the ribosomal proteins and any sensitive intermediates from inactivation by heat, and so we performed all these operations in a cold room and kept everything on ice.

Our initial attempts were failures. We could find only very slight protein-synthesizing activity. It seemed that the reaction might not be possible. Then we realized that living *E. coli* cells multiply most rapidly at 37 degrees Celsius (body temperature) and not at all at freezing temperatures, and that multiplying cells must certainly be assembling ribosomes quite efficiently. We also recognized that the cytoplasm of living *E. coli* contains a rather high concentration of salts; the salts might discourage nonspecific RNA-protein aggregation and thereby promote the specific RNA-protein assembly reaction. We therefore attempted the reconstitution at 37 degrees and with a high concentration of a salt, potassium chloride. Success! The 30S ribosome could be self-assembled [*see illustration on page 120, bottom*]. We found, indeed, that the reconstitution of 30S subunits from RNA and proteins is independent of the order of addition of proteins, whether they are core proteins or split proteins. By simply mixing all the proteins prepared directly from the 30S subunit with 16S RNA and incubating them in an optimal ionic environment at about 40 degrees C. for 10 minutes we were able to convert almost all the 16S RNA in the mixture into 30S particles [*see top illustration on page 120*]. In protein-synthesizing activity, in protein content and in sedimentation behavior the reconstituted particles were almost identical with the original 30S particles.

From this point one could proceed in many directions. One basic goal was to determine the functional role of the 16S RNA and of each 30S ribosomal protein. With regard to the RNA, we first considered the specificity requirements. For example, is an RNA molecule

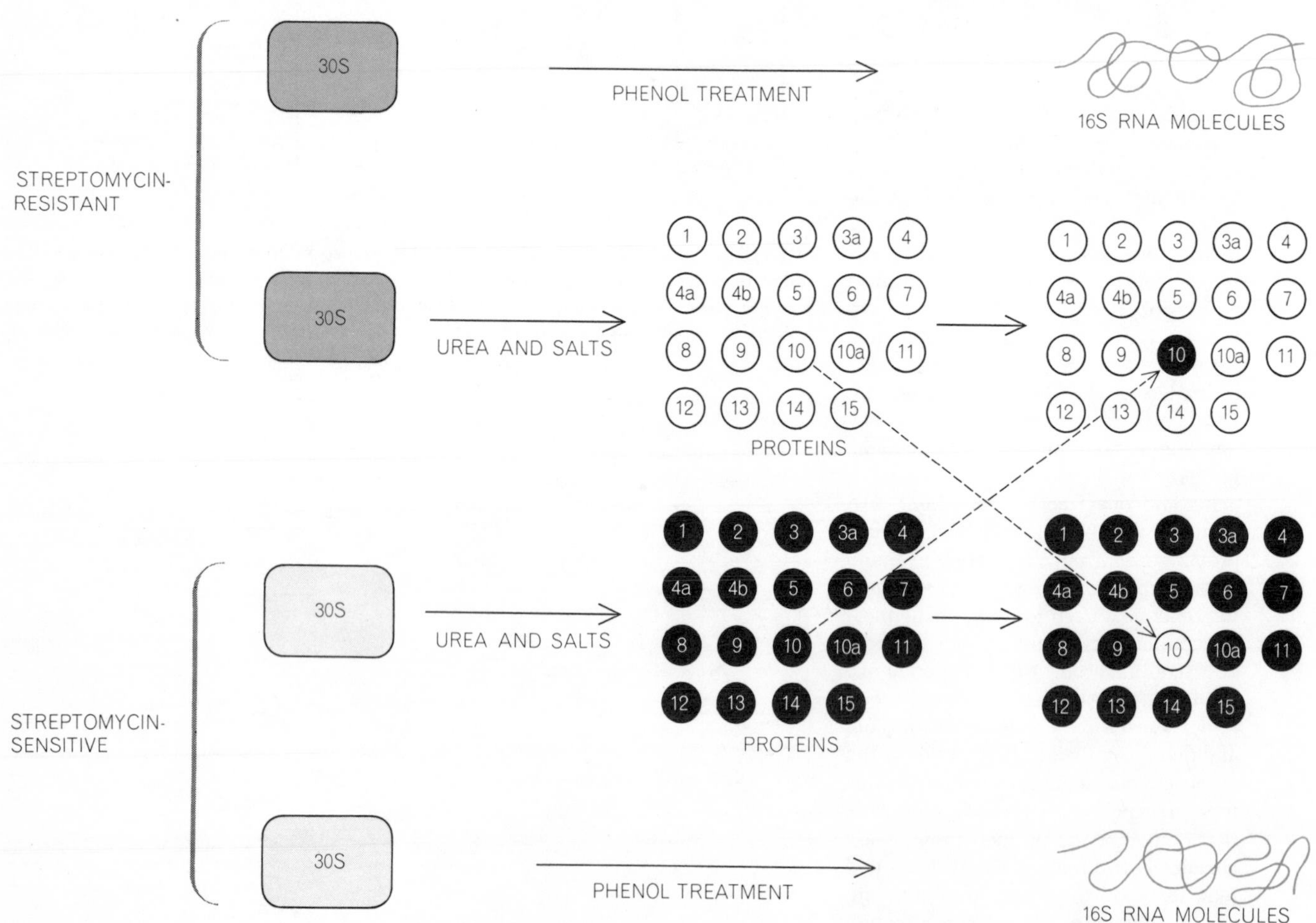

ONE PROTEIN in the 30S subunit is responsible for the effect of streptomycin on ribosome functioning. Proteins and RNA are separated from subunits derived from bacterial cells that are resistant to streptomycin and from other cells that are susceptible to its effect. The proteins are isolated. When the protein designated *P*10 from susceptible cells is combined with all the other proteins from resistant cells and with the RNA from resistant cells, the 30S ribosomes that result turn out to be susceptible to streptomycin. On the

that is merely similar in size to *E. coli* 16S RNA competent to reconstitute a physically and functionally intact 30S subunit? No. Neither 16S ribosomal RNA from yeast cells nor *E. coli* 23S RNA degraded to a 16S-sized fragment was active in reconstitutions with *E. coli* 30S proteins. In fact, the inactive products of such combinations did not even resemble 30S ribosomes physically, judging by their sedimentation behavior. This finding certainly came as no surprise; one would expect the requirements for a functional ribosome to be stringent. How stringent? When we performed the reconstitution with 16S RNA from one bacterial species and 30S ribosomal proteins from another species that is distantly related, we found that in many cases such "artificial" ribosomes were as active as the respective homologous RNA-protein combinations. We conclude from this that although there is definitely a specificity requirement for the RNA, the requirement is not absolute.

In determining the roles of the various ribosomal proteins our basic approach was to perform the reconstitution with one component omitted or specifically modified and then see if physically intact 30S particles were formed and, if so, whether or not they were functionally active. We first had to separate the 30S protein mixture into each of its 19 components. The fractionation of 30S ribosomal proteins had already been achieved by research groups at the University of Geneva, the Max Planck Institute for Molecular Genetics in Berlin, the University of Wisconsin and the University of Illinois. We employed methods similar to theirs, relying mainly on various types of column chromatography.

We then did the reconstitution with 16S RNA and 19 purified proteins rather than with the unfractionated protein mixture used in our earlier experiments. The extent of the reconstitution was not as good as it was with unfractionated proteins (and we therefore could not exclude the possibility that there are some ribosomal components other than the RNA and 19 protein molecules), but the reasonably high efficiency of the reconstitution made it possible to undertake the functional analysis of the separated protein components.

The first protein we studied in detail was the one responsible for sensitivity or resistance to the antibiotic streptomycin. Earlier studies had indicated that the drug's primary site of action is the bacterial ribosome, specifically the 30S subunit. When streptomycin is added to a cell-free system containing ribosomes from a streptomycin-sensitive strain, it inhibits protein synthesis. Streptomycin also causes the misreading of certain synthetic messenger RNA's, that is, it induces the incorporation into proteins of amino acids other than the ones dictated by the genetic code. This misreading effect of streptomycin was discovered first by Julian E. Davies, Walter Gilbert and Luigi Gorini at Harvard [see "Antibiotics and the Genetic Code," by Luigi Gorini; SCIENTIFIC AMERICAN Offprint 1041].

Bacteria can become resistant to streptomycin through mutation, and streptomycin does not inhibit synthesis or cause misreading of messages in a cell-free system if 30S ribosomal particles from streptomycin-resistant mutants are used. Traub and I showed that the component altered by the mutation was not the RNA but had to be in the protein fraction; 30S particles reconstituted from the protein of a resistant mutant and the RNA of a susceptible strain were resistant to streptomycin in cell-free protein-synthesizing systems, whereas the reverse combination produced 30S particles that were susceptible to streptomycin. Makoto Ozaki and Shoji Mizushima took over the job of identifying the altered protein. We purified 30S ribosomal proteins from both susceptible and resistant bacteria, systematically substituted single proteins from a resistant strain in a mixture of proteins from a susceptible strain [see *bottom illustration at left*] and assayed the reconstituted ribosomes for their response to streptomycin. In this way we established that a single protein, one we had designated *P*10, determines the susceptibility of the entire 30S ribosomal particle to the inhibitory action of streptomycin, its susceptibility to streptomycin-induced misreading and its ability to bind the antibiotic.

Having learned how an alteration in a given protein can affect the function of the ribosome, we investigated what happens to the ribosome when this protein is simply left out. Is the ribosome still able to assemble itself and, if so, how are its functional capabilities altered? We found that in the absence of *P*10, RNA and the other ribosomal proteins can still assemble into particles that sediment at 30S. These *P*10-deficient particles have several interesting properties, however. Under the conditions of the assays these particles show high activity when a synthetic messenger RNA is used as a template, but their activity is weak when directed by RNA from a natural source. It is known that a special mechanism for initiating protein synthesis is needed in the system directed by natural messenger RNA's but not in the system directed by synthetic messenger RNA's [see "How Proteins Start," by Brian F. C. Clark and Kjeld A. Marcker, Offprint 1093]; the *P*10-deficient particles cannot carry out this special initiating function.

The other interesting finding is that the frequency of translation errors with the *P*10-deficient particles is much reduced not only in the presence of streptomycin but also in the presence of certain other antibiotics, of ethyl alcohol or of high concentrations of magnesium ions, all of which are known to induce translation errors. In fact, the deficient particles read synthetic messenger RNA more accurately in the presence of error-inducing agents than normal ribosomes do even in the absence of such agents. In other words, it appears that protein *P*10 plays a role in increasing the frequency of errors in the translation of the genetic message. The inherent ability of ribosomes to make mistakes may be ad-

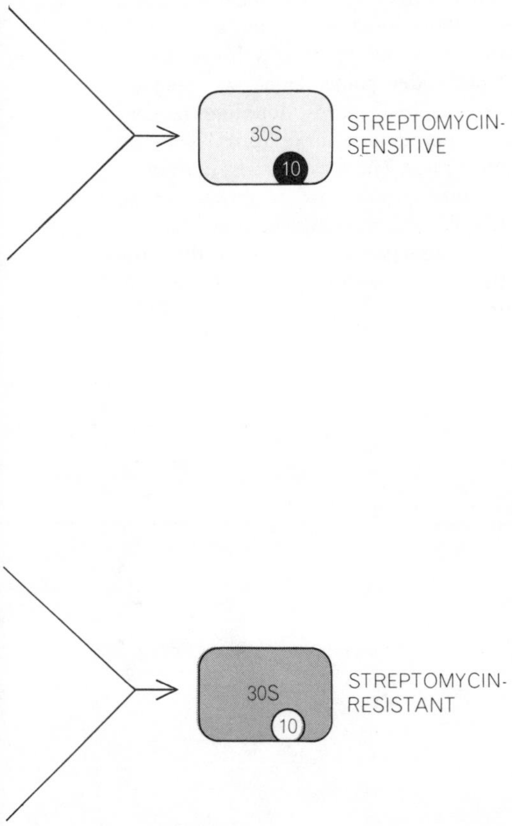

other hand, when *P*10 is derived from resistant cells and all the other components are taken from susceptible cells, the resulting subunit is resistant to streptomycin.

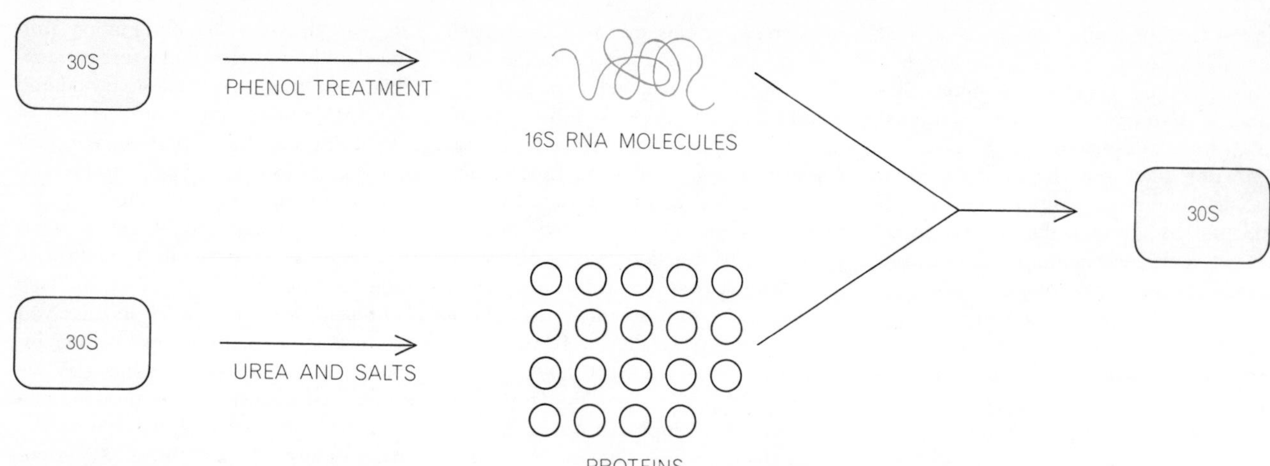

CORE AND SPLIT PROTEINS need not be assembled separately. Reconstitution can be accomplished by mixing all ribosomal proteins from 30*S* subunits with 16*S* RNA in the correct ionic environment and incubating at 37 to 40 degrees C. for 10 to 20 minutes.

vantageous to bacterial cells, as Gorini suggested, since it can suppress the effects of harmful mutations. On the other hand, the property may simply be an unavoidable consequence of the complexity of the machinery.

Although we have not yet completed similar detailed analyses of all the proteins, preliminary experiments with them lead to certain general conclusions. Omitting any one of several proteins affects a number of known 30*S* ribosomal functions. Conversely, several different 30*S* ribosomal functions are affected by the omission of any of a number of proteins. That is, these functions seem to require the presence of more than one protein component, and so one can say that the 30*S* ribosomal proteins function cooperatively. We have also found that the omission of some proteins drastically affects physical assembly. Particles formed in the absence of one of these proteins are deficient in several other proteins, including some that were present in the reconstitution mixture. In other words, the presence of certain proteins is essential for other proteins to be bound. In this sense the assembly process itself is cooperative.

One of the most effective tools that are available for the study of a reaction is chemical kinetics, the study of the rate at which a reaction proceeds. Most chemical reactions have a distinct kinetic mode, or "order," which is determined experimentally from the reaction rate's dependence on the concentration of the reactants. Since the 30*S* reconstitution reaction involves at least 20 components, one might expect the rate to have a very high order of dependence on their concentration—so that doubling the concentration would increase the rate by as much as 2^{20}, or more than a million times! To be sure, this would be an absurdity, since it would mean that it was necessary for all the components to collide simultaneously in order to form a complete subunit, and subunit formation would be an incredibly rare event. As a matter of fact, most chemical reactions, even those involving a number of reactants, turn out to have a first-order or second-order dependence on reactant concentration (that is, doubling the concentration doubles or quadruples the reaction rate). The reason is that most complex reactions proceed in steps, with one unimolecular rearrangement or bimolecular interaction being slower than the others and hence determining the overall rate.

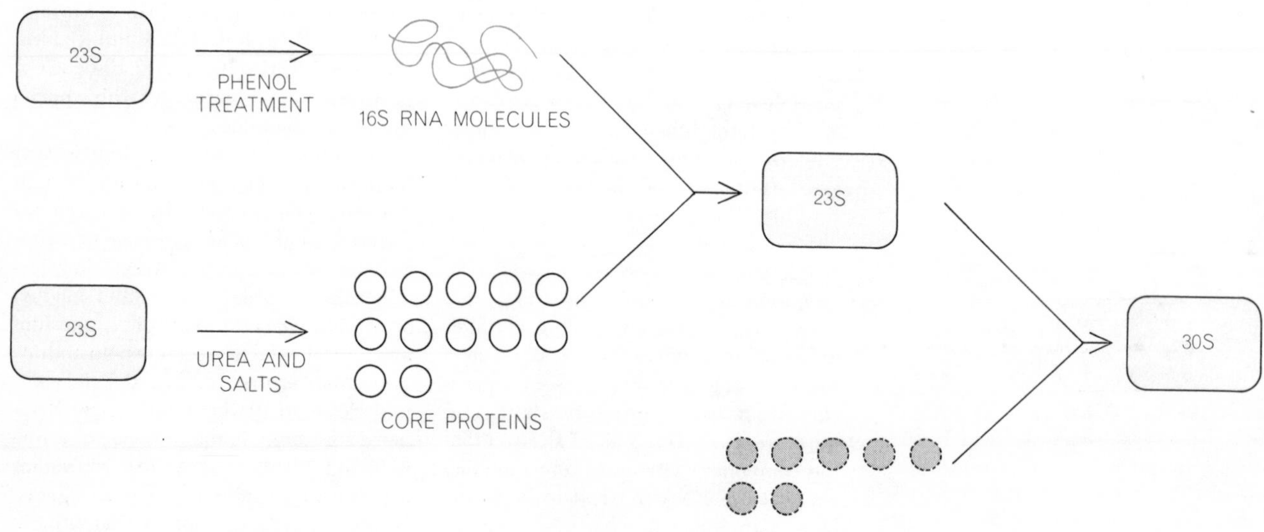

RECONSTITUTION of subunits was first accomplished as shown here. Core particles were treated with phenol to prepare 16*S* RNA and with urea and salts to yield the core proteins. The RNA and proteins were combined at 37 degrees Celsius to form 23*S* core particles. Then the reaction was completed by the addition of the split proteins, and in this way the 30*S* subunits were formed.

This generalization, however, has in the past applied to reactions involving far fewer components than are needed for reconstitution, and so we expected to find a somewhat higher order of reaction. We were rather surprised to observe that our assembly was in fact a first-order reaction; whether we double or halve the concentration of the reactants, the time it takes for all the components to assemble themselves into completed ribosomes is roughly the same—about five minutes under optimal conditions in the test tube. When one observes first-order kinetics in a reaction involving more than a single component, one can conclude that there is an intermediate step, involving a relatively slow rearrangement of a single component, that must take place before the reaction can be completed. This is exactly what we observe in the reconstitution reaction. The slow step in the assembly process may be occurring at any time before the binding of the 20 proteins to the RNA molecule is complete—after none of them or only some of them are bound—or after all are bound; in any case, our observation would be the same.

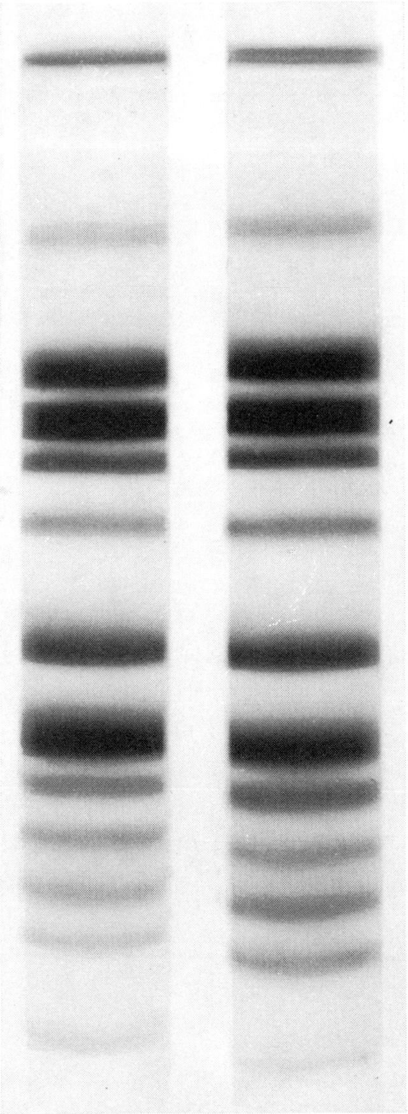

ALL THE PROTEINS normally found in a 30*S* subunit are found in a reconstituted 30*S* subunit, as shown by a comparison of electrophoresis results for the natural (*left*) and the reconstituted (*right*) particles. The protein mixtures are layered onto the top of a polyacrylamide gel column. When an electric current is applied, the proteins migrate down the column, each protein forming a band that moves at a different rate depending on the charge and the size of the molecule. Staining visualizes each protein.

We had noted with interest that the reaction is extremely dependent on temperature. It turned out that a considerable amount of heat energy is required to effect the rearrangement of the unknown intermediate product—about 40,000 calories per mole of ribosomes. On the other hand, many of the proteins attach themselves in an ice-cold solution. We were therefore able to isolate the intermediate, activate it by warming the solution and then observe the almost instantaneous binding of the rest of the proteins to form a completed ribosomal subunit. And so we can describe the general nature of the pathway of self-assembly: a rapid binding of some of the proteins to the RNA, a slow structural rearrangement of this intermediate that requires thermal energy and then a rapid binding of the rest of the proteins.

After obtaining all this information on assembly in the test tube, one comes to an obvious question: Do the same principles operate in the living cell? The problem of ribosome synthesis in living cells was being attacked long before test-tube assembly was even seriously considered. The early work was done by the group including Britten, Brian J. McCarthy and Roberts at the Carnegie Institution, and they were followed by a number of groups, notably Shozo Osawa's in Japan and David Schlessinger's at Washington University. In a series of intricate experiments they delivered short pulses of radioactive components to growing bacterial cells and monitored the flow of the labeled components into ribosomes. It was possible in this way to postulate the presence of several classes of precursor particles, but it was difficult to isolate and analyze them. The obvious limitations of such an experimental approach encouraged investigators to seek other directions. One was genetics. Genetics has been a powerful tool for identifying the flow of intermediates in numerous biosynthetic pathways because one of the easiest ways to find out how something works is to see what happens when it does not work. For example, it should be possible to isolate mutants that are defective in a specific step in the biosynthesis of ribosomes; the step reveals itself by the accumulation of the precursor whose conversion is blocked by the mutational defect.

What was needed was a systematic method of isolating mutants defective in ribosome assembly. The trouble is that since ribosomes are essential for growth such mutants are ordinarily inviable and cannot be cultured. They can be isolated only as "conditionally lethal mutants": cells with defects such that the organism is inviable under one condition but functions normally under some other condition. As I have mentioned, in our detailed study of the test-tube reaction we had been struck by the remarkable dependence of the reaction rate on temperature. If the same principle operated in living cells, we reasoned, then many mutational defects in ribosomes or in related components should manifest themselves more severely at lower temperatures, and so some assembly-defective mutants should be conditionally lethal—viable at high temperatures but inviable at lower temperatures. They could therefore be isolated as cold-sensitive mutants.

Our reasoning proved to be correct. Christine Guthrie, Hiroko Nashimoto and I have isolated a large number of cold-sensitive mutants of *E. coli*, a significant fraction of which appear to be defective specifically in ribosome assembly. (Independently John L. Ingraham and his co-workers at the University of California at Davis have found abnormal ribosome biosynthesis at low temperatures in cold-sensitive mutants of the related bacterium *Salmonella typhimurium.*) By sucrose-gradient sedimentation we have already identified three distinct classes of particles, from three different mutants, that accumulate in cells grown at 20 degrees C. Two of these particles appear to be precursors of 50*S* subunits; the third appears to be a 30*S* precursor. While proceeding with the biochemical characterization of these particles we are also conducting genetic analyses of the various mutants in the hope of obtaining information on the genetic organization and genetic control of the ribosome and of ribosome assembly. We hope that through the coupling of genetic techniques with the biochemical techniques of test-tube reconstitution this sophisticated and complex cellular instrument will soon be understood on a truly molecular level.

III

MEMBRANE STRUCTURE AND CELLULAR ENERGETICS

MEMBRANE STRUCTURE AND CELLULAR ENERGETICS III

INTRODUCTION

As discrete units of life, all living cells are separated from each other and from their external environment by a membranous structural boundary composed principally of lipids and proteins. Lipids are water-insoluble substances that include a number of important classes, such as triglycerides, phospholipids, and cholesterol. Other lipids of more specialized function found in higher eukaryotes include the steroid hormones (progesterone, testosterone), the vision pigment vitamin A-retinal, and the prostaglandins. The primary pigment for photosynthesis, chlorophyll, is necessarily present in any cell that carries out this function. Of the various classes, only phospholipids are present in all cellular membranes, and cholesterol is found specifically in the outer membrane of eukaryotic cells. Different biological membranes differ in their relative content of lipids and proteins as well as in the types of lipids and the particular proteins embedded in the lipid matrix. In addition to the outer membrane boundary that all cells have, eukaryotes possess an intricate internal network of membranes, including the endoplasmic reticulum, that subdivide the cells into compartments and organelles. The subcellular organelles, such as mitochondria and chloroplasts, carry out various specialized biochemical tasks, of which energy transduction is most notable. The lysosome, a sort of molecular stomach in the cell, encloses enzymes for digesting incorporated foodstuffs. Other enclosures, such as the Golgi apparatus, are centers for the production and release of secretory materials (e.g., digestive enzymes in certain specialized cells). Membranes serve as selective barriers that permit the interior composition of the cell and its organelles to be controlled over a wide range of external environments. Salts, nutrients, toxic by-products, and especially water are moved in or out of the cell in response to its metabolic needs and the osmotic conditions in the cellular environment. Although a few of the transport properties of the membrane are passive, the movement of most solutes is mediated by protein "carriers" that ferry selected hydrophilic molecules through the hydrophobic barrier. Many of these transport processes require the expenditure of metabolic energy. The regulation of the transport properties of membranes is complex and a marvel in its own right.

Membranes are much more than selective barriers, however. They have profound roles in cellular organization and intracellular transport, and the outer cell membranes carry unique protein identification markings, as already discussed in Mazia's article in the first section. Many of the ribosomes are organized along the endoplasmic reticulum, and even DNA replication is thought to occur at sites on membranes. Certain classes of metabolic activity are catalyzed by organized complexes of enzymes that have assembled in the hydrophobic environment provided by the lipid sheath. In particular, those reactions involving electron transfer, as in the oxidation of nutrients for energy production, occur at sites on membranes. The membrane serves as a matrix to organize the enzyme components in appropriate juxtaposition for the stepwise series of reactions in which energy in useable form is extracted from foodstuffs. The energy currency of life, ATP, is synthesized on the inner membranes of mitochondria. Although it is very difficult to study the enzymatic reactions that occur in close association with membranes, this field of so-called "solid-state" biochemistry has become a thriving area of investigation. The compartmentalization provided by membranes reduces the distances that intracellular substances must move by free diffusion, and it permits "local" high concentrations of specific molecules to exist at particular sites where they are needed in the cell. Furthermore, the existence of extensive membrane surfaces may reduce the traffic problem of random three-dimensional diffusion to a problem in only two dimensions and thus further accelerate the shuffling of metabolites to appropriate intracellular sites. We still have much to learn about the intrinsic transport properties and metabolic organization facilitated by membranes in living cells.

Most of the lipids in membranes are amphipathic; that is, they contain a polar "head group" as well as a hydrophobic "tail." In aqueous solutions these molecules naturally assemble in such a way that the hydrophobic regions come together and the polar groups are exposed to the solvent. (This you will recognize as a general feature of molecular behavior in polar solvents.) These molecules will also float on the surface of water, where they will orient with their tails up and their heads down. The notion that lipid elements might form a double layer around cells was first suggested in 1925, when it was found that the surface area of lipid films prepared from red blood cell membranes amounted to roughly twice the total area of the original cell surface. (Ironically, the experiment that gave this result was filled with errors and artifacts that cancelled one another.) Various models for biological membranes have been proposed. In the first to include protein as a constituent, the protein molecules were believed to form a coating on each side of the lipid bilayer (the unit-membrane hypothesis). With the development of more sophisticated electron microscopy, it was shown that in fact the proteins generally lie within the lipid bilayer, some with exposed portions but others totally submerged. The current "fluid-mosaic model" additionally takes into account the mobility of proteins and other components within the membrane. The recent spectacular advances in our understanding of membrane structure and function have been made possible by the development of certain methods of preparing tissue for electron microscopy, such as freeze-fracturing, as well as the application of new techniques for isolating intact hydrophobic proteins. Additionally, the application of the biophysical techniques of microcalorimetry and electron-spin resonance using spin-label probes have been essential to our current understanding of membrane dynamics.

In "A Dynamic Model of Cell Membranes," Roderick A. Capaldi demonstrates that the sac that surrounds the cell substance is as much a part of the living entity as is the interior. Just as proteins continuously change configuration to carry out their tasks, membranes are also constantly in motion. The lipids in membranes are by no means solid, but maintain the "consistency of light oil." Lateral diffusion of some hydrophobic proteins can therefore occur within the membrane, although others appear to be held in fixed sites by lattice structures. Capaldi describes the principal factors that influence the fluidity of membranes—namely, the degree to which the carbon atoms in the lipid "tails" are saturated with hydrogen and, not surprisingly, the temperature. It is possible to grow cells with different membrane fluidities by varying the saturation of the fatty acid precursors that they are fed.

The discussion of membrane structure is carried further by Harvey F. Lodish and James E. Rothman in their article on "The Assembly of Cell Membranes," in which they show that the inside and the outside of biological membranes are distinguishable. The distribution of proteins in membranes is controlled in different ways, and no general rules seem to apply. One factor that may be important in the asymmetric arrangement of some proteins in the outer membrane is that they apparently enter it from inside the cell after they are synthesized. When added from the outside, a protein will assume the reverse orientation; this is observed experimentally when proteins are added to reconstituted membranes. How does a protein "know" that it should be incorporated into a membrane? Lodish and Rothman describe evidence for a signal sequence of amino acids in the polypeptide that draws it to the membrane as it emerges from the ribosome. It is embedded in lipid even before its synthesis is complete. Not surprisingly these signal sequences, which are sometimes removed after their task is completed, contain a preponderance of hydrophobic amino acids. This mechanism may operate in the case of lysozyme, as we discussed in Section II. The article by Lodish and Rothman brings us back to the subject of viruses, as their research on cell membranes was done with the vesticular stomatitis virus, which conveniently buds off a portion of the host membrane to form its outer coat. The utility of viruses as simple test systems is illustrated once again; in this case, advantage is taken of the relatively few viral-specified membrane proteins rather than attempting analysis of the thousands of different proteins associated with the host membrane. The detailed example of the synthesis of the viral G protein provides a fascinating story of the necessary sequence of steps for properly placing a particular protein in a particular membrane. These steps include a complexity of coordinated reactions, some of which are spontaneous whereas others require enzyme catalysis. Finally, the authors discuss how and where the lipids themselves are synthesized. We learn that the lipids are synthesized within the membrane itself, using enzymes that have already been incorporated into the existing membrane structure.

The way in which electron transfer is coupled to the synthesis of ATP in living cells is discussed by Peter C. Hinkle and Richard E. McCarty in "How Cells Make ATP." This is one of the more sophisticated articles in this collection—a challenge to the more eager students. According to Peter Mitchell's chemiosmotic theory, ATP is generated by a gradient in proton concentration across a membrane. Essential tools for the study of chemiosmotic schemes are the ionophores, molecules originally identified by their uncoupling of phosphorylation from respiration, but now known to disrupt proton gradients by transporting ions through the membrane. In the final article in this section, "The Photosynthetic Membrane," Kenneth R. Miller takes us back to the primary source of all energy for the biosphere—the radiations from the sun. The energy in light is harnessed to transfer hydrogen atoms from water to carbon dioxide with a net release of oxygen. The sunlight is captured by chlorophyll and by accessory "antennae pigments" (e.g., carotenoids) to raise electrons to energetically excited states. In chloroplasts the pigments that absorb the light are located in the membrane, and it is the special nature of this thylakoid membrane that enables it to harness the energy of the excited electrons. The article includes excellent diagrammatic representations of both the light and dark reactions of photosynthesis. The technique of freeze-fracturing for the preparation of membranes for analysis by electron microscopy is also described in intricate detail with elegant illustrations.

In this section we have considered in molecular detail the biochemical processes for converting the radiant energy of sunlight into ATP for use in metabolic activity. But what of the energy needs of a living cell? How is the ATP utilized? Hinkle and McCarty discuss the studies of Howard Berg at the University of Colorado, which show that the "motor" for the motility of bacteria is driven directly by a proton gradient without the intermediate of an ATP battery. However, most cellular activity requires energy that is obtained directly or indirectly through coupled reactions in which one or both of the terminal phosphates are removed from ATP. Some of the ATP energy is channeled to specific biosynthetic pathways by using it to charge other nucleotide batteries. Thus uridine diphosphate is charged by ATP to yield uridine triphosphate (UTP), which in turn is coupled energetically to the synthesis of polysaccharides from simple sugars. Likewise, cytidine triphosphate (CTP) is generated for an analogous role in the synthesis of such lipids as lecithin, and guanosine triphosphate (GTP) and ATP are both utilized in protein synthesis. Of course all four of these compounds, ATP, UTP, CTP, and GTP, are also the activated precursors for RNA synthesis, and corresponding activated monomers are utilized in DNA synthesis, as detailed in Section V. It is evident that very intricate control systems must operate to channel the energy to those biochemical processes where it is needed while also maintaining adequate pools of the respective nucleotide precursors for nucleic acid synthesis. A growing bacterium maintains a pool of about a million ATP molecules, but it needs twice that many *per second* for protein synthesis alone. Thus the turnover rate of ATP must be extremely high. Since the addition of each amino acid to a polypeptide chain requires the expenditure of four high-energy phosphate bonds, our typical protein of 200 amino acids requires the energy supplied by 800 ATP molecules. That in turn could be furnished by the systematic oxidation of nearly 30 glucose molecules. The underlying principle in all biosynthetic processes is essentially the same: the terminal phosphate groups of ATP are used to activate each building-block molecule so that it may react with the next building block in a thermodynamically favorable manner.

No new physical laws are required to explain the buildup of molecular order that characterizes living systems. How, then, do we reconcile this statement with an apparent contradiction with the second law of thermodynamics, which asserts that the universe must be tending toward maximum disorder or entropy? The resolution of the conflict lies in the fact that the second law of thermodynamics deals with systems at equilibrium, whereas the living cell is in a steady state condition: there is a net flow of both matter and energy through the living cell. This flow of energy is in fact a self-organizing principle at both macroscopic and molecular levels according to Morowitz (1968). When the flow of matter and/or energy stops, the cell dies and entropy does increase as processes of molecular dissolution methodically dismantle the complex components of life.

REFERENCES CITED AND SUGGESTED FURTHER READING

Boyer, P. D., B. Chance, L. Ernster, P. Mitchell, E. Racker, and E. C. Slater. 1977. "Oxidative phosphorylation and photophosphorylation." *Ann. Rev. Biochem.* **46**:955–1026.

Bretscher, M. S. 1973. "Membrane structure: Some general principles." *Science* **181**:622–629.

Capaldi, R. A. 1974. "The structure of mitochondrial membranes." In G. Jamieson and D. Robinson, *Mammalian Cell Membranes* (Vol II). Butterworth's, London.

Fox, C. F. 1972. "The structure of cell membranes." *Scientific American*, February. (Offprint No. 1241.)

Hall, D. O., and K. K. Rao. 1972. *Photosynthesis: Studies in Biology*. E. Arnold, Ltd., London.

Jost, P. C., O. H. Griffith, R. A. Capaldi, and G. Vanderkooi. 1973. "Evidence for boundary lipid in membranes." *Proc. Nat. Acad. Sci.* **70**(2):480–484.

Lehninger, A. L. 1975. *Biochemistry* (2nd ed.). Worth, New York.

Miller, K. R. 1976. "A particle spanning the photosynthetic membrane." *J. Ultrastruc. Res.* **54**(1):159–167.

Morowitz, H. J. 1968. *Energy Flow in Biology*. Academic Press, New York.

Morowitz, H. J. 1978. *Foundations of Bioenergetics*. Academic Press, New York.

Quinn, P. J. 1976. *The Molecular Biology of Cell Membranes.* University Park Press, Baltimore, MD.

Racker, E. 1976. *A New Look at Mechanisms in Bioenergetics.* Academic Press, New York.

Rothman, J. E., and E. P. Kennedy. 1977. "Rapid transmembrane movement of newly synthesized phospholipids during membrane assembly." *Proc. Nat. Acad. Sci.* **74**(5):1821–1825.

Rothman, J. E., and J. Lenard. 1977. "Membrane asymmetry." *Science* **195:**743–753.

Rothman, J. E., and H. F. Lodish. 1977. "Synchronized transmembrane insertion and glycosylation of a nascent membrane protein." *Nature* **269:**775–780.

Singer, S. J., and G. Nicolson. 1972. "The fluid mosaic model of the structure of cell membranes." *Science* **175:**720–731.

Singer, S. J. 1974. "The molecular organization of membranes." *Ann. Rev. Biochem.* **41:**731–752.

Staehelin, L. A., P. A. Armond, and K. R. Miller. 1976. "Chloroplast membrane organization at the supramolecular level and its functional implications." *Brookhaven Symp. Biol.* **28:**278–315.

Tanford, C. 1973. *The Hydrophobic Effect: Formation of Micelles and Biological Membranes.* Wiley-Interscience, New York.

Wickner, W. 1979. "The assembly of proteins into biological membranes: The membrane trigger hypothesis." *Ann. Rev. Biochem.* **48:**23–45.

10

A Dynamic Model of Cell Membranes

by Roderick A. Capaldi
March 1974

The envelopes that surround entire cells, cell nuclei and the various cell organelles are thin assemblies of lipid and protein molecules. Their functions depend on how the membrane proteins are linked

The fundamental unit of living tissue is the cell. In recent years it has become plain that one cellular component, the membrane, plays a crucial role in almost all cellular activity. Cytoplasmic membrane, the outer envelope surrounding the cell, acts to regulate the internal environment of the cell and to transport substances into and out of it. Internal membranes, which enclose the nucleus of the cell and such cell organelles as microsomes, mitochondria and the chloroplasts of plants, play an equally important role. For example, the mitochondrial membrane is where adenosine triphosphate (ATP) is manufactured; hence this membrane supplies the fuel for all the cell's metabolic processes. Similarly, the chloroplast membrane is the site of photosynthesis, the process by which energy from the sun is trapped in a form that can be used by living cells. How then are membranes built to accomplish so many different tasks?

A good deal of information now exists concerning the basic structure of membranes. One fact to emerge recently is that cytoplasmic and internal membranes are essentially alike; both are composed of proteins and the fatty substances called lipids. In mammalian cells small amounts of carbohydrate are also present, associated either with protein as glycoproteins, that is, carbohydrate-bearing proteins, or with lipid as glycolipids.

Lipids account for about half of the mass of most membranes. In internal membranes the lipid is almost exclusively phospholipid. Cytoplasmic membranes, in contrast, contain both glycolipid and neutral, or uncharged, lipid in addition to phospholipid. For example, as much as 30 percent of the lipid in the membrane of red blood cells consists of one type of neutral lipid: cholesterol.

Individual lipid molecules have a head and two tails [*see illustration on page 132*]. At the point where the head and the tails meet, which C. Fred Fox of the University of California at Los Angeles calls the backbone, is a glycerol group. The tails that descend from the backbone are extended chains of fatty acids. The structure of these chains is quite similar to that of oil molecules and, just as oil and water tend to separate into different layers when they are mixed, so do the tails of phospholipid molecules tend to point away from water. Hence they are said to be hydrophobic. The heads of the phospholipid molecules, on the other hand, are soluble in water and are said to be hydrophilic. Molecules of this kind, with one hydrophobic end and one hydrophilic, are called amphipathic. Glycolipids and to some extent neutral lipids are also amphipathic.

The lipids in membranes are arranged so as to accommodate their amphipathic character. They form a bilayer, two layers back to back, so that their hydrophilic heads constitute the top and bottom surfaces of the membrane and their hydrophobic tails are buried in the membrane interior [*see upper illustration on page 133*]. The lipid bilayer is a sheet about 45 angstroms thick. It is the structural framework of the membrane. It is also the anchorage for the other major component of membranes: protein.

Proteins and glycoproteins play a variety of roles in membranes. They can contribute to the structural integrity of the membrane, they can act as enzymes or they can function as pumps, moving material into and out of cells and organelles. It is the diversity of its protein activity that gives each particular membrane its distinctive character.

Just which proteins are present in which membranes can be determined in various ways. One is to assay the membrane's various enzymic activities. Another is to identify proteins by molecular weight, utilizing the technique of gel electrophoresis. Proteins are made up of the long chains of amino acids called polypeptides. Some proteins have only a single polypeptide chain; others have many polypeptide chains tightly associated with one another. In preparation for gel electrophoresis a protein is broken down into its component polypeptide chains by exposure to a detergent, sodium dodecyl sulfate. The chains are then transferred to a polyacrylamide gel with an electric potential across it. They migrate through the gel in response to the potential at a rate proportional to their molecular weight; the lower the weight of the polypeptide, the farther it moves. The gel is then stained with a protein-specific substance, for example coomassie brilliant blue, and is scanned for absorbance. When, for example, the proteins associated with the membrane of the red blood cell are identified in this fashion, the scanning trace reveals polypeptides with molecular weights ranging from 255,000 to 12,500.

The two heaviest polypeptide components, with molecular weights of 255,000 and 220,000, are collectively known as spectrin. (Vincent T. Marchesi of the Yale University School of Medicine chose the name because he first isolated the components from "ghosts," the membranes of red blood cells that have been chemically deprived of their hemoglobin.) Spectrin accounts for about a third of all the protein in the red-cell membrane. Another third of the protein lies in a diffuse absorption band with a molecular weight of about 90,000. This band contains a number of different polypeptides, including a component with a molecular weight of 87,000 pres-

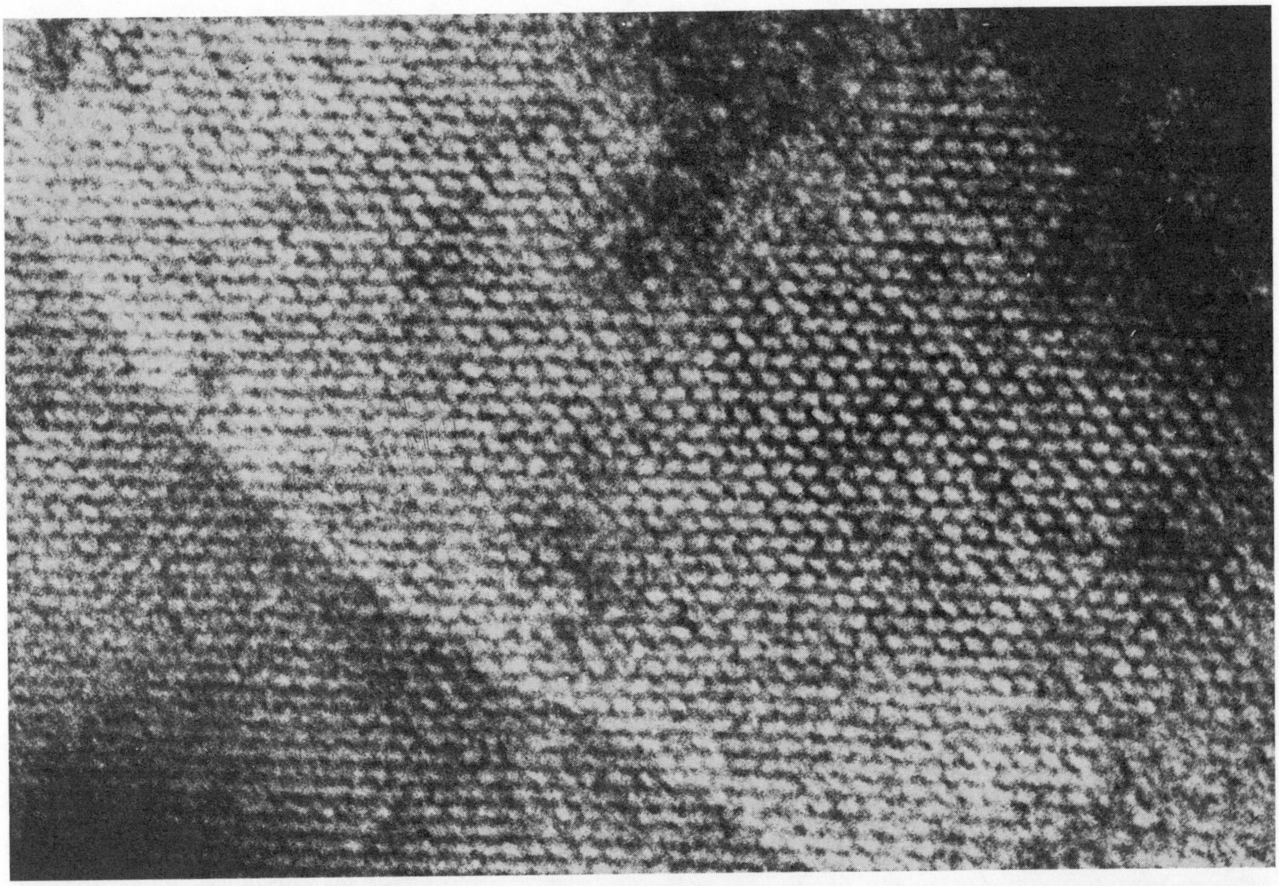

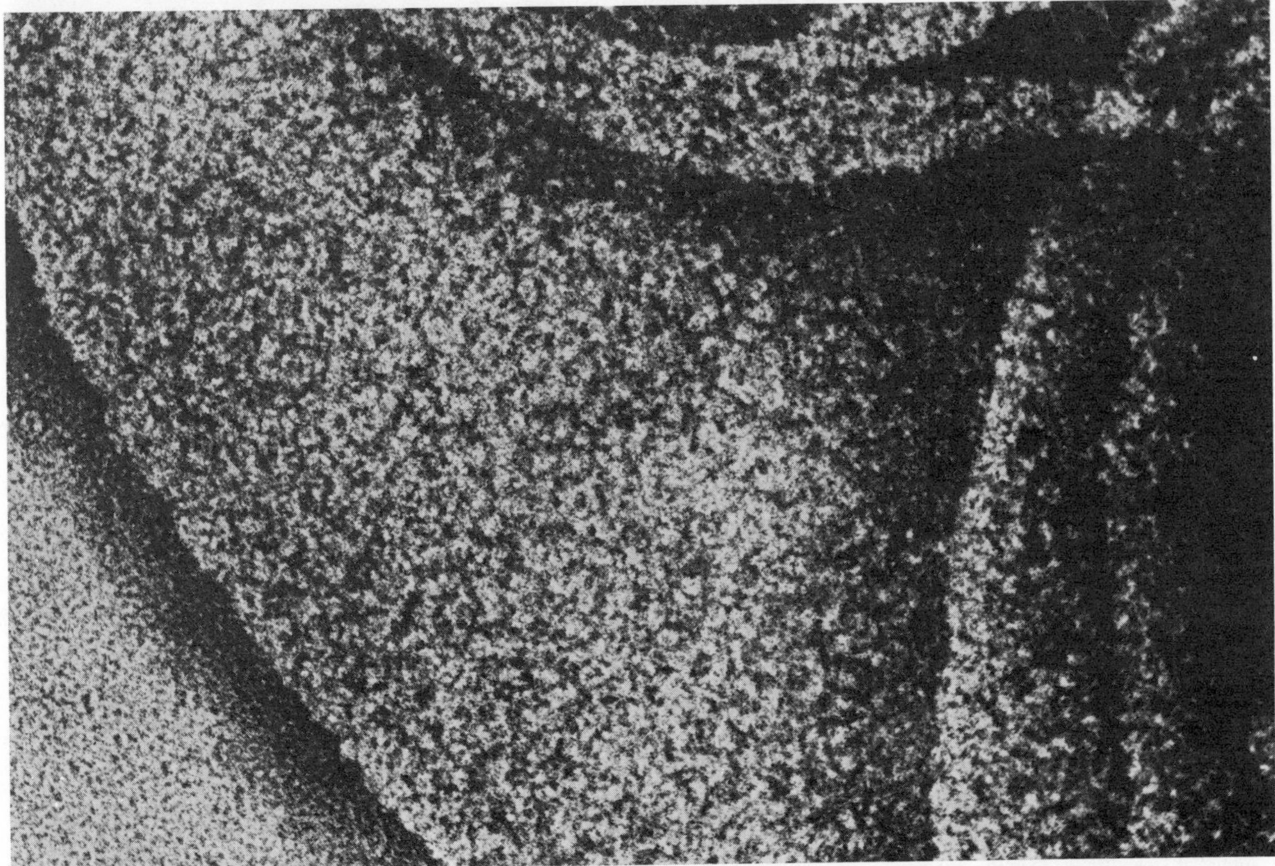

CYTOCHROME OXIDASE, a protein present in the membrane of mitochondria, can be isolated for examination under the electron microscope, as seen here, and for X-ray-diffraction studies. In its oxidized state (*top*) the protein is organized into a crystalline lattice. The lattice structure is lost when the protein is in its reduced state (*bottom*), but the structure reappears when the reductant is removed. Measurements show that the cytochrome oxidase molecule is oval in cross section and its main axis is 80 to 85 angstroms long.

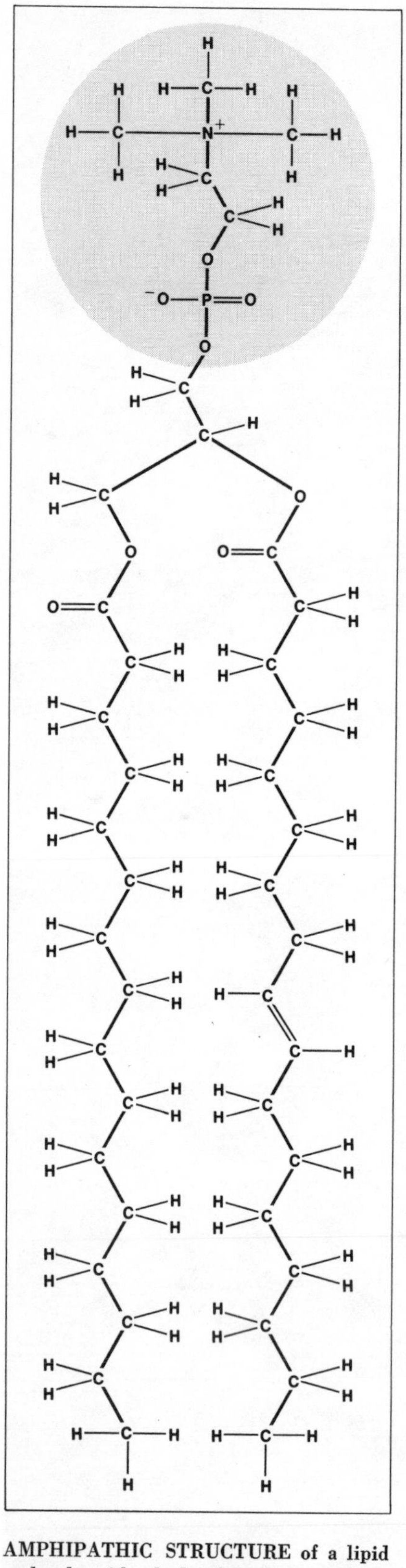

AMPHIPATHIC STRUCTURE of a lipid molecule, with a hydrophilic head and twin hydrophobic tails, is exemplified by this typical phospholipid, specifically a molecule of phosphatidylcholine. Various lipid molecules comprise about half of the mass of mammalian membrane, forming the membrane's structural framework. Their fatty-acid tails may be saturated (*left*), with a hydrogen atom linked to every carbon bond, or unsaturated (*right*), with carbons free.

ent in equal copies with each spectrin molecule. Proteins of molecular weight lower than 70,000 make up the remaining third of the protein in the membrane. The red-cell membrane may be unusual in the large amount of protein of high molecular weight that it incorporates. By way of comparison, almost all the polypeptides in mitochondrial membrane are below 70,000 in weight.

Membrane proteins can be divided into two classes depending on their location with respect to the lipids of the membrane framework. One class consists of those protein molecules that are associated only with the membrane surface. These "extrinsic" proteins are located adjacent to either the outer or the inner surface of the membrane. The second class is made up of proteins that actually penetrate the membrane surface. These "intrinsic" proteins enter the lipid bilayer and sometimes extend all the way through it [*see upper illustration on opposite page*].

Whether a membrane protein should be assigned to one or the other of the two classes can be determined on the basis of its chemical properties or on the basis of other kinds of analysis, such as X-ray diffraction or electron microscopy. For example, extrinsic proteins are relatively easy to remove from membranes by chemical dissociation methods, whereas intrinsic proteins form an integral part of the membrane continuum and are much more difficult to dislodge.

Two extrinsic proteins that are visible in electron micrographs are the enzyme ATPase, found in mitochondrial membrane, and spectrin, the polypeptide in red-cell membrane. Objects termed "headpieces" are visible sticking up from the membrane surface in electron micrographs of mitochondrial membrane; they are ATPase molecules. Similarly, in electron micrographs of red-cell ghosts the "fuzz" lining the inside of the membrane is composed of spectrin polypeptides.

One intrinsic protein that has been studied in detail is rhodopsin, the only protein present in the membranes of the disks that occupy the outer segments of the rod cells of the retina. J. Kent Blasie and his colleagues at the University of Pennsylvania School of Medicine, working with X-ray-diffraction techniques, have found that the rhodopsin molecule is globular and some 42 angstroms in diameter. When the retinal rods are in darkness, the rhodopsin molecules of the disk membrane are submerged for about a third of their diameter in the membrane's outer surface. When the rods are illuminated, the rhodopsin molecules sink deeper into the membrane until they are about half-submerged. Even then, however, the molecules have penetrated less than halfway through the bilayer.

Working in David E. Green's laboratory at the Institute for Enzyme Research at the University of Wisconsin, my colleagues and I have examined the organization of another intrinsic protein: cytochrome oxidase, an enzyme in mitochondrial membrane that is the terminal member of the electron-transfer chain involved in the synthesis of ATP. Now, one stumbling block in the path of membrane-protein research is the fact that most membranes contain a heterogeneous mixture of proteins, including both extrinsic and intrinsic proteins. Most of the methods that can be harnessed to examine the structure of membranes, such as X-ray diffraction, are averaging methods; the resulting data give only the average properties of all the proteins in the sample, whereas we really want to know the characteristics of individual membrane proteins. This is one reason why retinal-disk membrane, with its single protein rhodopsin, is a popular subject of investigation.

Fortunately for us it is possible to separate cytochrome oxidase from the other proteins in mitochondrial membrane. When the separated enzyme is placed in suspension with lipids, the lipids and the enzyme interact and gather in saclike vesicles that are in effect man-made membranes. The molecules of cytochrome oxidase in the artificial vesicles have the same enzymic properties that they show in normal mitochondrial membrane, and so it seems a good bet that the vesicles have the same structure as normal membrane. The advantage here is that a heterogeneous array of proteins has been reduced to a single protein.

We have used these membrane vesicles as a model system for the study of the isolated protein in its geometrical relations with the lipid bilayer. In its capacity as an electron-transfer substance cytochrome oxidase exists in one of two states, either oxidized or reduced. In the oxidized state (and in a narrow range of lipid-to-protein ratios) the enzyme is organized into a crystalline lattice that is visible in the electron microscope and can also be analyzed by X-ray diffraction [*see illustration on page 131*]. Utilizing both kinds of data, we found that individual molecules of cytochrome oxidase are some 55 angstroms long, 60 angstroms wide and 80 to 85 angstroms deep. This is enough depth to allow the molecule to penetrate the 45-angstrom bilayer completely, leaving one end jut-

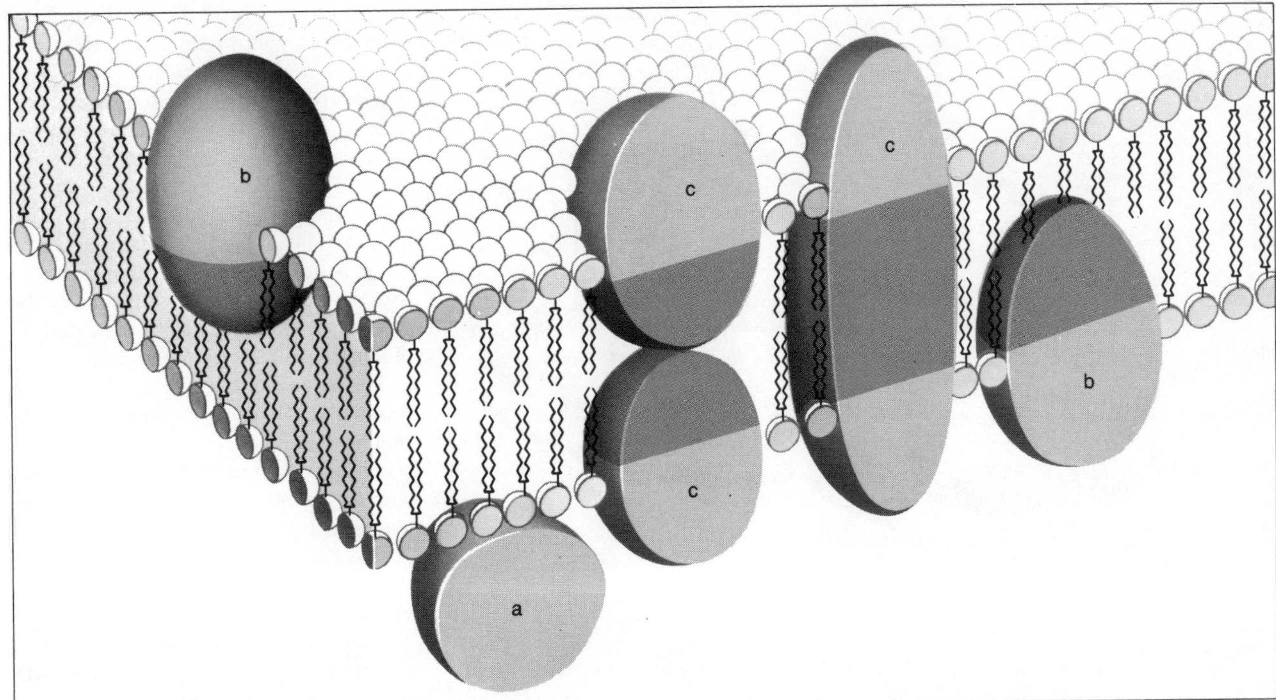

STRUCTURAL FRAMEWORK typical of cell membranes is made up of a bilayer of lipids with their hydrophilic heads forming outer and inner membrane surfaces and their hydrophobic tails meeting at the center of the membrane; the bilayer is about 45 angstroms thick. Proteins, the other membrane constituents, are of two kinds. Some (*a*) lie at or near either membrane surface. The others penetrate the membrane; they may intrude only a short way (*b*) or may bridge the membrane completely (*c*), singly or in pairs.

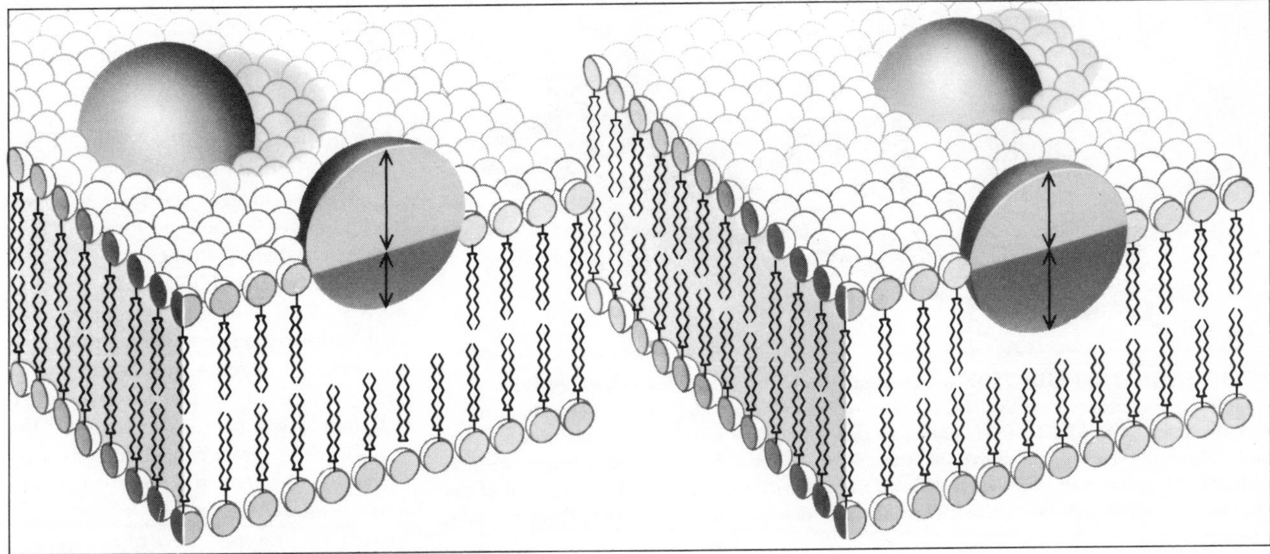

EXTENT OF PENETRATION can vary according to circumstance. Rhodopsin, the only protein in the membrane of retinal-rod disks, has been shown by J. Kent Blasie and his colleagues at the University of Pennsylvania School of Medicine to be a globular molecule 42 angstroms in diameter. In darkness a third of it is submerged in the membrane lipids (*left*); illuminated, it is half-submerged.

ting up from the exterior surface of the membrane and the other similarly exposed on the interior surface. Here is a membrane-penetrating protein that spans the bilayer by itself.

Intrinsic proteins have an environment quite unlike that of extrinsic or cytoplasmic proteins that are essentially surrounded by water. Only part of an intrinsic protein is exposed to water; the rest is essentially immersed in oil. In order to be stable in this unusual dual environment the protein must be amphipathic, as phospholipid molecules are. The parts of the molecule exposed to water must hold the majority of the seven hydrophilic amino acids: lysine, histidine, arginine, aspartic acid, glutamic acid, serine and threonine. The parts of the molecule buried in the lipid bilayer, in turn, must be mainly covered with hydrophobic amino acids.

The more deeply buried a protein is in the bilayer, the less water-surrounded surface there is to accommodate the hydrophilic amino acids in the molecule. It is therefore not surprising that many intrinsic proteins contain relatively few hydrophilic amino acids; instead they tend to be hydrophobic in character. For example, Alexander Tzagoloff of the Public Health Research Institute of the City of New York has isolated one in-

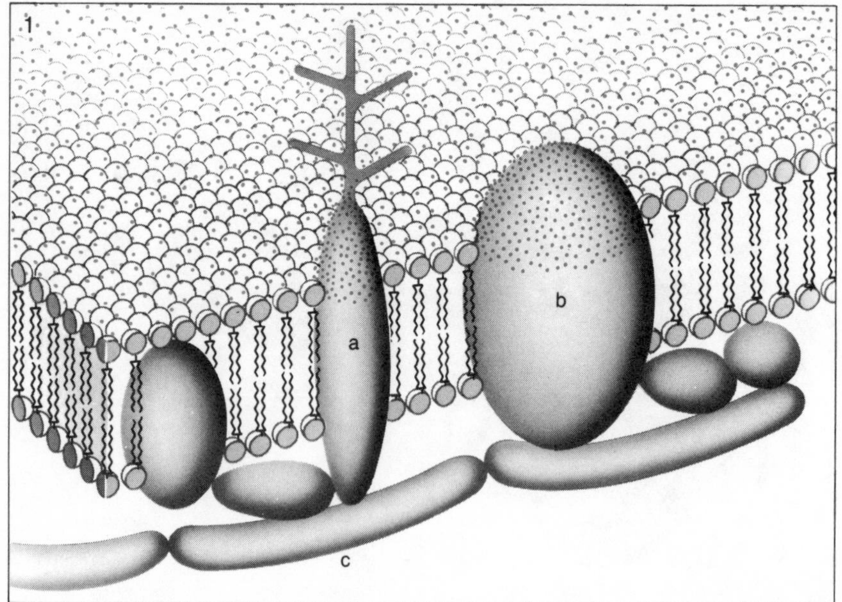

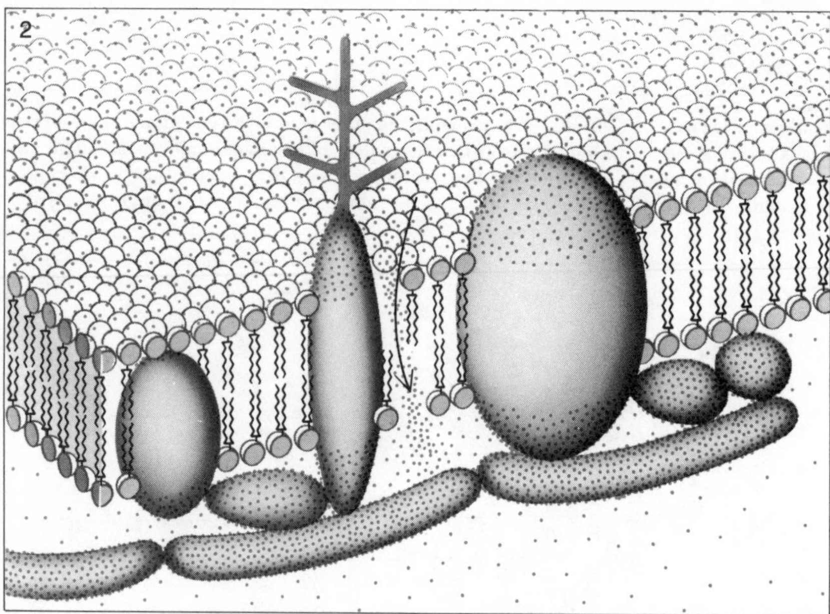

ASYMMETRIC DISTRIBUTION of the protein molecules associated with red-blood-cell membrane is demonstrated by labeling the proteins. When a chemical reagent that cannot pass through the lipid bilayer is applied to the cell surface (*1*), only two penetrating proteins with ends that extend above the surface (*a, b*) are labeled. When a lysing agent is added, making the membrane leaky (*2*), all the molecules on or near the other side of the membrane are labeled, including the spectrin (*c*) that probably holds the others in place.

trinsic protein of the inner membranes of mitochondria with a molecular weight of 10,000. Its amino acid content is 20 percent hydrophilic and 80 percent hydrophobic. This is in sharp contrast to cytoplasmic and extrinsic membrane proteins, which have on the average 47 percent hydrophilic amino acids and 53 percent hydrophobic amino acids. Another very hydrophobic intrinsic protein is the Folch-Lees protein, which can be isolated from the myelin covering of sciatic nerve; its amino acids are 29 percent hydrophilic and 71 percent hydrophobic. The intrinsic proteins rhodopsin and cytochrome oxidase are both hydrophobic; the amino acids of rhodopsin are 36 percent hydrophilic and 64 percent hydrophobic and those of cytochrome oxidase are 37 percent hydrophilic and 63 percent hydrophobic.

Two interesting intrinsic proteins have recently been characterized that, while they are not excessively hydrophobic in overall composition, have polypeptide-chain regions that are very rich in hydrophobic amino acids. One is cytochrome b_5, a protein isolated from the microsomal membrane of liver cells. Phillip Strittmatter and Lawrence Spatz of the University of Connecticut have shown that this protein, a single polypeptide, is folded at one end into a globular portion that is exposed at the surface of the membrane and is covered predominantly with hydrophilic amino acids. The polypeptide chain continues out of the globular portion into a "tail" of about 60 amino acids, almost all of them hydrophobic. The tail penetrates into the bilayer and serves to anchor the globular and enzymically active portion of the molecule to the membrane.

The second intrinsic protein with an unusual structure is the major glycoprotein found in the membrane of red blood cells. This molecule has been closely studied by Marchesi and various colleagues, first at the National Institutes of Health and more recently at Yale. Again it consists of a single polypeptide chain. One end of the chain, which holds all the carbohydrate associated with the molecule, consists predominantly of hydrophilic amino acids; this end is exposed to the water at the outer surface of the cell membrane. The other end of the chain, which also incorporates hydrophilic amino acids, extends into the watery interior of the red cell. The middle of the chain consists of some 30 amino acids. They are almost exclusively hydrophobic and lie inside the lipid bilayer of the membrane.

Because all the carbohydrate of the glycoprotein molecule is exposed at the outer surface of the red-cell membrane the membrane is asymmetric. Furthermore, labeling studies, using chemical reagents to label "available" proteins, show that the asymmetry in the red-cell membrane extends beyond carbohydrate imbalance. Reagents that cannot penetrate the lipid bilayer of the membrane will label two of the proteins in the membrane of intact red blood cells. One is the glycoprotein; the other is the protein of molecular weight 87,000; only those two proteins are exposed at the outer surface of the membrane. When the red cell is lysed and thereby made leaky to the labeling reagent, however, all the proteins in the membrane are labeled, indicating that the majority of proteins in the red-cell membrane are localized on the membrane's interior surface.

The red-blood-cell membrane is not the only one with an asymmetric organization. Labeling techniques have shown that the mitochondrial inner membrane is organized in a similar fashion. The protein molecules known as headpieces, actually ATPases, are located exclusively on the matrix, or inner, side of the membrane and the cytochrome *c* molecules are found only on the intracristal, or outer, surface. As one would expect, the molecules of cytochrome oxidase, which are exposed on both the intracristal and the matrix surfaces of the membrane, can be labeled from either side.

To review the membrane picture as it has been described so far, we see that the lipids, which account for roughly half of the mass of a membrane, are organized into a thin bilayer and that the membrane proteins are either perched on or near the two sides of the bilayer or penetrate into or completely through it. This picture is accurate but incomplete. One of the major advances in cell studies during the past few years has been the realization that membranes are by no means static. Both the lipids and the proteins have considerable freedom of movement.

Considering the mobility of the lipids first, whether a lipid bilayer is fluid or rigid depends on two factors: first, the extent of saturation of the lipid tails (that is, the extent to which all the available carbon bonds carry hydrogen atoms), and second, the ambient temperature. Now, a considerable proportion of the lipid in mammalian-cell membrane is unsaturated, so that the melting temperature for the bilayer is below the normal mammalian body temperature. Thus the bilayer is fluid and the fatty-acid tails of the lipid molecules are free to move. The freedom of movement has been studied in some detail by Harden M. McConnell and his colleagues at Stanford University and by O. Hayes Griffith and his colleagues at the University of Oregon. Both groups have used the same analytical technique: electron spin-resonance spectroscopy. The method involves attaching a "reporter" group, usually a nitroxide group that has an unpaired electron, to one of the carbons of a test molecule's fatty-acid tail.

The test molecule is either stearic acid or a phospholipid and the reporter group is usually attached to the fifth, 12th or 16th carbon atom along the tail. Several labeled test molecules are then inserted in the membrane bilayer that is to be examined; they fit in much as ordinary lipid molecules do. The inserted group is very sensitive to motion, however, and the spectroscopic trace unmistakably records any movement of the portion of the fatty-acid tail where the reporter group is attached. By fixing reporter groups at different positions along the tail one can determine just what degree of mobility exists where in the bilayer.

That is essentially what McConnell's

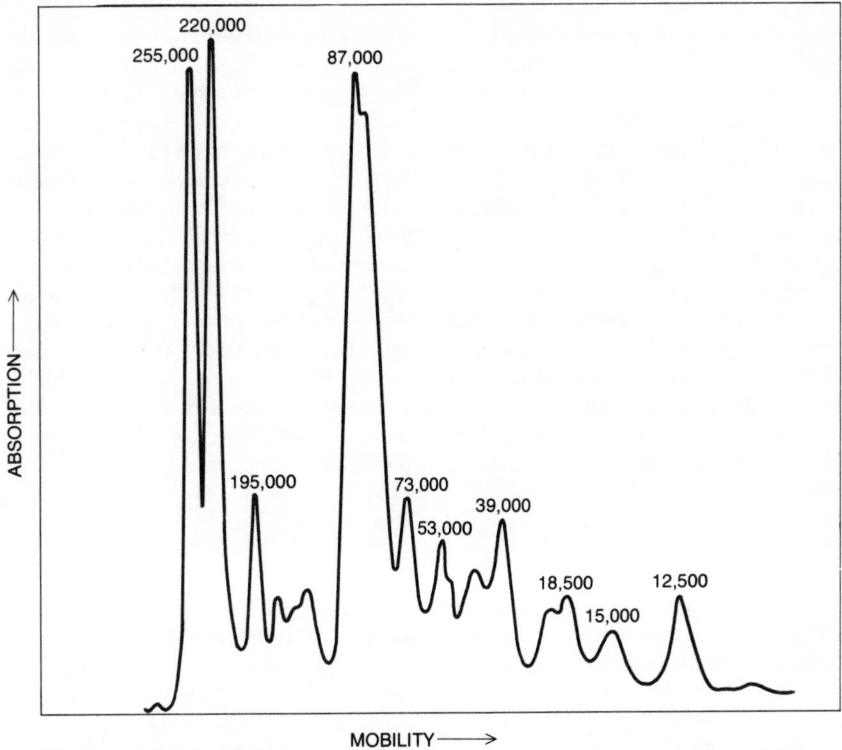

IDENTITY OF PROTEINS in the membrane of red blood cells can be determined by means of gel electrophoresis. In this instance the proteins have been stained and scanned, using a densiometer, after those with the lowest molecular weights have migrated farthest through the gel in response to an electric potential. The two proteins with the highest molecular weight (*left*) form the dimer collectively called spectrin. The scan shows at least 10 more absorption bands signifying the presence of other proteins even lower in weight.

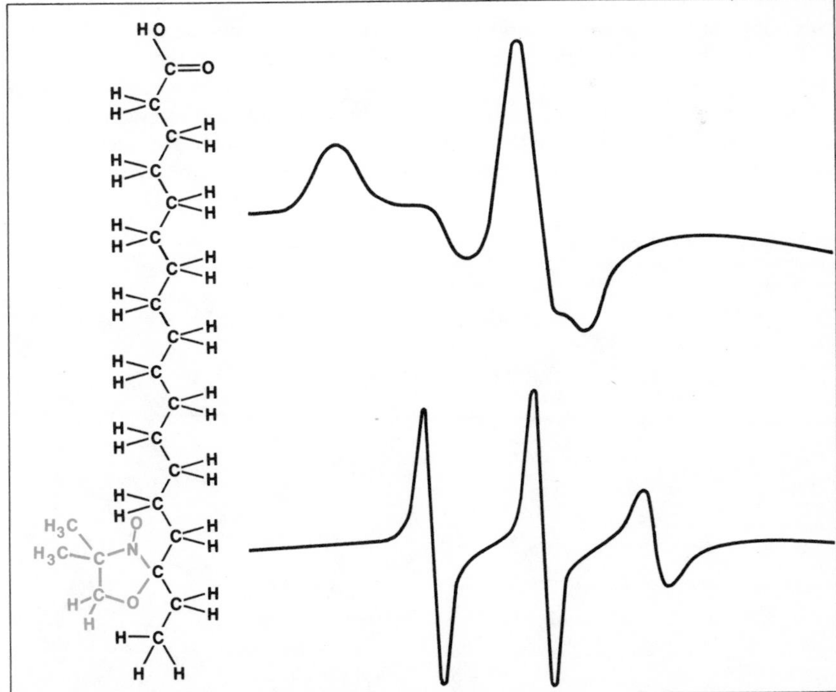

MOBILITY OF LIPIDS in the membrane bilayer can be determined by electron spin-resonance spectroscopy. A motion-sensitive "reporter" group (*color*) is attached to a carbon atom in the tail of a test molecule (*left*). When the test molecule is inserted in a bilayer, the spectrum of the reporter group will vary in accordance with the degree of bilayer mobility. If low temperature has "frozen" the bilayer so that the lipids are immobile, a characteristic spectrum results (*top right*). If the bilayer is mobile, however, the reporter group in the tail of the lipid produces a distinctly different spectrum (*bottom right*).

and Griffith's groups have done, and they have shown that there is a flexibility gradient in the bilayer. The part of the tail that is closest to the head of the lipid molecule and therefore nearest the surface of the bilayer is the least flexible. Conversely, the greatest flexibility is found at the tips of the lipids' tails and thus in the membrane zone nearest the center of the bilayer.

Not every lipid molecule in the bilayer is mobile. For example, penetrating protein molecules affect the mobility of the lipid molecules adjacent to them. In collaboration with Griffith and Patricia C. Jost, we have used spin-labeled stearic acids to probe the lipid environment immediately adjacent to molecules of cytochrome oxidase in our model membrane system. We find that the cytochrome oxidase molecules effectively immobilize sufficient lipid to coat themselves with a single layer of lipid molecules. We have given the name "boundary lipid" to this tightly associated layer. It is interesting that the amount of boundary lipid is just the amount of lipid required for full activity of this enzyme. Exactly how much of the total lipid content of a membrane will be immobilized by serving as boundary lipid depends on the number of penetrating protein molecules a particular membrane contains. In mitochondrial membrane, which contains many penetrating molecules, perhaps as much as 30 percent of the membrane lipid is so engaged.

Since a major proportion of the membrane bilayer is fluid, the membrane as a whole has the consistency of a light oil. It is therefore not surprising that, when constraints do not prevent it, both lipid and protein molecules are free to move about within the sheetlike structure. Such movement is more likely to be from side to side than up and down; the stability of the asymmetric configurations in membrane testifies to that. Indeed, it would be energetically extravagant for an intrinsic protein molecule to force its highly hydrophilic end down through the hydrophobic interior of the bilayer in order to reach the other side of the membrane. The waste of energy would be less acute in the case of a lipid molecule, but even so studies of spin-labeled phospholipids indicate that the flipping of a lipid from one side of the bilayer to the other is rare.

The lateral movement of molecules through the bilayer was first demonstrated in experiments conducted by David Frye and Michael Edidin at Johns Hopkins University in 1970. They were investigating the property of the membrane in what are known as cell-fusion heterokaryons, that is, "supercells" produced by the forced fusion of a number of individual cells under the influence of Sendai virus, a well-known fusion agent.

Frye and Edidin induced the fusion of human and mouse cells in culture and then studied the distribution of certain intrinsic membrane proteins: the antigenic components of the two kinds of cells. Just where which of the proteins was present could be determined by tagging antibodies with fluorescent dyes; the differently dyed antibodies were directed either to the mouse antigen or to the human one. The investigators found that shortly after the fusion of two cells the mouse and human antigenic components were clearly segregated in what might be called the mouse and human "halves" of the supercell's membrane. After the supercell had been incubated for 40 minutes at 98.6 degrees Fahrenheit, however, the mouse and human proteins became substantially intermixed. Because the mixing had taken place in the absence of ATP synthesis the process was evidently not an energized one. Frye and Edidin could only conclude that the intermixing was a product of lateral diffusion through the bilayer. They found that when the culture was held at 34 degrees F., a temperature more than low enough to "freeze" the membrane, the two proteins did not mix.

In the years since this classic experiment was done many more antigens, and other protein molecules as well, have been found to move about within the membrane bilayer. Evidence that lipid molecules are equally mobile has come from spin-labeling studies conducted by McConnell and Philippe Devaux. They inserted patches of spin-labeled phospholipids into membranes and recorded the time required to disperse the labeled molecules by diffusion. They found that the lipids moved around at a higher rate than protein molecules; that is what one would expect, since lipid molecules are smaller than protein molecules.

Now, if lipids and proteins were all free to move around in the membrane bilayer, one would expect that the two components would be distributed at random throughout the membrane. For many membranes, however, this is not so. For example, in the membrane of cells that line the intestine the glycoproteins are concentrated at the surface end of the cell and the proteins that pump sodium are concentrated at the opposite end. Another example is the nerve cell; here a key protein, the enzyme acetylcholinesterase, is localized exclusively in the membrane at one end of the cell. Some membranes even show crystalline lattices of the kind found in the cytochrome oxidase model membrane system: the chromatophore membrane of the bacterium *Halobacterium halobium* is one example. This "purple membrane" contains only one protein in a well-defined lattice structure. "Gap junctions" between cells provide another example. Daniel A. Goodenough of the Harvard Medical School has isolated gap junctions between the cells of mouse liver. It is thought that the junctions aid the

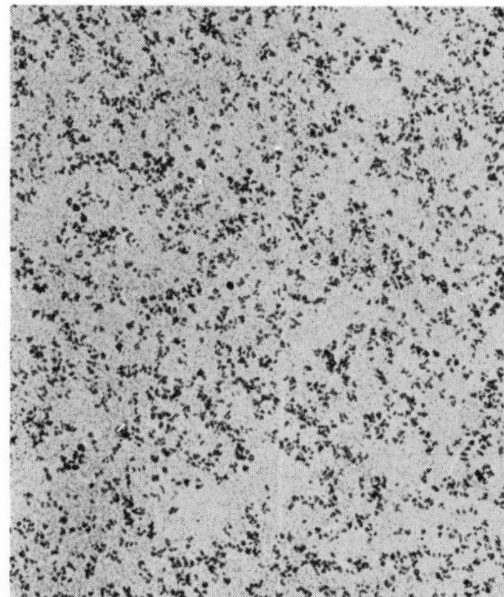

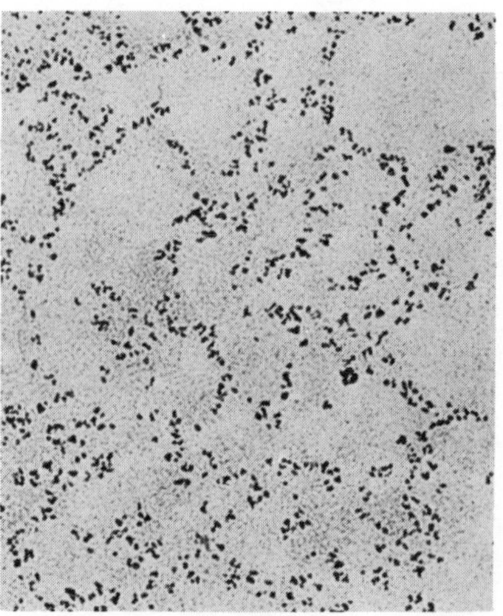

DISTRIBUTION PATTERN of constituent proteins in red-cell membrane appears to be controlled by the location of the spectrin molecules on its inner surface. These electron micrographs by Garth Nicolson of the Salk Institute for Biological Studies show the outer surface; molecules of glycoprotein that protrude from the surface have been stained. Their distribution (*left*) is more or less generalized. The second micrograph (*right*) shows the membrane after the spectrin is aggregated; movement has made glycoprotein move too.

transfer of substances from one cell to another. They contain one major protein and one major lipid, presumably donated from both cells, organized into a hexagonal lattice that is clearly visible in the electron microscope.

What influences affect the mobility and hence the distribution of the proteins in membranes? The information thus far available suggests that control is exercised in many different ways. For example, it seems likely that in the mitochondrial inner membrane a tight association between the various intrinsic membrane proteins is what maintains the spatial distribution. In this membrane almost all the proteins are associated with one or another of five complexes: four that are involved in electron transfer and one that synthesizes ATP. It seems likely that proportionate quantities of each of the five complexes combine to form supramolecular aggregates that lie in an orderly array throughout the membrane. Such aggregates would promote the efficiency of electron transfer and the coupling of electron transfer to ATP synthesis.

The distribution of proteins in the membrane of red blood cells appears to be maintained in a different way. Here the distribution pattern is evidently controlled by an interaction between the intrinsic proteins on the one hand and molecules of spectrin, the extrinsic protein located at the membrane's inner surface, on the other. Garth Nicolson of the Salk Institute for Biological Studies has provided a dramatic demonstration of the role played by spectrin in controlling the pattern. Nicolson has prepared an antibody to spectrin that, when it is applied to red-cell ghosts, causes the molecules of spectrin to aggregate abnormally. When this happens, the distribution of proteins intrinsic to the membrane, most noticeably the distribution of the major glycoprotein, is changed. The carbohydrate portion of this molecule contains many acidic groups, called sialic acids, that will bind an electron-dense material that is easily visible in the electron microscope. The material can therefore be used to "stain" the glycoprotein. When spectrin antibody is applied to a red-cell preparation, the position of the stained molecules is radically altered. Evidently the glycoprotein (or more probably a supramolecular aggregate that includes it) is so firmly linked to the spectrin on the inner surface of the membrane that when one moves, the other follows.

These are only two of the many systems that control the mobility and distribution of protein in membranes. A quite different system is involved with respect to the rhodopsin in retinal-rod membrane, for example, and still other systems, involving networks of microtubules and microfilaments located under the inner surface of cytoplasmic membrane, appear to control the mobility and distribution of the proteins in those membranes.

In summary, then, it appears likely that in some membranes the spatial distribution of proteins is fairly well fixed whereas in others this distribution is variable and may be controlled from within the cell. On the matter of the distribution of proteins and lipids in membranes no generalizations can be made. As was first realized by S. J. Singer of the University of California at San Diego, there is likely to be an interrelation between protein array and membrane function in cells with a variable distribution of proteins. When this is further understood, perhaps we shall be able to explain many of the key functions of membranes.

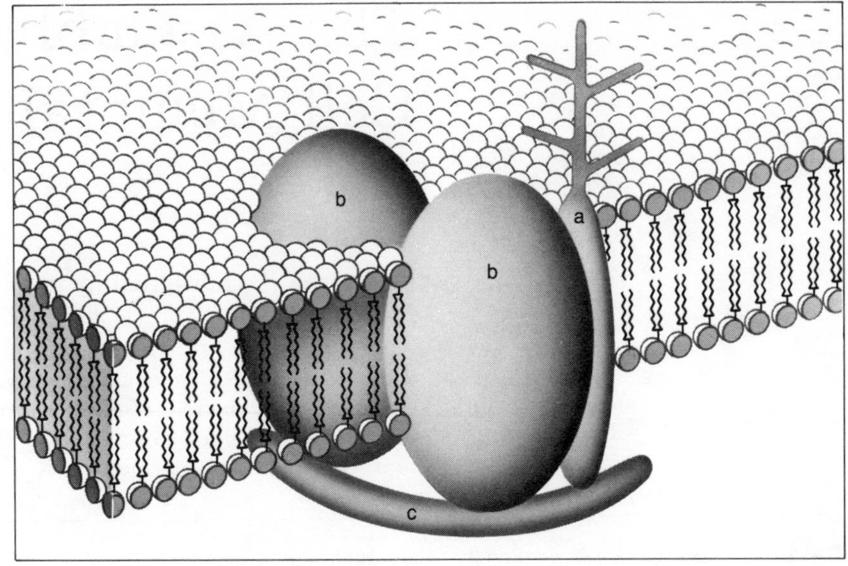

SUPRAMOLECULAR AGGREGATES in red-cell membrane include the two proteins that completely penetrate the membrane. One is a glycoprotein (*a*); the other is a protein with a molecular weight of 87,000 (*b*). The third protein in the hypothetical aggregate is spectrin (*c*). Evidently the three proteins are so linked that if one moves, the others follow.

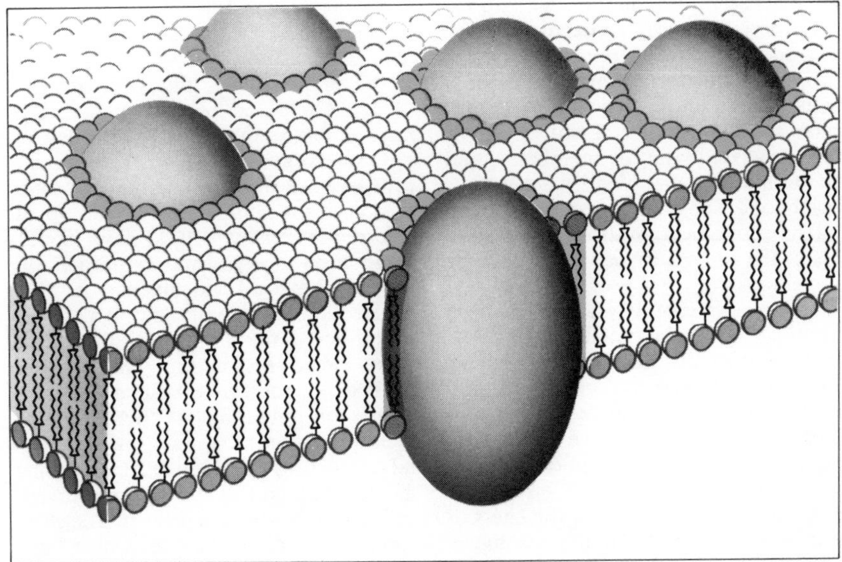

BOUNDARY LIPID is a one-molecule-thick "coating" of immobilized lipid molecules that surrounds penetrating protein molecules (*color*). Its existence was discovered in the course of electron spin-resonance spectroscopy studies of "model" membranes that contained only one protein: cytochrome oxidase. In membranes containing many penetrating proteins, such as mitochondrial membrane, perhaps 30 percent of all lipid is immobilized.

11

The Assembly of Cell Membranes

by Harvey F. Lodish and James E. Rothman
January 1979

The two sides of a biological membrane differ in structure and function. Studies of animal viruses and bacteria have helped to reveal how this asymmetry is preserved as the membrane grows

The membrane that surrounds a living cell is much more than a simple container or boundary; it not only defines the extent of the cell but also acts to maintain a distinction between inside and outside. Certain ions, for example, are pumped into the cell by large molecules embedded in the membrane and other ions are pumped out. Nutrients are taken up by the cell and concentrated inside by other components of the membrane.

For a membrane to preserve such concentration gradients one absolute requirement is that it form a closed vessel; otherwise a substance pumped in through the membrane would immediately leak back out. All known biological membranes form closed compartments. Another essential property is sidedness: the inner surface and the outer surface of a membrane must function differently. If they did not, an ion or a molecule pumped in at one point might be pumped out at another, with an expenditure of energy in each direction. Sidedness is also expressed in other properties of membranes. In animal cells, for example, receptors for hormones and for other chemical signals are associated with the membrane, and so are markers that identify a cell to its neighbors. These facilities for intercellular communication should be accessible at the outer surface; they would be useless inside.

In the past several years it has been established that the functional asymmetry of membranes reflects an underlying structural asymmetry. The protein molecules embedded in membranes or attached to them have a fixed orientation. Some of them are exposed only on the inner surface and some only on the outer surface. Others span the thickness of the membrane, but even they can invariably be regarded as having a fixed, asymmetric orientation. Every protein of the same type points in the same direction. The smaller phospholipid molecules that make up the structural matrix of the membrane have also been found to have an asymmetrical distribution, although it is a partial rather than an absolute asymmetry.

A major challenge to biologists is to understand how cells replicate and divide, and an important part of this problem is how cells form membranes. When a cell divides, new membrane constituents must be synthesized and then properly assembled. The new membrane must be laid down with great precision, and in particular the new components must have the same topographical arrangement as the old ones to ensure that the new membrane functions correctly.

In the past few years the outlines of the schemes cells use to solve these problems have been greatly clarified, although many biochemical details remain to be elucidated. The progress has resulted from the study of a few highly specialized kinds of membranes, which incorporate only a handful of molecular components and so are well suited for experimentation. It is these studies that we should like to discuss here.

The simplest cells, those of bacteria, have only one membrane, the plasma membrane that envelops the cell itself. Such organisms are prokaryotes, or cells without well-formed nuclei. Higher plants and animals are made up of eukaryotic cells, which do have nuclei. In addition to a plasma membrane eukaryotes have a great diversity of specialized internal membranes, which surround and define the subcellular structures called organelles. The nucleus itself is bounded by a system of membranes. In mitochondria there are two membranes, and the inner one has a vital role in the manufacture of adenosine triphosphate (ATP), the "energy currency" spent elsewhere in the cell. The chloroplasts of green plant cells also have two membranes, which may operate in a similar way. Another organelle of particular importance in the study of membranes is the endoplasmic reticulum, which is specialized for the fabrication of secreted proteins and new membrane material.

In our work we have been concerned mainly with the plasma membrane. (This should not be confused with the cell wall, which lies outside the plasma membrane in plant and bacterial cells and is not a membrane at all but a porous skeletal organ.) All cells have a plasma membrane, and in animals it is the only barrier separating the cell from its environment. There are intriguing differences between the plasma membrane and the various internal membranes of eukaryotic cells, but all biological membranes are remarkably similar in their basic structural features.

If a membrane is broken down into its molecular constituents, it is found to consist of three kinds of substance: lipids, proteins and carbohydrates. Lipid molecules are by far the most abundant; proteins, however, are much larger molecules, and in mass these two components are roughly equal. In any given membrane there are only a few species of lipid, each present in many copies; there is a vast assortment of proteins, on the other hand, some of them represented by no more than a few molecules per membrane. Carbohydrate is a minor constituent (in mass) and is mainly associated with the plasma membrane rather than with internal membranes. The carbohydrate in the membrane is always combined chemically with other membrane components; it binds to lipids to form glycolipids and to proteins to form glycoproteins.

It is the lipids that are responsible for the structural integrity of the membrane. Each lipid molecule is said to be amphipathic, which means it has one part that is hydrophobic and another part that is hydrophilic. If the hydrophobic part were separated from the rest of the molecule, it would be insoluble in water, just as salad oil is. The hydrophilic part, if it were isolated, would be soluble in water and insoluble in oil.

Two kinds of lipids are included in membranes, cholesterol and phospholipids. Cholesterol is found almost exclusively in the plasma membrane of mammalian cells and is absent entirely in

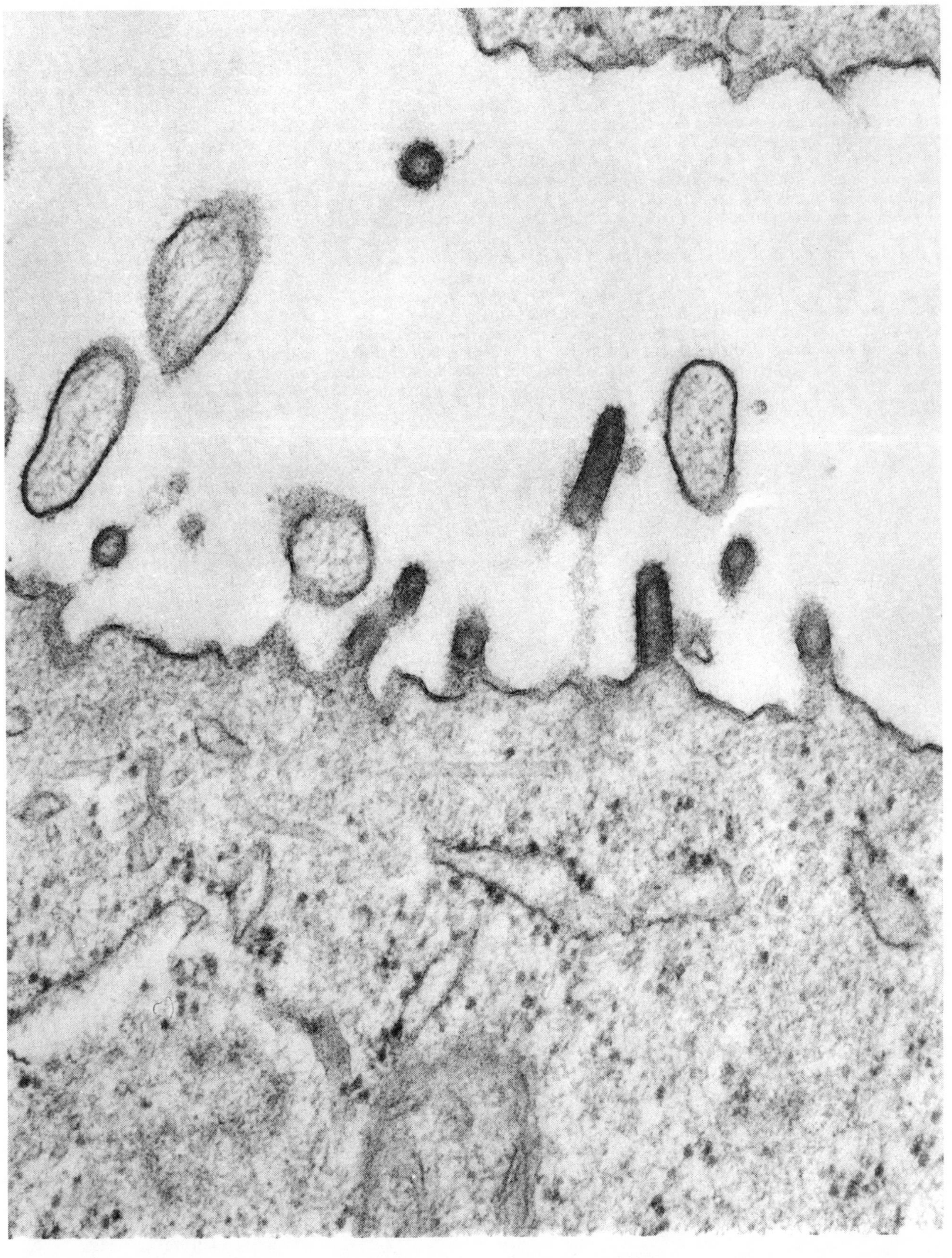

REGION OF CELL MEMBRANE envelops a virus particle as the virus buds off from an infected cell. The membrane appropriated by the virus incorporates only a single kind of surface protein, which makes the viral coat membrane a convenient experimental system for studies of how membranes are assembled. Each molecule of the surface protein is oriented in the membrane so as to form a spike on the surface of the virus. The cell is infected with vesicular stomatitis virus (VSV), which causes an influenzalike disease in farm animals. The virus particles are the small, dark, cigar-shaped objects emerging into the intercellular space at the top; a few are seen end on and appear round. The larger, irregular vesicles are cellular material isolated by the sectioning of the cell. This transmission electron micrograph, which magnifies the image some 60,000 times, was made by David M. Knipe of Massachusetts Institute of Technology.

bacteria. The hydrophilic head in cholesterol is a hydroxyl group (OH). Phospholipids are found in all membranes. The hydrophilic region includes a negatively charged phosphate group (PO_4^-); in many phospholipids a compensating positive charge is contributed by some other chemical entity, such as an amino group (NH_3^+).

Because the two parts of a membrane lipid have incompatible solubilities the molecules spontaneously organize themselves in the form of a bilayer, or double layer of molecules. In this way the hydrophobic portion of each molecule is shielded from water, whereas the hydrophilic heads are immersed in it. Such a bilayer sheet is a minimum-energy configuration for a suspension of lipids. The only place where the nonpolar tails must interact with water is at the edge of the sheet, and even that unfavorable contact can be eliminated: the growing bilayer simply folds over to form a closed vesicle, which has no edges.

The lipids of all membranes have the same organization, with two hydrophilic surfaces separated by a hydrophobic core. It is this lipid bilayer that determines the overall morphology of membranes and accounts for the fact that they form large sheets or vesicles. The structure of the bilayer also provides for one of the essential functions of the membrane: it is impermeable to most water-soluble molecules because they are insoluble in the oily core region.

The bilayer is by no means static; on the contrary, each monolayer is a two-dimensional fluid. The lipid molecules are free to diffuse laterally, like the molecules in a thin film of liquid. They exchange positions as often as a million times per second. In the third dimension, however, the mobility of the lipids is severely restricted. For a lipid molecule to jump from one monolayer to the other, a motion called a flip-flop, the polar head must pass through the hydrophobic core of the membrane, where it is insoluble. Recent measurements indicate that the flip-flop rate is so low that a given lipid molecule makes the passage no oftener than once a month.

Because the lipid bilayer is a two-dimensional fluid any protein molecules embedded in it can also diffuse laterally. The fluidity simplifies the task of membrane assembly, since both lipid molecules and protein molecules can be inserted anywhere in a monolayer with confidence that they will eventually reach any other point. At the same time the low rate of flip-flop could allow the two opposing monolayers to preserve different lipid and protein compositions. Indeed, the lipid bilayer has been found to be asymmetric in composition in every biological membrane that has been examined for this property. The function of the asymmetry is unclear.

The interactions of hydrophilic and hydrophobic chemical groups are important in proteins as well as in lipids. Proteins are polymers (polypeptides) made up of amino acids strung together in a linear sequence. Of the 20 amino acids specified by the genetic code six are strongly hydrophobic and a few others are weakly hydrophobic; the rest are hydrophilic. If a protein is considered as being merely a straight chain of amino acids, then a pattern can seldom be discerned in the sequence of hydrophilic and hydrophobic units. The native conformation of protein molecules, however, is not a straight chain but a densely folded one. In that form the soluble proteins of the cell cytoplasm generally have an excess of hydrophilic units at the surface of the molecule and an excess of hydrophobic ones inside.

Proteins associated with membranes fall into two classes. Integral membrane proteins are those with a portion of each molecule embedded in the lipid bilayer. All integral proteins that have been studied in detail have been found to span the full width of the bilayer and so to have regions exposed on both sides of the membrane. The other class is made up of peripheral membrane proteins, which are not inserted into the bilayer at all but reside at one surface or the other. Each peripheral protein molecule is bound to an integral protein.

In peripheral proteins the balance of hydrophilic and hydrophobic amino acids is much like that in cytoplasmic proteins. The peripheral proteins can be removed from the membrane by treatments that do not destroy the integrity of the lipid bilayer, and when they have

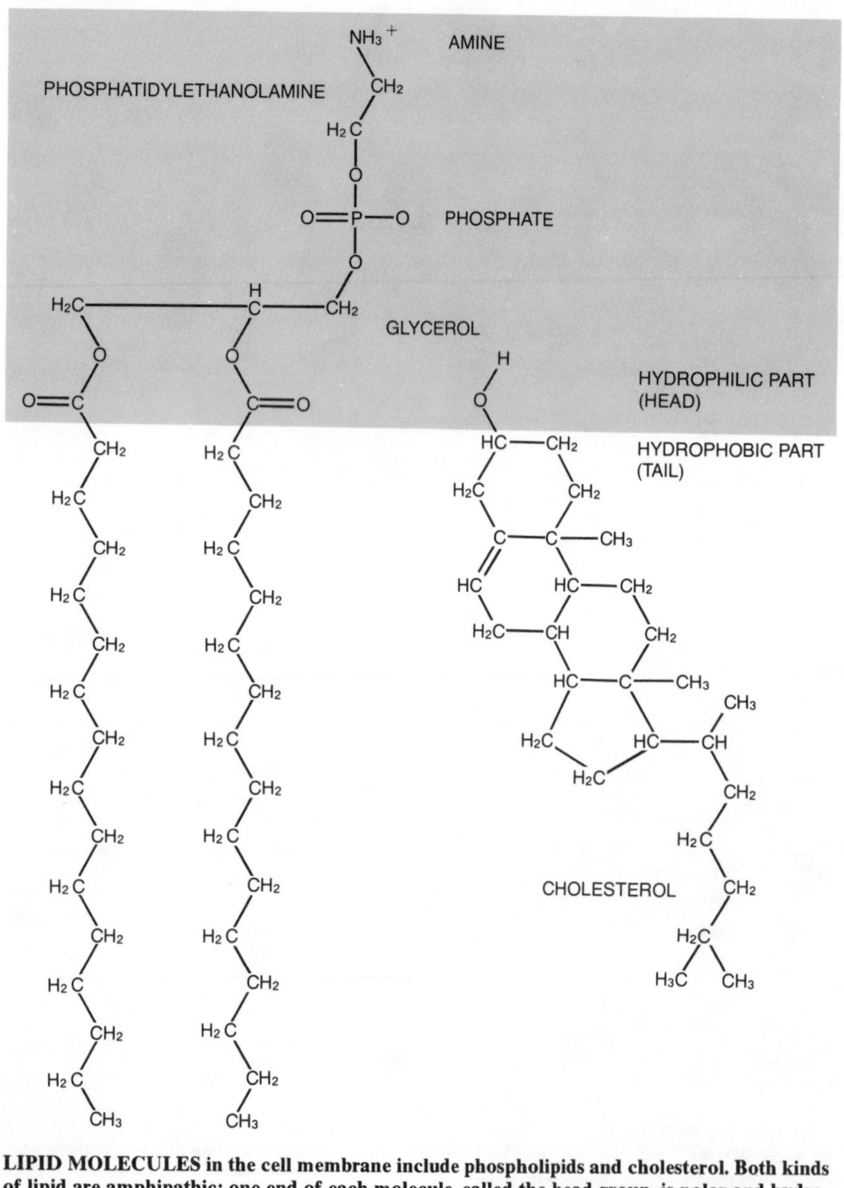

PHOSPHATIDYLETHANOLAMINE

AMINE

PHOSPHATE

GLYCEROL

HYDROPHILIC PART (HEAD)

HYDROPHOBIC PART (TAIL)

CHOLESTEROL

LIPID MOLECULES in the cell membrane include phospholipids and cholesterol. Both kinds of lipid are amphipathic: one end of each molecule, called the head group, is polar and hydrophilic; if it were isolated, it would be water-soluble. The other end, the tail, is nonpolar and hydrophobic, or insoluble in water. Cholesterol is found mainly in the outer, plasma membrane of mammalian cells; phospholipids of various kinds are components of all biological membranes.

been removed, they are fully soluble in water. Integral proteins, in contrast, are generally insoluble in water because they include substantial regions where most of the exposed amino acid units are hydrophobic. It is these regions that are embedded in the hydrophobic core of the membrane. They cannot be removed without disrupting the lipid bilayer itself.

The regions of an integral protein that extend into the external fluid or into the cytoplasm have the hydrophilic character common to soluble proteins. These hydrophilic regions essentially preclude the possibility of a flip-flop transition for membrane proteins. If the small polar head of a lipid molecule can rarely be forced through the core of the bilayer, then the much larger hydrophilic volume of a protein molecule is permanently excluded.

A cell that has proved useful for studies of membrane topography (although not for studies of membrane assembly) is the red blood cell. The plasma membrane of the red cell has only two main species of integral protein and several peripheral ones, and their orientation can readily be determined. From studies of the red cell several principles of membrane construction have emerged. All integral proteins are inserted asymmetrically, so that every molecule of a given protein species has the same orientation in the lipid bilayer. The lipid bilayer itself is asymmetric. Carbohydrates can be attached either to proteins or to lipids, but in each case they are invariably on the exterior surface rather than on the cytoplasmic surface. Peripheral proteins are usually found on the cytoplasmic side.

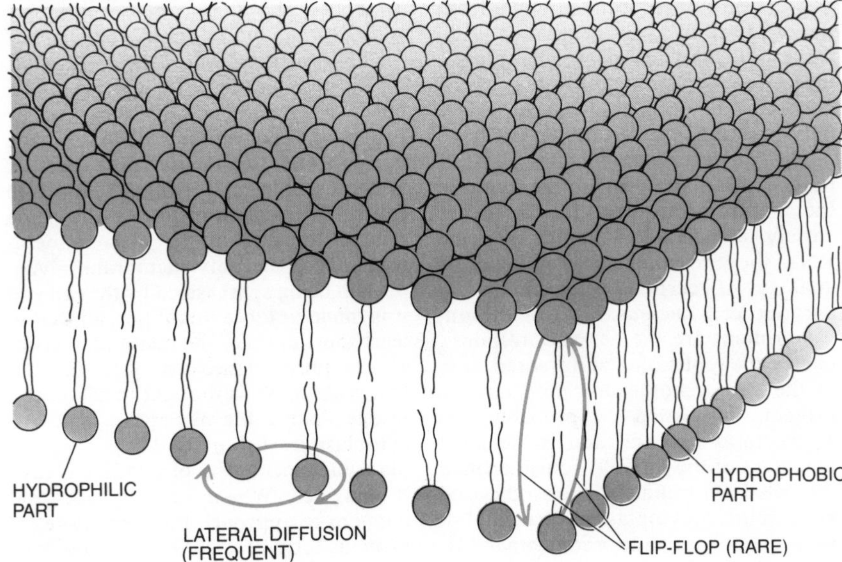

BILAYER OF PHOSPHOLIPIDS forms the structural matrix of a membrane. The lipids are arranged tail to tail so that only the hydrophilic heads are exposed to the aqueous medium on both sides of the membrane. This is the minimum-energy configuration for a suspension of lipids in water. Each monolayer is a two-dimensional fluid: lipid molecules can diffuse laterally as in a liquid film but can rarely execute a "flip-flop" transition from one layer to the other.

If suspended lipids can spontaneously form a closed vesicle, it is not unreasonable to suppose a complete and functional biological membrane might be assembled the same way. Certain viruses, muscle fibers and ribosomes spontaneously assemble themselves, but these structures are highly ordered and in some cases even crystalline, whereas membranes are not.

The self-assembly hypothesis can easily be tested. A membrane can be dispersed into individual protein and lipid molecules by disrupting the lipid bilayer with a high concentration of detergent. The detergent is a synthetic lipid that does not form a bilayer but instead forms droplets called micelles. The lipids and proteins of the membrane no longer need to stay together to avoid water; they can achieve the same thing inside the detergent micelles. With a sufficiently high concentration of detergent the components of the membrane can be completely separated, with no more than one membrane protein in each detergent micelle.

If the detergent is removed by any of several procedures, a membrane vesicle re-forms spontaneously, incorporating both lipids and proteins. The individual protein molecules are often functional, with the capability, for example, of transporting ions across the permeability barrier. There is just one flaw in such reconstituted membranes: they are almost always symmetrical. The proteins are embedded in the lipid bilayer, but no longer do all the molecules of a given species of protein point in the same direction. There are exceptions, which are interesting in themselves, but in most cases the absolute asymmetry of the original membrane is lost.

Why is the asymmetry not re-created in this experiment along with the rest of the membrane structure? In the self-assembly of a membrane the formation of the lipid bilayer and the insertion of the membrane proteins take place concurrently. Because there is no permeability barrier present during the assembly, proteins can be inserted from both sides, resulting in an approximately symmetric structure.

The alternative to the self-assembly hypothesis is that integral membrane proteins are given their orientation at the time of insertion. That mechanism requires that the protein always be inserted from the same side. For the sidedness to be defined and preserved at all times, the membrane must always have the form of a closed vessel. From these arguments a fundamental conclusion follows: Membranes can grow only by expansion, by the insertion of new material into a membrane that is already a closed vessel.

This mechanism also has a difficulty. If a protein (or a lipid) is to be inserted from one side of the membrane, then in many cases a hydrophilic region will have to be pushed through the core of the bilayer. It is the very inability of these regions to cross the membrane that preserves the asymmetry of the proteins once they are inserted. Our experiments have helped to identify the methods by which this intracellular rope trick can be performed.

It would be a formidable task to study the synthesis of membrane proteins in a typical eukaryotic cell or even in a prokaryotic one. There is too great a variety of proteins, each one present in too small a quantity. We and our colleagues have therefore turned to a simpler membrane system, that of a virus whose outer coat is a phospholipid membrane appropriated from the plasma membrane of an infected animal cell. The membrane segment is densely coated with integral membrane protein molecules, but they are all of one species, determined by the viral genome (the complete set of viral genes).

The virus is vesicular stomatitis virus (VSV), which infects mainly farm animals and elicits symptoms similar to those of influenza. When a VSV particle enters a cell, it halts the normal synthesis of the host cell's protein and commandeers enzymes and other components of the synthetic apparatus for its own purposes. In the subsequent period only viral proteins are assembled. At the same time the genome of the virus, which is a strand of RNA, is replicated many times by means of enzymes coded by the viral genome and carried in the virus particle. The life cycle is completed when the replicated RNA and the viral proteins come together at the plasma membrane. Each copy of the RNA, along with a set of proteins, is enveloped in a section of membrane and buds from the cell to form a new generation of virions, or virus particles. The viral membrane has the lipid composition of the host-cell membrane, but it incorporates only proteins specified by the virus.

The VSV genome encodes just five polypeptides, all of which are found in the mature virus particle. Three are closely associated with the genome RNA; they are called nucleoproteins, by analogy with certain proteins in the eukaryotic cell nucleus. Two of the nucleoproteins are the enzymes responsible for the replication and transcription of the viral genome; the third nucleoprotein, designated *N,* probably has a structural function and is the most abundant of the three. A fourth protein is labeled *M* for matrix; it is synthesized as a soluble cytoplasmic protein but subsequently is incorporated into the virion as a peripheral protein on the inner surface of the plasma membrane. The last protein, called *G* for glycoprotein, has been the major focus of our investigations. *G* is a membrane protein that spans the lipid bilayer; in electron micrographs it is visible as a pattern of spikes on the surface of the virion. It can also be detected in the plasma membrane of infected cells during budding. The amino acid sequence and three-dimensional structure of *G* have not been determined, but the protein is known to include about 550 amino acid units; there are also two carbohydrate side chains attached to it. The orientation of *G* in the bilayer is unquestionably asymmetrical: most of the polypeptide and all the carbohydrate is exposed on the external side of the membrane and only a stub of about 30 amino acids protrudes on the cytoplasmic side.

The steps involved in the budding of the virion are not known with certainty, but a likely sequence of events can be proposed. The genome RNA and the nucleoproteins evidently come together in the cytoplasm, forming a structure called the nucleocapsid. At the same time *G* appears in the plasma membrane. It is the only membrane glycoprotein being synthesized by the cell and it is made in large quantities, so that at least some parts of the membrane probably acquire a dense coat of it. The matrix protein, *M,* is thought to serve as a bridge linking the nucleocapsid to the cytoplasmic stub of *G.* If that is so, a plausible mechanism of budding is easily imagined. Whenever a nucleocapsid happens to approach the membrane, *M* proteins form cross-links at the point of first contact between the nucleoproteins and the cytoplasmic stub of *G.* As a result of the attractive interaction between the components, more of the membrane is then wrapped around the nucleocapsid and is stitched in place by the matrix protein. Any membrane proteins of the host cell that happen to be in the vicinity are shouldered aside. In the end the entire nucleocapsid is enveloped and the virion pinches off from the cell. The nucleocapsid has not been pushed through the membrane but rather is drawn into it by the cross-links of the matrix protein.

The viral proteins are made by the same process as normal host-cell proteins. The code specifying the amino acid sequence of each of the five viral proteins is carried by a length of messenger RNA, which is synthesized as a copy of a region of the viral genome RNA. The messenger RNA is translated into a polypeptide by the subcellular structure called a ribosome. The ribosome is made up of several dozen proteins and three or more kinds of RNA, organized into a large subunit and a small one. Messenger RNA is threaded through the ribosome, and the growing polypeptide is assembled in the large subunit. Each amino acid added to the polypeptide is carried by a molecule of transfer RNA, which recognizes a particular sequence of three nucleotide bases in the messenger RNA. Translation begins with the amino terminus and proceeds to the carboxyl terminus of the protein. The point where amino acids are joined to the growing chain is deeply embedded in the large subunit, and throughout translation about 40 amino acid units—the 40 most recently added—are buried inside the ribosome. As the chain grows, these units are successively pushed out by the new ones added behind them, and on completion of the polypeptide the entire protein molecule is released. In general a molecule of messenger RNA is translated by many ribosomes simultaneously, each ribosome generating a polypeptide chain.

It has often been pointed out that there are two kinds of ribosomes in the eukaryotic cell; some are free in the cytoplasm, whereas others are bound to membranes of the endoplasmic reticulum. It is now apparent that the ribosomes in these two classes are structurally identical and indeed are interchangeable. A membrane-bound ribosome can be released into the solution

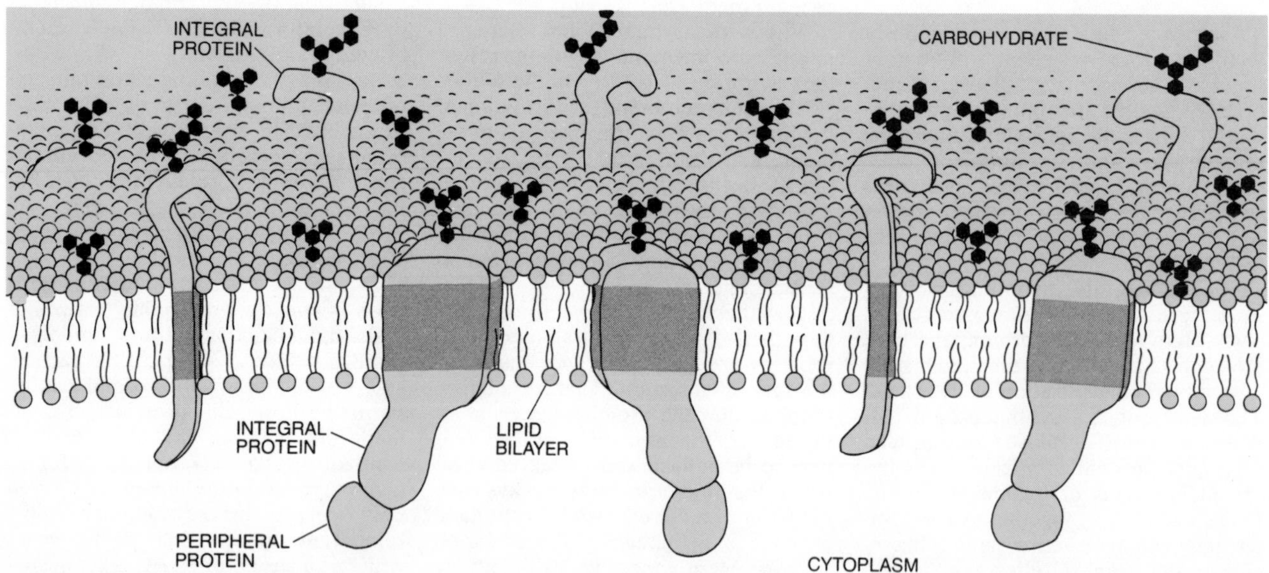

MODEL OF THE PLASMA MEMBRANE includes proteins and carbohydrates as well as lipids. Integral proteins are embedded in the lipid bilayer; peripheral proteins are merely associated with the membrane surface. The carbohydrate consists of monosaccharides, or simple sugars, strung together in chains that are attached to proteins (forming glycoproteins) or to lipids (forming glycolipids). The asymmetry of the membrane is manifested in several ways. Carbohydrates are always on the exterior surface and peripheral proteins are almost always on the cytoplasmic, or inner, surface. The two lipid monolayers include different proportions of the various kinds of lipid molecule. Most important, each species of integral protein has a definite orientation, which is the same for every molecule of that species.

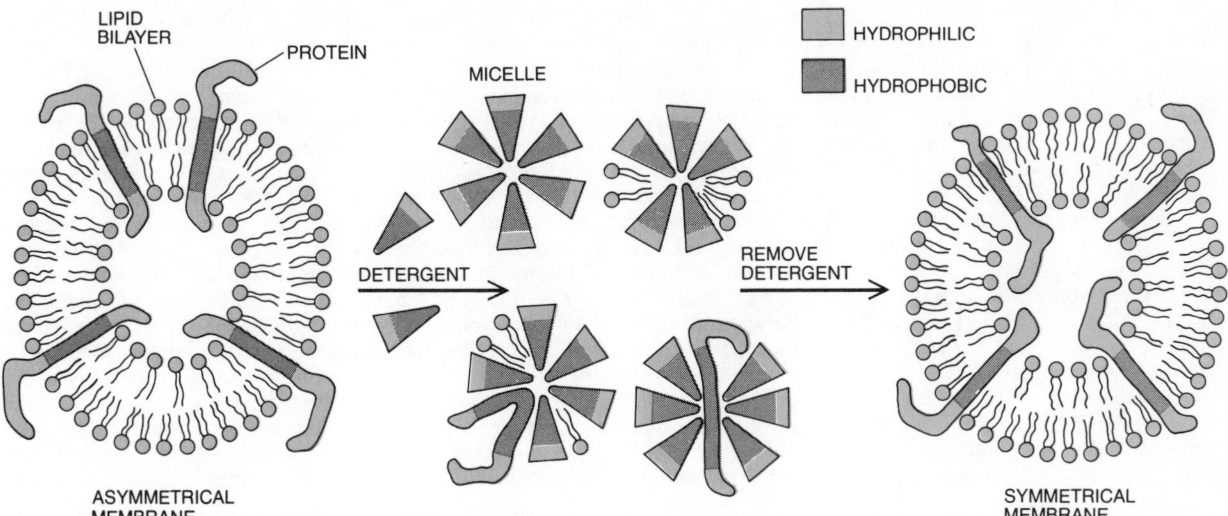

SELF-ASSEMBLY OF A MEMBRANE preserves its basic structure but not its asymmetry. A membrane can be disrupted by a high concentration of a detergent, which is an amphipathic molecule that forms the small droplets called micelles. The detergent dissolves the components of the membrane by enveloping the hydrophobic portions of both lipids and proteins in micelles, where they are protected from contact with water. If the detergent is then removed, the lipids spontaneously form a new bilayer, incorporating the integral proteins in it. The proteins, however, generally assume random orientations. Experiments such as this one have shown that membranes in the cell cannot be self-assembled; instead the integral proteins must be inserted in a membrane that already exists and has a defined sidedness.

and can later take up another place on the membrane. The two kinds of ribosomes do differ in function, however. Free ribosomes synthesize mainly soluble proteins, whereas membrane-bound ribosomes manufacture integral membrane proteins and proteins destined for secretion by the cell.

The VSV matrix protein is assembled on free ribosomes. That is not surprising, since immediately after synthesis the protein itself is free in the cytoplasm; it binds to the membrane (on the cytoplasmic side) only when budding begins. Other peripheral proteins have also been shown to originate as soluble proteins, and it is likely that all proteins attached to the cytoplasmic surface but not embedded in it are synthesized the same way.

The glycoprotein *G* follows a more complicated trajectory. It is synthesized by ribosomes on the endoplasmic reticulum and then transported to the plasma membrane, with at least one stop along the way. The endoplasmic reticulum is a network of membranes that is found near the cell nucleus and often partly surrounds it. The membranes are arranged in layers, and in electron micrographs each layer is seen to be a flattened, elongated vesicle, a cell within the cell.

The VSV glycoprotein first appears in the cell as a component of the rough endoplasmic reticulum, which is called rough because it is studded with ribosomes. *G* is embedded in the membrane of a reticulum vesicle, with the large spike portion, bearing the two carbohydrate chains, protruding into the lumen, or internal cavity, of the vesicle. The short stub at the carboxyl terminus of the protein faces the cytoplasm. After 20 to 30 minutes *G* can be found in the Golgi apparatus, another organelle made up of many stacked membrane vesicles. The Golgi apparatus is often called a smooth membranous organelle, because it lacks bound ribosomes. Only after the protein has passed through the Golgi apparatus does it reach the plasma membrane.

The means of transport between these organelles is obscure. The most likely possibility is that small vesicles bud off from one closed membrane and later fuse with another membrane, carrying the protein with them and preserving its orientation at all times. In this regard it is important to keep track of the orientation of the protein throughout its journey. At first it may seem that the orientation is reversed at some point in transit. In the endoplasmic reticulum the protein points inward, toward the lumen of the vesicle, but in the plasma membrane it points outward, toward the exterior of the cell. Actually this does not represent a reversal but the conservation of the same orientation. Note that the same portion of the protein (the stub) is always exposed to the cytoplasm and the same portion (the spike) is always outside the cytoplasm. Indeed, the lumen of an organelle such as the endoplasmic reticulum or the Golgi apparatus is topologically equivalent to the exterior of the cell, a fact that George E. Palade of the Yale University School of Medicine was the first to point out. In a topological sense the spike end of the protein is already outside the cell from the time it enters the endoplasmic reticulum.

The carbohydrate side chains of the glycoprotein are attached before it reaches the plasma membrane. In the virion the protein bears two identical carbohydrate structures, each made up of a dozen or so linked monosaccharides, or sugar units. The function of the sugar side chains in the VSV glycoprotein is not known, but in structure they resemble the carbohydrates of many normal glycoproteins found in the blood serum.

Donald F. Summers and his co-workers at the University of Utah have determined the sequence of monosaccharides in the carbohydrate chains and have shown that each chain can be considered to have two parts. The region nearer the protein and attached directly to it they designate the core, and they note that it consists exclusively of the monosaccharides mannose and *N*-acetylglucosamine. The outlying part of the carbohydrate constitutes the terminal region, and it is built up from three monosaccharides, *N*-acetylglucosamine, galactose and sialic acid. As we shall explain below, the core region is attached to the protein while it is still in the rough endoplasmic reticulum. In the Golgi apparatus a few of the core sugars are removed and the several new terminal monosaccharides are added. When the glycoprotein leaves the Golgi apparatus, it is complete.

We have studied the synthesis, glycosylation and insertion of the *G* protein in a cell-free system, a liquid medium that includes ribosomes, transfer RNA's, amino acids and various enzymes and other ingredients necessary for protein synthesis. To study the insertion and glycosylation of the protein, membranes from the endoplasmic reticulum that have been stripped of their endogenous ribosomes must also be present. Protein synthesis is initiated in this system simply by adding purified messenger RNA for the *G* protein, which can be obtained from infected cells. In order for newly made polypeptides to be identified later, one of the amino acids in the medium (methionine) includes an atom of a radioactive isotope of sulfur.

These experiments have been carried out at the Massachusetts Institute of Technology. Much of the work was done in collaboration with Flora N. Katz of M.I.T. and Vishu Lingappa and Gunter Blobel of Rockefeller University. Similar experiments have been conducted independently by Hara Ghosh and Frances Toneguzzo of McMaster University. We shall first present some of the conclusions derived from our findings and then discuss the experiments themselves.

The central thesis of our model is that an integral membrane protein is inserted through the lipid bilayer as the protein is synthesized and before it has assumed its ultimate, folded configuration. This mechanism accounts for two important observations. First, it helps to explain how the bulk of the protein can cross the hydrophobic core of the membrane: it passes through before the folding of the polypeptide makes the insertion even more difficult. Second, the model explains the asymmetry of the protein's orientation: the protein can be inserted in only one direction (the direction of polypeptide elongation) and from only one side of the membrane (the cytoplasmic side, where ribosomes are found).

The translation of *G*, like that of all proteins, begins with the binding of messenger RNA to the small subunit of a ribosome; then the large subunit is added. At this stage the ribosome is not associated with a membrane but is free in the cytoplasm. Some 40 amino acid units must be assembled before the polypeptide begins to emerge from the large subunit of the ribosome; these are the 40 units at the amino terminus of the molecule. After another 30 amino acids have been added to the growing chain the first 30 units are exposed. Within this section of the polypeptide is a group of amino acids, called the signal sequence, that is essential to the distinction be-

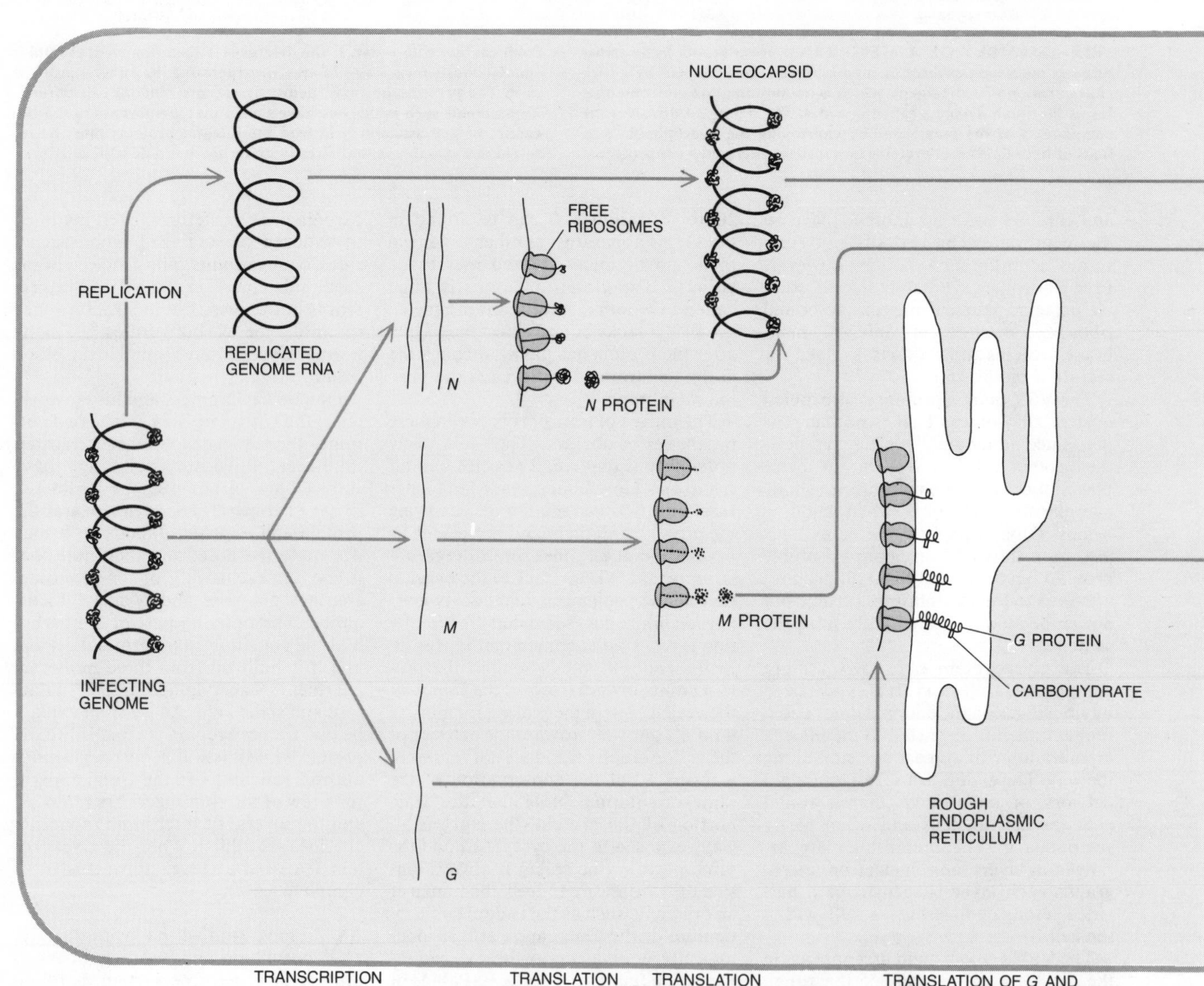

VIRAL PROTEINS are manufactured by the genetic machinery of the cell, but their structure is specified by the genome of the virus. The infectious cycle of VSV begins when a virus inserts its genetic material, a strand of RNA, into a cell. The genome RNA is transcribed by two viral enzymes into five molecules of messenger RNA. Later the genome RNA is replicated by similar viral enzymes. Each strand of messenger RNA is translated by the organelles called ribosomes into many identical protein molecules; since there are five messenger-RNA molecules, the viral genome encodes five proteins. Three are nucleoproteins; these include the two enzymes needed for transcription and replication and a third protein, designated *N*, whose main role is probably structural. The three nucleoproteins bind to the genome RNA to form a complex called the nucleocapsid. A fourth protein, designated *M* for matrix, is initially a soluble component of the cytoplasm. All four of these proteins are synthesized on ribosomes that are free in the cytoplasm. The last protein is designated *G* be-

tween membrane proteins and secreted proteins on the one hand and soluble proteins on the other. The signal sequence is evidently recognized by some component of the endoplasmic-reticulum membrane, presumably a membrane protein. This membrane transport site binds to the nascent polypeptide and helps to initiate its passage through the bilayer.

The existence of a signal sequence was first proposed by Cesar Milstein and George G. Brownlee and their colleagues at the Medical Research Council Laboratory of Molecular Biology in Cambridge, England, based on their studies of the synthesis of antibody proteins. Much additional evidence for the sequence has been provided by the experiments of Blobel and his co-workers, who have examined primarily proteins destined to be secreted by the cell. It appears that the cellular mechanisms for handling the two classes of proteins are similar in many ways. Both are extruded into the endoplasmic reticulum; the principal difference is that secreted proteins pass all the way through and are released into the lumen, whereas membrane proteins become embedded in the phospholipid bilayer before the polypeptide is complete. When a vesicle bearing such proteins fuses with the plasma membrane, the secreted proteins are released outside the cell, whereas the integral proteins remain bound to the membrane. In secreted proteins, once the signal sequence has piloted the polypeptide across the lipid bilayer it is clipped off by enzymes in the reticulum; it has recently been found that a signal sequence of the *G* protein, made up of 16 amino acids, is also removed. The signal sequences of the various secreted proteins and of *G* are not identical, but they share certain properties; in particular, they include a preponderance of hydrophobic amino acids.

It is only when the signal sequence draws the polypeptide to the endoplasmic reticulum that the ribosome becomes associated with a membrane. Initially the ribosome itself is probably not bound directly to the membrane but is merely tethered by the polypeptide chain. Later a weak electrostatic inter-

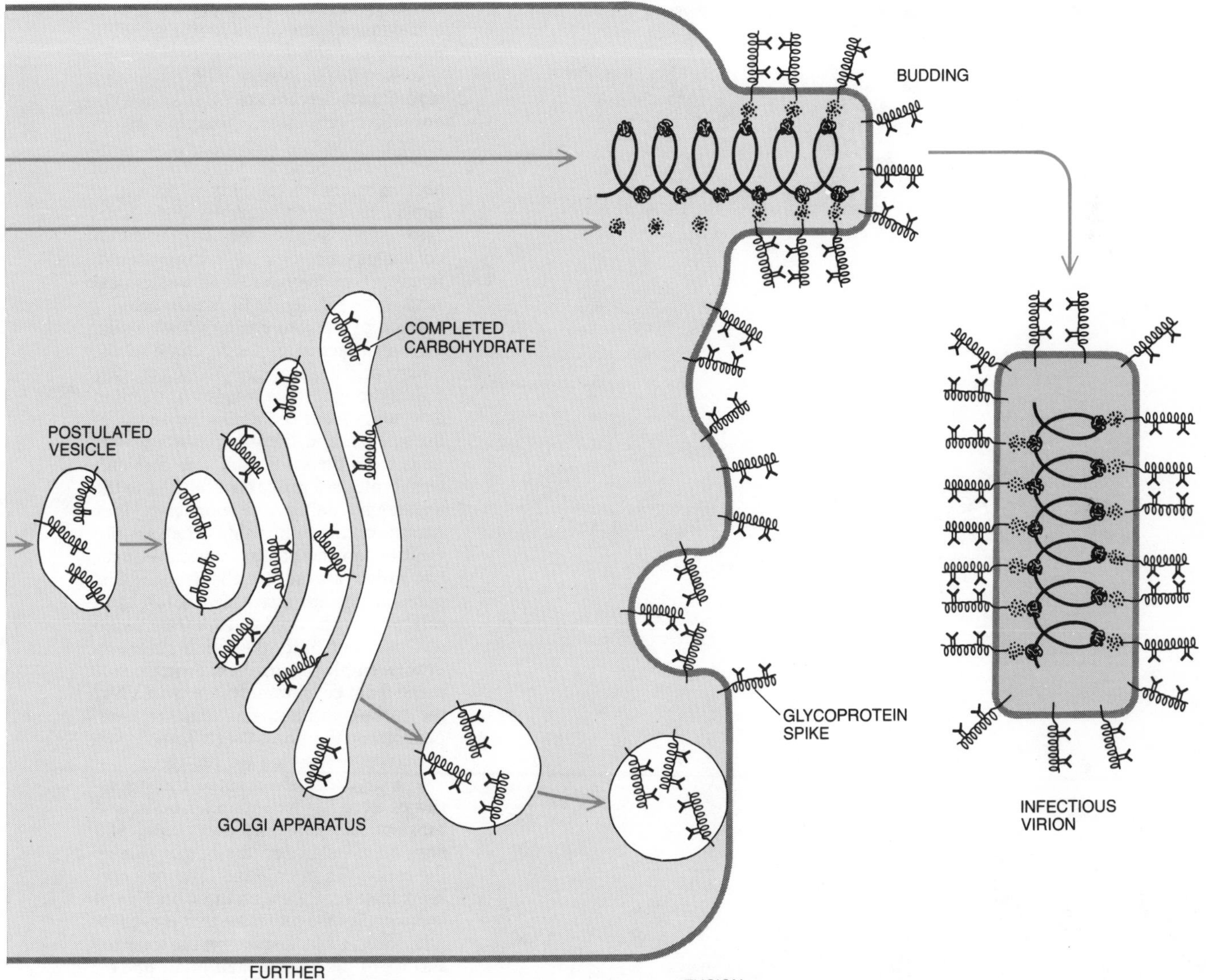

cause it is a glycoprotein, one with carbohydrate side chains (whose structure is not specified by the viral genome). *G* is assembled by ribosomes associated with membranes of the rough endoplasmic reticulum, and from the time the protein first appears it is bound to these membranes. Core monosaccharides are added to growing *G* molecules in the rough endoplasmic reticulum, and after 20 to 30 minutes the carbohydrate side chains are completed in the Golgi apparatus. The mechanism of transport between these organelles is not known, but it may depend on vesicles that bud off from one membrane and fuse with another. The glycoprotein could reach the plasma membrane, where it forms spikes on the surface of the cell, by the same means. The life cycle of the virus is completed when the nucleocapsid is wrapped in a piece of plasma membrane bearing the *G* protein; this newly assembled virus particle then buds from the cell. Budding is probably mediated by the matrix protein, which could stitch together nucleoproteins and a stub of *G* that protrudes into the cytoplasm.

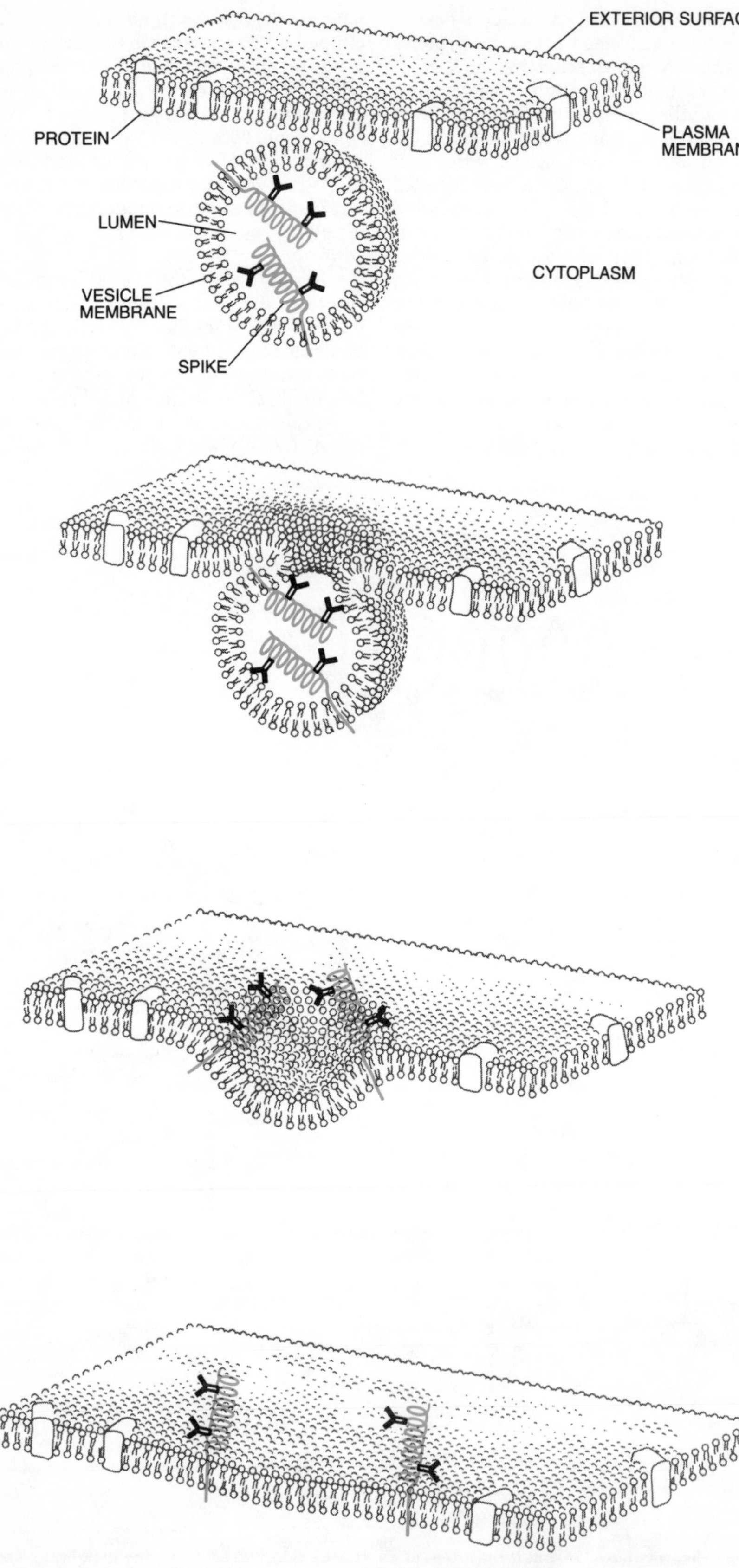

FUSION OF A VESICLE with the plasma membrane preserves the orientation of any integral proteins embedded in the vesicle bilayer. Initially the large spike end of the *G* protein faces the lumen, or inner cavity, of such a vesicle. After fusion the spike is on the exterior surface of the plasma membrane. That the orientation of the protein has not been reversed can be perceived by noting that the other end of the molecule, the small stub, is always immersed in the cytoplasm. The lumen of a vesicle and the outside of the cell are topologically equivalent.

action may develop between the ribosome and the membrane.

In all likelihood the protein does not penetrate the lipid bilayer directly but is aided by some other integral protein already in the membrane, which diffuses away when the insertion is complete. As more of the chain enters the lumen of the reticulum the polypeptide folds spontaneously, like a soluble protein, to achieve a configuration of minimum energy in an aqueous environment. Once in that state the protein cannot slide back through the bilayer; it is anchored like a rivet. What remains to be explained is why the small stub at the carboxyl terminus of the polypeptide does not slide through to the lumen side, as it does in secreted proteins. It has also not been established whether energy in addition to that normally expended in protein synthesis is required to transport the nascent protein across the membrane.

In a cell-free system where the only membranes present are those of the endoplasmic reticulum a complete glycoprotein cannot be constructed, because the modification of the carbohydrate portion is completed only in the Golgi apparatus. *G* protein grown in the laboratory does acquire the two chains of core sugars, however, and what is equally important, they appear to be identical with those formed in the intact cell.

Each of the two carbohydrate structures is attached to a side chain of the amino acid asparagine. One asparagine unit is thought to be roughly 150 amino acid units from the amino terminus of the protein and the other is about 400 units from the same end. The core carbohydrates are probably attached to the asparagine units very soon after they emerge on the lumen side. They are not built up on the protein sugar by sugar but rather are added all at once. The core regions are assembled beforehand on a membrane lipid and are transferred to *G* as a unit. Since the reticulum membranes employed in these experiments come from cells not infected with VSV, the carbohydrate chains must be components of the normal cell rather than structures assembled for the virus.

The glycoprotein created by this series of steps has its amino terminus and most of its mass, including about 500 amino acid units, on the lumen side of the reticulum membrane. The two carbohydrate chains are on the same side; it is this end of the molecule that is eventually observed as a spike on the exterior surface of the plasma membrane and on the virion. At least a few amino acid units must remain in the bilayer, and although their number is not known exactly, it is probably about 20 or 30. Only a stub made up of the last 30 units or so, including the carboxyl terminus, protrudes on the cytoplasmic side of the membrane.

How was this model of protein inser-

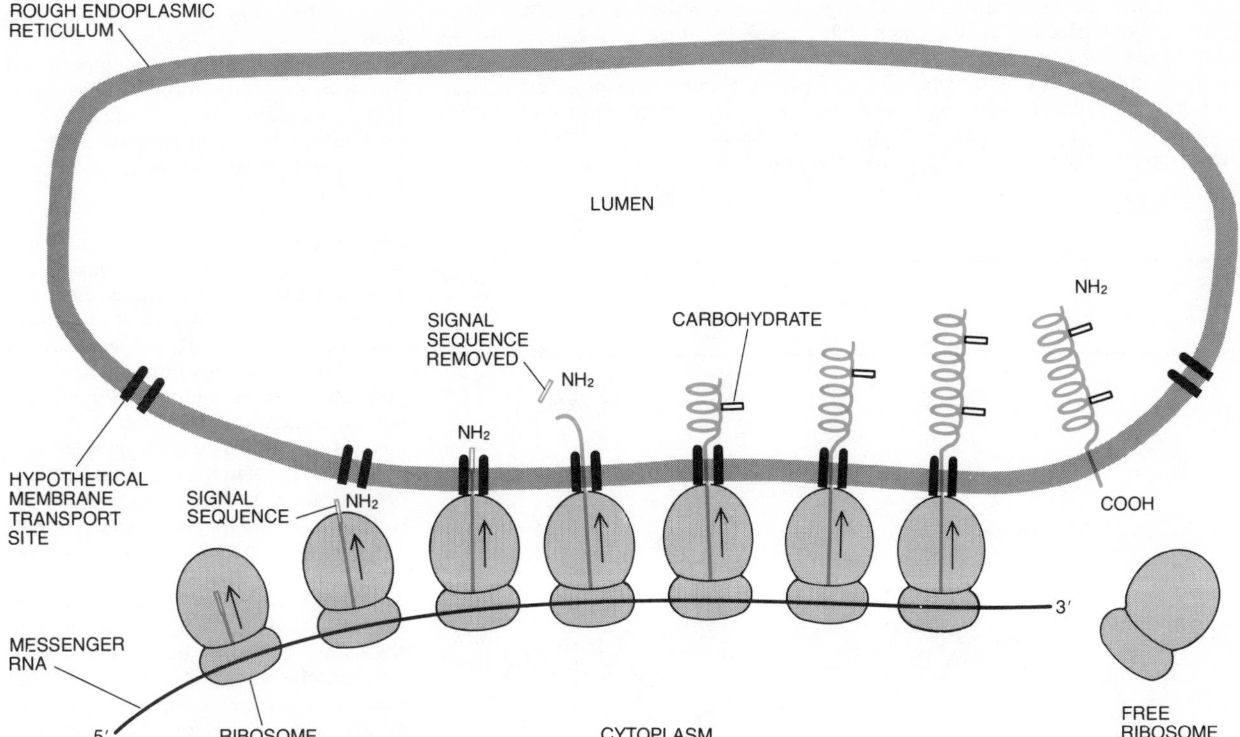

SYNTHESIS, INSERTION AND GLYCOSYLATION of the *G* protein are closely coupled. The protein, or polypeptide, is built up from amino acids linked together in a linear sequence specified by the messenger RNA. Synthesis begins with the amino terminus (NH_2) of the polypeptide; amino acids are added one at a time behind that terminus. Among the first 30 amino acid units of *G* is a "signal sequence" that identifies the protein as one destined to be inserted into the membrane of the rough endoplasmic reticulum. Because some 40 amino acids remain buried in the ribosome, the signal sequence does not emerge until the polypeptide is about 70 amino acid units long. At that time the signal sequence is recognized by some molecule, presumably a protein, in the membrane of the endoplasmic reticulum. This hypothetical protein is thought to facilitate the passage of the polypeptide through the lipid bilayer. Once in the lumen of the reticulum the signal sequence is removed. The protein continues to elongate, and as it grows it is extruded through the membrane and folds up in the lumen. As it enters, two identical, preformed carbohydrate side chains are fastened to it. Proteins secreted by the cell pass all the way through the membrane in this manner, but for reasons that are not fully understood *G* becomes stuck at about the time translation is completed, with some 30 amino acids remaining in the cytoplasm. Thus the completed glycoprotein has its amino terminus, most of its bulk and all its carbohydrate in the lumen of the reticulum, and a short stub that includes the carboxyl terminus (COOH) on the cytoplasmic side. Once the protein has folded it cannot be pulled out of the membrane, nor can it execute a flip-flop; it is anchored in an asymmetric orientation. Components of the cell are not shown to scale: the ribosome is some 50 times larger than *G* protein.

tion constructed and tested? A technique that had a part in almost all our experiments was gel electrophoresis, which separates proteins according to their size. A typical experiment yields a mixture of proteins, some in their native form, others reduced to fragments by enzymatic digestion, others with only a partial complement of carbohydrate. The mixture is placed on a porous gel and an electric field draws the proteins through the gel. Small molecules migrate through the pores of the gel faster than large ones, and so the various fractions are separated and can be identified.

In one experiment all the enzymes and other molecules required for the normal synthesis of *G* were provided except for reticulum membranes. Under these conditions the polypeptide can be assembled by the ribosome, but of course it cannot be inserted into a membrane. The resulting protein appeared to be normal in all respects except two: it had no carbohydrate, since that can be added only by enzymes of the reticulum, and it retained the signal sequence of 16 amino acid units at the amino terminus. Such nude protein is designated G_0. Protein with only core carbohydrates (and lacking the signal sequence) is G_1, and the finished protein with both core and terminal carbohydrate is G_2. Although G_1 and G_2 are normal intermediates found in infected cells, G_0 is not formed during infection.

If endoplasmic-reticulum membranes are added to the medium before translation begins, the protein is fabricated in the G_1 form, with both core chains of carbohydrate but without the signal sequence. A telling result was obtained, however, when the membranes were added while the synthesis was under way. The model presented above predicts that unless the membranes are added very early in the synthesis the emerging protein will not be inserted into the membrane. Part of the polypeptide will already have folded in the cytoplasm and will be unable to interact with receptors on the membrane or to cross the permeability barrier of the lipid bilayer. In fact, adding membranes well after synthesis had begun accomplished nothing; when it was subsequently completed, the protein remained in the G_0 form.

In order to check the orientation of the molecules, *G* protein was synthesized in the presence of membranes and after completion of protein synthesis the enzyme trypsin was added to the solution. Trypsin is a digestive enzyme that can cut a polypeptide at many points, but because it is a soluble protein it cannot cross a lipid membrane. Hence it can digest only those portions of the protein molecule that lie on the cytoplasmic side of the membrane. All molecules of *G* subjected to this procedure yielded a large fragment of *G* about 30 amino acids shorter than the normal G_1. We were able to show that only the carboxyl terminus of the polypeptide had been digested and that the large protected fragment carried both carbohydrate chains. These findings indicated that only the 30-unit carboxyl stub projects into the cytoplasm and that the carbohydrate is in the lumen of the vesicle. Moreover, the fact that all copies of *G* were reduced to the fragment indicated that all the G molecules in the membrane had the same orientation.

In order to find out exactly when insertion into the bilayer must begin, we carried out a series of experiments in which progressively longer segments of protein were synthesized before membranes were added to the medium. We found that the polypeptide was correctly inserted and went on to form a normal G_1 if the membranes were added before the first 70 amino acids had been polymerized. If the membranes were added later, only the sugarless G_0 was formed, and this polypeptide was not properly embedded in the membrane but was probably loose in the cytoplasm. Since about 40 of the first 70 amino acids are buried in the ribosome, the first 30 units must bear the crucial signal sequence.

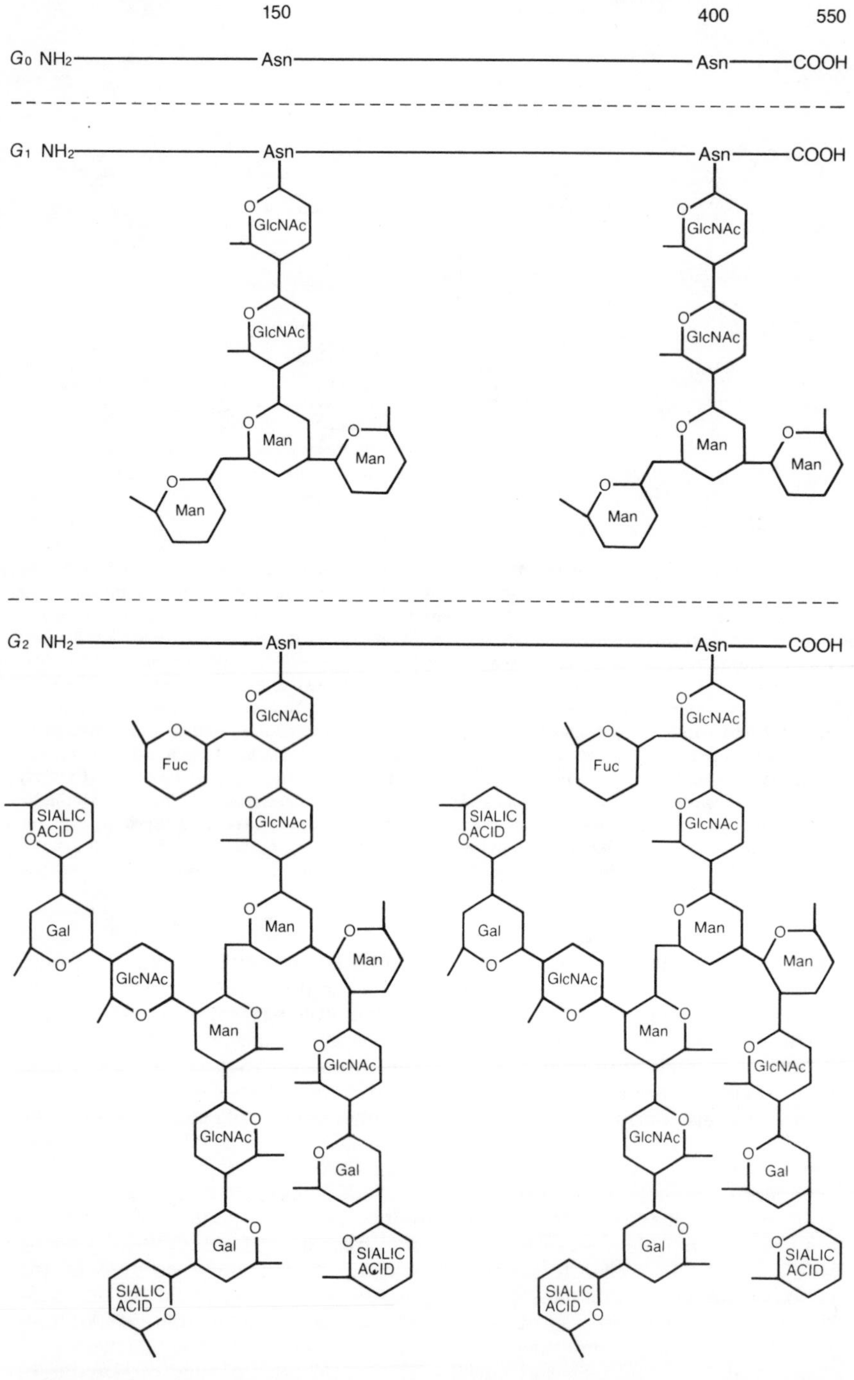

An experiment that was in some ways the obverse of this one tested the assertion that the protein is extruded through the membrane as it elongates. The interpretation of the experiment depended on the assumption that carbohydrate can be added only in the lumen of the endoplasmic reticulum.

Various lengths of polypeptide were allowed to grow in the presence of membranes; then the membrane was destroyed by dissolving it with a detergent. The detergent has no effect on protein synthesis, but it does halt the attachment of carbohydrates, probably because the enzymes that transfer the sugar side chains are bound to the lumen side of the endoplasmic-reticulum membrane and are separated from the ribosomes by detergent treatment. We found that when detergent was added before the polypeptide was about 150 amino acids long, and the protein was then completed, only the G_0 protein was observed. Thus no carbohydrates were attached before the polypeptide was 150 amino acids long. If the membrane was maintained intact until after some 400 amino acids had been incorporated in the chain, then the completed glycoprotein was the G_1 form, with two core carbohydrates.

This meant that all core sugars are added during protein synthesis, while the growing chain of G is still attached to the ribosome. It was a crucial finding. If the carbohydrate is added to G on the lumen side while the protein is still growing on a ribosome on the cytoplasmic side, then G must be extruded through the membrane during protein synthesis.

An intriguing result was obtained when we disrupted the membrane after 150 amino acids had been linked to the polypeptide but before 400 amino acids had been. The resulting protein was an intermediate form, never observed in nature, with just one carbohydrate chain. The creation of this novel intermediate provides strong evidence that the two carbohydrate structures are added in sequence as the polypeptide passes through the membrane.

The asymmetry of membrane lipids differs in detail from that of the proteins, but the same fundamental principle applies: The membrane can grow only by expansion; it cannot be broken open for the insertion of new material. In the case of the phospholipids this requirement has been met in a simple way:

CARBOHYDRATE SIDE CHAINS are constructed on the G protein in two steps. In the rough endoplasmic reticulum the growing protein receives the core regions of the two carbohydrate chains, which are made up exclusively of the monosaccharides mannose (Man) and *N*-acetylglucosamine (GlcNAc). In the Golgi apparatus the core regions are modified (additional core mannose units not shown here are removed) and the terminal sugar units are attached; these include additional *N*-acetylglucosamine as well as the monosaccharides galactose (Gal) and sialic acid. Fucose (Fuc) is also added. The completed glycoprotein has two identical carbohydrate chains, each bonded to the polypeptide through a side chain of the amino acid asparagine (Asn). The two asparagine units are thought to be about 150 and about 400 amino acid units from the amino terminus of the polypeptide. There is evidence that the core carbohydrates are attached to the protein as a unit, having been assembled on a glycolipid.

PROTEIN SYNTHESIZED WITHOUT MEMBRANES

PROTEIN SYNTHESIZED WITH MEMBRANES

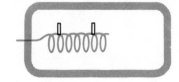

TRYPSIN TREATMENT

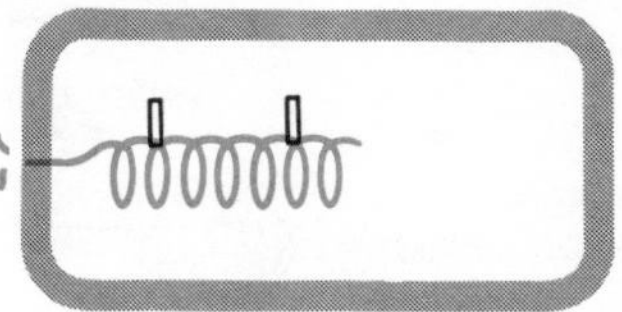

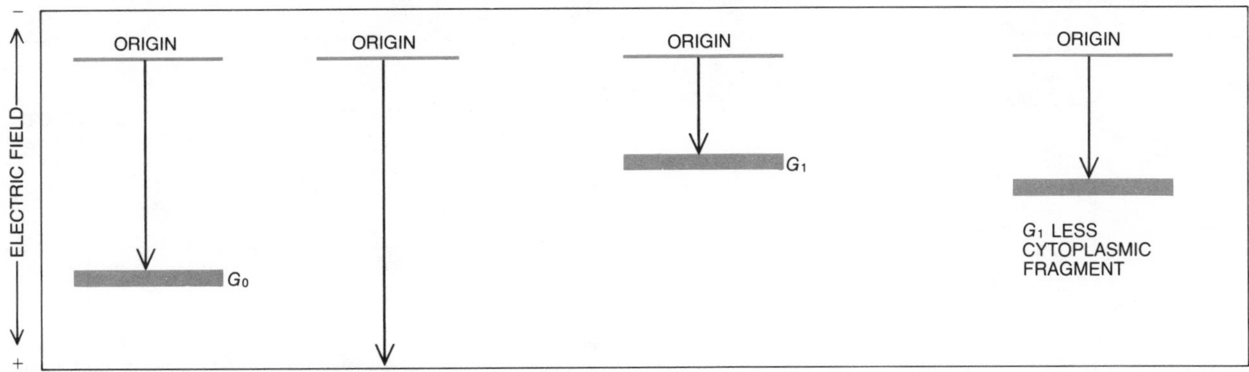

CELL-FREE SYSTEM for synthesizing *G* was employed to test the hypothesis that the protein is inserted into the membrane of the endoplasmic reticulum as the polypeptide elongates. The system is a liquid medium containing ribosomes, amino acids, various enzymes and other molecules necessary for protein growth. Synthesis is initiated by adding messenger RNA for *G* to the medium. In this system the protein can be grown under different conditions and the results assayed by gel electrophoresis, which separates molecules according to their size; small molecules move faster through the gel than large ones. If *G* is made in the absence of membranes, the product is a protein labeled G_0, which is normal except that it has no carbohydrate and retains the amino-terminal signal sequence. Repeating the procedure and adding the digestive enzyme trypsin after completion of protein synthesis reduces the protein to a multitude of small fragments, which migrate off the gel. When the membranes of the endoplasmic reticulum are added to the medium before protein synthesis begins, a different product is obtained: the protein G_1, which has the core regions of both carbohydrate chains and lacks the signal sequence. When trypsin is added in this procedure, only the cytoplasmic stub of the protein is degraded because trypsin cannot cross the lipid bilayer. The results of these experiments demonstrate that *G* acquires carbohydrate chains only if it enters endoplasmic reticulum.

the lipids are synthesized within the membrane itself.

An example is the lipid phosphatidylethanolamine, or PE, whose synthesis in the bacterium *Escherichia coli* follows a biochemical pathway that has been traced mainly by Eugene P. Kennedy and his colleagues at the Harvard Medical School. The synthesis begins with two molecules of a fatty acid, which are transferred from a donor to a molecule of glycerol phosphate, forming the most primitive phospholipid, phosphatidic acid. The two fatty acids are the hydrophobic tails of the molecule and the glycerol phosphate is the hydrophilic head. In subsequent steps a phosphorylated nucleotide, CMP, is grafted onto the head of phosphatidic acid, then the CMP is replaced by the amino acid serine. Finally, a molecule of carbon dioxide is extracted from the serine, yielding PE. Each step in the synthesis is catalyzed by an enzyme, and it has been found that all the enzymes except one are integral proteins of the bacterial plasma membrane. The lipid substrates and products of the reactions must therefore also be components of the bilayer. The membrane grows by fabricating lipids in situ.

Phospholipid synthesis is fundamentally similar in eukaryotic cells, the chief difference being that the events take place not in the plasma membrane but in the endoplasmic reticulum. Again the enzymes concerned are themselves membrane proteins, indicating that newly made lipids and intermediates are part of the bilayer. Hence the endoplasmic reticulum is a true membrane factory: it is the site where both the lipid and most of the protein components of membranes are assembled before they are dispatched to other parts of the cell.

It is easy to understand in a general way how lipid membranes can grow by expansion, but the analysis presented above overlooks the one feature of membrane structure we set out to explain, namely that bilayers have two sides, which are not identical. Are new lipids formed on only one side of the bilayer or on both sides? Many of the precursors in lipid synthesis and one of the enzymes are soluble molecules in the cytoplasm; the fact that these substances cannot cross the hydrophobic core of the membrane argues that lipid synthesis should take place only at the cytoplasmic surface. If that is the case, however, some means must be provided for ferrying lipid molecules to the external surface.

The mechanism of lipid synthesis must also be consistent with the compositional asymmetry of the bilayer. This asymmetry is not an absolute one, as it is for proteins, where each species of protein has a designated orientation observed by every molecule of that species. The asymmetry of lipid composition is better described as a statistical bias. There are no lipids found exclusively in one monolayer or the other, but most lipids do seem to favor one surface. In certain bacteria, for example, PE is roughly twice as abundant on the cytoplasmic side of the membrane as it is on the external side. Other phospholipids are found predominantly in the external monolayer.

Asymmetries of lipid composition have been investigated at the Harvard Medical School by one of us (Rothman) in collaboration with Kennedy. The membrane employed in several of these experiments was that of *Bacillus megaterium*. A bacterium of this type, called gram-positive, was chosen because the presence of only one membrane per cell simplifies the analysis of lipid composition. This particular organism was selected in part because its membrane includes large quantities of PE, which makes up some 70 percent of the total membrane lipid. Most of the remainder is phosphatidylglycerol, in which the ethanolamine group of PE is replaced by a second glycerol. In other gram-positive bacteria phosphatidylglycerol is the major lipid constituent of the membrane. The abundance of PE in *B. megaterium* is an experimental convenience because PE is easily labeled.

The distribution of PE in the membrane of *B. megaterium* was measured by

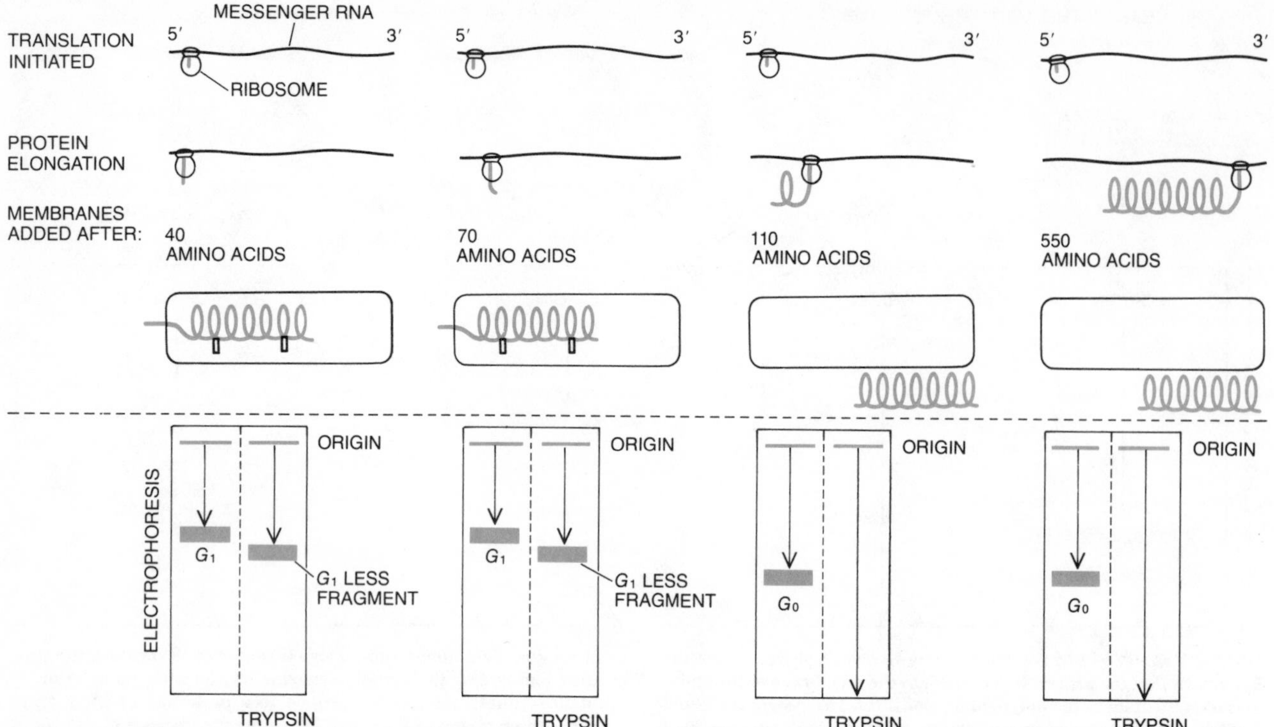

ADDITION OF MEMBRANES after protein synthesis has already begun demonstrates that insertion must be simultaneous with synthesis if it is to succeed at all. If the polypeptide is allowed to reach a length of 70 amino acid units before reticulum membranes are introduced, the protein can still penetrate the endoplasmic reticulum membrane, and it subsequently develops normally. Electrophoresis shows that it acquires the two core carbohydrates and that only a small fragment (the stub) is degraded by trypsin. With any longer delay between the initiation of synthesis and the introduction of membranes, however, the folding of the protein has progressed too far for it to enter the bilayer. Because the folding exposes many hydrophilic amino acids the polypeptide is permanently excluded from the reticulum. When proteins are analyzed by electrophoresis, they are found to be without carbohydrate and to be completely digested by trypsin.

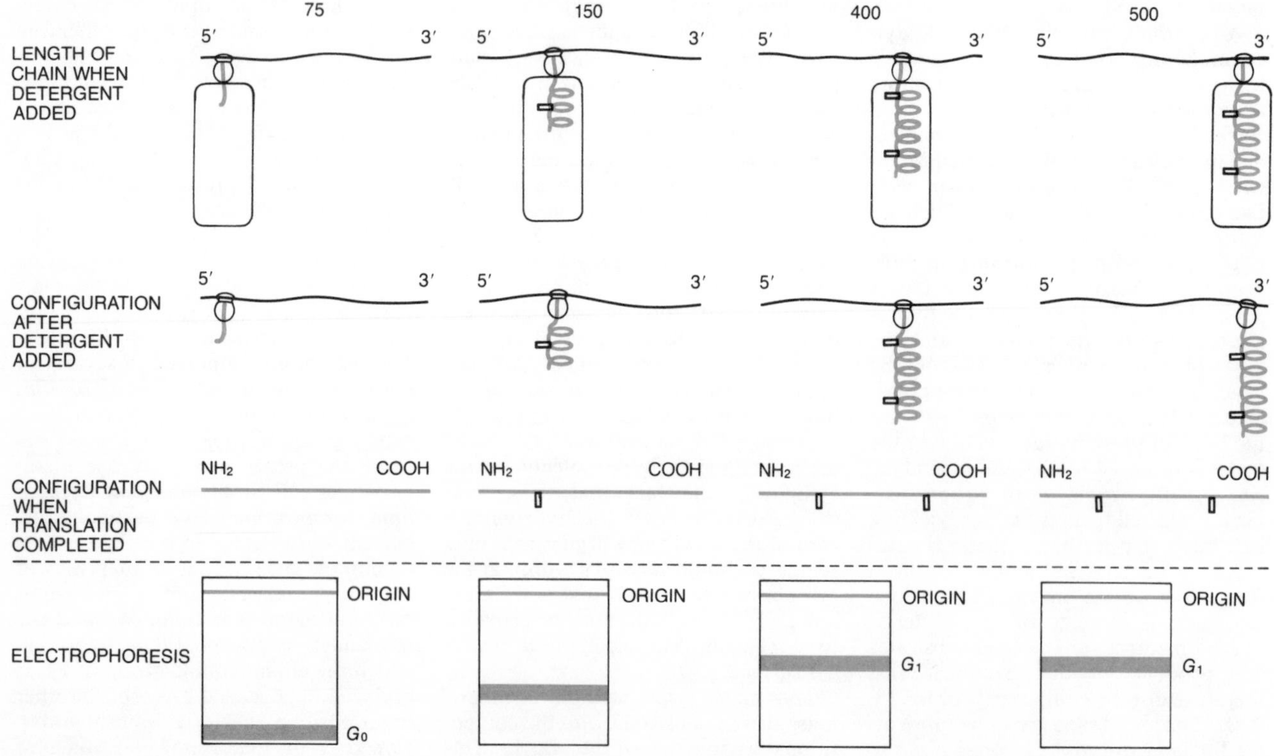

DESTRUCTION OF MEMBRANES after protein synthesis has begun proves that core carbohydrates can be joined to the polypeptide only within membranes of the endoplasmic reticulum. The translation of many copies of *G* is begun simultaneously in a medium with abundant reticulum membranes and is allowed to continue for various periods before a detergent is added to the medium, destroying the membranes. Elongation of the polypeptide is then allowed to proceed to completion in the absence of membranes. When the membranes are disrupted before the chain is about 150 amino acid units long, the only protein observed is G_0, which lacks all carbohydrate. If detergent is not added until some 400 or more amino acids are completed, then normal G_1, with two carbohydrate chains, is produced. Applying detergent at intermediate intervals yields a protein that is never observed in nature, with only the first carbohydrate chain.

labeling all the PE molecules exposed at the exterior surface but none of those at the cytoplasmic surface. The proportion of the PE on each side could then be determined by comparing the number of labeled and unlabeled molecules.

The labeling reagent was trinitrobenzenesulfonic acid, or TNBS, which reacts with the amino group of the PE molecule but cannot combine with phosphatidylglycerol or other lipids that lack an amino group. If TNBS is present in sufficient concentrations, it will tag all molecules of PE to which it can gain access; the accessible molecules are precisely those of the exterior membrane surface. TNBS cannot reach the cytoplasmic side of the membrane because it is water-soluble and cannot cross the permeability barrier. This simple but important principle of labeling membranes with nonpenetrating reagents was devised by Mark S. Bretscher of the Medical Research Council laboratories in Cambridge, who applied it in pioneering studies of lipids and proteins in the membrane of the red blood cell.

When the labeling was completed, the bacterial membranes were broken down by dissolving the lipids in an organic solvent. It was then necessary to separate the labeled and unlabeled PE, which was done by the technique of thin-layer chromatography. A sample of the extracted lipids was applied to the end of a thin sheet of a silica gel and a mixture of organic solvents was allowed to rise through the gel by capillary action, sweeping past the lipid extracts. PE labeled with TNBS is more soluble in these solvents than unlabeled PE, and so the labeled lipids were carried farther through the gel. In a sense the two halves of the phospholipid bilayer were separated by this method. When the amount of each material was determined, it was found that only about 30 percent of the PE had been labeled. It follows that about 30 percent of the PE is on the outer surface of *B. megaterium* and the other 70 percent is inside. Since the total amount of lipid in the two monolayers is roughly equal, the other major lipid, phosphatidylglycerol, must have the opposite distribution.

Through a somewhat more elaborate procedure we were able to approach the question not only of where the lipids reside in the membrane but also of where they are made. The idea of the experiment was to employ two independent labels: one would mark only the newly synthesized lipids and the other would distinguish inner-surface lipids from outer-surface ones. In this way the distribution of the new lipid material would be revealed.

In order to label freshly synthesized PE we incubated bacterial cells with inorganic phosphate prepared with the radioactive isotope phosphorus 32. The radioactive phosphate was incorporated

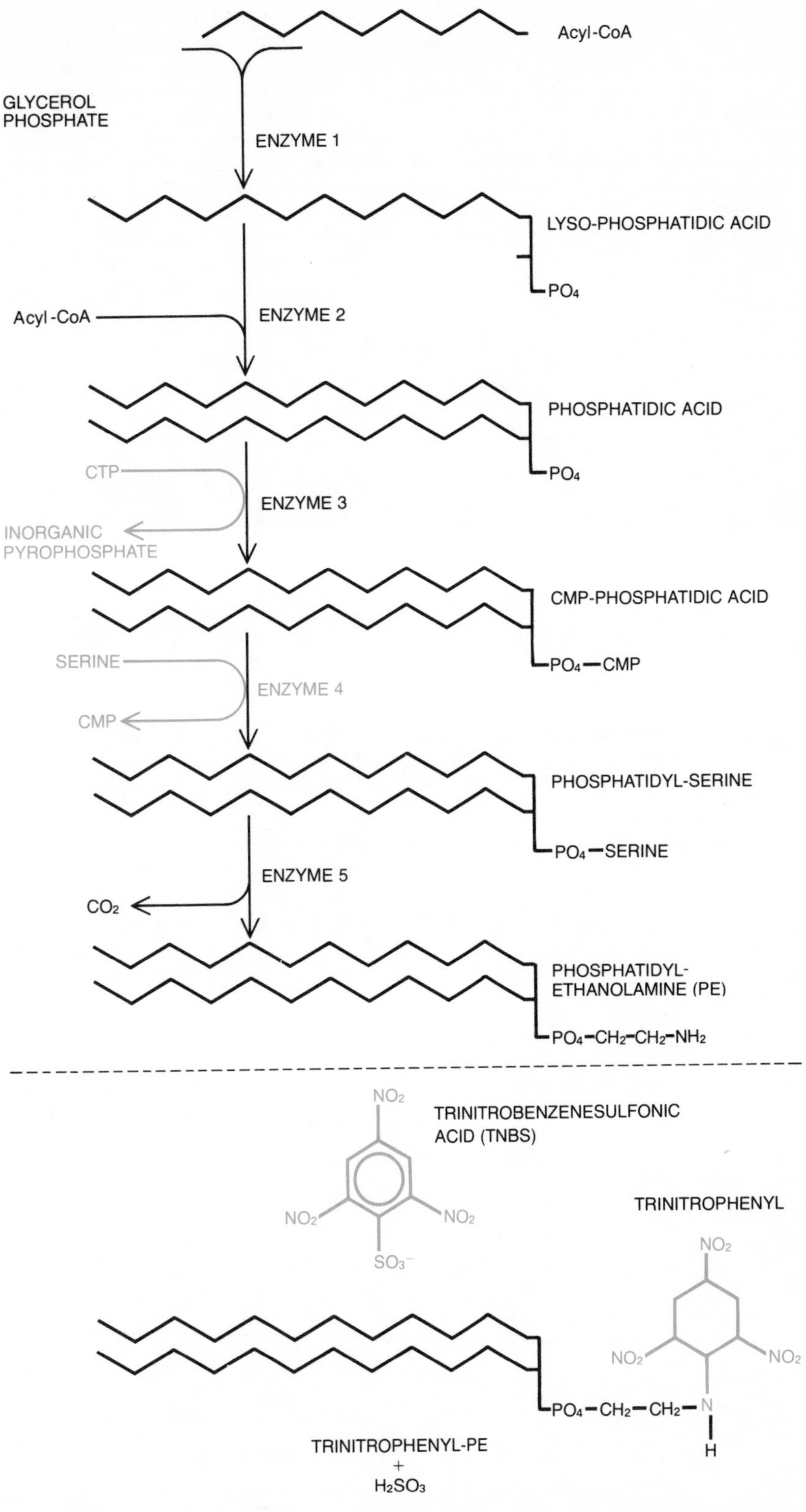

SYNTHESIS OF A LIPID relies on several enzymes that are integral membrane proteins. The lipid is phosphatidylethanolamine, or PE. Enzymes, substrates and precursors that are soluble in the cytoplasm are shown in color; all other molecules are oil-soluble and are confined to the lipid bilayer. The synthesis begins with the transfer of two fatty acids from donor molecules (Acyl-CoA) to glycerol phosphate. A molecule of cytosine monophosphate (CMP) is attached to the glycerol phosphate and is then replaced by the amino acid serine. In the last step of the synthesis PE is made by extracting a molecule of carbon dioxide (CO_2) from the serine. Because four of the five enzymes required for this procedure can be found only in the lipid bilayer, it appears that lipids are formed within the membrane itself. Because the one remaining enzyme and some substrates are present only in the cytoplasm, the lipid synthesis must take place on the cytoplasmic side of the bilayer. PE can be selectively labeled with trinitrobenzenesulfonic acid (TNBS), which bonds to the amino group in the hydrophilic head of the lipid.

into glycerol phosphate and hence into membrane lipids. The incubation lasted for only about a minute, and so only the lipids made during that period would exhibit radioactivity. TNBS was then added to the suspension of cells and allowed to combine with PE molecules on the exterior surface of the membrane. The membrane was next dissolved and the lipids were separated by chromatography and tested for radioactivity. If both the TNBS-labeled PE and the unlabeled PE were radioactive, the lipid must be synthesized in both monolayers, since these fractions come from opposite sides of the membrane. We found, however, that radioactivity could be detected only in the PE that had not combined with TNBS, indicating that synthesis of this lipid is confined to the cytoplasmic surface, which TNBS cannot reach.

This finding is consistent with the cytoplasmic origin of the lipid precursors, but it also raises a troubling question. If lipids are made at the cytoplasmic surface and immediately become part of the cytoplasmic monolayer, how do they ever get to the other side of the membrane? As we pointed out above, the rate of spontaneous flip-flops across the bilayer is exceedingly low, so that a given lipid could be expected to make the transition only about once a month. Yet a bacterium can double in mass and divide in less than an hour, thereby doubling the surface area of the plasma membrane. If lipids could not cross from the inner surface to the outer one, the bilayer would soon become a lipid monolayer.

Another experiment has shown that lipids do in fact cross the membrane, although it has not revealed how they do so. Bacteria were again briefly incubated with radioactive phosphate, but instead of their being labeled immediately with TNBS another reagent was added that inhibits further lipid synthesis. The inhibitor, hydroxylamine (NH_2OH), inhibits the last enzyme in the sequence leading to PE, the enzyme that extracts a molecule of carbon dioxide from serine. TNBS could then be added after various intervals and would show how much lipid had migrated across the membrane in the period after PE synthesis was halted.

When TNBS was added immediately after the inhibitor, essentially all the radioactivity was found in the inner monolayer, again showing it to be the site of lipid synthesis. If the labeling with TNBS was delayed by 30 minutes, however, we found that the distribution of new, radioactive PE was the same as the distribution of the old, nonradioactive lipid, with about 30 percent of both on the exterior surface. Evidently the new PE molecules were able to cross the membrane rapidly enough to reach their natural asymmetric distribution.

SITE OF LIPID SYNTHESIS was determined by independently labeling newly made PE and the fraction of the PE that is exposed on the outer surface of a bacterial cell. The newly formed PE is labeled by briefly incubating the cell with radioactive phosphate, which is incorporated in all phospholipids. The external PE is marked with TNBS, which combines with every PE molecule it can reach but which cannot cross the bilayer. When the membrane is dissolved, the lipids can be segregated according to whether or not they are radioactive and whether or not they are labeled with TNBS. (The latter determination is made by thin-layer chromatography.) Almost none of the radioactive PE molecules—the recently synthesized ones—are labeled with TNBS, indicating that the lipid is made only on the cytoplasmic side of the membrane.

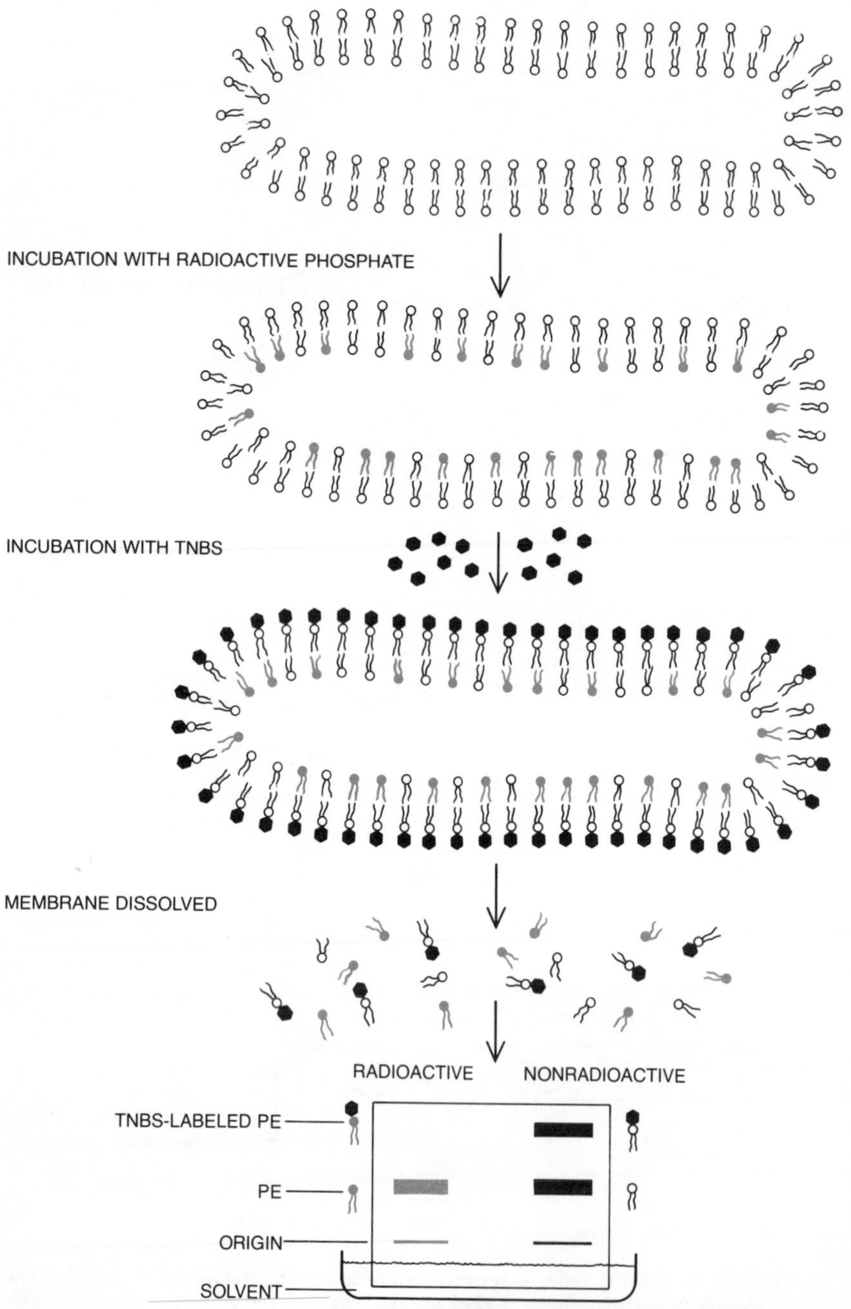

From a series of such measurements we calculated that at physiological temperature the time required for a batch of new lipids to return halfway to their final distribution is about five minutes. That is almost 100,000 times faster than the rate of flip-flop observed in other membrane systems, and it cannot plausibly be attributed to spontaneous flip-flops. We have therefore proposed that growing membranes include proteins that facilitate the equilibration of lipids across the membrane, a notion first put forward by Bretscher. Such a protein need not influence the direction of lipid movement; it might merely provide a channel. The proteins would be needed only where membranes are actively growing, such as the plasma membrane of bacteria and the endoplasmic reticulum of eukaryotes; their absence elsewhere might explain the much lower rate of interchange between the layers of other membranes. It should be emphasized, however, that no such proteins have yet been shown to exist.

In this account of the lipid bilayer one further perplexity arises: If lipids can pass freely from one side of the membrane to the other in a growing membrane such as that of *B. megaterium*, how can the asymmetry of lipid composition be maintained? The explanation now favored by many investigators is that the asymmetry is not actively maintained at all but reflects a thermodynamic equilibrium between the molecules at the two surfaces. This hypothesis assumes that some species of lipids have a lower free energy on one side of the membrane than they do on the other. The difference in energy need not be great to explain the observed asymmetries, which are almost always modest compared with those of proteins. It is not hard to imagine a mechanism that might give rise to such a small energy difference. For example, if proteins found only at the cytoplasmic surface or ions in the cytoplasm bound PE with greater affinity than the proteins outside the cell, then each PE molecule might spend more of its time on the cytoplasmic side. In this way an asymmetric distribution would arise even though the lipid molecules were not physically restrained.

This hypothesis makes the asymmetries of proteins and lipids fundamentally different. Protein asymmetry is a nonequilibrium state enforced at the time of synthesis; if the proteins are allowed to come to equilibrium (by disrupting and reconstituting the membrane), they assume a more random configuration. Lipid asymmetry, according to the hypothesis, results from a somewhat biased equilibrium and does not depend on how or where the lipids are synthesized; if they were inserted in the outer monolayer, their ultimate distribution would be the same.

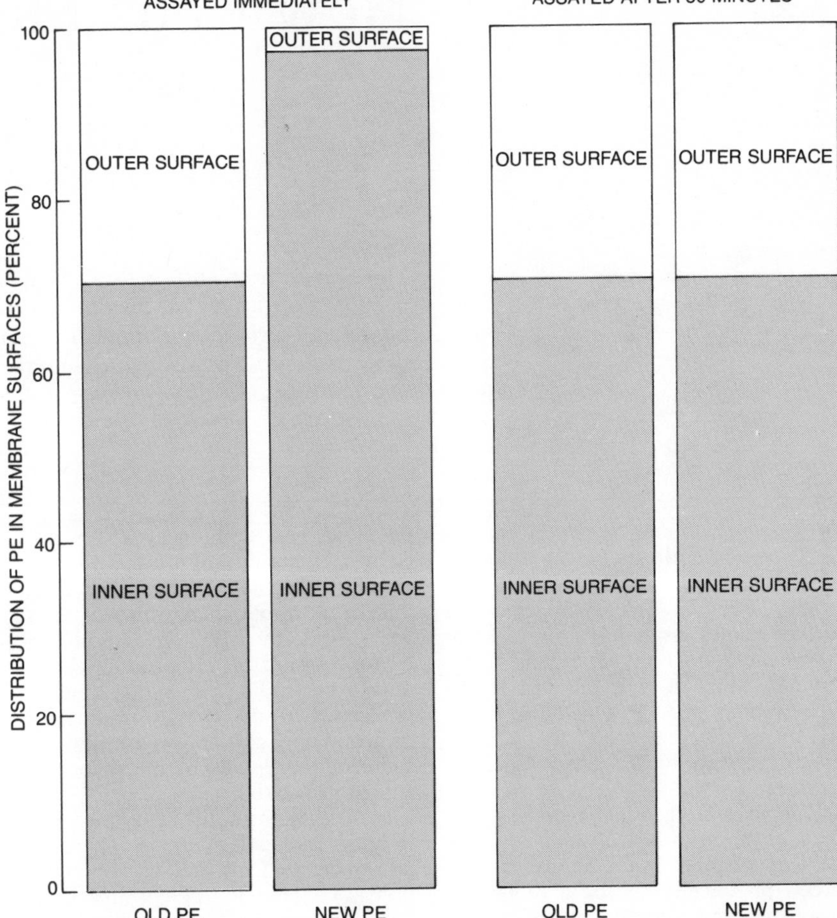

TRANSPORT OF LIPIDS across the bilayer is required if new lipid is formed at only one surface. In the bacterium *Bacillus megaterium* the distribution of PE in the two layers of the membrane is asymmetric: about 70 percent of it is found at the cytoplasmic surface. When the distribution of newly synthesized PE is measured, however, as in the experiment diagrammed on the opposite page, almost all of it is found at the inner surface. In order to maintain the integrity of the bilayer some of the new material must be ferried to the outer surface. In fact, within just 30 minutes the distribution of newly made PE matches that of the older material. It has been proposed that proteins in the growing bacterial membrane facilitate the movement of lipids across the bilayer, perhaps allowing them to reach thermodynamic equilibrium.

Our conclusions about the synthesis of lipids in bacterial membranes have recently been extended by other investigators who have found similar mechanisms at work in the endoplasmic reticulum of eukaryotic cells. Rosalind Coleman and Robert M. Bell of Duke University have shown that the enzymes responsible for the synthesis of several phospholipids all have their active sites on the cytoplasmic side of the reticulum membrane. Donald B. Zilversmit and his colleagues at Cornell University have measured transmembrane movement in the reticulum. They find that the lipids reach an equilibrium across the membrane of the endoplasmic reticulum in minutes, just as they do in bacteria. Therefore our model of how the lipid bilayer is assembled appears to be of general validity.

The studies described here, although they are confined to a few simple experimental systems, have brought to light several principles of membrane synthesis that seem to be widely applicable. It is only fair to point out, however, that other aspects of membrane assembly remain obscure.

One aspect is the transport of assembled membrane material from the endoplasmic reticulum to other sites in the cell. Each of the various organelle membranes, and the plasma membrane, has its own characteristic suite of lipids and proteins; to preserve these distinctions newly assembled membrane materials must somehow be sorted out as they leave the endoplasmic reticulum or the Golgi apparatus to ensure that they reach the proper destinations. There are also proteins peculiar to the endoplasmic reticulum itself, and it is not clear what prevents them from being exported along with all the rest. The supposed mechanism of intracellular transport—vesicles that bud off from the reticulum and fuse with another membrane—may well serve to ferry material to the plasma membrane, but it meets with difficulties elsewhere. Certain membrane proteins of mitochondria, for example, must cross both of the membranes surrounding that organelle.

Another deficiency of our models is that they cannot yet describe in satisfactory detail the one event that is crucial to the emplacement of both proteins and lipids: transport across the bilayer. For proteins this happens only once, at the time of synthesis and insertion. For lipids there is probably a rapid but undirected interchange for as long as new material is being added to the membrane. In both cases a membrane-bound protein probably serves as a gatekeeper, but the identity of these proteins and how they work are not known. It is in these areas, where present understanding is most vague and tentative, that the results of future investigations should prove most illuminating.

12 How Cells Make ATP

by Peter C. Hinkle and Richard E. McCarty
March 1978

The prevailing theory is the "chemiosmotic" one. Light or oxidation drives protons across a membrane; then the energy-rich compound ATP is formed as the protons flow back through a complex of enzymes

Energy acquired by living cells is conserved in useful form mainly as molecules of adenosine triphosphate, abbreviated ATP. Whether the energy comes ultimately from light or from the oxidation of organic compounds, most of it is invested in the manufacture of ATP, which then serves as an "energy currency" that can be spent in powering the other functions of the cell.

The overall chemical reactions that lead to the synthesis of ATP are now well understood. In the chloroplasts of green plant cells hydrogen atoms—or electrons and protons—are extracted from water. The protons are released into solution and the electrons are driven by the energy of light through a sequence of carrier molecules. Eventually the electrons and the protons combine with carbon dioxide to form organic molecules. In the mitochondria of all cells with nuclei electrons donated by organic molecules are passed along a similar chain of carrier molecules and are ultimately accepted by oxygen, forming water. Many of the intermediate stages in these energy transformations have been worked out, but one crucial step has remained puzzling. It has not been clear how the transfer of electrons through the series of carrier molecules is coupled to the synthesis of ATP. Enough energy for ATP formation is made available by the electron transfer, but the mechanism of ATP synthesis has proved difficult to characterize.

A hypothetical mechanism for coupling electron transfer to ATP synthesis was proposed in 1961 by Peter Mitchell of the Glynn Research Laboratories in England. Mitchell suggested that the flow of electrons through the system of carrier molecules drives positively charged hydrogen ions, or protons, across the membranes of chloroplasts, mitochondria and bacterial cells. As a result an electrochemical proton gradient is created across the membrane. The gradient consists of two components: a difference in hydrogen ion concentration, or *p*H, and a difference in electric potential. The synthesis of ATP is driven by a reverse flow of protons down the gradient. Mitchell's proposal is called the chemiosmotic theory. In the past 15 years work by Mitchell and his colleague Jennifer M. Moyle and by many other workers has shown that the basic postulates of the chemiosmotic theory are almost certainly right, although some of the details of the theory are still controversial.

In the chemiosmotic theory there is no isolated molecular engine where the flow of energy from light or from oxidation is coupled to ATP synthesis. Instead the crucial role is played by a membrane that divides one region from another. The membrane provides much more than shelter, confinement and a controlled internal milieu. It is an asymmetrical arrangement of the carrier molecules across the membrane that allows the proton gradient to be established. Even the topological properties of the membrane are important: it must form a closed envelope if the proton gradient is to be maintained, so that in energy metabolism the most important distinction is the one between inside and outside.

An essential concept for an understanding of ATP synthesis is that of free energy, which measures the amount of energy in a chemical system available for doing work. Any chemical reaction, whether or not it takes place in the living cell, can proceed only in the direction of lower free energy.

The reaction of carbohydrate with oxygen (to yield carbon dioxide and water) releases a large quantity of free energy. Hence the reaction, which takes place in the physiological process called respiration, is thermodynamically favored. In photosynthesis the same overall reaction proceeds in reverse: carbon dioxide and water are combined to yield carbohydrates and molecular oxygen (O_2). By itself this reaction would call for an increase in free energy, and so it can take place only when energy is supplied from an external source. The energy, of course, comes from sunlight; several photons, or quanta of light, must be absorbed for every molecule of carbon dioxide converted into carbohydrate. The photons are absorbed by pigment molecules, which are thereby promoted to an excited state of high free energy. The overall reaction, in which the excited pigments are considered among the reactants, proceeds as required by the laws of thermodynamics in the direction of lower free energy.

Energy constraints of the same kind apply to the synthesis of ATP. The ATP molecule consists of a nitrogen-containing organic base (adenine), a sugar (ribose) and a string of three phosphate groups. It is one of the class of molecules called nucleotides. In most of the reactions where ATP serves as an energy currency only the terminal phosphate group is involved. ATP is formed by attaching a third phosphate to adenosine diphosphate (ADP) with the elimination of water. The reaction does not proceed spontaneously; on the contrary, a substantial quantity of energy must be supplied. Much of that energy can then be recovered in the reverse reaction, in which ATP is split into ADP and inorganic phosphate. By donating the terminal phosphate group to other molecules, ATP creates phosphorylated species of high free energy, which can participate in reactions that otherwise could not take place.

The energy transactions of biochemistry are conveniently measured in units of kilocalories per mole. One kilocalorie is the quantity of energy needed to raise the temperature of a kilogram of water one degree Celsius. A mole is the amount of a substance in grams that is numerically equal to its molecular weight; a mole of the sugar glucose, for example, weighs 180 grams. The complete oxidation of glucose yields about 700 kilocalories per mole. The amount of energy needed to form ATP depends on the chemical environment, but it is never more than about 15 kilocalories per mole. Hence each mole of glucose oxidized provides enough energy for the

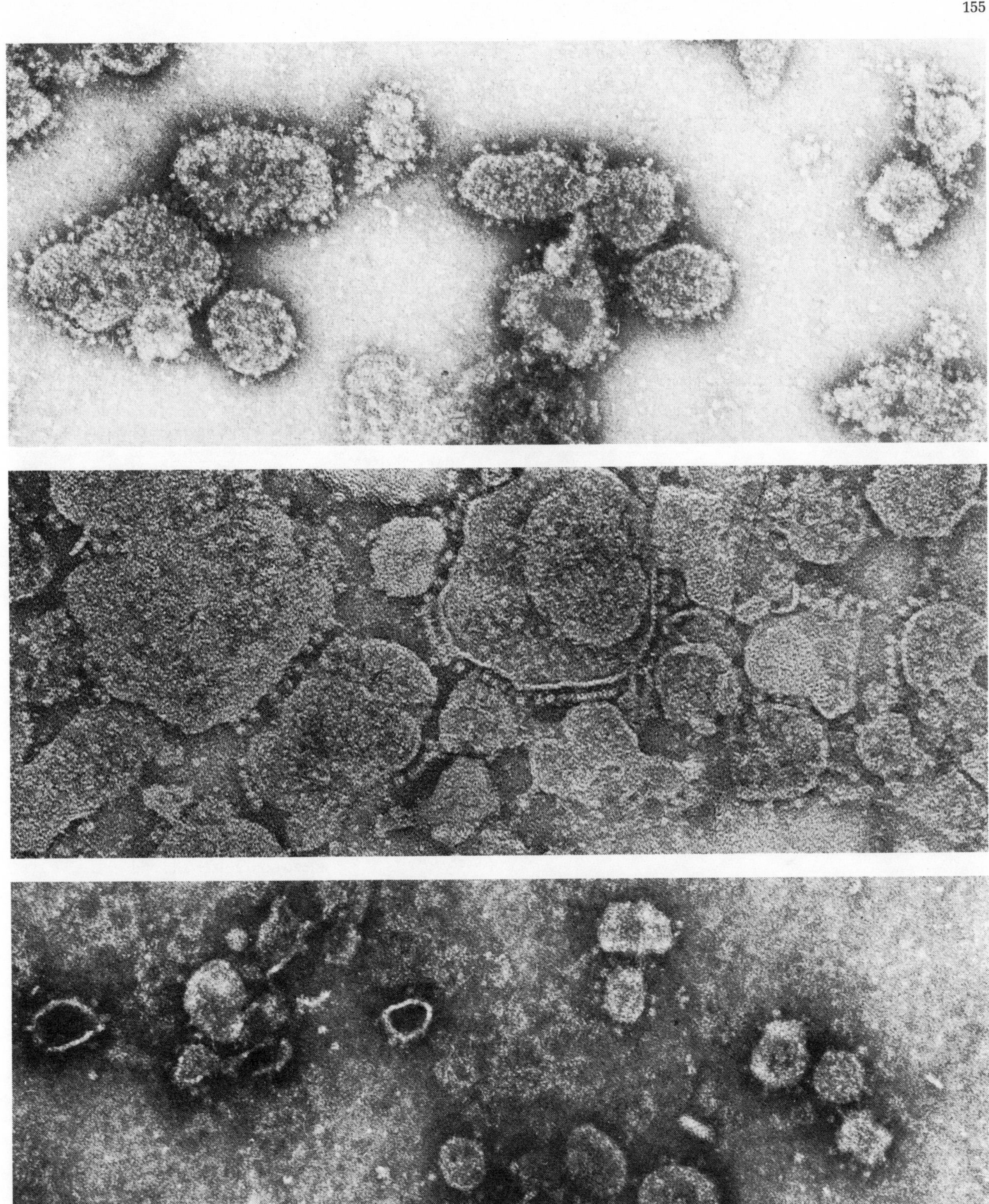

MEMBRANE VESICLES display "knobs" on their surface that have been identified as part of an enzyme complex capable of synthesizing adenosine triphosphate, or ATP. The vesicles at the top were prepared from the inner membranes of the subcellular organelles called mitochondria; those in the middle are from the inner membranes of chloroplasts, the photosynthetic organelles found in green plant cells; the vesicles at the bottom are from membranes of the bacterium *Escherichia coli*. All the membrane systems are capable of making ATP, and their similarity in appearance suggests that they employ similar mechanisms. The knobs are made up of several protein subunits, known collectively as F_1; they are attached to a membrane protein called F_0. In the vesicles shown here the F_1 protein protrudes from the outer surface of the membrane, but that is its natural orientation only in chloroplasts. In mitochondria and bacteria the protein is ordinarily inside; it is outside here because the disrupted membranes form inside-out vesicles. The mitochondrial and chloroplast vesicles were prepared by the authors and the bacterial vesicles by E. Hertzberg. The micrographs were made by John N. Telford of Cornell University.

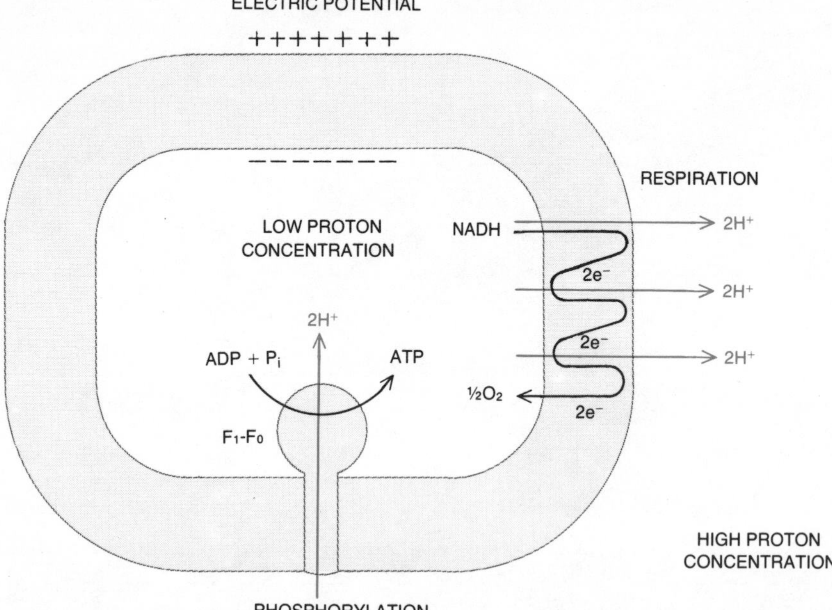

OXIDATIVE PHOSPHORYLATION couples the release of energy by the oxidation of molecules derived from carbohydrates and fats to the synthesis of ATP. In cells with nuclei oxidative phosphorylation takes place only in mitochondria. The basic events of the mitochondrial process are depicted here as they are interpreted in the chemiosmotic theory. Electrons and protons (or hydrogen atoms) from carbohydrates and fats are conveyed by molecules of the hydrogen carrier NADH to a system of enzymes in the mitochondrial membrane. In respiration pairs of electrons cross the membrane three times, each time transporting two protons from inside the mitochondrion to outside. The result is a gradient in proton concentration and electric potential that tends to force the protons back through the membrane. The energy of the gradient drives the process of ATP synthesis. For each two protons traversing the F_1-F_0 complex one molecule of ATP is formed from adenosine diphosphate (ADP) and inorganic phosphate (P_i).

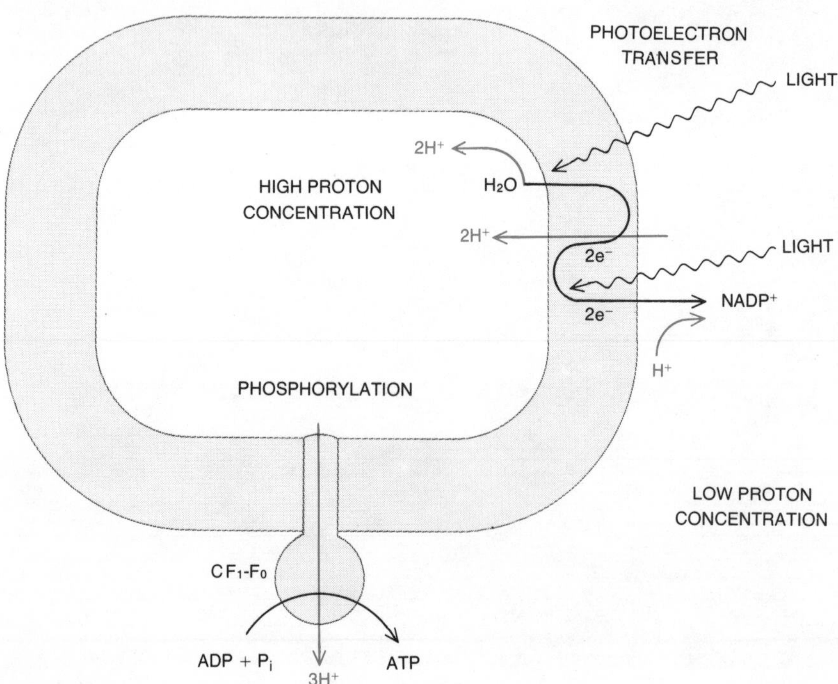

PHOTOSYNTHETIC PHOSPHORYLATION in chloroplasts derives the energy needed for making ATP from light. As in oxidative phosphorylation, hydrogen ions are transported across the membrane to create a proton gradient, and ATP is synthesized as the protons flow back across the membrane down the gradient. In chloroplasts, however, the direction of proton flow is reversed: the light-driven movement of electrons pumps protons inward, making the interior acidic, and phosphorylation is driven by an outward flow. Moreover, the stoichiometry, or ratio of reactant molecules and ions, is different from that in mitochondria. Each two electrons cross the membrane only twice, translocating only four protons, and for each molecule of ATP formed three protons must pass through the enzyme complex, which is designated CF_1-F_0.

synthesis of about 46 moles of ATP. Actually no more than 36 moles of ATP, and perhaps as few as 25, are formed for each mole of glucose oxidized. The coupling of these reactions is clearly allowed thermodynamically.

If a chemical process leads to a state of lower free energy, the reaction can proceed, but it should not be assumed that it necessarily will. Glucose and oxygen, for example, are quite stable at room temperature; they react (the glucose burns) only when they are heated. The applied heat represents an activation energy, which is recovered along with the 700 kilocalories per mole evolved by the oxidation. In biological systems the need for activation energy is reduced by the catalytically active proteins called enzymes. Because virtually all biochemical reactions are mediated by enzymes they can take place at physiological temperatures and pressures and they are confined to particular chemical pathways. It is important to note that enzymes do not affect the direction in which a reaction proceeds, only the rate of the reaction. The action of an enzyme is somewhat like that of a lubricant: it cannot make the reaction go uphill, but it can make it roll downhill faster.

In oxidative metabolism almost 95 percent of the energy captured in ATP is derived from the transfer of electrons from carbohydrates or other substrates to oxygen. In photosynthesis all the energy is derived from the transport of electrons in the other direction, from water to carbon dioxide. The chemiosmotic theory is concerned with those stages of metabolism that begin with electron transfer and end with ATP synthesis. In mitochondria and bacteria these processes are called oxidative phosphorylation; in chloroplasts and photosynthetic bacteria they are called photosynthetic phosphorylation. Many of the details of both systems remain uncertain. What we shall present here is the theory of phosphorylation that now seems to be most nearly in accord with experimental observations.

The overall chemistry of oxidation and of photosynthesis consists in a transfer of hydrogen, but it is not necessary at each stage of the process to transport complete hydrogen atoms. Indeed, in the chemiosmotic theory hydrogen carriers alternate with molecules that carry only electrons. The use of electron carriers is possible because protons are soluble in water and in the aqueous medium of the cell, whereas electrons are not. When a molecule that bears a whole hydrogen atom interacts with another molecule that accepts only electrons, the proton is released into solution. When an electron carrier then donates its electron to a hydrogen carrier, the hydrogen atom is reconstituted by withdrawing a proton from the medium.

The transfer of an electron or a hydrogen atom from one molecule to another is called an oxidation-reduction reaction. The molecule that donates the electron or the hydrogen is said to be oxidized by the molecule that accepts it; the electron acceptor or hydrogen acceptor, conversely, is said to be reduced. Whenever one substance is oxidized, another must be reduced. In this context the term oxidation sometimes causes confusion. Oxygen is indeed a strong oxidizing agent, but the chemical process of oxidation is a general one that can proceed in the absence of oxygen. Any acceptor of electrons can be regarded as an oxidant.

In the mitochondrion hydrogen is extracted from carbohydrates through the complicated series of chemical transformations called the citric acid cycle. The details of the cycle will not concern us here; it suffices to report that its overall effect is to break down the carbon chain of glucose to carbon dioxide and to deliver the liberated hydrogen atoms to the molecule nicotinamide adenine dinucleotide, or NAD^+. The plus sign has been included in the abbreviation to show that the molecule ordinarily carries a positive electric charge. Each molecule of NAD^+ accepts two electrons and one proton. The proton and one electron bind to a carbon atom in the NAD^+ molecule; the other electron neutralizes the positive charge. This reduced form of NAD^+ is designated NADH. NADH is the principal intermediary between the citric acid cycle and the enzymes in the inner mitochondrial membrane that eventually deliver electrons to oxygen, forming water. The overall process is called respiration, and the hydrogen carriers and electron carriers in the membrane are called the respiratory chain.

In the version of the chemiosmotic theory of mitochondrial respiration that now seems most plausible, each pair of electrons transferred from NADH to oxygen results in the outward translocation of six protons across the membrane. The ratio is expressed in terms of pairs of electrons because the electrons appear in pairs both at the beginning of the respiratory chain (at NADH) and at the end (where they reduce an oxygen atom to water). Within the respiratory chain the electrons are transported one at a time by some carriers and in pairs by others.

NADH donates its two electrons and one proton to a carrier group called flavin mononucleotide, or FMN. In this process NADH is oxidized, or restored to the form NAD^+, and FMN, having accepted two electrons and a proton, takes up an additional proton from the medium inside the membrane and is thereby reduced to $FMNH_2$. The FMN molecule is attached to a large protein, which is embedded in the mitochondrial membrane and probably extends all the way across it.

By a mechanism that has not yet been elucidated $FMNH_2$ transfers the two hydrogen atoms from the interior surface of the membrane to the exterior surface. There the atoms are ionized and the protons are released in the extramitochondrial medium, so that the first two protons have been transported across the membrane. In one hypothetical transport mechanism the protons are released by the flavin group inside the membrane, and they reach the external

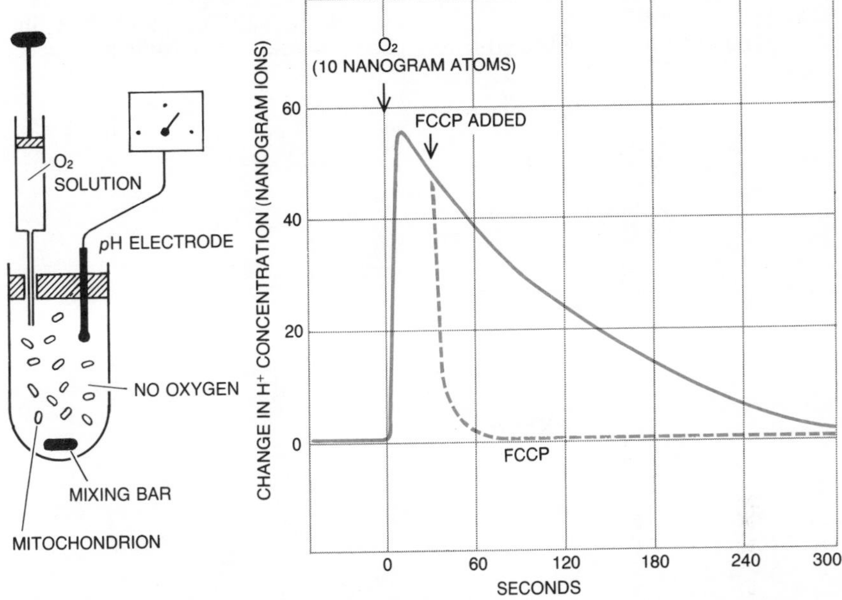

PROTON TRANSPORT by active mitochondria can be observed as a change in the concentration of hydrogen ions in the medium surrounding the organelles. The mitochondria were suspended in a medium that contained a source of electrons for respiration but had been depleted of oxygen. When a pulse of oxygen was introduced into the suspension, the *p*H of the medium declined sharply, a change that corresponds to an increase in the concentration of protons. When the oxygen was used up, protons slowly leaked back into the mitochondria. FCCP, a molecule that makes membranes permeable to protons, dissipated the proton gradient quickly.

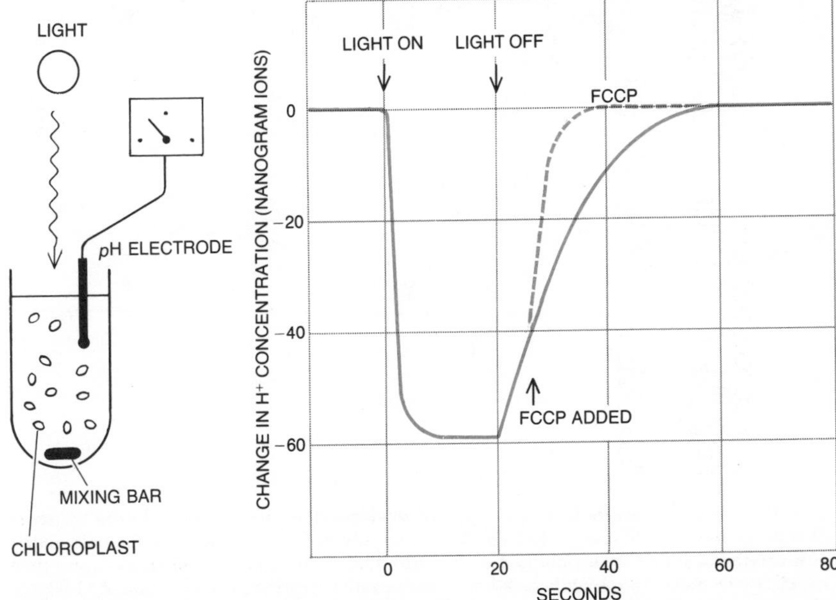

CHANGE IN *p*H in a suspension of chloroplasts is in the direction opposite to that observed with mitochondria. The chloroplasts were suspended in a medium containing an acceptor of hydrogen atoms and were kept in the dark. When the suspension was illuminated, the concentration of hydrogen ions declined (indicating that protons were being taken up by the organelles), and when the light was turned off, the gradient decayed slowly. Again FCCP accelerated the leakage of protons, which in this case was out of the organelle and into the medium.

medium through a channel in the protein molecule.

As the two protons are released by FMN the two electrons are passed on to other proteins, which because of the elements associated with them are called iron-sulfur proteins, or FeS. Having given up two protons and two electrons, $FMNH_2$ is restored to its original form of FMN and can be reduced again by NADH.

Unlike FMN, the iron-sulfur proteins carry only electrons, not complete hydrogen atoms; moreover, they carry the electrons one at a time rather than in pairs. There are several types of iron-sulfur proteins in this part of the respiratory chain, and they may be arranged across the membrane so that they can transfer the electrons back to the interior side. The electrons then enter the most complicated part of the respiratory chain, and the part where the chemiosmotic theory is the most speculative. Mitchell has recently proposed a mechanism for electron transfer through this section of the chain, which we shall present, although definitive experimental support for the idea is still lacking.

The carrier to which the iron-sulfur proteins donate the pair of electrons is a small molecule called ubiquinone, or coenzyme Q; it is usually abbreviated Q. Mitchell calls his proposed mechanism the Q cycle. Ubiquinone is a ring-shaped molecule with two oxygen atoms attached to it and with three possible oxidation states. In the fully oxidized, or quinone, form both oxygen atoms are connected to the ring by double bonds. Attaching a hydrogen atom to one of the oxygens creates the semiquinone form, or QH•. In the fully reduced form, called hydroquinone or QH_2, both oxygens have hydrogen atoms attached.

In the proposed Q cycle the iron-sulfur proteins donate an electron to each of two ubiquinone molecules, which take up protons from the internal medium to form two molecules of the semiquinone, QH•. Two more electrons are then supplied by molecules of another protein in the respiratory chain, cytochrome *b*. (The source of these electrons will be discussed below.) By taking up two more protons from inside the mitochondrion two molecules of fully reduced QH_2 are formed.

Ubiquinone is soluble in the lipid matrix of the membrane, and it therefore may be mobile. Unlike the other carriers, which presumably remain stationary and serve merely as conduits for hydrogen atoms or electrons, ubiquinone may migrate as a molecule from one side of the membrane to the other. In support of this hypothesis is the observation that there are about 10 times as many ubiquinone molecules in the membrane as there are molecules of other components of the respiratory chain. In the proposed Q cycle the two molecules of hydroquinone cross from the inner surface of the membrane, where they acquired a total of four hydrogen atoms, to the outer surface. There each QH_2 molecule gives up an electron to the next protein in the respiratory chain, cytochrome c_1, and releases a proton outside the mitochondrion. Hence the total number of protons transported so far is four.

Each of the ubiquinone molecules is now in the semiquinone, or QH , state. They complete the cycle by returning to the fully oxidized form. Each gives up its remaining proton to the external medium and transfers the associated electron to cytochrome *b*. All six protons have now been moved across the membrane and expelled from the mitochondrion, but the fate of four electrons remains to be determined: two were deposited with cytochrome c_1 and two

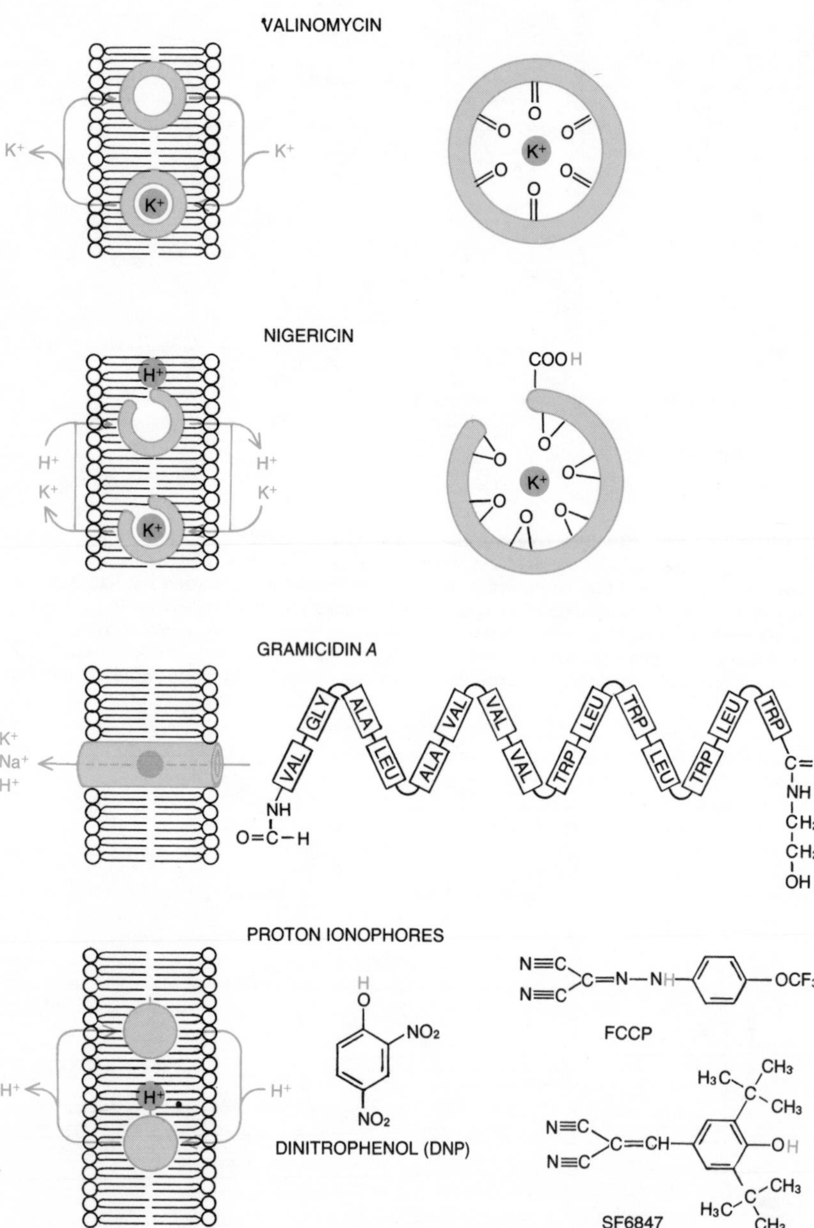

IONOPHORES, molecules that carry ions across membranes, are essential tools in the study of chemiosmotic mechanisms. Most simple ions are soluble in water but not in the lipid matrix of membranes, and so the membranes are impermeable to them. The ionophore valinomycin is a cyclic molecule that is soluble in membranes and selectively carries potassium ions (K^+). Nigericin also transports potassium, but only in exchange for protons; it is insoluble in the membrane unless one ion or the other is bound. Gramicidin *A* forms a pore in the membrane, made up of two helical molecules, that is permeable to ions with one positive charge. The proton ionophores were first studied as molecules classified functionally as "uncouplers": they inhibit phosphorylation but stimulate respiration. That mode of action is explained by their ability to dissipate a proton gradient. The proton ionophores are all weak acids that readily gain or lose a proton (*at the positions shown in color*) and can diffuse across the membrane in either state.

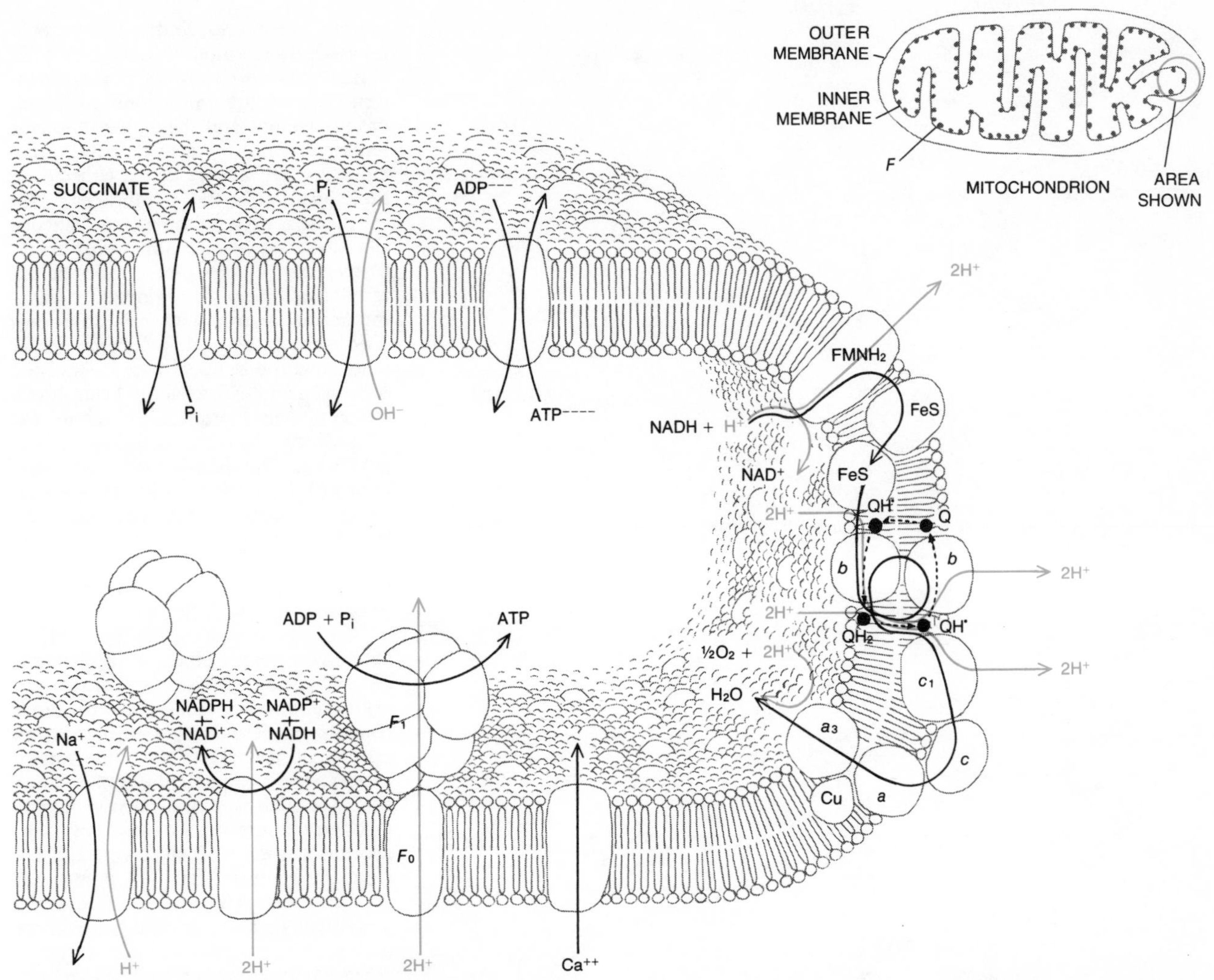

MEMBRANE OF THE MITOCHONDRION has embedded in it the enzymes and other components of the respiratory chain. The arrangement of the molecules, however, is not yet certain, and the model presented here is somewhat conjectural. Respiration begins with NADH, which gives up two electrons and a proton to flavin mononucleotide (FMN); another proton is picked up from the internal medium, so that the reduced form of the molecule ($FMNH_2$) carries two complete hydrogen atoms. The protons are expelled and the electrons return through an iron-sulfur protein (FeS) to the inner surface of the membrane. There the two electrons are donated to two molecules of ubiquinone (Q), each of which acquires a proton to form the semiquinone (QH•). Unlike the other components of the respiratory chain the quinones probably migrate as molecules through the membrane (*broken lines*). The semiquinone takes on two more electrons from cytochrome *b* and with two more protons from inside the mitochondrion is converted into the fully reduced hydroquinone (QH_2). Each hydroquinone gives up one electron to cytochrome c_1 and releases the corresponding proton outside. The remaining two electrons are then returned to the cycle through cytochrome *b* and the last two of six protons are released. Finally the two electrons deposited with cytochrome c_1 pass through cytochromes *c*, *a* and a_3 to oxygen, which is thereby reduced to water. The proton circuit is completed by the F_1-F_0 complex, where each two protons driven inward bring about the synthesis of one ATP molecule. Other processes are also powered by the proton gradient. They include the reduction of $NADP^+$ by NADH, the transport of calcium (Ca^{++}) and sodium (Na^+) ions and the exchange of ADP for ATP. The illustrations on this page and the next two pages are based on drawings made by Maija V. Hinkle.

with cytochrome *b*. The cytochromes are all proteins with an embedded heme group: a large ring structure with a central iron atom that can readily accept or donate an electron. The two electrons on cytochrome *b* molecules are returned across the membrane to the Q cycle: they ultimately reduce two more molecules of QH• to QH_2. It was from this closed loop that the electrons were obtained to reduce the semiquinones discussed above.

The two electrons deposited with molecules of cytochrome c_1 continue through the respiratory chain to its end. They are passed from cytochrome c_1, which is embedded in the exterior surface of the membrane, to cytochrome *c*, which lies on the exterior surface. The two electrons then go to cytochrome *a* and, crossing the membrane for the last time, are transferred to cytochrome a_3. Finally, cytochrome a_3 is oxidized by molecular oxygen. The two electrons are donated to an oxygen atom and two protons are picked up from the interior medium of the mitochondrion to form a water molecule.

In this long sequence of oxidation-reduction reactions the pair of electrons makes three round-trip crossings of the membrane and exports two protons each time. The first passage is made through reduced flavin mononucleotide, $FMNH_2$. After returning through the iron-sulfur proteins the electrons traverse to the exterior side again with hydroquinone. The second return trip is made through cytochrome *b*, where the electrons reduce another pair of ubiquinone molecules and migrate to the outside surface for the third and last time. Finally they return to the inside of the mitochondrion through the system of cytochromes (c_1, *c*, *a* and a_3) and are delivered to oxygen. The flow of electrons in the mitochondrion is an electric current. The potential difference between NADH and oxygen is about 1.2 volts, and the total current in the mitochondria of a human being at rest is

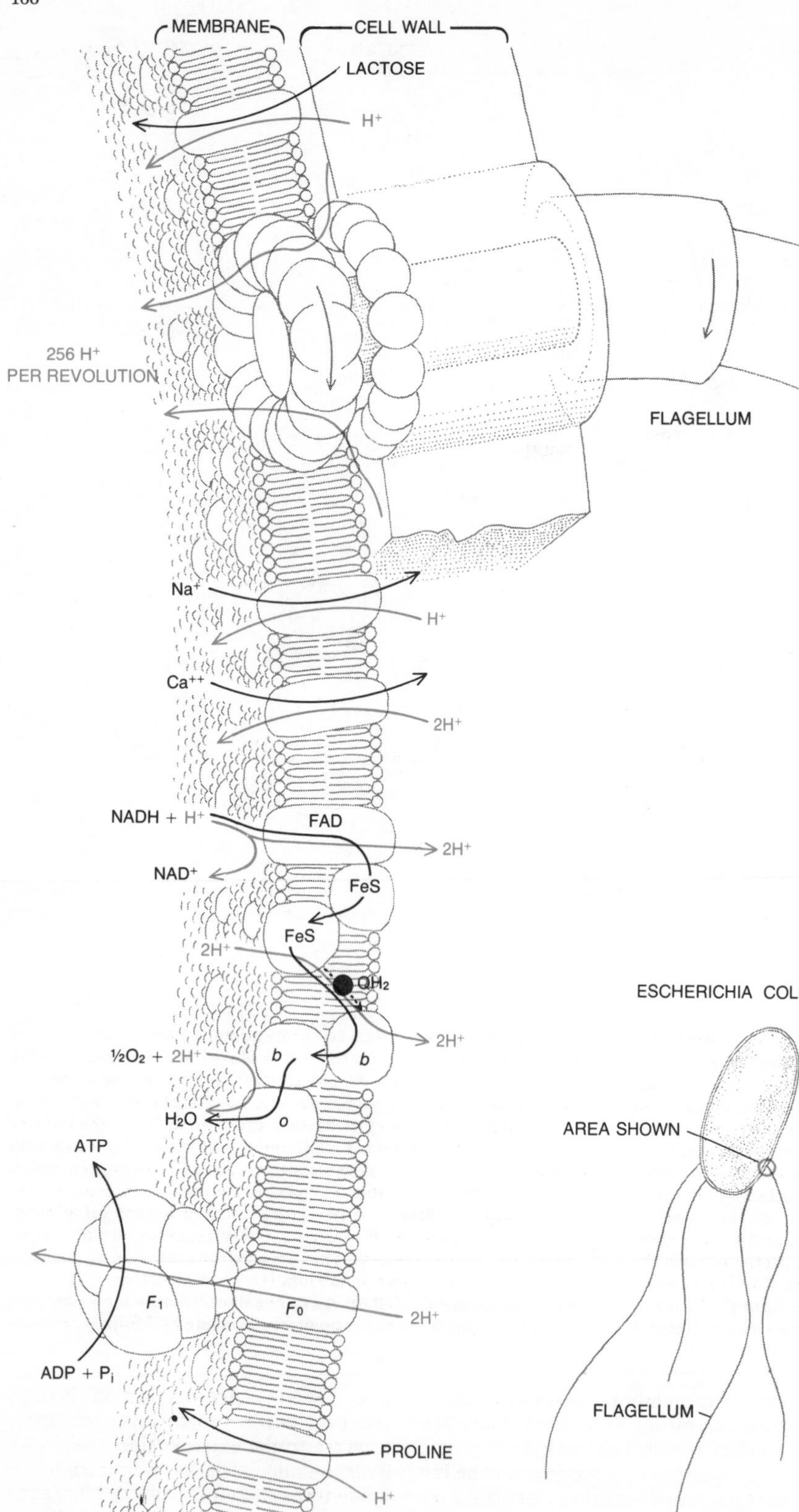

MEMBRANE OF THE BACTERIUM *E. coli* **has a system of respiratory proteins similar to the system in mitochondria. Electrons and protons are donated to flavin adenine dinucleotide ($FADH_2$), which exports the two protons; the electrons are returned to the inner surface through iron-sulfur proteins (FeS). The electrons, together with two protons from the internal medium, then reduce a single molecule of ubiquinone to hydroquinone (QH_2), which diffuses across the membrane and releases the protons outside. Finally the electrons proceed through cytochromes *b* and *o* to water. Just four protons are translocated for each electron pair instead of six. As in mitochondria, phosphorylation requires two protons for each molecule of ATP. Sodium and calcium transport are coupled to the movement of protons down the gradient, and so is the uptake of certain sugars, such as lactose, and amino acids, such as proline. The rotation of the flagellum is also powered by the influx of protons. At the root of the flagellum is a ring of 16 proteins, apposed to a similar ring in the cell wall. If a proton must pass through each protein to rotate the flagellum a sixteenth of a turn, 256 protons would be consumed in each revolution.**

about 100 amperes, so that 120 watts of power are generated.

It must be emphasized that this reconstruction of the mitochondrial membrane proteins and their interactions is tentative. Only for a few of the molecules is there any evidence bearing on their position in the membrane. FMN must be exposed on the interior surface since NADH reacts with it only there. Conversely, cytochrome *c* must be external since it can be removed by washing mitochondria with salt solutions. Two complexes of cytochrome molecules, one made up of *b* and *c* and the other of *a* and a_3, have both been shown to span the membrane, although the exact position of each cytochrome has not yet been established. Other arrangements of the same molecules, however, could lead to the same result: that the transfer of two electrons from NADH to oxygen is coupled to the export of six protons from inside the mitochondrion to outside it.

The respiratory chain of the bacterium *Escherichia coli* incorporates several components analogous to those found in mitochondrial membranes and operates according to the same principles. There is a major difference, however: for each pair of electrons passing through the respiratory chain only four protons (instead of six) are exported from the bacterial cell. The number of protons transported also varies with the conditions under which the bacteria are grown.

Similarities between bacteria and mitochondria are not unexpected. Both mitochondria and chloroplasts are believed to have evolved from bacteria that may have entered cells as parasites or symbionts early in the history of the nucleated cells and only later became captive organelles. Mitochondria and chloroplasts have their own genetic material, which in form resembles that of bacteria. Many proteins of mitochondria and chloroplasts are more like bacterial proteins than they are like cellular proteins of equivalent function.

In *E. coli* the respiratory chain is apparently simpler than it is in mitochondria. Electrons are carried to the respiratory proteins in the cell membrane by NADH and are donated to a flavin group analogous to FMN; the group is flavin adenine dinucleotide, or FAD. Like FMN, FAD is associated with a protein, and it carries two hydrogen atoms to the outer surface of the membrane, where the protons are released. The electrons are passed on to iron-sulfur proteins, which transport them back across the membrane. As in mitochondrial respiration, the second hydrogen carrier is ubiquinone, but at present there is no evidence for a Q cycle in *E. coli*. The two electrons are donated to a single quinone molecule, which takes up two protons from inside the cell to form

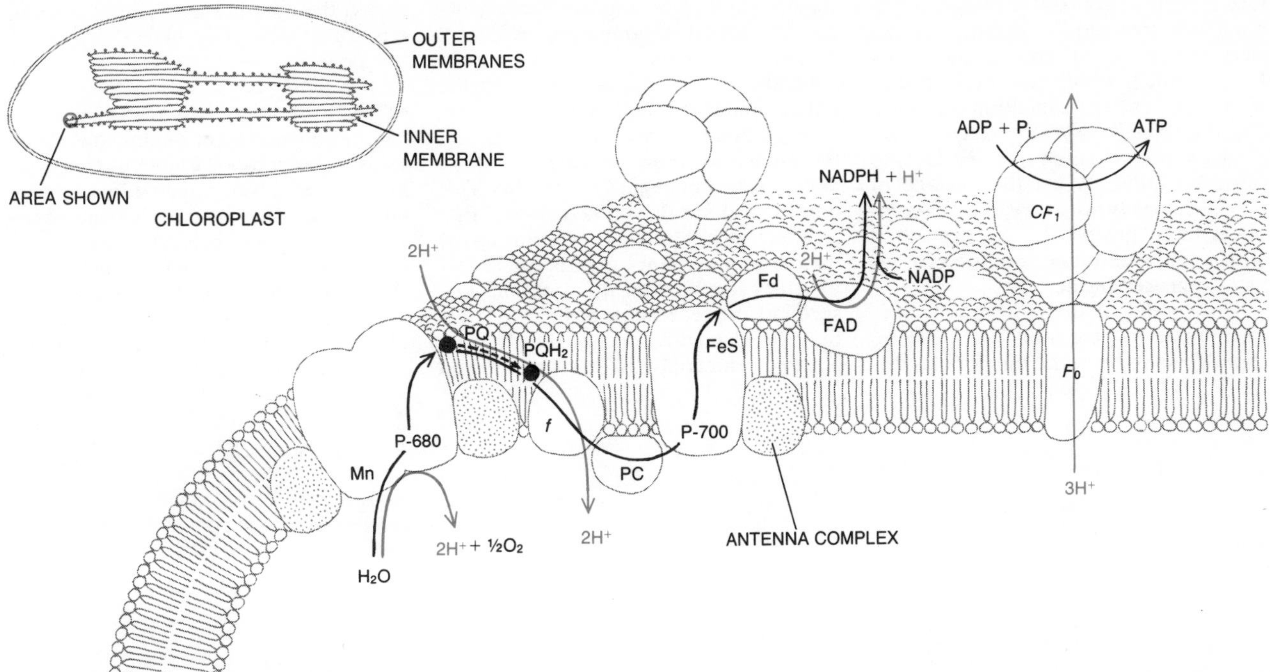

MEMBRANE OF THE CHLOROPLAST incorporates a light-driven system of pigments and other molecules that translocates protons inward. Two photons, or quanta of light, must be absorbed for each electron carried from water to the ultimate acceptor of electrons, $NADP^+$. For each pair of electrons passing through the chain of carriers three protons are taken up outside the chloroplast and four appear inside. The first photon is absorbed by an array of chlorophyll molecules, the antenna complex associated with a specialized chlorophyll designated P-680. Two electrons freed from P-680 molecules cross the membrane and are replaced by electrons taken from a molecule of water. The electrons from P-680, with two protons from outside the membrane, reduce plastoquinone (PQ) to PQH_2. The protons are released inside when the electrons are transferred to cytochrome *f*. The electrons then proceed through plastocyanin (PC) to a second photocenter, P-700. With the absorption of additional photons the electrons complete their journey through an iron-sulfur protein (FeS), ferredoxin (Fd) and flavin adenine dinucleotide ($FADH_2$) to $NADP^+$. The synthesis of ATP in the CF_1-F_0 complex is the only process in chloroplasts known to require a proton gradient; three protons apparently cross the membrane for each ATP molecule formed.

the hydroquinone. After it has migrated across the membrane the hydroquinone releases the two protons outside and passes the two electrons on to two cytochrome *b* molecules. Finally the electrons flow back across the membrane through cytochrome *b* to cytochrome *o*, which is oxidized by molecular oxygen. In summary, the pair of electrons crosses the membrane twice in each direction and translocates four protons outward. In other bacteria the respiratory chain takes other forms, and in some it is similar to the mitochondrial chain. It should also be pointed out that several compounds in addition to NADH can be oxidized by mitochondria and bacteria.

In chloroplasts the electron-transfer chain has some carrier molecules similar to those of the respiratory chains of mitochondria and bacteria, but the flow of electrons is in the opposite direction. The chain begins with water and ends with a phosphorylated form of NAD^+, nicotinamide adenine dinucleotide phosphate, or $NADP^+$. What is even more remarkable, the direction of proton translocation is also opposite to that of mitochondria. Chloroplasts have a triple system of membranes: the outer two define the organelle, and the inner one, which is very convoluted, contains the apparatus of photosynthetic phosphorylation. In chloroplasts protons are translocated inward across the inner membrane and are accumulated inside.

For electrons to flow from water to $NADP^+$ the free energy of the initial state must be raised. The required energy is supplied by the absorption of light by pigments stationed at two points in the electron-transfer chain. Each photon absorbed drives an electron across the inner membrane.

The first photon is absorbed by an array of molecules that includes several hundred chlorophyll molecules and other pigments, all of which are bound to proteins. This dense collection of light-absorbing substances, called the antenna complex, is embedded in the inner surface of the membrane. The energy of excitation is communicated rapidly from the antenna complex to a specialized chlorophyll molecule called P-680. ("P" stands for "pigment" and the number refers to the wavelength of light, expressed in nanometers, that the molecule absorbs.) Chlorophylls have a ringlike structure similar to that of the heme group, except that the central cavity is occupied by a magnesium atom instead of an iron atom. In P-680 light energy alters the distribution of electrons in the ring and makes one electron available for transfer.

The electron from P-680 is apparently transferred to an electron acceptor on the outer surface of the membrane immediately after a photon is absorbed. On the absorption of another photon the process is repeated. The two oxidized P-680 molecules are reduced by two electrons derived from water with the help of an enzyme that contains manganese. The oxygen atom from the water molecule is released and diffuses out of the chloroplast and the two protons go into solution inside the inner membrane.

The two electrons that cross the membrane from P-680 are picked up at the outer surface by a hydrogen carrier similar in structure to ubiquinone but called plastoquinone, or PQ. The electrons, with two protons extracted from the external solution, reduce PQ to PQH_2, which migrates back across the membrane to the inner surface. There the PQH_2 releases the protons internally and transfers the electrons to cytochrome *f*, a protein of the same type as cytochrome *c*.

The next electron carrier in the chain is the copper-containing protein plastocyanin, and it turns the electrons over to the second photochemically active site. The most important component of this system is another specialized chlorophyll molecule, designated P-700. When P-700 is excited by light absorption in the antenna complex, the electrons are once again moved to the outer surface of the membrane, where they are accepted by an iron-sulfur protein. The route then continues along the outer surface from that iron-sulfur protein to another one, ferredoxin. From there the electrons go to a proton containing FAD, which withdraws two protons from the external medium to form $FADH_2$. Finally $NADP^+$ acquires the electrons and a proton from $FADH_2$ to form NADPH. When the passage of two electrons through the chain is complete, three protons have disappeared outside the membrane and four protons have appeared inside.

This model of the chloroplast inner membrane, like the model of the mitochondrial and bacterial membranes discussed above, is a tentative reconstruction based on fragmentary evidence. Structural studies of the membrane have shown that ferredoxin is on the outside and that functional plastocyanin is on the inner surface. Other features of the model are more speculative. H. T. Witt and his colleagues at the Max Volmer Institute in Berlin have shown that an electric potential is created across the chloroplast inner membrane in less than 10^{-8} second after illumination. The speed of the reaction is strong evidence that the electrons released by the specialized chlorophylls are transferred across the membrane; less direct methods for creating a potential would be too slow.

The proton gradient established by photochemical or oxidative electron transport represents a store of free energy, much like water pumped up a gravitational gradient to an elevated reservoir. The energy can be recovered by allowing the protons to flow back across the membrane through an appropriate "mill." The main work done by the flow of protons down the gradient is the phosphorylation of ADP to form ATP,

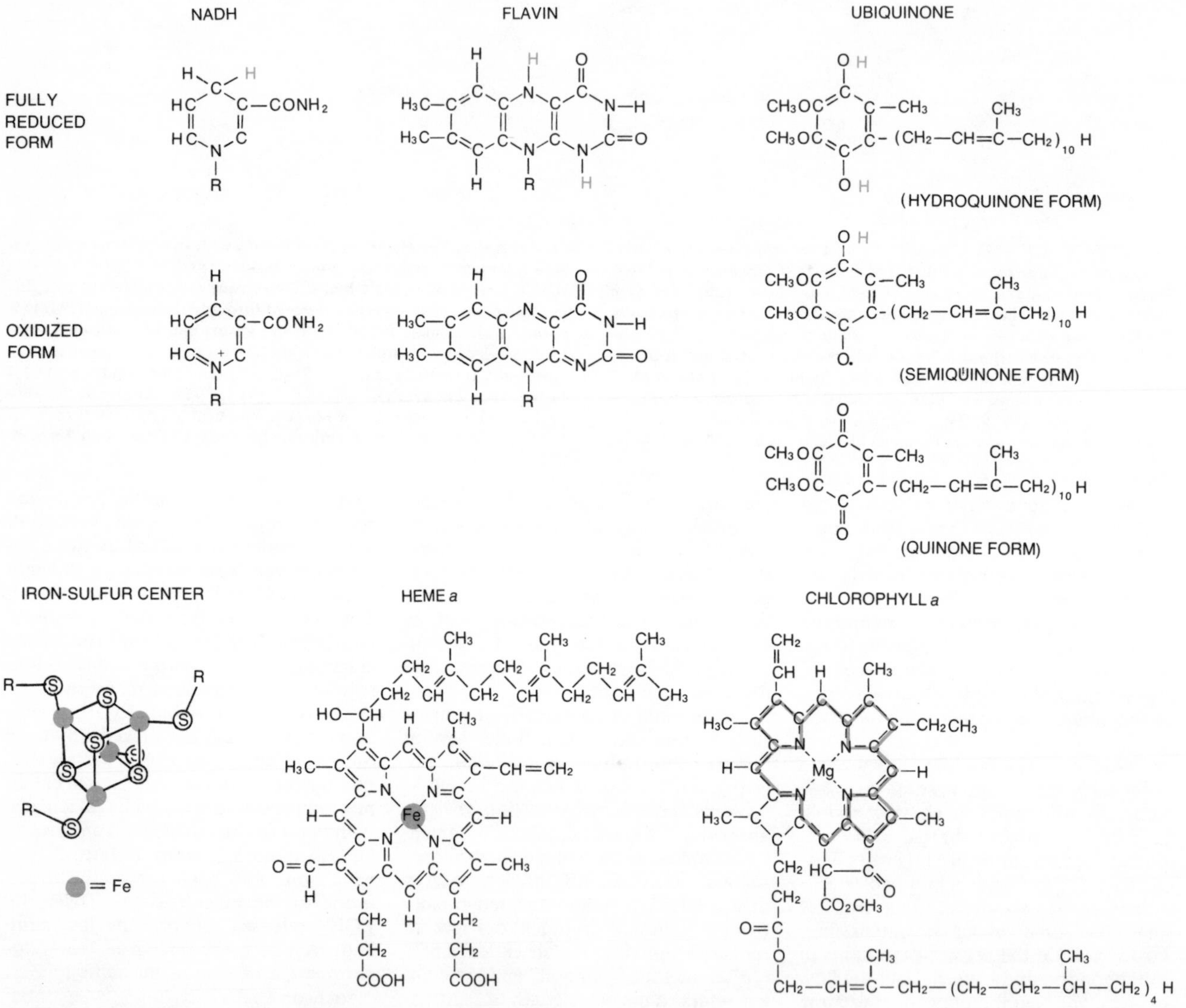

NADH FLAVIN UBIQUINONE

FULLY REDUCED FORM

(HYDROQUINONE FORM)

OXIDIZED FORM

(SEMIQUINONE FORM)

(QUINONE FORM)

IRON-SULFUR CENTER HEME *a* CHLOROPHYLL *a*

= Fe

CARRIER MOLECULES in both respiratory and photosynthetic chains are specialized for the transport of electrons or hydrogen atoms. The active centers of several carriers are shown; *R* stands for the rest of the molecules. The hydrogen atom (or atoms) donated to the oxidized forms of the carriers to generate the reduced forms are in color. NAD^+ accepts two electrons and a proton to form NADH. One electron neutralizes the charge on the ring nitrogen atom, and the other combines with a proton to hydrogenate a carbon atom. A very similar molecule, $NADP^+$, is a carrier in chloroplasts. The flavins FAD and FMN are the active parts of flavoproteins. Two hydrogen atoms are donated to FAD and FMN to form $FADH_2$ and $FMNH_2$. Iron-sulfur proteins are electron carriers rather than hydrogen carriers; the active center of one type is a cagelike structure in which the electron's charge is distributed among the iron and sulfur atoms. Some of the sulfur atoms are sulfide (S^{--}), and the remainder are from the amino acid cysteine in the protein (*R*). Ubiquinone is a hydrogen carrier in mitochondrial and bacterial membranes. In its fully reduced and protonated form the oxygen atoms on the benzene ring have hydrogen atoms attached to them. In the fully oxidized state the oxygen atoms are linked to the carbon ring by double bonds. Ubiquinone can also exist in a half-reduced or semiquinone form, in which one oxygen is protonated. The semiquinone is a free radical, and the unpaired electron is delocalized by resonance over the carbon ring. Plastoquinone, a hydrogen carrier in chloroplasts, is very similar to ubiquinone. Chlorophyll and the heme group of cytochromes are also similar in structure. In cytochromes electron is carried primarily by iron atom (Fe) at the center of the large heme ring; the positive charge of oxidized chlorophyll is probably distributed throughout the ring.

but other processes such as the active transport of some ions are also powered by the gradient.

Energy is stored in the proton gradient in two forms, or in other words the gradient has two components. One component is the difference in concentration or chemical activity of protons on opposite sides of the membrane. The protons tend to diffuse from a region of high concentration to a region of low concentration when a pathway through the membrane is provided. Proton concentration is measured in terms of *p*H, which is defined as the negative logarithm of the hydrogen ion concentration. The energy of the concentration gradient is determined by the difference in *p*H across the membrane and is independent of the absolute magnitude of the *p*H.

The electric charge carried by the proton contributes the second component to the energy of the gradient. The net movement of charges across the membrane creates a difference in electric potential, and all charged particles are affected by the resulting electrostatic field. The total energy of the proton gradient is the sum of the concentration (or osmotic) component and the electric component.

Because of the difference in concentration and in electric potential a proton that has been expelled from a mitochondrion experiences a force tending to draw it back across the membrane. The movement of the proton in response to that force can be made to do work, such as the work of phosphorylation.

The enzymes that couple the diffusion of protons back through the membrane to the synthesis of ATP appear to be remarkably similar in mitochondria, bacteria and chloroplasts. They are conspicuous in electron micrographs as globular bodies protruding from the surface of the membrane. The protruding "knob," designated F_1, was isolated in 1960 by Efraim Racker, Maynard E. Pullman and Harvey Penefsky of the Public Health Research Institute of the City of New York. It is a soluble protein made up of five kinds of subunit, some of which are present in multiple copies. F_1 is attached to the membrane through another set of proteins designated F_0, which is embedded in the membrane and is thought to pass all the way through it. F_1 is readily removed from the membrane, but F_0 can be removed only when the membrane itself is destroyed with detergents. The entire system of enzymes is called the F_1-F_0 complex.

In mitochondria and in bacteria the F_1-F_0 complex is oriented with the knobs protruding into the interior matrix from the inner surface of the membrane. The evidence indicates that for every two protons passing inward through the complex one molecule of ADP is combined with inorganic phosphate to form ATP. An important observation is that the reaction is reversible: under appropriate circumstances the F_1-F_0 complex can split ATP molecules and employ the energy made available to pump protons out of the mitochondrion or bacterial cell. Like all enzymes, the F_1-F_0 complex controls the rate of the reaction, but the direction of the reaction is determined by the balance of free energy.

In chloroplasts the equivalent enzyme system is called the CF_1-F_0 complex, and it is oriented in the opposite direction, that is, the knob protrudes from the outer surface of the membrane. The orientation is exactly what would be expected, since the direction of the proton gradient in chloroplasts is also opposite to that in mitochondria and bacteria. Protons flow outward through the CF_1-F_0 complex, and the ATP is formed in the space between the inner and the outer chloroplast membranes. Another difference is more difficult to account for: the chloroplast CF_1-F_0 complex apparently requires three protons per ATP synthesized instead of two.

The mechanism of ATP synthesis at the active site of the F_1-F_0 complex is not understood. Several hypotheses have been formulated, but the evidence from studies of the enzyme is not yet sufficient to decide among them.

The active site is apparently associated with the F_1 part of the complex, the knobs seen in electron micrographs. Even when the F_1 components are isolated and in solution, they catalyze the splitting of ATP to yield ADP and phosphate, the reverse of the phosphorylation reaction driven by a proton gradient. The forward reaction cannot be observed with isolated F_1 molecules, of course, because a proton gradient cannot be established in solution. Of the five kinds of protein subunit in F_1, only the two largest are required for ATP-splitting activity. The smallest of the subunits is an inhibitor of catalytic activity that presumably serves as a check valve: it prevents the reaction from proceeding at all when the proton gradient is low, since under those circumstances the reaction would proceed in reverse, consuming ATP.

The F_0 part of the complex, which is left behind in the membrane when F_1 is removed, exhibits no catalytic activity with ATP. On the other hand, the removal of F_1 makes the membrane highly permeable to protons, and these leaks can be plugged by rebinding F_1 or by certain inhibitors. These observations have led to the view that F_0 serves as a channel through which protons cross the membrane to F_1.

One of the hypothetical catalytic mechanisms was proposed by Mitchell. In this scheme a phosphate group binds to the enzyme at an active site within the F_1 part of the complex but near the end of the proton channel through F_0. Two protons driven down the channel by the membrane potential and the *p*H gradient attack one of the phosphate oxygens, pulling it loose to form a molecule of water. The unattached phosphorus bond created in this way converts the phosphate group into a highly reactive species that can bind directly to ADP.

In the main alternative hypothesis protons have a less direct role. This indirect hypothesis actually embraces several possible mechanisms. The idea common to all of them is that the passage of protons through the F_1 part of the complex could change the conformation of protein. For example, the binding of a proton to a group in the protein could cause it to move as a result of electrical interactions with other charged regions. Paul D. Boyer of the University of California at Los Angeles has proposed that proton-induced changes in protein conformation near the active site could result in ATP synthesis. In one such mechanism the dissociation of the completed ATP molecule from the enzyme is the crucial, energy-requiring step. Although free ADP cannot combine with phosphate without an input of energy, the reaction might take place spontaneously if both molecules were bound to a protein. The resulting ATP would remain bound to the enzyme and could be dissociated from it only by the addition of energy. That energy would be supplied by a change of conformation, driven in turn by the translocation of protons.

In 1971 I. J. Ryrie and André T. Jagendorf of Cornell University observed changes in the conformation of CF_1 bound to chloroplasts when the chloroplasts were illuminated or when the *p*H of the medium in which they were suspended was rapidly changed from acidic to basic, creating a momentary *p*H gradient. R. P. Magnusson, J. Fagan and one of us (McCarty) showed that the reactivity of a chemical group in one subunit of CF_1 is enhanced when chloroplasts are illuminated. Penefsky and others have shown that the F_1 protein of mitochondria also undergoes structural changes when it is exposed to an electrochemical proton gradient. In 1973 E. C. Slater and his colleagues at the University of Amsterdam found that CF_1 and F_1 have ADP and ATP firmly bound to them. In chloroplasts the ADP bound to CF_1 can be exchanged for nucleotide molecules in the medium when a proton gradient is formed. The bound ADP can be phosphorylated, but only at a low rate. It is probably not bound at a catalytic site in the enzyme but may regulate the structure or the functioning of F_1 molecules.

These findings can be interpreted as evidence for an indirect mechanism of phosphorylation, in which conformational changes have an essential role. Mitchell's direct hypothesis cannot be ruled out, however. Conformational changes in enzymes are hardly unusual and may be universal in enzymes. The changes in F_1 could be associated with a direct mechanism or they could be incidental to the catalytic activity.

The movement of molecules or ions across a membrane can proceed spontaneously only in a direction that tends to reduce a concentration gradient or a gradient in electric potential. In the cell and in subcellular organelles many substances must be assimilated or expelled against a gradient, and so most transport requires energy. The movement of some ions and molecules is driven directly by the proton gradient.

An example of proton-linked transport is the flow of sodium ions (Na^+) out of the mitochondrion in exchange for hydrogen ions. In this system one sodium ion is expelled from the mitochondrion for each proton that crosses the membrane in the opposite direction. The exchange is assumed to be mediated by a membrane protein.

Sodium transport is driven by only one component of the electrochemical proton gradient, namely the difference in proton concentration, or *p*H. The membrane potential has no effect, since while it accelerates the proton's inward passage it retards the outward movement of the sodium ion and the two effects exactly cancel. The reaction comes to equilibrium when the sodium gradient is equal to the *p*H gradient in proton concentration. An equivalent transport system powers the uptake of inorganic

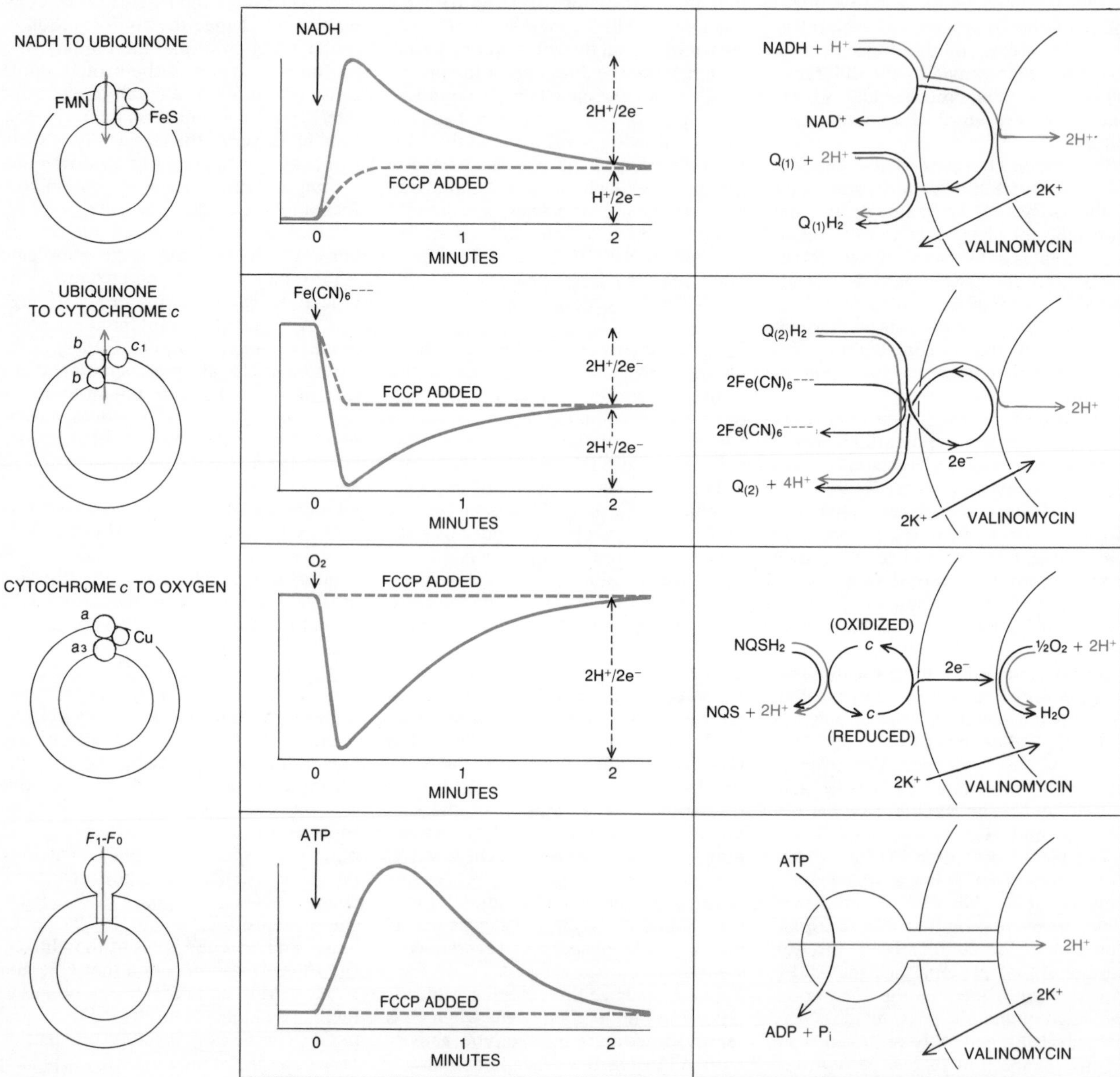

SEGMENTS OF THE ENZYME SYSTEM from mitochondrial membranes can be isolated and assembled in artificial membrane vesicles; when each segment is supplied with appropriate substrates, it is seen to transport protons across the membrane. The first segment consists of those components of the respiratory chain extending from NADH to ubiquinone, including FMN and the iron-sulfur proteins. The second segment includes all the molecules from ubiquinone to cytochrome *c*. The third segment completes the respiratory chain, extending from cytochrome *c* to oxygen. The fourth isolated system is the F_1-F_0 complex, which is operated in reverse, so that it splits ATP to form a proton gradient. The respiratory systems are driven by adding a donor and an acceptor of electrons; they can be natural substrates, such as NADH or oxygen, or artificial ones, such as ferricyanide ($Fe(CN^{---})_6$), ubiquinone analogues (Q_1 or Q_2) or the reductant naphthoquinol sulfonate ($NQSH_2$). Proton transport is measured by monitoring the *p*H of the external medium. In some cases the reaction causes a permanent change in the *p*H, which is independent of the gradient, but since that offset is not affected by the proton ionophore FCCP it can be measured separately. When corrections are applied for the permanent offset, so that only the effects of the proton gradient are measured, it is seen that each segment of the respiratory chain transports two protons for each electron pair.

phosphate, which enters the mitochondrion as a negatively charged ion. The movement of the phosphate ions is balanced by a counterflow of hydroxyl ions (OH^-).

The other component of the proton gradient, the membrane potential, can also drive active transport. The uptake of calcium ions (Ca^{++}) by the mitochondrion, for example, is powered by the membrane potential and is independent of the *p*H gradient. Calcium probably enters without a countercurrent of any other ion, its transport being motivated entirely by electrostatic forces that pull it toward the negatively charged inner surface of the membrane. The membrane potential also provides the motive force for the exchange of ADP and ATP molecules, ensuring a continuing supply of substrate for phosphorylation and exporting the product. ADP carries a charge of minus 3 and ATP a charge of minus 4, so that the exchange is equivalent to the outward movement of one negative charge or the inward movement of one positive charge.

The proton gradient in chloroplasts is apparently not coupled to any function other than phosphorylation, with the possible exception of magnesium ion transport. In bacteria, on the other hand, a great variety of metabolic processes seem to derive their energy from the proton gradient. Indeed, Mitchell was studying transport across bacterial membranes when he proposed the chemiosmotic theory in 1961.

The mechanism of sodium transport in *E. coli* is apparently similar to that in mitochondria. Calcium transport, however, is in the direction opposite to that in mitochondria, that is, *E. coli* excretes calcium. The bacterial calcium transport is probably an electrically neutral exchange of two protons for each Ca^{++} ion, a movement powered by the *p*H gradient.

In 1963 Mitchell proposed that uptake of the sugar lactose by *E. coli* could be coupled to an influx of protons, even though lactose is an electrically neutral molecule rather than an ion. Subsequent studies have shown that lactose transport can be driven by an artificially formed proton gradient and, conversely, that the influx of lactose into starved cells is accompanied by an influx of protons. The transport system appears to be symmetrical: the sugar can be concentrated on either side of the membrane, depending on the polarity of the proton gradient. Similar systems for the uptake of certain amino acids have been discovered in *E. coli*.

It should be pointed out that these are not the only systems of active transport in bacteria; there are at least two others that are not directly coupled to the proton gradient. Glucose and some other sugars are brought into the bacterial cell

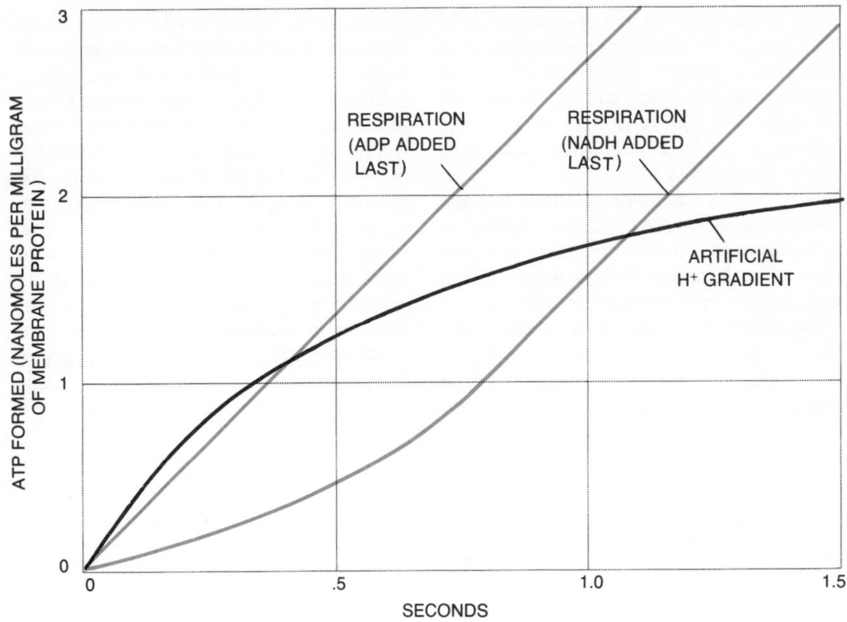

RATE OF PHOSPHORYLATION appears to be limited by the magnitude of the proton gradient. The rate of ATP synthesis was measured in vesicles made from mitochondrial membranes. When an artificial electrochemical proton gradient was imposed, the initial rate of synthesis was even greater than that observed under natural conditions, but the rate declined as the gradient decayed. When the gradient was created by respiration, a constant rate of synthesis was eventually attained, but the initial rate depended on the experimental conditions. If all the reactants needed for respiration were supplied but ADP was withheld, phosphorylation began immediately at a high, steady rate. If the missing reactant was NADH, however, there was a lag of about a second before the maximum rate of synthesis was achieved. Presumably the time was required for the respiratory chain to build up a proton gradient. These findings indicate that a proton gradient can supply energy fast enough to meet the demands of phosphorylation.

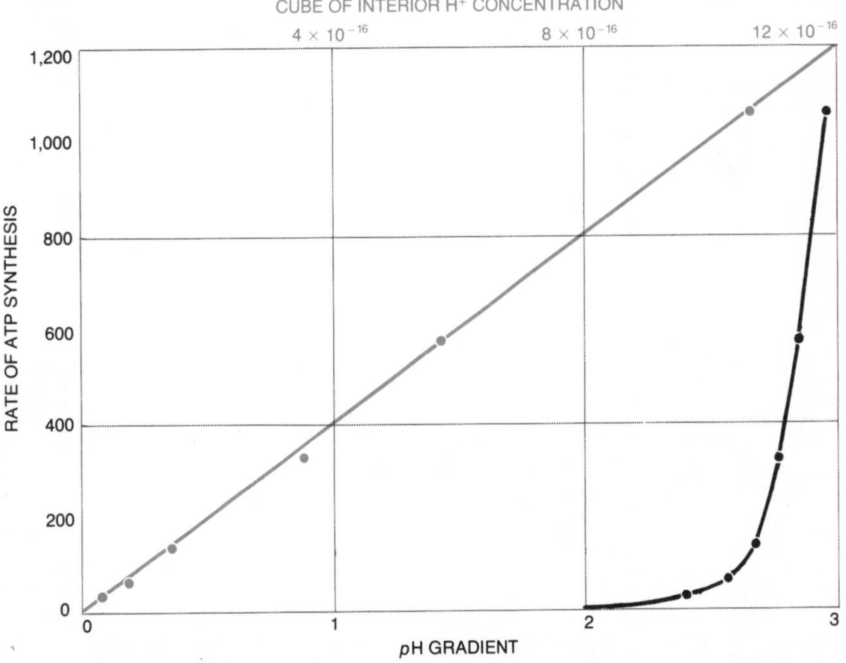

KINETICS OF PHOTOPHOSPHORYLATION were studied in functioning chloroplasts. The rate of ATP synthesis and the magnitude of the proton gradient were determined separately but under similar conditions in chloroplasts illuminated with various intensities of light. The relation derived from these measurements (*black curve and scale*) shows that the rate of phosphorylation is strongly dependent on the *p*H gradient. In particular no ATP is made until the difference in *p*H reaches about two *p*H units, presumably because protons moving down a smaller gradient have too little energy to drive the reaction. The rate measurements can be made to fall along a straight line by replotting them as a function of the cube of the internal proton concentration (*colored curve and scale*). This third-power relation supports other evidence suggesting three protons are required for each molecule of ATP formed in chloroplasts.

by a system in which the essential preliminary event is the phosphorylation of the sugar. Other molecules cross the membrane by first binding to soluble proteins in a system that derives the energy needed for active transport from the splitting of ATP.

An extraordinary mechanical linkage driven by the proton gradient is the flagellum responsible for motility in *E. coli* and other bacteria. It now appears that the flagellum moves by rotating like a propeller. The shaft at the root of the flagellum passes through a bearing in the rigid cell wall and terminates in a rotor in the cell membrane. The rotor appears to be a ring of 16 proteins, and it is apposed to another protein ring that may act as a "stator" in the cell wall.

Howard C. Berg and his co-workers at the University of Colorado have recently determined that the flagellar rotor is powered by the electrochemical proton gradient. Berg's calculation of the energy required for rotation suggests that about 250 protons would have to flow inward to power one revolution if the gradient is a typical one for *E. coli*. The protons presumably flow through the protein ring that forms the rotor, generating force against the adjacent ring in the cell wall. An obvious hypothesis is that the passage of one proton through each of the 16 subunits in the ring rotates the flagellum through a sixteenth of a turn. A complete revolution would then require 16 × 16, or 256, protons.

When Mitchell proposed the chemiosmotic theory, the translocation of protons in mitochondria and chloroplasts was unknown. (Indeed, he first suggested that respiration would pump protons into mitochondria rather than out of them.) The existence of a proton-transport system coupled to respiration was established by Mitchell and Moyle, who employed a comparatively

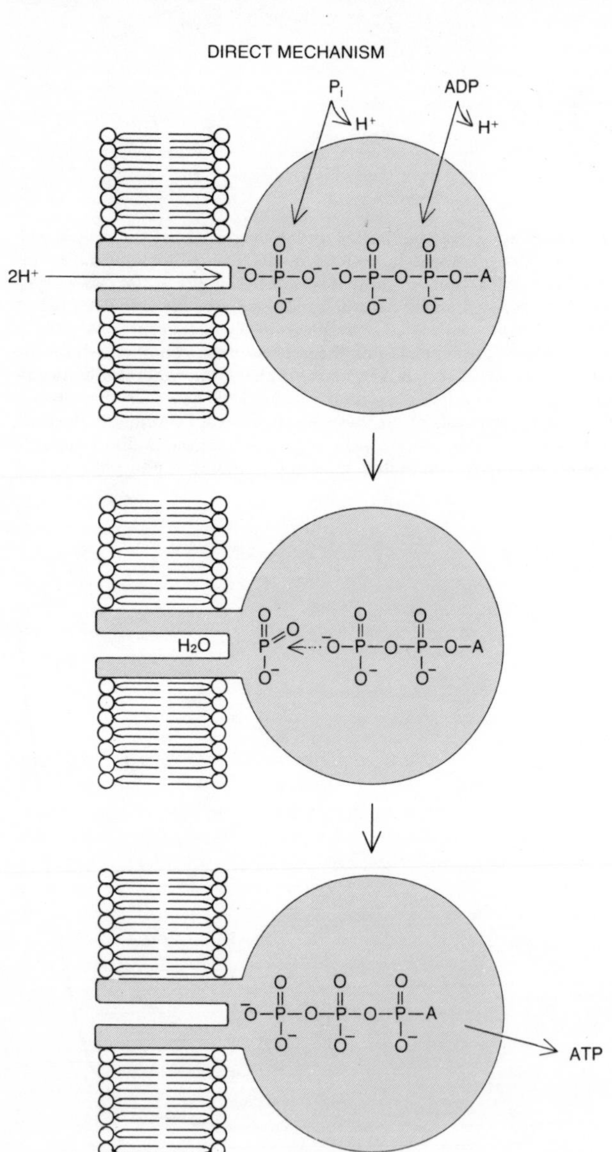

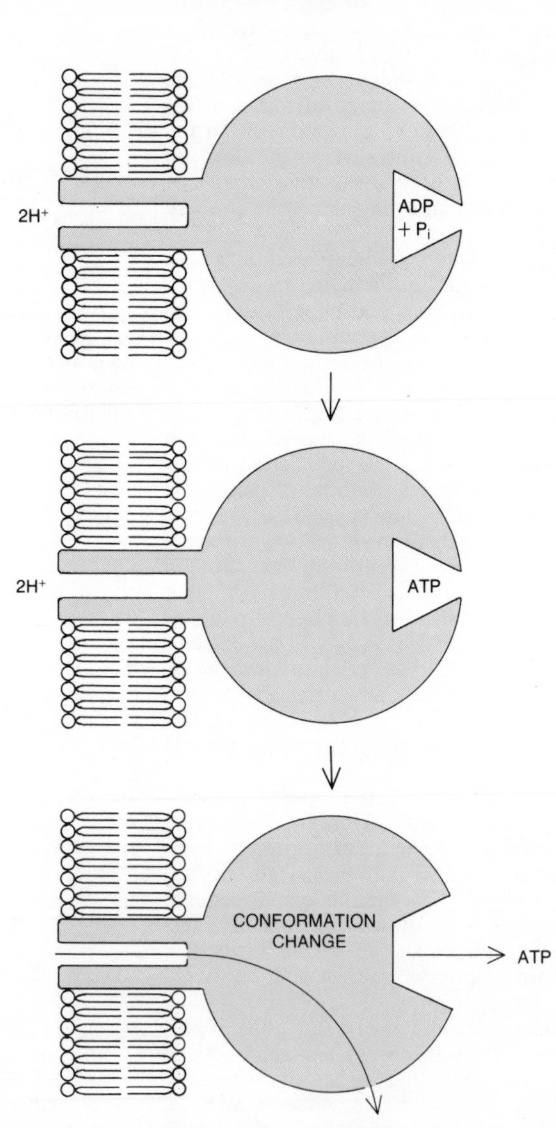

POSSIBLE MECHANISMS of phosphorylation in the F_1-F_0 complex fall into two categories, called direct and indirect. In the direct mechanism (*left*) first event is binding of a phosphate ion and ADP to the F_1 part of the enzyme complex. Protons move through a channel in the F_0 part and then attack one of the phosphate oxygens, which is removed as a molecule of water. Finally an oxygen of ADP attacks the phosphorus atom, forming ATP. The ATP molecule then dissociates from the enzyme. A variety of indirect methods are possible; one example is shown here (*right*). At the active site of the enzyme it might be possible for ADP and inorganic phosphate to combine spontaneously, without an input of free energy. The resulting ATP molecule would be securely bound to the enzyme, however, and energy would be needed to release it. The energy might be supplied by protons that bind at a position other than the active site and that thereby change the conformation of the enzyme. The protons would then be released into the solution on the F_1 side of the membrane.

simple technique to measure the number of protons transported for each pair of electrons. Mitochondria are suspended in a medium that contains no oxygen; then the concentration of hydrogen ions is monitored as a known quantity of oxygen is injected into the solution. The proton concentration is observed to rise immediately and rapidly; when the oxygen is exhausted, the concentration slowly returns to normal as protons leak back across the membrane. Adding a substance called a proton ionophore immediately dissipates the gradient; such ionophores are known to make membranes permeable to protons. The inverse behavior can be demonstrated in chloroplasts. After a brief flash of light the proton concentration in the external medium falls rapidly and then slowly regains its original level as protons leak out of the chloroplasts. Again proton ionophores collapse the gradient.

Measuring the magnitude of the gradient is complicated by the need to determine the difference in *p*H and the membrane potential separately. The gradient in proton concentration can be determined by measuring the uptake of weak acids into mitochondria or bacteria; in chloroplasts a weak base is employed instead of an acid. The membrane potential is revealed by the uptake of an ion (other than H^+ or OH^-) that is free to pass through the membrane and reach an equilibrium concentration gradient.

In mitochondria the proton gradient has been found to consist of a *p*H difference of about 1.4 *p*H units (acid outside) and a membrane potential of some 140 millivolts (positive outside). Vesicles prepared from *E. coli* membranes have a *p*H gradient of two units and a membrane potential of about 70 millivolts, with the same polarity as in mitochondria. The way in which the electrochemical gradient is partitioned into its two components depends on what ions are present in the solution and on the external *p*H. In chloroplasts, of course, the gradient is oriented in the opposite direction. What is more, in the steady state almost the entire gradient in chloroplasts is expressed as a difference in proton concentration. Under these conditions the membrane potential is negligible because the membrane is permeable to chloride ions, but the *p*H gradient reaches a magnitude of about 3.5 units. As a result the interior matrix of chloroplasts can become strongly acidic, reaching a *p*H of about 4. In effect the chloroplast accumulates hydrochloric acid inside. It should be noted in passing that the enzymes of carbon dioxide fixation in chloroplasts are all between the inner and outer membranes and thus are not exposed to this harsh environment. The interior of mitochondria and bacteria, on the other hand, is the site of many enzymatic reactions. Mitochondria and bacteria have mechanisms, such as the exchange of protons for sodium ions, that maintain the internal *p*H close to neutrality.

The entire electrochemical gradient, regardless of how it is partitioned into components of *p*H difference and membrane potential, can be measured in terms of energy. It is the energy acquired by a proton when it is transported across the membrane against the gradient, or the energy given up by a proton when it is carried back across the membrane with the gradient. In chloroplasts the electrochemical gradient is equal to about 4.8 kilocalories per mole of protons; in vesicles made from *E. coli* membranes it is 4.4 kilocalories per mole. Mitochondria have the largest total gradient, about 5.3 kilocalories per mole of protons.

The detection of a proton gradient generated by respiration is by no means a demonstration that phosphorylation is driven by the gradient. The proton ionophores, however, provide evidence bearing on this point. The first of the ionophores to be discovered, dinitrophenol, was found to have an effect on oxidative phosphorylation long before its mode of action was understood. In 1948 William F. Loomis and Fritz A. Lipmann of the Harvard Medical School showed that dinitrophenol inhibits ATP synthesis but stimulates electron transport; in other words, it uncouples respiration from phosphorylation. In 1963 Mitchell proposed that dinitrophenol acts by ferrying protons across membranes, and he showed that the addition of dinitrophenol to bacteria equilibrates a *p*H gradient across the membrane. It has since been shown that the effectiveness of various uncouplers is correlated with their capacity to act as ionophores. One alternative hypothesis is that the uncouplers act by interfering with the F_1-F_0 complex, but the most potent uncoupler, a molecule designated SF6847, is completely effective at a concentration of about one molecule for every five F_1-F_0 complexes.

For complete uncoupling both the *p*H gradient and the membrane potential must be dissipated. The proton ionophores can accomplish this because they carry the protons themselves across the membrane. Each of the two components of the gradient can also be abolished individually by other ionophores. Valinomycin, an ionophore that carries potassium ions (K^+), dissipates only the membrane potential. In the presence of valinomycin a countercurrent of potassium ions neutralizes the difference in electric potential, but with continued respiration there is a compensating increase in the *p*H gradient. Nigericin, another ionophore, brings about an electrically neutral exchange of protons for potassium ions. It thereby acts to eliminate the *p*H gradient, so that continued respiration enhances the membrane potential. Phosphorylation can continue in the presence of either ionophore. In the presence of both ionophores, however, both components of the proton gradient are collapsed and phosphorylation is strongly inhibited.

When a proton ionophore is present, the energy of respiration is dissipated as heat. In the mitochondria of brown fat tissue this effect is employed to maintain body temperature. In these mitochondria a protein in the membrane short-circuits the proton gradient; fatty acids may also act as proton ionophores.

A persuasive demonstration of the role of the proton gradient in phosphorylation was achieved in 1966 by Jagendorf and Ernest G. Uribe of Johns Hopkins University. They showed that imposing an artificial *p*H gradient can result in ATP synthesis. Chloroplast inner membranes were suspended in a solution at *p*H 4 (moderately acidic), then the solution was rapidly brought to *p*H 8 (slightly basic), so that protons were driven outward. ATP was formed in amounts corresponding to 100 molecules for each CF_1 complex. Mitchell later reported the synthesis of ATP in mitochondria exposed to a transition in the opposite direction, from basic to acidic. William S. Thayer of Cornell and one of us (Hinkle) have also elicited ATP synthesis by imposing artificial gradients on inverted vesicles formed from mitochondrial membranes. The maximum yield of about 10 ATP for each F_1 was obtained when a *p*H gradient was combined with a membrane potential.

If an electrochemical proton gradient is to serve as the intermediate state coupling respiration to phosphorylation, then it must be energetically and kinetically competent to do so. Enough protons must be translocated to account for the amount of ATP synthesized, the protons must have sufficient energy to provide the free energy of synthesis, and the rates of proton translocation and phosphorylation must correspond.

A basic assumption of the membrane models presented above was that two protons are translocated for each "coupling site," or each passage of a pair of electrons across the membrane. Thus in the mitochondrion, which is thought to have three coupling sites, six protons are thought to be translocated for each pair of electrons passing from NADH to oxygen. The mitochondrion is said to have an H^+/O ratio, or an $H^+/2e^-$ ratio, of 6. *E. coli* and chloroplasts, which are thought to have two coupling sites, have an $H^+/2e^-$ ratio of 4.

The study of such ratios in chemical

reactions is called stoichiometry. The stoichiometry of proton translocation can be determined experimentally in mitochondria and bacteria by the oxygen-pulse method and in chloroplasts by measuring proton translocation following a light flash. Recently Martin D. Brand, Baltazar Reynafarje and Albert L. Lehninger of the Johns Hopkins School of Medicine have reported experimental findings that more than six protons are translocated by mitochondria for each electron pair passing from NADH to oxygen. If their findings turn out to be correct, significant modifications of Mitchell's theory will be necessary. One of us (Hinkle) has studied proton translocation in vesicles formed from mitochondrial membranes and in artificial membrane vesicles incorporating mitochondrial respiratory enzymes. The $H^+/2e^-$ ratio measured in these systems agrees with the findings of Mitchell and Moyle, but the controversy is not yet resolved.

The stoichiometry of proton translocation by the F_1-F_0 complex is most conveniently measured by driving the process in reverse, that is, by splitting ATP to create a proton gradient. What is measured is the number of protons translocated for each ATP molecule split, which is presumably equal to the number of protons required to synthesize a molecule of ATP. Through an ATP-pulse method, analogous to the oxygen-pulse method, Mitchell determined that mitochondria translocate approximately two protons for each ATP, a finding confirmed by Thayer and one of us (Hinkle) in inverted mitochondrial vesicles. As we noted above, the corresponding enzyme of chloroplasts, CF_1-F_0, seems to require three protons per ATP, but direct measurements of this ratio are difficult to make. Values ranging from 2 to 4 have been reported, and the ratio of 3 was deduced indirectly.

From these ratios the overall stoichiometry of oxidative and photosynthetic phosphorylation can be estimated. If six protons are translocated for each pair of electrons in mitochondria, and two protons are required to flow back across the membrane for each ATP formed, then no more than three ATP molecules can be synthesized for each electron pair. The ratio is only an upper limit, since fewer ATP molecules would be produced if some protons were diverted into other processes, such as active transport. The overall ratio has been determined experimentally many times by measuring the number of ATP molecules synthesized for each oxygen atom reduced to water. In mitochondria the experimental evidence is generally considered to indicate a ratio of three ATP molecules for each pair of electrons passing from NADH to oxygen. Several recent studies, however, have suggested that the ratio may be as low as two. A possible mechanism that might account for the lower ratio, based on the need for energy to transport ADP and ATP, will be discussed below.

In chloroplasts, where only four protons are translocated for each electron pair and where three protons are required for each ATP synthesized by CF_1-F_0, the maximum stoichiometry of synthesis should be four ATP molecules for every three electron pairs. Most observed ratios are consistent with that value, although ratios greater than 4/3 have occasionally been reported.

From these stoichiometries an energy balance sheet for ATP synthesis can be drawn up. As we calculated above, the electrochemical proton gradient in chloroplasts represents an energy store of about 4.8 kilocalories per mole of protons. Since three protons flow back across the membrane for each ATP formed, an energy of 3×4.8, or 14.4, kilocalories is available for phosphorylation. Other estimates suggest that the maximum free energy required for ATP synthesis in chloroplasts is 14.5 kilocalories per mole, and so the energy budget is very close to balancing.

The energetic demands of mitochondrial phosphorylation are somewhat more complicated. The protons crossing the mitochondrial membrane, it will be remembered, have an energy of 5.3 kilocalories per mole, but only two protons are consumed for each ATP formed. The free energy made available, then, is 10.6 kilocalories per mole, which is insufficient since a maximum of about 15 kilocalories is required for phosphorylation. There is an important difference, however, between phosphorylation in chloroplasts and in mitochondria, since in the latter the reaction takes place inside the inner membrane. In that environment the high concentration of reactants (ADP and phosphate) and the low concentration of product (ATP) shifts the equilibrium point of the reaction and effectively reduces the free energy required. Inside the mitochondrion ATP synthesis demands only about 11 kilocalories per mole, which is within range of the calculated energy available.

The reduction in the energy requirement of phosphorylation in mitochondria does not come without a cost. The cost is the energy needed to concentrate ADP and phosphate in the interior cavity and to remove ATP. From a study of the transport reactions Martin Klingenberg and his colleagues at the University of Munich have concluded that the counterflow of ADP and ATP ions involves the net movement of charge across the membrane. The exchange of reactants and products could therefore be powered by the membrane potential. Similarly, phosphate is absorbed in exchange for OH^- ions and its transport is therefore driven by the pH gradient. The combined effect of these transactions is to transport an additional proton for each ATP synthesized and exported. Hence a total of three protons is expended for each ATP appearing outside the mitochondrion, and the total energy available is 3×5.3, or 15.9, kilocalories per mole. The stoichiometry of mitochondrial phosphorylation is also changed: the ratio of phosphate ions consumed to oxygen atoms consumed is reduced from 3 to 2.

If a proton gradient is to be a satisfactory intermediate between electron flow and phosphorylation, then it not only must provide sufficient free energy but also must be kinetically competent. The rate at which protons are produced must at least equal the rate at which they are consumed.

Thayer and one of us (Hinkle) made a stringent test of the chemiosmotic theory by comparing the rate of ATP synthesis in inverted mitochondrial vesicles driven by respiration with that driven by an artificial electrochemical proton gradient. In the first tenth of a second the artificial gradient drove ATP synthesis faster than the oxidation of NADH did, but thereafter synthesis driven by the artificial gradient declined as the gradient decayed. The implication is that the rate of phosphorylation is controlled by the magnitude of the gradient. A difference in the initial rate of synthesis was also observed in two systems of vesicles with naturally generated gradients. One suspension of vesicles had all the reactants needed for oxidative phosphorylation except NADH, which was added at the last moment to start the reaction; in the other suspension the one missing component was ADP. The system triggered by adding NADH started much slower, presumably because time was required for the proton gradient to build up. In the system lacking ADP respiration could establish the gradient in advance and phosphorylation could begin immediately when ADP was added.

More recently Boyer and his colleagues have shown that the rate of ATP synthesis in chloroplasts exposed to an acid-to-base transition is equal to the maximum rate driven by light. A. R. Portis, Jr., and one of us (McCarty) have found that the rate of phosphorylation in chloroplasts is critically dependent on the magnitude of the pH gradient. Indeed, the rate is proportional to the cube of the internal proton concentration, a finding consistent with other indications that three protons pass through the CF_1-F_0 complex for each molecule of ATP formed.

One of the most powerful techniques of biochemistry consists in taking apart a complex system of enzymes and reassembling selected parts of it so that the operation of each part can be examined in isolation. This method can be applied to the study of oxidative and photosynthetic phosphorylation by embedding selected proteins in an artificial membrane.

In 1971 Yasuo Kagawa and Racker, who had by then moved to Cornell, developed a method of incorporating isolated transport enzymes into artificial vesicles made of phospholipids, the fatlike molecules that form the basic matrix of all biological membranes. The vesicles were made by mixing the phospholipids and the selected proteins with a detergent and then removing the detergent by dialysis. The molecules assemble themselves spontaneously into closed vesicles.

The first vesicles made by Kagawa and Racker incorporated the mitochondrial F_1-F_0 complex. They were inverted with respect to normal mitochondria, that is, the F_1 knobs were on the outer surface rather than the inner one. As a result the splitting of ATP by the F_1-F_0 complex transported protons inward.

Since then one of us (Hinkle) and his co-workers have employed the same method to study each of the three segments of the mitochondrial respiratory chain separately. Each of the systems could be driven by adding a different combination of oxidants and reductants. In each case the resulting electron flow was found to be coupled to the transport of protons with a stoichiometry similar to that observed in mitochondria. Both a proton concentration gradient and a membrane potential could be formed.

Racker and his co-workers have constructed vesicles incorporating both the F_1-F_0 complex and segments of the respiratory chain. Care must be taken in preparing such particles to see that all the components in a single vesicle have the same sidedness. When all the functional proteins are properly oriented, the vesicles are capable of oxidative phosphorylation. With Walther Stoeckenius of the University of California at San Francisco, Racker has also combined the mitochondrial F_1-F_0 complex with a light-driven proton pump, called bacteriorhodopsin, from a salt-loving bacterium. The hybrid vesicles exhibit light-driven phosphorylation.

These reconstituted systems offer persuasive evidence that each segment of the respiratory chain transports protons and that a proton gradient alone, generated by any means, will drive phosphorylation by the F_1-F_0 complex. They strongly suggest that a proton gradient is not only necessary for ATP synthesis but also sufficient. The investigation of the chemiosmotic theory can thus move on from testing basic postulates to examining detailed mechanisms.

13 The Photosynthetic Membrane

by Kenneth R. Miller
October 1979

The conversion of light energy into chemical energy by green plants is accomplished in the thylakoid membrane of the plant cell. Electron microscopy reveals the asymmetry that makes the conversion possible

The earth is a planet bathed in light. It is therefore not surprising that many of the living organisms that have evolved on the earth have developed the capacity to trap light energy. Of all the ways in which life interacts with light the most fundamental is photosynthesis, the biological conversion of light energy into chemical energy. In an energetic sense all living things are ultimately dependent on photosynthesis, which is the source of all forms of food and even of the oxygen in the earth's atmosphere. The vast majority of living cells, from simple algae to large and complex terrestrial plants, are photosynthetic. Much has recently been learned about the design and function of the biological structure in which light energy is initially trapped: the photosynthetic membrane.

The overall chemistry of photosynthesis can be expressed in a deceptively simple equation. Six molecules of carbon dioxide are taken up from the environment together with six molecules of water. In the presence of light these molecules are converted into a single molecule of the six-carbon sugar glucose and six molecules of oxygen are released. The products of the reaction hold more chemical energy than the reactants do, and so in a sense the energy of sunlight is captured in the glucose and oxygen that are produced in the process.

For purposes of analysis photosynthesis is most conveniently divided into two steps: a light reaction and a dark reaction. The light reaction captures energy from the sun in two comparatively unstable molecules: adenosine triphosphate (ATP) and nicotinamide adenine dinucleotide phosphate (NADPH). In the dark reaction ATP and NADPH supply the energy needed to form glucose from carbon dioxide.

Both of these reactions, the light and the dark, take place in regions of the plant cell known as chloroplasts. The chloroplast is a complex organelle that has its own genetic material and is able to synthesize at least a few of the proteins needed for its own functioning. It is filled with membranous sacs called thylakoids, which in many chloroplasts are piled on top of one another in stacks called grana. Thylakoid membranes carry out the light reaction of photosynthesis; the dark reaction takes place in the soluble, or nonmembranous, component of the chloroplast.

Because the enzymes that carry out the conversion of carbon dioxide into glucose are soluble and can be easily studied by the biochemist, the dark reaction was the part of photosynthesis that yielded to experimentation first. Melvin Calvin and his associates at the University of California at Berkeley unraveled the enzymatic steps of the dark reaction some 20 years ago. The dark reaction actually occurs most often in the light (when leaves are illuminated); it is called the dark reaction because it does not require the direct participation of light. The reaction can proceed in total darkness if the appropriate compounds are supplied.

The mechanism by which the light reaction achieves the initial conversion of light into chemical energy is the inner sanctum of photosynthesis. Only very recently have workers in several fields of biology and biochemistry begun to arrive at a comprehensive picture of how the light reaction is organized in the photosynthetic membrane.

The green pigment chlorophyll is the central molecule of photosynthesis. Chlorophyll and accessory pigments such as beta-carotene and xanthophyll absorb light over much of the visible spectrum. The absorption of a single photon of light by a chlorophyll molecule gives rise to a separation of positive and negative charges in a special region of the membrane known as a reaction center. After the absorption event the reaction center, which has lost an electron, has a positive charge, and the free electron, now the recipient of much of the energy of the absorbed photon, is capable of reducing (that is, adding an electron to) an appropriate molecule in the membrane.

The ability to maintain this charge separation is one of the critical features of the system. Although positive and negative charges are plentiful in the membrane, a mechanism exists to prevent the energized electron from recombining with positive charges and giving up its energy as heat or fluorescence. As we shall see, the thylakoid membrane is specifically designed to harness the energy available in the excited electrons.

An analogous process of excitation and charge separation takes place in a man-made solar cell based on crystalline silicon. In such a photovoltaic cell p (positive) and n (negative) regions are created in the silicon crystal by the controlled addition of small amounts of impurities. In effect the n region is made more hospitable to negative charges and the p region more hospitable to positive ones. If the photovoltaic cell is wired into a circuit, the accumulation of excited electrons can be tapped to yield a small electric current.

The chloroplast lacks the luxury of a wiring system, but it does have a special array of molecules near the reaction center that prevent the energy stored in excited electrons from being wasted. These molecules, some protein and some lipid, form an electron-transport chain that carries electrons from the reaction center and ultimately utilizes them to convert $NADP^+$ into NADPH. $NADP^+$ is the oxidized form of NADPH. On the addition of a pair of energized electrons and a proton $NADP^+$ is reduced to NADPH, one of the molecules required for the dark reaction. Since electrons cannot flow continuously to $NADP^+$ from the reaction center without being replenished, the membrane incorporates a system to reduce the reaction center once it has been oxidized (that is, depleted of an electron). The system removes electrons from water and in the process releases from the membrane protons (hydrogen ions) and oxygen molecules. Electron flow in the thylakoid therefore follows a linear pathway from water to NADPH.

One of the essential features of this linear flow is that it is vectorial, or unidirectional. The movement of electrons is from inside the thylakoid sac to the

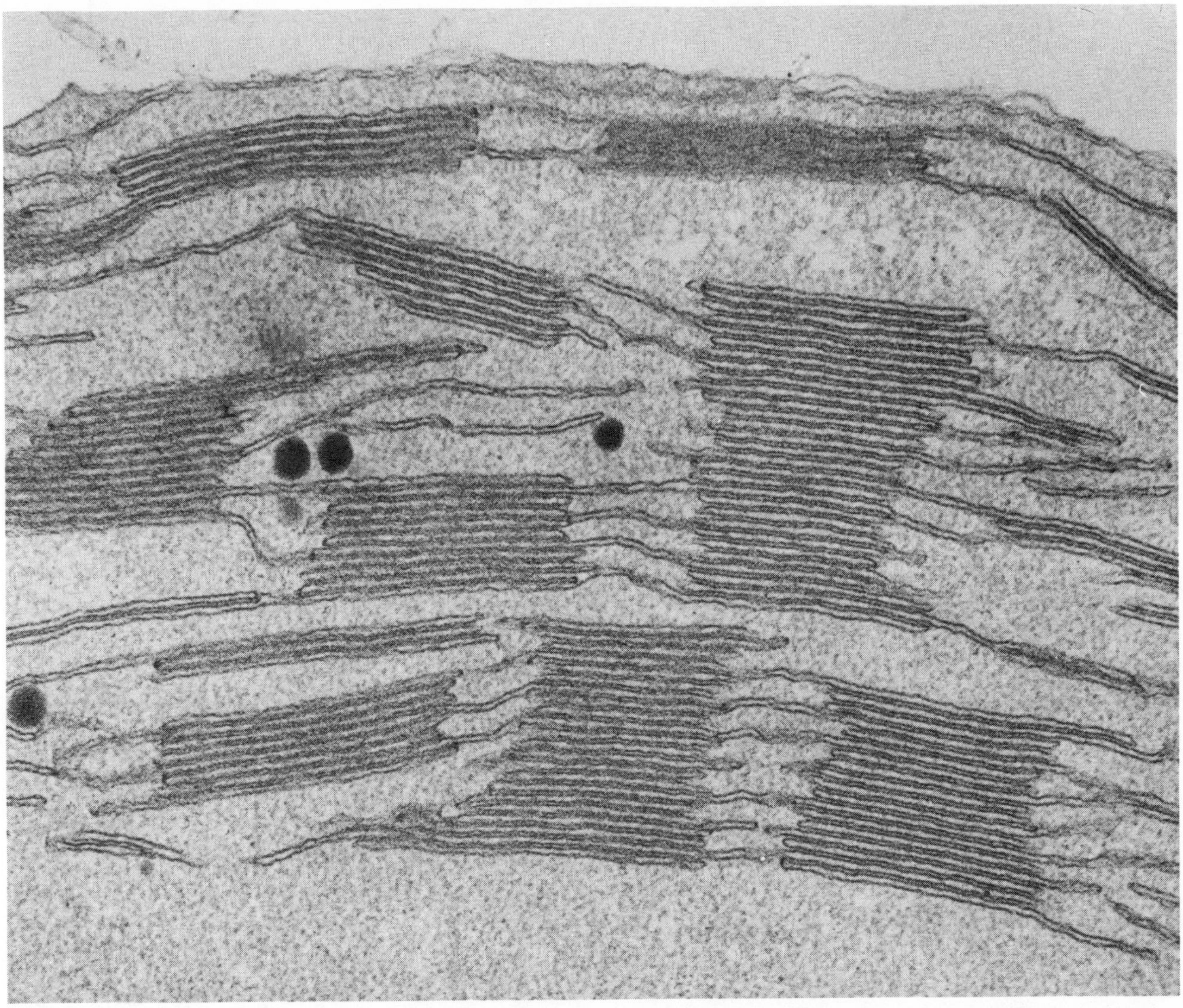

SITE OF PHOTOSYNTHESIS in the green-plant cell is the organelle known as the chloroplast. This electron micrograph shows part of a chloroplast from a leaf cell of a corn plant magnified 70,000 diameters. The closely spaced membranes, which in many cells are piled into stacks called grana, conduct the initial step in photosynthesis: the "light reaction" that traps the photons of light. The capture is carried out by pigment-protein complexes embedded in the membranes. They can be revealed by a technique in which the bilayer membrane structure is split into two halves by freezing and fracturing. Splitting one of the membranes of a closely spaced pair exposes large particles of the kind identified in the micrograph below. Unless it is otherwise indicated all the micrographs in this article were made by the author.

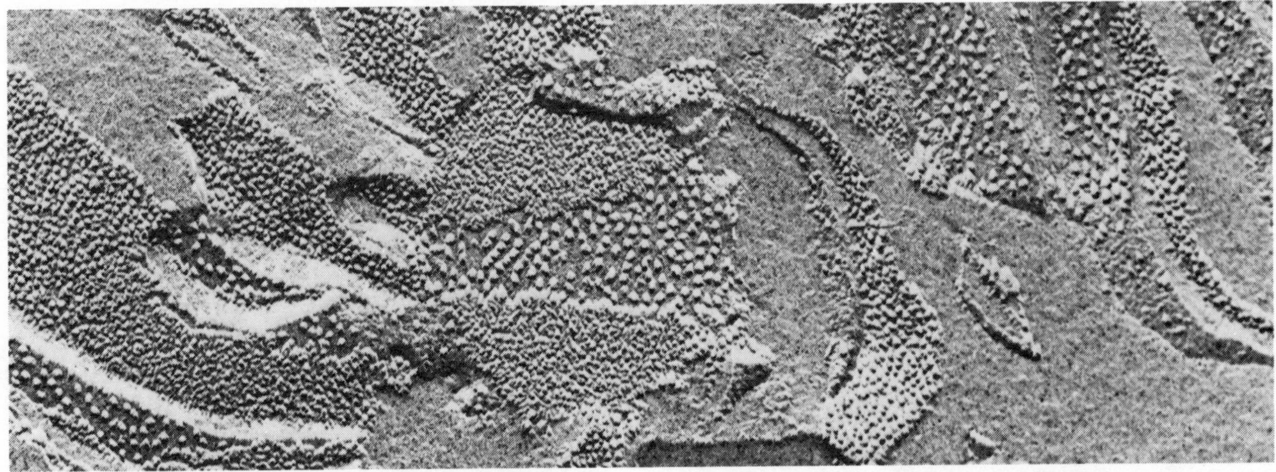

VARIETY OF EMBEDDED PARTICLES are exposed when grana are frozen and fractured. The fractured surface is "shadowed" with platinum and coated with carbon, producing a replica that can be separated from the frozen sample and examined in the electron microscope. The complex contours arise because the fracture tends to jump from one level to another as it passes through the membranes. Large particles in middle are exposed when one membrane of a close pair in a granum is fractured. The magnification is 70,000 diameters.

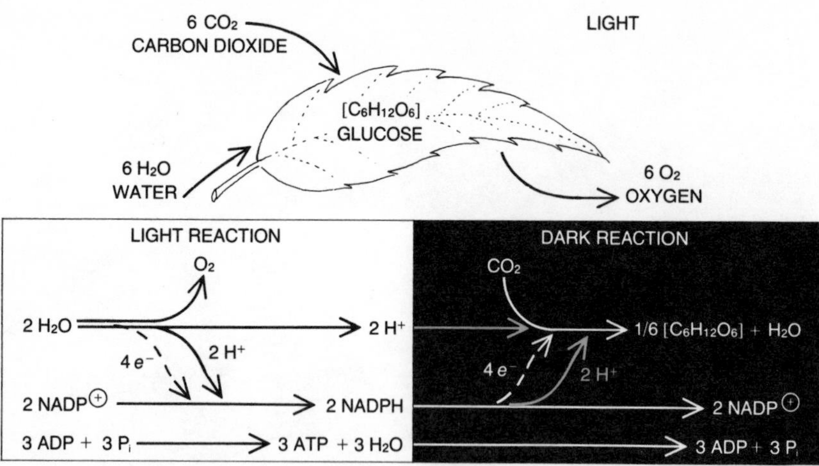

OVERALL CHEMISTRY OF PHOTOSYNTHESIS is simply expressed (*top*) as the synthesis of glucose (a six-carbon sugar) from water and carbon dioxide with the release of oxygen. The products of the reaction store more energy than the reactants do, so that the energy of light is captured chemically. The process has two distinct steps (*bottom*). In the first step (the light reaction) photons captured by chlorophyll and other pigments supply the energy to split two molecules of water into a molecule of oxygen, four protons (or hydrogen nuclei, denoted H^+) and four electrons (e^-). Simultaneously the four electrons and two of the protons convert two molecules of nicotinamide adenine dinucleotide phosphate ($NADP^+$) into its reduced form (NADPH). Another portion of the light energy helps to add inorganic phosphate (P_i) to adenosine diphosphate (ADP) to form a molecule that stores more energy: adenosine triphosphate (ATP). In the dark reaction (so named because it can proceed in the absence of light) the energy stored in NADPH and ATP is extracted to convert carbon dioxide and hydrogen into glucose and water. Half of the hydrogen and all the electrons that are needed are supplied by NADPH.

STRUCTURAL SIMILARITIES are evident in the two molecular systems that provide temporary chemical storage for the light energy trapped by the photopigments. In the ADP-ATP system energy is stored in bond of the added phosphate group. In $NADP^+$-NADPH system energy storage is achieved by the addition of a proton and two electrons to nicotinamide group.

outside. Therefore the inside of the sac quickly becomes positively charged and the outside becomes negatively charged. This has two important consequences. First, the displacement of electrons creates an electric potential across the membrane, similar in many respects to the potential developed across the *p-n* junction of a silicon photovoltaic cell. Second, the concentration of protons is now dramatically different on the two sides of the membrane, with the interior being much the richer. Because the random forces of diffusion tend to drive molecules in such a way as to equalize their concentrations on both sides of a barrier, the difference in concentration, or proton gradient, is a source of potential energy. The larger the proton gradient, the greater the amount of potential energy.

One can now begin to see why membranes are associated with the photosynthetic process. It is likely that the light reaction is associated with a membrane precisely in order to hold the reaction centers and electron-transport systems in an asymmetrical orientation capable of establishing a proton gradient. The membrane also provides a tough barrier to maintain the difference in charge that has built up, and to store that energy for a very brief period.

Although there is (just barely) enough energy in a single absorbed quantum of light to drive electron flow uphill from water to $NADP^+$, the thylakoid membrane evidently requires the absorption of two quanta at separate reaction centers, linked in series, to accomplish the task. The two reaction centers, generally called photosystem I and photosystem II, were discovered because the chlorophyll molecules at the two centers absorb light of slightly different wavelengths. Why does the plant expend an additional photon on the transport of electrons? The answer may lie in the way the other molecule required by the dark reaction—ATP—is synthesized by the thylakoid.

The transport of electrons is accompanied by an acidification of the internal space of the thylakoid sac. The hydrogen ions responsible for the acidification seem to come from two sources: the protons released during the removal of electrons from water and the protons actually pumped across the thylakoid membrane in the course of electron transport. Much of the energy trapped in electrons excited by the absorption of light seems to be consumed in this shuttling of protons across the membrane, with much of the remaining energy accounted for by the formation of NADPH.

The British biochemist Peter Mitchell was the first to realize that the buildup of protons in the thylakoid sac might serve as a source of energy to drive phosphorylation: the adding of a phos-

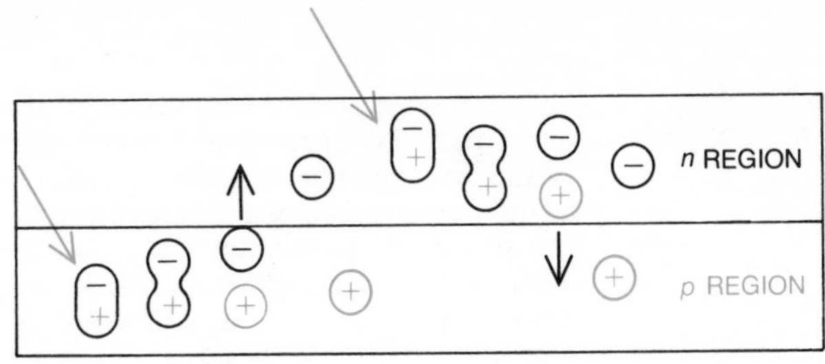

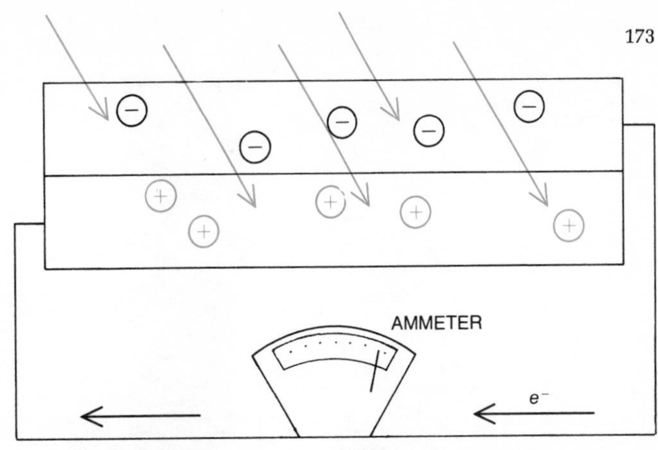

MECHANISM OF CHARGE SEPARATION is exploited in man-made solar cells. Such photovoltaic cells consist of crystalline silicon grown in the form of a junction in which two adjacent layers are separately doped with trace amounts of other elements. These "impurities" make one layer hospitable to positive charges (*p*) and the other layer hospitable to negative charges (*n*). When a photon enters the *n* region of the cell, an energized positive charge (actually a "hole," or the absence of an electron) passes into the *p* region. Similarly, the absorption of a photon in the *p* region releases a negative charge into the *n* region (*left*). Under steady illumination (*right*) the buildup of positive and negative charges on opposite sides of the *p-n* junction can eventually reach a level as high as .6 volt. The effectiveness of the photovoltaic cell rests on its inherent asymmetry and the ability of the junction to keep the positive and negative charges separate.

phate group to adenosine diphosphate (ADP) to form ATP. The phosphorylation of ADP to ATP calls for a source of energy and therefore does not occur spontaneously. Mitchell recognized that an electrochemical gradient created across a membrane by the transport of protons might serve as just such a source. He suggested that specialized enzymes in the thylakoid membrane might be able to extract the energy inherent in the gradient to drive the synthesis of ATP. At the time he proposed his ideas few biochemists were willing to consider them. Now the essential elements of Mitchell's proposals have been accepted by most investigators.

The key predictions of Mitchell's model, which is often called the chemiosmotic theory, were confirmed by experiments in many laboratories. Among them was an experiment by André T. Jagendorf of Cornell University and Ernest G. Uribe of Washington State University showing that ATP could be synthesized in total darkness if a proton gradient was artificially created across the thylakoid membrane. Work in Efraim Racker's laboratory at Cornell demonstrated that a large protein molecule at the outer surface of the membrane was critical to photophosphorylation. Although electron transport was not disturbed when the molecule was removed from the membrane, ADP could no longer be phosphorylated. When the molecule was added back to the membrane, photophosphorylation could again proceed. Because this molecule couples electron transport to photophosphorylation it is generally called the coupling factor.

It should be noted that one of Mitchell's critical predictions is that the thylakoid membrane should be asymmetri-

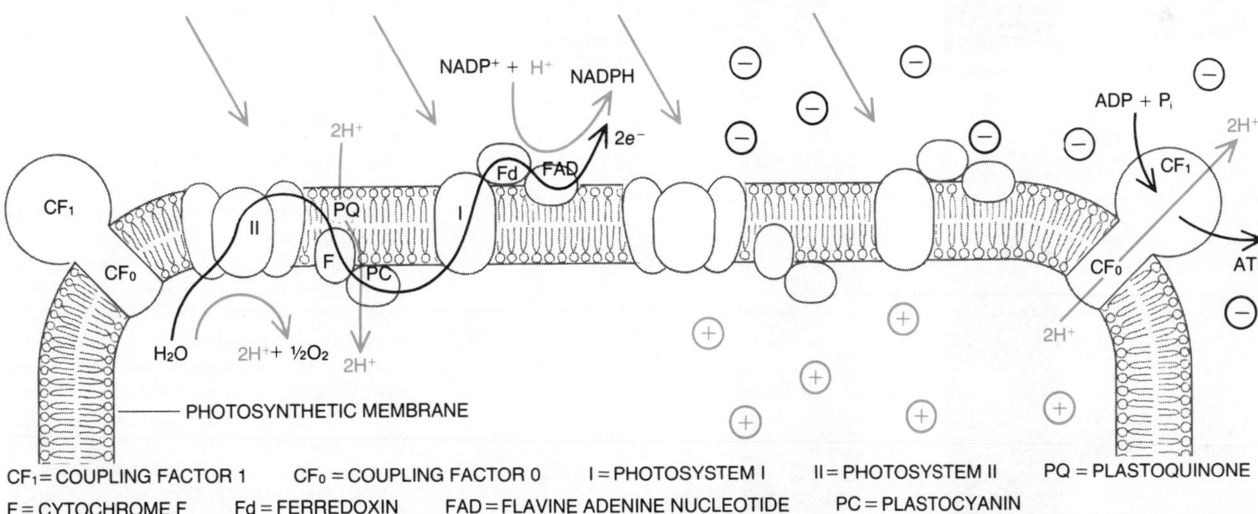

CHAIN OF REACTIONS powered by photons is exploited by the photosynthetic membrane inside the chloroplast to accomplish a separation of charges. In the process NADP⁺ is reduced to NADPH. The voltage differential that results from the separation of charges also helps to drive the conversion of ADP into ATP. The two processes, which constitute the light reaction, proceed simultaneously in the membranous sacs called thylakoids. For the purposes of clarity the processes are depicted here as taking place separately within a thylakoid that is part of a large granum. At the left in the illustration two reaction centers, designated photosystem I and photosystem II, absorb light of slightly different wavelengths. In photosystem II water is dissociated into two protons (H^+), oxygen and two electrons (e^-). The electrons are shuttled through an electron-transport chain involving plastoquinone, plastocyanin, photosystem I, ferredoxin and flavine adenine dinucleotide, which finally enables the electrons, with the aid of a proton, to reduce $NADP^+$ to NADPH at the outside surface of the thylakoid membrane. At the right positive charges have accumulated on the inside of the membrane from two sources: the protons that are left behind after dissociation of water and positive charges that seem to be shuttled across the membrane from the exterior in electron transport. The outflow of protons through the coupling factors (CF_0 and CF_1) drives the synthesis of ATP from ADP.

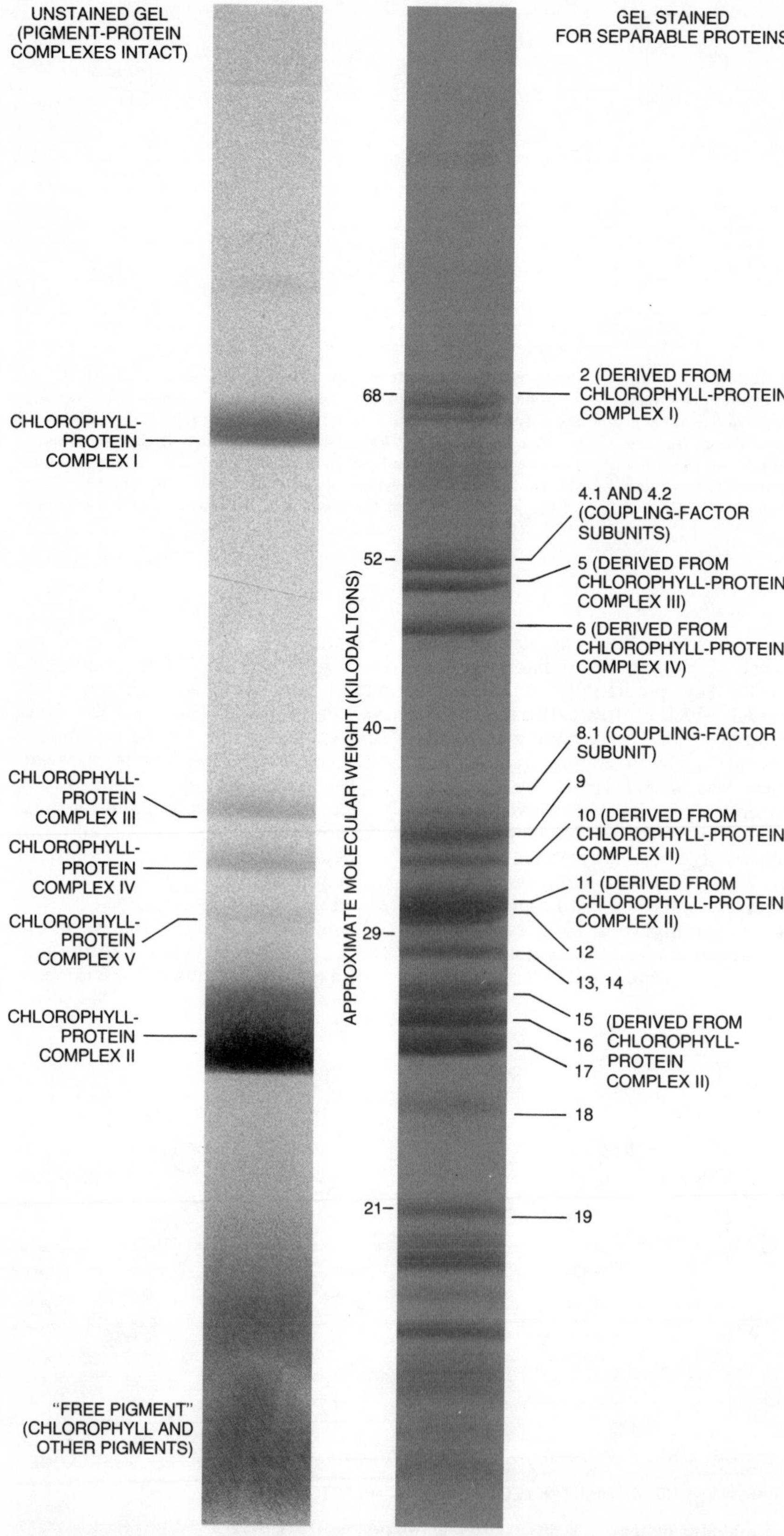

COMPONENTS OF THE PHOTOSYNTHETIC MEMBRANE can be separated by dissolving the membrane in a detergent (lithium dodecyl sulfate) and allowing the protein components to migrate through a polyacrylamide gel under the influence of an electric field, the technique known as electrophoresis. The rate of migration of a molecule is determined largely by its size. The green color of several of the fractions in the chromatogram at the left shows that chlorophyll forms complexes with several different proteins in the membrane. The chromatogram at the right has been stained with a blue dye that brings out all the polypeptide components (proteins and protein subunits) of the membrane. Bands are more numerous and appear at positions different from those in the gel at the left because the pigment-protein complexes were dissociated by heating. The chromatograms were made by Nam-Hai Chua of Rockefeller University.

cal, both structurally and biochemically. A perfectly symmetrical membrane could not build up the differences of proton concentration that are required for photophosphorylation, nor could it effectively utilize such gradients. As we shall see, this prediction has been dramatically realized in studies of the structural organization of the thylakoid membrane.

The purple membrane of the salt-loving bacterium *Halobacterium halobium,* the simplest energy-transducing membrane, has only a single protein for shuttling protons in response to light. The thylakoid membrane of the chloroplast is far more complex. In electrophoretic gels, where proteins and shorter chains of amino acids (polypeptides) can be separated into bands according to their molecular weight, the proteins and polypeptides of the thylakoid membrane can be resolved into at least 40 molecular species, of which only a few have been identified biochemically. The identified bands include some of the components of the coupling factor and also a few of the electron-transport components.

One of the main questions engaging the interest of investigators working on the thylakoid membrane is how the green pigment chlorophyll is dispersed within it. The membrane basically consists of a bilayer of lipid molecules, with protein molecules either embedded in it or associated with its surface. Chlorophyll can freely associate with a lipid bilayer, so that there is at least a possibility that some chlorophyll molecules are freely mobile in the membrane. Workers in several laboratories have now found, however, that most of the pigment seems to be associated with particular proteins. When the membrane is carefully dissolved in certain detergents, several proteins bind chlorophyll and appear as green bands in an electrophoretic gel. Intensive work is now under way in several laboratories in the effort to learn more about such chlorophyll-protein complexes. Several of the complexes have been shown to be associated directly with either photosystem I or photosystem II.

As I have mentioned, the two photosynthetic reaction centers are driven by slightly different wavelengths of light. Does this imply that the two centers exist in separate structural centers within the membrane? Apparently it does. When thylakoid membranes are dissolved in very mild detergent, large complexes can be recovered from the membrane that contain chlorophyll, together with several polypeptides, and that display activity characteristic of either photosystem I or photosystem II. Experiments such as this one, which have been done with a wide variety of detergent agents, are evidence for two physically distinct reaction centers.

In my laboratory at Harvard University we have been concerned with developing a structural description of the thylakoid membrane that will deepen understanding of the light reaction of photosynthesis. Although conventional techniques of preparing specimens for electron microscopy are useful, they have not enabled us to investigate the nature of structures within the photosynthetic membrane itself. We have therefore resorted to the technique of freeze-fracturing, in which a sample of photosynthetic membranes is rapidly frozen and then fractured in a vacuum chamber. The surface exposed by the fracture turns out to be of considerable interest for studying the organization of a biological membrane. A replica is made of the surface by covering it with a thin layer of platinum and carbon, which conforms to the shape of the fractured surface and can be removed from it and placed in the electron microscope. Such replicas reproduce very fine detail from the original specimen. Moreover, by placing the platinum source to one side of the sample, shadows are created that indicate the relative heights of projecting structures as if they were seen in three dimensions.

Freeze-fracture replicas can be prepared in two slightly different ways. When a membrane specimen is simply fractured, the lipid bilayers of membranes in the path of the fracture split along a central plane, thereby separating the two layers. For certain purposes the freshly fractured surface can be "etched" by allowing ice to sublime away from the frozen membrane before the replica is made. Etched preparations are useful for revealing the outer surfaces of a membrane that would otherwise be covered with ice.

Whereas natural or artificial membranes with little or no protein associated with them exhibit large, smooth surfaces when they are frozen and fractured, the photosynthetic thylakoid membrane reveals a bewildering variety of particles of different sizes and shapes. Indeed, the complexity is so great that several years passed before those of us involved in freeze-fracture studies were able to agree on what we were seeing. The complexity of the structure is indicative of the complexity of the membrane's function. It has turned out, for example, that the internal organization of the membrane is different in regions where two membranes are stacked on top of each other. The stacking, which is characteristic of the thylakoid membrane, gives rise to the piles of grana visible in the intact chloroplast.

In regions where membranes are stacked one can see a dense concentration of large particles on one of the fracture faces. In adjacent regions of the same membrane that are not stacked the density of the same type of particle is

PARTICLES OF VARIOUS TYPES are visible in photosynthetic membranes of barley that have been exposed by the freeze-fracture technique and magnified 100,000 diameters. The fracture path jumped from the middle of one membrane to the middle of an adjacent closely stacked membrane, so that four different "fracture faces" were exposed, designated EF_u, EF_s, PF_s and PF_u. EF and PF stand for exoplasmic face and periplasmic face; s and u, for stacked and unstacked. An explanation of how the different faces are formed is diagrammed below.

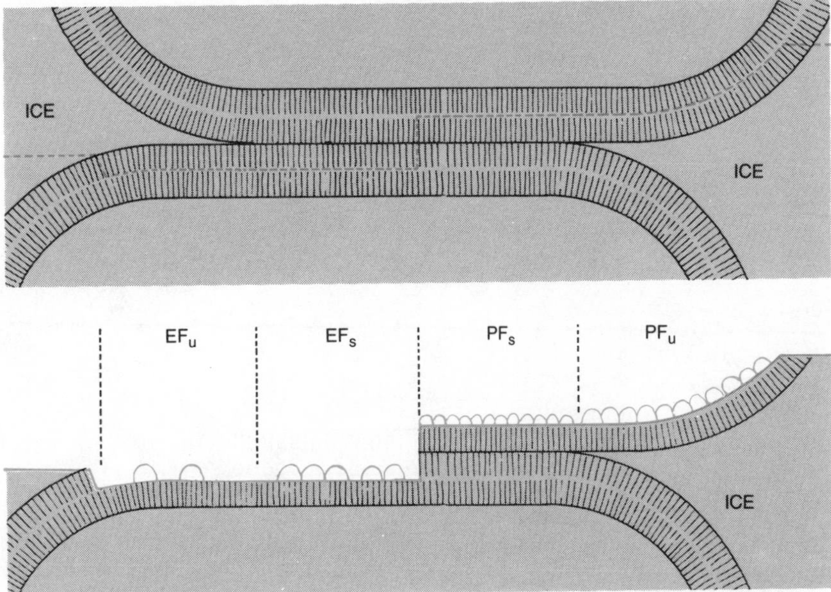

FRACTURE PATH that exposed the four types of fracture face in the micrograph at the top of the page is depicted schematically. The largest particles (EF_s) are on the inner half of a thylakoid membrane that closely abuts another membrane. Where the two membranes are not in contact, at the left, the large particles are reduced in size and number; the exposed face is one designated EF_u. Where the fracture has split the upper membrane of the adjacent pair, at the right, the exposed faces are different again. Here the exposed surface is the inside of the outer half of the thylakoid membrane rather than the inner half. Exposed faces are those designated PF_s in the region where the membranes touch each other and PF_u where membranes do not.

much lower. In 1966 Seikichi Izawa and Norman E. Good of Michigan State University noted that when thylakoid membranes are suspended in a solution of low ionic strength, the stacked regions disappear and the membranes appear freely unfolded. When such membranes are studied by freeze-fracturing, internal surfaces that have dense populations of large particles are no longer seen.

Thylakoid membranes that have been studied by the etching technique show that Mitchell's prediction of a highly asymmetrical photosynthetic membrane is borne out. The inner and outer surfaces of the membrane are remarkably different. The outer surface is covered by a mixture of large particles (about 12 nanometers across) and smaller ones (about eight nanometers across). The inner surface is densely covered by boxcar-like particles with an overall diameter of about 18 nanometers, seemingly composed of four or more subunits.

Several years ago, while I was working in L. Andrew Staehelin's laboratory at the University of Colorado, I removed the coupling factor, which synthesizes ATP from ADP, from the photosynthetic membrane in order to see how its absence might change the appearance of the freeze-etched specimens. Our results and those from several other laboratories showed that the large particles on the outer surface of the membrane are indeed molecules of coupling factor. In studying the effect of

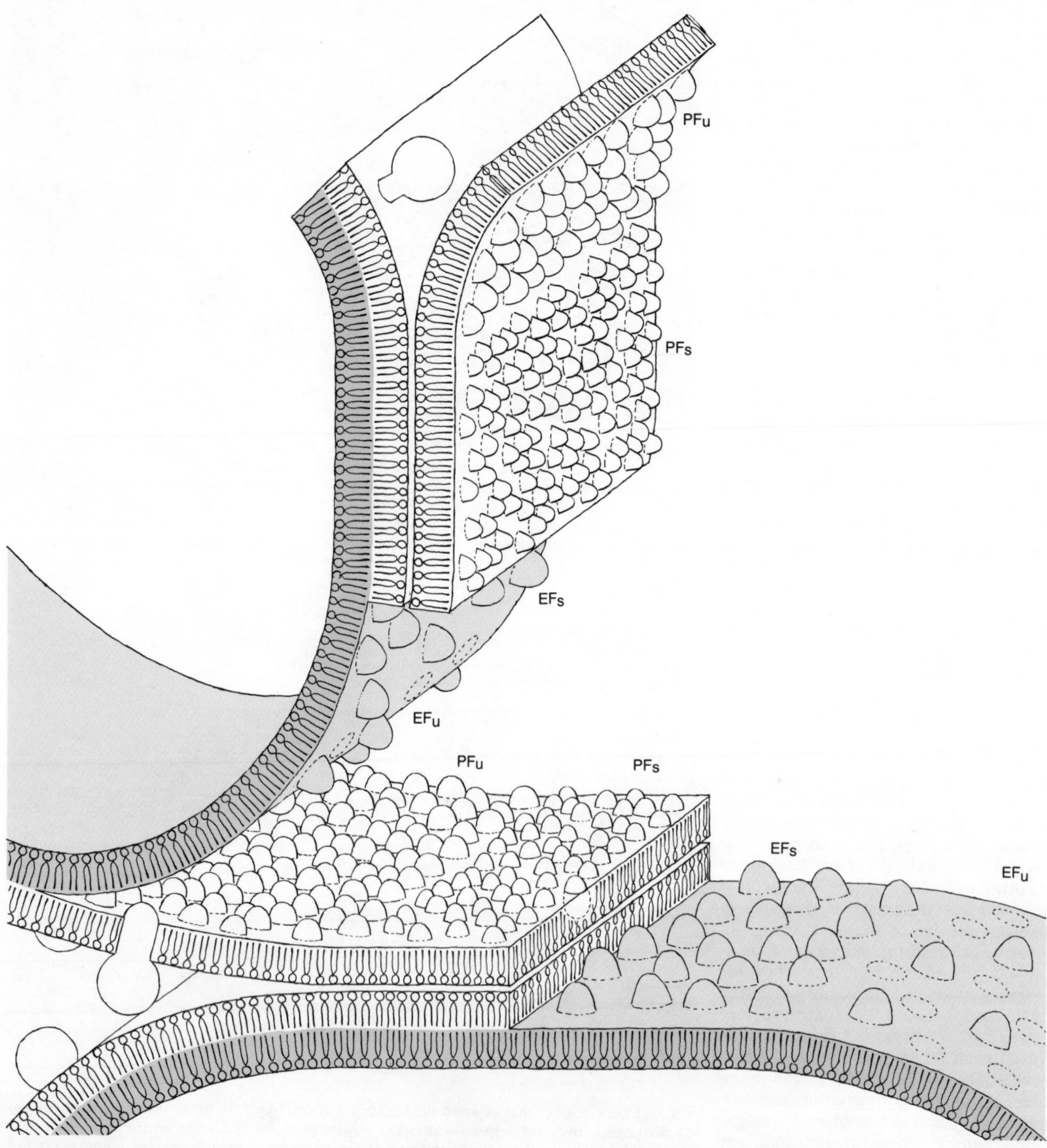

ASYMMETRY OF THYLAKOID MEMBRANE is demonstrated by analysis of the faces exposed by freeze-fracturing. The particles on the exposed inner half of the fractured membrane (*color*) are different in appearance from the particles on the exposed outer half (*black*). In other words, the membrane has a distinct inside-outside polarity. Large particles on the EF_s and EF_u faces seem to be multiunit structures associated with photosystem II. Some PF_s and PF_u particles may be associated with photosystem I. Light-harvesting components, portions of electron-transport chain and membrane-bound elements of coupling factor may also form parts of the small PF particles.

membrane stacking and unstacking on the density of these particles we found, to our great surprise, that they seemed to be excluded from the outer surface of the thylakoid, where the membranes came together to form part of a granum; they were found only on surfaces in unstacked regions. This result is further support for Mitchell's chemiosmotic theory because it shows that the ATP-synthesizing system need not be directly connected to an electron-transport chain in order to take advantage of the light-induced proton gradient.

The inner surface of the membrane does not show coupling-factor molecules but rather is populated by the curious particle with four or more subunits. Under certain conditions, which we have only begun to understand, these particles seem to "crystallize" into a regular two-dimensional lattice. Such lattices are found not only at the inner surface of the membrane but also at the outer one and in fractured preparations. The particle is visible at both surfaces of the thylakoid and in the interior membrane as well, so that one must conclude that the particle spans the membrane. What is the nature of this unusual structure, and what is its association with photosynthesis?

Independent studies conducted in my laboratory and in the laboratory of Charles J. Arntzen at the University of Illinois at Urbana gave the first evidence that a particular chlorophyll-protein complex (CP II) is associated with the membrane-spanning structure. Each study demonstrated that the absence of CP II resulted in a reduction in the diameter of the large membrane-spanning particle; each group therefore concluded that this particle must be the reaction complex of photosystem II. Evidently the photosystem II reaction center is at the core of the particle and CP II molecules are arrayed around it.

More direct evidence has recently been obtained in a study of a mutant tobacco plant that is deficient in photosystem II activity. Robert Cushman and I found that the thylakoid membranes of the mutant deficient in photosystem II are almost completely devoid of the large particle and that all the other structures of the membrane are apparently unaltered by the mutation. The fact that the large membrane-spanning particles seem to be concentrated in stacked regions of the thylakoid-membrane system is also consistent with the idea that they represent photosystem II reaction complexes. Studies in which the membrane system is broken up into grana and stroma (nongrana) fractions show that photosystem II activity is much higher in the grana membranes, which contain large numbers of the particles.

Recent studies suggest that membrane stacking may also be mediated by CP II

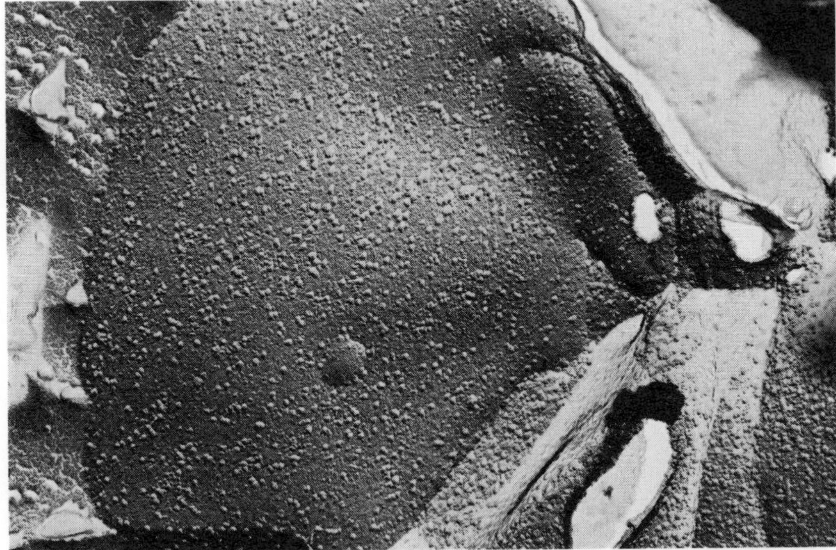

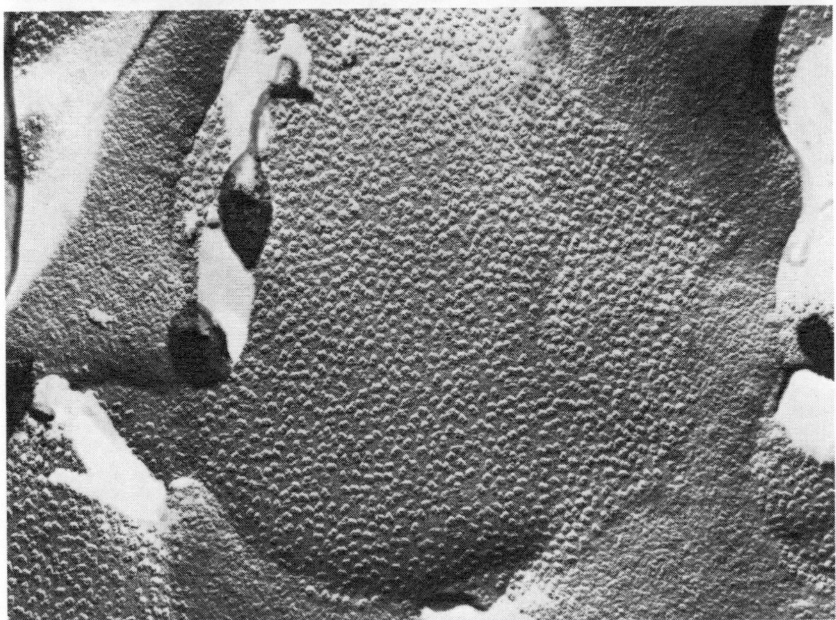

EXTERIOR SURFACES OF THE THYLAKOID MEMBRANE are made visible by freeze-etching. In this method surfaces of membranes that have not been fractured but that were originally covered by ice are exposed by allowing the ice to sublime away. The exposed surfaces are then shadowed with platinum and replicated as usual. The outer surface of the photosynthetic membrane (*top*) is covered by a mixture of large and small particles, 12 nanometers and eight nanometers in diameter. The larger particles have been identified as coupling-factor molecules. Inside surface of membrane (*bottom*) is studded with a boxcar-like particle that seems to have four or more subunits. Magnification in both micrographs is 55,000 diameters.

MAGNIFIED VIEW OF INSIDE SURFACE of the thylakoid membrane shows the boxcar-like particles in greater detail. The magnification here is 135,000 diameters. The particles penetrate the membrane and can also be recognized on the outside surface. These membrane-spanning structures correspond to the photosystem II particle EF_s in illustration on opposite page.

molecules associated with this particle. Evidently some interaction between inorganic salts and CP II is responsible for the stacking process; it may also help to arrange the particles into regular lattices. For example, when purified molecules of CP II are incorporated into artificial membrane systems, the molecules arrange themselves into regular lattices.

What is the location of photosystem I in the thylakoid membrane? Does it also appear as a distinct particle? And if it does, how does it interact with photosystem II? Preliminary studies in our laboratory of mutants deficient in photosystem I indicate that it is associated with a subset of the smaller particles found in the photosynthetic membrane after freeze-fracturing. Other studies support this idea. For example, in hybrid systems consisting of a synthetic lipid bilayer to which the purified photosystem I reaction center has been added, particles with a diameter matching that of a class of particles found in the natural membrane are observed.

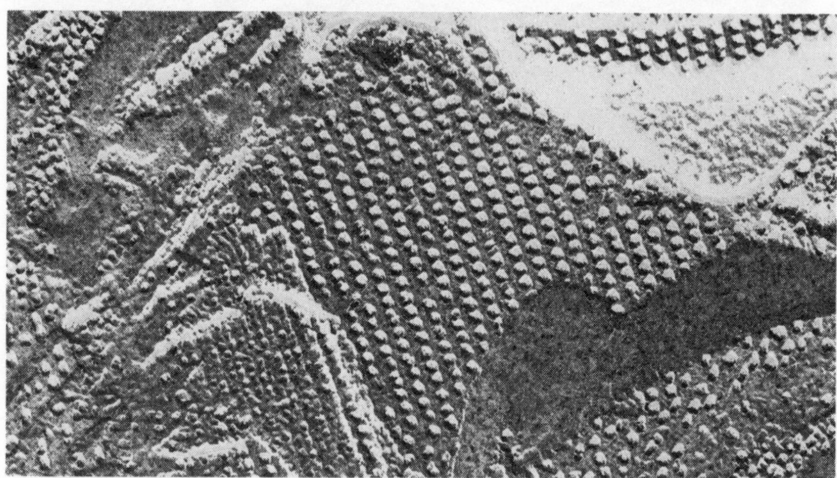

PARTICLES FORM REGULAR LATTICES under certain conditions. These three micrographs, all at a magnification of 135,000 diameters, depict such lattices inside a fractured membrane (*top*), on the inner surface of the intact membrane (*middle*) and on the outer surface of the intact membrane (*bottom*). The lattice has the same dimensions in all three pictures. It is least evident in bottom micrograph but can be seen if image is viewed at a shallow angle. Lattices are apparently formed by membrane-spanning particle associated with photosystem II.

It is clear that a number of distinct substructures within the thylakoid membrane have important roles to play in the light reaction of photosynthesis. One of the disappointing aspects of this story is that little can yet be said about the details of molecular organization within the complex particles revealed by freeze-fracturing. One would like to know, for example, the detailed arrangement of the components of the electron-transport chain, the arrangement of chlorophyll molecules around the two reaction centers and the arrangement of the special proteins that seem to be important to various activities of the membrane.

At present detailed molecular structures are known only for two systems that are related to the photosynthetic membrane. One of these systems is the proton-pumping protein in the purple membrane of the bacterium *Halobacterium halobium*. The bacteriorhodopsin molecules in this membrane are arranged in a regularly repeating hexagonal lattice. That feature enabled Richard Henderson and P. N. T. Unwin of the Medical Research Council Laboratory of Molecular Biology at Cambridge in England to combine electron microscopy, electron diffraction and X-ray diffraction to map the structure of the protein to a resolution of .7 nanometer, which is sufficient to reveal such molecular details as the foldings of amino acid chains.

The second molecular structure now known in detail is that of the water-soluble bacteriochlorophyll-protein complex found in the photosynthetic bacterium *Chlorobium limicola*. Roger E. Fenna and Brian W. Matthews of the University of Oregon have studied crystals of this pigment-protein complex by X-ray diffraction. Their detailed picture of the bacteriochlorophyll-protein complex provides an exciting glimpse of the intricate arrangement of chlorophyll molecules and amino acid chains needed to capture solar energy at the submicroscopic level.

There is reason to believe, however, that neither of these studies will be able to tell us much about the photosynthetic membrane, because in each case the work was done on a system quite distinct from a true photosynthetic one. The purple membrane of *Halobacterium* participates in the light-activated pumping of protons, but it does so without electron transport and without the release of molecular oxygen. Further, although the structure of the chlorophyll protein from *Chlorobium* is most interesting, the protein is a water-soluble one

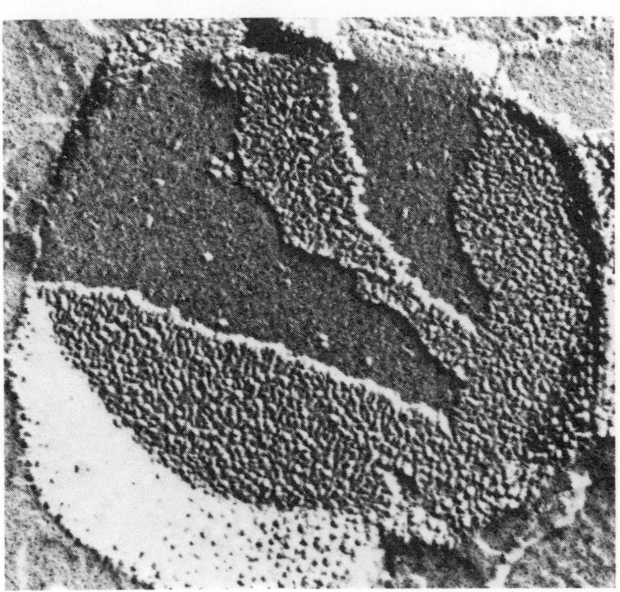

MEMBRANE-SPANNING PARTICLES are the large particles at the left in the freeze-fracture replica of a photosynthetic membrane from a normal tobacco plant. In a mutant deficient in photosystem II a comparable replica of photosynthetic membrane shows no large particles (*right*). All the other structures of the mutant membrane seem to be unaltered. The replicas are magnified 120,000 diameters.

and not a true membrane protein such as the proteins associated with the thylakoid membrane proper. In all probability the structures of the chlorophyll-binding proteins of the photosynthetic membrane will be quite different.

My colleagues and I hope to learn more about the photosynthetic membrane by applying recently developed techniques in which diffraction images of ordered molecular systems are subjected to Fourier analysis. The techniques seem capable of yielding high-resolution images of how the membrane systems are put together. In this article I have described two examples of such ordered systems that should be amenable to the new analytical approach: the ordered lattices that occasionally appear in the thylakoids of higher plants and the regular lattices formed by purified CP II molecules in synthetic lipid membranes.

We have also begun to investigate regular structures in the photosynthetic membrane of the bacterium *Rhodopseudomonas viridis*. With these analytical methods we have obtained images of this membrane with a resolution of about three nanometers. It is the hope of an impatient investigator that not too many years will pass before these approaches and others will reveal the structure of the photosynthetic membrane at the molecular level. For now one is left to marvel at the intricate workmanship that has shaped this membrane and equipped it to operate with such exquisite efficiency. As remarkable and as varied as other living systems have become, it is always worth remembering that the photosynthetic membrane is the engine that has made such extraordinary diversity possible.

ARTIFICIAL MEMBRANE was prepared in the author's laboratory by adding photosystem I complexes, which were purified by Michael Newman, to bilayers formed from a commercially available lipid. When the membrane was frozen and fractured, particles identical in size with certain particles that are present in natural membranes were visible on the fracture face. Magnification is 68,000 diameters.

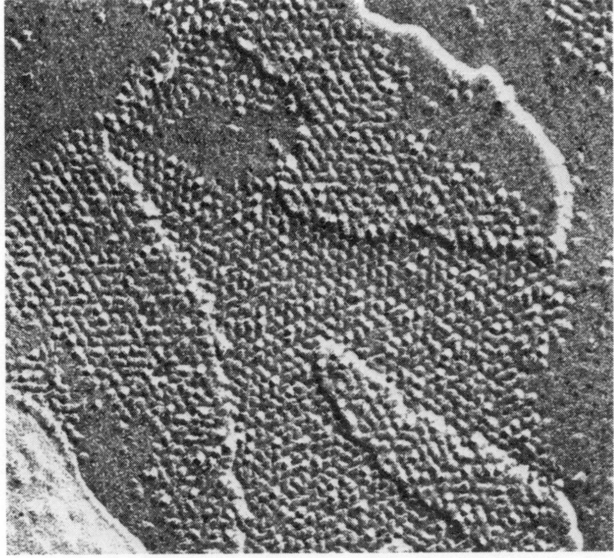

CHLOROPHYLL-PROTEIN COMPLEXES (CP II) that are associated with photosystem II form regular lattices in this micrograph of a freeze-fracture replica of an artificial membrane prepared by Annette McDonnel in the laboratory of L. Andrew Staehelin at the University of Colorado. The replica is magnified 125,000 diameters. The chlorophyll-protein complex also seems to promote grana formation.

IV

NUCLEIC ACID STRUCTURE: EXPRESSION OF THE GENETIC MESSAGE

IV NUCLEIC ACID STRUCTURE: EXPRESSION OF THE GENETIC MESSAGE

INTRODUCTION

The nucleic acids are responsible for genetic continuity in all living cells, and they are also essential to the expression of the encoded hereditary message. Nucleic acids are composed of long chains of nucleotides linked together by phosphodiester bonds. Each nucleotide consists of a sugar, a phosphate group, and a nitrogenous base. The alternating sugars and phosphates form the monotonous backbone of a chain (analogous to the polypeptide backbone in proteins); the bases attached to the sugars may be considered as "side groups." Recall that in proteins the backbone carries no net charge, although some of the amino acid side chains may be charged, depending upon the pH. In nucleic acids at neutral pH, the backbone carries one negative charge per phosphate. This large charge mitigates against other factors that might lead to compact folding of the chain. Thus, whereas polypeptide chains tend to coil, polynucleotide chains tend to remain open and extended. The specificity of a nucleic acid resides in the sequence of bases, and only four types of bases are normally found in DNA: the two pyrimidines, thymine and cytosine; and the two purines, adenine and guanine. RNA differs from DNA in that the sugar in the backbone structure is ribose instead of deoxyribose. Another significant difference is that RNA normally contains the base uracil instead of the thymine (5-methyl uracil) found in DNA. (There are important exceptions to this rule, however.) It is clear that a polynucleotide chain would have to contain many more monomer units than would a corresponding polypeptide chain bearing the same amount of information. A sequence of 200 nucleotides in a strand of DNA could be arranged in only 4^{200} different ways, compared to the 20^{200} possible unique polypeptides that can be assembled as a sequence of 200 amino acids. As we have learned, either DNA or RNA can serve as the primary genome in sub-living entities, but in living cells this responsibility is always entrusted to double-stranded DNA. One might wonder why there must be two distinct types of nucleic acid, so alike and yet so profoundly different.

The general features of RNA and DNA are quite similar, particularly with regard to their ability to form stable helical configurations in which two strands are paired. The pairing of nucleic acid strands is highly specific and depends upon the base sequences. Thus guanine in one strand always pairs with cytosine in the other; likewise, adenine always pairs with thymine (in DNA) or uracil (in RNA). Note that in general a purine in one strand will be found in juxtaposition to a pyrimidine in the other. In addition, the two paired strands are antiparallel, making it possible for a strand to fold back on itself in places to form a hairpin structure, as we have already noted in the case of TMV RNA. In fact, any stretch of nucleotides in either an RNA or a DNA strand may pair with the complementary sequence in another RNA or DNA strand or within the same strand. Self-pairing of a nucleic acid strand is particularly notable in the case of transfer RNA, but it also occurs in two-stranded structures in which palindromic sequences can loop out to form cruciform configurations where the respective strands are self-paired. Within cells RNA and DNA are distinguished by their functional roles, which in turn are as determined by the enzymes that act upon them. Thus DNA polymerases promote the duplication and repair of DNA while RNA polymerases produce the necessary RNA transcripts from DNA templates.

We learned of the essential relationships between genes and their expression in metabolic activity long before we knew anything about nucleic acids except that they were found in chromosomes. Archibald Garrod in 1908 developed the concept of "inborn errors of metabolism" to explain his observations on certain heritable defects in biochemical reactions in humans. His analysis of family pedigrees led to the conclusion that the disease alkaptonuria, in which the urine is darkened by excretion of an abnormal product of nitrogen metabolism, involves the expression of a recessive gene. This first connection between genes and enzymes was far ahead of its time. It remained for George Beadle, Boris Ephrussi, and Edward Tatum to restate and amplify these ideas three decades later as model systems became available for more detailed genetic and biochemical analysis.

Noble Laureate George W. Beadle, in his 1948 article on "The Genes of Men and Molds," suggested that "Genes are probably nucleoproteins that serve as patterns in a model-copy process by which new genes are copied from old ones and by which nongenic proteins are produced with configurations that correspond to those of the gene templates." This classic account of the beginnings of biochemical genetics appeared just five years before James Watson and Francis Crick announced their double-helical model for the structure of DNA. In an analysis of the formation of eye pigments in the fruit fly, *Drosophila melanogaster*, George Beadle and Boris Ephrussi had discovered some precursor-product relationships that suggested to them the possibility of understanding metabolic pathways through use of mutants blocked at different steps in those pathways. Beadle then turned to the simple fungus *Neurospora*, which grows on a chemically defined medium and which has a haploid vegetative phase, so that recessive mutations could be readily identified. Working together at Stanford University, Beadle, a geneticist, and Edward Tatum, a biochemist, were among the first to appreciate the value of combining biochemistry with formal genetics. As a result of their efforts, the "gene" advanced from its status as a hypothetical construct to that of a defined chemical entity. Beadle reasoned that "one ought to be able to discover what genes do by making them defective" (i.e., by studying mutant forms), and he developed the idea that "each gene controls a single protein." That idea was to be proved essentially correct some years later when Charles Yanofsky and his co-workers (working in Tatum's former basement laboratory in Stanford's Jordan Hall) demonstrated a linear correspondence between the amino acids in a specific polypeptide and the sequence of nucleotides in the gene that coded for that polypeptide.

The first four articles in this section describe important historic landmarks in the coupling of genetic analysis to biochemical structure determination. The precision of genetics is dramatically illustrated in Seymour Benzer's article "The Fine Structure of the Gene," which describes how the unit of genetic variability was pushed to the limit of the nucleotide. Different classes of mutation are defined, including localized point mutations and deletions (in which nucleotides are missing from a region of the genome). In addition to his demonstration that recombination of DNA molecules could occur at sites within genes, Benzer was able to infer from his studies that "nonsense" sequences of nucleotides may exist that don't specify any amino acid but rather terminate the corresponding chain of amino acids. While Benzer genetically dissected the rII region of the T4 bacterial virus, F. H. C. Crick focused on the nature of the code itself and by further clever genetic analysis concluded that it must require three nucleotides to specify an amino acid. (Accordingly, each triplet of nucleotides is termed a "codon.") In "The Genetic Code," Crick describes another type of mutant, the reading frame-shift class produced when a foreign molecule, such as the dye acridine orange, is wedged between the stacked bases in the duplex DNA molecule. The resulting misalignment causes the message to be shifted one nucleotide out of register during replication, so that a garbled sequence of amino acids is ultimately obtained when the frame-shift mutation is expressed. Crick's results also confirmed the existence of certain triplets of nucleotides that terminate the growing amino acid chain. The genetic code itself was finally cracked through intensive biochemical efforts in the laboratories of Marshall Nirenberg, Severo Ochoa, and Gobind Khorana after Nirenberg's discovery that defined polypeptides were produced in an *in vitro* protein-synthesizing system when fed synthetic polynucleotide "messenger RNA" of known composition. (See Table 1.) Charles Yanofsky and his associates confirmed the codon assignments in living bacteria and demonstrated the linear correspondence between the sequence of triplet codons in DNA and the amino acid sequence in a protein, thus proving the central dogma of molecular biology. In "Gene Structure and Protein Structure" Yanofsky details the experiments by which this important correspondence was proved in his laboratory; he also describes the parallel studies of Sydney Brenner and co-workers, who established a colinear relation between the genetic map

TABLE 1. THE GENETIC CODE

First position (5′ end)	Second position				Third position (3′ end)
	U	C	A	G	
U	Phe	Ser	Tyr	Cys	U
	Phe	Ser	Tyr	Cys	C
	Leu	Ser	Stop	Stop	A
	Leu	Ser	Stop	Trp	G
C	Leu	Pro	His	Arg	U
	Leu	Pro	His	Arg	C
	Leu	Pro	Gln	Arg	A
	Leu	Pro	Gln	Arg	G
A	Ile	Thr	Asn	Ser	U
	Ile	Thr	Asn	Ser	C
	Ile	Thr	Lys	Arg	A
	Met	Thr	Lys	Arg	G
G	Val	Ala	Asp	Gly	U
	Val	Ala	Asp	Gly	C
	Val	Ala	Glu	Gly	A
	Val	Ala	Glu	Gly	G

Source: From *Biochemistry* by Lubert Stryer. W. H. Freeman and Company. Copyright © 1975.

Note: Given the position of the bases in a codon, it is possible to find the corresponding amino acid. For example, the codon 5′ AUG 3′ on mRNA specifies methionine, whereas CAU specifies histidine. UAA, UAG, and UGA are termination signals. AUG is part of the initiation signal, in addition to coding for internal methionines.

and a structural protein from the T4 bacteriophage capsid. We now know of an amazing complication in the one gene-one protein relationship in which a given stretch of DNA may in fact include overlapping sequence information for more than one polypeptide chain, and that information may be read in different reading frames for the different proteins. This kind of efficiency in the utilization of a limited size genome is illustrated in the expression of the tiny DNA bacteriophage ϕX174.

The crucial elements in the translation of the genetic code are the enzymes (called aminoacyl-tRNA synthetases) that link the amino acids to the appropriate tRNA molecules, which then convey them to the ribosomal assembly sites. Each tRNA contains an "anticodon" triplet of nucleotides that is capable of recognizing (i.e., pairing with) a complementary codon sequence on a mRNA template. Thus, if a faulty aminoacyl-tRNA synthetase incorrectly links the amino acid glycine to a tRNA for histidine, then that glycine may ultimately appear in a polypeptide at a site designated for histidine by the mRNA. However, the accurate translation of the message also involves other factors, including the ribosome itself. For example, errors in translation are caused by the antibiotic streptomycin, which interacts with the 30S ribosomal subunit; sometimes the effect is to "suppress" a nonsense codon and promote the incorporation of an amino acid in response to that codon instead of terminating the chain (Gorini, 1966).

As with proteins, the first determination of the primary sequence of a nucleic acid was tediously worked out many years before the elucidation of its tertiary configuration. Robert Holley achieved the first sequencing of a tRNA over 12 years before Alexander Rich and Sung Hou Kim revealed the structure of yeast phenylalanine tRNA by means of x-ray diffraction analysis. The determination of base sequences in DNA is now even easier than that for bases in RNA or amino acids in proteins, thanks to a surprisingly simple technique developed by Allan Maxam and Walter Gilbert (1977).

Rich and Kim report in fascinating detail on their work in "The Three-Dimensional Structure of Transfer RNA." Finally, we have a structure for the "key" that deciphers the genetic code. Unfortunately, the revealed structure does not appear to answer many of the questions about the diverse regulatory roles of tRNA in addition to its decoding function. The authors discuss some of these regulatory roles and predict that this is an area in which the next significant discoveries will be made.

Regulation of gene expression occurs at many different levels. The number of DNA copies of a given gene in the cell may be varied, the frequency of transcription into mRNA copies can be regulated, and the persistence of those mRNA copies can be controlled. Further control exists at the level of translation. Finally, control of the action of the protein products is achieved through activation and inhibition reactions as well as through ultimate degradation of proteins themselves. A negative-feedback mechanism for regulation of transcription—the operon model—was originally postulated by Francois Jacob and Jacques Monod purely on the basis of genetic analysis. It was over ten years later that their model was confirmed by parallel biochemical studies on two different gene systems by Mark Ptashne and Walter Gilbert in their separate laboratories at Harvard University. The putative repressor protein for regulating genes that provide for the metabolism of lactose in *E. coli* was isolated by Gilbert and Benno Müller-Hill while Mark Ptashne and co-workers identified the phage lambda repressor that is required to maintain the viral genome in a dormant prophage state in the host chromosome. The article on "A DNA Operator-Repressor System," by Tom Maniatis and Mark Ptashne, presents the essentials of regulation at the level of gene expression and then describes in great experimental detail the elucidation of relevant structures in the lambda regulatory scheme. The article includes a description of the DNA sequencing method developed in Frederick Sanger's laboratory, a procedure somewhat more complicated than that devised by Maxam and Gilbert (1977).

Our understanding of the processes of transcription and translation seemed quite complete just a few years ago. Transcription begins at a promoter site at the beginning of a gene and ends at a termination site to produce a defined length of mRNA. That RNA is then threaded through a ribosome to achieve translation into at least one but sometimes several polypeptides. We now know of a number of complications in this simple scheme. For example, even in bacteria the transcription process does not always produce the same size messenger RNA. Sometimes transcription stops at an earlier site termed the "attenuator," so that an early part of the message may be expressed to a greater extent than a later part. In eukaryotic cells the surprising new concept of messenger RNA splicing has appeared. Sometimes a gene coding for a particular protein has an intervening sequence of "nonsense" DNA. The cell produces an mRNA that includes these stretches of nonsense DNA, but then the nonsense is spliced out before the RNA is sent to the cytoplasm for translation. Thus there may be long sequences of nucleotides within a gene that do not appear in the final RNA transcript. Even in the simplest eukaryotes, such as yeasts, over a dozen nucleotides are deleted from the middle of a precursor tRNA, and the end pieces are then spliced to form the mature tRNA. Such basic differences in the mechanisms of RNA formation in eukaryotes as compared to prokaryotes have strengthened the notion that these higher forms have been evolving independently for a very long time. There are also a number of other significant differences in the way mRNA is constructed in eukaryotes as compared to that in prokaryotes. Most eukaryotic mRNA's contain at one end (the 5′ terminus) a modified methylated structure termed a "cap." At the other end (the 3′ terminus), a stretch of adenine nucleotides is added after transcription to essentially all eukaryotic mRNA's, with the exception of the messenger RNA that codes for the histones (basic proteins in chromatin). Furthermore, some of the adenines are methylated after the mRNA has been synthesized. Thus processing of the mRNA chain after synthesis is an important and nearly universal characteristic in eukaryotes. We have yet to figure out the rationale for the seemingly complex manner in which gene expression operates in eukaryotes. However, we can be certain that it has evolved that way for some purposes of efficient design. James Darnell (1978) has prepared a very provocative discussion of the implications of RNA:RNA splicing for evolution and regulation in eukaryotic cells. Yet another notable feature of the organization of the nucleotide sequences in eukaryotes is the existence of a large degree of redundancy of some sequences. Over half of the DNA in some eukaryotic systems consists of multiple copies of a few of the genes, and some short nucleotide sequences may exist in as many as a million copies. Eric Davidson and Roy Britten (1979) have considered possible regulatory roles for these repetitive sequences. The regulatory complexities of embryonic development in humans and other multicellular organisms undoubtedly include a wealth of additional principles to be discovered.

REFERENCES CITED AND SUGGESTED FURTHER READING

Britten, R. J., and D. E. Kohne. 1970. "Repeated segments of DNA." *Scientific American*, April. (Offprint No. 1173.)

Campbell, A. 1979. Structure of complex operons." In R. F. Goldberger (editor), *Biological Population and Development.* Plenum Press, New York.

Changeux, J. 1965. "The control of biochemical reactions." *Scientific American*, April. (Offprint No. 1008.)

Crick, F. 1979. "Split genes and RNA splicing." *Science* **204**:264–271.

Darnell, J. E., Jr. 1978. "Implication of RNA-RNA splicing in evolution of eukaryotic cells." *Science* **202:**1257–1260.

Davidson, E. H., and R. J. Britten. 1979. "Regulation of gene expression: Possible role of repetitive sequences." *Science* **204:**1052–1059.

Dulbecco, R. 1979. "Contributions of microbiology to eukaryotic cell biology: New directions for microbiology." *Microbiol. Rev.* **43:**443–452.

Gilbert, W., and A. Maxam. 1973. "The nucleotide sequence of the lac operator." *Proc. Nat. Acad. Sci.* **70**(12):3581–3584.

Gorini, L. 1966. "Antibiotics and the genetic code." *Scientific American*, April.

Hurwitz, J., and J. J. Furth. 1962. "Messenger RNA." *Scientific American*, February. (Offprint No. 119.)

Kim, S. H., G. J. Quigley, F. L. Suddath, A. McPherson, D. Sneden, J.-J. P. Kim, J. Weinzierl, and A. Rich. 1973. "Three-dimensional structure of yeast phenylalanine transfer RNA: Folding of the polynucleotide chain." *Science* **179:**285–288.

Kim, S. H. 1976. "Three-dimensional structure of transfer RNA." *Prog. Nucleic Acid Res. Mol. Biol.* **17:**181–216.

Maniatis, T., M. Ptashne, and R. Maurer. 1974. "Control elements in the DNA of bacteriophage λ." *Cold Spring Harbor Symp. Quant. Biol.* **38:**857–869.

Maxam, A. M., and W. Gilbert. 1977. "A new method for sequencing DNA." *Proc. Nat. Acad. Sci.* **74:**560–564.

Miller, J. H., and W. S. Rexnikoff (editors). 1978. *The Operon.* Cold Spring Harbor Laboratory, Long Island.

Ptashne, M., and W. Gilbert. 1970. "Genetic repressors." *Scientific American*, February. (Offprint No. 1133.)

Ptashne, M. 1975. "Repressor, operators, and promoters in bacteriophage lambda." The Harvey Lectures, Series 69. Academic Press, New York.

Rich, A., and U. L. Raj Bhandary. 1976. "Transfer RNA: Molecular structure, sequence and properties." *Ann. Rev. Biochem.* **45:**805–860.

Sanger, F. et al. 1978. "Nucleotide sequence of the DNA of φX174cs70 and the amino acid sequence of the protein for which it codes." In D. T. Denhardt, D. Dressler, and D. S. Ray (editors), *The Single-stranded DNA Phages.* Cold Spring Harbor Laboratory, Long Island.

Sarabhai, A. S., A. O. W. Stretton, S. Brenner, and A. Bolle. 1964. "Co-linearity of the gene with the polypeptide chain." *Nature* **201:**13–17.

Stent, G., and R. Calendar. 1978. *Molecular Genetics. An Introductory Narrative* (2nd ed.). W. H. Freeman and Company, San Francisco.

Watson, J. 1976. *Molecular Biology of the Gene* (3rd ed.). Benjamin, Menlo Park, CA.

Yanofsky, C., G. R. Drapeau, J. R. Guest, and B. C. Carlton. 1967. "The complete amino acid sequence of the tryptophan synthetase A protein (α subunit) and its colinear relationship with the genetic map of the A gene." *Proc. Nat. Acad. Sci.* **57:**296–298.

14

The Genes of Men and Molds

by George W. Beadle
September 1948

The study of the red fungus Neurospora crassa *sheds light on exactly how the units of heredity determine the characteristics of all living things*

EIGHTY-FIVE years ago, in the garden of a monastery near the village of Brünn in what is now Czechoslovakia, Gregor Johann Mendel was spending his spare moments studying hybrids between varieties of the edible garden pea. Out of his penetrating analysis of the results of his studies there grew the modern theory of the gene. But like many a pioneer in science, Mendel was a generation ahead of his time; the full significance of his findings was not appreciated until 1900.

In the period following the "rediscovery" of Mendel's work biologists have developed and extended the gene theory to the point where it now seems clear that genes are the basic units of all living things. They are the master molecules that guide the development and direct the vital activities of men and amoebas.

Today the specific functions of genes in plants and animals are being isolated and studied in detail. One of the most useful genetic guinea pigs is the red bread mold *Neurospora crassa*. Its genes can conveniently be changed artificially and the part that they play in the chemical alteration and metabolism of cells can be analyzed with considerable precision. We are learning what sort of material the genes are made of, how they affect living organisms and how the genes themselves, and thereby heredity, are affected by forces in their environment. Indeed, in their study of genes biologists are coming closer to an understanding of the ultimate basis of life itself.

It seems likely that life first appeared on earth in the form of units much like the genes of present-day organisms. Through the processes of mutation in such primitive genes, and through Darwinian natural selection, higher forms of life evolved—first as simple systems with a few genes, then as single-celled forms with

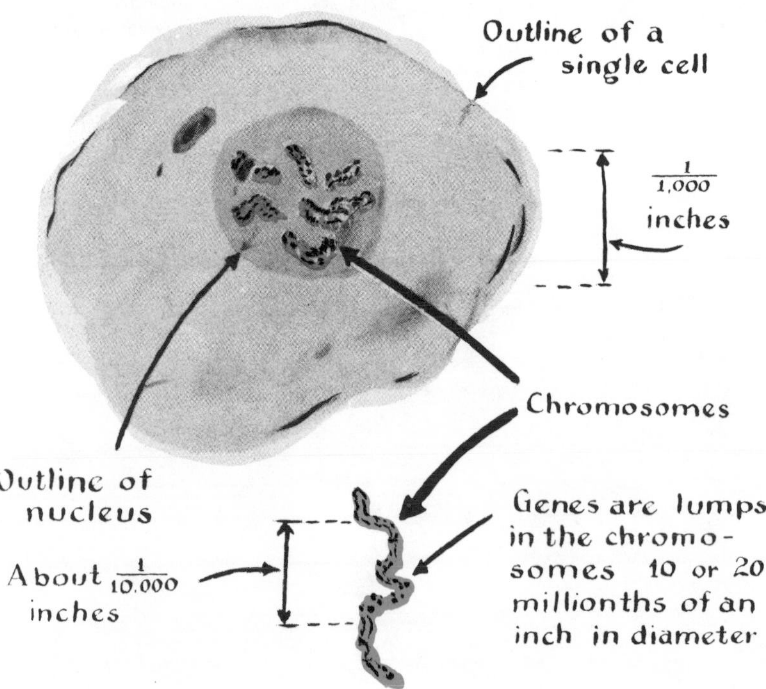

THE CELL is the site of nearly all the interactions between the gene and its environment. The genes themselves are located in the chromosomes, shown above in the stage before cell divides, duplicating each gene in the process.

THE MOLD *Neurospora* is an admirable organism for the study of genes, mainly because of its unusually simple reproductive apparatus. This may be neatly dissected to isolate a single complete set of genes. The sequence of steps in the drawing at the right shows how the tiny fruiting body of the mold is taken apart in the laboratory. With the aid of a microscope, the laboratory worker is able to spread out a set of spore sacs, each containing eight spores. One spore sac may then be separated from the others, and its spores carefully removed. The individual spores are lined up on a block of agar and finally planted in a test tube which contains all the substances that are normally required for the mold to grow.

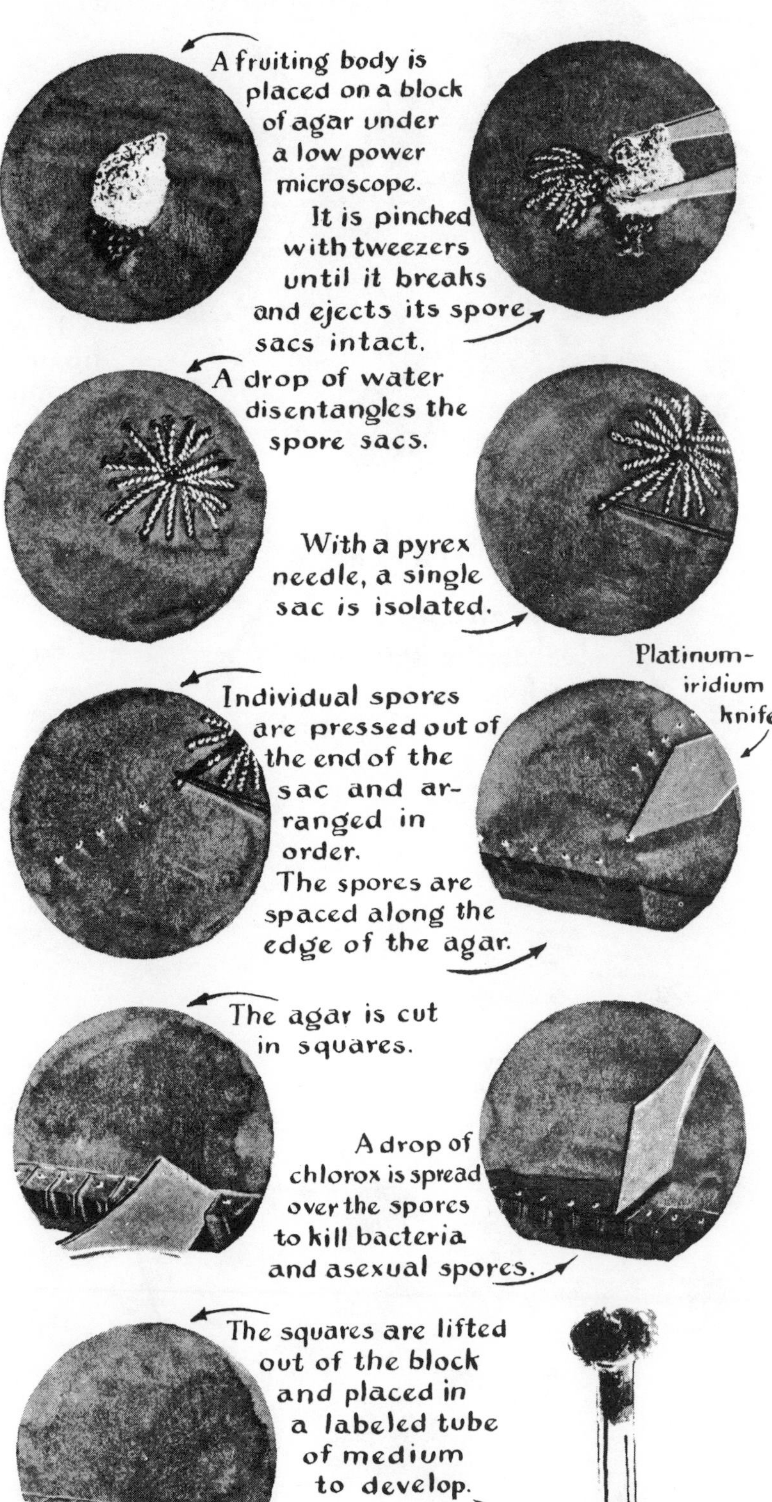

many genes, and finally as multicellular plants and animals.

A fruiting body is placed on a block of agar under a low power microscope.
It is pinched with tweezers until it breaks and ejects its spore sacs intact.
A drop of water disentangles the spore sacs.
With a pyrex needle, a single sac is isolated.
Platinum-iridium knife
Individual spores are pressed out of the end of the sac and arranged in order.
The spores are spaced along the edge of the agar.
The agar is cut in squares.
A drop of chlorox is spread over the spores to kill bacteria and asexual spores.
The squares are lifted out of the block and placed in a labeled tube of medium to develop.

What do we know about these genes that are so all-important in the process of evolution, in the development of complex organisms, and in the direction of those vital processes which distinguish the living from the non-living worlds?

In the first place, genes are characterized by students of heredity as the units of inheritance. What is meant by this may be illustrated by examples of some inherited traits in man.

Blue-eyed people may differ by a single gene from those with brown eyes. This eye-color gene exists in two forms, which for convenience may be designated *B* and *b*.

Every person begins as a single cell a few thousandths of an inch in diameter—a cell that comes into being through the fusion of an egg cell from the mother and a sperm cell from the father. This fertilized egg carries two representatives of the eye-color gene, one from each parent. Depending on the parents, there are therefore three types of individuals possible so far as this particular gene is concerned. They start from fertilized eggs represented by the genetic formulas *BB*, *Bb* and *bb*. The first two types, *BB* and *Bb*, will develop into individuals with brown eyes. The third one, *bb*, will have blue eyes. You will note that when both forms of the gene are present the individual is brown-eyed. This is because the form of the gene for brown eyes is *dominant* over its alternative form for blue eyes. Conversely, the form for blue eyes is said to be *recessive*.

During the division of the fertilized egg cell into many daughter cells, which through growth, division and specialization give rise to a fully developed person, the genes multiply regularly with each cell division. As a result each of the millions of cells of a fully developed individual carries exact copies of the two representatives of the eye-color gene

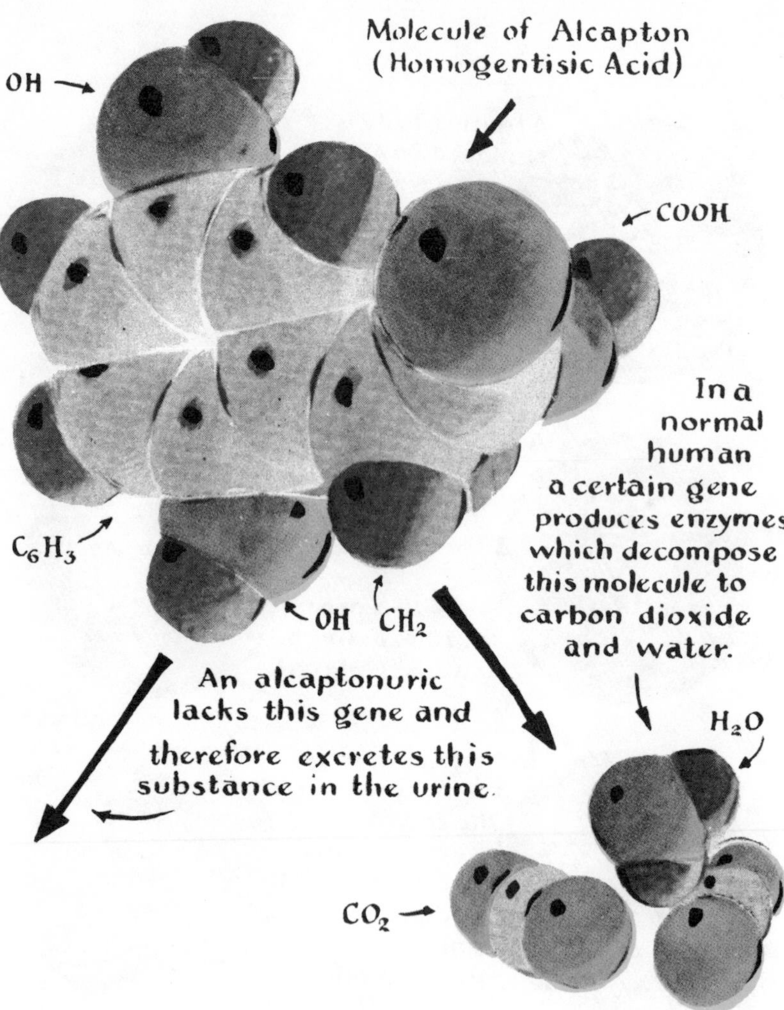
Molecule of Alcapton
(Homogentisic Acid)
OH
COOH
C_6H_3
OH
CH_2
In a normal human a certain gene produces enzymes which decompose this molecule to carbon dioxide and water.
An alcaptonuric lacks this gene and therefore excretes this substance in the urine.
H_2O
CO_2
Phenylpyruvic acid
Normal human beings oxidize this substance to form this.
A phenylketonuric idiot, lacking one gene, excretes this substance in the urine:
p-Hydroxy-phenylpyruvic acid

DEFECTIVE GENES in man can cause serious hereditary disorders. The chemical basis of two such disorders is shown in the drawings on this page. The large molecule in the drawing at the left is homogentisic acid, or alcapton. In most human beings a single gene produces an enzyme which is capable of breaking alcapton down to carbon dioxide and water. When the gene that produces the enzyme is faulty, however, alcapton is not decomposed. It must be eliminated in the urine, to which it gives a dark color. This excretion of alcapton in the urine is called alcaptonuria. The drawing at the bottom of this page shows the basis of a much more serious genetic disorder. The biochemical apparatus of most human beings, again, is able to transform phenylpyruvic acid into p-hydroxy phenylpyruvic acid. Those who cannot transform it are called phenylketonurics. Phenylketonuria is characterized by extreme feeblemindedness. Most phenyketonurics are imbeciles or idiots; a few are low-grade morons. The faulty genes that are responsible for both are recessive. This means that they are expressed only when two such genes are paired in the union of an egg and sperm cell. Thus most of the genes responsible for these disorders are carried by normal people without being expressed.

which has been contributed by the parents.

In the formation of egg and sperm cells, the genes are again reduced from two to one per cell. Therefore a mother of the type *BB* forms egg cells carrying only the *B* form of the gene. A type *bb* mother produces only *b* egg cells. A *Bb* mother, on the other hand, produces both *B* and *b* egg cells, in equal numbers on the average. Exactly corresponding relations hold for the formation of sperm cells.

With these facts in mind it is a simple matter to determine the types of children expected to result from various unions. Some of these are indicated in the following list:

Mother	*Father*	*Children*
BB (brown)	*BB* (brown)	All *BB* (brown)
Bb (brown)	*Bb* (brown)	¼ *BB* (brown) ½ *Bb* (brown) ¼ *bb* (blue)
BB (brown)	*bb* (blue)	All *Bb* (brown)
Bb (brown)	*bb* (blue)	½ *Bb* (brown) ½ *bb* (blue)
bb (blue)	*bb* (blue)	All *bb* (blue)

This table shows that while it is expected that some families in which both parents have brown eyes will include blue-eyed children, parents who are both blue-eyed are not expected to have brown-eyed children.

LIFE CYCLE of the mold *Neurospora* is illustrated in the drawing at the right. The hyphal fusion of Sex A and Sex a at the bottom of the page is taken as a starting point. *Neurospora* enters a sexual stage rather similar to the union of sperm and egg cells in higher organisms. The union produces a fertile egg, in which two complete sets of genes are paired. The fertile egg cell then divides (*center of drawing*), and divides again. This produces four nuclei, each of which has only a single set of genes. Lined up in a spore sac, the four nuclei divide once more to produce four pairs of nuclei that are genetically identical. A group of spore sacs is gathered in a fruiting body. The sacs and the spores may then be dissected by the technique outlined on page 189. Following this, the germinating spores (*top of page*) may be planted in test tubes containing the necessary nutrients. It is at this point that genetic defects can be exposed by changing the constitution of the medium. Here also *Neurospora* may be allowed to multiply by asexual means. This makes it possible to grow large quantities of the mold without genetic change for convenient chemical analysis. The entire life cycle of the mold takes only 10 days, another reason why *Neurospora* is an exceptionally useful experimental organism.

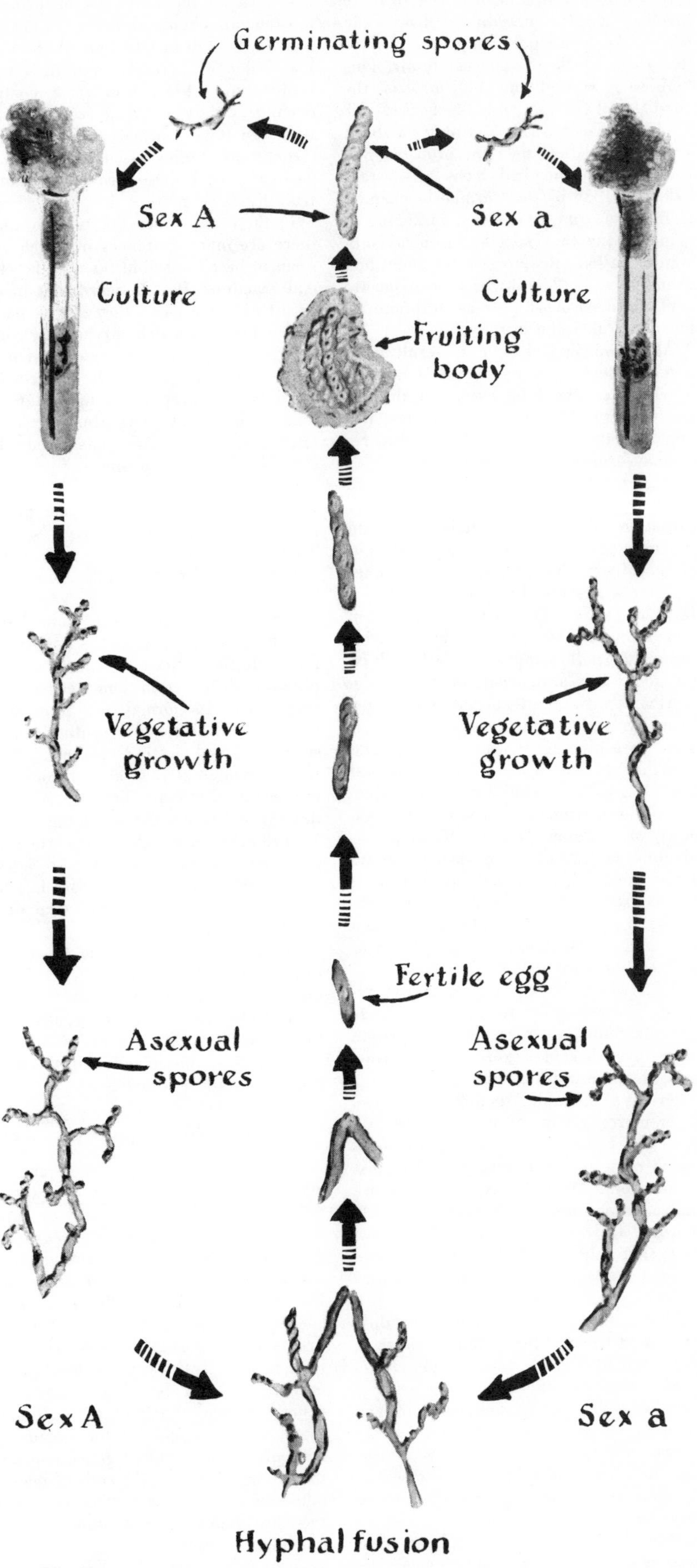

It is important to emphasize conditions that may account for apparent exceptions to the last rule. The first is that eye-color inheritance in man is not completely worked out genetically. Probably other genes besides the one used as an example here are concerned with eye color. It may therefore be possible, when these other genes are taken into account, for parents with true blue eyes to have brown-eyed children. A second factor which accounts for some apparent exceptions is that brown-eyed persons of the *Bb* type may have eyes so light brown that an inexperienced observer may classify them as blue. Two parents of this type may, of course, have a *BB* child with dark brown eyes.

Another example of an inherited trait in man is curly hair. Ordinary curly hair, such as is found frequently in people of European descent, is dominant to straight hair. Therefore parents with curly hair may have straight-haired children but straight-haired parents do not often have children with curly hair. Again there are other genes concerned, and the simple rules based on a one-gene interpretation do not always hold.

Defective Genes

Eye-color and hair-form genes have relatively trivial effects in human beings. Other known genes are concerned with traits of deeper significance. One of these involves a rare hereditary disease in which the principal symptom is urine that turns black on exposure to air. This "inborn error of metabolism," as the English physician and biochemist Sir Archibald Garrod referred to it, has been known to medical men for probably 300 years. Its biochemical basis was established in 1859 by the German biochemist C. Bödeker, who showed that darkening of urine is due to a specific chemical substance called alcapton, later identified chemically as 2,5-dihydroxyphenylacetic acid. The disease is known as alcaptonuria, meaning "alcapton in the urine."

Alcaptonuria is known to result from a gene defect. It shows typical Mendelian inheritance, like blue eyes, but the defective form of the gene is much less frequent in the population than is the recessive form of the eye-color gene.

The excretion of alcapton is a result of the body's inability to break it down by oxidation. Normal individuals possess an enzyme (a protein-containing catalyst, often called a biocatalyst) which makes possible a reaction by which alcapton is further oxidized. This enzyme is absent in alcaptonurics. As a result alcaptonurics cannot degrade alcapton to carbon dioxide and water as normal individuals do.

Alcaptonuria is of special interest genetically and biochemically because it gives us a clue as to what genes do and how they do it. It is clear that the normal kind of gene is essential for the production of the enzyme necessary for the breakdown of alcapton. If the cells of an individual contain only the recessive or inactive form of the gene, no enzyme is formed, alcapton accumulates and is excreted in the urine. The relations between gene and chemical reaction are shown in the diagram at the top of page 190.

A hereditary error of metabolism related biochemically to alcaptonuria is phenylketonuria, a rare disease in which phenylpyruvic acid is excreted in the urine. Like alcaptonuria, this metabolic defect is inherited as a simple Mendelian recessive. It is more serious in its consequences, however, because it is invariably associated with feeble-mindedness of an extreme kind. Most phenylketonurics are imbeciles or idiots; a few are low-grade morons. It should be made clear, however, that only a small fraction of feebleminded persons are of this particular genetic type.

Phenylketonurics excrete phenylpyruvic acid because they cannot oxidize it, as normal individuals can, to a closely related derivative differing from phenylpyruvic acid by having one more oxygen atom per molecule (*see diagram at the bottom of page 190*). Again it is evident that the normal form of a gene is essential for the carrying out of a specific chemical reaction.

Man, however, is far from an ideal organism in which to study genes. His life cycle is too long, his offspring are too few, his choice of a mate is not often based on a desire to contribute to the knowledge of heredity, and it is inconvenient to subject him to a complete chemical analysis. As a result, most of what we have learned about genes has come from studies of such organisms as garden peas, Indian corn plants and the fruit fly *Drosophila*.

In these and other plants and animals there are many instances in which genes seem to be responsible for specific chemical reactions. It is believed that in most or all of these cases they act as pattern molecules from which enzymes are copied.

Many enzymes have been isolated in a pure crystalline state. All of them have proved to be proteins or to contain proteins as essential parts. Gene-enzyme relations such as those considered above suggest that the primary function of genes may be to serve as models from which specific kinds of enzyme proteins are copied. This hypothesis is strengthened by evidence that some genes control the presence of proteins that are not parts of enzymes.

For example, normal persons have a specific blood protein that is important in blood clotting. Bleeders, known as hemophiliacs, differ from non-bleeders by a single gene. Its normal form is presumed to be essential for the synthesis of the specific blood-clotting protein. Hemophilia, incidentally, is almost completely limited to the male because it is sex-linked; that is, it is carried in the so-called X chromosome, which is concerned with the determination of sex. As is well known, this hereditary disorder has been carried for generations by some of the royal families of Europe.

The genes that determine blood types in man and other animals direct the production of so-called antigens. These are giant molecules which apparently derive their specificity from gene models, and which are capable of inducing the formation of specific antibodies.

Neurospora

The hypothesis that genes are concerned with the elaboration of giant protein molecules has been tested by experiments with the red mold *Neurospora*. This fungus has many advantages in the study of what genes do. It has a short life cycle—only 10 days from one sexual spore generation to the next. It multiplies profusely by asexual spores. The result is that any strain can be multiplied a millionfold in a few days without any genetic change. Each of the cell nuclei that carry the genes of the bread mold has only a single set of genes instead of the two sets found in the cells of man and other higher organisms. This means that recessive genes are not hidden by their dominant counterparts.

During the sexual stage, in which molds of opposite sex reactions come together, there is a fusion comparable to that between egg and sperm in man. The fusion nucleus then immediately undergoes two divisions in which genes are reduced again to one per cell. The four products formed from a single fusion nucleus by these divisions are lined up in a spore sac. Each divides again so as to produce pairs of nuclei that are genetically identical. The eight resulting nuclei are included in eight sexual spores, each one-thousandth of an inch long. This life cycle of *Neurospora* is shown in the illustration on page 191.

Using a microscope, a skilled laboratory worker can dissect the sexual spores from the spore sac in orderly sequence. Each of them can be planted separately in a culture tube (*see illustration on page 189*). If the two parental strains differ by a single gene, four spores always carry descendants of one form of the gene and four carry descendants of the other. Thus if a yellow and a white strain are crossed, there occur in each spore sac four spores that will give white molds and four that will give yellow.

The red bread mold is almost ideally suited for chemical studies. It can be grown in pure culture on a chemically known medium containing only nitrate, sulfate, phosphate, various other inorganic substances, sugar and biotin, a vitamin of the B group. From these relatively

EXPERIMENT to determine the role of a single *Neurospora* gene essentially consists in disabling a gene and tracking down its missing biochemical function. Spores of the mold are first exposed to radiation that will cause mutation, *i.e.*, change in a gene. This culture is then crossed with another. The spores resulting from this union are then planted in a medium that contains all the substances that normal *Neurospora* needs for growth, plus a few that the mold normally manufactures for itself. All the spores, including those which may carry a defective gene, germinate on this medium. Spores from these same cultures are then planted in a medium that contains only the bare minimum of substances required by *Neurospora*. Four of the cultures fail to grow, indicating that they have lost the power to manufacture one substance that *Neurospora* normally synthesizes. In test tubes at the bottom of opposite page, the detailed identification of exactly what synthetic power has been lost is begun by planting the defective culture in media that contain (1) all substances required by the normal mold plus vitamins, and (2) all substances plus amino acids. When mold grows on first medium, it appears it has lost the power to synthesize vitamin.

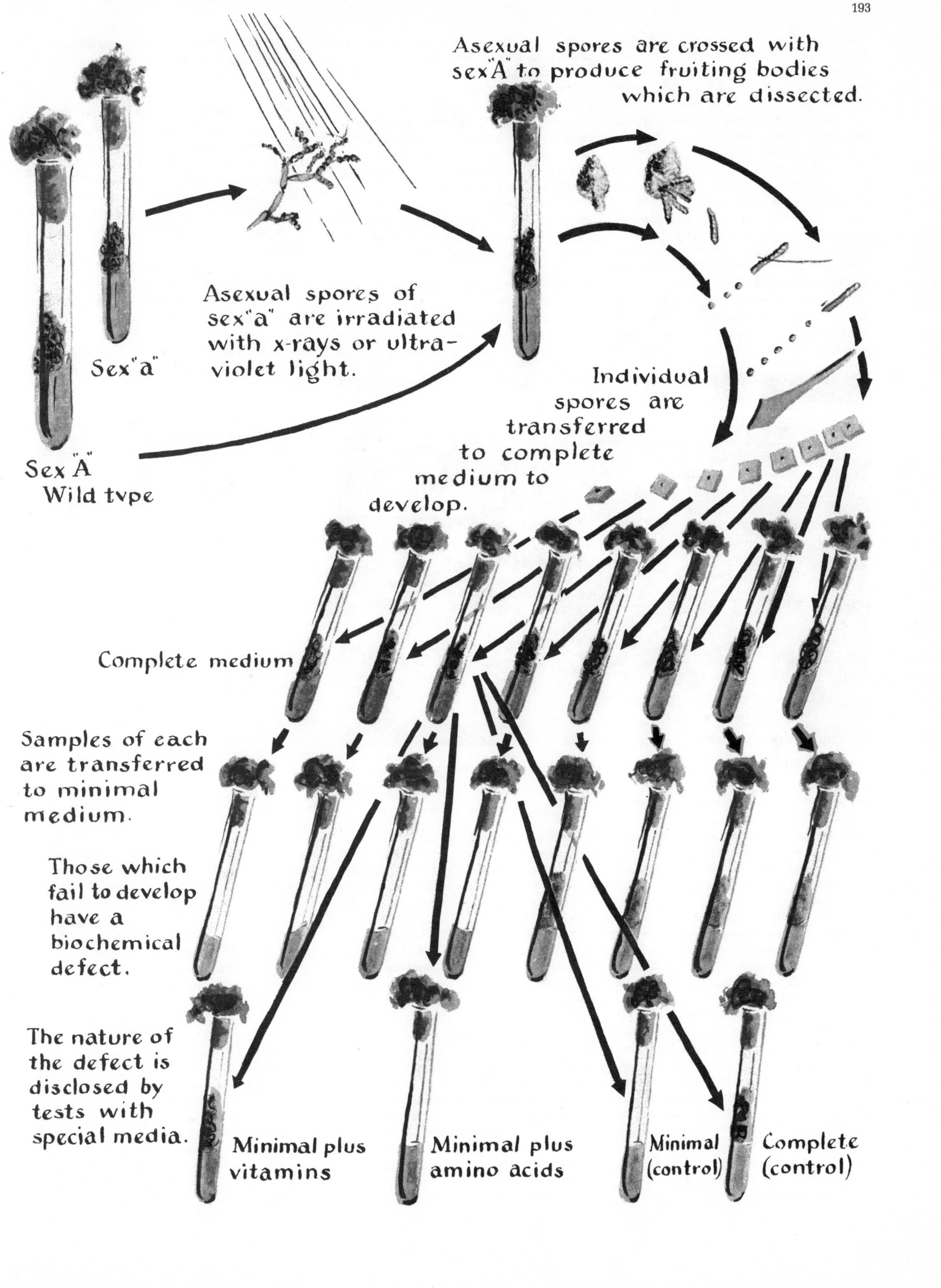

Asexual spores are crossed with sex"A" to produce fruiting bodies which are dissected.

Sex"a"

Sex"A" Wild type

Asexual spores of sex"a" are irradiated with x-rays or ultra-violet light.

Individual spores are transferred to complete medium to develop.

Complete medium

Samples of each are transferred to minimal medium.

Those which fail to develop have a biochemical defect.

The nature of the defect is disclosed by tests with special media.

Minimal plus vitamins

Minimal plus amino acids

Minimal (control)

Complete (control)

simple starting materials, the mold produces all the constituent parts of its protoplasm. These include some 20 amino acid building blocks of proteins, nine water-soluble vitamins of the B group, and many other organic molecules of vital biological significance.

To one interested in what genes do in a human being, it might at first thought seem a very large jump from a man to a mold. Actually it is not. For in its basic metabolic processes, protoplasm—Thomas Huxley's physical stuff of life—is very much the same wherever it is found.

If the many chemical reactions by which a bread mold builds its protoplasm out of the raw materials at its disposal are catalyzed by enzymes, and if the proteins of these enzymes are copied from genes, it should be possible to produce molds with specific metabolic errors by causing genes to mutate. Or to state the problem somewhat differently, one ought to be able to discover what genes do by making them defective.

The simplicity of this approach can be illustrated by an analogy. The manufacture of an automobile in a factory is in some respects like the development of an organism. The workmen in the factory are like genes—each has a specific job to do. If one observed the factory only from the outside and in terms of the cars that come out, it would not be easy to determine what each worker does. But if one could replace able workers with defective ones, and then observe what happened to the product, it would be a simple matter to conclude that Jones puts on the radiator grill, Smith adds the carburetor, and so forth. Deducing what genes do by making them defective is analogous and equally simple in principle.

It is known that changes in genes—mutations—occur spontaneously with a low frequency. The probability that a given gene will mutate to a defective form can be increased a hundredfold or more by so-called mutagenic (mutation producing) agents. These include X-radiation, neutrons and other ionizing radiations, ultraviolet radiation, and mustard gas. Radiations are believed to cause mutations by literally "hitting" genes in a way to cause ionization within them or by otherwise causing internal rearrangements of the chemical bonds.

A bread-mold experiment to test the hypothesis that genes control enzymes and metabolism can be set up in the manner shown in the diagrams on pages 193 and 195. Asexual spores are X-rayed or otherwise treated with mutagenic agents. Following a sexual phase of the life cycle, descendants of mutated genes are recovered in sexual spores. These are grown separately, and the molds that grow from them are tested for ability to produce the molecules out of which they are built.

If a gene essential for the production of vitamin B-1 by the mold is made defective, then B-1 must be supplied in the medium if a mold is to develop from a spore carrying the defective gene. But in the present state of our knowledge it is not possible to produce mutations in specific genes at will. By X-raying, for example, any one or more of several thousand genes may be mutated, or in many cases none at all will be changed. There is no known method of predicting which of the genes, if any, will be hit. It is therefore necessary to grow presumptive mutant spores on a medium supplemented with protoplasmic building blocks of which the formation could be blocked if defective genes were present.

Molds grown on such supplemented medium may grow normally either (1) by making a particular essential part themselves or (2) by taking it ready-made from the culture medium, as they must do if the gene involved in making it is defective. The two possibilities can be distinguished by trying to grow the mold on an unsupplemented medium and on media to which single supplements are added.

Following heavy ultraviolet treatment, about two sexual spores out of every hundred tested carry defective forms of those genes which are necessary for the production of essential substances supplied in the supplemented medium. For example, strain number 5531 of the mold cannot manufacture the B-vitamin pantothenic acid. For normal growth it requires an external supply of this vitamin just as human beings do.

How do we know that the inability of the mold to produce its own pantothenic acid involves a gene defect? The only way this question can be answered at present is by seeing if inability to make pantothenic acid behaves in crosses as a single unit of inheritance.

The answer is that it does. If the mold that cannot make pantothenic acid is crossed with a normal strain of the other sex, the resulting spore sacs invariably contain four spores that produce molds like one parent and four that produce

IN CONTINUATION of the experiment begun on page 193, the strain of *Neurospora* that carries a defective gene is put through another series of steps. On page 193 it had been determined that the strain in question did not grow in the absence of vitamins. This indicated that the defective gene was involved in the synthesis of a vitamin. Now the question is: exactly what vitamin? This may be found by planting the strain carrying the defective gene on a group of minimal media, each of which is supplemented by a single vitamin. The mold will then grow on the medium which contains the vitamin that it has lost the power to synthesize. In the experiment outlined on the opposite page, the missing vitamin turns out to be pantothenic acid, a vitamin of the B group. When this has been established, further experiments must be run to determine whether the deficiency of the strain involves a single gene. This is done by crossing the strain bearing the defective gene with a normal strain. All the spores from the union flourish in a medium supplemented with pantothenic acid. When they are planted in a medium that does not contain pantothenic acid, however, only four cultures grow. This is proof that one gene is involved.

PHOTOMICROGRAPH of *Neurospora* shows the sturcture of its fine red tendrils. This photograph, supplied through the courtesy of Life Magazine, was made by Herbert Gehr in the genetics laboratory of E. L. Tatum at Yale.

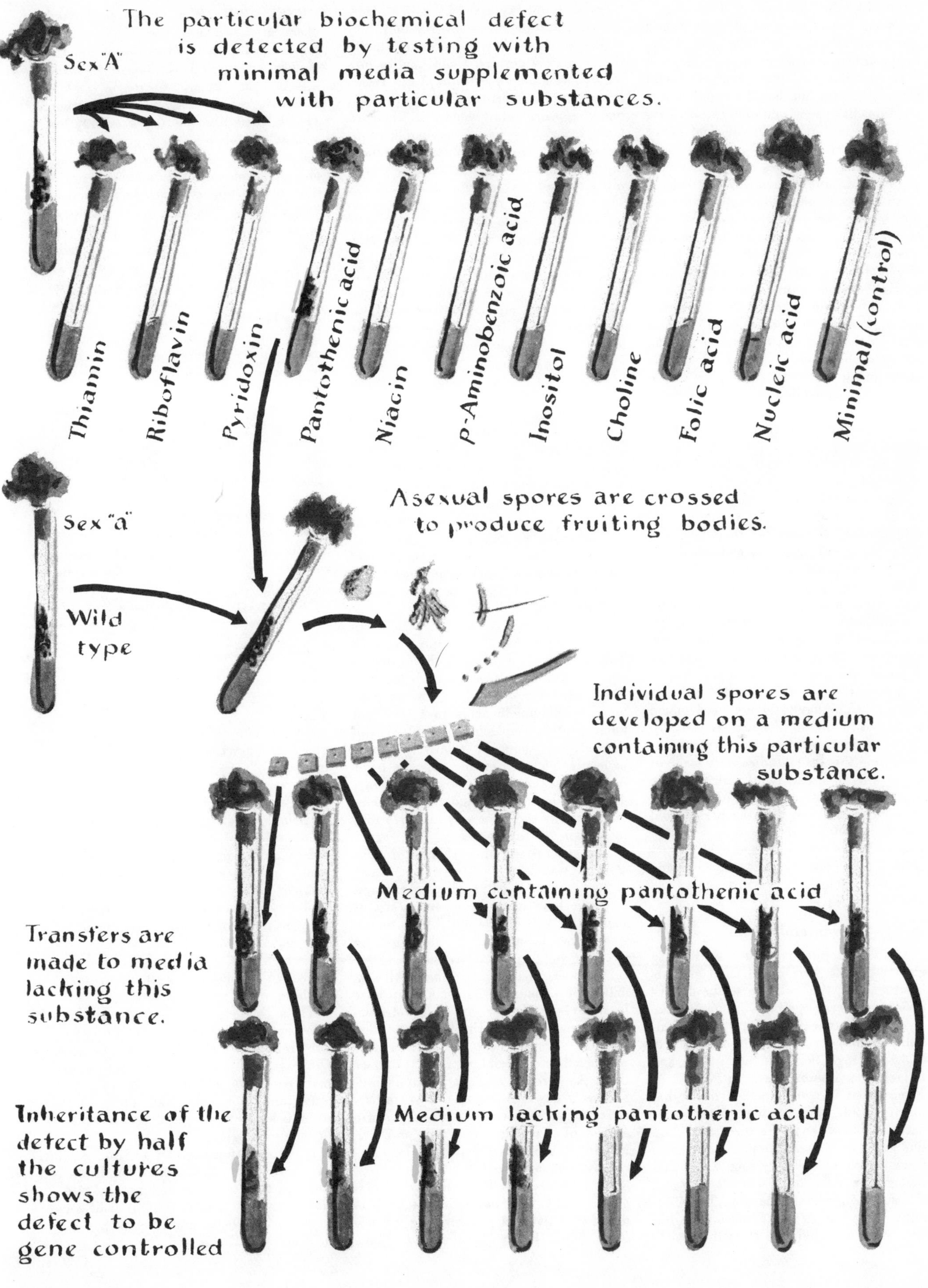
The particular biochemical defect is detected by testing with minimal media supplemented with particular substances.
Sex "A"
Thiamin
Riboflavin
Pyridoxin
Pantothenic acid
Niacin
p-Aminobenzoic acid
Inositol
Choline
Folic acid
Nucleic acid
Minimal (control)
Sex "a"
Wild type
Asexual spores are crossed to produce fruiting bodies.
Individual spores are developed on a medium containing this particular substance.
Medium containing pantothenic acid
Transfers are made to media lacking this substance.
Medium lacking pantothenic acid
Inheritance of the defect by half the cultures shows the defect to be gene controlled

strains like the other parent. Four daughter molds out of each set of eight from a spore sac are able to make pantothenic acid, and four are not (*see page 195*).

In a similar way, genes concerned with many other specific bread-mold chemical reactions have been mutated. In each case that has been studied in sufficient detail to be sure of the relation, it has been found that single genes are directly concerned with single chemical reactions.

An example that illustrates not only that genes are concerned with specific chemical reactions but also how mutant types can be used as tools for the study of metabolic processes involves the production of the amino acid tryptophane and the vitamin niacin (also known as nicotinic acid) by bread mold. Several steps in the synthesis of tryptophane, an indispensable component of the protoplasm of all organisms, have been shown to be gene-controlled. These have been used to show that bread mold forms this component by combining indole and the amino acid serine.

It has been found that indole, in turn, is made from anthranilic acid. If the second gene in the series in the accompanying diagram is made defective, anthranilic acid cannot be converted to indole, and if the mold carrying this gene in defective form is grown on a small amount of tryptophane it accumulates anthranilic acid in much the same way as an alcaptonuric accumulates alcapton. The accumulated anthranilic acid has been chemically identified in the culture medium of such a defective strain.

A recent report that rats fed on diets rich in tryptophane did not need niacin suggested to animal biochemists that possibly niacin is made from tryptophane. Following this lead, studies were made of the strains of bread mold which require ready-made tryptophane and niacin. They gave clear evidence that the bread mold does indeed derive its niacin from tryptophane. Intermediates in the chain of reactions by which the conversion is made were then identified (*see drawing on the opposite page*).

Men and Molds

The tryptophane-niacin relation so clearly disclosed by bread mold mutants has an interesting relation to the dietary deficiency disease pellagra in man. In the past this disease has been variously attributed to poor quality of dietary proteins, to a toxic factor in Indian corn, and to lack of a vitamin. When, in 1937, C. A. Elvehjem of Wisconsin demonstrated that niacin would bring about spectacular cures of black tongue, a disease of dogs like pellagra in man, the problem seemed to be solved. It was very soon found that pellagra in man, too, is cured by small amounts of niacin in the diet. The alternative hypotheses were promptly forgotten, even though the facts that led to them were not explained by niacin alone.

The tryptophane-niacin relation now makes it clear that the protein quality theory also is correct. Good quality proteins contain plenty of tryptophane. If this is present in sufficient amounts in the diet, niacin appears not to be needed. The corn toxin theory also has a reasonable basis. There appear to be chemical substances in this grain that interfere with the body's utilization of tryptophane and niacin in such a way as to increase the requirements of those two materials.

Another point of interest in connection with the tryptophane-niacin story is that it illustrates again that, in terms of basic protoplasmic reactions, pretty much the same things go on in men and molds. It is supposed that in much the same way as a single gene is in control of the enzyme by which alcaptonuria is broken down in man, genes of the bread mold guide chemical reactions indirectly through their control of enzyme proteins. In most instances the enzymes involved have not yet been studied directly.

Bread-mold studies have contributed strong support to the hypothesis that each gene controls a single protein. But they have not proved it to the satisfaction of all biologists. There remains a possibility that some genes possess several distinct functions and that such genes were automatically excluded by the experimental procedure followed.

What is the process by which genes direct the formation of specific proteins? This is a question to which the answer is not yet known. There is evidence that genes themselves contain proteins combined with nucleic acids to form giant nucleoprotein molecules hundreds of times larger than the relatively simple molecules pictured on the opposite page. And it has been suggested that genes direct the building of non-genic proteins in essentially the same way in which they form copies of themselves.

The general question of how proteins are synthesized by living organisms is one of the great unsolved problems of biology. Until we have made headway toward its solution, it will not be possible to understand growth, normal or abnormal, in anything but superficial terms.

Do all organisms have genes? All sexually reproducing organisms that have been investigated by geneticists demonstrably possess them. Until recently there was no simple way of determining whether bacteria and viruses also have them. As a result of very recent investigations it has been found that some bacteria and some bacterial viruses perform a kind of sexual reproduction in which hereditary units like genes can be quite clearly demonstrated.

By treatment of bacteria with mutagenic agents, mutant types can be produced that parallel in a striking manner those found in the bread mold. These make it almost certain that bacterial genes are functionally like the genes of molds.

So we can sum up by asserting that genes are irreducible units of inheritance in viruses, single-celled organisms and in many-celled plants and animals. They are organized in threadlike chromosomes which in higher plants and animals are carried in organized nuclei. Genes are probably nucleoproteins that serve as patterns in a model-copy process by which new genes are copied from old ones and by which non-genic proteins are produced with configurations that correspond to those of the gene templates.

Through their control of enzyme proteins many genes show a simple one-to-one relation with chemical reactions. Other genes appear to be concerned primarily with the elaboration of antigens—giant molecules which have the property of inducing antibody formation in rabbits or other animals.

It is likely that life first arose on earth as a genelike unit capable of multiplication and mutation. Through natural selection of the fittest of these units and combinations of them, more complex forms of life gradually evolved.

GENES DIRECT a sequence of vital chemical reactions in *Neurospora.* Each of the molecules shown in the models on the opposite page is made up of the atoms hydrogen (*white spheres*), oxygen (*light color*), carbon (*black*) and nitrogen (*dark color*). Reactions involving the genes switch these atoms around to manufacture one molecule out of another. Beginning at the upper left, a single gene is known to be involved in the synthesis of anthranilic acid. Two genes are then involved in making anthranilic acid into indole, with an unknown intermediate indicated by a question mark. Indole is combined with serine to make the amino acid tryptophane, with water left over. Tryptophane is made into kynurenine. Two genes transform kynurenine into 3-hydroxy-anthranilic acid, again with an unknown intermediate molecule. Two genes finally synthesize the last product of the chain: niacin, the B vitamin that is an essential of both plant and animal life. This sequence of events is also involved in the human nutritional disease pellagra. A diet poor in the amino acid tryptophane obviously will lead to a deficiency of niacin, which causes the symptoms of pellagra. Therefore supplying either tryptophane or niacin to patient will alleviate disease.

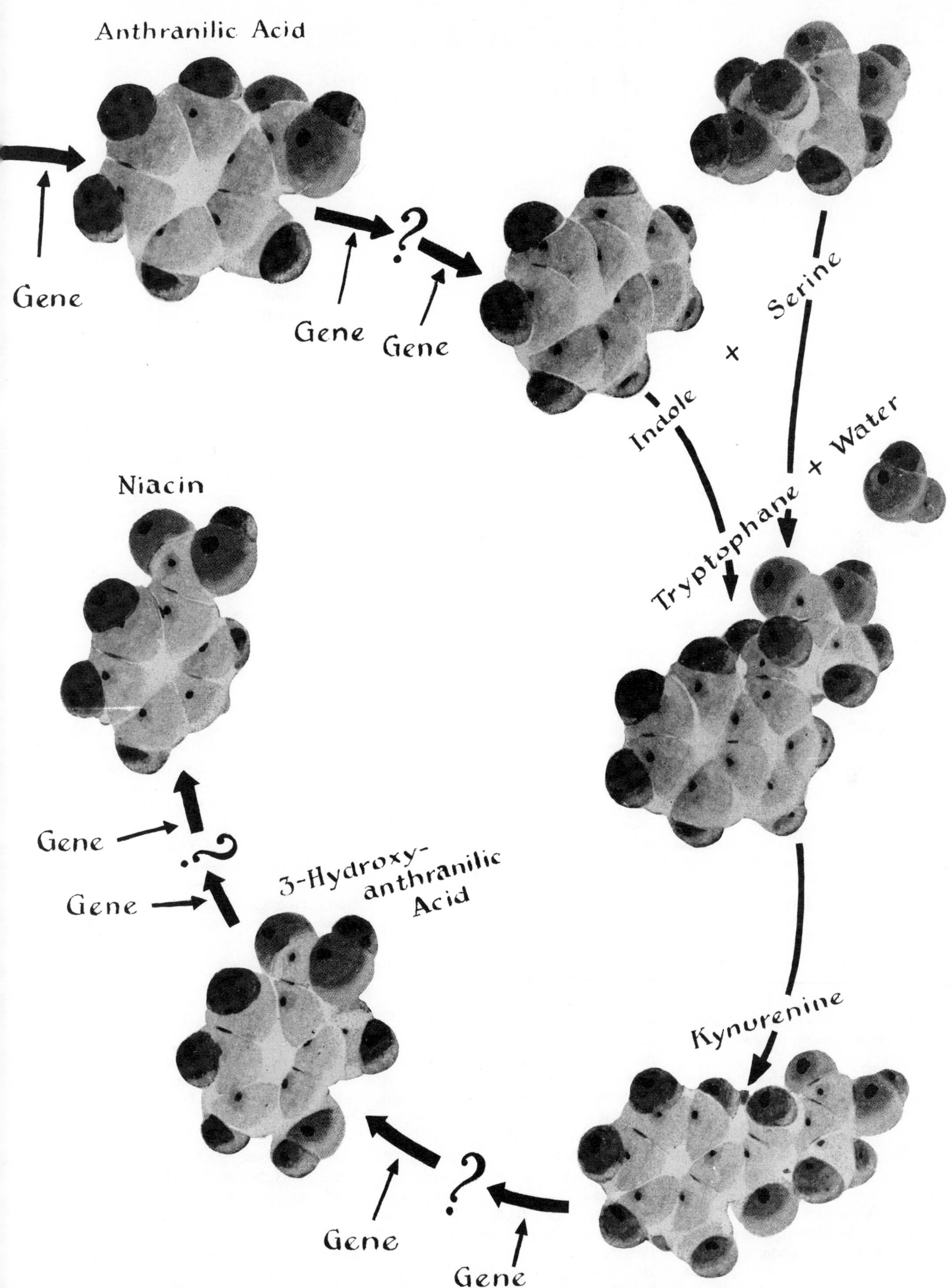

Anthranilic Acid
Gene
Gene
Gene
?
Serine
+
Indole
Tryptophane + Water
Niacin
Gene
?
Gene
3-Hydroxy-
anthranilic
Acid
Kynurenine
?
Gene
Gene

15

The Fine Structure of the Gene

by Seymour Benzer
January 1962

The question "What is a gene?" has bothered geneticists for fifty years. Recent work with a small bacterial virus has shown how to split the gene and make detailed maps of its internal structure

Much of the work of science takes the form of map making. As the tools of exploration become sharper, they reveal finer and finer details of the region under observation. In the December, 1961 issue of *Scientific American* John C. Kendrew of the University of Cambridge described the mapping of the molecule of the protein myoglobin, revealing a fantastically detailed architecture. A living organism manufactures thousands of different proteins, each to precise specifications. The "blueprints" for all this detail are stored in coded form within the genes. In this article we shall see how it is possible to map the internal structure of a single gene, with the revelation of detail comparable to that in a protein.

It has been known since about 1913 that the individual active units of heredity—the genes—are strung together in one-dimensional array along the chromosomes, the threadlike bodies in the nucleus of the cell. By crossing such organisms as the fruit fly *Drosophila*, geneticists were able to draw maps showing the linear order of various genes that had been marked by the occurrence of mutations in the organism. Most geneticists regarded the gene as a more or less indivisible unit. There seemed to be no way to attack the questions "Exactly what is a gene? Does it have an internal structure?"

In recent years it has become apparent that the information-containing part of the chromosomal chain is in most cases a giant molecule of deoxyribonucleic acid, or DNA. (In some viruses the hereditary material is ribonucleic acid, or RNA.) Indeed, the threadlike molecule of DNA can be seen in the electron microscope [*see bottom illustration on opposite page*]. For obtaining information about the fine structure of DNA, however, modern methods of genetic analysis are a more powerful tool than even the electron microscope.

It is important to understand why this fine structure is not revealed by conventional genetic mapping, as is done with fruit flies. Genetic mapping is possible because the chromosomes sometimes undergo a recombination of parts called crossing over. By this process, for example, two mutations that are on different chromosomes in a parent will sometimes emerge on the same chromosome in the progeny. In other cases the progeny will inherit a "standard" chromosome lacking the mutations seen in the parent. It is as if two chromosomes lying side by side could break apart at any point and recombine to form two new chromosomes, each made up of parts derived from the original two. As a matter of chance two points far apart will recombine frequently; two points close together will recombine rarely. By carrying out many crosses in a large population of fruit flies one can measure the frequency—meaning the ease—with which different genes will recombine, and from this one can draw a map showing the parts in correct linear sequence. This technique has been used to map the chromosomes of many organisms. Why not, then, use the technique to map mutations inside the gene? The answer is that points within the same gene are so close together that the chance of detecting recombination between them would be exceedingly small.

In the study of genetics, however, everything hinges on the choice of a suitable organism. When one works with fruit flies, one deals with at most a few thousand individuals, and each generation takes roughly 20 days. If one works with a microorganism, such as a bacterium or, better still, a bacterial virus (bacteriophage), one can deal with billions of individuals, and a generation takes only minutes. One can therefore perform in a test tube in 20 minutes an experiment yielding a quantity of genetic data that would require, if humans were used, the entire population of the earth. Moreover, with microorganisms special tricks enable one to select just those individuals of interest from a population of a billion. By exploiting these advantages it becomes possible not only to split the gene but also to map it in the utmost detail, down to the molecular limits of its structure.

Replication of a Virus

An extremely useful organism for this fine-structure mapping is the T4 bacteriophage, which infects the colon bacillus. T4 is one of a family of viruses that has been most fruitfully exploited by an entire school of molecular biologists founded by Max Delbrück of the California Institute of Technology. The T4 virus and its relatives each consist of a head, which looks hexagonal in electron micrographs, and a complex tail by which the virus attaches itself to the bacillus wall [*see top illustration on opposite page*]. Crammed within the head of the virus is a single long-chain molecule of DNA having a weight about 100 million times that of the hydrogen atom. After a T4 virus has attached itself to a bacillus, the DNA molecule enters the cell and dictates a reorganization of the cell machinery to manufacture 100 or so copies of the complete virus. Each copy consists of the DNA and at least six distinct protein components. To make these components the invading DNA specifies the formation of a series of special enzymes, which themselves are proteins. The entire process is controlled by the battery of genes that constitutes the DNA molecule.

According to the model for DNA de-

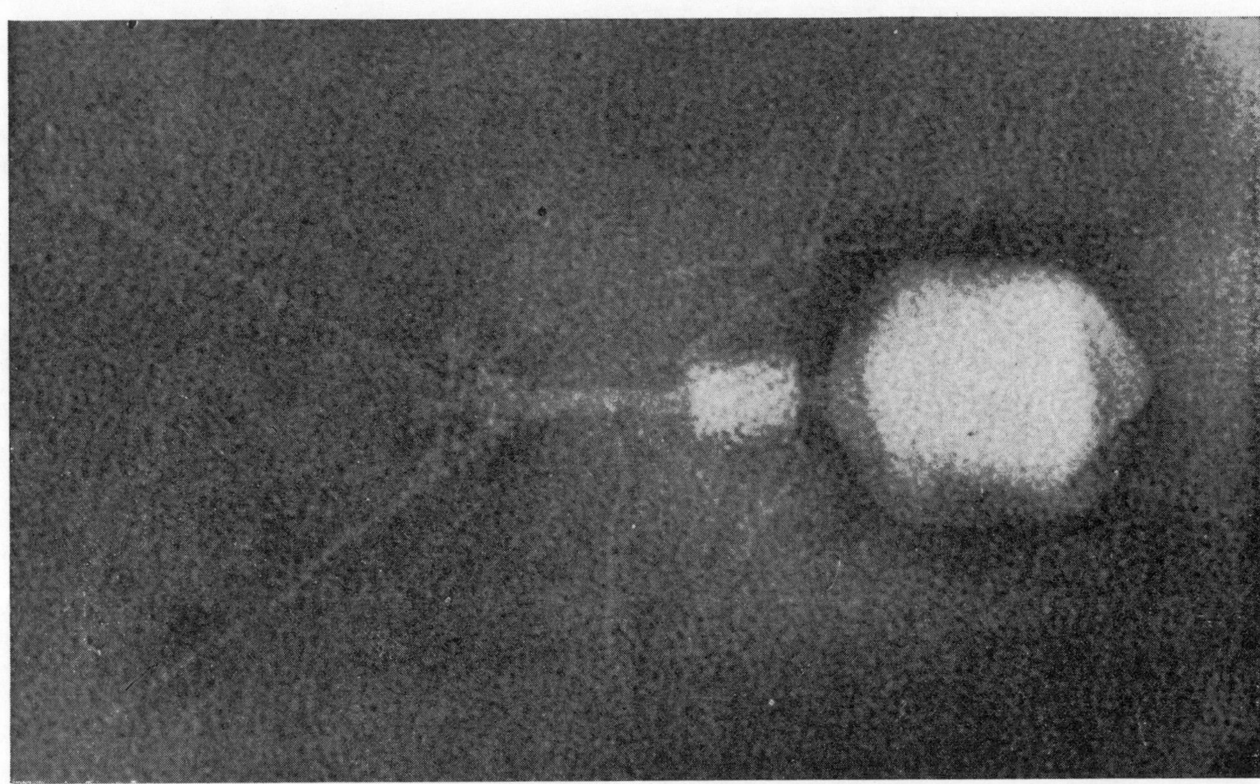

T2 BACTERIOPHAGE, magnified 500,000 diameters, is a virus that contains in its head complete instructions for it own replication. To replicate, however, it must find a cell of the colon bacillus into which it can inject a giant molecule of deoxyribonucleic acid (DNA). This molecule, comprising the genes of the phage, subverts the machinery of the cell to make about 100 copies of the complete phage. The mutations that occasionally arise in the DNA molecule during replication enable the geneticist to map the detailed structure of individual genes. The electron micrograph was made by S. Brenner and R. W. Horne at the University of Cambridge.

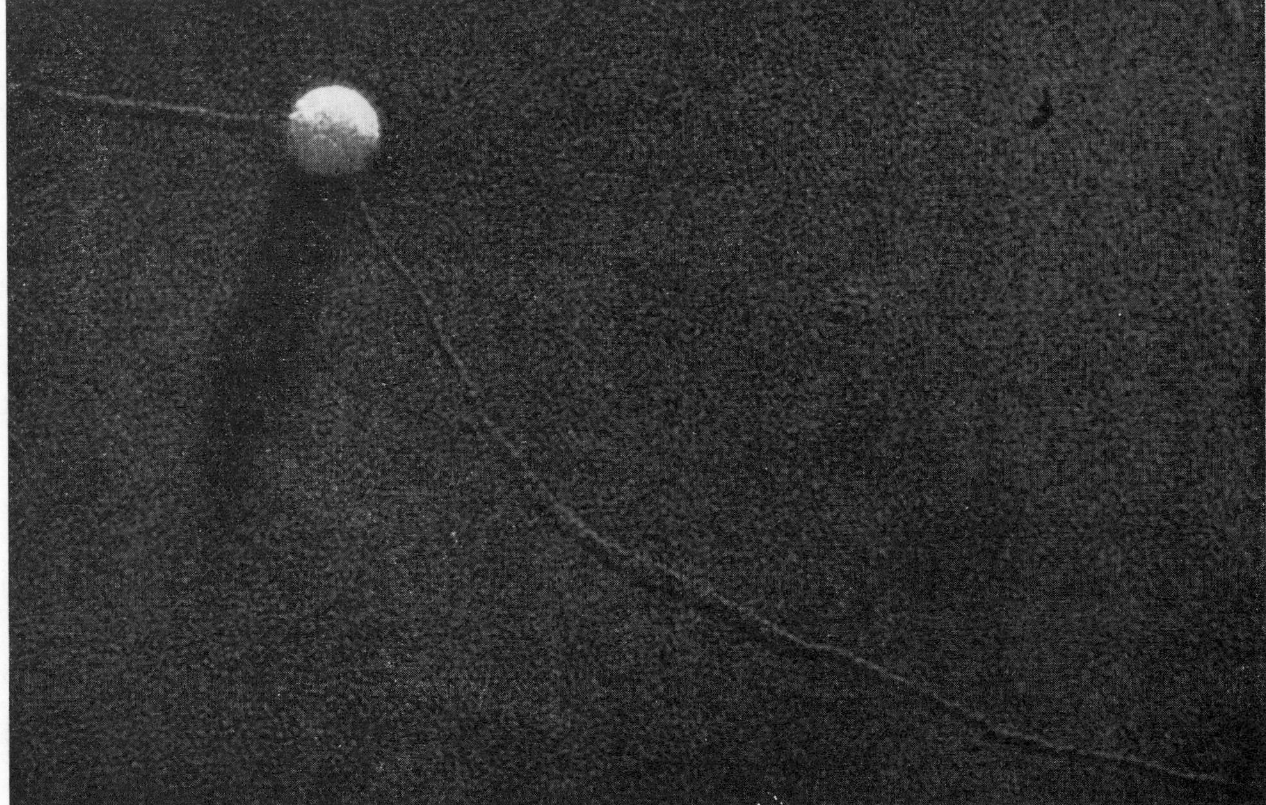

MOLECULE OF DNA is the fundamental carrier of genetic information. This electron micrograph shows a short section of DNA from calf thymus; its length is roughly that of the *r*II region in the DNA of T4 phage studied by the author. The DNA molecule in the phage would be about 30 feet long at this magnification of 150,000 diameters. The white sphere, a polystyrene "measuring stick," is 880 angstrom units in diameter. The electron micrograph was made by Cecil E. Hall of the Massachusetts Institute of Technology.

vised by James D. Watson and F. H. C. Crick, the DNA molecule resembles a ladder that has been twisted into a helix. The sides of the ladder are formed by alternating units of deoxyribose sugar groups and phosphate groups. The rungs, which join two sugar units, are composed of pairs of nitrogenous bases: either adenine paired with thymine or guanine paired with cytosine. The particular sequence of bases provides the genetic code of the DNA in a given organism. The DNA in the T4 virus contains some 200,000 base pairs, which, in amount of information, corresponds to much more than that contained in this article. Each base pair can be regarded as a letter in a word. One word (of the DNA code) may specify which of 20-odd amino acids is to be linked into a polypeptide chain. An entire paragraph might be needed to specify the sequence of amino acids for a polypeptide chain that has functional activity. Several polypeptide units may be needed to form a complex protein.

One can imagine that "typographical" errors may occur when DNA molecules are being replicated. Letters, words or sentences may be transposed, deleted or even inverted. When this occurs in a daily newspaper, the result is often humorous. In the DNA of living organisms typographical errors are never funny and are often fatal. We shall see how these errors, or mutations, can be used to analyze a small portion of the genetic information carried by the T4 bacteriophage.

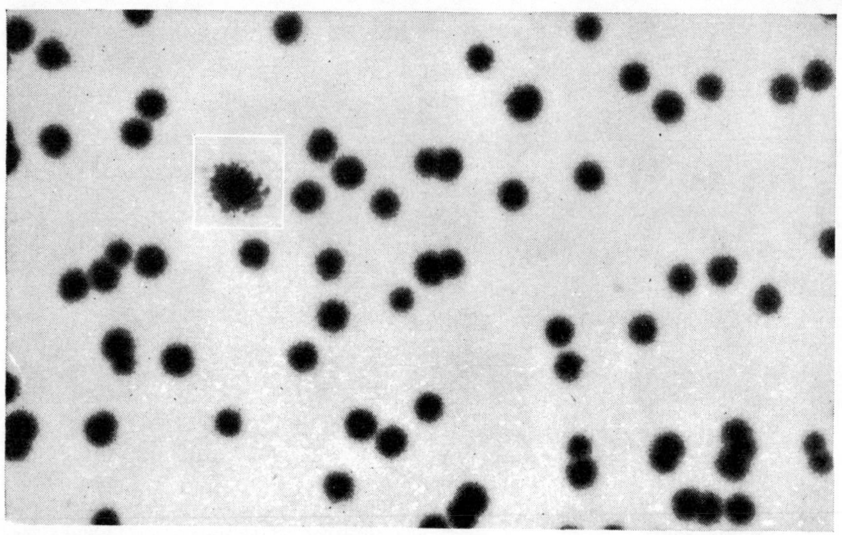

SPONTANEOUS MUTATIONAL EVENT is disclosed by the one mottled plaque (*square*) among dozens of normal plaques produced when standard T4 phage is "plated" on a layer of colon bacilli of strain B. Each plaque contains some 10 million progeny descended from a single phage particle. The plaque itself represents a region in which cells have been destroyed. Mutants found in abnormal plaques provide the raw material for genetic mapping.

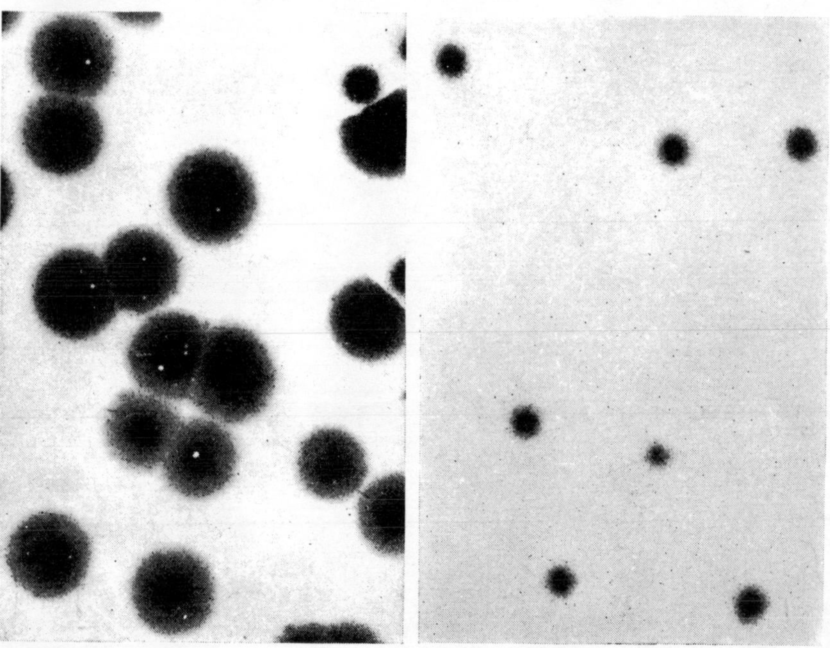

DUPLICATE REPLATINGS of mixed phage population obtained from a mottled plaque, like that shown at top of page, give contrasting results, depending on the host. Replated on colon bacilli of strain B (*left*), *r*II mutants produce large plaques. If the same mixed population is plated on strain K (*right*), only standard type of phage produce plaques.

Genetic Mapping with Phage

Before examining the interior of a gene let us see how genetic experiments are performed with bacteriophage. One starts with a single phage particle. This provides an important advantage over higher organisms, where two different individuals are required and the male and female may differ in any number of respects besides their sex. Another simplification is that phage is haploid, meaning that it contains only a single copy of its hereditary information, so that none of its genes are hidden by dominance effects. When a population is grown from a single phage particle, using a culture of sensitive bacteria as fodder, almost all the descendants are identical, but an occasional mutant form arises through some error in copying the genetic information. It is precisely these errors in reproduction that provide the key to the genetic analysis of the structure [*see upper illustration on pages 202 and 203*].

Suppose that two recognizably different kinds of mutant have been picked up; the next step is to grow a large population of each. This can be done in two test tubes in a couple of hours. It is now easy to perform a recombination experiment. A liquid sample of each phage population is added to a culture of bacterial cells in a test tube. It is arranged that the phage particles outnumber the bacterial cells at least three to one, so that each cell stands a good chance of being infected by both mutant forms of phage DNA. Within 20 minutes about 100 new phage particles are formed within each cell and are released when the cell bursts. Most of the progeny will resemble one or the other parent. In a few of them, however, the genetic information from the two parents may have been recombined to form a DNA molecule that is not an exact copy of the molecule possessed by either parent but a combination of the two. This new recombinant phage particle can carry both mutations or neither of them [*see lower illustration on next two pages*].

When this experiment is done with various kinds of mutant, some of the mutant genes tend to recombine almost independently, whereas others tend to be tightly linked to each other. From such experiments Alfred D. Hershey and Raquel Rotman, working at Washington University in St. Louis, were able to construct a genetic map for phage showing an ordered relationship among the various kinds of mutation, as had been done earlier with the fruit fly *Drosophila* and other higher organisms. It thus appears that the phage has a kind of chromosome —a string of genes that controls its hereditary characteristics.

One would like to do more, however, than just "drosophilize" phage. One would like to study the internal structure of a single gene in the phage chromosome. This too can be done by recombination experiments, but instead of choosing mutants of different kinds one chooses mutants that look alike (that is, have modifications of what is apparently the same characteristic), so that they are likely to contain errors in one or another part of the same gene.

Again the problem is to find an experimental method. When looking for mutations in fruit flies, say a white eye or a bent wing, one has to examine visually every fruit fly produced in the experiment. When working with phage, which reproduce by the billions and are invisible except by electron microscopy, the trick is to find a macroscopic method for identifying just those individuals in which recombination has occurred.

Fortunately in the T4 phage there is a class of mutants called *r*II mutants that can be identified rather easily by the appearance of the plaques they form on a given bacterial culture. A plaque is a clear region produced on the surface of a culture in a glass dish where phage particles have multiplied and destroyed the bacterial cells. This makes it possible to count individual phage particles without ever seeing them. Moreover, the shape and size of the plaques are hereditary characteristics of the phage that can be easily scored. A plaque produced in several hours will contain about 10 million phage particles representing the progeny of a single particle. T4 phage of the standard type can produce plaques on either of two bacterial host strains, B or K. The standard form of T4 occasionally gives rise to *r*II mutants that are easily noticed because they produce a distinctive plaque on B cultures. The key to the whole mapping technique is that

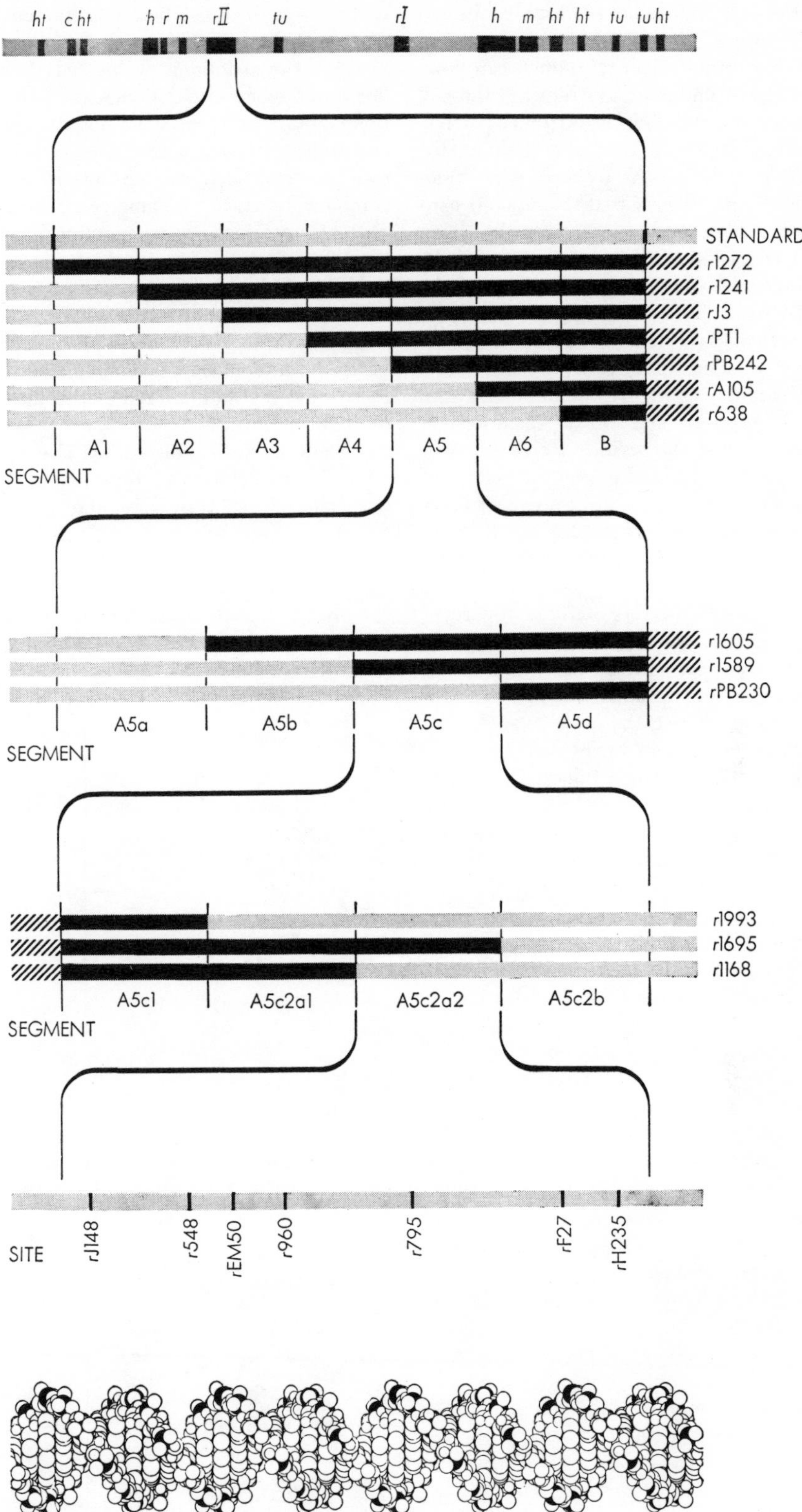

MAPPING TECHNIQUE localizes the position of a given mutation in progressively smaller segments of the DNA molecule contained in the T4 phage. The *r*II region represents to start with only a few percent of the entire molecule. The mapping is done by crossing an unknown mutant with reference mutants having deletions (*dark gray tone*) of known extent in the *r*II region (*see illustration of method on page 204*). The order and spacing of the seven mutational sites in the bottom row are still tentative. Each site probably represents the smallest mutable unit in the DNA molecule, a single base pair. The molecular segment (*extreme bottom*), estimated to be roughly in proper scale, contains a total of about 40 base pairs.

these mutants do not produce plaques on K cultures.

Nevertheless, an *r*II mutant can grow normally on bacterial strain K if the cell is simultaneously infected with a particle of standard type. Evidently the standard DNA molecule can perform some function required in K that the mutants cannot. This functional structure has been traced to a small portion of the DNA molecule, which in genetic maps of the T4 phage is designated the *r*II region.

To map this region one isolates a number of independently arising *r*II mutants (by removing them from mutant plaques visible on B) and crosses them against one another. To perform a cross, the two mutants are added to a liquid culture of B cells, thereby providing an opportunity for the progeny to recombine portions of genetic information from either parent. If the two mutant versions are due to typographical errors in different parts of the DNA molecule, some individuals of standard type may be regenerated. The standards will produce plaques on the K culture, whereas the mutants cannot. In this way one can easily detect a single recombinant among a billion progeny. As a consequence one can "resolve" two *r*II mutations that are extremely close together. This resolving power is enough to distinguish two mutations that are only one base pair apart in the DNA molecular chain.

What actually happens in the recombination of phage DNA is still a matter of conjecture. Two defective DNA molecules may actually break apart and rejoin to form one nondefective molecule, which is then replicated. Some recent evidence strongly favors this hypothesis. Another possibility is that in the course of replication a new DNA molecule arises from a process that happens to copy only the good portions of the two mutant molecules. The second process is called copy choice. An analogy for the two different processes can be found in the methods available for making a good tape recording of a musical performance from two tapes having defects in different places. One method is to cut the defects out of the two tapes and splice the good sections together. The second method (copy choice) is to play the two tapes and record the good sections on a third tape.

Mapping the *r*II Mutants

A further analogy with tape recording will help to explain how it has been established that the *r*II region is a simple linear structure. Given three tapes, each with a blemish or deletion in a different place, labeled *A*, *B* and *C*, one can imagine the deletions so located that deletion *B* overlaps deletion *A* and deletion *C*, but that *A* and *C* do not overlap each other. In such a case a good performance can be re-created only by recombining *A* and *C*. In mutant forms of phage DNA containing comparable deletions the existence of overlapping can be established by recombination experiments of just the same sort.

To obtain such deletions in phage one looks for mutants that show no tendency to revert to the standard type when they reproduce. The class of nonreverting mutants automatically includes those in which large alterations or deletions have occurred. (By contrast, *r*II mutants that revert spontaneously behave as if their alterations were localized at single points). The result of an exhaustive study covering hundreds of nonreverting *r*II mutants shows that all can be represented as containing deletions of one size or another in a single linear structure. If the structure were more complex, containing, for example, loops or branches, some mutations would have been expected to overlap in such a way as to make it impossible to represent them in a linear map. Although greater complexity cannot be absolutely excluded, all observations to date are satisfied by the postulate of simple linearity.

Now let us consider the *r*II mutants that do, on occasion, revert spontaneously when they reproduce. Conceivably they arise when the DNA molecule of the phage undergoes an alteration of a single base pair. Such "point" mutants are those that must be mapped if one is to probe the fine details of genetic structure. However, to test thousands of point mutants against one another for recombination in all possible pairs would

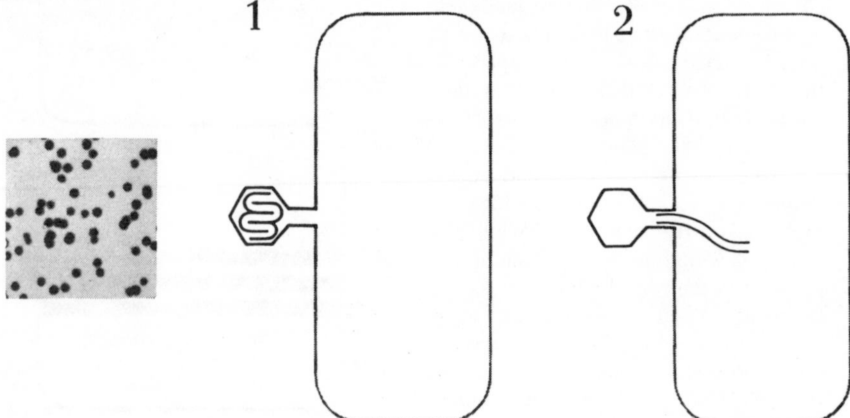

REPLICATION AND MUTATION occur when a phage particle infects a bacillus cell. The experiment begins by isolating a few standard particles from a normal plaque (*photograph at far left*) and growing billions of progeny in a broth culture of strain B colon bacilli. A sample of the broth is then spread on a Petri dish containing the same strain, on which the

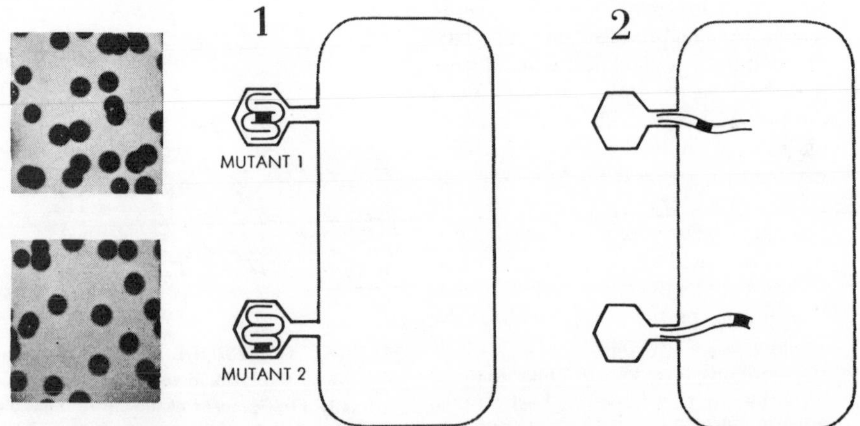

PROCESS OF RECOMBINATION permits parts of the DNA of two different phage mutants to be reassembled in a new DNA molecule that may contain both mutations or neither of them. Mutants obtained from two different cultures (*photographs at far left*) are introduced into a broth of strain B colon bacilli. Crossing occurs (*1*) when DNA from each mutant type

require millions of crosses. Mapping of point mutations by such a procedure would be totally impracticable.

The way out of this difficulty is to make use of mutants of the nonreverting type, whose deletions divide up the *r*II region into segments. Each point mutant is tested against these reference deletions. The recombination test gives a negative result if the deletion overlaps the point mutation and a positive result (over and above the "noise" level due to spontaneous reversion of the point mutant) if it does not overlap. In this way a mutation is quickly located within a particular segment of the map. The point mutation is then tested against a second group of reference mutants that divide this segment into smaller segments, and so on [*see illustration on pages 206 and 207*]. A point mutation can be assigned by this method to any of 80-odd ordered segments.

The final step in mapping is to test against one another only the group of mutants having mutations within each segment. Those that show recombination are concluded to be at different sites, and each site is then named after the mutant indicating it. (The mutants themselves have been assigned numbers according to their origin and order of discovery.) Finally, the order of the sites within a segment can be established by making quantitative measurements of recombination frequencies with respect to one another and neighbors outside the segment.

The Functional Unit

Thus we have found that the hereditary structure needed by the phage to multiply in colon bacilli of strain K consists of many parts distinguishable by mutation and recombination. Is this region to be thought of as one gene (because it controls one characteristic) or as hundreds of genes? Although mutation at any one of the sites leads to the same observed physiological defect, it does not necessarily follow that the entire structure is a single functional unit. For instance, growth in strain K could require a series of biochemical reactions, each controlled by a different portion of the region, and the absence of any one of the steps would suffice to block the final result. It is therefore of interest to see whether or not the *r*II region can be subdivided into parts that function independently.

This can be done by an experiment known as the *cis-trans* comparison. It will be recalled that the needed function can be supplied to a mutant by simultaneous infection of the cell with standard phage; the standard type supplies an intact copy of the genetic structure, so that

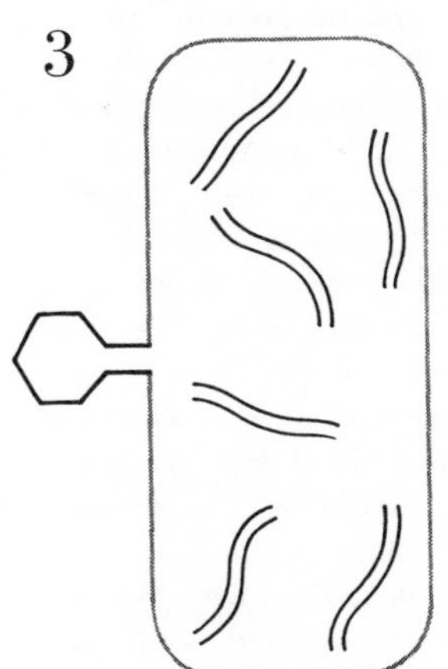

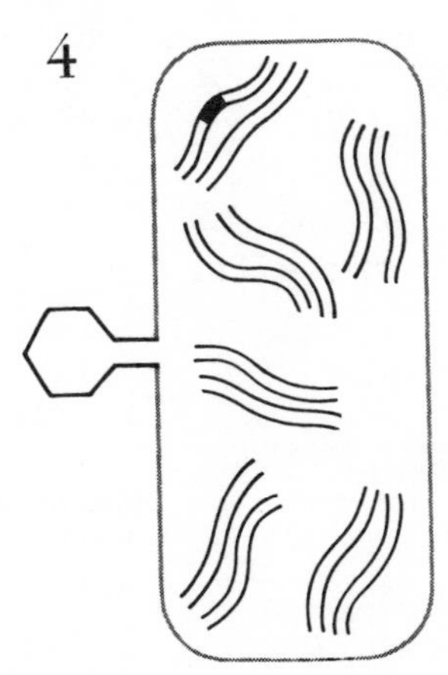

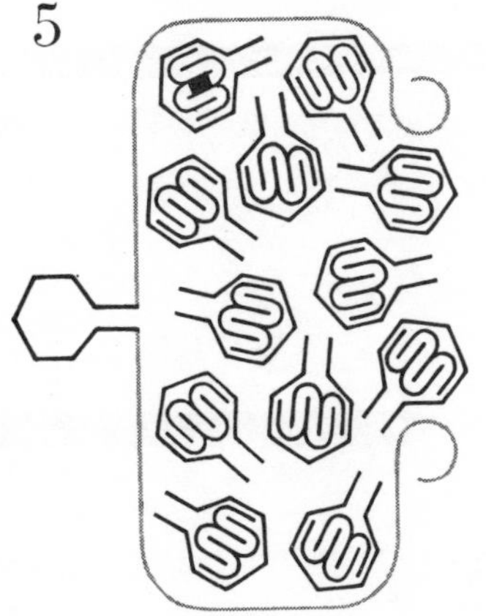

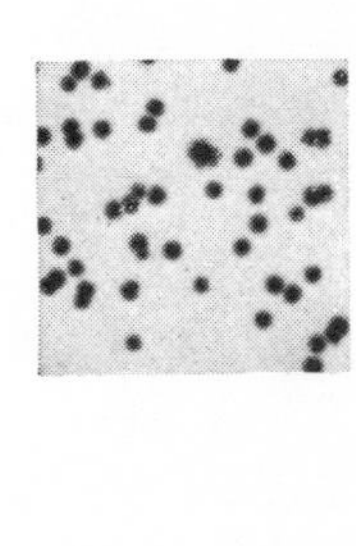

mutants and standard phage produce different plaque types. The diagrams show a bacillus infected by a single standard phage. The DNA molecule from the phage enters the cell (2) and is replicated (*3 and 4*). Among scores of perfect replicas, one may contain a mutation (*dark patch*). Encased in protein jackets, the phage particles finally burst out of the cell (5). When a mutant arises during development of a plaque, the mixture of its mutant progeny and standard types makes plaque look mottled (*photograph at right*).

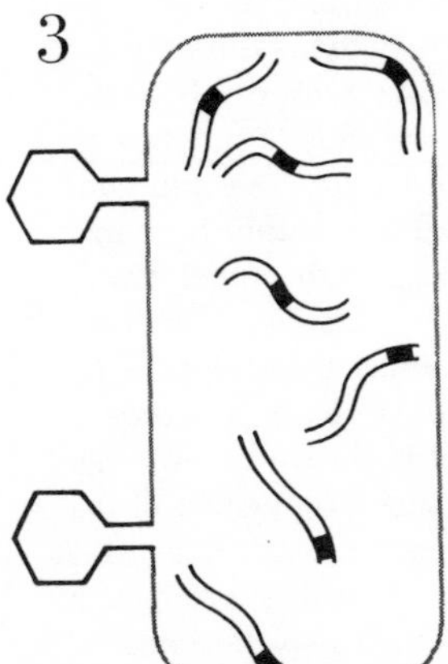

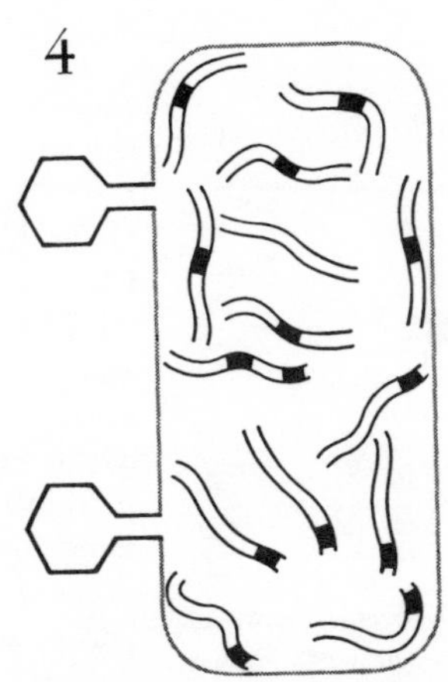

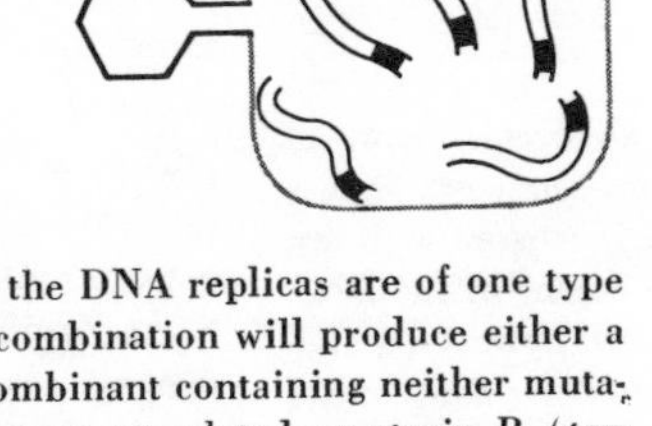

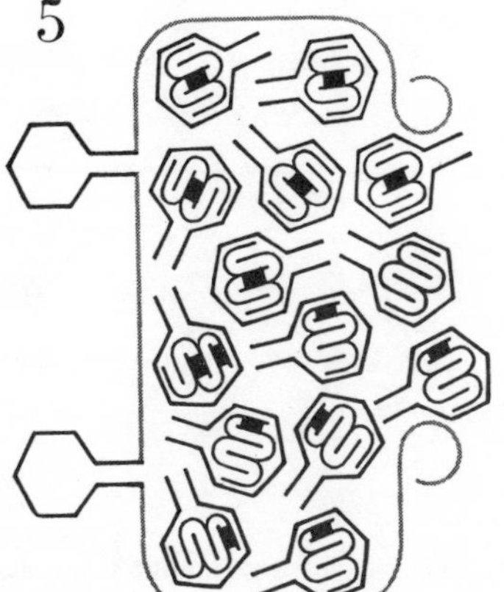

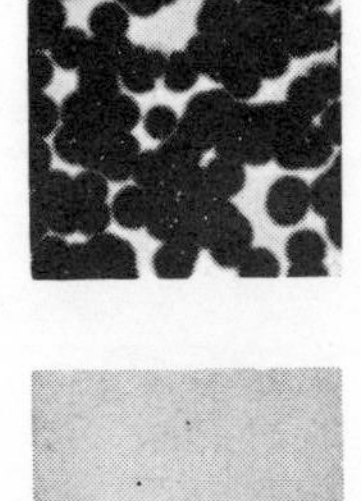

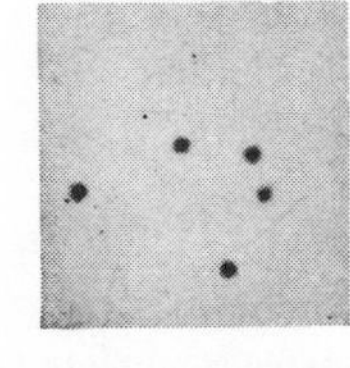

infects a single bacillus. Most of the DNA replicas are of one type or the other, but occasionally recombination will produce either a double mutant or a standard recombinant containing neither mutation. When the progeny of the cross are plated on strain B (*top photograph at far right*), all grow successfully, producing many plaques. Plated on strain K, only the standard recombinants are able to grow (*bottom photograph at right*). A single standard recombinant can be detected among as many as 100 million progeny.

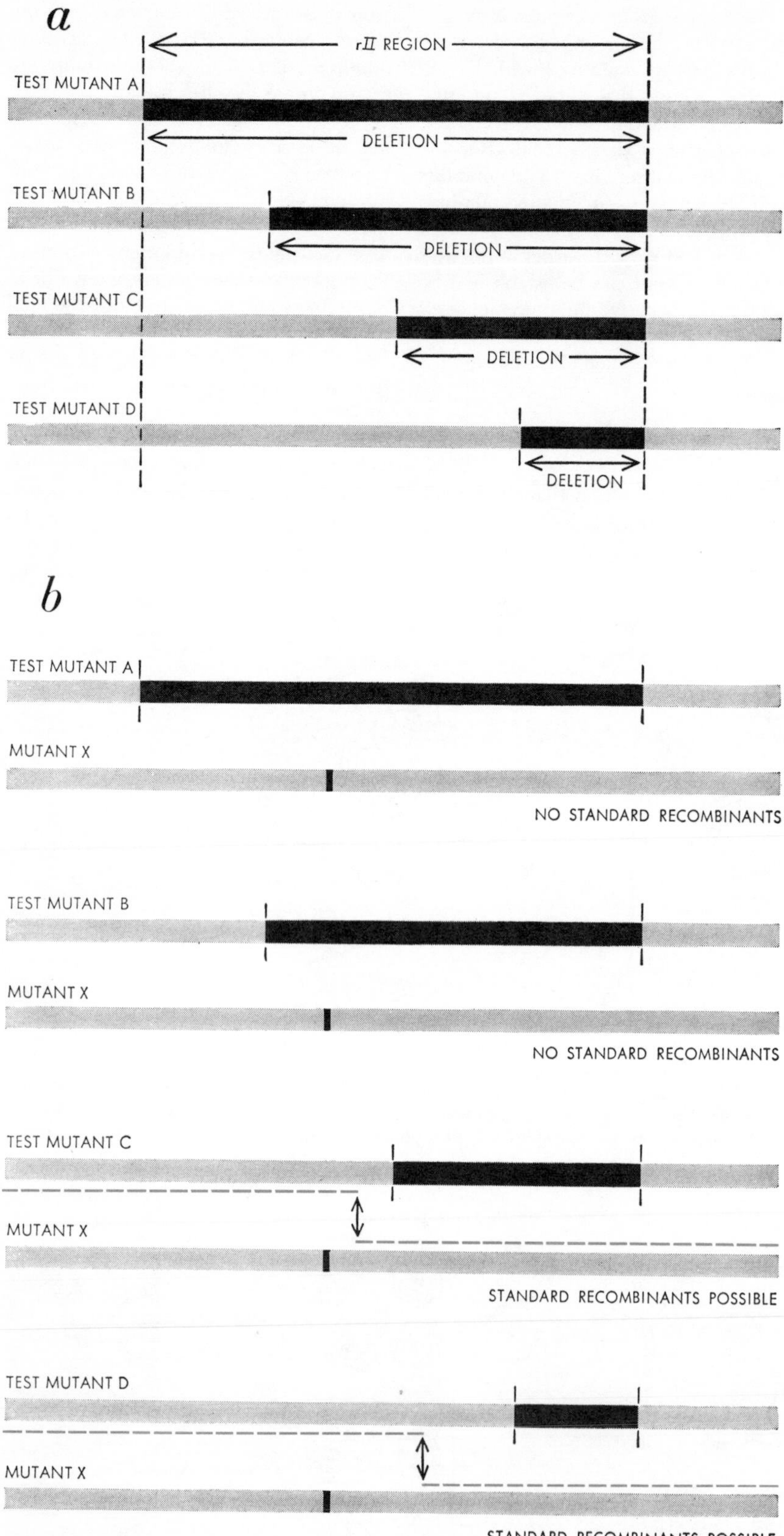

DELETION MAPPING is done by crossing an unknown mutant with a selected group of reference mutants (*four at top*) whose DNA molecules contain deletions—or what appear to be deletions—of known length in the rII region. Thus when mutant *X* is crossed with test mutants *A* and *B*, no standard recombinants are observed because both copies of the DNA molecule are defective at the same place. When *X* is crossed with *C* and *D*, however, standard recombinants can be formed, as indicated by broken lines and arrows. By using other reference mutants with appropriate deletions the location of *X* can be further narrowed.

it does not matter what defect the *r*II mutant has and both types are enabled to reproduce. Now suppose the intact structure of the standard type could be split into two parts. If this were to destroy the activity, the two parts could be regarded as belonging to a single functional unit. Although the experiment as such is not feasible, one can do the next best thing. That is to supply piece *A* intact by means of a mutant having a defect in piece *B,* and to use a mutant with a defect in piece *A* to supply an intact piece *B*. If the two pieces *A* and *B* can function independently, the system should be active, since each mutant supplies the function lacking in the other. If, however, both pieces must be together to be functional, the split combination should be inactive.

The actual experimental procedure is as follows. Let us imagine that one has identified two mutational sites in the *r*II region, *X* and *Y*, and that one wishes to know if they lie within the same functional unit. The first step is to infect cells of strain K with the two different mutants, *X* and *Y;* this is called the *trans* test because the mutations are borne by different DNA molecules. Now in K the decision as to whether or not the phage will function occurs very soon after infection and *before* there is any opportunity for recombination to take place. To carry out a control experiment one needs a double mutant (obtainable by recombination) that contains both *X* and *Y* within a single phage particle. When cells of strain K are infected with the double mutant and the standard phage, the experiment is called the *cis* test since one of the infecting particles contains both mutations in a single DNA molecule. In this case, because of the presence of the standard phage, normal replication is expected and provides the control against which to measure the activity observed in the *trans* test. If, in the *trans* test, the phage fails to function or shows only slight activity, one can conclude that *X* and *Y* fall within the same functional unit. If, on the other hand, the phage develops actively, it is probable (but not certain) that the sites lie in different functional units. (Certainty in this experiment is elusive because the products of two defective versions of the same functional unit, tested in a *trans* experiment, will sometimes produce a partial activity, which may be indistinguishable from that produced by a *cis* experiment.)

As applied to *r*II mutants, the test divides the structure into two clear-cut parts, each of which can function independently of the other. The functional units have been called cistrons, and we say that the rII region is composed of an A cistron and a B cistron.

We have, then, genetic units of various sizes: the small units of mutation and recombination, much larger cistrons and finally the rII region, which includes both cistrons. Which one of these shall we call the gene? It is not surprising to find geneticists in disagreement, since in classical genetics the term "gene" could apply to any one of these. The term "gene" is perfectly acceptable so long as one is working at a higher level of integration, at which it makes no difference which unit is being referred to. In describing data on the fine level, however, it becomes essential to state unambiguously which operationally defined unit one is talking about. Thus in describing experiments with rII mutants one can speak of the rII "region," two rII "cistrons" and many rII "sites."

Some workers have proposed using the word "gene" to refer to the genetic unit that provides the information for one enzyme. But this would imply that one should not use the word "gene" at all, short of demonstrating that a specific enzyme is involved. One would be quite hard pressed to provide this evidence in the great majority of cases in which the term has been used, as, for example, in almost all the mutations in *Drosophila*. A genetic unit should be defined by a genetic experiment. The absurdity of doing otherwise can be seen by imagining a biochemist describing an enzyme as that which is made by a gene.

We have seen that the topology of the rII region is simple and linear. What can be said about its topography? Are there local differences in the properties of the various parts? Specifically, are all the subelements equally mutable? If so, mutations should occur at random throughout the structure and the topography would then be trivial. On the other hand, sites or regions of unusually high or low mutability would be interesting topographic features. To answer this question one isolates many independently arising rII mutants and maps each one to see if mutations tend to occur more frequently at certain points than at others. Each mutation is first localized into a main segment, then into a smaller segment, and finally mutants of the same small segment are tested against each other. Any that show recombination are said to define different sites. If two or more reverting mutants are found to show no detectable recombination with each other, they are considered to be

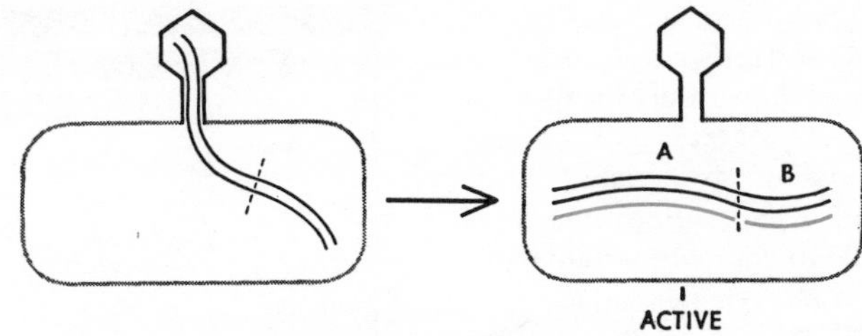

PHAGE ACTIVITY requires that the coded information inside functional units of the DNA molecule be available intact. The rII region consists of two functional units called A cistron and B cistron. When both are present intact (*right*), the phage actively replicates inside colon bacillus of strain K. Colored lines indicate effective removal of coded information.

a

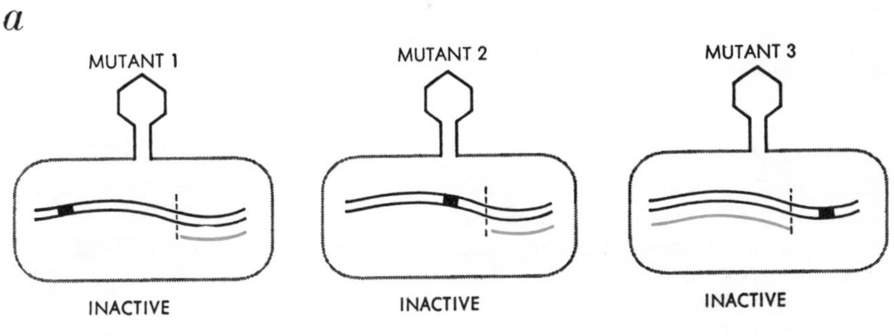

b

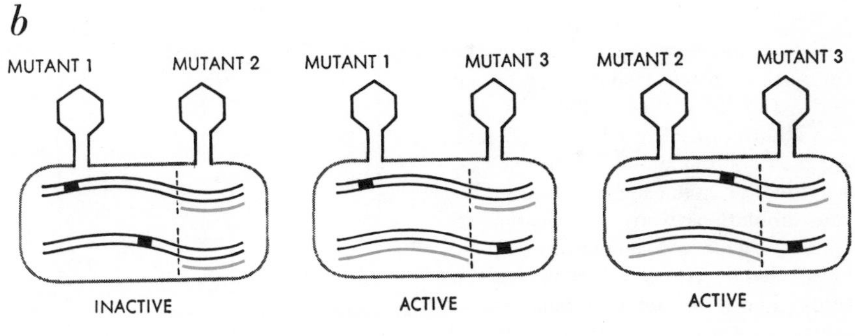

c

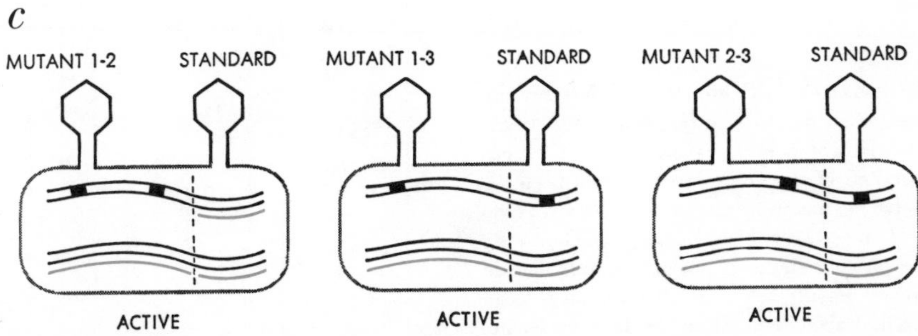

***CIS-TRANS* TEST determines the size of functional units. In bacillus of strain K, T4 phage is active only if both A and B cistrons are provided intact; hence mutants *1*, *2* and *3* are inactive. (The sites of mutation have been previously established.) Tests with the three mutants taken two at a time (*b*) show that sites *1* and *2* must be in the same cistron. A test of each mutant with standard phage (*c*) provides a control; in this case all are active.**

repeats, and one of them is chosen to represent the site in further tests. A set of distinct sites is thereby obtained, each with its own group of repeats. The designation of a mutant as a repeat is, of course, tentative, since in principle it remains possible that a more sensitive test could show some recombination.

The illustration on pages 208 and 209 shows a map of the *r*II region with each occurrence of a spontaneous mutation indicated by a square. These mutations, as well as other data from induced mutations, subdivide the map into more than 300 distinct sites, and the distribution of repeats is indeed far from random. The topography for spontaneous mutation is evidently quite complex, the structure consisting of elements with widely different mutation rates.

Spontaneous mutation is a chronic disease; a spontaneous mutant is simply one for which the cause is unknown. By using chemical mutagens such as nitrous acid or hydroxylamine, or physical agents such as ultraviolet light, one can alter the DNA in a more controlled manner and induce mutations specifically. A method of inducing specific mutations has long been the philosophers' stone of genetics. What the genetic alchemist desired, however, was an effect that could be directed at the gene controlling a particular characteristic. Chemical mutagenesis is highly specific but not in this way. When Rose Litman and Arthur B. Pardee at the University of California discovered the mutagenic effect of 5-bromouracil on phage, they regarded it as a nonspecific mutagen because mutations were induced that affected a wide assortment of different phage characteristics. This nonspecificity resulted because each functional gene is a structure with many parts and is bound to contain a number of sites that are responsive to any particular mutagen. Therefore the rate at which mutation is

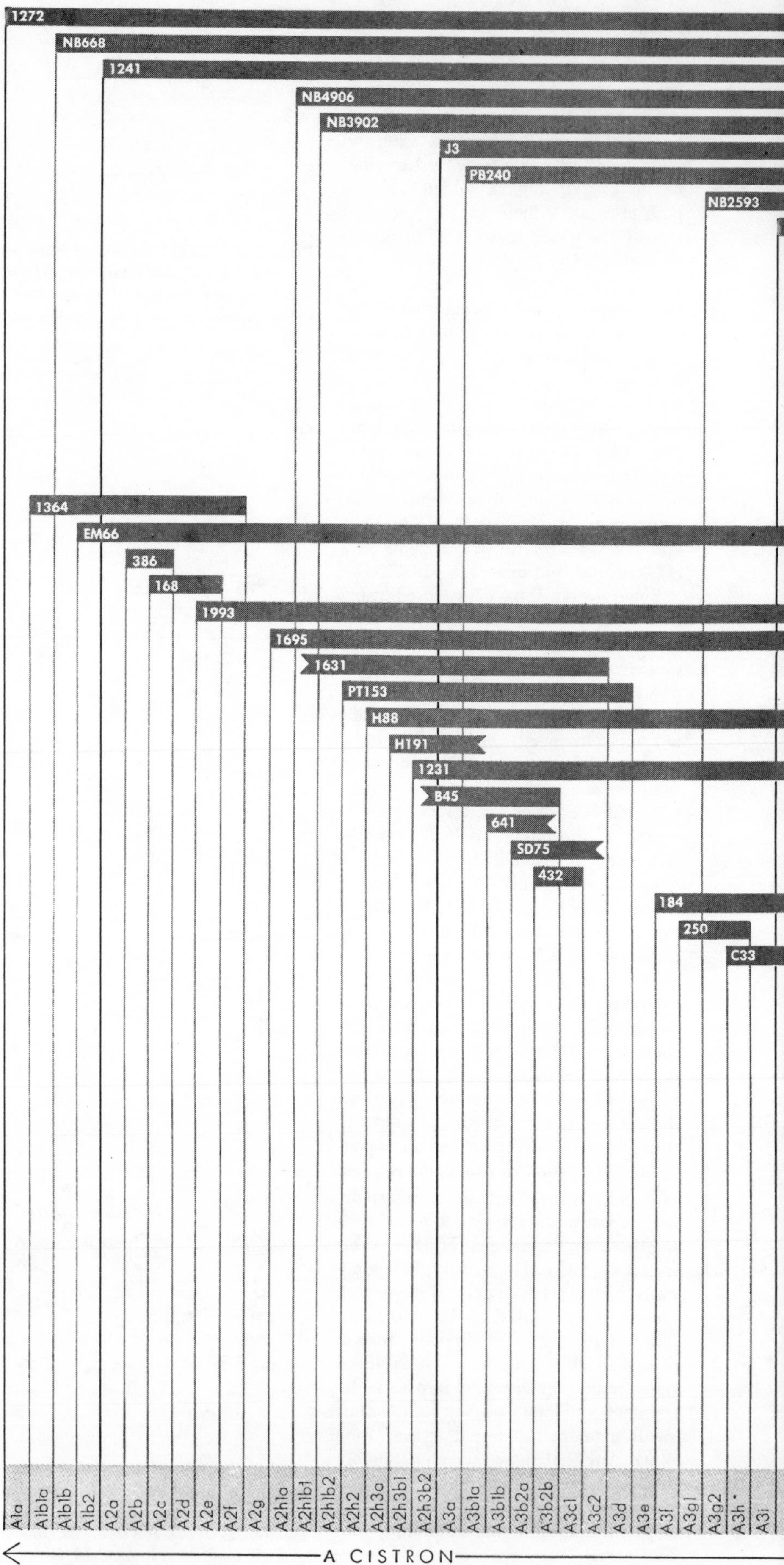

DELETION MAP shows the reference mutants that divided the *r*II region into 80 segments. These mutants behave as if various sections of the DNA molecule had been deleted or inactivated, and as a class they do not revert, or back-mutate, spontaneously to produce standard phage. Mutants that do revert usually act as if the mutation is localized at a single point on the DNA molecule. Where this point falls in the *r*II region is determined by systematically crossing the revertible mutant with these reference deletion mutants, as illustrated on page 204. The net result is to assign the point mutation to smaller and smaller segments of the map.

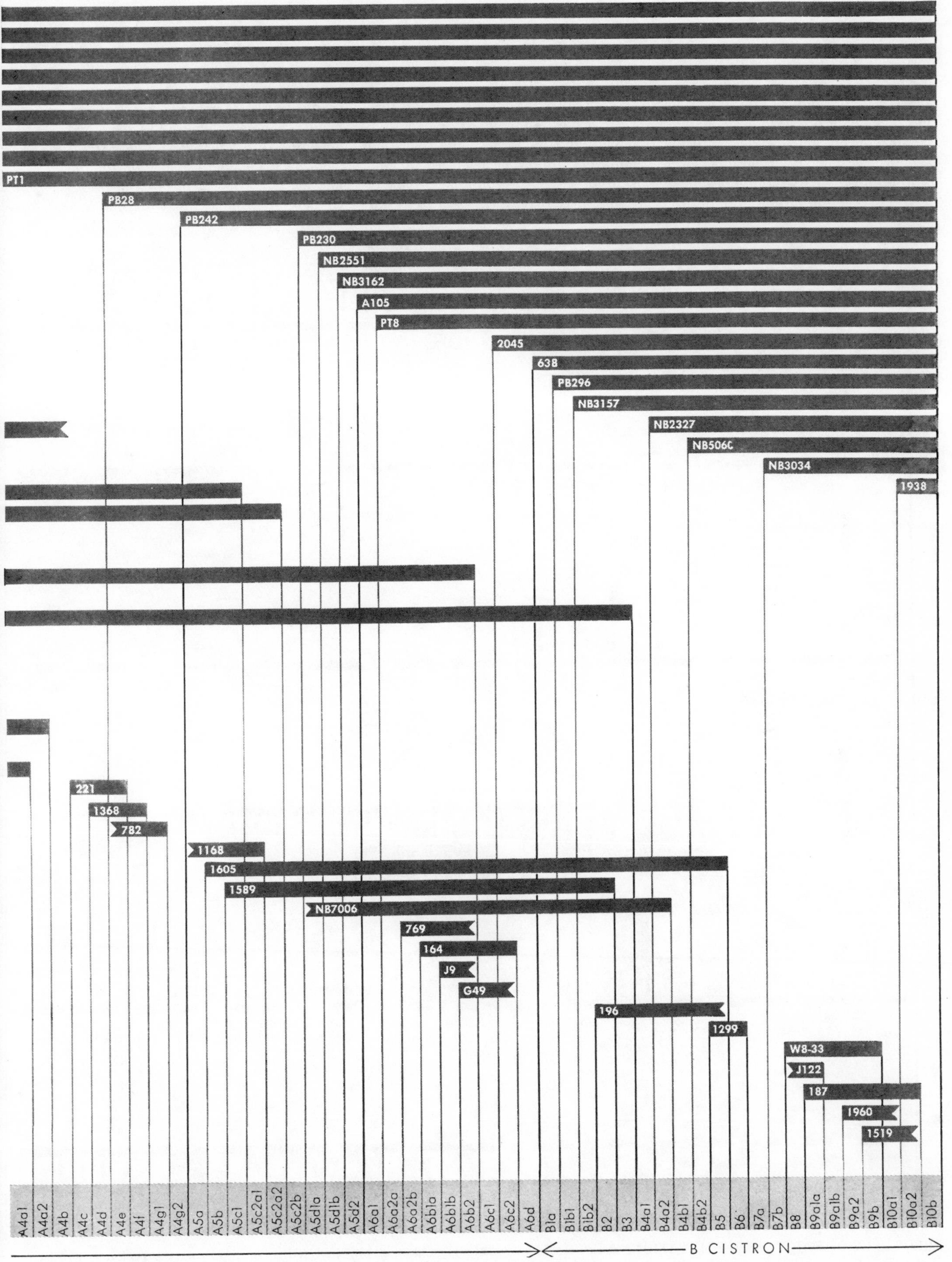
PT1
PB28
PB242
PB230
NB2551
NB3162
A105
PT8
2045
638
PB296
NB3157
NB2327
NB5060
NB3034
1938
221
1368
782
1168
1605
1589
NB7006
769
164
J9
G49
196
1299
W8-33
J122
187
1960
1519
A4a1
A4a2
A4b
A4c
A4d
A4e
A4f
A4g1
A4g2
A5a
A5b
A5c1
A5c2a1
A5c2a2
A5c2b
A5d1a
A5d1b
A5d2
A6a1
A6a2a
A6a2b
A6b1a
A6b1b
A6b2
A6c1
A6c2
A6d
B1a
B1b1
B1b2
B2
B3
B4a1
B4a2
B4b1
B4b2
B5
B6
B7a
B7b
B8
B9a1a
B9a1b
B9a2
B9b
B10a1
B10a2
B10b
B CISTRON

induced in various genes is more or less the same. By fine-structure genetic analysis, however, Ernst Freese and I, working in our laboratory at Purdue University, have found that 5-bromouracil increases the mutation rate at certain sites by a factor of 10,000 or more over the spontaneous rate, while producing no noticeable change at some other sites. This indicates a high degree of specificity indeed, but at the level within the cistron. Furthermore, other mutagens specifically alter other sites. The response of part of the B cistron to a variety of mutagens is shown in the illustration on the following two pages.

Each site in the genetic map can, then, be characterized by its spontaneous mutability and by its response to various mutagens. By this means many different kinds of site have been found. Some response patterns are represented at only a single site in the entire structure; for example, the prominent spontaneous hot spot in segment B4. This is at first surprising, because according to the Watson-Crick model for DNA the structure should consist of only two types of element, adenine-thymine (AT) pairs and guanine-hydroxymethylcytosine (GC) pairs. One possible explanation for the uneven reactivity among various sites is that the response may depend not

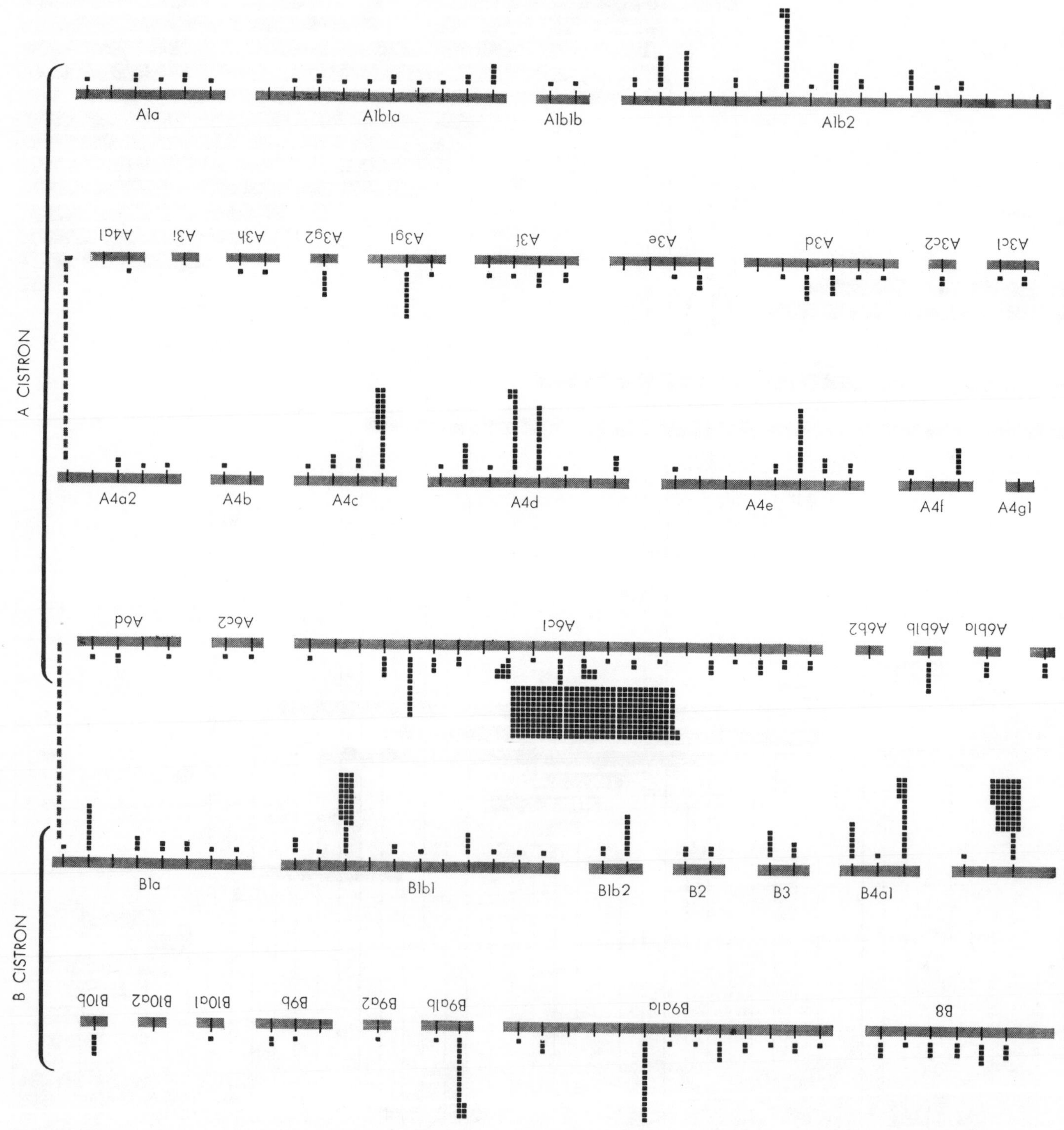

FREQUENCY OF SPONTANEOUS MUTATIONS at various sites is shown in this complete map of the *r*II region. Alternate rows have been deliberately inverted to indicate that the region is a continuous molecular thread. Each spontaneous mutation at a site

only on the particular base pair at a site but also very much on the type and arrangement of neighboring base pairs.

Once a site is identified it can be further characterized by the ease with which a particular mutagen makes reverse mutations produce phage of standard type. Combining such studies with studies of the chemical mechanism of mutagenesis, it may be possible eventually to translate the genetic map, bit by bit, into the actual base sequence.

Saturation of the Map

How far is the map from being run into the ground? Since many of the sites are represented by only one occurrence of a mutation, it is clear that there must still exist some sites with zero occurrences, so that the map as it stands is not saturated. From the statistics of the distribution it can be estimated that there must exist, in addition to some 350 sites now known, at least 100 sites not yet discovered. The calculation provides only a minimum estimate; the true number is probably larger. Therefore the map at the present time cannot be more than 78 per cent saturated.

Everything that we have learned about the genetic fine structure of T4 phage is compatible with the Watson-

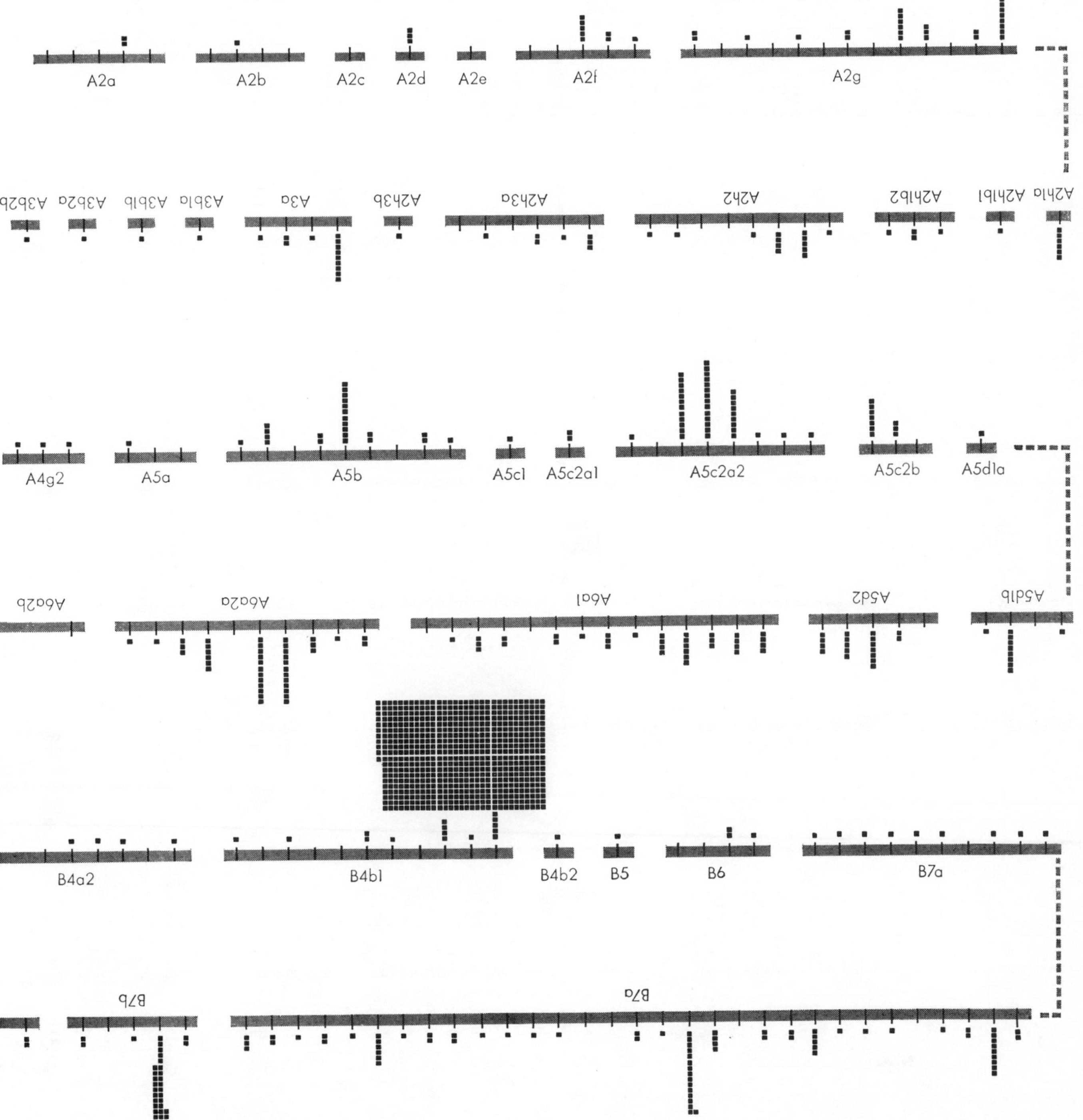

is represented by a small black square. Sites without squares are known to exist because they can be induced to mutate by use of chemical mutagens or ultraviolet light (*see illustration on next two pages*), but they have not been observed to mutate spontaneously.

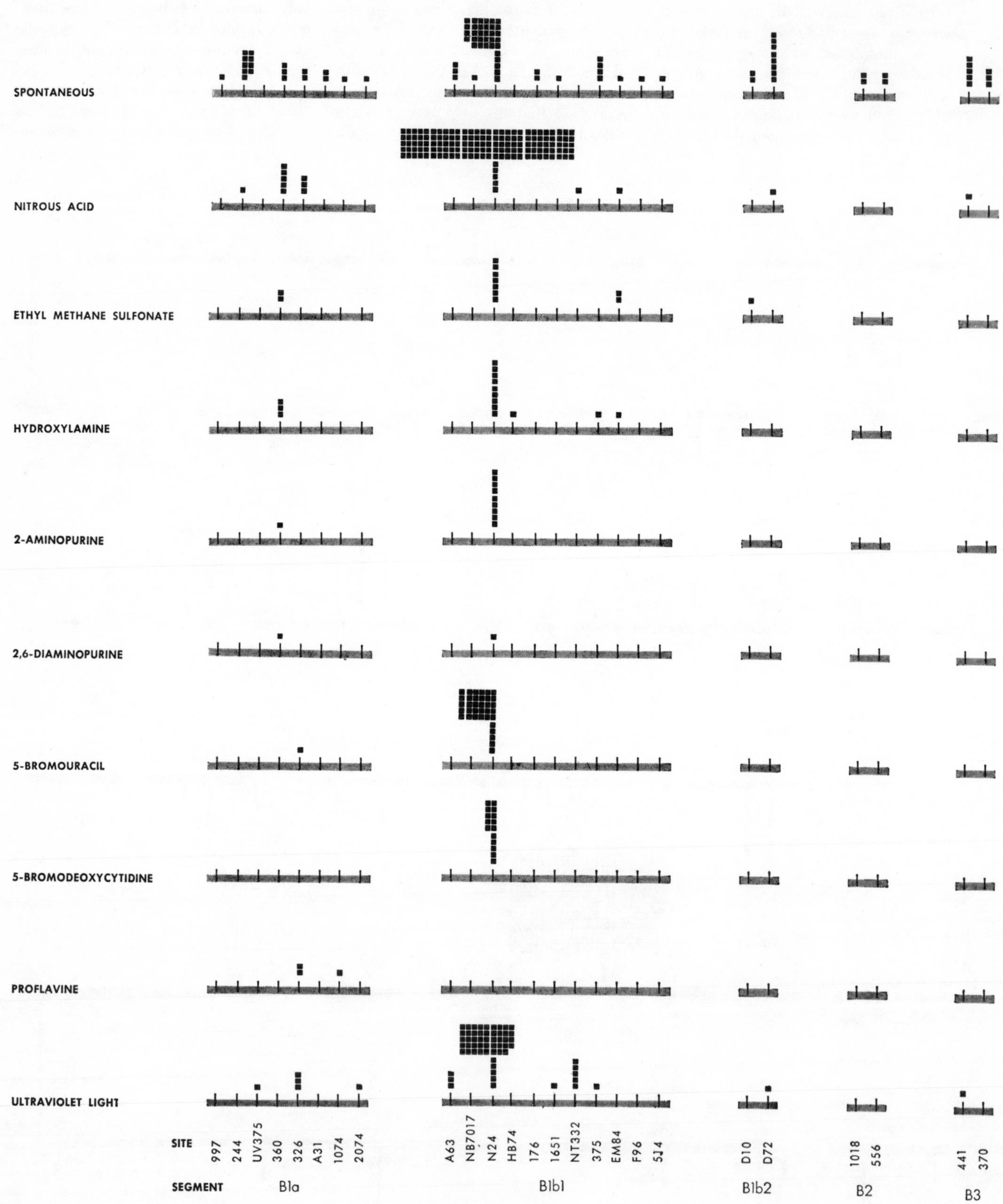

RESPONSE OF PHAGE TO MUTAGENS is shown for a portion of the B cistron. The total number of mutations studied is not the same for each mutagen. It is clear, nevertheless, that mutagenic action is highly specific at certain sites. For example, site EM26,

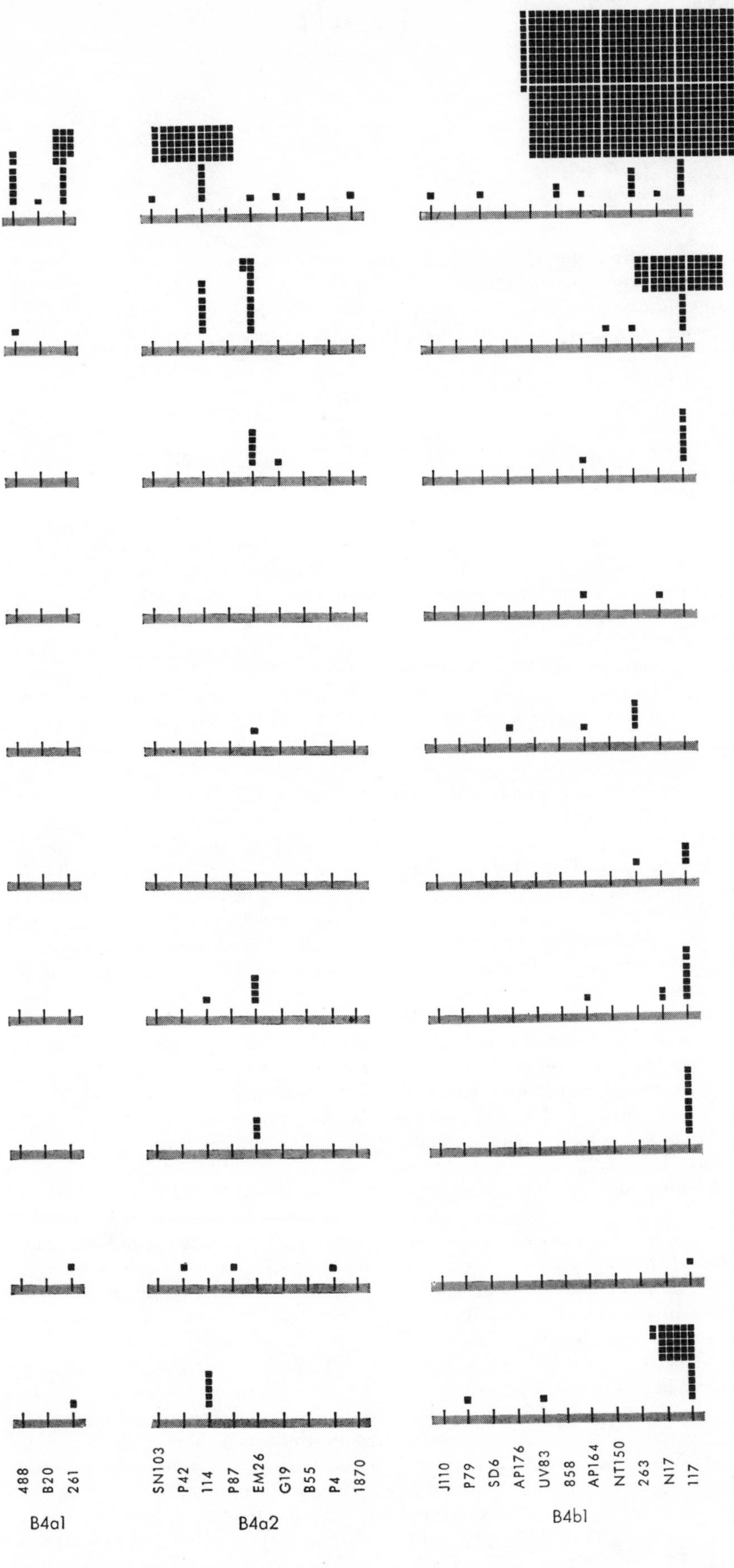

which resists spontaneous mutation, responds readily to certain mutagens. However, site 117 in segment B4b1 is more apt to mutate spontaneously than in response to a mutagen.

Crick model of the DNA molecule. In this model the genetic information is contained in the specific order of bases arranged in a linear sequence. The four-letter language of the bases must somehow be translated into the 20-letter language of the amino acids, so that at least several base pairs must be required to specify one amino acid, and an entire polypeptide chain should be defined by a longer segment of DNA. Since the activity of the resulting enzyme, or other protein, depends on its precise structure, this activity should be impaired by any of a large number of changes in the DNA base sequence leading to amino acid substitutions.

One can also imagine that certain changes in base sequence can lead to a "nonsense" sequence that does not specify any amino acid, and that as a result the polypeptide chain cannot be completed. Thus the genetic unit of function should be vulnerable at many different points within a segment of the DNA structure. Considering the monotonous structure of the molecule, there is no obvious reason why recombination should not be possible at every link in the molecular chain, although not necessarily with the same probability. In short, the Watson-Crick model leads one to expect that the functional units—the genes of traditional genetics—should consist of linear segments that can be finely dissected by mutation and recombination.

Mapping Other Genes

The genetic results fully confirm these expectations. All mutations can in fact be represented in a strictly linear map, the functional units correspond to sharply defined segments, and each functional unit is divisible by mutation and recombination into hundreds of sites. Mutations are induced specifically at certain sites by agents that interact with the DNA bases. Although the data on mutation rates are complex, it is quite probable that they can be explained by interactions between groups of base pairs.

In confining this investigation to *r*II mutants of T4, attention has been focused on a tiny bit of hereditary material constituting only a few per cent of the genetic structure of a virus, enabling the exploration to be carried almost to the limits of the molecular structure. Similar results are being obtained in many other microorganisms and even in higher organisms such as corn. Given techniques for handling cells in culture in the appropriate way, man too may soon be a subject for genetic fine-structure analysis.

16 The Genetic Code

by F. H. C. Crick
October 1962

How does the order of bases in a nucleic acid determine the order of amino acids in a protein? It seems that each amino acid is specified by a triplet of bases, and that triplets are read in simple sequence

Within the past year important progress has been made in solving the "coding problem." To the biologist this is the problem of how the information carried in the genes of an organism determines the structure of proteins.

Proteins are made from 20 different kinds of small molecule—the amino acids—strung together into long polypeptide chains. Proteins often contain several hundred amino acid units linked together, and in each protein the links are arranged in a specific order that is genetically determined. A protein is therefore like a long sentence in a written language that has 20 letters.

Genes are made of quite different long-chain molecules: the nucleic acids DNA (deoxyribonucleic acid) and, in some small viruses, the closely related RNA (ribonucleic acid). It has recently been found that a special form of RNA, called messenger RNA, carries the genetic message from the gene, which is located in the nucleus of the cell, to the surrounding cytoplasm, where many of the proteins are synthesized [see "Messenger RNA," by Jerard Hurwitz and J. J. Furth; SCIENTIFIC AMERICAN Offprint 119].

The nucleic acids are made by joining up four kinds of nucleotide to form a polynucleotide chain. The chain provides a backbone from which four kinds of side group, known as bases, jut at regular intervals. The order of the bases, however, is not regular, and it is their precise sequence that is believed to carry the genetic message. The coding problem can thus be stated more explicitly as the problem of how the sequence of the four bases in the nucleic acid determines the sequence of the 20 amino acids in the protein.

The problem has two major aspects, one general and one specific. Specifically one would like to know just what sequence of bases codes for each amino acid. Remarkable progress toward this goal was reported early this year by Marshall W. Nirenberg and J. Heinrich Matthaei of the National Institutes of Health and by Severo Ochoa and his colleagues at the New York University School of Medicine. [Editor's note: Brief accounts of this work appeared in "Science and the Citizen" for February and March 1962. This article is a companion to one by Nirenberg, which deals with the biochemical aspects of the genetic code].

The more general aspect of the coding problem, which will be my subject, has to do with the length of the genetic coding units, the way they are arranged in the DNA molecule and the way in which the message is read out. The experiments I shall report were performed at the Medical Research Council Laboratory of Molecular Biology in Cambridge, England. My colleagues were Mrs. Leslie Barnett, Sydney Brenner, Richard J. Watts-Tobin and, more recently, Robert Shulman.

The organism used in our work is the bacteriophage T4, a virus that infects the colon bacillus and subverts the biochemical machinery of the bacillus to make multiple copies of itself. The infective process starts when T4 injects its genetic core, consisting of a long strand of DNA, into the bacillus. In less than 20 minutes the virus DNA causes the manufacture of 100 or so copies of the complete virus particle, consisting of a DNA core and a shell containing at least six distinct protein components. In the process the bacillus is killed and the virus particles spill out. The great value of the T4 virus for genetic experiments is that many generations and billions of individuals can be produced in a short time. Colonies containing mutant individuals can be detected by the appearance of the small circular "plaques" they form on culture plates. Moreover, by the use of suitable cultures it is possible to select a single individual of interest from a population of a billion.

Using the same general technique, Seymour Benzer of Purdue University was able to explore the fine structure of the A and B genes (or cistrons, as he prefers to call them) found at the "*r*II" locus of the DNA molecule of T4 [see the article "The Fine Structure of the Gene," by Seymour Benzer, beginning on page 198]. He showed that the A and B genes, which are next to each other on the virus chromosome, each consist of some hundreds of distinct sites arranged in linear order. This is exactly what one would expect if each gene is a segment, say 500 or 1,000 bases long, of the very long DNA molecule that forms the virus chromosome [*see illustration on opposite page*]. The entire DNA molecule in T4 contains about 200,000 base pairs.

The Usefulness of Mutations

From the work of Benzer and others we know that certain mutations in the A and B region made one or both genes inactive, whereas other mutations were only partially inactivating. It had also been observed that certain mutations were able to suppress the effect of harmful mutations, thereby restoring the function of one or both genes. We suspected that the various—and often puzzling—consequences of different kinds of mutation might provide a key to the nature of the genetic code.

We therefore set out to re-examine the effects of crossing T4 viruses bearing mutations at various sites. By growing two different viruses together in a common culture one can obtain "recombinants" that have some of the properties

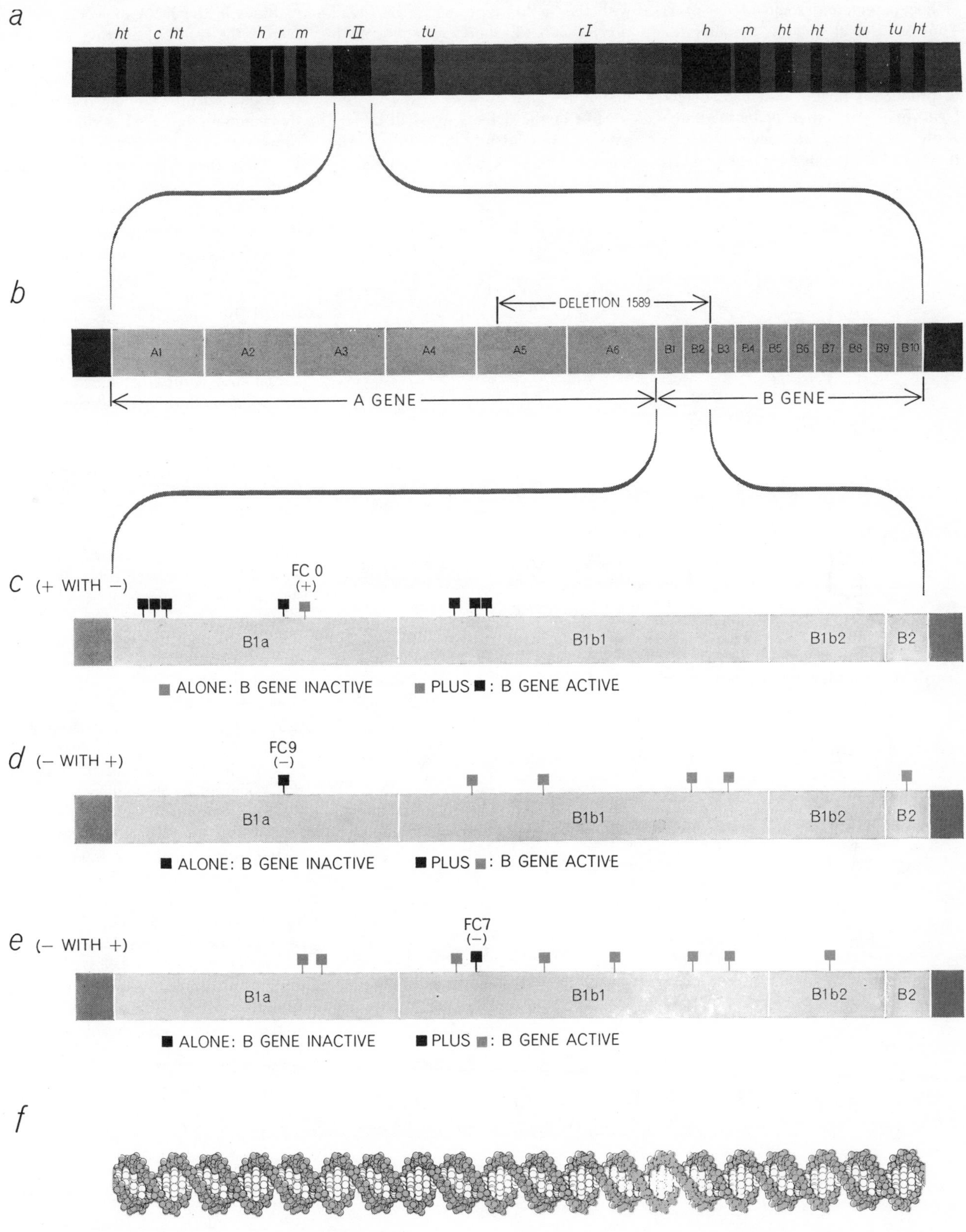

rII REGION OF THE T4 VIRUS represents only a few per cent of the DNA (deoxyribonucleic acid) molecule that carries full instructions for creating the virus. The region consists of two genes, here called A and B. The A gene has been mapped into six major segments, the B gene into 10 (*b*). The experiments reported in this article involve mutations in the first and second segments of the B genes. The B gene is inactivated by any mutation that adds a molecular subunit called a base (*colored square*) or removes one (*black square*). But activity is restored by simultaneous addition and removal of a base, as shown in *c*, *d* and *e*. An explanation for this recovery of activity is illustrated on page 216. The molecular representation of DNA (*f*) is estimated to be approximately in scale with the length of the B1 and B2 segments of the B gene. The two segments contain about 100 base pairs.

of one parent and some of the other. Thus one defect, such as the alteration of a base at a particular point, can be combined with a defect at another point to produce a phage with both defects [*see upper illustration below*]. Alternatively, if a phage has several defects, they can be separated by being crossed with the "wild" type, which by definition has none. In short, by genetic methods one can either combine or separate different mutations, provided that they do not overlap.

Most of the defects we shall be considering are evidently the result of adding or deleting one base or a small group of bases in the DNA molecule and not merely the result of altering one of the bases [*see lower illustration on this page*]. Such additions and deletions can be produced in a random manner with the compounds called acridines, by a process that is not clearly understood. We think they are very small additions or deletions, because the altered gene seems to have lost its function completely; mutations produced by reagents capable of changing one base into another are often partly functional. Moreover, the acridine mutations cannot be reversed by such reagents (and vice versa). But our strongest reason for believing they are additions or deletions is that they can be combined in a way that suggests they have this character.

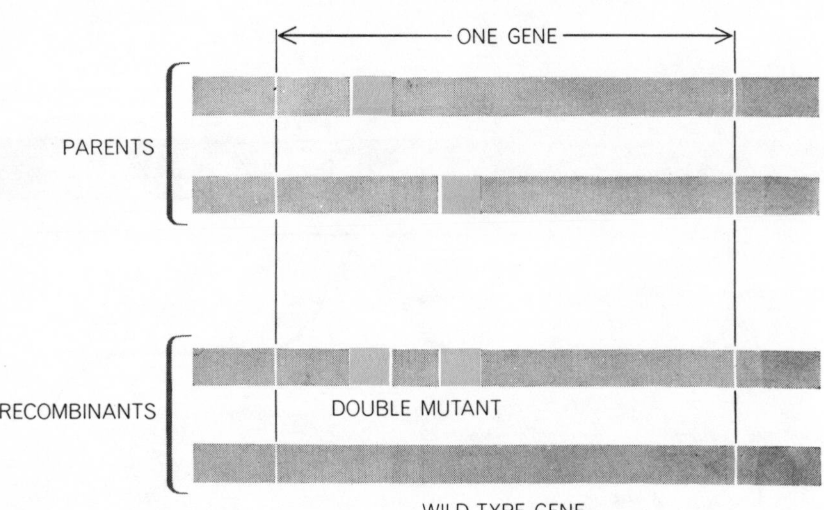

GENETIC RECOMBINATION provides the means for studying mutations. Colored squares represent mutations in the chromosome (DNA molecule) of the T4 virus. Through genetic recombination, the progeny can inherit the defects of both parents or of neither.

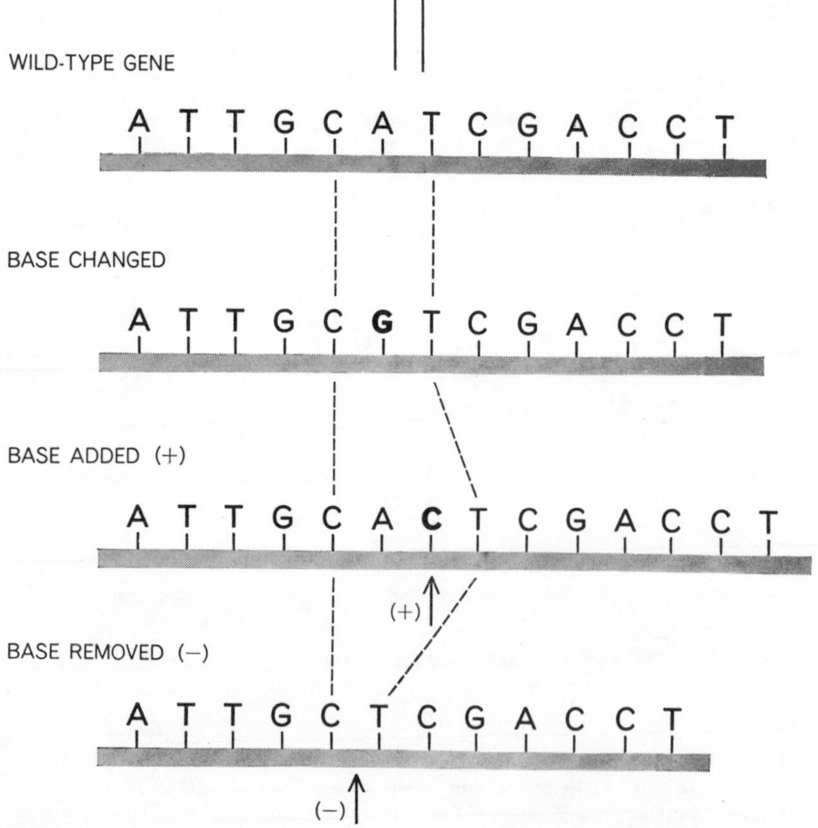

TWO CLASSES OF MUTATION result from introducing defects in the sequence of bases (A, T, G, C) that are attached to the backbone of the DNA molecule. In one class a base is simply changed from one into another, as A into G. In the second class a base is added or removed. Four bases are adenine (A), thymine (T), guanine (G) and cytosine (C).

To understand this we shall have to go back to the genetic code. The simplest sort of code would be one in which a small group of bases stands for one particular acid. This group can scarcely be a pair, since this would yield only 4 × 4, or 16, possibilities, and at least 20 are needed. More likely the shortest code group is a triplet, which would provide 4 × 4 × 4, or 64, possibilities. A small group of bases that codes one amino acid has recently been named a codon.

The first definite coding scheme to be proposed was put forward eight years ago by the physicist George Gamow, now at the University of Colorado. In this code adjacent codons overlap as illustrated on the following page. One consequence of such a code is that only certain amino acids can follow others. Another consequence is that a change in a single base leads to a change in three adjacent amino acids. Evidence gathered since Gamow advanced his ideas makes an overlapping code appear unlikely. In the first place there seems to be no restriction of amino acid sequence in any of the proteins so far examined. It has also been shown that typical mutations change only a single amino acid in the polypeptide chain of a protein. Although it is theoretically possible that the genetic code may be partly overlapping, it is more likely that adjacent codons do not overlap at all.

Since the backbone of the DNA molecule is completely regular, there is nothing to mark the code off into groups of three bases, or into groups of any other size. To solve this difficulty various ingenious solutions have been proposed. It was thought, for example, that the code might be designed in such a way that if the wrong set of triplets were chosen, the message would always be complete nonsense and no protein would be produced. But it now looks as if the most obvious solution is the correct one. That is, the message begins at a fixed starting point, probably one end of the gene, and is simply read three bases at a time. Notice that if the reading started at the wrong point, the message would fall into the wrong sets of three and would then be hopelessly incorrect. In fact, it is easy to see that while there is only one correct reading for a triplet code, there are two incorrect ones.

If this idea were right, it would immediately explain why the addition or the deletion of a base in most parts of the gene would make the gene completely nonfunctional, since the reading of the genetic message from that point onward would be totally wrong. Now, although our single mutations were always without function, we found that if we put certain pairs of them together, the gene would work. (In point of fact we picked up many of our functioning double mutations by starting with a nonfunctioning mutation and selecting for the rare second mutation that restored gene activity, but this does not affect our argument.) This enabled us to classify all our mutations as being either plus or minus. We found that by using the following rules we could always predict the behavior of any pair we put together in the same gene. First, if plus is combined with plus, the combination is nonfunctional. Second, if minus is combined with minus, the result is nonfunctional. Third, if plus is combined with minus, the combination is nonfunctional if the pair is too widely separated and functional if the pair is close together.

The interesting case is the last one. We could produce a gene that functioned, at least to some extent, if we combined a plus mutation with a minus mutation, provided that they were not too far apart.

To make it easier to follow, let us assume that the mutations we called plus really had an extra base at some point and that those we called minus had lost a base. (Proving this to be the case is rather difficult.) One can see that, starting from one end, the message would be read correctly until the extra base was reached; then the reading would get out of phase and the message would be wrong until the missing base was reached, after which the message would come back into phase again. Thus the genetic message would not be wrong over a long stretch but only over the short distance between the plus and the minus. By the same sort of argument one can see that for a triplet code the combination plus with plus or minus with minus should never work [*see illustration on following page*].

We were fortunate to do most of our work with mutations at the left-hand end of the B gene of the *r*II region. It appears that the function of this part of the gene may not be too important, so that it may not matter if part of the genetic message in the region is incorrect. Even so, if the plus and minus are too far apart, the combination will not work.

Nonsense Triplets

To understand this we must go back once again to the code. There are 64 possible triplets but only 20 amino acids to be coded. Conceivably two or more triplets may stand for each amino acid. On the other hand, it is reasonable to expect that at least one or two triplets may not represent an amino acid at all but have some other meaning, such as "Begin here" or "End here." Although such hypothetical triplets may have a meaning of some sort, they have been named nonsense triplets. We surmised that sometimes the misreading produced in the region lying between a plus and a minus mutation might by chance give rise to a nonsense triplet, in which case the gene might not work.

We investigated a number of plus-with-minus combinations in which the distance between plus and minus was relatively short and found that certain combinations were indeed inactive when we might have expected them to function. Presumably an intervening nonsense triplet was to blame. We also found cases in which a plus followed by a minus worked but a minus followed by a plus did not, even though the two mutations appeared to be at the same sites, although in reverse sequence. As I have indicated, there are two wrong ways to read a message; one arises if the plus is to the left of the minus, the other if the plus is to the right of the minus. In cases where plus with minus gave rise to an active gene but minus with plus did not, even when the mutations evidently occupied the same pairs of sites, we concluded that the intervening misreading produced a nonsense triplet in one case but not in the other. In confirmation of this hypothesis we have been able to modify such nonsense triplets by mutagens that turn one base into another, and we have thereby restored the gene's activity. At the same time we have been able to locate the position of the nonsense triplet.

Recently we have undertaken one

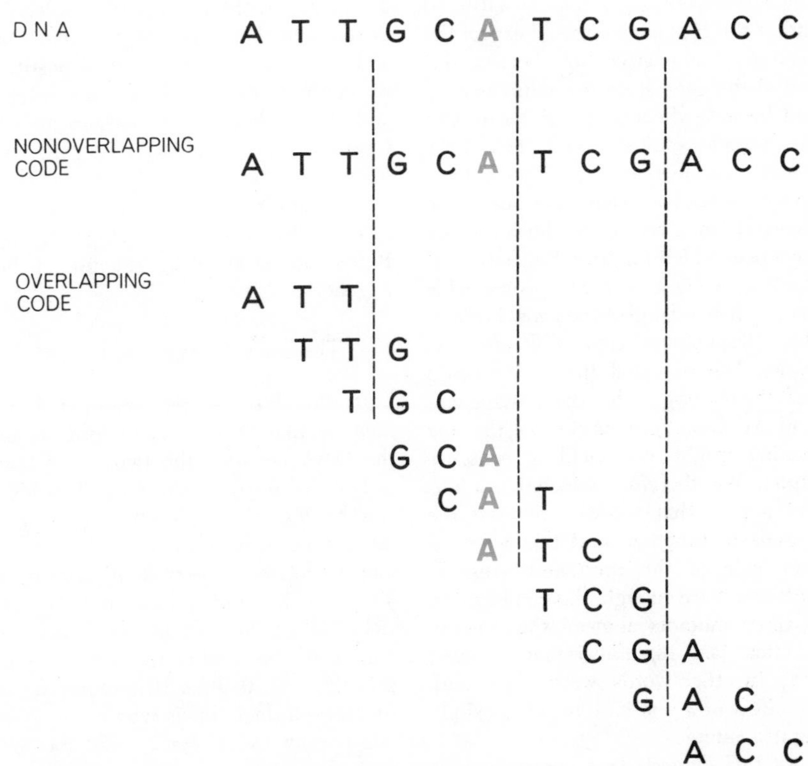

PROPOSED CODING SCHEMES show how the sequence of bases in DNA can be read. In a nonoverlapping code, which is favored by the author, code groups are read in simple sequence. In one type of overlapping code each base appears in three successive groups.

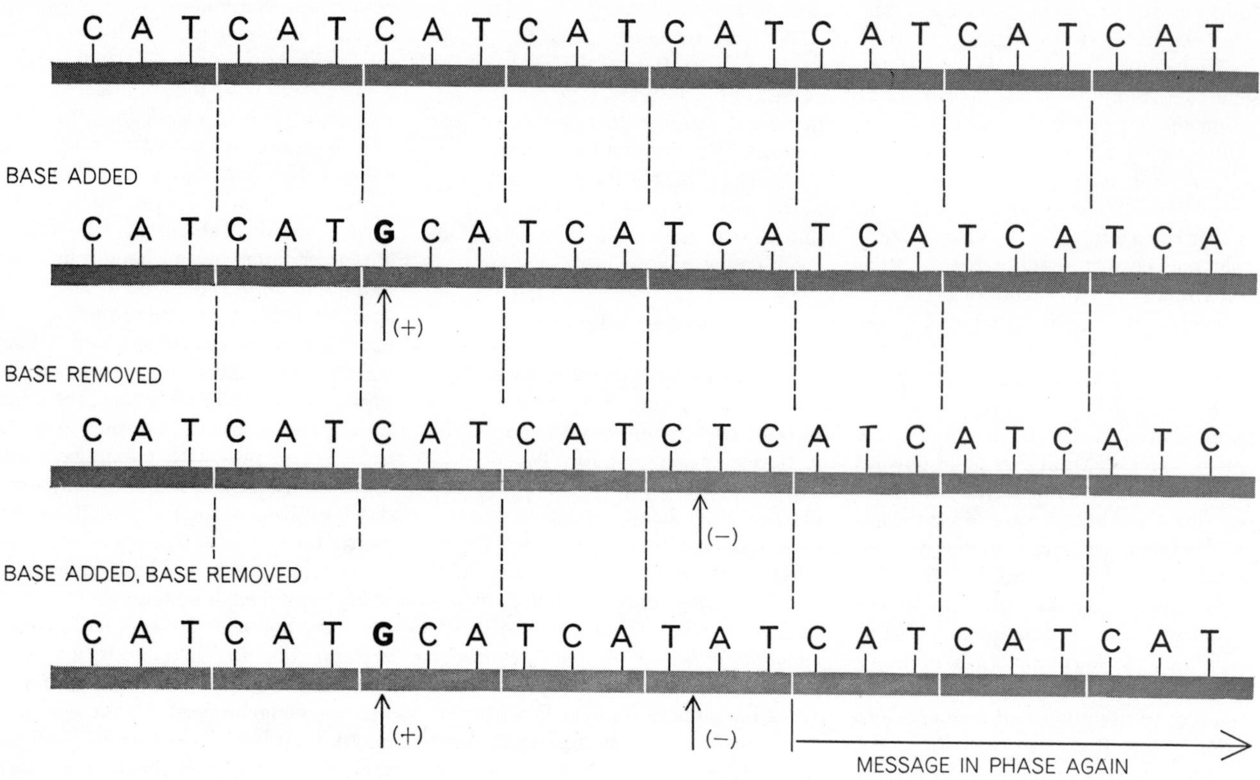

EFFECT OF MUTATIONS that add or remove a base is to shift the reading of the genetic message, assuming that the reading begins at the left-hand end of the gene. The hypothetical message in the wild-type gene is CAT, CAT... Adding a base shifts the reading to TCA, TCA... Removing a base makes it ATC, ATC... Addition and removal of a base puts the message in phase again.

other rather amusing experiment. If a single base were changed in the left-hand end of the B gene, we would expect the gene to remain active, both because this end of the gene seems to be unessential and because the reading of the rest of the message is not shifted. In fact, if the B gene remained active, we would have no way of knowing that a base had been changed. In a few cases, however, we have been able to destroy the activity of the B gene by a base change traceable to the left-hand end of the gene. Presumably the change creates a nonsense triplet. We reasoned that if we could shift the reading so that the message was read in different groups of three, the new reading might not yield a nonsense triplet. We therefore selected a minus and a plus that together allowed the B gene to function, and that were on each side of the presumed nonsense mutation. Sure enough, this combination of three mutants allowed the gene to function [*see top illustration on page 218*]. In other words, we could abolish the effect of a nonsense triplet by shifting its reading.

All this suggests that the message is read from a fixed point, probably from one end. Here the question arises of how one gene ends and another begins, since in our picture there is nothing on the backbone of the long DNA molecule to separate them. Yet the two genes A and B are quite distinct. It is possible to measure their function separately, and Benzer has shown that no matter what mutation is put into the A gene, the B function is not affected, provided that the mutation is wholly within the A gene. In the same way changes in the B gene do not affect the function of the A gene.

The Space between the Genes

It therefore seems reasonable to imagine that there is something about the DNA between the two genes that isolates them from each other. This idea can be tested by experiments with a mutant T4 in which part of the *r*II region is deleted. The mutant, known as T4 1589, has lost a large part of the right end of the A gene and a smaller part of the left end of the B gene. Surprisingly the B gene still shows some function; in fact this is why we believe this part of the B gene is not too important.

Although we describe this mutation as a deletion, since genetic mapping shows that a large piece of the genetic information in the region is missing, it does not mean that physically there is a gap. It seems more likely that DNA is all one piece but that a stretch of it has been left out. It is only by comparing it with the complete version—the wild type—that one can see a piece of the message is missing.

We have argued that there must be a small region between the genes that separates them. Consequently one would predict that if this segment of the DNA were missing, the two genes would necessarily be joined. It turns out that it is quite easy to test this prediction, since by genetic methods one can construct double mutants. We therefore combined one of our acridine mutations, which in this case was near the beginning of the A gene, with the deletion 1589. Without the deletion present the acridine mutation had no effect on the B function, which showed that the genes were indeed separate. But when 1589 was there as well, the B function was completely destroyed [*see top illustration on next page*]. When the genes were joined, a change far away in the A gene knocked out the B gene completely. This strongly suggests that the reading proceeds from one end.

We tried other mutations in the A

gene combined with 1589. All the acridine mutations we tried knocked out the B function, whether they were plus or minus, but a pair of them (plus with minus) still allowed the B gene to work. On the other hand, in the case of the other type of mutation (which we believe is due to the change of a base and not to one being added or subtracted) about half of the mutations allowed the B gene to work and the other half did not. We surmise that the latter are nonsense mutations, and in fact Benzer has recently been using this test as a definition of nonsense.

Of course, we do not know exactly what is happening in biochemical terms. What we suspect is that the two genes, instead of producing two separate pieces of messenger RNA, produce a single piece, and that this in turn produces a protein with a long polypeptide chain, one end of which has the amino acid sequence of part of the presumed A protein and the other end of which has most of the B protein sequence—enough to give some B function to the combined molecule although the A function has been lost. The concept is illustrated schematically at the bottom of this page. Eventually it should be possible to check the prediction experimentally.

How the Message Is Read

So far all the evidence has fitted very well into the general idea that the message is read off in groups of three, starting at one end. We should have got the same results, however, if the message had been read off in groups of four, or indeed in groups of any larger size. To test this we put not just two of our acridine mutations into one gene but three of them. In particular we put in three with the same sign, such as plus with plus with plus, and we put them fairly close together. Taken either singly or in pairs, these mutations will destroy the function of the B gene. But when all three are placed in the same gene, the B function reappears. This is clearly a remarkable result: two blacks will not make a white but three will. Moreover, we have obtained the same result with several different combinations of this type and with several of the type minus with minus with minus.

The explanation, in terms of the ideas described here, is obvious. One plus will put the reading out of phase. A second plus will give the other wrong reading. But if the code is a triplet code, a third plus will bring the message back into phase again, and from then on to the end it will be read correctly. Only between

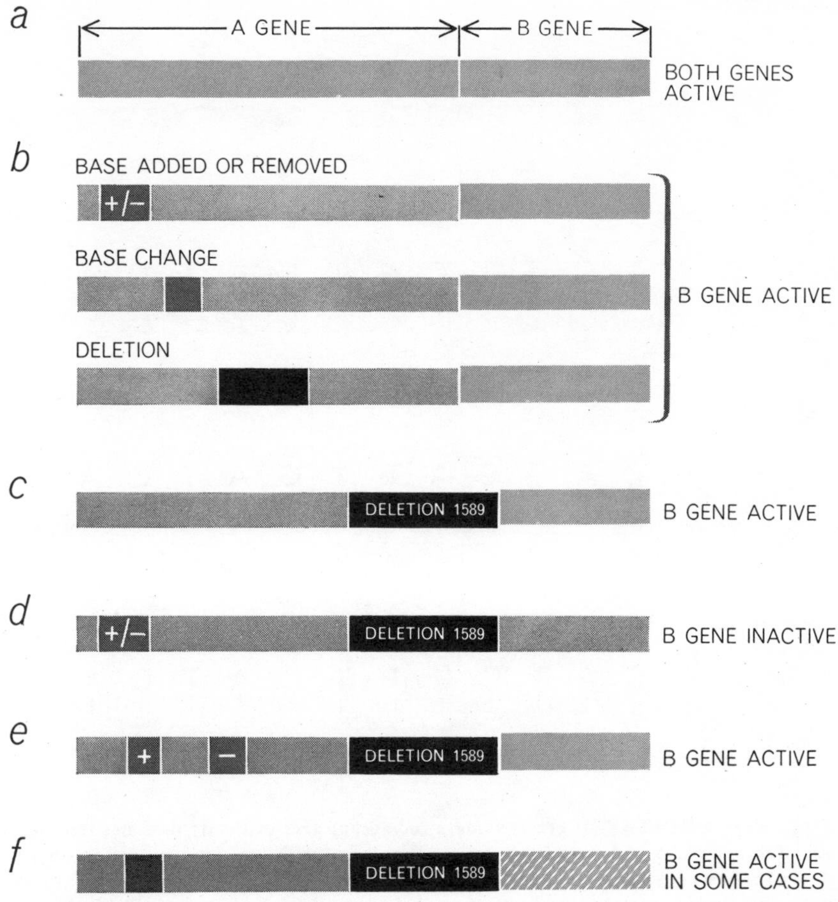

DELETION JOINING TWO GENES makes the B gene vulnerable to mutations in the A gene. The messages in two wild-type genes (*a*) are read independently, beginning at the left end of each gene. Regardless of the kind of mutation in A, the B gene remains active (*b*). The deletion known as 1589 inactivates the A gene but leaves the B gene active (*c*). But now alterations in the A gene will often inactivate the B gene, showing that the two genes have been joined in some way and are read as if they were a single gene (*d, e, f*).

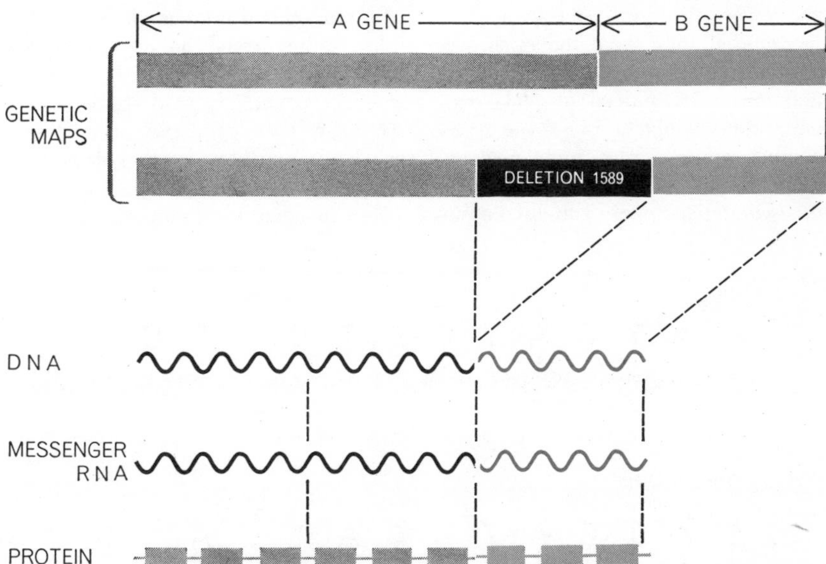

PROBABLE EFFECT OF DELETION 1589 is to produce a mixed protein with little or no A-gene activity but substantial B activity. Although the conventional genetic map shows the deletion as a gap, the DNA molecule itself is presumably continuous but shortened. In virus replication the genetic message in DNA is transcribed into a molecule of ribonucleic acid, called messenger RNA. This molecule carries the message to cellular particles known as ribosomes, where protein is synthesized, following instructions coded in the DNA.

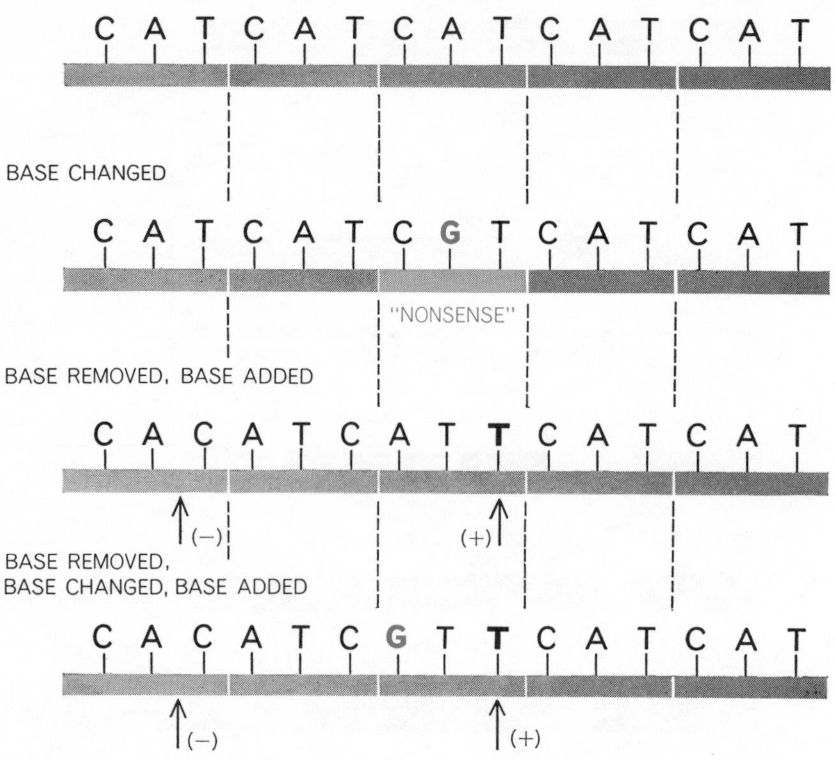

NONSENSE MUTATION is one creating a code group that evidently does not represent any of the 20 amino acids found in proteins. Thus it makes the gene inactive. In this hypothetical case a nonsense triplet, CGT, results when an A in the wild-type gene is changed to G. The nonsense triplet can be eliminated if the reading is shifted to put the G in a different triplet. This is done by recombining the inactive gene with one containing a minus-with-plus combination. In spite of three mutations, the resulting gene is active.

the pluses will the message be wrong [*see illustration below*].

Notice that it does not matter if plus is really one extra base and minus is one fewer; the conclusions would be the same if they were the other way around. In fact, even if some of the plus mutations were indeed a single extra base, others might be two fewer bases; in other words, a plus might really be minus minus. Similarly, some of the minus mutations might actually be plus plus. Even so they would still fit into our scheme.

Although the most likely explanation is that the message is read three bases at a time, this is not completely certain. The reading could be in multiples of three. Suppose, for example, that the message is actually read six bases at a time. In that case the only change needed in our interpretation of the facts is to assume that all our mutants have been changed by an even number of bases. We have some weak experimental evidence that this is unlikely. For instance, we can combine the mutant 1589 (which joins the genes) with medium-sized deletions in the A cistron. Now, if deletions were random in length, we should expect about a third of them to allow the B function to be expressed if the message is indeed read three bases at a time, since those deletions that had lost an exact multiple of three bases should allow the B gene to function. By the same reasoning only a sixth of them should work (when combined with 1589) if the reading proceeds six at a time. Actually we find that the B gene is active in a little more than a third. Taking all the evidence together, however, we find that although three is the most likely coding unit, we cannot completely rule out multiples of three.

There is one other general conclusion we can draw about the genetic code. If we make a rough guess as to the actual size of the B gene (by comparing it with another gene whose size is known approximately), we can estimate how many bases can lie between a plus with minus combination and still allow the B gene to function. Knowing also the frequency with which nonsense triplets are created in the misread region between the plus and minus, we can get some idea whether there are many such triplets or only a few. Our calculation suggests that nonsense triplets are not too common. It seems, in other words, that most of the 64 possible triplets, or codons, are not nonsense, and therefore they stand for amino acids. This implies that probably more than one codon can stand for one amino acid. In the jargon of the trade, a code in which this is true is "degenerate."

In summary, then, we have arrived at three general conclusions about the genetic code:

1. The message is read in nonover-

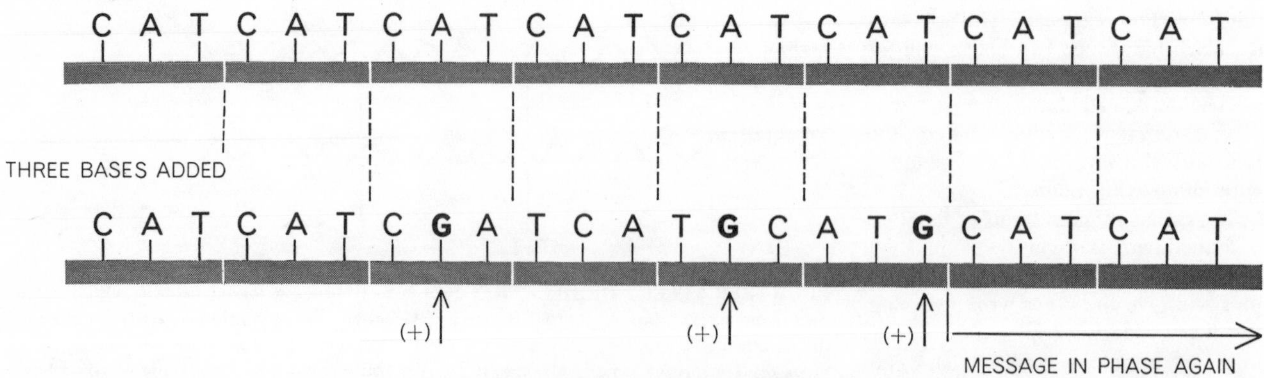

TRIPLE MUTATION in which three bases are added fairly close together spoils the genetic message over a short stretch of the gene but leaves the rest of the message unaffected. The same result can be achieved by the deletion of three neighboring bases.

lapping groups from a fixed point, probably from one end. The starting point determines that the message is read correctly into groups.

2. The message is read in groups of a fixed size that is probably three, although multiples of three are not completely ruled out.

3. There is very little nonsense in the code. Most triplets appear to allow the gene to function and therefore probably represent an amino acid. Thus in general more than one triplet will stand for each amino acid.

It is difficult to see how to get around our first conclusion, provided that the B gene really does code a polypeptide chain, as we have assumed. The second conclusion is also difficult to avoid. The third conclusion, however, is much more indirect and could be wrong.

Finally, we must ask what further evidence would really clinch the theory we have presented here. We are continuing to collect genetic data, but I doubt that this will make the story much more convincing. What we need is to obtain a protein, for example one produced by a double mutation of the form plus with minus, and then examine its amino acid sequence. According to conventional theory, because the gene is altered in only two places the amino acid sequences also should differ only in the two corresponding places. According to our theory it should be altered not only at these two places but also at all places in between. In other words, a whole string of amino acids should be changed. There is one protein, the lysozyme of the T4 phage, that is favorable for such an approach, and we hope that before long workers in the U.S. who have been studying phage lysozyme will confirm our theory in this way.

The same experiment should also be useful for checking the particular code schemes worked out by Nirenberg and Matthaei and by Ochoa and his colleagues. The phage lysozyme made by the wild-type gene should differ over only a short stretch from that made by the plus-with-minus mutant. Over this stretch the amino acid sequence of the two lysozyme variants should correspond to the same sequence of bases on the DNA but should be read in different groups of three.

If this part of the amino acid sequence of both the wild-type and the altered lysozyme could be established, one could check whether or not the codons assigned to the various amino acids did indeed predict similar sequences for that part of the DNA between the base added and the base removed.

17

Gene Structure and Protein Structure

by Charles Yanofsky
May 1967

A linear correspondence between these two chainlike molecules was postulated more than a dozen years ago. Here is how the correspondence was finally demonstrated

The present molecular theory of genetics, known irreverently as "the central dogma," is now 14 years old. Implicit in the theory from the outset was the notion that genetic information is coded in linear sequence in molecules of deoxyribonucleic acid (DNA) and that the sequence directly determines the linear sequence of amino acid units in molecules of protein. In other words, one expected the two molecules to be colinear. The problem was to prove that they were.

Over the same 14 years, as a consequence of an international effort, most of the predictions of the central dogma have been verified one by one. The results were recently summarized in these pages by F. H. C. Crick, who together with James D. Watson proposed the helical, two-strand structure for DNA on

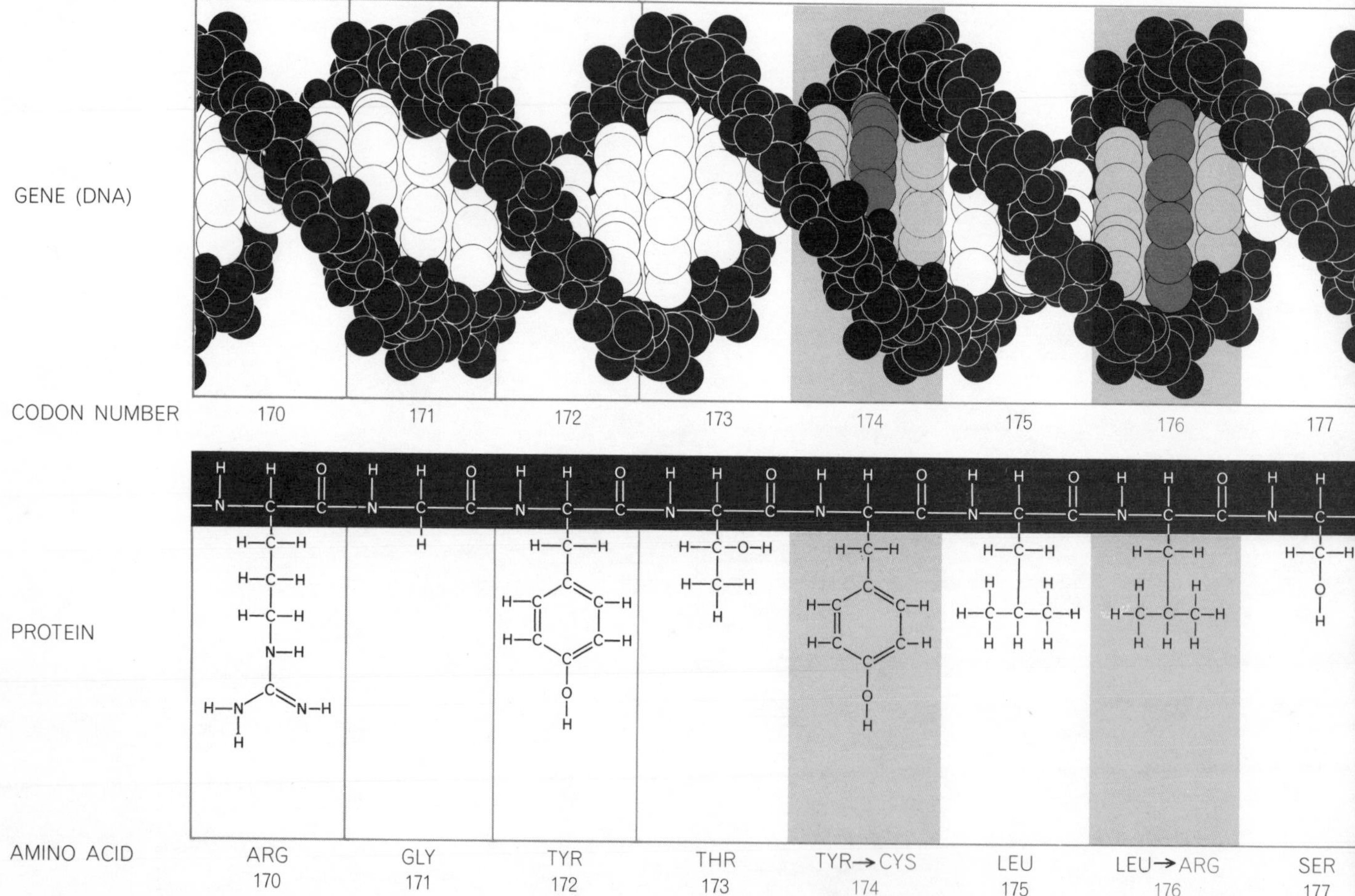

STRUCTURES OF GENE AND PROTEIN have been shown to bear a direct linear correspondence by the author and his colleagues at Stanford University. They demonstrated that a particular sequence of coding units (codons) in the genetic molecule deoxyribonucleic acid, or DNA (*top*), specifies a corresponding sequence of amino acid units in the structure of a protein molecule (*bottom*). In the DNA molecule depicted here the black spheres represent repeating units of deoxyribose sugar and phosphate, which form the helical backbones of the two-strand molecule. The white spheres connecting the two strands represent complementary pairs of the four kinds of base that provide the "letters" in which the genetic message is written. A sequence of three bases attached to one strand of DNA

which the central dogma is based [see "The Genetic Code: III," by F. H. C. Crick; SCIENTIFIC AMERICAN Offprint 1052]. Here I shall describe in somewhat more detail how our studies at Stanford University demonstrated the colinearity of genetic structure (as embodied in DNA) and protein structure.

Let me begin with a brief review. The molecular subunits that provide the "letters" of the code alphabet in DNA are the four nitrogenous bases adenine (A), guanine (G), cytosine (C) and thymine (T). If the four letters were taken in pairs, they would provide only 16 different code words—too few to specify the 20 different amino acids commonly found in protein molecules. If they are taken in triplets, however, the four letters can provide 64 different code words, which would seem too many for the efficient specification of the 20 amino acids. Accordingly it was conceivable that the cell might employ fewer than the 64 possible triplets. We now know that nature not only has selected the triplet code but also makes use of most (if not all) of the 64 triplets, which are called codons. Each amino acid but two (tryptophan and methionine) are specified by at least two different codons, and a few amino acids are specified by as many as six codons. It is becoming clear that the living cell exploits this redundancy in subtle ways. Of the 64 codons, 61 have been shown to specify one or another of the 20 amino acids. The remaining three can act as "chain terminators," which signal the end of a genetic message.

A genetic message is defined as the amount of information in one gene; it is the information needed to specify the complete amino acid sequence in one polypeptide chain. This relation, which underlies the central dogma, is sometimes expressed as the one-gene-one-enzyme hypothesis. It was first clearly enunciated by George W. Beadle and Edward L. Tatum, as a result of their studies with the red bread mold *Neurospora crassa* around 1940. In some cases a single polypeptide chain constitutes a complete protein molecule, which often acts as an enzyme, or biological catalyst. Frequently, however, two or more polypeptide chains must join together in order to form an active protein. For example, tryptophan synthetase, the enzyme we used in our colinearity studies, consists of four polypeptide chains: two alpha chains and two beta chains.

How might one establish the colinearity of codons in DNA and amino acid units in a polypeptide chain? The most direct approach would be to separate the two strands of DNA obtained from some organism and determine the base sequence of that portion of a strand which is presumed to be colinear with the amino acid sequence of a particular protein. If the amino acid sequence of the protein were not already known, it too would have to be established. One could then write the two sequences in adjacent columns and see if the same codon (or its synonym) always appeared adjacent to a particular amino acid. If it

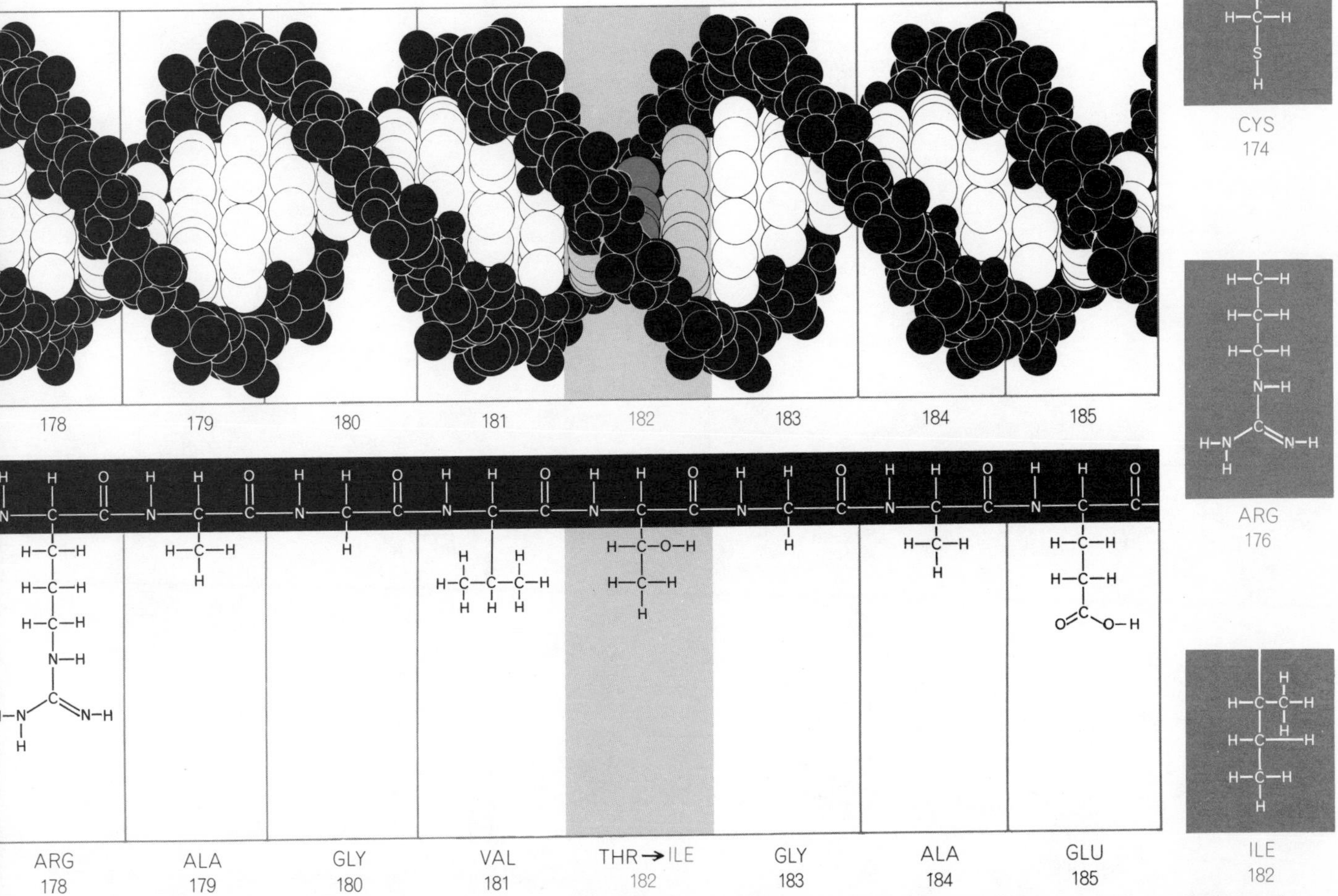

is a codon and specifies one amino acid. The amino acid sequence illustrated here is the region from position 170 through 185 in the A protein of the enzyme tryptophan synthetase produced by the bacterium *Escherichia coli*. It was found that mutations in the A gene of *E. coli* altered the amino acids at three places (*174, 176 and 182*) in this region of the A protein. (A key to the amino acid abbreviations can be found on page 223). The three amino acids that replace the three normal ones as a result of mutation are shown at the extreme right. Each replacement is produced by a mutation at one site (*dark color*) in the DNA of the *A* gene. In all, the author and his associates correlated mutations at eight sites in the *A* gene with alterations in the *A* protein.

did, a colinear relation would be established. Unfortunately this direct approach cannot be taken because so far it has not been possible to isolate and identify individual genes. Even if one could isolate a single gene that specified a polypeptide made up of 150 amino acids (and not many polypeptides are that small), one would have to determine the sequence of units in a DNA strand consisting of some 450 bases.

It was necessary, therefore, to consider a more feasible way of attacking the problem. An approach that immediately suggests itself to a geneticist is to construct a genetic map, which is a representation of the information contained in the gene, and see if the map can be related to protein structure. A genetic map is constructed solely on the basis of information obtained by crossing individual organisms that differ in two or more hereditary respects (a refinement of the technique originally

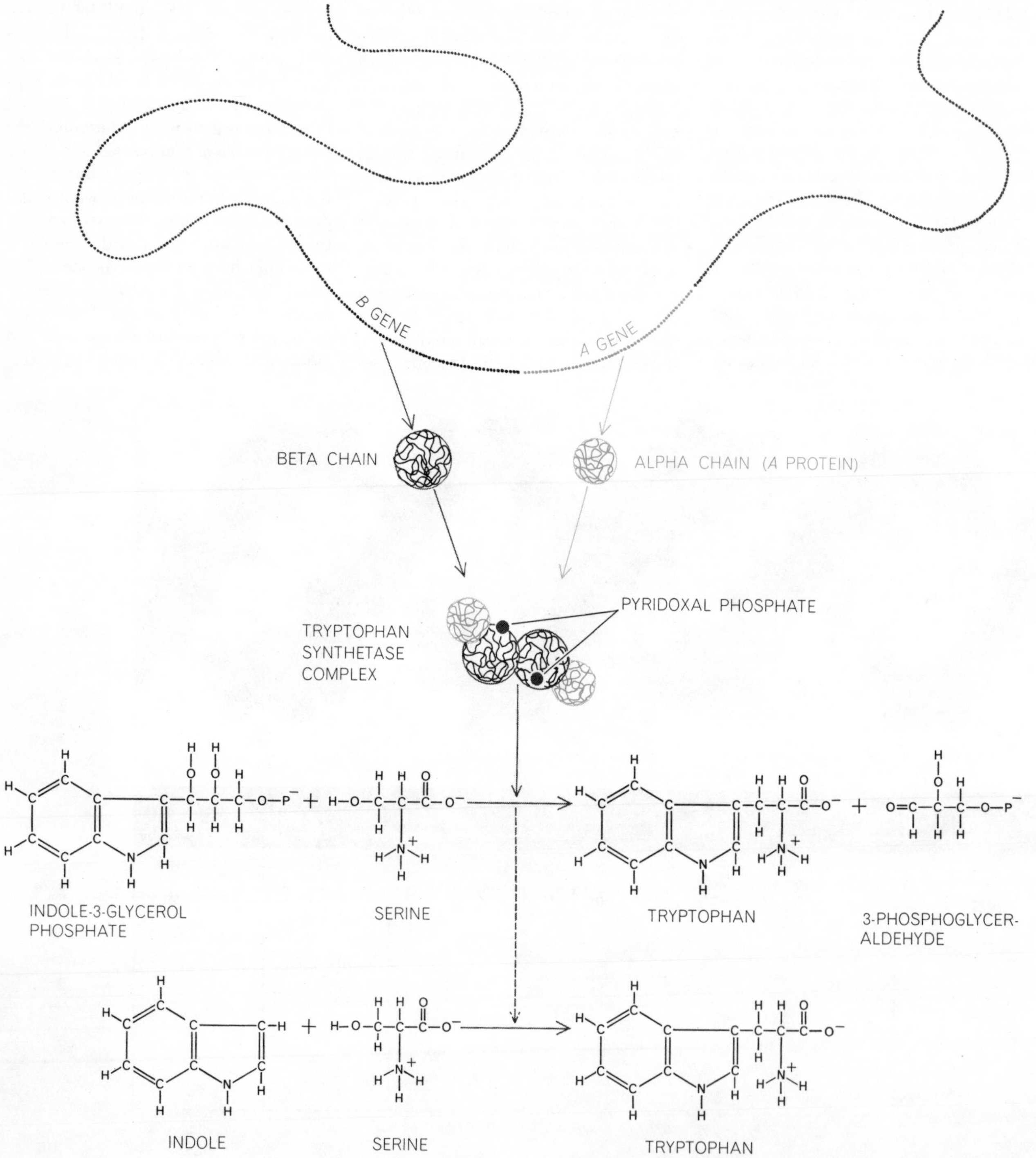

GENETIC CONTROL OF CELL'S CHEMISTRY is exemplified by the two genes in *E. coli* that carry the instructions for making the enzyme tryptophan synthetase. The enzyme is actually a complex of four polypeptide chains: two alpha chains and two beta chains. The alpha chain is the *A* protein in which changes produced by mutations in the *A* gene have provided the evidence for gene-protein colinearity. One class of *A*-protein mutants retains the ability to associate with beta chains but the complex is no longer able to catalyze the normal biochemical reaction: the conversion of indole-3-glycerol phosphate and serine to tryptophan and 3-phosphoglyceraldehyde. But the complex can still catalyze a simpler nonphysiological reaction: the conversion of indole and serine to tryptophan.

used by Gregor Mendel to demonstrate how characteristics are inherited).

By using bacteria and bacterial viruses in such studies one can catalogue the results of crosses involving millions of individual organisms and thereby deduce the actual distances separating the sites of mutational changes in a single gene. The distances are inferred from the frequency with which parent organisms, each with at least one mutation in the same gene, give rise to offspring in which neither mutation is present. As a result of the recombination of genetic material the offspring can inherit a gene that is assembled from the mutation-free portions of each parental gene. If the mutational markers lie far apart on the parental genes, recombination will frequently produce mutation-free progeny. If the markers are close together, mutation-free progeny will be rare [*see bottom illustration on next page*].

In his elegant studies with the "*r*II" region of the chromosome of the bacterial virus designated T4, Seymour Benzer, then at Purdue University, showed that the number of genetically distinguishable mutation sites on the map of the gene approaches the estimated number of base pairs in the DNA molecule corresponding to that gene. (Mutations involve pairs of bases because the bases in each of the two entwined strands of the DNA molecule are paired with and are complementary to the bases in the other strand. If a mutation alters one base in the DNA molecule, its partner is eventually changed too during DNA replication.) Benzer also showed that the only type of genetic map consistent with his data is a map on which the sites altered by mutation are arranged linearly. Subsequently A. D. Kaiser and David Hogness of Stanford University demonstrated with another bacterial virus that there is a linear correspondence between the sites on a genetic map and the altered regions of a DNA molecule isolated from the virus. Thus there is direct experimental evidence indicating that the genetic map is a valid representation of DNA structure and that the map can be employed as a substitute for information about base sequence.

This, then, provided the basis of our approach. We would pick a suitable organism and isolate a large number of mutant individuals with mutations in the same gene. From recombination studies we would make a fine-structure genetic map relating the sites of the mutations. In addition we would have to be able to isolate the protein specified by that gene and determine its amino acid sequence. Finally we would have to analyze the protein produced by each mutant (assuming a protein were still produced) in order to find the position of the amino acid change brought about in its amino acid sequence by the mutation. If gene structure and protein structure were colinear, the positions at which amino acid changes occur in the protein should be in the same order as the positions of the corresponding mutationally altered sites on the genetic map. Although this approach to the question of colinearity would require a great deal of work and much luck, it was logical and experimentally feasible. Several research groups besides our own set out to find a suitable system for a study of this kind.

The essential requirement of a suitable system was that a genetically

ALA	ALANINE	GLY	GLYCINE	PRO	PROLINE
ARG	ARGININE	HIS	HISTIDINE	SER	SERINE
ASN	ASPARAGINE	ILE	ISOLEUCINE	THR	THREONINE
ASP	ASPARTIC ACID	LEU	LEUCINE	TRP	TRYPTOPHAN
CYS	CYSTEINE	LYS	LYSINE	TYR	TYROSINE
GLN	GLUTAMINE	MET	METHIONINE	VAL	VALINE
GLU	GLUTAMIC ACID	PHE	PHENYLALANINE		

AMINO ACID ABBREVIATIONS identify the 20 amino acids commonly found in all proteins. Each amino acid is specified by a triplet codon in the DNA molecule (*see below*).

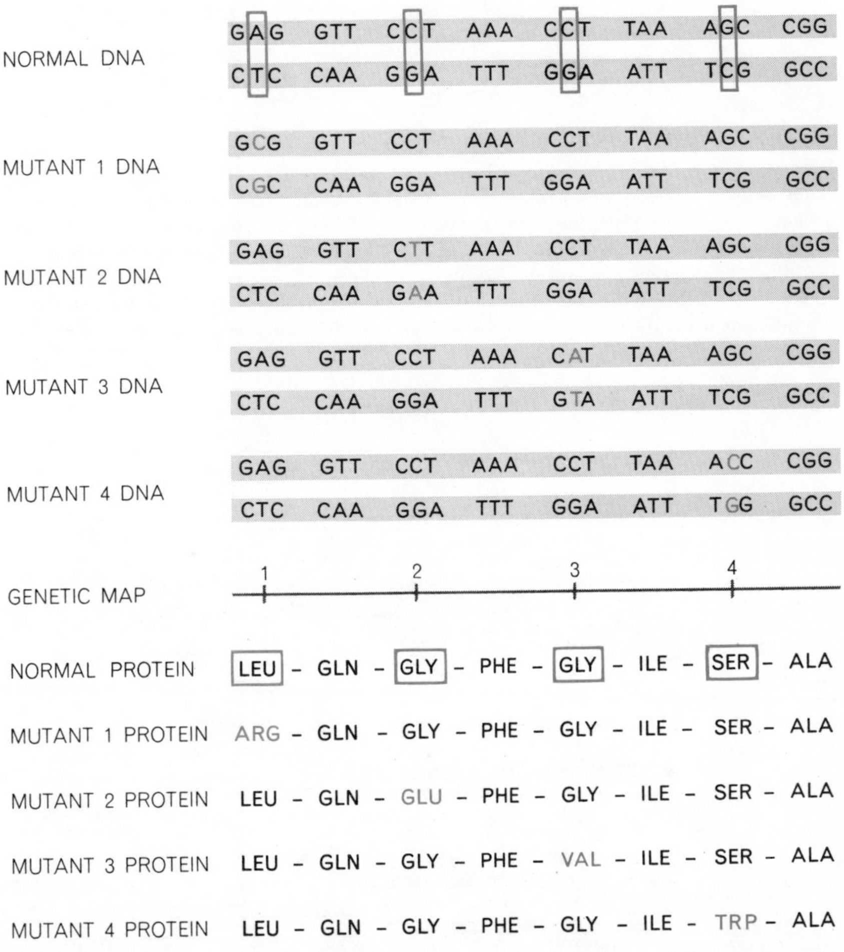

NORMAL DNA	GAG	GTT	CCT	AAA	CCT	TAA	AGC	CGG
	CTC	CAA	GGA	TTT	GGA	ATT	TCG	GCC
MUTANT 1 DNA	GCG	GTT	CCT	AAA	CCT	TAA	AGC	CGG
	CGC	CAA	GGA	TTT	GGA	ATT	TCG	GCC
MUTANT 2 DNA	GAG	GTT	CTT	AAA	CCT	TAA	AGC	CGG
	CTC	CAA	GAA	TTT	GGA	ATT	TCG	GCC
MUTANT 3 DNA	GAG	GTT	CCT	AAA	CAT	TAA	AGC	CGG
	CTC	CAA	GGA	TTT	GTA	ATT	TCG	GCC
MUTANT 4 DNA	GAG	GTT	CCT	AAA	CCT	TAA	ACC	CGG
	CTC	CAA	GGA	TTT	GGA	ATT	TGG	GCC
GENETIC MAP	1		2		3		4	
NORMAL PROTEIN	LEU	GLN	GLY	PHE	GLY	ILE	SER	ALA
MUTANT 1 PROTEIN	ARG	GLN	GLY	PHE	GLY	ILE	SER	ALA
MUTANT 2 PROTEIN	LEU	GLN	GLU	PHE	GLY	ILE	SER	ALA
MUTANT 3 PROTEIN	LEU	GLN	GLY	PHE	VAL	ILE	SER	ALA
MUTANT 4 PROTEIN	LEU	GLN	GLY	PHE	GLY	ILE	TRP	ALA

GENETIC MUTATIONS can result from the alteration of a single base in a DNA codon. The letters stand for the four bases: adenine (A), thymine (T), guanine (G) and cytosine (C). Since the DNA molecule consists of two complementary strands, a base change in one strand involves a complementary change in the second strand. In the four mutant DNA sequences shown here (*top*) a pair of bases (*color*) is different from that in the normal sequence. By genetic studies one can map the sequence and approximate spacing of the four mutations (*middle*). By chemical studies of the proteins produced by the normal and mutant DNA sequences (*bottom*) one can establish the corresponding amino acid changes.

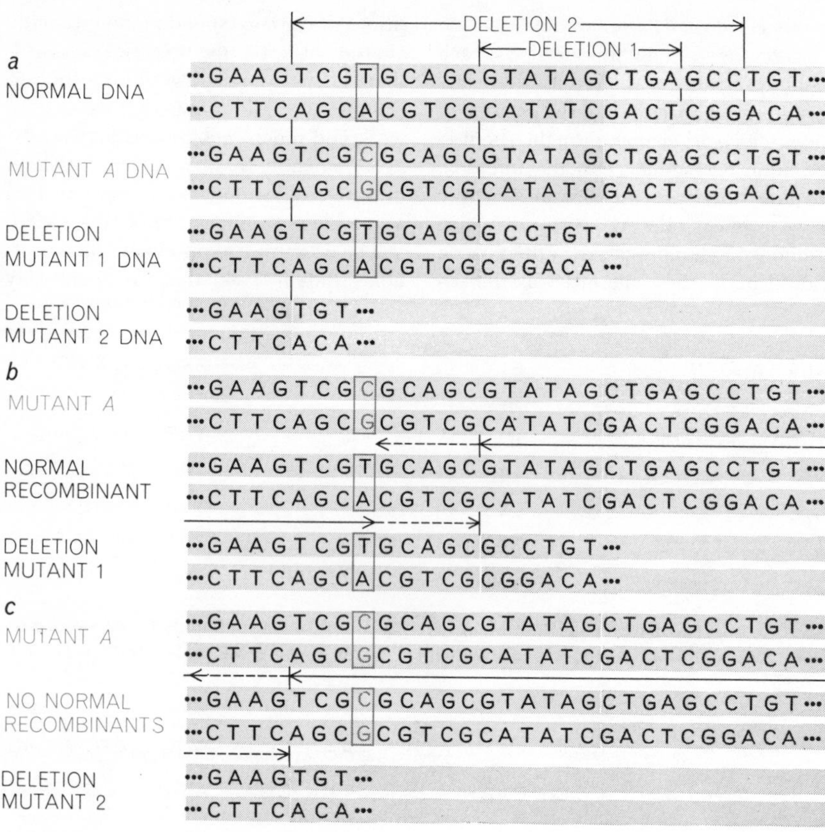

"DELETION" MUTANTS provide one approach to making a genetic map. Here (*a*) normal DNA and mutant *A* differ by only one base pair (*C–G* has replaced *T–A*) in a certain portion of the *A* gene (*colored area*). In deletion mutant 1 a sequence of 10 base pairs, including six pairs from the *A* gene, has been spontaneously deleted. In deletion mutant 2, 22 base pairs, including 15 pairs from the *A* gene, have been deleted. By crossing mutant *A* with the two different deletion mutants in separate experiments (*b*, *c*), one can tell whether the mutated site (*C–G*) in the *A* gene falls inside or outside the deleted regions. A normal-type recombinant will appear (*b*) only if the altered base pair falls outside the deleted region.

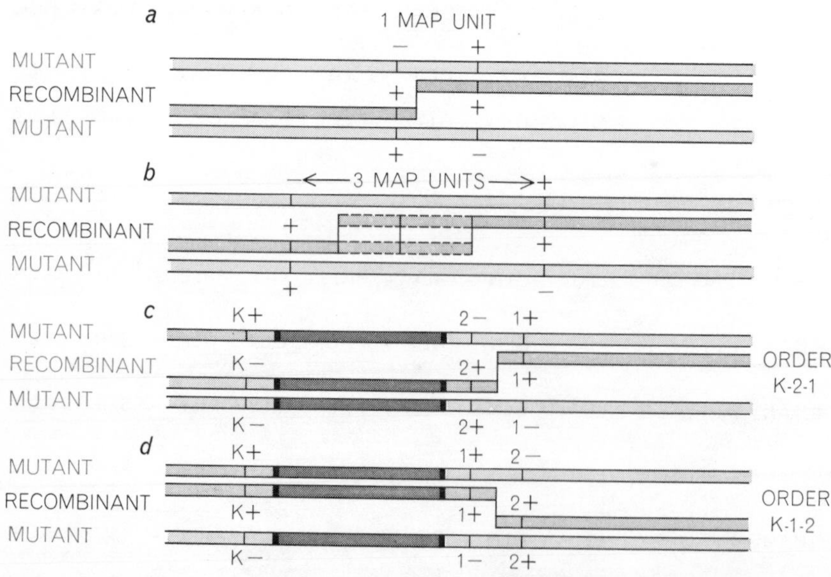

OTHER MAPPING METHODS involve determination of recombination frequency (*a*, *b*) and the distribution of outside markers (*c*, *d*). The site of a mutational alteration is indicated by "—," the corresponding unaltered site by "+." If the altered sites are widely spaced (*b*), normal recombinants will appear more often than if the altered sites are close together (*a*). In the second method the mutants are linked to another gene that is either normal (K^+) or mutated (K^-). Recombinant strains that contain 1^+ and 2^+ will carry the K^- gene if the correct order is *K*–2–1. They will carry the K^+ gene if the order is *K*–1–2.

mappable gene should specify a protein whose amino acid sequence could be determined. Since no such system was known we had to gamble on a choice of our own. Fortunately we were studying at the time how the bacterium *Escherichia coli* synthesizes the amino acid tryptophan. Irving Crawford and I observed that the enzyme that catalyzed the last step in tryptophan synthesis could be readily separated into two different protein species, or subunits, one of which could be clearly isolated from the thousands of other proteins synthesized by *E. coli.* This protein, called the tryptophan synthetase *A* protein, had a molecular weight indicating that it had slightly fewer than 300 amino acid units. Furthermore, we already knew how to force *E. coli* to produce comparatively large amounts of the protein—up to 2 percent of the total cell protein—and we also had a collection of mutants in which the activity of the tryptophan synthetase *A* protein was lacking. Finally, the bacterial strain we were using was one for which genetic procedures for preparing fine-structure maps had already been developed. Thus we could hope to map the *A* gene that presumably controlled the structure of the *A* protein.

To accomplish the mapping we needed a set of bacterial mutants with mutational alterations at many different sites on the *A* gene. If we could determine the amino acid change in the *A* protein of each of these mutants, and discover its position in the linear sequence of amino acids in the protein, we could test the concept of colinearity. Here again we were fortunate in the nature of the complex of subunits represented by tryptophan synthetase.

The normal complex consists of two *A*-protein subunits (the alpha chains) and one subunit consisting of two beta chains. Within the bacterial cell the complex acts as an enzyme to catalyze the reaction of indole-3-glycerol phosphate and serine to produce tryptophan and 3-phosphoglyceraldehyde [*see illustration, page 222*]. If the *A* protein undergoes certain kinds of mutations, it is still able to form a complex with the beta chains, but the complex loses the ability to catalyze the reaction. It retains the ability, however, to catalyze a simpler reaction when it is tested outside the cell: it will convert indole and serine to tryptophan. There are still other kinds of *A*-gene mutants that evidently lack the ability to form an *A* protein that can combine with beta chains; thus these strains are not able to catalyze even the simpler reaction. The first class of mutants—those that produce an *A* protein that is still able to combine with beta chains and exhibit catalytic activity when they are tested outside the cell—proved to be the most important for our study.

A fine-structure map of the *A* gene was constructed on the basis of genetic crosses performed by the process called transduction. This employs a particular bacterial virus known as transducing phage *P1kc*. When this virus multiplies in a bacterium, it occasionally incorporates a segment of the bacterial DNA within its own coat of protein. When the virus progeny infect other bacteria, genetic material of the donor bacteria is introduced into some of the recipient cells. A fraction of the recipients survive the infection. In these survivors segments of the bacterium's own genetic material pair with like segments of the "foreign" genetic material and recombination between the two takes place. As a result the offspring of an infected bacterium can contain characteristics inherited from its remote parent as well as from its immediate one.

In order to establish the order of mutationally altered sites in the *A* gene we have relied partly on a set of mutant bacteria in which one end of a deleted segment of DNA lies within the *A* gene. In each of these "deletion" mutants a segment of the genetic material of the bacterium was deleted spontaneously. Thus each deletion mutant in the set retains a different segment of the *A* gene. This set of mutants can now be crossed with any other mutant in which the *A* gene is altered at only a single site. Recombination can give rise to a normal gene only if the altered site does not fall within the region of the *A* gene that is missing in the deletion mutant [*see top illustration on opposite page*]. By crossing many A-protein mutants with the set of deletion mutants one can establish the linear order of many of the mutated sites in the *A* gene. The ordering is limited only by the number of deletion mutants at one's disposal.

A second method, which more closely

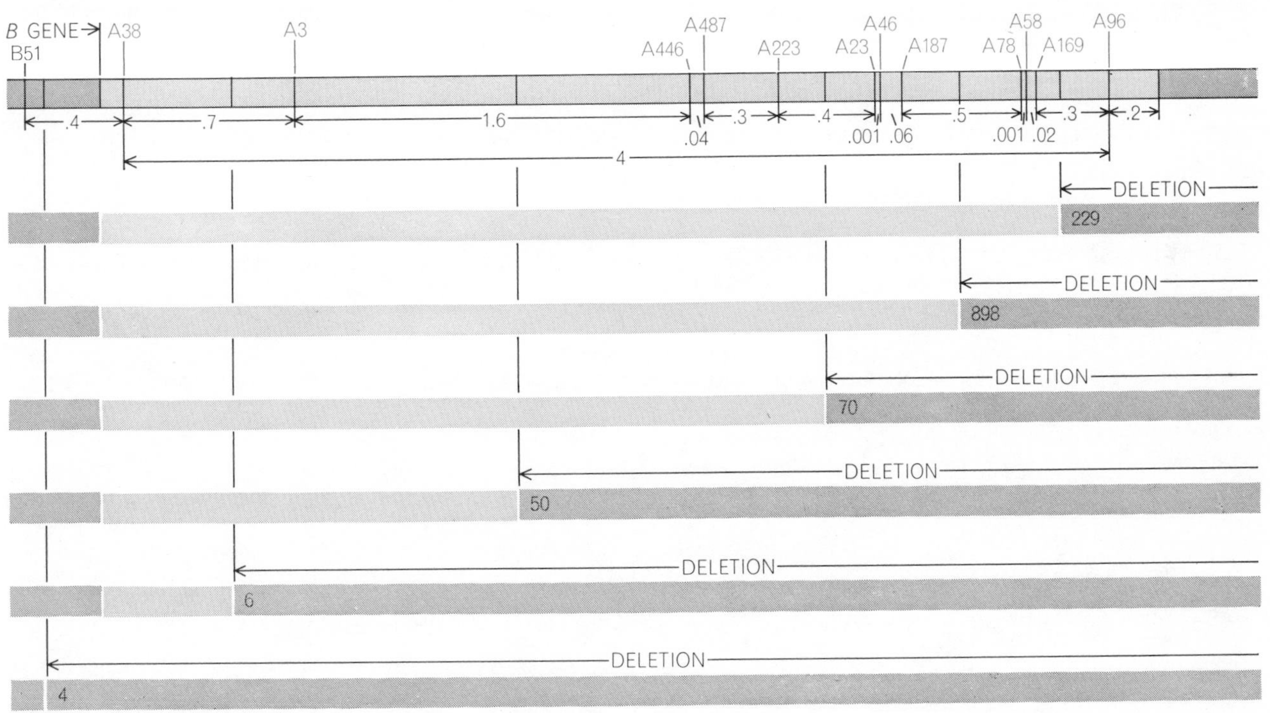

MAP OF *A* GENE shows the location of mutationally altered sites, drawn to scale, as determined by the three genetic-mapping methods illustrated on the opposite page. The total length of the *A* gene is slightly over four map units (probably 4.2). Below map are six deletion mutants that made it possible to assign each of the 12 *A*-gene mutants to one of six regions within the gene. The more sensitive mapping methods were employed to establish the order of mutations and the distance between mutation sites within each region.

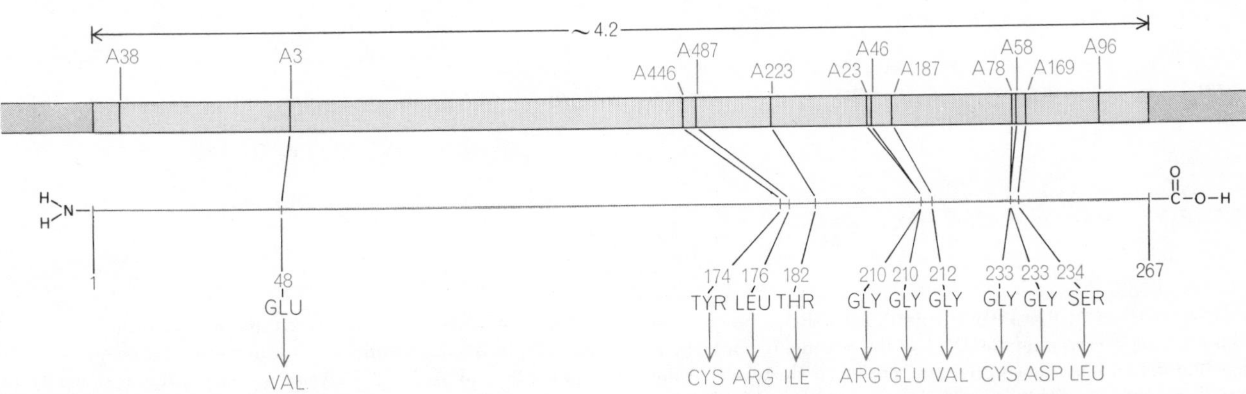

COLINEARITY OF GENE AND PROTEIN can be inferred by comparing the *A*-gene map (*top*) with the various amino acid changes in the *A* protein (*bottom*), both drawn to scale. The amino acid changes associated with 10 of the 12 mutations are also shown.

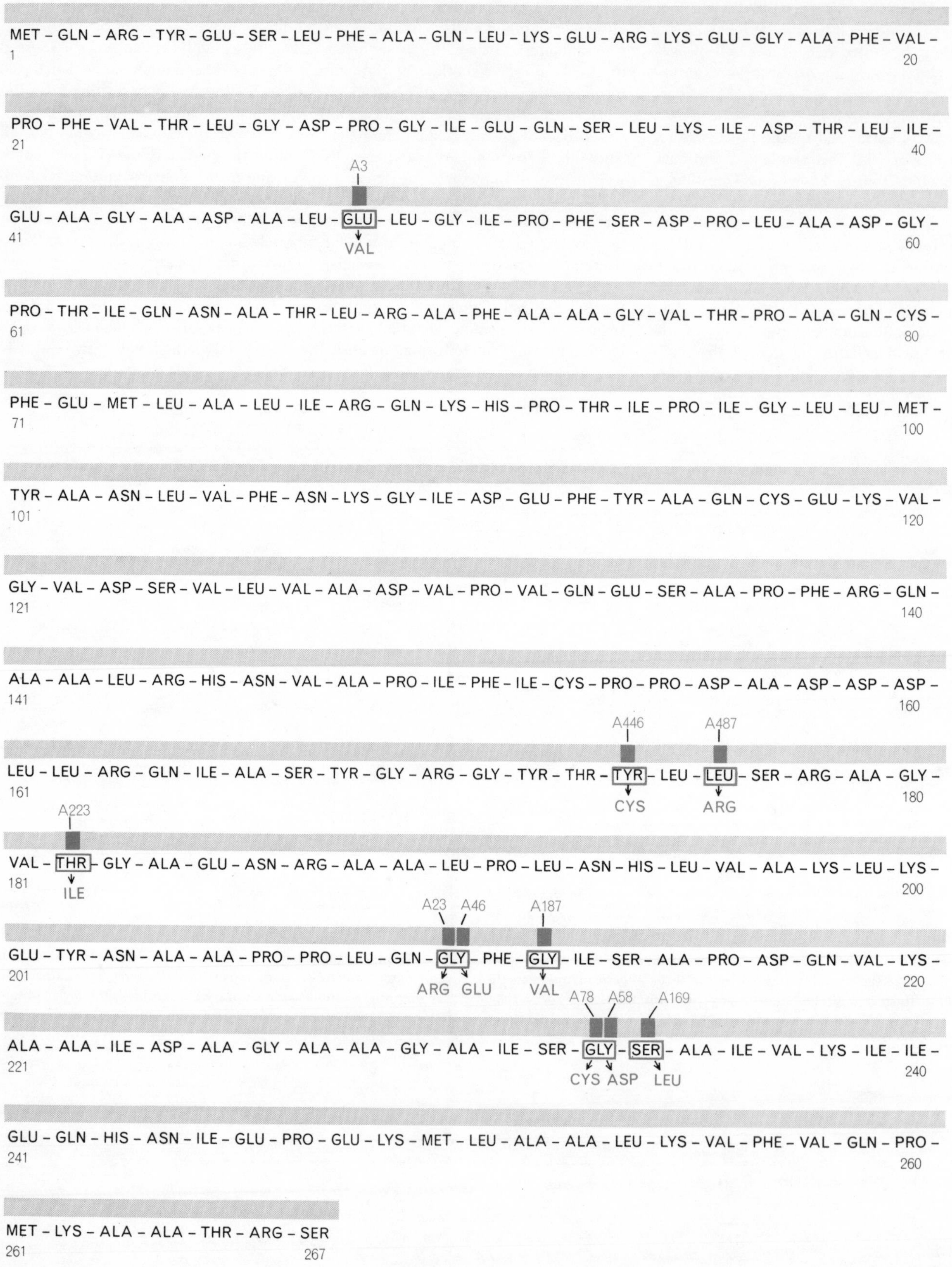

MET – GLN – ARG – TYR – GLU – SER – LEU – PHE – ALA – GLN – LEU – LYS – GLU – ARG – LYS – GLU – GLY – ALA – PHE – VAL –
1 20

PRO – PHE – VAL – THR – LEU – GLY – ASP – PRO – GLY – ILE – GLU – GLN – SER – LEU – LYS – ILE – ASP – THR – LEU – ILE –
21 40

A3

GLU – ALA – GLY – ALA – ASP – ALA – LEU – GLU – LEU – GLY – ILE – PRO – PHE – SER – ASP – PRO – LEU – ALA – ASP – GLY –
41 VAL 60

PRO – THR – ILE – GLN – ASN – ALA – THR – LEU – ARG – ALA – PHE – ALA – ALA – GLY – VAL – THR – PRO – ALA – GLN – CYS –
61 80

PHE – GLU – MET – LEU – ALA – LEU – ILE – ARG – GLN – LYS – HIS – PRO – THR – ILE – PRO – ILE – GLY – LEU – LEU – MET –
71 100

TYR – ALA – ASN – LEU – VAL – PHE – ASN – LYS – GLY – ILE – ASP – GLU – PHE – TYR – ALA – GLN – CYS – GLU – LYS – VAL –
101 120

GLY – VAL – ASP – SER – VAL – LEU – VAL – ALA – ASP – VAL – PRO – VAL – GLN – GLU – SER – ALA – PRO – PHE – ARG – GLN –
121 140

ALA – ALA – LEU – ARG – HIS – ASN – VAL – ALA – PRO – ILE – PHE – ILE – CYS – PRO – PRO – ASP – ALA – ASP – ASP – ASP –
141 160

A446 A487

LEU – LEU – ARG – GLN – ILE – ALA – SER – TYR – GLY – ARG – GLY – TYR – THR – TYR – LEU – LEU – SER – ARG – ALA – GLY –
161 CYS ARG 180

A223

VAL – THR – GLY – ALA – GLU – ASN – ARG – ALA – ALA – LEU – PRO – LEU – ASN – HIS – LEU – VAL – ALA – LYS – LEU – LYS –
181 ILE 200

A23 A46 A187

GLU – TYR – ASN – ALA – ALA – PRO – PRO – LEU – GLN – GLY – PHE – GLY – ILE – SER – ALA – PRO – ASP – GLN – VAL – LYS –
201 ARG GLU VAL 220

A78 A58 A169

ALA – ALA – ILE – ASP – ALA – GLY – ALA – ALA – GLY – ALA – ILE – SER – GLY – SER – ALA – ILE – VAL – LYS – ILE – ILE –
221 CYS ASP LEU 240

GLU – GLN – HIS – ASN – ILE – GLU – PRO – GLU – LYS – MET – LEU – ALA – ALA – LEU – LYS – VAL – PHE – VAL – GLN – PRO –
241 260

MET – LYS – ALA – ALA – THR – ARG – SER
261 267

AMINO ACID SEQUENCE OF *A* PROTEIN is shown side by side with a ribbon representing the DNA of the *A* gene. It can be seen that 10 different mutations in the gene produced alterations in the amino acids at only eight different places in the *A* protein. The explanation is that at two of them, 210 and 233, there were a total of four alterations. Thus at No. 210 the mutation designated A23 changed glycine to arginine, whereas mutation A46 changed glycine to glutamic acid. At No. 233 glycine was changed to cysteine by one mutation (A78) and to aspartic acid by another mutation (A58). On the genetic map A23 and A46, like A78 and A58, are very close.

resembles traditional genetic procedures, relies on recombination frequencies to establish the order of the mutationally altered sites in the *A* gene with respect to one another. By this method one can assign relative distances—map distances—to the regions between altered sites. The method is often of little use, however, when the distances are very close.

In such cases we have used a third method that involves a mutationally altered gene, or genetic marker, close to the *A* gene. This marker produces a recognizable genetic trait unrelated to the *A* protein. What this does, in effect, is provide a reading direction so that one can tell whether two closely spaced mutants, say No. 58 and No. 78, lie in the order 58–78, reading from the left on the map, or vice versa [*see bottom illustration on page 224*].

With these procedures we were able to construct a genetic map relating the altered sites in a group of mutants responsible for altered *A* proteins that could themselves be isolated for study. Some of the sites were very close together, whereas others were far apart [*see upper illustration, page 225*]. The next step was to determine the nature of the amino acid changes in each of the mutationally altered proteins.

It was expected that each mutant of the *A* protein would have a localized change, probably involving only one amino acid. Before we could hope to identify such a specific change we would have to know the sequence of amino acids in the unmutated *A* protein. This was determined by John R. Guest, Gabriel R. Drapeau, Bruce C. Carlton and me, by means of a well-established procedure. The procedure involves breaking the protein molecule into many short fragments by digesting it with a suitable enzyme. Since any particular protein rarely has repeating sequences of amino acids, each digested fragment is likely to be unique. Moreover, the fragments are short enough—typically between two and two dozen amino acids in length—so that careful further treatments can release one amino acid at a time for analysis. In this way one can identify all the amino acids in all the fragments, but the sequential order of the fragments is still unknown. This can be established by digesting the complete protein molecule with a different enzyme that cleaves it into a uniquely different set of fragments. These are again analyzed in detail. With two fully analyzed sets of fragments in hand, it is not difficult to find short sequences of amino acids that are grouped together in the fragment of one set but that are divided between two fragments in the other. This provides the clue for putting the two sets of fragments in order. In this way we ultimately determined the identity and location of each of the 267 amino acids in the unmutated *A* protein of tryptophan synthetase.

SEGMENT OF PROTEIN	MUTANT H11	C140	B17	B272	H32	B278	C137	H36	A489	C208	NORMAL
I	+	+	+	+	+	+	+	+	+	+	+
II	–	+	+	+	+	+	+	+	+	+	+
III	–	–	+	+	+	+	+	+	+	+	+
IV	–	–	–	+	+	+	+	+	+	+	+
V	–	–	–	–	+	+	+	+	+	+	+
VI	–	–	–	–	–	+	+	+	+	+	+
VII	–	–	–	–	–	–	+	+	+	+	+
VIII	–	–	–	–	–	–	–	+	+	+	+
IX	–	–	–	–	–	–	–	–	+	+	+
X	–	–	–	–	–	–	–	–	–	+	+
XI	–	–	–	–	–	–	–	–	–	–	+

GENETIC MAP: H11 C140 B17 B272 H32 B278 C137 H36 A489 C208

INDEPENDENT EVIDENCE FOR COLINEARITY of gene and protein structure has been obtained from studies of the protein that forms the head of the bacterial virus T4D. Sydney Brenner and his co-workers at the University of Cambridge have found that mutations in the gene for the head protein alter the length of head-protein fragments. In the table "+" indicates that a given segment of the head protein is produced by a particular mutant; "–" indicates that the segment is not produced. When the genetic map was plotted, it was found that the farther to the right a mutation appears, the longer the fragment of head protein.

Simultaneously my colleagues and I were examining the mutants of the *A* protein to identify the specific sites of mutational changes. For this work we used a procedure first developed by Vernon M. Ingram, now at the Massachusetts Institute of Technology, in his studies of naturally occurring abnormal forms of human hemoglobin. This procedure also uses an enzyme (trypsin) to break the protein chain into peptides, or polypeptide fragments. If the peptides are placed on filter paper wetted with certain solvents, they will migrate across the paper at different rates; if an electric potential is applied across the paper, the peptides will be dispersed even more, depending on whether they are negatively charged, positively charged or uncharged under controlled conditions of acidity. The former separation process is chromatography; the latter, electrophoresis. When they are employed in combination, they produce a unique "fingerprint" for each set of peptides obtained by digesting the *A* protein from a particular mutant bacterium. The positions of the peptides are located by spraying the filter paper with a solution of ninhydrin and heating it for a few minutes at about 70 degrees centigrade. Each peptide reacts to yield a characteristic shade of yellow, gray or blue.

When the fingerprints of mutationally altered *A* proteins were compared with the fingerprint of the unmutated protein, they were found to be remarkably similar. In each case, however, there was a difference. The mutant fingerprint usually lacked one peptide spot that appears in the nonmutant fingerprint and exhibited a spot that the nonmutant fingerprint lacks. The two peptides would presumably be related to each other with the exception of the change resulting from the mutational event. One can isolate each of the peptides and compare their amino acid composition. Guest, Drapeau, Carlton and I, together with D. R. Helinski and U. Henning, identified the amino acid substitutions in each of a variety of altered *A* proteins.

The final step was to compare the locations of these changes in the *A* protein with the genetic map of the mutationally altered sites. There could be no doubt that the amino acid sequence of the *A* protein and the map of the *A* gene are in fact colinear [*see lower illustration on page 225*].

One can also see that the distances between mutational sites on the map of the *A* gene correspond quite closely to the distances separating the corresponding amino acid changes in the *A* protein. In two instances two separate mutational changes, so close as to be almost at the same point on the genetic map, led to changes of the same amino acid in the unmutated protein. This is to be expected if a codon of three bases in DNA is required to specify a single amino acid in a protein. Evidently the most closely spaced mutational sites in our genetic map represent alterations in two bases within a single codon.

Thus our studies have shown that each

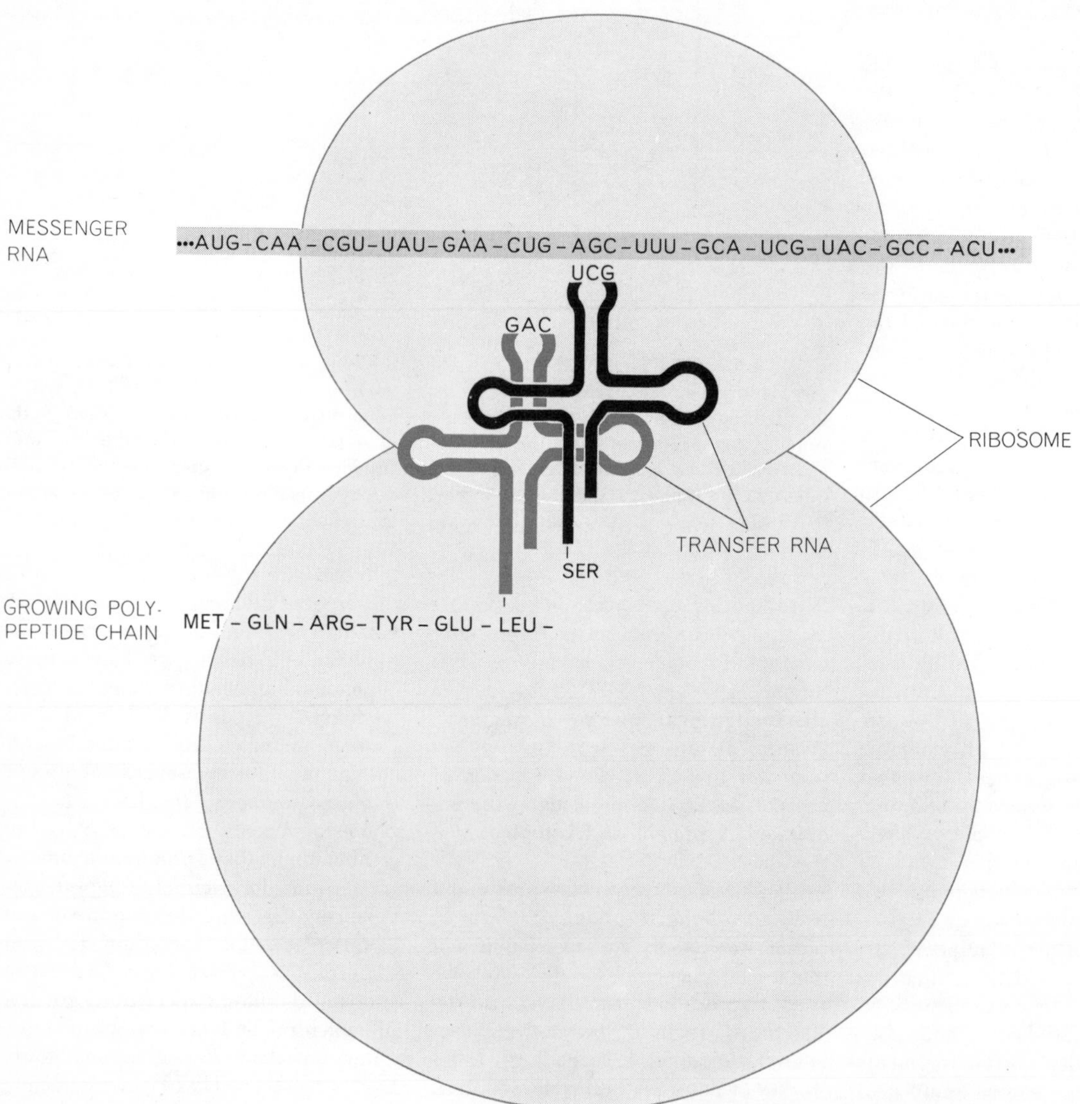

SCHEME OF PROTEIN SYNTHESIS, according to the current view, involves the following steps. Genetic information is transcribed from double-strand DNA into single-strand messenger ribonucleic acid (RNA), which becomes associated with a ribosome. Amino acids are delivered to the ribosome by molecules of transfer RNA, which embody codons complementary to the codons in messenger RNA. The next to the last molecule of transfer RNA to arrive (*color*) holds the growing polypeptide chain while the arriving molecule of transfer RNA (*black*) delivers the amino acid that is to be added to the chain next (serine in this example). The completed polypeptide chain, either alone or in association with other chains, is the protein whose specification was originally embodied in DNA.

AGY AGX AGY AGY

CGZ GGZ UAX ACZ [UAX] XUZ [CUZ] UCZ CGZ GCZ GGZ GUZ [ACW] GGZ GCZ GAY AAX CGZ GCZ GCZ XUZ

– ARG – GLY – TYR – THR – [TYR] – LEU – [LEU] – SER – ARG – ALA – GLY – VAL – [THR] – GLY – ALA – GLU – ASN – ARG – ALA – ALA – LEU –

170 190

CCZ XUZ AAX CAX XUZ GUZ GCZ AAY XUZ AAY GAY UAX AAX GCZ GCZ CCZ CCZ XUZ CAY [GGA]

PRO – LEU – ASN – HIS – LEU – VAL – ALA – LYS – LEU – LYS – GLU – TYR – ASN – ALA – ALA – PRO – PRO – LEU – GLN – [GLY] –

191 210

AGX

UUX [GGZ] AUW UCZ GCZ CCZ GAX CAY GUZ AAY GCZ GCZ AUW GAX GCZ GGZ GCZ GCZ GGZ GCZ

PHE – [GLY] – ILE – SER – ALA – PRO – ASP – GLN – VAL – LYS – ALA – ALA – ILE – ASP – ALA – GLY – ALA – ALA – GLY – ALA –

211 230

AGX

AUW UCZ [GGX] [UCZ] GCZ AUW GUZ AAY AUW AUW GAY CAY CAX AAX AUW GAY CCZ GAY AAY AUG

ILE – SER – [GLY] – [SER] – ALA – ILE – VAL – LYS – ILE – ILE – GLU – GLN – HIS – ASN – ILE – GLU – PRO – GLU – LYS – MET –

231 250

W = U, C or A X = U or C Y = A or G Z = U, C, A or G

PROBABLE CODONS IN MESSENGER RNA that determines the sequence of amino acids in the *A* protein are shown for 81 of the protein's 267 amino acid units. The region includes seven of the eight mutationally altered positions (*colored boxes*) in the *A* protein. The codons were selected from those assigned to the amino acids by Marshall Nirenberg and his associates at the National Institutes of Health and by H. Gobind Khorana and his associates at the University of Wisconsin. Codons for the remaining 186 amino acids in the *A* protein can be supplied similarly. In most cases the last base in the codon cannot be specified because there are usually several synonymous codons for each amino acid. With a few exceptions the synonyms differ from each other only in the third position.

unique sequence of bases in DNA—a sequence constituting a gene—is ultimately translated into a corresponding unique linear sequence of amino acids—a sequence constituting a polypeptide chain. Such chains, either by themselves or in conjunction with other chains, fold into the three-dimensional structures we recognize as protein molecules. In the great majority of cases these proteins act as biological catalysts and are therefore classed as enzymes.

The colinear relation between a genetic map and the corresponding protein has also been convincingly demonstrated by Sydney Brenner and his co-workers at the University of Cambridge. The protein they studied was not an enzyme but a protein that forms the head of the bacterial virus T4. One class of mutants of this virus produces fragments of the head protein that are related to one another in a curious way: much of their amino acid sequence appears to be identical, but the fragments are of various lengths. Brenner and his group found that when the chemically similar regions in fragments produced by many mutants were matched, the fragments could be arranged in order of increasing length. When they made a genetic map of the mutants that produced these fragments, they found that the mutationally altered sites on the genetic map were in the same order as the termination points in the protein fragments. Thus the length of the fragment of the head protein produced by a mutant increased as the site of mutation was displaced farther from one end of the genetic map [*see illustration on page 227*].

The details of how the living cell translates information coded in gene structure into protein structure are now reasonably well known. The base sequence of one strand of DNA is transcribed into a single-strand molecule of messenger ribonucleic acid (RNA), in which each base is complementary to one in DNA. Each strand of messenger RNA corresponds to relatively few genes; hence there are a great many different messenger molecules in each cell. These messengers become associated with the small cellular bodies called ribosomes, which are the actual site of protein synthesis [*see illustration on page 228*]. In the ribosome the bases on the messenger RNA are read in groups of three and translated into the appropriate amino acid, which is attached to the growing polypeptide chain. The messenger also contains in code a precise starting point and stopping point for each polypeptide.

From the studies of Marshall Nirenberg and his colleagues at the National Institutes of Health and of H. Gobind Khorana and his group at the University of Wisconsin the RNA codons corresponding to each of the amino acids are known. By using their genetic code dictionary we can indicate approximately two-thirds of the bases in the messenger RNA that specifies the structure of the *A*-protein molecule. The remaining third cannot be filled in because synonyms in the code make it impossible, in most cases, to know which of two or more bases is the actual base in the third position of a given codon [*see illustration above*]. This ambiguity is removed, however, in two cases where the amino acid change directed by a mutation narrows down the assignment of probable codons. Thus at amino acid position 48 in the *A*-protein molecule, where a mutation changes the amino acid glutamic acid to valine, one can deduce from the many known changes at this position that of the two possible codons for glutamic acid, GAA and GAG, GAG is the correct one. In other words, GAG (specifying glutamic acid) is changed to GUG (specifying valine). The other position for which the codon assignment can be made definite in this way is No. 210. This position is affected by two different mutations: the amino acid glycine is replaced by arginine in one case and by glutamic acid in the other. Here one can infer from the observed amino acid changes that of the four possible codons for glycine, only one—GGA—can yield by a single base change either arginine (AGA) or glutamic acid (GAA).

Knowledge of the bases in the messenger RNA for the *A* protein can be translated, of course, into knowledge of the base pairs in the *A* gene, since each base pair in DNA corresponds to one of the bases in the RNA messenger. When the ambiguity in the third position of most of the codons is resolved, and when we can distinguish between two quite different sets of codons for arginine, leucine and serine, we shall be able to write down the complete base sequence of the *A* gene—the base sequence that specifies the sequence of the 267 amino acids in the *A* protein of the enzyme tryptophan synthetase.

The Three-Dimensional Structure of Transfer RNA

18

by Alexander Rich and Sung Hou Kim
January 1978

This nucleic acid plays a key role in translating the genetic code into the sequence of amino acids in a protein. The determination of its structure has clarified the mechanism of protein synthesis

It is now widely known that the instructions for the assembly and organization of a living system are embodied in the DNA molecules contained within the living cell. The sequence of nucleotide bases along the linear chain of the DNA molecule specifies the structure of the thousands of proteins that are the construction materials of the cell and the catalysts of its intricate biochemical reactions. By itself, however, a DNA molecule is rather like a strip of magnetic recording tape: the information embodied in its structure cannot be expressed without a decoding mechanism.

The development of such a decoding mechanism was one of the crucial events in the origin of life some four billion years ago. A basic biochemical system gradually evolved in which the nucleotide sequence of DNA is first transcribed into the complementary sequence of messenger RNA (abbreviated mRNA). The messenger RNA then directs the assembly of amino acids into the specific linear sequence characteristic of a given protein, a process called translation.

A central role in translation is played by another kind of RNA: transfer RNA (tRNA). The molecules of transfer RNA form a class of small globular polynucleotide chains (as distinct from fibrous polynucleotide chains such as DNA and mRNA) about 75 to 90 nucleotides long. They act as vehicles for transferring amino acids from the free state inside the cell into the assembled chain of the protein. This vital function as an intermediary between the nucleic acid language of the genetic code and the amino acid language of the working cell has made transfer RNA a major subject of research in molecular biology. Recently, in an important step toward the goal of understanding the process of translation in precise molecular terms, the three-dimensional structure of a tRNA molecule has been worked out at high resolution.

The translation of the nucleotide sequence of messenger RNA into protein proceeds in two major steps. First an amino acid molecule is attached to a particular transfer-RNA molecule, a reaction catalyzed by a large enzyme called an aminoacyl-tRNA synthetase. There are many different types of synthetase in living cells, each specific for one of the 20 different amino acids found in proteins. For example, leucyl-tRNA synthetase selectively binds to itself both the amino acid leucine and

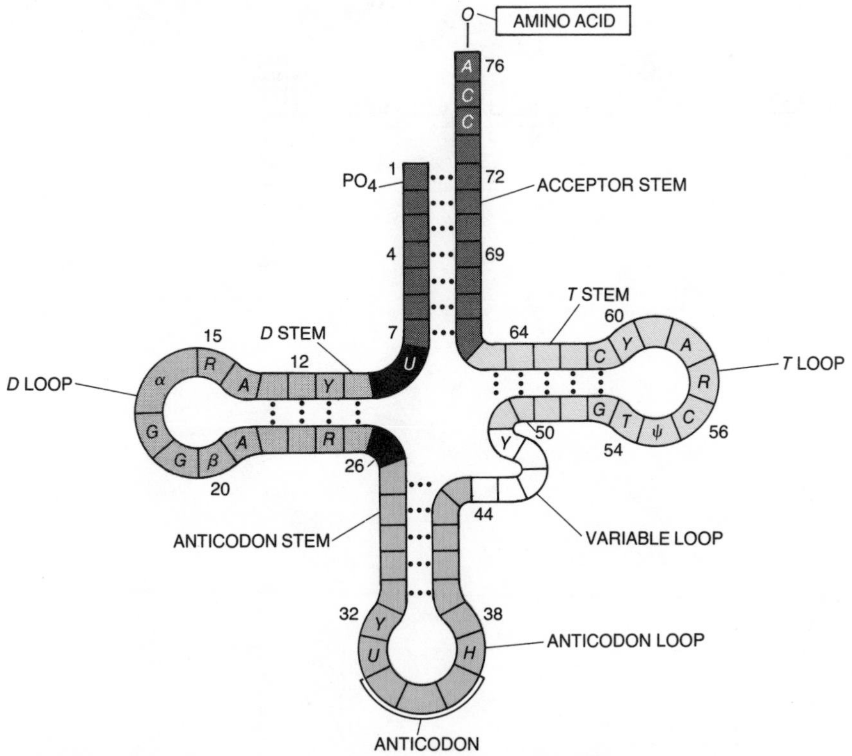

CLOVERLEAF DIAGRAM is the two-dimensional folding pattern of the transfer-RNA (tRNA) molecule, which was first deduced in 1965 from the sequence of nucleotide building blocks in yeast alanine tRNA. Since then the diagram has been found to fit the nucleotide sequences of about 100 tRNA's isolated from plant, animal and bacterial cells. Nucleotide bases found in the same positions in all tRNA sequences are indicated. The ladderlike stems are made up of complementary bases in different parts of the polynucleotide chain that pair up and form hydrogen bonds, causing the chain to fold back on itself. The number of nucleotides in the various stems and loops is generally constant except for two parts of the *D* loop designated α and β (which consist of from one to three nucleotides in different tRNA's) and the variable loop (which usually has four or five nucleotides but may have as many as 21). Abbreviations are *A* (adenosine), *G* (guanosine), *C* (cytidine), *U* (uridine), *R* (adenosine or guanosine), *Y* (cytidine or uridine), *T* (ribothymidine), ψ (pseudouridine), *H* (modified adenosine or guanosine).

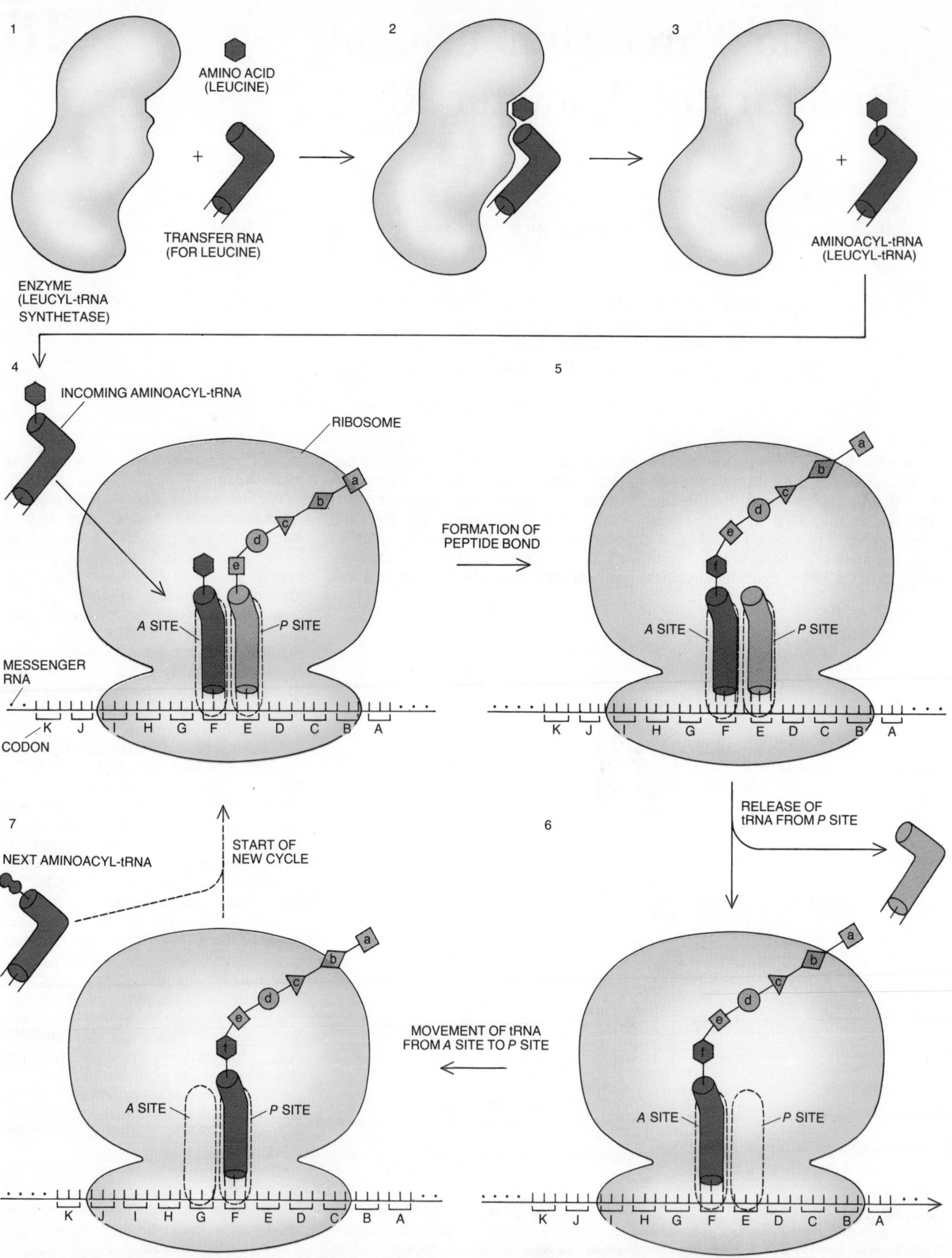

FUNCTION OF TRANSFER RNA in the synthesis of a protein molecule is to make a chain of amino acids that reflects the nucleotide sequence of the template represented by messenger RNA. First a large enzyme called an aminoacyl-tRNA synthetase joins a specific transfer-RNA molecule to its corresponding amino acid with a covalent bond (*1–3*). The transfer RNA with the amino acid attached to it binds at the *A* site to the ribosome: the organelle where the amino acids are linked into the polypeptide chain of a protein. This interaction requires specific hydrogen bonding between the three codon bases on the messenger-RNA strand that specify an amino acid and the three anticodon bases of the transfer RNA (*4*). A transfer-RNA molecule in the adjacent *P* site then transfers the growing polypeptide chain to the tRNA in the *A* site (*5*). The "empty" tRNA leaves the *P* site and the ribosome moves along the messenger RNA a distance of one codon, so that the transfer RNA carrying the polypeptide chain is shifted from the *A* site to the *P* site (*6, 7*). Then the cycle begins anew.

the tRNA for leucine; a complex of leucine and leucine tRNA is then formed and released. Once a tRNA has an amino acid attached to it, it is ready to participate in the second major step of protein synthesis.

This second step, the joining of the amino acids into a chain, is carried out inside the cellular organelle known as the ribosome, an aggregate of more than 50 different protein molecules and three RNA molecules. The ribosome is an intricate piece of molecular machinery designed to help translate the polynucleotide sequence of messenger RNA into the polypeptide sequence of protein. Although the exact details of the process have not been worked out, its general features are known.

Each amino acid in a protein is specified by a group of three adjacent nucleotide bases, designated a codon, on the messenger-RNA strand. There are four kinds of nucleotide base in messenger RNA, and so there is a total of 4^3, or 64, possible codons. The relation between the codons and the amino acids they specify is the genetic code. The fact that the code appears to be the same in all living organisms is a remarkable proof of the unity of life at the molecular level.

Inside the ribosome are two sites that are involved in translation. One of them is the *A* site, which stands for aminoacyl-tRNA binding site. It is at this position that the transfer-RNA molecule and its attached amino acid are bound to the ribosome. The tRNA is positioned there partly by a set of specific interactions with the messenger RNA, which has already become associated with the ribosome. Three special nucleotide bases in the transfer-RNA molecule, designated the anticodon, interact with three complementary codon bases in the messenger RNA. The interaction involves the weak directional bonds known as hydrogen bonds, in which a hydrogen atom with a slight positive charge is shared by two other atoms with a slight negative charge. Hydrogen bonding is also the force that holds together the complementary nucleotide bases in the double helix of DNA: the base guanine on one strand of the helix is always paired with the base cytosine on the other strand, and the base adenine is always paired with the base thymine.

Immediately adjacent to the *A* site in the ribosome is the peptidyl-tRNA binding site, or *P* site. The transfer-RNA molecule with the growing chain of amino acids attached to it is bound to this site and specifically interacts with the next codon triplet of bases on the messenger-RNA chain. In the course of protein synthesis the growing polypeptide chain is cleaved from the tRNA molecule in the *P* site and is transferred to the end of the single amino acid attached to

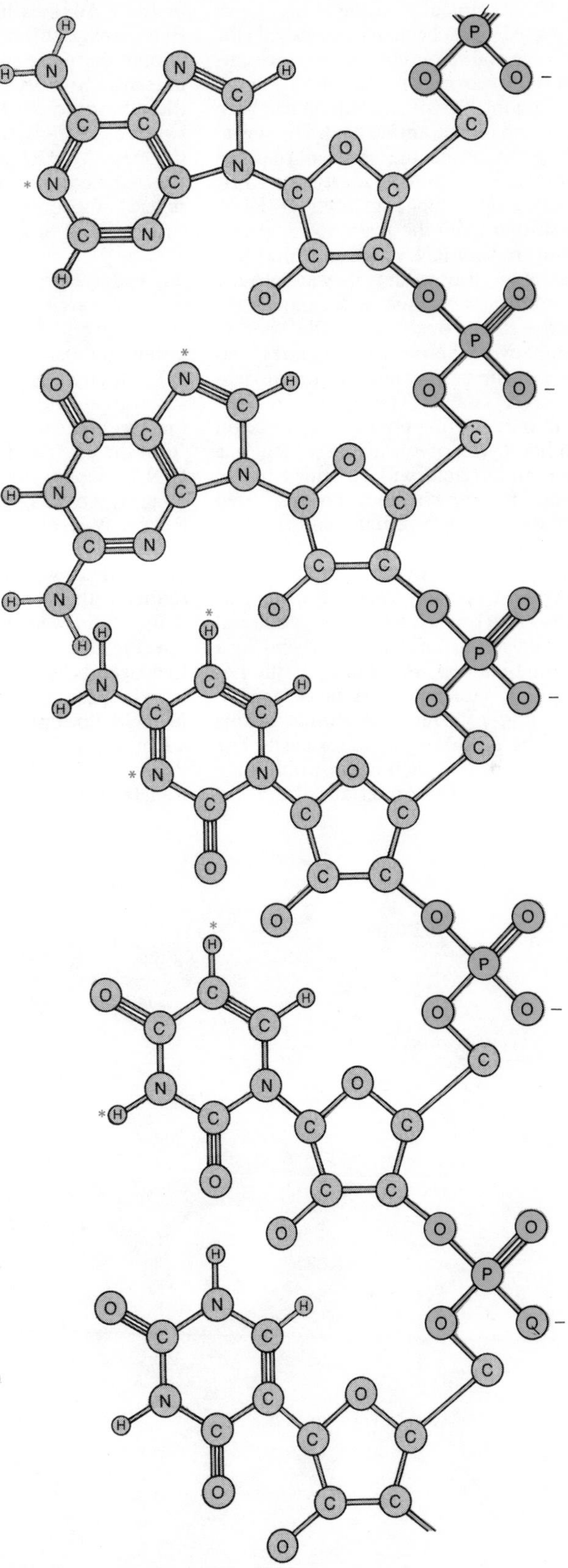

NUCLEOSIDES, consisting of a nucleotide base attached to the sugar ribose, are joined by negatively charged phosphate (PO_4) groups to form the polynucleotide chain of transfer RNA. The four major nucleosides in the molecule are adenosine, guanosine, cytidine and uridine. Transfer RNA also incorporates many modified nucleosides, more than 50 of which have been identified. The commonest modification is the replacement of a hydrogen atom by a methyl group (CH_3). This reaction is catalyzed by special enzymes and occurs at the sites indicated by an asterisk. Other structural modifications also occur. For example, the nucleoside pseudouridine (ψ) has its base attached to the ribose through a carbon atom instead of a nitrogen atom.

the tRNA molecule in the *A* site. Once the transfer has been accomplished (the cleavage and rejoining reactions are carried out by an enzyme in the ribosome) the growing polypeptide chain has been elongated by one amino acid. The "empty" tRNA molecule is then released from the *P* site, and the messenger RNA and the newly elongated peptidyl-tRNA are shifted from the *A* site to the *P* site. A new transfer RNA with an amino acid attached to it now finds its way into the ribosome and becomes lodged in the vacated *A* site through the specific interaction between its anticodon bases and those of the next codon on the messenger-RNA strand. The system is now back to its starting point, ready to begin another cycle of events in which one more amino acid will be added to the chain. This stepwise addition is repeated until the complete protein has been synthesized.

The process of polypeptide-chain elongation is fairly rapid: it occurs as many as 20 times a second in a bacterial cell and about once every second in a mammalian cell. For example, the hemoglobin molecule is a large protein consisting of four polypeptide chains with about 140 amino acids each. The synthesis of one such chain would take seven seconds in a bacterial cell and two or three minutes in a mammalian cell. Even though this rate of synthesis is fairly high there are surprisingly few errors in translation, because the machinery of the ribosome ensures a careful fit between each transfer-RNA molecule and the messenger RNA. The process is also very efficient, because there are usually several ribosomes at work translating a single strand of messenger RNA.

In order to understand how transfer RNA carries an amino acid into the ribosome and transfers it to the growing polypeptide chain it is essential to have a knowledge of the three-dimensional structure of the tRNA molecule. One of the first clues to that structure emerged from the nucleotide sequence of a yeast tRNA specific for the amino acid alanine, which was determined in 1965 by Robert W. Holley and his colleagues at Cornell University. These workers noted that there were certain regions of the sequence that would be complementary if the chain were folded back on itself. Specifically, these regions could form hydrogen bonds with each other, much like the base pairing in the double helix of DNA (except that in RNA's the base adenine is paired with uracil instead of thymine). The polynucleotide chain of transfer RNA could thus be arranged in such a way that it would contain hydrogen-bonded double-strand regions called stems and nonbonded regions called loops. The postulated combination of stems and loops resembled a four-leaf clover, and so it became known as the cloverleaf diagram.

One feature of the nucleotide sequence of transfer RNA is that it includes many unusual bases, most of them common RNA bases that have been modified by the addition of one or more methyl groups (CH_3). Because of this feature some parts of the cloverleaf diagram have been named for the modified bases that occur in them. For example, the *T* loop is so named because it includes thymine (*T*), which is found in DNA but is not found in any RNA species other than transfer RNA. Similarly, the *D* loop usually includes the modified base dihydrouracil (*D*). Other regions of the cloverleaf are the variable loop, which in different tRNA's has different numbers of nucleotides (ranging from four to 21), the anticodon loop, which includes the three bases of the anticodon, and the acceptor stem, which accepts the amino acid specific to that particular tRNA.

An interesting feature of the cloverleaf diagram is the presence of nucleotide sequences that are constant in all 100 of the tRNA sequences that have been determined so far. The number of base pairs in the stem regions is also constant: seven in the acceptor stem, five in the *T* stem, five in the anticodon stem and three or four in the *D* stem. These features are maintained in tRNA molecules from plants, animals, bacteria and viruses. Indeed, the pattern of stems, loops and constant nucleotides found in the tRNA cloverleaf appears to have the same universality as the genetic code. Much of the explanation for this constancy was later provided by the three-dimensional structure of tRNA.

Today the three-dimensional structure of large biological molecules is commonly determined by means of the X-ray-diffraction analysis of molecular crystals. A molecular crystal is an assembly of molecules packed together in a regular three-dimensional array. When X rays with a wavelength comparable to the distance between atoms are directed into the crystal, they are diffracted, or scattered, in a variety of directions by the electron clouds of the atoms in the crystal lattice. The diffraction pattern of the crystal can be detected as a series of spots on a piece of X-ray film, with the blackening of the emulsion being proportional to the intensity of each scattered beam.

This pattern contains a great deal of information about the structure of the crystal. For one thing, the amplitude of the wave scattered by an atom is proportional to the number of electrons in the atom, so that a carbon atom will scatter

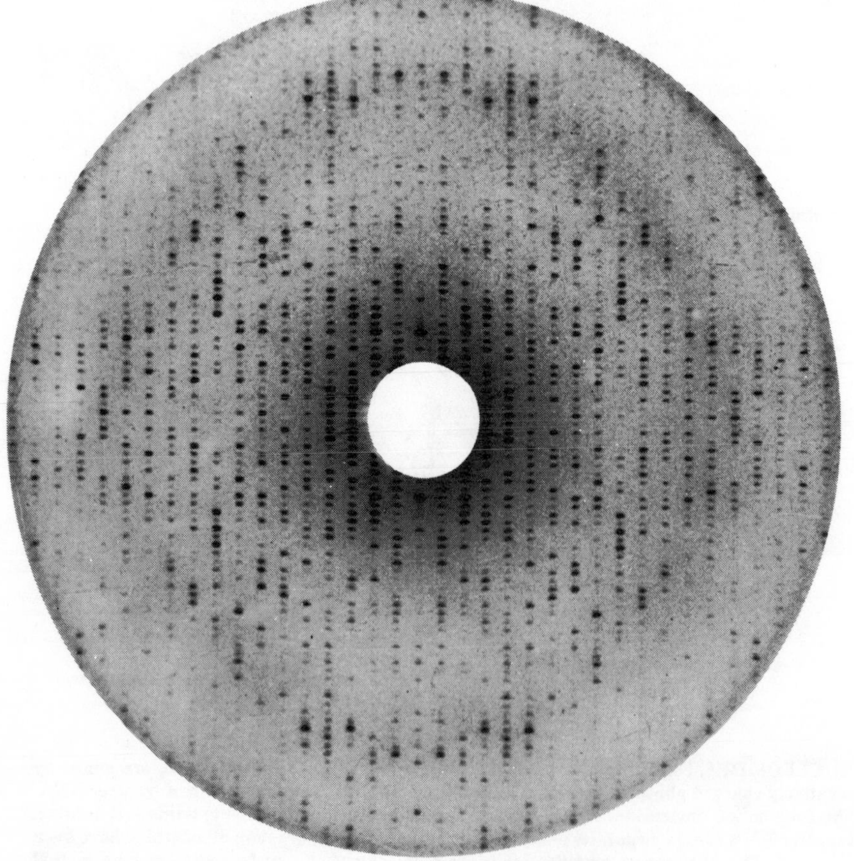

X-RAY-DIFFRACTION PATTERN was one of patterns utilized by the authors to deduce the three-dimensional structure of transfer RNA. The pattern was created by directing X rays into a crystalline array of transfer-RNA molecules and capturing the scattered beams on a piece of film. The spots contain information about distribution of electrons within the crystal.

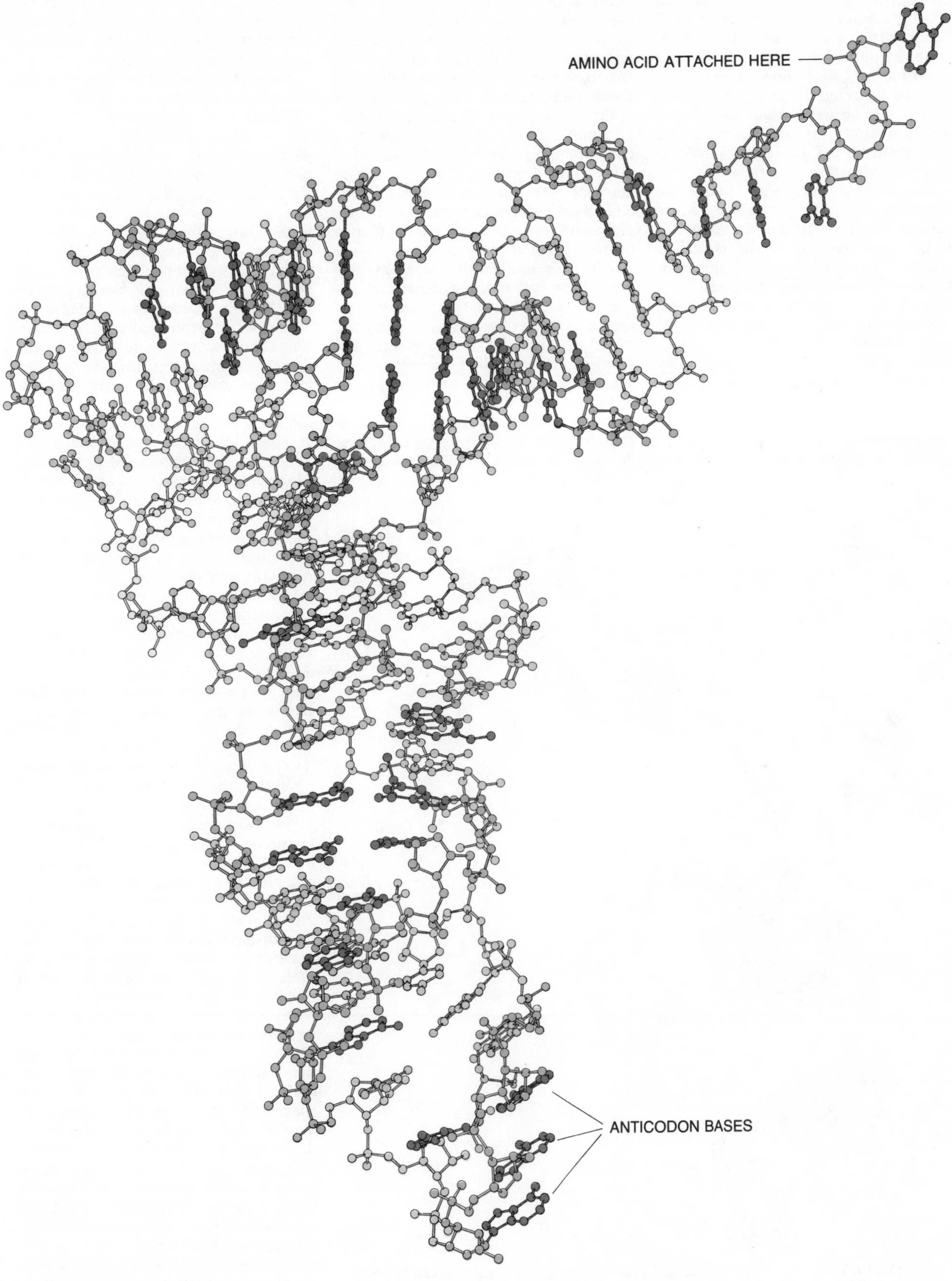

DETAILED SKELETAL MODEL of yeast phenylalanine tRNA shows the hydrogen-bonding interactions between the nucleotide bases. It was derived in 1974 from an X-ray-crystallographic study at a resolution of three angstroms. Projection shown here was generated on a computer by one of the authors (Kim). Ribose-phosphate backbone of the molecule is shaded in color; the bases are shaded in gray.

X rays six times more strongly than a hydrogen atom. Secondly, the scattered waves recombine inside the crystal lattice; depending on whether they are in phase or out of phase, they will either reinforce or cancel one another. The way the scattered waves recombine depends only on the arrangement of the atoms in the crystal, and so it is possible to reconstruct the image of a molecule from its diffraction pattern.

To analyze the three-dimensional structure of a large protein or nucleic acid molecule a crystal of the substance is first prepared. Then the crystal is mounted in a capillary tube and positioned in a precise orientation with respect to the X-ray beam and the film. The crystal is rotated along each of its axes to yield a series of X-ray photographs in which there is a regular array of spots of various intensities. Each of these photographs is actually a two-dimensional section through a three-dimensional array of spots.

Next the intensities of all the spots in the diffraction patterns are measured, either from the film or through the use of a Geiger counter. Additional information is needed, however, before one can establish the three-dimensional structure, namely the phases of the scattered X-ray beams with respect to an arbitrary fixed point in the crystal. This information is obtained by inserting heavy-metal atoms such as those of platinum or gold into the crystal lattice as markers. The addition of these atoms changes the diffraction pattern slightly and enables one to calculate the phases of the diffracted beams.

With this information in hand it is possible to calculate the density of the electrons at a large number of regularly spaced points in the crystal, making use of a Fourier series: a sum of sine and cosine terms. A high-speed computer is needed to handle the enormous number of terms (more than a billion) involved in determining the structure of a large protein or nucleic acid molecule. The first such molecule whose structure was determined in this way was the protein myoglobin; the feat was accomplished in 1958. Today the technique is almost routinely exploited for the structural analysis of large molecules.

The end product of the technique is a three-dimensional map showing the distribution of the electrons in the crystal. The map is usually drawn as a series of parallel sections stacked on top of one another, each section being a transparent plastic sheet on which the electron-density distribution is represented by black contour lines resembling those of a topographical map. The critical factor in the interpretation of the electron-density map is its resolution, which is determined by the number of scattered-beam intensities incorporated in the Fourier series. For example, a map at a resolution of six angstroms, derived from the innermost spots of the diffraction pattern, may reveal the general shape of the molecule but few additional structural details. (An angstrom is 10^{-10} meter, about the diameter of a hydrogen atom.) Maps of higher resolution are needed to delineate groups of atoms, which may be three to four angstroms apart, or individual atoms, which are from one to two angstroms apart. A large molecule is usually analyzed at different levels of resolution, making it possible to visualize different features of the structure. The ultimate resolution of an X-ray analysis, however, is determined by the degree of perfection of the crystal. For large biological molecules the best resolution one can usually obtain is about two angstroms.

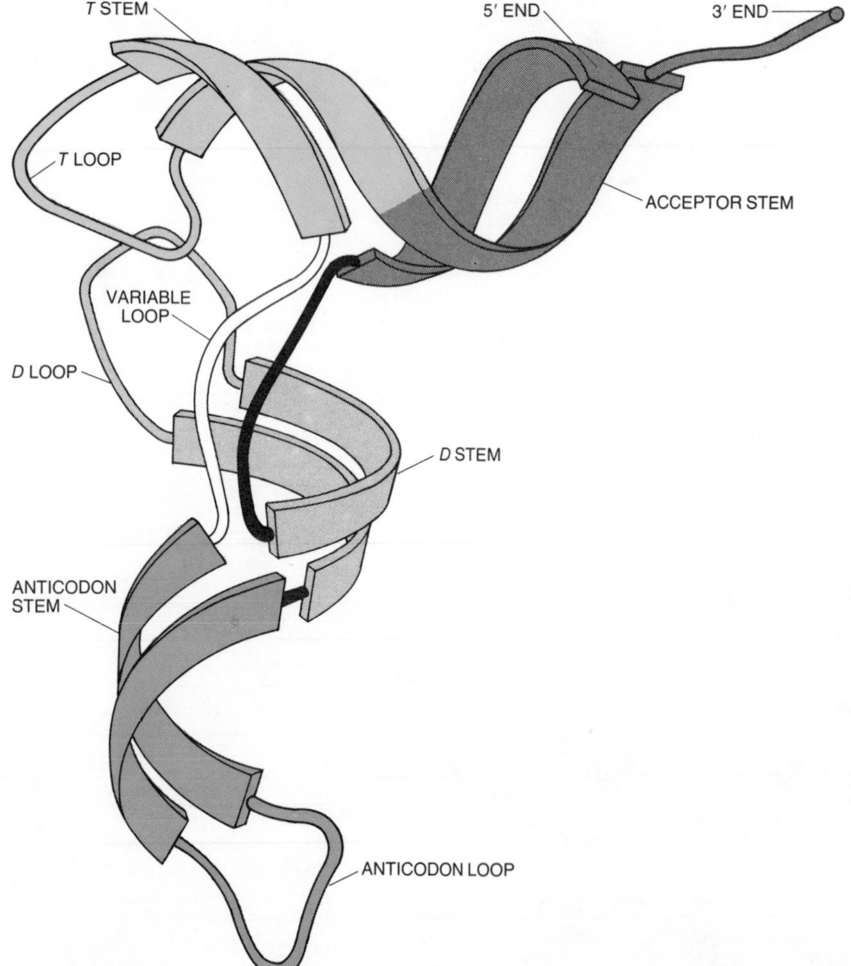

HELICAL SEGMENTS of the tRNA molecule, corresponding to the four stems of the cloverleaf diagram, are represented by ribbons in this schematic view. The two helical regions are arranged at right angles to provide the structural framework for the *L*-shaped folding pattern. Each region consists of about 10 base pairs, corresponding to roughly one turn of the double helix. The helix in these regions is similar to the double helix of DNA, except that in transfer RNA the two strands of the helix are formed by different parts of same polynucleotide chain.

With transfer RNA the first step of the process—crystallizing the molecule—turned out to be a major hurdle. In 1968 our group at the Massachusetts Institute of Technology and workers in five other laboratories discovered that it was possible to crystallize different species of tRNA by dissolving them in various mixtures of solvents and allowing the solvents to evaporate slowly. This advance caused great excitement among molecular biologists, since it seemed that the major hurdle had been overcome and that the three-dimensional structure of transfer RNA was within reach. Our elation was soon followed by some degree of despair when it was realized that although many different species of tRNA had been crystallized, most of the crystals were quite disordered. As a result the crystals provided diffraction patterns with very low resolution (usually between 10 and 20 angstroms) and hence could reveal little of the detailed structure of the molecule. Although it was exciting to discover that tRNA was crystallizable, it was frustrating to realize that further work had to be done before suitable material was available for X-ray-diffraction analysis.

Together with Gary J. Quigley and Fred L. Suddath we made a concerted effort to find conditions where tRNA would form a well-ordered crystal that would produce an X-ray-diffraction pattern with sufficient resolution to reveal the three-dimensional structure of the molecule. For two years we surveyed a large number of different tRNA species and crystallizing conditions. Finally we made an important discovery: the addition of spermine, a small positively charged molecule, resulted in the formation of a highly ordered crystal of a tRNA extracted from yeast cells that was specific for the amino acid phenylalanine. The spermine-stabilized crystal had a diffraction pattern that extended out to a resolution of nearly two angstroms.

Late in 1972, working with Alexander McPherson, Daryll Sneden, Jung-Ja Park Kim and Jon Weinzierl, we obtained an electron-density map of the crystal in which we were able to trace the backbone of the polynucleotide chain of the tRNA at a resolution of four angstroms. At that resolution it was not possible to perceive the individual bases of the polynucleotide chain, but the electron-dense phosphate (PO_4) groups along the backbone of the molecule could be seen as a string of beads coiled in three-dimensional space. To our great surprise the polynucleotide chain was organized in such a way that the molecule was shaped like an *L*, with one arm of the *L* made up of the acceptor stem and the *T* stem and the other arm made up of the *D* stem and the anticodon stem. The complementary hydrogen-bonded sequences that had been identified in the cloverleaf diagram were clearly seen as RNA double helixes. The various loops occupied strategic positions either at one end of the molecule or at the corner of it, where the *T* and *D* loops were coiled together in a complex manner.

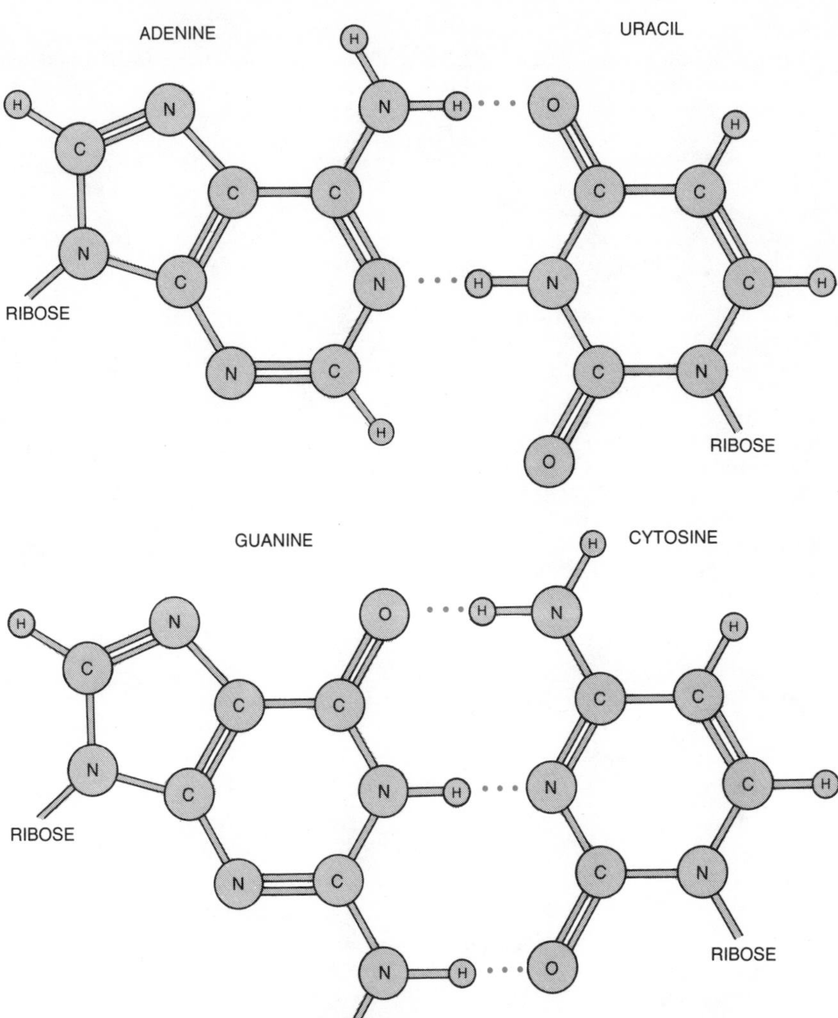

COMPLEMENTARY HYDROGEN BONDING between bases in the helical regions of the transfer-RNA molecule follows the pattern first proposed by James D. Watson and Francis H. C. Crick for double helix of DNA (except that in tRNA uracil replaces thymine). Adenine and uracil pair with two hydrogen bonds, whereas guanine and cytosine pair with three.

This folding of the molecule was entirely unexpected. Over the preceding few years a number of investigators had recognized the features common to the cloverleafs of all transfer RNA's and had tried to predict how the tRNA molecule might be folded. As is so often the case, however, nature proved to be subtler than had been imagined. The *L*-shaped folding served to explain a number of chemical observations that had accumulated, and it also made people wonder what functional purpose was served by this unusual shape.

By mid-1974, together with Joel L. Sussman, Andrew H.-J. Wang and Nadrian C. Seeman, we had interpreted the electron-density map at a resolution of three angstroms. The overall form of the molecule was the same as the one apparent at four angstroms, but now many more details were visible, including the positions of most of the nucleotide bases. At about this time Jon Robertus, Brian F. C. Clark, Aaron Klug and their colleagues at the British Medical Research Council Laboratory of Molecular Biology in Cambridge described their X-ray-crystallographic analysis of a transfer RNA at a resolution of three angstroms. Their tRNA was the same spermine-stabilized yeast phenylalanine tRNA, but it was in a different crystal form. Even though the molecule was packed differently in the crystal lattice, comparison of the two three-dimensional structures resulting from the analyses showed that the structures were virtually identical. This agreement between the findings of the two groups provided important evidence that the structure of the tRNA molecule is independent of how it is packed in a crystal.

The map of the tRNA molecule at a resolution of three angstroms confirmed our earlier finding that it is organized into two columns of nucleotide bases stacked at right angles to each other. These columns have both helical and nonhelical regions corresponding to the stems and loops of the cloverleaf diagram. The high-resolution map further revealed that the two helical regions each consist of about 10 base pairs, corresponding to one turn of the double helix, and possess the same type of hydrogen bonding between complementary nucleotide bases as that found in the double helix of DNA.

In the nonhelical parts of the tRNA molecule many of the nucleotide bases are oriented with their hydrogen-bonding groups pointed toward the interior of the molecule, where they participate in a variety of unusual hydrogen-bonding interactions known as tertiary interactions. Such bonds may occur between two or three bases that are not usually considered complementary, between a base and the ribose-phosphate backbone of the transfer-RNA chain or even between different parts of the backbone itself. The fact that several tertiary interactions in tRNA involve the hydroxyl (OH) groups of the sugar ribose is of particular interest, because hydroxyl groups are absent from the sugar molecules of DNA. Such tertiary interactions

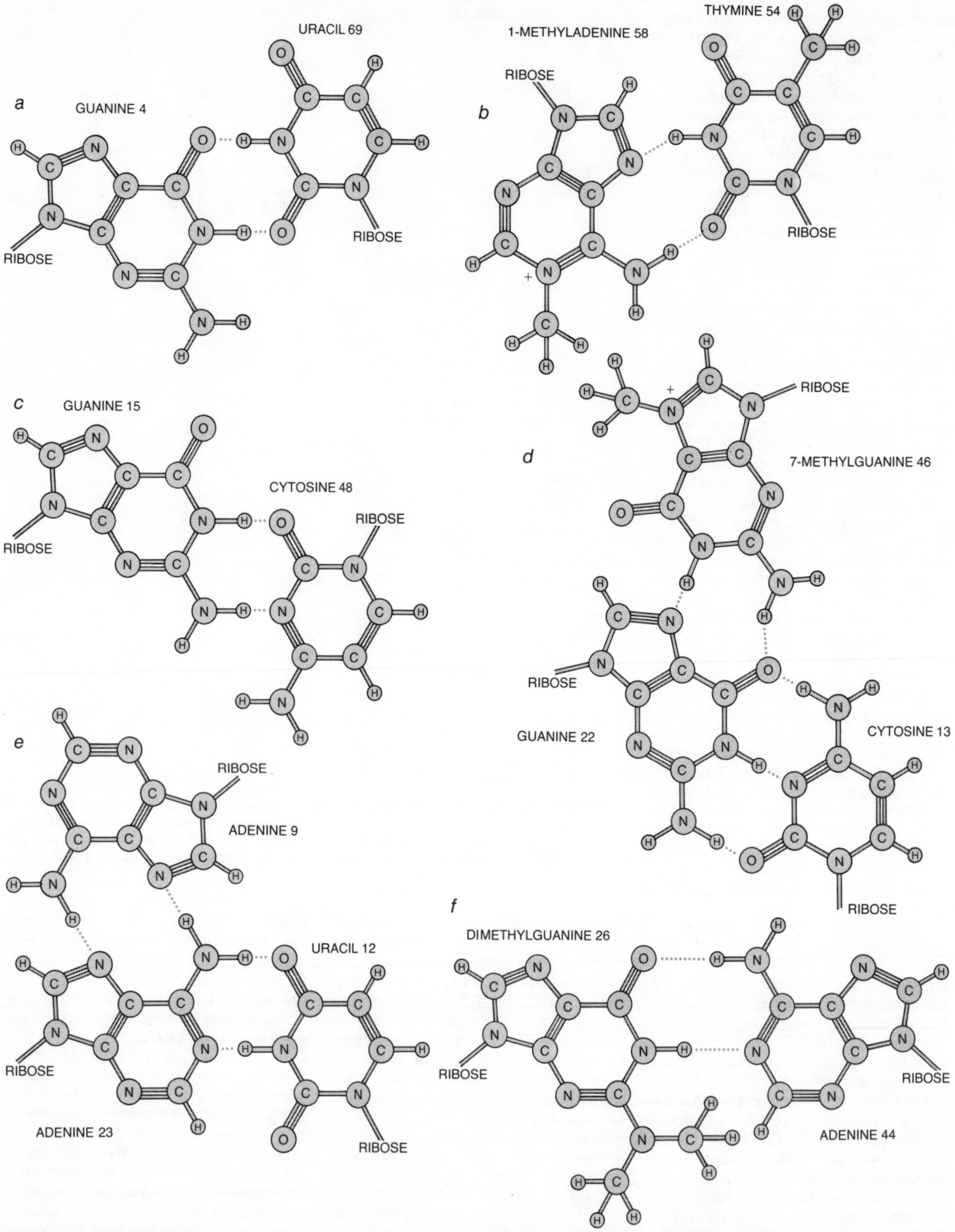

UNUSUAL INTERACTIONS between bases stabilize the folding pattern of the transfer-RNA molecule. In the acceptor stem the normally noncomplementary bases guanine and uracil are held together by two hydrogen bonds as the result of a slight lateral "wobble," or displacement, in one of the bases (*a*). In the *T* loop 1-methyladenine is paired with thymine, a modified form of uracil that has an added methyl group (*b*). In the core region of the molecule, immediately below the corner, guanine and cytosine are paired, but with two hydrogen bonds instead of the usual three. This pairing is of the *trans* type because the ribose groups fall on opposite sides of the pair (*c*). Also in the core region are two complex systems of hydrogen bonding involving three bases in the same plane (*d*, *e*). In the region joining the *D* stem and the anticodon stem a dimethylated guanine is paired with an adenine by two hydrogen bonds (*f*). Because of the bulky methyl groups on the guanine this base pair is not planar; the two bases are tilted about 25 degrees away from each other like the blades of a propeller. The dimethyl guanine is stacked at the bottom of the *D* stem and the adenine is stacked at the top of the anticodon stem, an arrangement that stabilizes the junction between the two stems. For a more schematic view of these interactions see the diagram on page 241.

are simply not needed in a regular linear nucleotide chain such as that of DNA, but they are essential for stabilizing the complex coiling of the polynucleotide chain in tRNA.

One unusual hydrogen-bonding arrangement was found in the acceptor stem, where the pair of nucleotide bases guanine-uracil occurs in place of the normal pair guanine-cytosine or adenine-uracil. The possibility of such a pairing had been suggested several years earlier when Francis H. C. Crick made the observation that it was likely certain additional types of base pairing would be found at the position of the third base in the interaction between the messenger-RNA codon and the transfer-RNA anticodon. One of the "unconventional" arrangements Crick had postulated was a guanine-uracil pair that would be connected by two hydrogen bonds as a result of a "wobble," or slight lateral displacement, in one of the bases. Continued analysis and refinement of the electron-density map at a resolution of 2.5 angstroms confirmed the wobble type of pairing between guanine and uracil in the acceptor stem.

Several other novel arrangements of hydrogen bonds have been discovered among the tertiary base-base interactions in the transfer-RNA molecule [*see illustration on opposite page*]. The variety of these interactions was one of the most surprising findings to emerge from our structure-determination work.

Most of the flat nucleotide bases in transfer RNA are organized in two stacked columns that form the arms of the *L*-shaped molecule. This arrangement explains the unusual stability of tRNA. If one heats a solution containing tRNA molecules, they will denature, that is, the polynucleotide chain will unravel and assume random conformations in the solution. As soon as the solution cools, however, the molecule will immediately snap back to its native conformation. This behavior is quite different from that exhibited by most proteins, which denature irreversibly; egg albumin, for example, turns white and opaque when the egg is boiled and stays that way when the egg is cooled.

Why does the transfer-RNA molecule revert so readily to its native structure? It is known that the stacking interaction between the adjacent nucleotide bases in the interior of the DNA double helix is one of the major stabilizing features of that molecule. Similarly, the bases of tRNA are predominantly hydrophobic (water-repelling), so that they retreat from the surrounding solvent into the interior of the folded polynucleotide chain; this behavior helps to return the tRNA molecule to its native—and stablest—conformation. In proteins there is usually no comparable interaction that will make the polypeptide chain refold spontaneously. Thus it appears that the structure of tRNA is organized to preserve the stabilizing feature of the stacking interactions between bases. At the same time some very complex molecular architecture holds the two stacked columns at right angles to each other.

An important aspect of the tertiary interactions found in yeast phenylalanine tRNA is the fact that many of them involve bases that are the same in the polynucleotide sequences of all tRNA's. Moreover, bases occurring in regions of the polynucleotide chain that have variable numbers of nucleotides are usually unstacked and located in loops that protrude from the surface of the tRNA molecule. These findings suggest that the structural framework of yeast phenylalanine tRNA may accommodate the nucleotide sequences found in other tRNA's. For example, in yeast phenylalanine tRNA one variable region of the *D* loop contains two nucleotides, and this segment of the polynucleotide chain arches away from the molecule and returns. If there were more nucleotides in this region, it is likely that the bulge would be somewhat larger; conversely, if there were fewer nucleotides, it would be smaller. The size of such variable loops, however, would not affect the overall folding pattern of the molecule.

A number of important problems concerning the three-dimensional structure of transfer RNA's in general remain unsolved. It is not clear, for example, what the detailed structure will be for tRNA's with very large variable loops. The structure of "initiator" tRNA's, which start the synthesis of proteins by laying down the first amino acid, is also of interest. Some initiator tRNA's have polynucleotide sequences that depart somewhat from the sequences common to other tRNA's, particularly in the *T* loop. It is quite likely that these differences are associated with a structure slightly different from that of yeast phenylalanine tRNA.

Our crystals of yeast phenylalanine tRNA contain almost 75 percent water. It is important to ask whether the molecule has the same form in solution (where it is biologically active) that it has in the crystal. Fortunately there have been numerous investigations of yeast phenylalanine tRNA in solution. These studies make it possible to correlate the structure observed in the crystal with various chemical characteristics of the molecule. For example, one of the features of yeast phenylalanine tRNA in solution is that some nucleotides seem to be readily available for chemical modification when chemical reagents are added to the solution, whereas other nucleotides are not. This disparity was puzzling until the structure of the molecule in the crystal emerged. Then it became apparent that only certain nucleotides, such as those that protrude from the molecule in the crystalline state, are readily available for chemical modification. In general there is an excellent correlation between the susceptibility of a region of the tRNA molecule to chemical modification and the accessibility of that region of the molecule in the crystalline state.

Several other types of experiments carried out in solution can be interpreted in the light of the three-dimensional structure, including experiments based on nuclear magnetic resonance, which is sensitive to the three-dimensional structure of a molecule. Several investigators have found a good correlation between the nuclear-magnetic-resonance signals obtained from transfer-RNA molecules in solution and the three-dimensional structure deduced from X-ray-diffraction analysis of yeast phenylalanine tRNA in the crystal. These and other findings provide convincing evidence that the structure of the tRNA molecule in the crystal is the structure of the biologically active form of the molecule.

Such correlations are important because the principal reason for determining the structure of a biological molecule is to perceive how it functions in a biological system. What has the structure of transfer RNA taught us about how the molecule works? Here one can speak with considerably less confidence because the necessary experiments have not yet been carried out. First one would like to know how the enzyme aminoacyl-tRNA synthetase recognizes and selects only the correct tRNA for attachment to a specific amino acid. If this process is to be understood fully, it will be necessary to determine the three-dimensional structure of the synthetase when it is complexed with the tRNA, so that the nature of the specific interactions between the enzyme and the nucleic acid can be perceived. Studies of this kind are now under way in many laboratories, and the answers should be forthcoming in the near future. Already some experiments suggest that in the recognition process certain regions of the tRNA molecule are more important than others.

Another major question is: Why does the tRNA molecule have an *L*-shaped form in which the anticodon is more than 76 angstroms away from the attached amino acid? The definitive answer has not yet been obtained, but it is quite likely that the shape of tRNA is related to its essential transfer function inside the ribosome. For two adjacent tRNA's in the *A* and *P* sites to be brought close together on the messenger-RNA strand so that the growing polypeptide chain can be transferred from one to the other, it may have been necessary for the cell to have evolved a tRNA molecule bent in this peculiar fashion. Perhaps the acceptor arm of the *L* rotates inside the ribosome so that the protein chain can be transferred to the tRNA bound to the next codon on

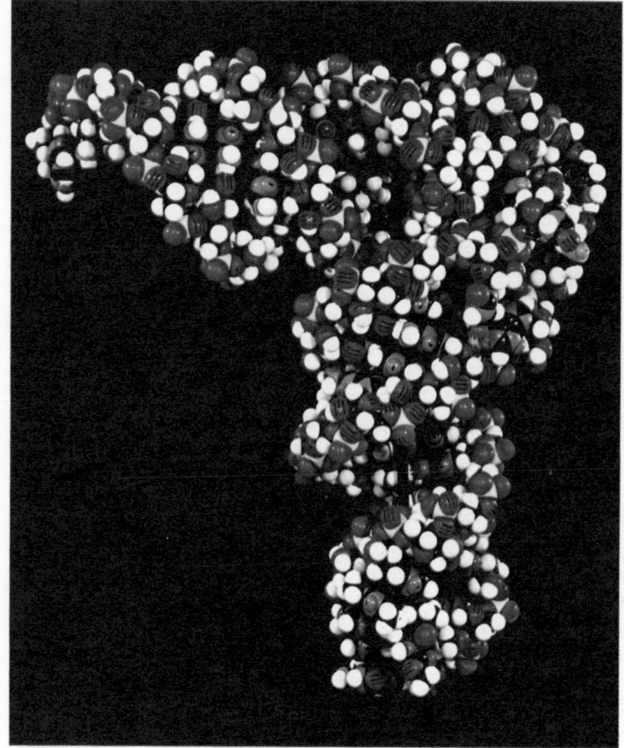
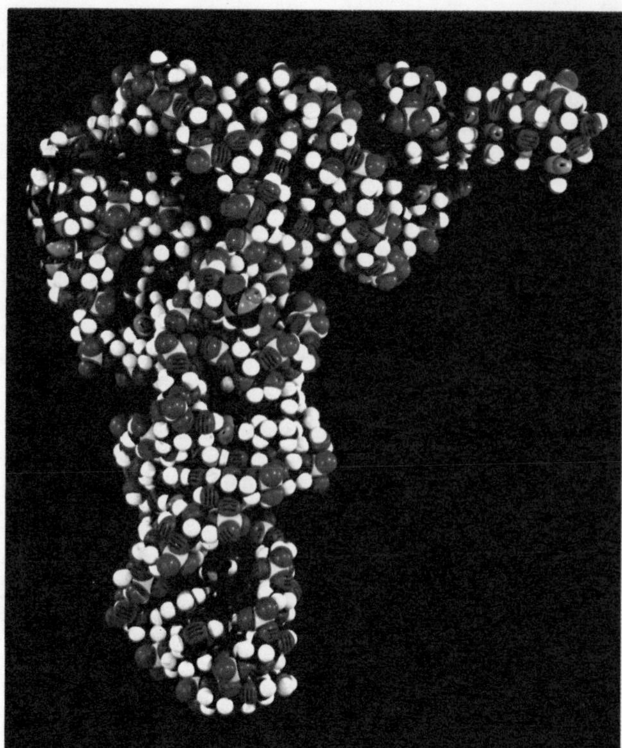

SPACE-FILLING MODEL of yeast phenylalanine tRNA approximates the actual shape of the molecule. It was constructed on the basis of X-ray-diffraction analyses conducted in the authors' laboratories at the Massachusetts Institute of Technology and the Duke University School of Medicine. The polynucleotide chain of tRNA is folded into a compact *L*-shaped structure. During protein synthesis the amino acid phenylalanine is joined to the end of the horizontal arm of the *L*. Three nucleotide bases at the end of the vertical arm then recognize the genetic code for phenylalanine on the strand of messenger RNA (mRNA). Finally the amino acid is transferred to the growing protein chain. In this molecular model carbon is black, oxygen red, nitrogen blue, phosphorus yellow and hydrogen white.

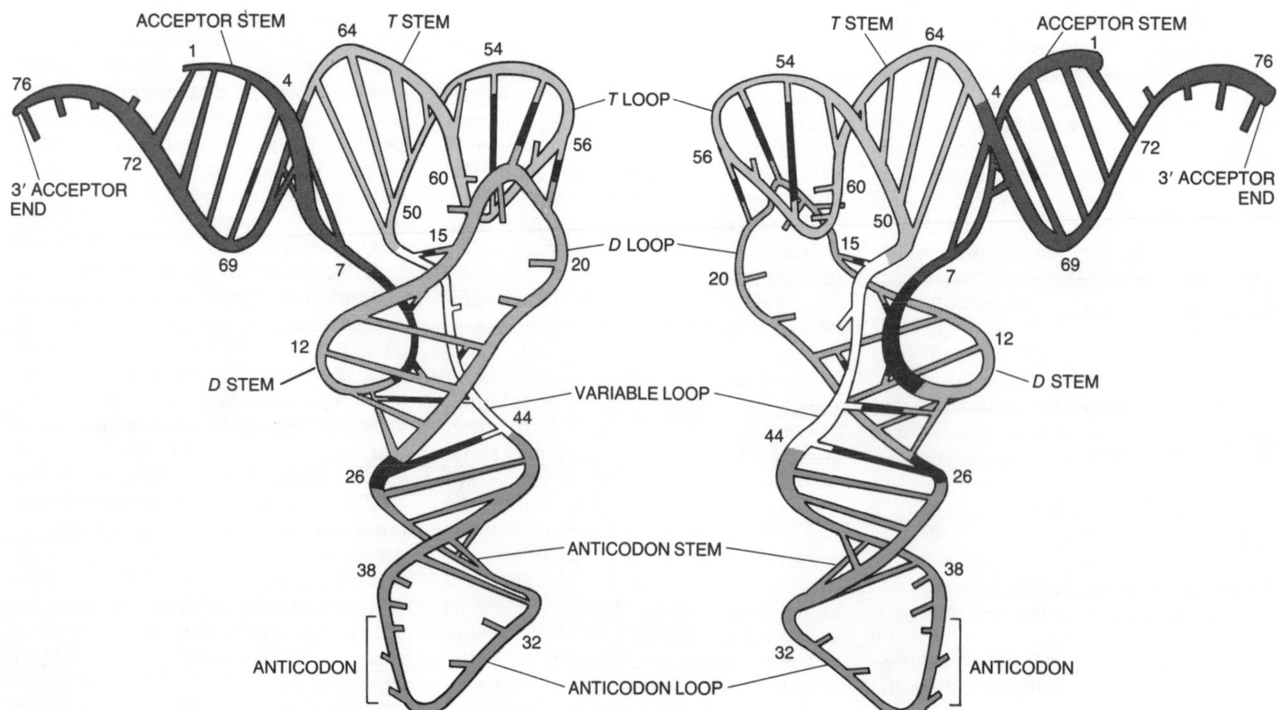

FOLDING PATTERN of the polynucleotide chain in yeast phenylalanine transfer RNA is diagrammed. The sugar-phosphate backbone of the molecule is represented as a coiled tube, with the cross rungs standing for the nucleotide base pairs in the stem regions. The short rungs indicate bases that are not involved in base-base hydrogen bonding. The shading refers to cloverleaf diagram on page 231.

the messenger-RNA chain. This second tRNA would then take the chain onto its own amino acid and in turn pass the chain along. The considerable distance between the end of the acceptor stem and the anticodon loop of the tRNA molecule may also be functionally important in that different ribosomal proteins can simultaneously interact with several regions of the tRNA in order to help maintain the precision of protein synthesis.

The view of the tRNA molecule that has been obtained from X-ray-diffraction analyses of molecules in a crystal is essentially a static one. In its natural environment within the cell the molecule may undergo conformational changes, particularly when it interacts with large molecular structures such as the ribosome. Recent experiments suggest that inside the ribosome the *D* loop and the *T* loop of the tRNA molecule may move away from each other when the molecule shifts from the *A* site to the *P* site. It is also possible that the shape of the anticodon loop is modified when it comes in contact with the messenger RNA inside the ribosome. A fuller evaluation of these proposals will have to await the results of further research.

At the beginning of this article we described the role of transfer RNA in protein synthesis in some detail because that is the molecule's most essential role in biological systems. Without the tRNA molecule genetic information could not be expressed in the synthesis of proteins. In addition tRNA participates in a variety of other processes, some of which are of great importance. For example, tRNA molecules can donate amino acids to preformed protein molecules or to the molecular structure of the cell wall in bacteria independently of the ribosome.

Another process in which tRNA participates is the control of gene expression. Certain tRNA's with an amino acid attached to them are known to determine whether or not genes, that is, segments of DNA, will be expressed by regulating their transcription into messenger RNA. The detailed mechanism is not known, but in some systems the control function is associated with a particular modified nucleotide in the tRNA molecule, for example a uracil that has been converted into a pseudouracil. It is thought that tRNA helps to control the expression of many different genes, although the exact number is not known. The instances that have been most intensively studied are those of the genes that regulate the synthesis of amino acids, where tRNA-mediated regulation plays a major role.

Other observations suggest that transfer RNA may be involved in still more types of biochemical regulation. For example, in the course of embryonic development one kind of modification of certain nucleotides in a tRNA gives way to another kind. Similarly, when a normal cell becomes cancerous, the kinds of modifications of nucleotides in its tRNA molecules change substantially. It is not yet known whether these transformations are associated with the regulatory functions of tRNA.

Another mysterious area concerns the relatively high number of modifications in the nucleotide sequences of the *D* loop as well as in those of the variable loop. Why has nature gone to such trouble to vary the nucleotides that project from the surface of the molecule? It is generally believed these sequences are not required for the specificity of protein synthesis; instead they may be involved in the regulatory functions of tRNA molecules, since the variable regions could provide sites for specific recognition by other molecules.

Finally, tRNA is associated not only with the synthesis of polypeptide chains but also with that of polynucleotide chains. This synthesis is carried out by special enzymes such as reverse transcriptase, which was discovered a few years ago as a constituent of several tumor viruses. Reverse transcriptase synthesizes a strand of DNA from a template of single-strand RNA, a direction of information flow that is the reverse of the normal one. Surprisingly, a specific type of tRNA first binds to the enzyme and to the viral RNA and signals the synthesis of the DNA copy to begin. Why a tRNA serves this purpose is completely unknown.

It is probable that tRNA-like molecules were an essential component of the earliest living systems. Once these molecules were formed their unusual stability may have resulted in their gradually being utilized to serve purposes other than their main function in protein synthesis. Although the elucidation of the three-dimensional structure of tRNA has been an important step forward, a great deal remains to be learned about this versatile molecule and its many roles in the living cell.

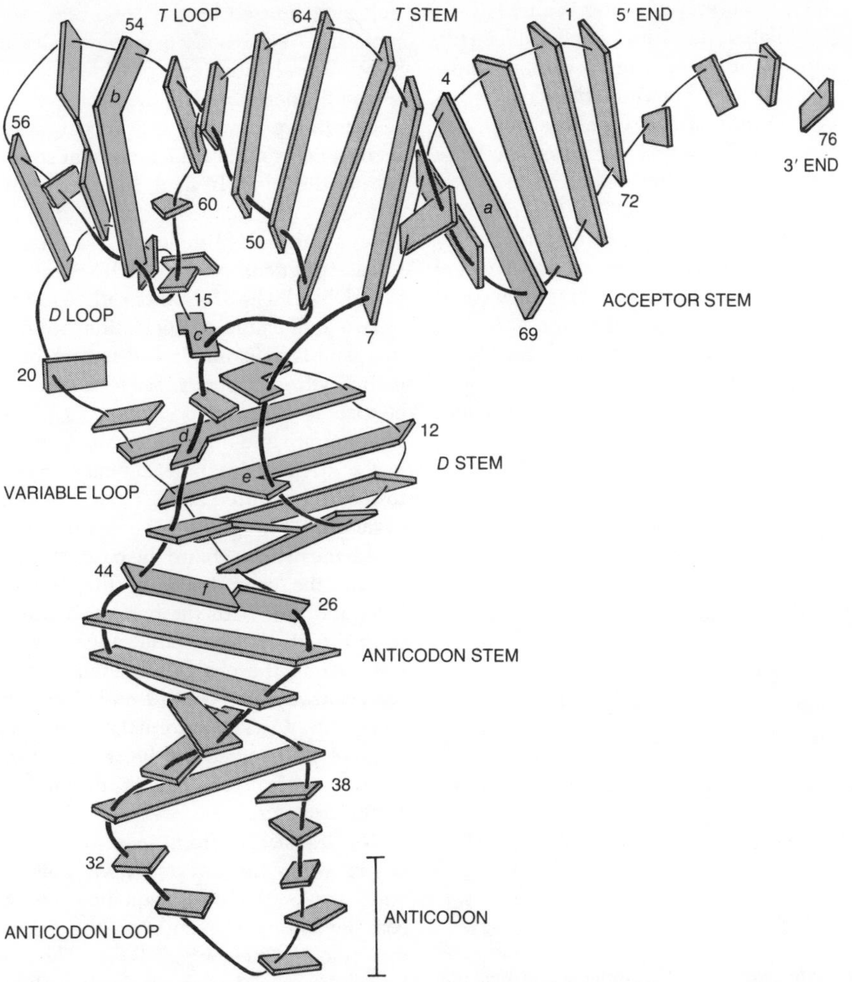

ORIENTATION OF BASES in the transfer-RNA molecule is shown in this schematic view. The polynucleotide backbone is reduced to a thin line, with the short boardlike structures representing unpaired bases and the longer boards representing base pairs. The letters refer to the molecular diagrams on page 238. Note the presence of tertiary interactions between three bases in the core region of the molecule below the corner of the *L*. Overall the molecule is composed of two stacked columns of bases at right angles to each other. The stacking interactions between the parallel bases in the interior of the molecule provide a major stabilizing force.

19

A DNA Operator-Repressor System

by Tom Maniatis and Mark Ptashne
January 1976

An operator is a segment of DNA adjacent to a gene; a repressor is a protein that binds to the operator and controls the expression of the gene. How such a system works in a virus is explored in detail

A fundamental property of living cells is their ability to turn their genes on and off in response to extracellular signals. In the human body, for example, every cell (with the exception of a few cell types such as the red blood cell) has the same set of genes, yet in the course of embryonic development cells take on different shapes and functions as their genes are selectively switched on and off. How are the genes regulated? Is there a common mechanism underlying such regulation in different organisms?

Through the study of gene regulation in bacteria and viruses it has been learned in recent years that a fundamental mechanism of gene control depends on the interaction of protein molecules with specific regions on the long-chain molecule of DNA, the material that embodies the genetic instructions of all organisms from bacteria to man. As a result of this interaction genes are switched on or off. In the best-understood instances genes are switched off by controlling molecules named repressors. The existence of repressors was first hypothesized in 1960 by François Jacob and Jacques Monod of the Pasteur Institute in Paris. Seven years later Walter Gilbert (in collaboration with Benno Müller-Hill) and one of us (Ptashne), working independently at Harvard University, succeeded in isolating repressors from bacteria [see "Genetic Repressors," by Mark Ptashne and Walter Gilbert; Scientific American Offprint 1179]. Later it was shown that repressors could bind tightly and specifically to sites on DNA called operators and that in so doing they could prevent genes adjacent to the operators from being transcribed and translated into proteins.

Since these early discoveries we and many others have pursued the molecular details of gene repression. This article is a brief progress report on some of the things that have been learned. We now know, for example, the sequence of bases, or code units, in DNA that constitutes the operators to which a repressor binds. In the case we shall discuss here the operators have several nearly identical binding sites, each capable of being recognized by the same repressor molecule. We are only beginning to learn why several sites are provided when seemingly one would do the job.

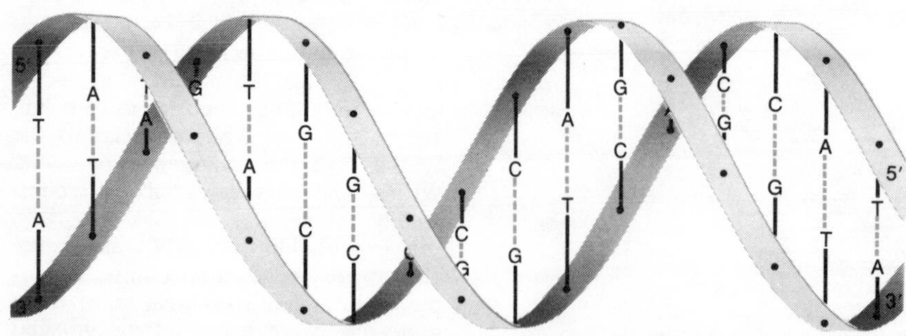

DOUBLE HELIX OF DNA encodes the genetic information of all cellular organisms and most viruses. (In some viruses the genetic material is RNA.) The genetic code is written in the particular sequences of nitrogenous bases that connect the two strands of the DNA molecule. The bases are of four kinds: adenine (*A*), thymine (*T*), guanine (*G*) and cytosine (*C*). *A* always pairs with *T* and *G* always pairs with *C*. The strands of the double helix consist of alternating subunits of phosphate and ribose, a five-carbon sugar. In one strand the phosphate links the No. 5 carbon in one sugar to the No. 3 carbon in the adjacent sugar, creating what is denoted a 5′–3′ linkage. In the opposing strand, proceeding in the same direction, the linkage is 3′–5′. Thus each strand of DNA molecule has a 5′ end and a 3′ end.

Before we describe this recent work let us quickly review the molecular structure of the gene. In man, as in bacteria, a gene can be defined as a sequence of bases along a DNA molecule. (In certain viruses the gene consists of RNA rather than DNA.) The DNA molecule consists of two long chains of nucleotides wound in a double helix and linked to each other by hydrogen bonds. Each nucleotide consists of a deoxyribose sugar, a phosphate group and one of four nitrogenous bases: adenine, guanine, thymine or cytosine (abbreviated *A*, *G*, *T* and *C*). The sugar and phosphate groups form the backbone of each chain; the bases extend toward the central axis of the double helix and pair with the bases extending from the other chain. The sequences of bases along the chains are complementary: *A* always pairs with *T* and *G* always pairs with *C*. The information content of DNA is specified by the sequence of bases. A typical gene consists of roughly 1,000 base pairs.

The translation from gene to protein begins when the enzyme RNA polymerase copies the base sequence into a complementary sequence on the linear molecule of "messenger" RNA. The intracellular translating machines called ribosomes attach themselves to the messenger RNA and translate its base sequence into a sequence of amino acids, which are linked to form a protein molecule. Since there are only four different bases and 20 different amino acids, a sequence of three bases is needed to specify one amino acid. (For example, *ACA* specifies the amino acid threonine.) It is clear that the translation of a gene into a protein molecule might conceivably be repressed, or blocked, at any one of several stages along this complex pathway. It turns out in the case we have studied that repression takes place directly at the DNA molecule, so that the genetic information is not transcribed into messenger RNA unless the repressor is inactivated. Just how is this repression achieved?

Bacteriophage Lambda

The repressor we have studied is a protein molecule manufactured by bacteriophage lambda, a virus that infects the common colon bacterium *Escherichia coli.* The bacteriophage, or phage, particle consists of a single DNA molecule with about 47,000 base pairs, enough for some 50 genes, enclosed in a protein coat equipped with a tail through which the DNA is injected into the bacterial cell. Once the DNA is inside the cell it can follow either of two pathways. It can complete the course of infection by causing the machinery of the bacterial cell to translate the phage genes into proteins. Some of the proteins are enzymes that replicate several hundred

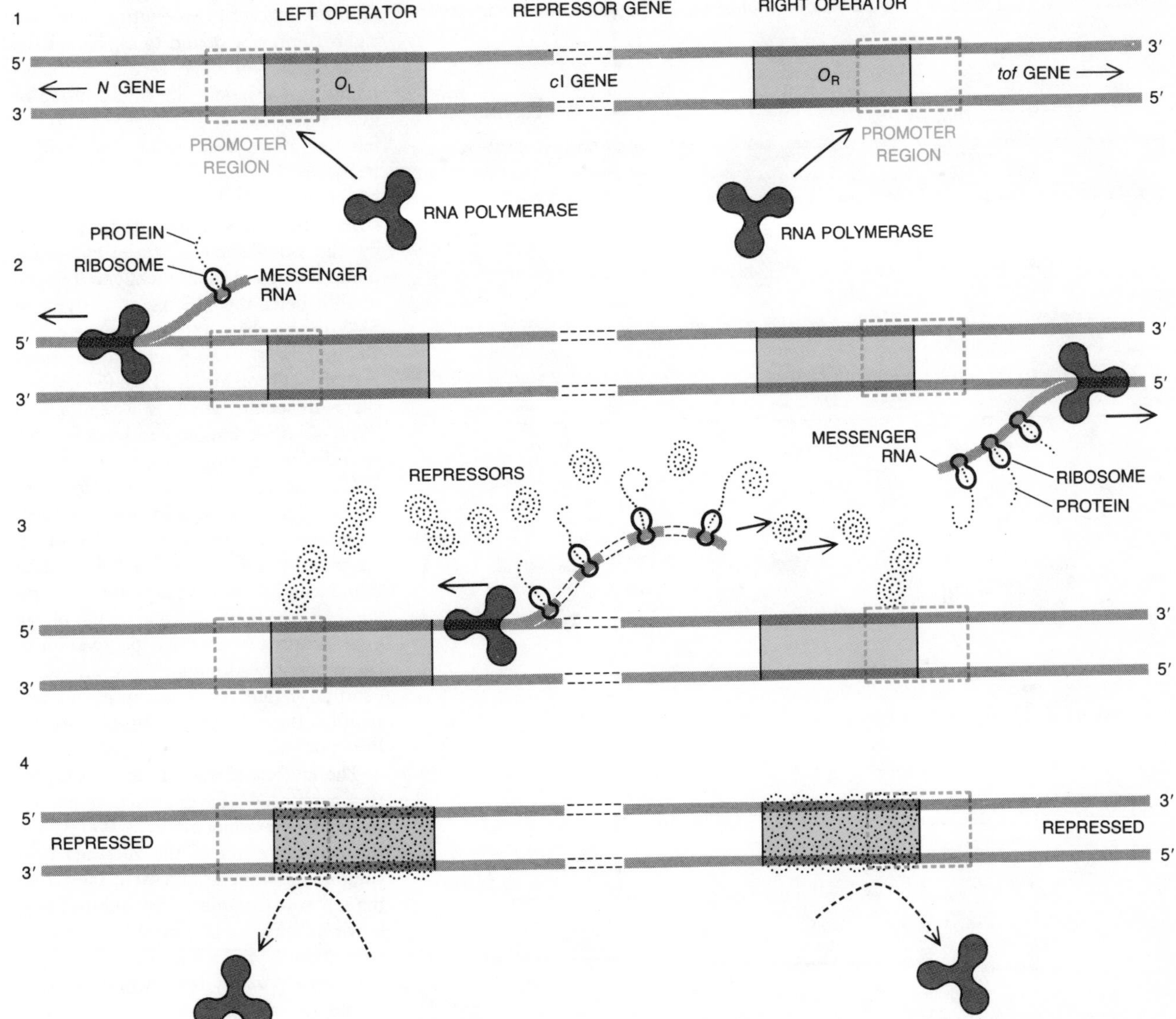

EXPRESSION AND REPRESSION OF GENES is conveniently studied in bacteriophage lambda, a virus that infects the bacterium *Escherichia coli.* When the 50 or so genes of the phage-lambda DNA are translated into protein inside a cell of *E. coli,* the phage multiplies and kills the cell in about 45 minutes. In some cases, however, repressor proteins specified by a particular gene in the lambda DNA, the *c*I gene, get the upper hand and block the transcription of genes on either side, forcing the lambda DNA into a state of dormancy. Normal transcription of lambda DNA into "messenger" RNA and its translation into protein molecules are depicted in *1* and *2*. The transcription is effectuated by the enzyme RNA polymerase, which attaches itself to a promoter region and assembles an RNA chain in the 5′–3′ direction by copying a DNA strand of the opposite polarity. Thus the RNA-polymerase molecules travel in opposite directions, copying the different strands of the DNA as they transcribe into messenger RNA the complete instructions for replicating the phage, beginning with the *N* gene on the left and the *tof* gene on the right. Transcription of these two genes begins just outside the operator. The ribosomes fasten themselves to the emerging "tape" of messenger RNA and translate the encoded message into the protein molecule. (In this diagram the structures are not drawn to scale; the ribosomes in particular are much larger than they are shown.) Under other conditions (*3*) the *c*I repressor gene is transcribed by the same process and translated into repressor molecules. (Specific terminator signals in the DNA prevent RNA polymerase from continuing on into left operator.) Perhaps singly as monomers, but more likely as dimers depicted in *4* or even as tetramers, repressor molecules migrate to binding sites in two operator regions of lambda DNA, blocking access of RNA polymerase to promoter regions of *N* and *tof* genes.

copies of the phage-DNA molecule; other phage proteins package each DNA copy in a protein coat, thus creating multiple copies of the original phage particle. Typically within 45 minutes the bacterial cell, swollen with phage particles, bursts.

The other pathway is the more interesting one for our purposes. Occasionally after the phage genes enter the bacterial cell they are switched off and the phage DNA becomes integrated into the DNA of the host cell. There it remains dormant, replicating with the bacterial DNA at every cell division and giving rise to a population of *E. coli* cells each of which contains a chain of phage genes. The dormant phage genes are called a prophage.

It has been known for some years that the phage genes are turned off by a specific repressor molecule specified by one of the phage's own genes, the *c*I gene. The repressor actually binds to two separate operators on the phage-DNA molecule, thereby blocking the transcription of two different sets of genes. The turning off of these two sets of genes is sufficient to cause the 40-odd remaining genes, with the exception of *c*I, to stop functioning. The dormant phage genes can be switched on again by a suitable inducing agent such as a low dose of ultraviolet radiation, which causes the repressor to be inactivated.

The two operators, which are separated by some 2,000 base pairs (including the *c*I gene), are designated O_L and O_R, the subscripts denoting left and right. Repressor bound to O_L blocks the transcription of gene *N* and repressor bound to O_R blocks the transcription of gene *tof*. Gene *N* is transcribed to the left beginning near O_L and gene *tof* is transcribed to the right from the opposite DNA strand, beginning near O_R. (DNA chains have a polarity determined by the orientation of the sugar-phosphate linkage in their backbone. In each double helix the linkage is designated 5′–3′ in one chain and 3′–5′ in the other. The numbers 3 and 5 refer to the third and fifth carbon atoms in the five-carbon sugar molecule. RNA is assembled in the 5′–3′ direction, copied from a DNA strand of the opposite polarity.)

One can speculate that it is clearly to the phages' advantage for some of them to go into the prophage, or dormant, stage and to be "revived" at a later time under conditions that may be more favorable for multiplication. The phage genes are reactivated by inactivation of the repressor. Although there are many conditions that result in repressor inactivation, the details of the process are not fully understood.

The lambda-phage repressor is a protein composed of subunits that have a molecular weight of about 27,000. When the concentration of the subunits in a suitable medium is increased, they form dimers and tetramers: two-subunit and four-subunit associations. When the concentration is reduced, they dissociate. Dimers or possibly tetramers must form before the repressor is able to interact strongly with DNA. We shall return to the possible significance of this fact.

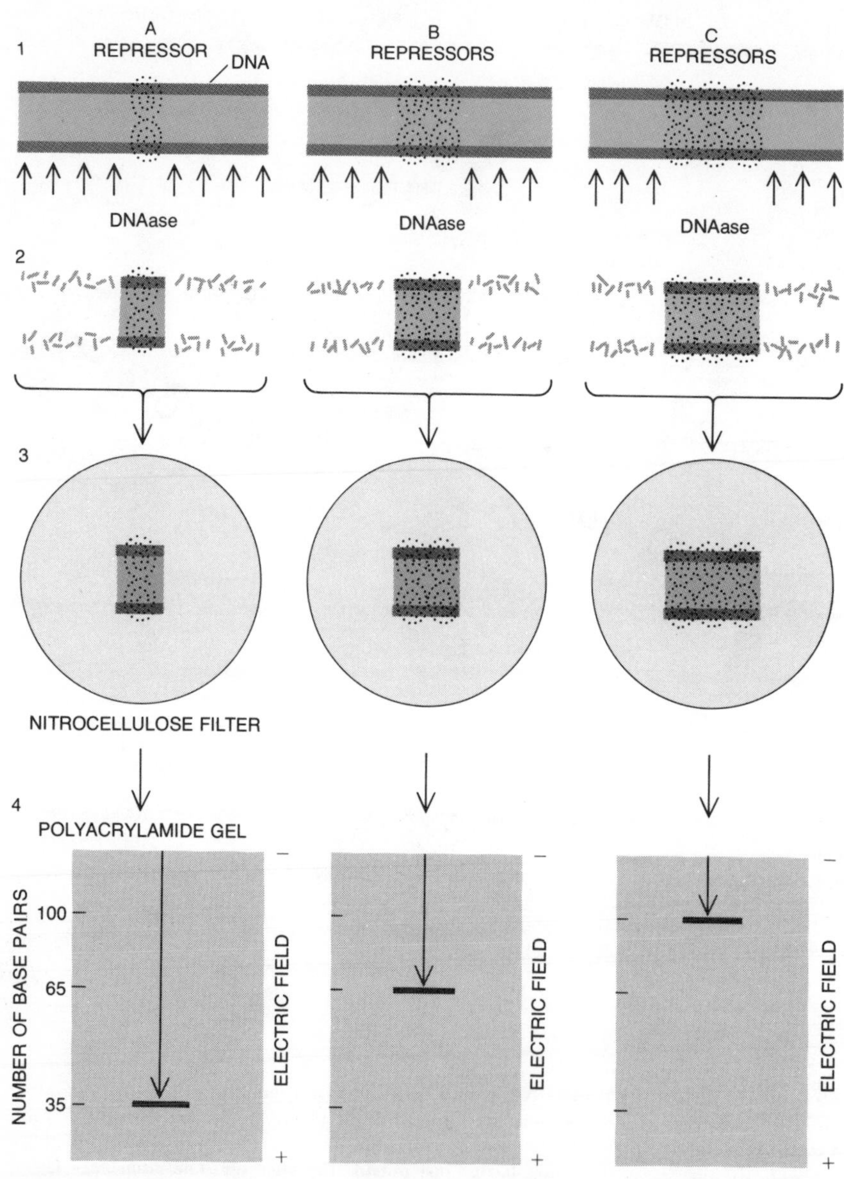

SIZE OF REPRESSOR BINDING SITES in operator regions of lambda DNA was established by mixing the DNA with repressor molecules. Wherever repressors bind to the DNA they protect it from digestion when the enzyme DNAase is added to the mixture (*1, 2*). As the ratio of repressor to DNA is increased (*A, B, C*) the segment of DNA protected increases in length, thus indicating that repressor binds to several adjacent sites. The pieces of protected DNA are collected on nitrocellulose filters (*3*), extracted and subjected to electrophoresis in a polyacrylamide gel (*4*). The fewer base pairs present in a fragment, the farther it will migrate when it is subjected to an electric field. Evidently this repressor can cover from 30 to 100 base pairs. If the DNA has been labeled with atoms of a radioactive isotope such as phosphorus 32, location of fragments can be visualized with autoradiogram.

The Isolation of Lambda Operators

Regions on the DNA molecule that are bound to specific proteins can be isolated by virtue of the fact that the enzyme DNAase digests any naked DNA it encounters but leaves intact any DNA that is covered by a protein. This property of DNAase was utilized by Allan Maxam and Gilbert to isolate an operator region from the DNA of *E. coli* and was later utilized by Vincenzo Pirrotta, then a postdoctoral fellow working with one of us (Ptashne), to isolate the operators on the DNA of the lambda phage to which the repressor becomes attached.

By growing phage particles in a nutrient medium of *E. coli* cells containing the radioactive isotope phosphorus 32 Pirrotta obtained molecules of phage DNA in which the radioactive atoms replaced many nonradioactive phosphorus atoms. The highly radioactive DNA was mixed with purified repressor and then with DNAase. The protected segments of operator DNA were recovered simply by passing the mixture through a nitrocellulose filter. The repressor, like many other proteins, binds tightly to the filter whereas free DNA and DNA-digestion products pass through it. The operator fragment bound to repressor is retained in the filter and can be washed out with a detergent solution. The operator fragment isolated in this way was found to be surprisingly large. A protein the size of the lambda-repressor dimer or tetramer should cover only 15 to 30 base pairs of DNA, but Pirrotta found that roughly 85 base pairs were protected from DNAase digestion by repressor.

We continued these studies of the lambda operators by trying to discover why such an unexpectedly large stretch of DNA is covered by repressor. One possibility we considered (the right one, it turned out) was that the lambda operators may have more than one repressor binding site. We reasoned that if the operators do have more than one binding site, the size of the fragment protected from DNAase digestion should depend on the ratio of repressor to operator in the digestion mixture.

We tested this possibility as follows. We repeated Pirrotta's procedure but varied the ratio of repressor to operator in the DNAase digestion mixture. We then determined the size of the operator fragments by subjecting them to electrophoresis in a polyacrylamide gel. The gel acts as a molecular sieve; under the influence of an electric field smaller DNA fragments migrate through the gel faster than larger ones. Since the operator fragments were labeled with radioactive phosphorus their position at the end of electrophoresis could be determined by placing the gel on photographic film and making an autoradiograph.

We discovered not only that the size of the operator fragment protected by repressor increases as the ratio of repressor is increased but also that the size increases in discrete steps. At low ratios of repressor to operator a single DNA fragment about 30 base pairs long is recovered. At the highest ratios a single fragment about 100 base pairs long is obtained. At intermediate ratios the fragments are of several intermediate lengths. Moreover, this interesting result was obtained whether we used lambda DNA containing only the left operator or lambda DNA containing only the right operator. These experiments and related ones led us to conclude that each lambda operator does in fact have more than one binding site for repressor. Apparently repressor molecules can line up adjacent to each other on the operator, covering a minimum of 30 base pairs and a maximum of 100.

An analysis of the complexity of the DNA sequence in the various operator fragments identified in these experiments revealed two important additional facts about the structure of the two operators. First, the repressor does not bind randomly to any site within the 100-base-pair operator sequence. Rather, it binds initially at a site adjacent to the *N* gene (in the case of O_L) and at one or two sites adjacent to the *tof* gene (in the case of O_R). As the repressor-to-operator ratio is increased secondary sites adjacent to the first sites are filled. Second, the base sequences of the various repressor binding sites are similar but not identical. We obtained strong evidence supporting these two facts by certain experiments we need not review here and confirmed them by the work we shall now describe.

Host-Restriction Endonucleases

Experiments on the properties of DNA molecules have been revolutionized in recent years by the use of the enzymes known as host-restriction endonucleases. Most of these enzymes, which are widely distributed in the bacterial kingdom, have the remarkable property of recognizing certain base sequences in DNA molecules and cutting the DNA within those sequences. For example, an enzyme (abbreviated *Hin*) isolated from

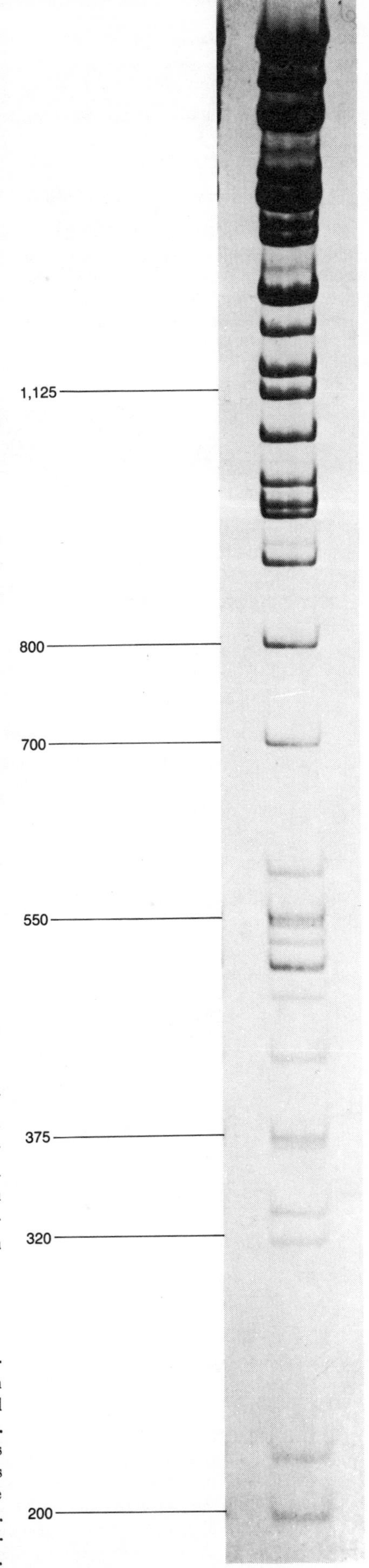

AUTORADIOGRAM shows the result of digesting lambda DNA with a host-restriction endonuclease, an enzyme (*Hin*) isolated from the bacterium *Haemophilus influenzae*. The enzyme cuts DNA when it encounters a particular sequence of six bases. It cuts lambda DNA into about 50 fragments whose sizes are established by electrophoresis. Numbers beside autoradiogram show number of base pairs in representative fragments.

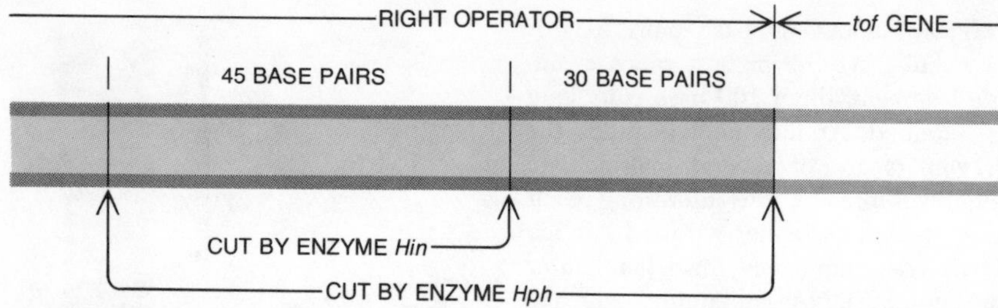

HOST-RESTRICTION ENDONUCLEASES were used to cut lambda-phage DNA in various ways until fragments were obtained in which the base sequence of the operator could be determined. An enzyme (*Hph*) that was isolated from the bacterium *Haemophilus parahaemolyticus* yields a DNA fragment consisting of 75 base pairs that include most of the right operator, which lies next to the *tof* gene. The fragment can be cut in two by enzyme *Hin*.

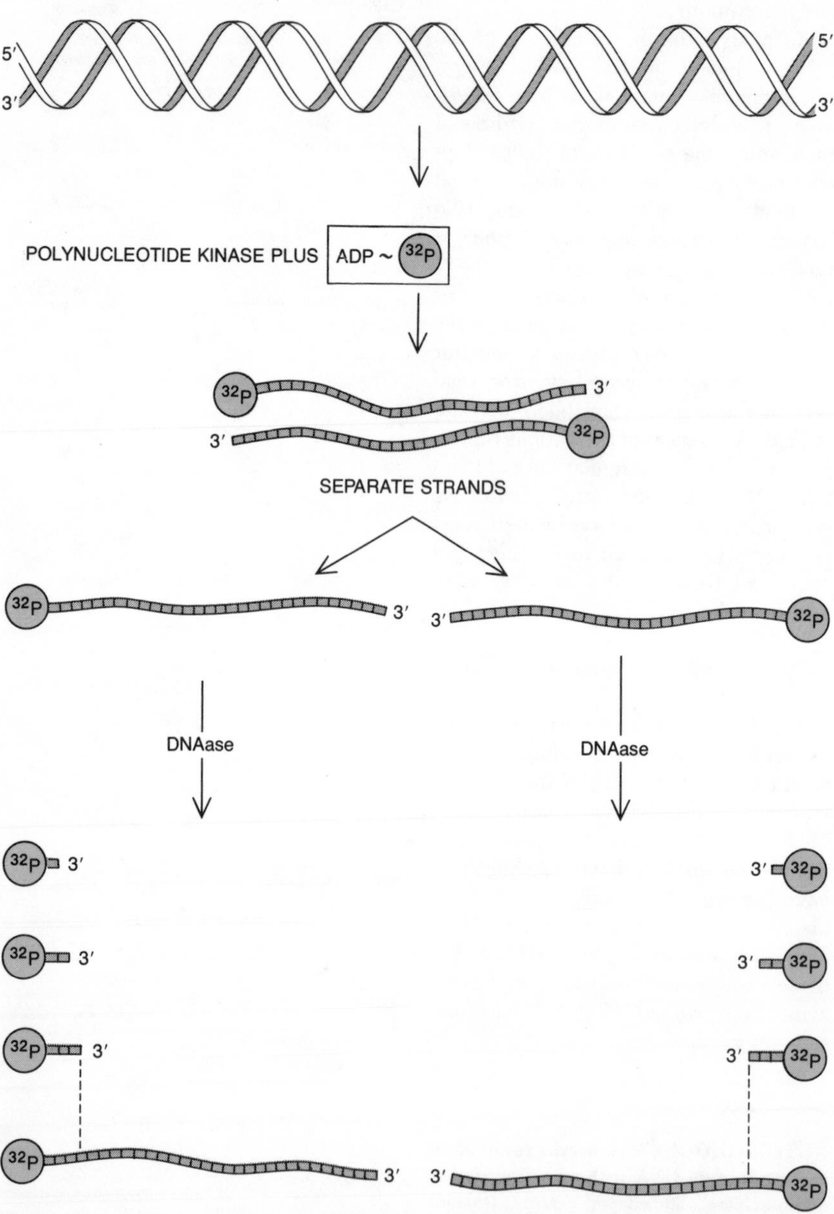

DETERMINATION OF BASE SEQUENCE in segments of operator DNA is established with the help of enzymes. First, polynucleotide kinase is used to attach an atom of phosphorus 32 (^{32}P) to the 5′ end of each strand in an operator segment including perhaps 30 bases. The strands are separated and mixed with exonuclease, a DNAase that degrades DNA starting from 3′ ends. By removing samples at intervals fragments from one base to 30 bases in length are obtained. Subsequent steps are described in illustration on opposite page.

the influenza bacterium *Haemophilus influenzae* cuts DNA in the middle of the sequence *GTTGAC* paired (on the other chain of the double helix) with *CAACTG*. Another enzyme (*Hpa*) isolated from the bacterial strain *Haemophilus parainfluenzae* cuts DNA in the middle of the sequence *CCGG* paired with *GGCC*. We need not be concerned here that such enzymes may help to protect bacterial species from foreign DNA's (hence the term restriction enzymes) or that some restriction endonucleases act in a more complex way than *Hin* and *Hpa*. The important point is that because the recognition sequences are short they tend to appear at many specific sites on DNA molecules. For example, the enzymes *Hin* and *Hpa* both cut lambda DNA at about 50 sites, generating about 50 specific DNA fragments, ranging in length from fewer than 100 base pairs to several thousand. (Remember that the total length of the lambda DNA is about 47,000 base pairs.)

We designed an experiment with lambda DNA and host-restriction enzymes that we hoped would yield two specific fragments: one containing the left operator and the other the right operator. We reasoned that since the enzymes *Hin* and *Hpa* cut lambda DNA into segments with an average length of 1,000 base pairs and since the operators are separated by some 2,000 base pairs, the enzymes should cut at least once between the operators and, of course, somewhere on either side of the operators. The result should be two fragments with one operator in each. We further reasoned that we could isolate the two fragments by mixing the digestion mixture with repressor and passing the mixture through a nitrocellulose filter. Presumably only those fragments bearing an operator would bind to the repressor molecules and would be trapped in the filter.

We performed this experiment using the enzyme *Hpa*. When we examined the trapped pieces of the lambda-DNA molecule, we found, as we had hoped, two specific fragments, one bearing the left operator and the other the right operator. When we repeated the experiment with the enzyme *Hin*, however, the results were strikingly different. Now we recovered not two fragments bound to repressor molecules but four. We interpreted this result to mean that *Hin*, unlike *Hpa*, cuts the lambda DNA within each operator, thereby splitting each operator into two fragments that independently bind repressor. Our conclusion implied that the largest fragment of operator, 100 base pairs long, isolated in the DNAase digestion experiment would be cut by the *Hin* enzyme. That prediction was soon verified.

The experiments with host-restriction enzymes enabled us to state conclusively not only that each lambda operator has multiple repressor binding sites but also that the sites can function independently. The latter conclusion was demonstrated by the sites' ability to bind repressor even when they are separated with the cutting of the lambda DNA by *Hin*.

The Base Sequence of Operators

The ability of the lambda repressor to recognize and to bind to a number of different sites within each of the two lambda-DNA operators provides an unusual opportunity for studying the molecular basis of specific interactions between protein and DNA. With the hope of being able to identify the sequence of bases within each operator that interact with repressor, we traveled to the Medical Research Council Laboratory of Molecular Biology in Cambridge, England, in the spring of 1973. There Frederick Sanger, George Brownlee, Bart Barrell and their co-workers had developed novel methods for determining the sequence of bases in RNA molecules. Just before our visit Sanger and his co-workers reported their first success at developing methods for doing the same thing with DNA molecules. In collaboration with Barrell and John Donelson, a visiting American postdoctoral fellow, we determined the base sequence of the binding site in the left operator of the lambda DNA that had the highest affinity for the repressor protein.

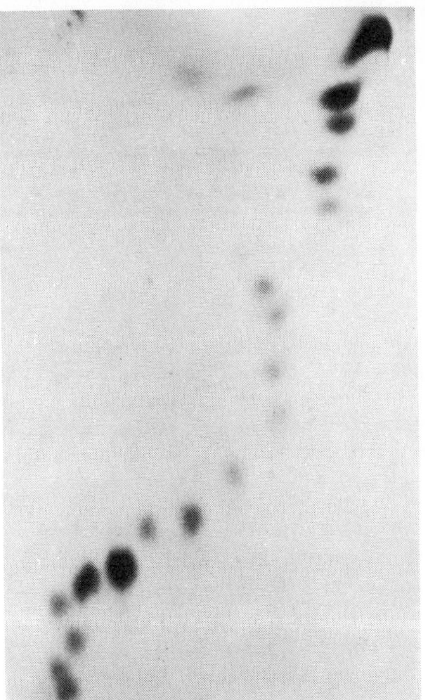

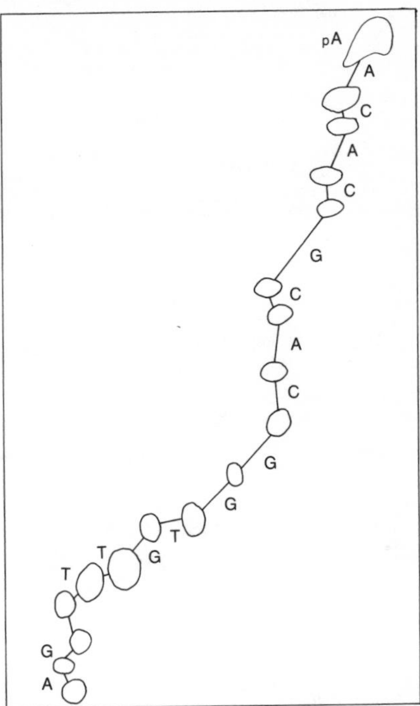

DIFFERENTIAL MOBILITY OF FRAGMENTS is exploited to separate the DNA pieces containing from one base to 30 or 40 bases produced by the method described in illustration on opposite page. The mixture of fragments is first subjected to a special type of electrophoresis (*not shown*), which separates them in one dimension according to base composition, with fragments rich in *T*'s traveling fastest and fragments rich in *C*'s traveling slowest. The separated fragments are then carefully transferred to one edge of a glass plate coated with an ionic resin to be separated in a second dimension by homochromatography. The edge of the plate is dipped in a solution containing RNA molecules of assorted lengths, which compete with the DNA fragments for ionic sites. The shorter DNA fragments are bumped along faster than the longer ones. The result of the two-dimensional separation is shown in the autoradiogram at the left. Reading upward, each spot represents a DNA fragment one base shorter than the fragment responsible for the spot below it. The angle of displacement between any two spots reveals which base is present in the lower fragment but absent in the upper one and hence also reveals the sequence in which bases were removed when the strands were digested one base at a time with DNAase. Thus the letters in the drawing at the right represent sequence of bases in a central region of right operator; if bases are read downward, they represent sequence along one strand in the 5′–3′ direction. This method of determining base sequence was devised by Frederick Sanger and co-workers at Medical Research Council Laboratory of Molecular Biology in Cambridge, England.

On returning to Harvard some eight months later one of us (Maniatis) established the base sequence of most of the right operator with the help of Andrea Jeffrey, a research assistant, and Dennis Kleid, a postdoctoral fellow. The methods used to determine the base sequences can be described very briefly. First, we took advantage of the fact that Richard J. Roberts and his colleagues at the Cold Spring Harbor Laboratory had collected and characterized a substantial number of host-restriction endonucleases, samples of which they had isolated from various strains of bacteria. With these enzymes, which were generously provided by Roberts, we proceeded to dissect lambda DNA in and around the right operator. We found that *Hph*, an enzyme isolated from *Haemophilus parahaemolyticus*, neatly excises a 75-base-pair fragment that incorporates most of the right operator. As one would expect, the fragment is itself cleaved by the enzyme *Hin*. The cleavage specifically yields two DNA fragments, one of which is 45 base pairs in length and the other 30, each incorporating part of the right operator.

By combining some new methods of our own with methods developed by workers in Sanger's laboratory (including Edward Ziff and John Sedat, who were then visiting American postdoctoral fellows) we soon developed a fast and accurate technique for determining the base sequence in small lengths of DNA. Remember that DNA chains have a polarity depending on whether the backbone linkage is 5′–3′ or 3′–5′. Our technique involves the use of the enzyme polynucleotide kinase to attach a radioactive phosphorus atom to the 5′ end of each DNA chain. The chains are then separated from each other and subjected to electrophoresis under conditions such that the base composition of each chain determines its relative mobility in the electric field.

We now determine the sequence of bases along each chain with an ingenious method developed by Sanger's group. Each chain is mixed with exonuclease, a DNAase that degrades DNA one base at a time starting from the 3′ end, the end that does not bear the radioactive phosphorus atom. Samples are removed from the digestion mixture at regular intervals and the enzyme is inactivated; hence we recover all the partial products of the degradation. Starting with a chain 30 bases long we end up with as many as 30 different DNA fragments having from

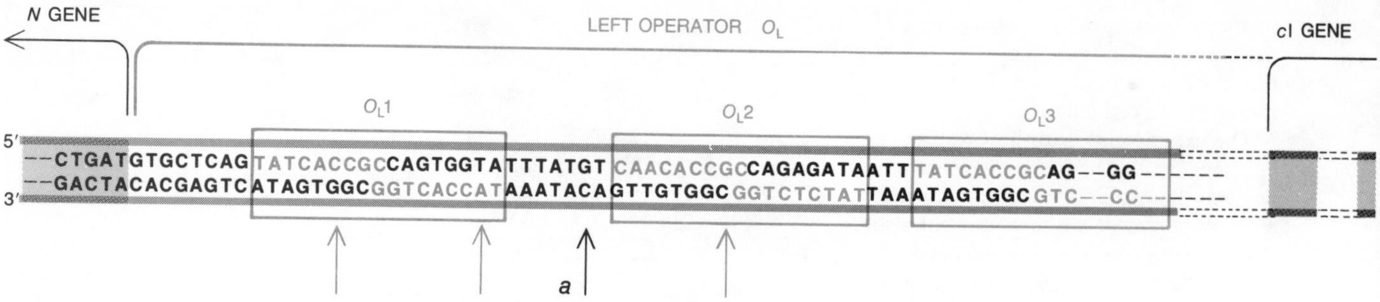

N GENE

LEFT OPERATOR O_L

cI GENE

O_L1 O_L2 O_L3

5′ --CTGATGTGCTCAGTATCACCGCCAGTGGTATTTATGTCAACACCGCCAGAGATAATTTATCACCGCAG--GG-----

3′ --GACTACACGAGTCATAGTGGCGGTCACCATAAATACAGTTGTGGCGGTCTCTATTAAATAGTGGCGTC--CC-----

a

SEQUENCES OF BASES have now been determined for 60-odd base pairs in the left operator of the lambda-phage DNA and 70-odd base pairs in the right operator. The presumed repressor binding sites, each 17 base pairs long, are in colored rectangles. Base sequence of the O_L3 site has not been completely determined. Binding sites are similar in base sequence; moreover, each site has considerable rotational symmetry around its central base pair. This is most readily seen by reading the base sequence on opposite strands starting from opposite ends of a binding site and proceeding toward the middle. Within each site one set of these "half-site" sequences

one to 30 bases, all labeled at the 5′ end with radioactive phosphorus.

The fragments are now separated from one another by electrophoresis; they are placed along one edge of a sheet of cellulose acetate and subjected to the electric field. Under the influence of the field the fragments migrate at a speed influenced by their base composition. *T*'s give rise to the fastest migration and *C*'s to the slowest; the overall order is *T*, *G*, *A*, *C*. Thus at the end of the electrophoresis molecules rich in *T*'s, regardless of their size, will have traveled farther than those rich in *G*'s, and so on.

The distributed molecules are carefully transferred to one edge of a glass slide coated with a thin layer of ionic resin. There they are further separated along an axis at right angles to the electrophoresis axis by the procedure called homochromatography. The plate, clamped vertically with the fractionated DNA fragments lined up along its lower edge, is dipped in a solution containing RNA molecules of various lengths. As the RNA molecules compete with the DNA molecules for sites in the layer of ionic resin, the DNA molecules are displaced upward; the shorter the DNA molecule is, the farther it travels. The final position of each DNA fragment is revealed on photographic film by autoradiography.

Since the fragments are labeled with radioactive phosphorus, they appear as dark spots running up the film. Remarkably, from the angle of displacement of each fragment from its neighbor one can deduce the identity of the base removed by each step in the DNAase digestion [*see illustration on preceding page*]. By using both chains of the double-chain DNA fragment, each labeled at its 5′ end, we can determine the probable sequence of 30 to 40 base pairs quite quickly. Although the sequence assignments made in this way are not completely reliable, they can be verified and the ambiguities can be resolved by further manipulations. The important general point is that these methods enable us to determine the base sequence of any double DNA chain up to about 40 base pairs in length.

Our efforts in Cambridge and at Harvard have now yielded the base sequence of a large portion of the left and right operators of phage-lambda DNA [*see illustration above*]. The sequence of the right operator was also determined by Pirrotta at the University of Basel, who used a method different from ours. It is gratifying that the two approaches yield the same sequence.

What do we learn by determining the sequence of bases that constitutes the lambda operators? As was first noticed by Keith C. Backman, a graduate student working with us at Harvard, there are base sequences in both operators that are strikingly similar. Presumably they are the sites recognized by the repressor. Each of the sites is exactly 17 base pairs long; moreover, they are separated from one another by "spacers," strings of three to seven bases that contain only, or nearly only, the base pair *AT*.

Mutations in Binding Sites

One of the strongest indications that the closely related 17-base-pair sequences are indeed the sites recognized by repressor comes from a study of mutations that change a base pair in the operator region. Kleid, Zafri Humayun and Stuart M. Flashman, working with one of us (Ptashne), have now determined the sequence of 12 such mutants, many of which were selected and studied by Flashman. Ten of the mutants change the sequence of bases in the sites we have called O_L1, O_L2, O_R1 and O_R2; all of them decrease the affinity of that portion of the operator for repressor. Two mutations located in two different spacer regions, and a third probably located in another spacer, have no effect on repressor binding. Instead they drastically decrease the efficiency with which the enzyme RNA polymerase initiates transcription of the adjacent genes.

An interesting feature of the 17-base-pair repressor binding sites is that each has a partial internal symmetry. What is meant by this is as follows. One reads off the sequence of bases on opposite strands of the binding site starting at opposite ends and proceeding toward the middle. One finds that the two sequences of eight bases on either side of the central base pair are more similar than would be expected if the sequences were random. This partial twofold rotational symmetry, as it can be called, is most apparent in the site O_L1, where six of the eight positions are occupied by the identical base. If the sequence were random, one would expect only two of the eight bases to be the same. The other 17-base-pair binding sites are also more symmetrical than would be expected by chance, although they are less symmetrical than the O_L1 site. That site, perhaps significantly, is also the site with the highest affinity for repressor.

One way to compare the various repressor binding sequences with one another is to consider only the "half-site" sequences obtained by reading the opposing chains in the 5′–3′ direction, beginning at the ends of each 17-base sequence and proceeding toward the middle. If any site were perfectly symmetrical, the two half-site sequences compared in this way would be identical. In fact, certain symmetrically arranged positions in every half-site are identical [*see top illustration on page 250*]. For example, at position No. 2 we always find the base *A*; at positions No. 4 and No. 6 we always find the base *C*. Positions No. 5 and No. 8 are usually occupied by an *A* and a *G* respectively. We infer that the identity of the bases at these positions cannot be changed without strongly af-

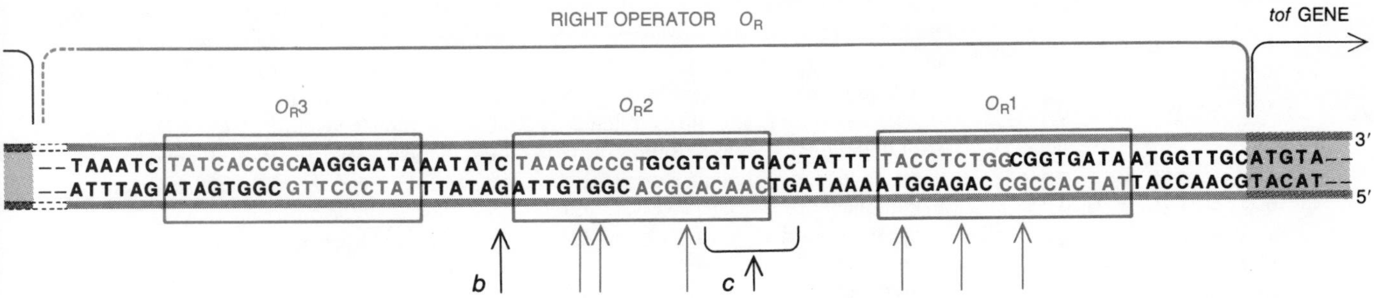

RIGHT OPERATOR O_R

tof GENE

O_R3 O_R2 O_R1

3′ --TAAATC TATCACCGCAAGGGATA AATATC TAACACCGTGCGTGTTG ACTATTT TACCTCTGGCGGTGATA ATGGTTGCATGTA--

--ATTTAG ATAGTGGCGTTCCCTAT TTATAG ATTGTGGCACGCACAAC TGATAAA ATGGAGACCGCCACTAT TACCAACGTACAT-- 5′

b *c*

is identified by letters in color. Nine colored arrows identify base pairs within repressor binding sites that have been altered in mutant lambda DNA. In every case the mutation abolishes the affinity of the repressor for that binding site. Three black arrows (*a*, *b*, *c*) show location of mutations that do not affect repressor binding but do interfere markedly with the ability of RNA polymerase to recognize promoter regions for gene *N* and genes *c*I and *tof* respectively. (Exact location of the mutation indicated by the arrow *c* is uncertain.) The base sequence of the *c*I gene, which codes for the repressor, has not been determined; it is about 1,000 base pairs long.

fecting the affinity of the site for repressor. Evidently the requirements of the other positions are not so stringent, although the favored sequence is *TATCACCG*, with either *C* or *G* as the ninth base (that is, the middle base of the 17-base sequence). The 10 operator mutations that have been found, each of which abolishes affinity for repressor, are located at positions No. 2, No. 5, No. 6 and No. 8.

Is there any reason to expect that the base sequences recognized by repressor would be symmetrical? Some years ago Walter Gierer of the Max Planck Institute for Biophysical Chemistry in Göttingen suggested that symmetrical sequences in DNA might form looped structures and that the loops might be recognized by repressors. We now have good reason to believe repressors bind to DNA in linear form; thus we prefer to think that symmetry of the operator reflects a symmetry in the structure of the repressor. As we have mentioned, the lambda repressor binds to DNA as a dimer or tetramer made up of identical subunits. On the basis of general considerations of protein architecture we suspect the repressor is largely, if not entirely, symmetrical; therefore we would not be surprised to find symmetry in the operator to which it binds.

As was first noted by Hamilton O. Smith and Kent W. Wilcox of Johns Hopkins University, the sites recognized by some restriction endonucleases, although much shorter than those recognized by repressors, also have a twofold symmetry. Moreover, as Gilbert, Maxam and their colleagues have shown, the operator recognized by a different repressor, the *lac* repressor, is highly symmetrical. (The *lac* repressor turns off the genes in *E. coli* that specify enzymes for the utilization of the sugar lactose.)

Our analysis of the lambda operators has emphasized the internal symmetries in the repressor binding sites. It must be said, however, that there is no direct evidence that the repressor actually interacts with the DNA symmetrically. Moreover, we know of cases where proteins recognize DNA sequences without there being any apparent symmetries. In fact, there may be a systematic asymmetry in the various lambda-repressor binding sites. Each 17-base-pair site has on one side the sequence *TATCACCGC* or a sequence closely related to it, together with another half-site where the sequence is more variable. It is possible that repressor protein bound to its operator DNA is somewhat deformed and not perfectly symmetrical.

An appraisal of the role of symmetry in operator-repressor interaction may call for the chemical synthesis of various operator sequences and a study of their interactions with repressor. In any case we believe each repressor will recognize its own favored DNA sequence; we have no reason to believe the sequences of operators recognized by different repressors will be similar to the sequences of the lambda operators. In particular, the base sequence of the *lac* operator, the only other operator whose sequence is known, bears no obvious relation to the sequences of the lambda operators. Moreover, we cannot yet describe in any detail how the repressors recognize their operator targets in DNA. For example, the 17 base pairs of a binding site occupy nearly two full turns of the double-chain DNA helix. One can only guess at what the repressor protein "sees." Perhaps some simple rules govern the recognition of base sequences by proteins, but if they do, they remain to be discovered.

How the Repressor Works

Our experiments with lambda repressors and operators, taken together with the results from several other laboratories, have revealed how the repressor turns off its target genes. At the University of Basel, Alfred Walz and Pirrotta have shown that in the absence of repressor the enzyme RNA polymerase binds tightly to, and protects from DNAase digestion, a 45-base-pair sequence that includes most of one of the repressor binding sites. The same sequence includes about 20 base pairs of the beginning of the gene the RNA polymerase transcribes into messenger RNA. Many other cases have been found where RNA polymerase covers about 20 base pairs on either side of the beginning of a gene. (The covered region does not include all the bases required for RNA-polymerase recognition because, as was found by Russell A. Maurer, a graduate student working with us, and by others, mutations some seven bases to the left of that region severely impede the action of RNA polymerase. The entire DNA region required for RNA-polymerase binding is called the promoter, and its exact extent is not known.) RNA polymerase apparently recognizes some aspect of the base sequence near the beginning of many genes. When the enzyme is supplied with the appropriate substances on which to act, it copies one chain of the DNA sequence into messenger RNA.

The RNA-polymerase molecule is some 20 times heavier than the repressor molecule and is therefore several times larger in volume. From the DNA region covered by RNA polymerase one can see that the enzyme competes with the repressor for its binding site. Therefore when repressor is bound to the operator, RNA polymerase cannot bind to the promoter region. The same effect is seen with the *lac* repressor.

Why does each lambda operator have more than one repressor binding site? The *lac* operator has only one repressor binding site, so that multiple binding sites do not seem to be the general rule. One possibility is that multiple sites allow for graded control. For example, if

	WILD-TYPE SITES									MUTANT BASES
O_L1	T	A	T	C	A	C	C	G	C	A OR T
	T	A	C	C	A	C	T	G	G	G
O_R1	T	A	T	C	A	C	C	G	C	T
	T	A	C	C	T	C	T	G	G	G, A
O_L2	C	A	A	C	A	C	C	G	C	T
	T	A	T	C	T	C	T	G	G	
O_R2	C	A	A	C	A	C	G	C	A	A
	T	A	A	C	A	C	C	G	T	G, A
O_R3	T	A	T	C	C	C	T	T	G	
	T	A	T	C	A	C	C	G	C	
O_L3	T	A	T	C	A	C	C	G	C	
	T_9	A_{11}	T_6	C_{11}	A_8	C_{11}	C_6	G_9	C_5	
	C_2		A_3		T_2		T_4	C_1	G_4	
			C_2		C_1		G_1	T_1	A_1	
									T_1	

OPERATOR HALF-SITES are compared to reveal their similarity. The 11 sequences are those designated by colored letters in illustration on preceding two pages. Each sequence is written left to right in the 5′–3′ direction. Letters in color at left indicate nine sites where 10 different mutations have been found. Mutant bases that replace normal bases and abolish repressor affinity are listed in the column at the right. Tabulation at lower left summarizes frequency with which each base appears at each position in the 11 half-sites. One can see that the second, fourth and sixth positions are invariant.

only one operator site is occupied by repressor, some gene transcription might occur, whereas if all sites were filled, transcription would be abolished. Flashman and Barbara Meyer, another graduate student, have recently shown that the maximum repression of transcription of the *tof* gene requires that repressor be bound to both O_R1 and O_R2. Thus mutation of either O_R1 or O_R2 decreases the effect of repressor on the transcription of *tof*, and mutation of both sites has a stronger effect. Significantly, two mutations at one site do not have as strong an effect as two mutations at two sites. Similar experiments show that the maximum repression of gene *N* requires repressor bound to both O_L1 and O_L2. We imagine that occasional unnecessary expression of the *lac* genes is not harmful to the cell, whereas occasional expression of the lambda genes could be lethal. We may therefore speculate that the lambda system, involving multiple repressor binding sites at each operator, has evolved stricter controls than the *lac* system, involving only a single repressor binding site.

Recently Meyer has discovered a most remarkable function for O_R3: the third repressor binding site in the right operator. She found that RNA polymerase recognizes not one promoter region in the right operator but two regions. Moving to the right from one promoter region, RNA polymerase transcribes the *tof* gene. Moving to the left from the second promoter region, it transcribes the *cI* gene, the gene that codes for the repressor. Although the exact starting point for the transcription of the *cI* gene is not known, it is probably just to the left of O_R3. We have mentioned the fact that transcription of *tof* begins just to the right of O_R1. Therefore we have two genes, *cI* and *tof*, transcribed in opposite directions and separated by three repressor binding sites.

Meyer has found that repressor not only turns off the transcription of *tof* but also turns off transcription of *cI*. In other words, repressor regulates its own level in the cell. From analysis of the effects of mutations in the right operator we deduce that repressor turns off the transcription of *cI* primarily by binding to O_R3. Because repressor binds more weakly to O_R3 than to O_R1 and O_R2 it allows a higher level of *cI* transcription (and hence a higher level of repressor protein) than it does of *tof* transcription [*see illustration below*].

Even this description of the role of the reiterated repressor binding sites is probably incomplete. For example, Meyer has found that relatively large amounts of repressor turn off the transcription of both *tof* and *cI*, whereas smaller amounts suffice to turn off the transcription of *tof*. On the other hand, the smaller amounts of repressor actually enhance the transcription of *cI*, the repressor's own gene. This positive effect of repressor on the transcription of its own gene had been predicted by work on whole bacterial cells infected with bacteriophage lambda; the work was done by, among others, Louis Reichardt, who is now at the Harvard Medical School. We do not yet know what molecular mechanisms are involved.

We suspect that as investigations progress even more sophisticated roles in gene regulation will be assigned to the system we have been describing. We have described some features of the interaction of two proteins, lambda repressor and RNA polymerase, with the sequences in and around the lambda operators. We know, however, that at least two other proteins, the products of genes *N* and *tof*, are themselves regulatory proteins that almost certainly recognize sequences in the operators. Although our understanding is far from complete, these studies have begun to reveal how complex patterns of gene regulation can be described in terms of specific interactions between proteins and DNA.

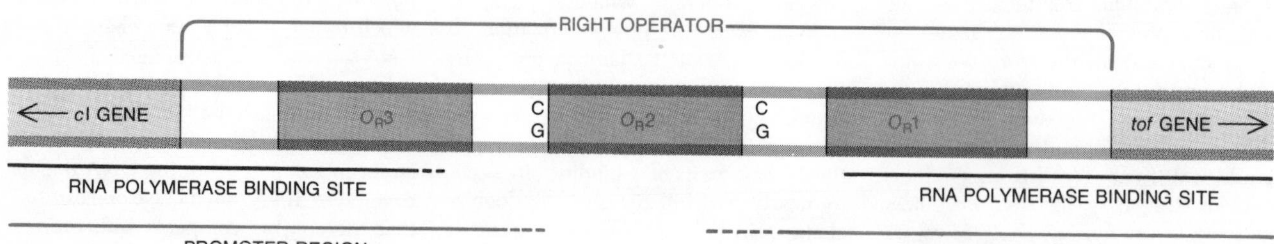

RNA-POLYMERASE BINDING SITES overlap at one end of the left lambda operator and both ends of the right operator. For simplicity only the right operator is shown. Repressors produced by the *cI* gene bind preferentially to O_R1 and O_R2, blocking expression of the *tof* gene. As long as no repressor is bound to O_R3, RNA-polymerase molecules have access to the binding site next to the *cI* gene and keep generating more repressors. (The exact location of the RNA-polymerase binding site next to the *cI* gene is not known.) Eventually, however, O_R3 is occupied by repressors and access to the *cI* gene is blocked. The ability of RNA polymerase to bind to lambda DNA is influenced by the integrity of bases lying somewhat outside the binding site. Thus mutations in the base pair *CG* to the left of O_R2 make it impossible for polymerase to transcribe the *cI* gene. Location of this gene within right operator was first noted by Gerald Smith and his colleagues at University of Geneva. A similar mutation in a *CG* pair between O_L1 and O_L2 (*not shown*) prevents transcription of *N*. A mutation believed to affect a *CG* pair to right of O_R2 blocks transcription of *tof*. Total sequence of bases required for polymerase recognition is termed a promoter; exactly how many bases are involved is not established.

V

REPLICATION, PRESERVATION, AND MANIPULATION OF DNA

V

REPLICATION, PRESERVATION, AND MANIPULATION OF DNA

INTRODUCTION

In this final section we will examine some of the remarkable features of DNA as the master blueprint for all living cells. The essential continuity of life from generation to generation and, indeed, much of the history of evolution is carefully preserved in the sequences of nucleotides that constitute DNA molecules. This legacy is not taken lightly, and elaborate enzymatic repair systems have evolved to ensure that the hereditary information is not accidentally lost due to "mishandling" or attack by deleterious agents in the environment. We should also be reminded that the precious genetic blueprint is not used directly for the synthesis of proteins; rather, the information is first transcribed to produce expendable RNA "field copies" which may be destroyed later when no longer needed. A unique feature of cellular DNA as a blue print is that it incorporates the instructions for its own expression. Imagine a blueprint for a house so detailed that it includes information on how to make and use the tools, a time table for the construction, and even a maintenance plan for the completed structure.

One might predict that the master blueprint would be an extremely durable molecular structure, chemically inert and structurally sound. In fact, DNA is quite pliable, chemically reactive, and easily manipulated. The ease with which the interwoven chains can be taken apart and rejoined is, of course, crucial to the functions of the DNA molecule in replication and transcription. In addition, one or both of the strands can be easily cut out of one part of a genome and transposed to another part of the same genome or even to another genome in another cell. Such transpositions of DNA segments occur naturally, and they have had undoubtedly profound consequences for the evolution of the varied life forms we see today. The intrinsic features of complementary strand pairing and the phosphodiester linkage of monomer nucleotides facilitate the rearrangement of segments of DNA within a cell, genetic exchange with DNA from other cells, the integration of bits of DNA (e.g., viral chromosomes) into cellular genomes, and even the repair of damaged sections of one strand or the other. However, in spite of the ease with which DNA can be manipulated, it is clear that living systems are well buffered against the possible deleterious consequences of such rearrangements in their primary genomes—otherwise, life could not be sustained.

We begin the readings in this section with some early history, romantically set in a castle on the Neckar River in Germany, where in 1868 a hardworking postdoctoral student, Friedrich Miescher, extracted from the nuclei of white blood cells a new phosphorus-rich substance that he termed "nuclein." Several years later Miescher analyzed nuclein in salmon sperm and concluded that "If one wants to assume that a single substance . . . is the specific cause of fertilization, then one should undoubtedly first of all think of nuclein." However, as Alfred E. Mirsky points out in "The Discovery of DNA," Miescher unfortunately did not pursue this idea, and it remained for others to develop that concept further. The article gives some feeling for the way in which the field of molecular biology was developing, and some of the personalities involved, at this very early stage—long before the term "molecular biology" had been coined and before the catalytic merger of genetics and biochemistry. The article reproduces elegant early drawings from the microscopic observations of dividing cells in which the movements of chromosomes and the events following fertilization of *Ascaris* eggs are depicted. In examining these illustrations one can appreciate how the early cell biologists might well have become preoccupied with the observational approach to studying life rather than the analytical methods of biochemistry. One could be tempted to feel that somehow the detailed secrets of life might be unveiled simply through the eyepiece of a microscope. Of course, even today the microscopic visualization of biological structure is an essential complement to the genetic and biochemical approaches.

Many years elapsed between the discovery of DNA and the understanding of its three-dimensional structure. The purine and pyrimidine bases were chemically characterized in the early part of this century. The chemical analysis of DNA was carried a crucial step further by Erwin Chargaff (1950), who discovered that the amount of adenine corresponds to the amount of thymine and the amount of guanine always equals that of cytosine in any DNA sample. (These rules apply to double-stranded DNA.) The significance of this fact was fully appreciated when the master model-builders Watson and Crick (1953) proposed their helical "ladder of life" structure for DNA three years later. Each step in the ladder is a pair of bases, either adenine-thymine or guanine-cytosine. A plausible replication scheme was apparent to Watson and Crick as soon as they had completed their model. In their first publication they cautiously state, "It has not escaped our notice that the specific pairing we have postulated immediately suggests a possible copying mechanism for the genetic material." The complementary parental strands unwind and separate so that each may serve as the template for a new strand. This mode of replication is called semiconservative, because each double-stranded daughter molecule contains both an old, conserved strand and a newly synthesized strand. The actual proof of the semiconservative replication of DNA was provided in the elegant experiments of Matthew Meselson and Franklin Stahl. Meselson, then a graduate student with Linus Pauling, and Stahl, a postdoctoral fellow, in collaboration with Jerome Vinograd at the California Institute of Technology developed the important technique of equilibrium sedimentation of DNA in a density gradient of the heavy salt cesium chloride. By "density labeling" the daughter DNA strands using ^{15}N (the predominant isotope is ^{14}N) in growing bacteria, they were able to physically separate the replicated "hybrid" DNA from those fragments of parental DNA that had not yet replicated. (The ^{14}N:^{15}N hybrid DNA duplex has a greater buoyant density than the ^{14}N:^{14}N DNA and less than that of ^{15}N:^{15}N double-stranded DNA.) This technique has proved invaluable for research on nucleic acids generally, and it was in principle the method by which the repair replication of damaged DNA was first demonstrated in my laboratory.

New strands of DNA are synthesized sequentially by the addition of nucleotides as the parental template strands unwind. DNA polymerase, an enzyme that catalyzes the synthesis of DNA from nucleotide precursors, was first purified from *Escherichia coli* by Arthur Kornberg and his co-workers, a Nobel-Prize-winning achievement. At the time that this first *in vitro* synthesis of DNA was reported in 1958, it required a monumental amount of effort just to prepare the substrates for the test reaction and to label them radioactively so that the tiny amount of synthesis could be detected. In the ensuing years, as we learned more about the properties of DNA polymerases, several intrinsic problems became apparent for the sequential replication of double-stranded DNA templates. First, since the parental strands are antiparallel, and since DNA polymerases operate in a particular orientation with respect to the template strand, it wasn't clear how *both* strands could be replicated in the same direction at a growing fork. Furthermore, DNA polymerase was shown to require a "primer" with an exposed 3′OH end in order to initiate replication. Unlike RNA polymerase it cannot simply recognize a starting sequence in the DNA and then begin copying the template at that point. These problems have now been resolved as discussed below, but the overall process of replication still appears to be much more complex than we might have imagined. The polymerase first characterized by Kornberg is called pol I, and we now know of two additional DNA polymerases in *E. coli:* pol II, with no essential biological role yet established; and pol III, required for normal replication but in a complex with other proteins. Eukaryotic cells have also been found to contain multiple DNA polymerases to carry out replication and repair functions.

The polarity problem in DNA replication was resolved by Reiji Okazaki and his co-workers in Japan, who demonstrated that nascent DNA strands were synthesized discontinuously in short fragments (appropriately termed Okazaki Fragments) that could be stitched together later. While synthesis of one daughter strand could be continuous in the direction of overall fork progression, that of the other could produce short segments with the opposite polarity. In principal the initiation problem might be resolved in a number of ways. The end of a strand might fold around to form a hairpin loop, a primer DNA fragment might be added at an appropriate site, or an RNA polymerase might synthesize an RNA primer at that site. The latter is the usual mode, and we now know that in *E. coli* the RNA primers are later removed by means of a unique capability of pol I. In its single polypeptide chain, pol I possesses a number of distinct enzymatic activities. In addition to the polymerase activity, there is an exonuclease activity (5′ exo) that can remove nucleotides (e.g., RNA) ahead of the progressing chain elongation. Not only is this exonuclease activity an essential component in normal replication, but it also has an important role in the repair of damaged DNA, as discussed below.

A number of the important contributions of Arthur Kornberg and his associates are described in his article, "The Synthesis of DNA." One of the interesting properties of DNA pol I is that when presented with nucleotides containing only A and T bases, but no template, a polymer of alternating A and T nucleotides eventually materializes; thus the polymerase itself can exert some selective influence on the ordering of nucleotides. However, when a DNA template is provided, the enzyme faithfully copies it according to the base-pairing rules. It has more recently been shown in Kornberg's laboratory that the fidelity of replication is further ensured by the presence of another exonuclease activity (3′ exo) in the pol I molecule, a "proofreading" nuclease that senses whether the most recently incorporated nucleotide is appropriately paired with the corresponding template nucleotide. If not, the erroneous addition is removed and the polymerase backs up and tries again. Proofreading has been shown to be an intrinsic feature of most of the prokaryotic DNA polymerases (except, interestingly enough, those in the mycoplasma), but it has not yet been found in integral association with any eukaryotic DNA polymerase. Kornberg describes the experiments in which he and his associates Mehran Goulian and Robert Sinsheimer demonstrated *in vitro* synthesis of biologically active φX174. Sinsheimer at Caltech had originally characterized the φX174 DNA and had shown that it was single-stranded. The problem of priming was not appreciated at the time that this first *in vitro* replication of biologically active φX174 was carried out. A few years later Douglas Brutlag, Randy Schekman, and Kornberg showed that an RNA primer was involved in the initiation of DNA replication on these small single-strand virus templates. Kornberg's article ends on a prophetic note suggesting that an important future research goal should be to complement genetic deficiencies with the appropriate synthetic genes. The field of recombinant DNA and gene cloning developed into a flourishing technology within ten years of his article!

Our present concept of the replicating region of DNA in *E. coli* is schematically represented in Figure 1 (adapted from Kornberg's keynote address at the 1978 Cold Spring Harbor Symposium on DNA replication and recombination). By analogy with the ribosome for protein synthesis, Kornberg terms the growing-point complex in DNA the "replisome." Many different enzymes are required to facilitate the sequential progression of the replication fork. While the parental strands unwind at a rate in excess of 10,000 revolutions per minute, "gyrase" releases the overtwisting of strands in the region ahead of the advancing fork by periodically nicking and joining parental strands. The parental strands are melted apart at the fork by a type of protein called a "helicase." An activity of this type, the *rep* gene product, has been directly implicated in replication, and several other helicases have been reported by Hartmut Hoffman-Berling and co-workers in Heidelberg. The separated parental strands are protected from deleterious nuclease action by a single-strand-specific DNA-binding protein. As shown, replication proceeds in the 5′ to 3′ direction along each of the respective parental strand templates. The actual chain elongation is carried out by DNA pol III in a complex with six additional proteins to form the "holoenzyme." On the so-called lagging-strand side of the fork, initiation must occur at periodic intervals to generate the short Okazaki Fragments. This is in itself a very complex process, which requires a number of proteins in order to set the stage before the *dna* G protein, called "primase," is able to prepare an RNA primer (shown in Fig. 1 by the wavy line). Eventually, the pol III holoenzyme catalyzes DNA synthesis from the primer, and then pol I carries out the sequential deletion of the RNA

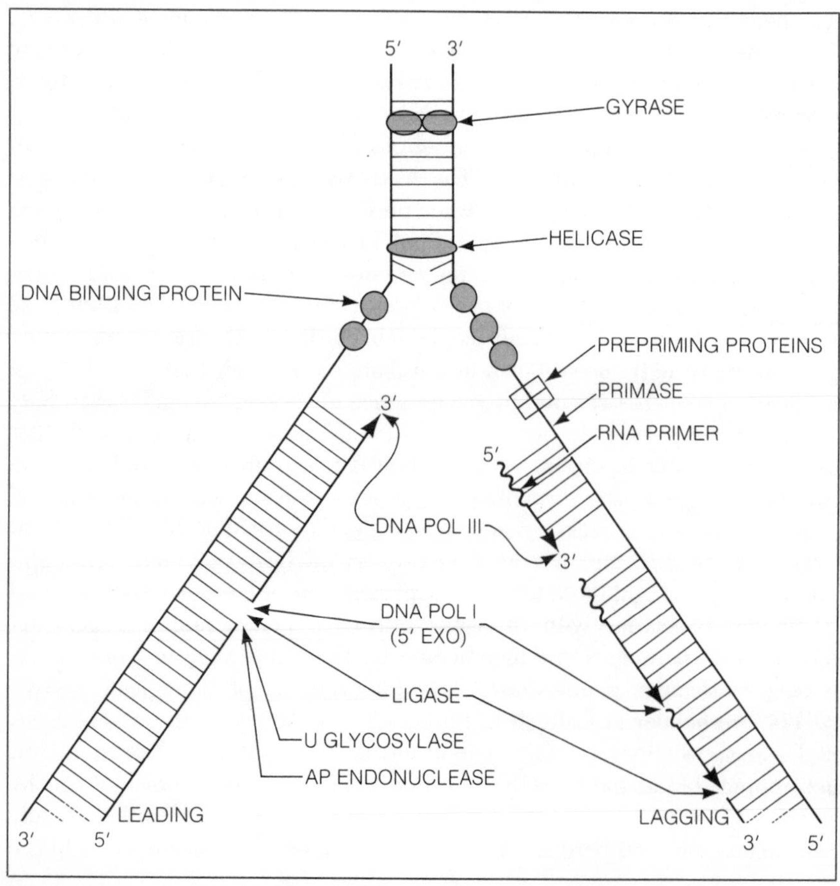

Figure 1. Schematic view of the replicating region of DNA in *E. coli*.

and its replacement with a continuous stretch of deoxyribonucleotides. Ultimately, the enzyme polynucleotide ligase joins the contiguous daughter strands to yield an intact daughter duplex of DNA. However, a complication illustrated on the leading-strand side sometimes results in the appearance of discontinuous synthesis on that side as well. Although the pool of deoxyuridine triphosphate (dUTP) that each cell contains is usually held to a low level by the enzyme deoxyuridine triphosphatase, some dUTP is periodically incorporated erroneously in place of the correct nucleotide, thymidine triphosphate. Since uracil pairs appropriately with adenine, the proofreading exonuclease does not detect the error. Fortunately, other enzymes monitor the DNA for mistakes after its synthesis. One of these enzymes is a uracil glycosylase that cleaves the uracil from the deoxyribose sugar. An AP endonuclease (AP stands for either "apyrimidinic" or "apurinic") then cuts the strand near the site of the missing base so that a repair process involving pol I and ligase can restore the strand to normal with the correct thymidine nucleotide.

As with other important macromolecules in cells, the DNA undergoes some processing after its synthesis; some of the bases are methylated to give rise to a unique "branding" pattern for each type of cell. This serves in some instances to protect the DNA from internal "restriction endonucleases," the molecular security police that normally cut up foreign DNA from an invading species. (The invading DNA is not appropriately branded, hence is seen as alien.) In some as-yet undefined manner these methylation patterns may also have roles in regulating the expression of the genome in eukaryotes.

Let us move now from the microtopology of replication to the macrotopology, the overall pattern of duplication of a chromosome. John Cairns first revealed the intact circular genome of *E. coli* by the technique of autoradiography combined with an impressive degree of patience and care. The fragility of this structure, 1 mm long, makes it nearly impossible to isolate without shearing into smaller fragments. Nevertheless, Cairns achieved this remarkable feat, and thus confirmed the evidence from genetic analysis that the *E. coli* genome exists in one circular linkage group of genes. At the time he was preparing his article, "The Bacterial Chromosome," replication was thought to be unidirectional in *E. coli*, hence that is the interpretation of the autoradiographic representation shown. However, we now know that replication is bidirectional from a unique origin, so that in fact both of the forks shown are growing points. Moreover, "topoisomerases," such as gyrase, were unknown at that time, and it was thought that the entire chromosome must rotate to facilitate unwinding of parental strands. Instead of one swivel in the chromosome, it now appears that there are many sites at which the parental strands can be untwisted as a result of the catalytic activity of topoisomerases. At 37°C it takes 40 minutes to replicate the bacterial chromosome with two growing points. However, the cells are capable of dividing nearly every 20 minutes when cultured in a rich growth medium. This is accomplished simply by initiating new replication forks at the origin before the primary ones have reached the terminus. Thus a dividing cell already contains more than two genome equivalents, and each daughter cell will receive a chromosome already in the process of replication. Developing embryonic cells of fruit flies with 40 times as much DNA as *E. coli* are able to replicate their entire genomes in only one minute! This remarkable feat is achieved by operating 21,000 replication forks at the same time! The frequency of replication forks is great enough and the multiple tandem replicons are small enough that it is possible to visualize their bidirectional replication in the electron microscope, as first demonstrated by H. J. Kriegstein and David Hogness, and illustrated in this volume in an electron micrograph prepared by David R. Wolstenholme (see p. 52). Multiple tandem replicons are characteristic of all eukaryotic genomes. Another characteristic feature of eukaryotic genomes is the packaging of the DNA in "nucleosome" structures in which successive stretches of DNA are wrapped about core units composed of histones, to give a beads-on-a-string appearance in electron micrographs of the intact chromatin. The role of the nucleosome structure in the regulation of replication and the expression of the eukaryotic genome is currently a thriving field of study.

An even more rapidly developing field right now is the study of how damaged DNA is repaired in living cells. In several previous *Scientific American* readers, Robert Haynes and I have included our own 1967 article, "The Repair of DNA," as an introduction to that topic. In this reader I have replaced our article with the one by Raymond Devoret, "Bacterial Tests for Potential Carcinogens," which updates some aspects of the field and discusses an important application of our understanding of DNA repair in relation to human health—namely, the detection of chemical carcinogens.

The DNA in living cells is exposed to a variety of physical and chemical agents in the environment that can alter its structure so that replication is inhibited or the essential genetic information is changed. For example, the very same ultraviolet radiations from the sun that may have helped to bring living order from molecular chaos 3 billion years ago can also destroy that order today. The damage produced in DNA by ultraviolet light interferes with replication, and a single lesion in the entire genome of a bacterial cell is sufficient to kill the cell. It is not surprising that one of the pressures in evolution has been to develop efficient schemes for dealing with such damage, which threatens all living forms with extinction. At the same time it is important that the developing repair systems not be too perfect, so that some errors can slip by and be perpetuated as mutations. The processes of natural selection that produce the best-adapted species require that there exist an adequate pool of mutant forms. One may suspect that the evolution of DNA repair itself has had a profound impact on the rates of evolution of different species.

One of the simplest DNA repair schemes is enzymatic photoreactivation, which deals specifically with the principal damage produced by ultraviolet light—a fusing of adjacent pyrimidines in a DNA strand to form a dimer. The photoreactivating enzyme binds to the dimer-containing region of DNA in the dark, but then in the presence of visible light photocatalytically cleaves the fused bases to reverse the damage. Although this process is widespread in the biosphere, it does not appear to be of predominant importance for the survival of most organisms in which it is found, including humans.

The most ubiquitous repair mode known is excision-repair, in which one strand of the DNA is used as a template to reconstruct the information lost when the complementary strand is damaged. The usefulness of the two-stranded redundant form of the primary genetic message for repair was perhaps the only feature of DNA overlooked when Watson and Crick reflected upon their double-helical structure over 25 years ago. The first experimental evidence for excision-repair was the discovery by my former mentor, Richard Setlow, and his associate William Carrier that ultraviolet-induced pyrimidine dimers were selectively excised from the DNA in *E. coli.* This finding was confirmed independently by Richard Boyce and Paul Howard-Flanders and by Emmanuel Riklis. My graduate student David Pettijohn and I first demonstrated in *E. coli* the resynthesis step in excision-repair, termed repair replication. Within a few years repair replication had also been demonstrated in mammalian cells and other eukaryotes. James Cleaver then read in the *San Francisco Chronicle* about a rare human hereditary disease that results in extreme sunlight sensitivity and skin carcinogenesis upon exposure to ultraviolet light. He suspected that a DNA repair deficiency might be responsible for this condition. Following up on his hunch he showed that this disease, known as xeroderma pigmentosum, does indeed involve a deficiency in an early step in the excision-repair of damaged DNA. Cleaver's work stands as an important example of the utilization of the discoveries in basic research for eventual understanding of problems in human health. The existence of repair-deficient conditions such as xeroderma pigmentosum, described in Devoret's article, specifically attests to the importance of excision-repair for the survival of humans in the presence of environmental carcinogens, such as ultraviolet light.

Many, if not most, of the chemical carcinogens in the environment have been shown to damage DNA directly or to be converted metabolically to active forms that damage DNA. Because of this fact the determined mutagenicity of a test compound has predictive value in estimating its carcinogenicity. The Ames test, which makes use of this correlation, is described by Raymond Devoret, who also promotes a prophage induction test as another sensitive indicator of potential carcinogenic activity.

Although we have learned much about the multiple pathways for the excision-repair of damaged DNA (see Fig. 2), we still know relatively little about the mechanism of genetic recombination and the inducible processes that may enable cells to tolerate unrepaired damage in their genomes. A key element in the so-called SOS scheme, as well as in the process of genetic recombination in *E. coli,* is the *recA* protein described in Devoret's article. We still do not know the actual mechanism of the SOS scheme, but the principal model proposed includes the suppression of the proofreading 3′ exonuclease activity so that a DNA polymerase may synthesize DNA on the damaged template, albeit with the expected high error frequency. It is important that we learn more about the ways in which living systems respond to and tolerate damage to their primary genomes with the possibility that the effective carcinogenicity of environmental agents may someday be reduced. Alternatively, the preferred solution is to minimize contact with those agents and to control their dispersal into the biosphere.

Not all cancers are caused by the action of carcinogens on DNA; in fact, some have been shown to be caused by viruses that can integrate their genomes into the chromosomes of the host cell. Again, the models derived from basic research on bacterial systems have proved of value. Allan M. Campbell

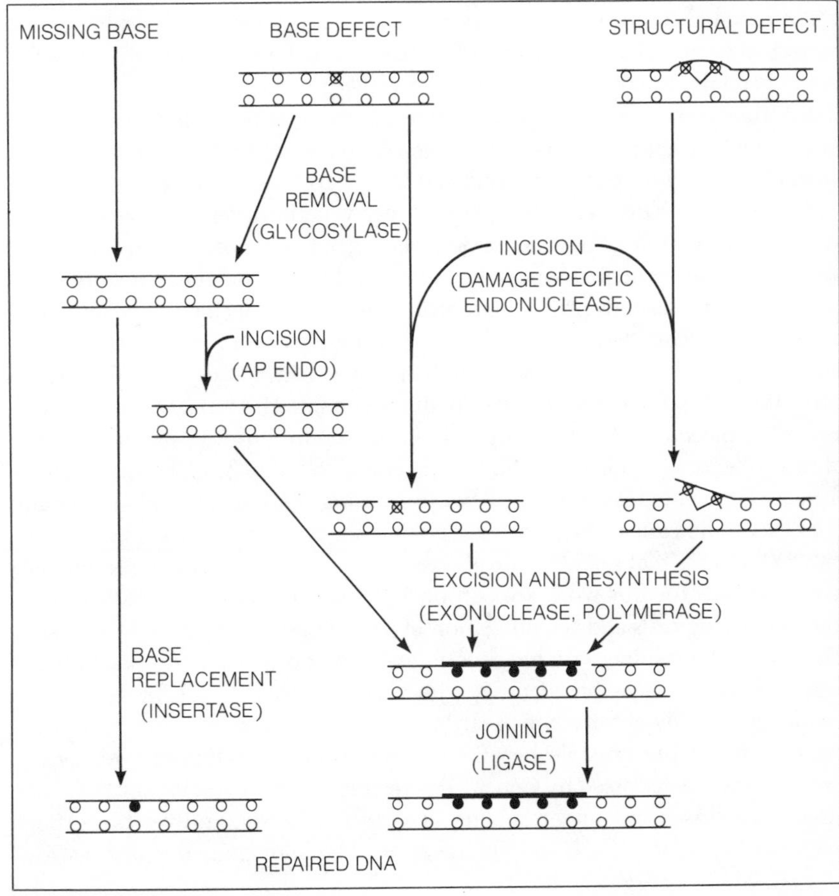

Figure 2. Schematic overview of excision repair pathways. [After P. C. Hanawalt, P. K. Cooper, A. K. Ganesan, and C. A. Smith, "DNA Repair in Bacteria and Mammalian Cells." ***Annual Review of Biochemistry*** **48:783-836, 1979.]**

originally proposed the mechanism by which the lambda virus is integrated in *E. coli.* In his article, "How Viruses Insert Their DNA into the DNA of the Host Cell," he describes the way in which this process can be utilized as a tool to probe the structural organization of the host genome. Electron micrographs are shown to detail the powerful technique of heteroduplex mapping used to visualize regions of DNA in which the two strands are not complementary; Campbell also explains the process of "transduction," one of the mechanisms by which DNA from one host may be transferred to another host and then recombined to alter the genotype of the recipient. In this case an integrated virus picks up adjacent segments of host DNA when it is subsequently excised from the genome. These host DNA sequences may then be transferred along with the viral DNA to which they are attached when a new host is infected. Bacterial DNA can also be transmitted from one cell to another by a "generalized transduction" mode in which fragments of the donor genome are accidentally packaged in viral capsids. Alternatively the donor genome can be transferred as free DNA fragments (bacterial transformation) or even through direct contact (conjugation) of the donor and recipient bacteria. These genetic transfer mechanisms are outlined in the article by Stanley N. Cohen and James A. Shapiro on "Transposable Genetic Elements." However, the principal emphasis in that article is upon recent discoveries in the fascinating field of "illegitimate" recombination, in which the combining DNA molecules lack a significant degree of homology with each other.

Transposable genetic elements ("transposons") are known to exist in plants and animals as well as in bacteria: they include intragenomic rearrangements of DNA in addition to the transfer of DNA from cell to cell. Cohen and Shapiro discuss the profund role that these transposons may have played in evolution. The manner in which these elements affect the regulation of genetic expression is also detailed. Transposons can operate as biological switches, turning genes on and off as the DNA sequences are inserted or removed from particular sites in the genome. Finally, the authors discuss the implications for public health of the transposon-mediated proliferation of genes responsible for resistance to antibiotics among bacterial strains. Resistance to antibiotics such as streptomycin or ampicillin can be transmitted from harmless bacteria to dangerous pathogens by means of "plasmids" that carry certain genes. This serves as yet another example of how our understanding of basic molecular mechanisms for the manipulation of DNA can be of great relevance to human health.

Viruses and plasmids provide important means for transferring genetic material from one cell to another. This is a natural process mediated entirely by the intrinsic properties of the biological systems involved. It is possible to carry this a step further and to artificially splice into the virus DNA "carrier" any other piece of DNA that can be isolated in pure form. In the final article in this collection, "The Manipulation of Genes," Stanley N. Cohen shows how this can be accomplished in the laboratory using viral or plasmid vectors for the DNA to be cloned. The procedures for manipulating genes are "artificial" only in the sense that scientists are choosing the pieces of DNA to be combined. The essentials for this work are certain DNA-specific enzymes that are produced in living cells and the utilization of the inherent features of DNA itself. The risks in tampering with the genetic makeup of organisms are carefully assessed by committees of scientists who review each research proposal that could conceivably entail some hazard before the researcher is allowed to proceed. Some of the possible benefits of genetic manipulation as well as the precautions are discussed by Cohen. The technology of recombinant DNA has already increased our understanding of gene regulation, and we can expect even more spectacular advances in the applied technology in the next decade.

REFERENCES CITED AND SUGGESTED FURTHER READING

Ames, B. N., J. McCann, and E. Yamasaki. 1975. "Methods for detecting carcinogens and mutagens with the *Salmonella*/Mammalian microsome mutagenicity test."

Mut. Res. **31:**347–364.

Berg, P. (Chmn., Committee on Recombinant DNA Molecules). 1974. "Potential biohazards of recombinant DNA molecules." *Proc. Nat. Acad. Sci.* **71**(7): 2593–2594.

Brown, D. D. 1973. "The isolation of genes." *Scientific American,* August. (Offprint No. 1278.)

Bukhari, A. I., J. A. Shapiro, and S. L. Adhya, (editors). 1977. *DNA Insertion Elements, Plasmids, and Episomes.* Cold Spring Harbor Laboratory, Long Island.

Chargaff, Erwin 1950. "Chemical specificity of nucleic acids and mechanisms of their enzymatic degradation." *Experientia* **6:**201–209.

Crick, F. 1974. "The double helix: A personal view." *Nature* **248:**766–769.

Elespuru, R. K., and M. B. Yarmolinsky. 1979. "A colorimetric assay of lysogenic induction designed for screening potential carcinogenic and carcinostatic agents." *Environ. Mut.* **1:**65–78.

Franklin, N. C. 1978. "Genetic fusions for operon analysis." *Ann. Rev. Genet.* **11:** 103–126.

Freifelder, D. 1979. *Recombinant DNA: Readings from Scientific American.* W. H. Freeman and Company, San Francisco.

Goulian, M., A. Kornberg, and R. Sinsheimer. 1967. "Enzymatic synthesis of DNA, XXIV: Synthesis of infectious phage φX174 DNA." *Proc. Nat. Acad. Sci.* **58**(6): 2321–2328.

Hanawalt, P. C., E. C. Friedberg, and C. F. Fox (editors). 1978. *DNA Repair Mechanisms.* Academic Press, New York.

Hanawalt, P. C., P. K. Cooper, A. K. Ganesan, and C. A. Smith. 1979. "DNA repair in bacteria and mammalian cells." *Ann. Rev. Biochem.* **48:**783–836.

Hershey, A. D. (editor). 1971. *The Bacteriophage Lambda.* Cold Spring Harbor Laboratory, Long Island.

Kornberg, A. 1980. *DNA Replication.* W. H. Freeman and Company, San Francisco.

Kornberg, R. 1977. "Structure of chromatin." *Ann. Rev. Biochem.* **46:**931–954.

Lehman, I. R., M. J. Bessman, E. S. Sims, and A. Kornberg. 1958. "Enzymatic synthesis of deoxyribonucleic acid." *J. Biol. Chem.* **233:**163–170.

Maxam, A. M., and W. Gilbert. 1977. "A new method for sequencing DNA." *Proc. Nat. Acad. Sci.* **74:**560–64.

Meselson, M., and F. W. Stahl. 1958. "The replication of DNA in *Escherichia coli.*" *Proc. Nat. Acad. Sci.* **44:**671–682.

Moreau, P., A. Bailone, and R. Devoret. 1976. "Prophage λ. Induction in *Escherichia coli* K12 *envA uvrB:* A Highly sensitive test for potential carcinogens." *Proc. Nat. Acad. Sci.* **73**(10):3700–3704.

Morgan, J., and W. J. Whelan (editors). 1979. *Recombinant DNA and Genetic Experimentation.* Pergamon Press, Elmsford, NY.

Morrow, J. F., S. N. Cohen, A. C. Y. Chang, H. W. Boyer, H. M. Goodman, and R. B. Helling. 1974. "Replication and transcription of eukaryotic DNA in *Escherichia coli.*" *Proc. Nat. Acad. Sci.* **71**(5):1743–1747.

Research with Recombinant DNA, An Academy Forum. 1977. U. S. National Academy of Sciences Publishing Office, Washington, D.C.

Sheinin, R., J. Humbert, and R. E. Pearlman. 1978. "Some aspects of eukaryotic DNA replication." *Ann. Rev. Biochem.* **45:**277–316.

Sinsheimer, R. L. 1977. "Recombinant DNA." *Ann. Rev. Biochem.* **46:**415–438.

Smith, H. O. 1979. "Nucleotide sequence specificity of restriction endonucleases." *Science* **205:**455–462.

Stahl, F. W. 1979. *Genetic Recombination: Thinking About It in Phage and Fungi.* W. H. Freeman and Company, San Francisco.

Starlinger, P. 1977. "DNA rearrangements in prokaryotes." *Ann. Rev. Genet.* **II:** 103–126.

Tomizawa, J., and G. Selzer. 1979. "Initiation of DNA synthesis in *E. coli.*" *Ann. Rev. Biochem.* **48:**999–1034.

Watson, J. D. 1968. *The Double Helix.* Atheneum, New York.

Watson, J. D., and F. H. C. Crick. 1953. "Molecular structure of nucleic acids." *Nature* **171:**737–738.

Wickner, S. H. 1978. "DNA replication proteins of *Escherichia coli.*" *Ann. Rev. Biochem.* **47:**1163–1194.

Witkin, E. M. 1976. "Ultraviolet mutagenesis and inductible DNA repair in Escherichia coli." *Bacteriol. Rev.* **40**(4): 869–907.

Wu, R. 1978. "DNA sequence analysis." *Ann. Rev. Biochem.* **47:**607–634.

20 The Discovery of DNA

by Alfred E. Mirsky
June 1968

In 1869 Friedrich Miescher found a substance in white blood cells that he called "nuclein." Cell biologists saw that it was a constituent of chromosomes and hence must play a major role in heredity

Deoxyribonucleic acid was discovered in 1869 by Friedrich Miescher. The reader may find it surprising that DNA, which has been the focus of so much recent work in biology, was isolated almost a century ago. If so, he will find it even more surprising that the function of DNA as the substance in chromosomes that transmits hereditary characteristics was recognized only a few years later by a number of biologists (not, however, by Miescher). Here I shall recall how it happened that Miescher discovered "nuclein" (as he called it) and what it meant to him and his contemporaries, and tell something of the investigative history of DNA until, some 25 years ago, it was conclusively shown to be the genetic material.

Miescher was the son of a Swiss physician who practiced in Basel and taught at the university there, and he followed his father into medicine. As a medical student he came under the influence of his uncle Wilhelm His, professor of anatomy and one of the outstanding investigators and teachers of his time. That influence was profound and lifelong. Miescher's later views on biology are often best understood in the light of His's attitudes and ideas; Miescher's writings are available today largely because His gathered and published the younger man's letters and papers after Miescher's death in 1895 at the age of 51.

His urged Miescher to go into histochemistry, the study of the chemical composition of tissues, because in "my own histological investigations I was constantly reminded that the ultimate problems of tissue development would be solved on the basis of chemistry." Miescher took the advice, and after receiving his degree in 1868 he went to the University of Tübingen first to learn organic chemistry and then to work in the laboratory of the biochemist F. Hoppe-Seyler. (The first laboratory devoted entirely to biochemistry, it was located incongruously in an ancient castle overlooking the Neckar River. In later years Miescher liked to tell his students how the narrow, deep-set windows and the dark vault of his room reminded him of a medieval alchemist's laboratory.) Within a few months, in the course of experiments on cells in pus, he discovered DNA.

Pus may seem an unlikely material for a study of cell composition, but Miescher considered the white blood cells present in pus to be among the simplest of animal cells, and there was a fresh supply of pus every day in the Tübingen surgical clinic. He was given the bandages removed from postoperative wounds, and he washed the white cells from the bandages for experiments. If the bandages were washed with ordinary saline solution, the cells swelled to form a gelatinous mass. In a dilute sodium sulfate solution, on the other hand, the cells were preserved and sedimented rapidly, making it easy to separate them from the blood serum and other material in pus.

Miescher undertook a general study of the chemical composition of the white cells. First he extracted them in various ways—with salt solutions, acid, alkali and alcohol. Earlier workers, including Hoppe-Seyler, had extracted pus cells with concentrated salt solutions and had obtained a gelatinous material that reminded them of myosin, the protein of muscle. Miescher got the same result. What is of interest to us is that the gelatinous substance consists largely of DNA! This did not become known until 1942, when it was demonstrated that concentrated neutral salt solutions are an exceedingly useful medium for the extraction of polymerized (and therefore gelatinous) DNA. To extract DNA under these conditions, however, one needs a centrifuge. If Miescher had had a centrifuge, it is quite possible that he would have obtained DNA in its natural form instead of in the depolymerized form he eventually discovered.

Miescher's route to DNA was necessarily more indirect, as is frequently the case for the pathfinder. When he extracted pus cells with dilute alkali, he obtained a substance that precipitated on the addition of acid and redissolved when a trace of alkali was added. At this point Miescher noted: "According to recognized histochemical data I had to ascribe such material to the nuclei... and I therefore tried to isolate the nuclei." The isolation of the nucleus—or of any other cell organelle—had not been attempted before, as far as we know, although the nucleus had been identified in 1831. Miescher's primary observation, on which the isolation of nuclei depended, was that dilute hydrochloric acid dissolves most of the materials of a cell, leaving the nuclei behind. (This observation is the basis of what is still a valuable procedure for isolating nuclei.)

When Miescher examined the isolated nuclei under his microscope, he could still see contamination, which he suspected was cell protein. To obtain clean nuclei he therefore added the protein-digesting enzyme pepsin to the dilute hydrochloric acid he was using. More precisely (since this was in 1868), he made a hydrochloric acid extract of pig's stomach and applied it to pus cells. The isolated nuclei so prepared were somewhat shrunken but were clean enough for chemical study. The next step was to extract the isolated nuclei (rather than the whole cells) with dilute alkali. The extracted material precipitated on the addition of acid and redissolved readily in alkali. Miescher analyzed this material into its elements (finding, for example, 14 percent nitrogen and 2.5 percent phosphorus) and studied some of its other properties. He came to the conclusion that it did not fit into any known group of substances: it was a substance "*sui generis*" and he called it nuclein. The analytical data indicate that somewhat less than 30 percent of Miescher's first nuclein preparation consisted of DNA.

The year with Hoppe-Seyler was now ended. Hoppe-Seyler, who agreed that a new substance had been discovered, was sufficiently interested to repeat the preparation of nuclein after Miescher's departure, and sufficiently cautious to delay the publication of Miescher's paper until he had satisfied himself that the work was sound. The paper was published in 1871.

Miescher left Tübingen in the fall of 1869 to spend his vacation at home in Basel. He decided that during the two-month vacation he would broaden his study of nuclein by looking for it in various cells. The first material he chose was the hen's egg, because His had recently claimed that the microscopic particles called yolk platelets were genuine cells of connective-tissue origin. This was a controversial claim, and Miescher believed a biochemical approach—the search for nuclein in the platelets—would help to establish its validity. Following the procedure he had worked out for pus cells, he soon isolated from yolk platelets what he took to be nuclein. The phosphorus content and some other properties of platelet nuclein did differ somewhat from those of pus nuclein, but Miescher (like many another investigator!) was determined to find what he was looking for, and he was convinced that this was nuclein. In spite of the curious appearance of the platelets, he wrote, "nobody would any longer deny that they have genuine nuclei, because it is not in the optical properties but in the chemical nature of a structure that its role in the molecular events of a cell's life is rooted."

Miescher wrote up his new work and sent it off to Hoppe-Seyler; it was published along with the Tübingen research. The paper on pus cells remains a classic but the one on yolk platelets is forgotten. It was wrong. The microscopists, considering what Miescher called the "optical properties" of platelets, never did accept the idea that they were cells, and in time detailed chemical analysis showed (and Miescher had to agree) that what he had taken for platelet nuclein in fact had a very different composition. It is only recently that careful microscopic observation has demonstrated that yolk platelets are derived from the subcellular particles called mitochondria; they contain only a trace of DNA, whereas Miescher thought they contained a large amount of nuclein.

Having learned biochemistry at Tübingen, Miescher next spent a year at the University of Leipzig in Carl Lud-

CASTLE ON THE NECKAR RIVER in Germany was where nuclein, or deoxyribonucleic acid (DNA), was discovered. The castle housed F. Hoppe-Seyler's biochemical laboratory at the University of Tübingen, where Friedrich Miescher was a postdoctoral student.

wig's laboratory, a world center of physiology. Here he became convinced of the central role of the physical sciences in biology and in particular learned the importance of developing new instrumentation for research.

Miescher returned to Basel in 1870 and soon began the investigation on which he did his finest work: an analysis of the nuclein and other components of salmon spermatozoa. Wilhelm His introduced him to the salmon fishery then flourishing along the banks of the Rhine at Basel. The salmon, having swum all the way up from the North Sea to spawn, were sexually mature, so that huge quantities of ripe sperm were available. The nucleus is extremely large in any sperm cell, and among spermatozoa the salmon's is remarkable: its nucleus accounts for more than 90 percent of its mass. For a young investigator who had recently discovered nuclein in pus cells washed out of bandages, the sperm of the Rhine salmon must have seemed to present a God-given opportunity, and Miescher seized it. He was an intense and rapid worker. Within a year and a half he had completed most of his investigation—in spite of the fact that his laboratory was so crowded with medical students that he could only do his chemical analyses, an essential part of the job, at night and on Sundays. His classic paper on the sperm cell of the salmon was published in 1874.

FRIEDRICH MIESCHER, the discoverer of DNA, was born in 1844 and died in 1895. This portrait is the frontispiece of a collection of his letters and papers published posthumously.

By acidifying a suspension of sperm cells, Miescher first caused the cells to aggregate and settle. Then he treated them with hydrochloric acid (omitting the pepsin he had used with pus cells) and extracted from the sperm an organic base with a high content of nitrogen. It accounted for 27 percent of the mass of the sperm, and he called it "protamine." Then he treated the residue of the sperm with dilute alkali. This extracted the nuclein, which accounted for 49 percent of the sperm's mass.

Miescher saw that nuclein was an acid containing a number of acid groups (a "polybasic" acid) and that it combined with protamine, a base, to form an insoluble salt in the nucleus. He experimented with modifying the chemical equilibrium of the nuclein-protamine combination, a problem that still interests biochemists today. He found, for example, that if sperm washed with acetic acid and alcohol were treated with a sodium chloride solution, much of the protamine was released from combination and passed into solution. If *fresh* sperm were treated with the same salt solution, however, the material became a lumpy gel that could almost be cut with a pair of scissors, as in the case of the pus cells treated with salt solution. The reason was that the DNA in the preparation was present as an extremely long linear polymer. Miescher, as I have mentioned, never did isolate DNA in its natural state and learn the importance of its fibrous structure. He did, however, get some idea that nuclein consisted of large molecules because he found that it would not pass through a parchment filter, whereas protamine would do so readily.

Although most of Miescher's work was done on salmon sperm, he also investigated the sperm of other species, notably the carp and the bull. He was disappointed not to find protamine in either, or even in unripe salmon sperm, and in a letter to His he referred to its presence in ripe salmon sperm as a "miserable special case." Ten years later, however, the biochemist Albrecht Kossel discovered histone, a base analogous to protamine, in the nucleus of red blood cells, and soon it was found also in the nucleus of lymphocytes, white cells from the thymus gland. Today it is known that either histones or protamines are present in saltlike combination with nuclein in the nuclei of all plant and animal cells.

Miescher's analyses of protamine and nuclein into their constituent elements were done with great care, and his results for nuclein are very close to those of later workers for what came to be called nucleic acid. (That term was introduced by the biochemist Richard Altmann in 1889. Altmann's method of preparation was different, but Miescher recognized that the substance was the same, and he did not object to the new name because he had been aware that nuclein was an acid.) To obtain good analytical results Miescher considered it essential to isolate his nuclein at low temperatures. He analyzed nuclein in the fall and winter, sometimes working from five in the morning until late at night in unheated rooms. (Later in his life, when he again took up his study of salmon sperm, he went back to spending long hours in the cold. His health, which had always been delicate, gave way and he died of a chest ailment.)

The process of elementary analysis was particularly important for Miescher because it was his main guide to the composition and identification of a substance. He could not know, for example, that protamine is a protein, and that it is basic and has a nitrogen content because it contains large amounts of the basic amino acid arginine. He knew that phosphoric acid was responsible for the acidity of sperm nuclein, and that nuclein was a substance *sui generis* quite distinct from proteins, but he did not know that nucleic acids had two kinds of subunit: purines and pyrimidines. The detailed chemistry of nucleic acids and the proteins associated with them in the cell nucleus was worked out by later biochemists, beginning with Kossel in 1884. It took longer still, as we shall see, to understand the biological function of these substances.

Miescher's work on nuclein led him to a deep interest in the life of the Rhine salmon. In conducting what he called his "sperm campaign" almost every year he was witness to one of the most impressive events of animal life: the migration of the salmon 500 miles from the sea on their way to spawn in the headwaters of the Rhine. Between 1874 and 1880 Miescher devoted himself to the physiology of the salmon, examining more than 2,000 fish. He learned that the salmon spent eight to 10 months in fresh water, and that during this period, although they were extremely active, they did not feed; they did not, indeed, secrete any digestive juices. And yet during this time an extraordinary metabolic rearrangement occurred: a massive increase in the size of the sex organs. Miescher found that when female salmon left the sea, their ovaries accounted for .5 percent of their body weight; by the time the fish reached Basel the ovaries represented 26 percent of their weight. The growth, he discovered, was at the expense of the large "lateral trunk" muscle; the loss of protein and phosphate by this muscle was sufficient to account for the growth of the ovaries. From physiology Miescher moved on to ecology, making a number of suggestions for the conservation of the Rhine salmon.

In 1872, at the age of 28, Miescher had been appointed professor of physiology at Basel, succeeding His, who had moved to Leipzig. Miescher became a leading figure at the university. In time he built a new physiological institute (which he named the Vesalianum in honor of Andreas Vesalius, the 16th-century Belgian anatomist who had spent six months in Basel while his *De humani corporis fabrica* was being printed there), stocked it with new precision instruments as they became available and guided it in researches on such classical problems of physiology as respiration, the circulation and the effects of high altitude. From time to time he undertook public service projects. The canton of Basel asked his advice on the nutrition of inmates of prisons and other institutions, and he also gave a series of public lectures on nutrition and home economics. Living in a country that produced milk but did not consume much of it, he was aware of milk's nutritional value, and in one lecture he heaped scorn on the "sordid avarice" of peasants who withheld milk from their children in order to make "the last drop" into salable cheese. "If you ask these pale and feeble people what they eat," he said, "they reply: potatoes, coffee, more potatoes and schnapps to keep down hunger." Miescher's social concern led a colleague to comment that he would surely have been active in politics had he not been hard of hearing.

All his life Miescher's primary and absorbing interest was nuclein. From time to time he went back to investigating its chemistry; he was also preoccupied with the question of its biological function. During his work on the white cells of pus he had made no mention, either in his letters or in the paper published in 1871, of the possible function of the nucleus. This is hardly surprising, since at that time and for some years to come the role of the nucleus in the life of the cell was simply not understood. As late as 1882 the leading cell biologist Walther Flemming, describing the latest work on the nucleus, conceded that "concerning the biological significance of the nucleus we remain completely in the dark."

When Miescher came to the investigation of sperm, however, he began to ask questions about the role of nuclein in fertilization and about the nature of fertilization. Does the sperm contain certain special substances that are effective in fertilization? Willy Kühne, the Heidelberg biochemist who had just introduced the word "enzyme," suggested that there might be enzymes in sperm. Since salmon sperm seemed to be a clean and uncontaminated material, Miescher looked in it for enzymes, but he failed to find anything that appeared to be promising. Then he went on to say: "If one wants to assume that a single substance . . . is the specific cause of fertilization, then one should undoubtedly first of all think of nuclein." Coming on this passage written in 1874, the reader holds his breath for a moment—but only for a moment, because Miescher turned away from the idea. He supposed, one must remember, that the egg contained a rich supply of nuclein in its yolk granules. And in his opinion there was no special characteristic that distinguished sperm nuclein from the great mass of egg nuclein. Indeed, he believed "the riddle of fertilization is not hidden in a particular substance"; the sperm is acting as a whole through the cooperation of all its parts, and if one considers the magnitude of the sperm's contribution to heredity, it must work in a most complex way. This line of thought led Miescher to think of fertilization as a physical procedure in which a certain movement (*Bewegung*) of the sperm is transmitted to the egg. In casting about for the nature of the movement, Miescher pointed to the "molecular process" that occurs when a nerve stimulates a muscle as perhaps being analogous to the effect of the sperm on the egg.

When Miescher invoked physical motion to explain fertilization, it is clear that he was thinking of kinetic theory, which at that time was in its formative period. By 1892 another physicochemical approach to fertilization appealed to

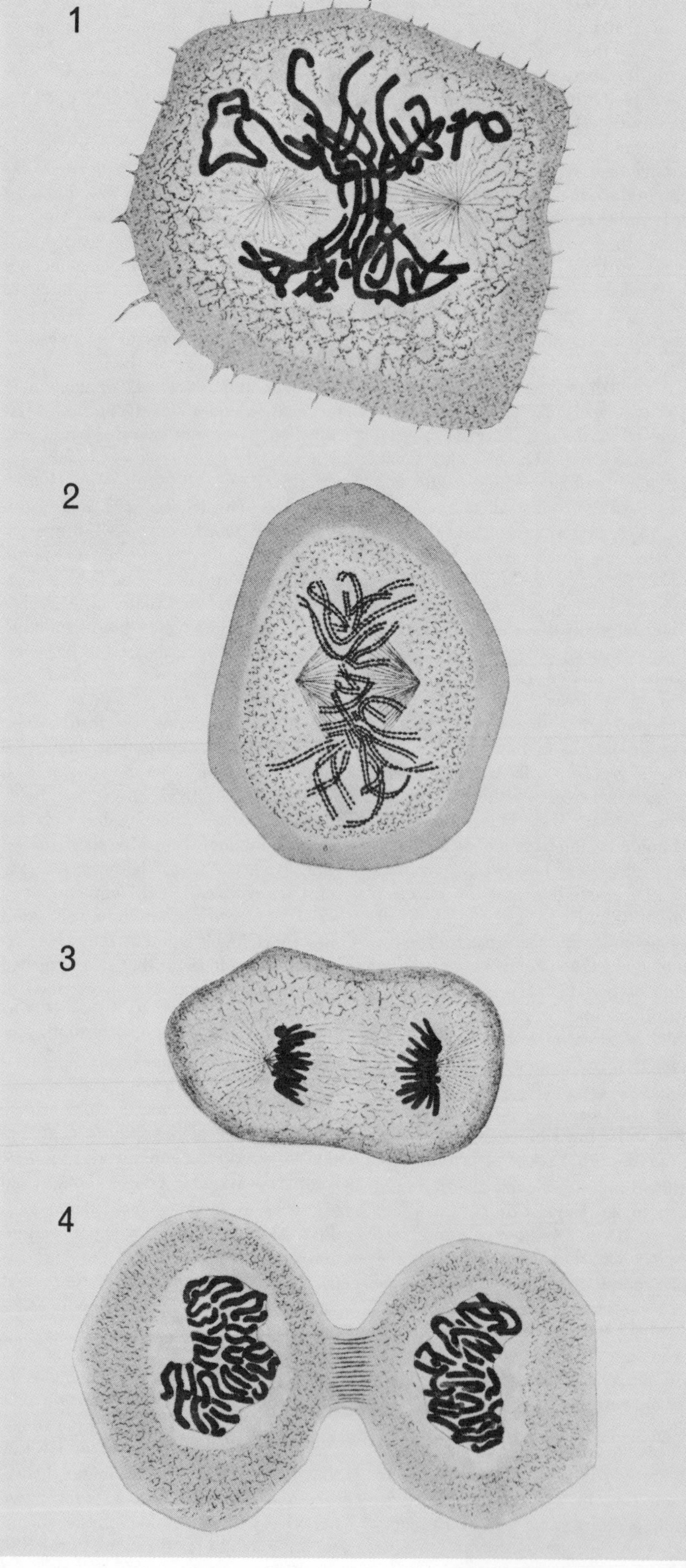

him and he wrote: "The key to sexuality lies for me in stereochemistry," that is, in the varying positions of the asymmetric carbon atoms in molecules. Miescher's desire to find explanations in physical science did not lead him to an understanding of the biological role of nuclein. At about the same time, however, another group of biologists—who were not interested in kinetic theory or stereochemistry—succeeded in laying some of the foundations of biology and at the same time recognizing the fundamental role of nuclein.

Miescher was nurtured in the great 19th-century school of "molecular biology." Both His and Carl Ludwig (in whose laboratory Miescher had spent a year) were leaders of the movement to analyze vital phenomena on the basis of physical science. Other outstanding figures in the movement were Claude Bernard and the neurophysiologists Emil du Bois-Reymond and Hermann von Helmholtz. The contributions of these men taken together constitute the foundation of modern physiology. Ludwig's work on kidney function, to take an example, is the basis of modern kidney physiology. In 1885 Michael Foster, professor of physiology at the University of Cambridge, spoke of "a new molecular physiology," maintaining that "the more these molecular problems of physiology ... are studied, the stronger becomes the conviction that the consideration of what we call 'structure' and 'composition' must, in harmony with the modern teachings of physics, be approached under the dominant conception of modes of motion.... The phenomena in question are the result not of properties of kinds of matter, in the vulgar sense of these words, but of kinds of motion." It was from this point of view that His considered fertilization to be a process in which the sperm communicates a mode

IN CELL DIVISION it is the chromosomes that provide continuity, as is shown in these drawings of salamander epithelial cells published in 1882 by Walther Flemming. In an early stage of cell division the chromosomes (which were known to contain nuclein) are ribbon-like and a "spindle" (two rayed structures) is beginning to form (*1*). The chromosomes seem to split (*2*); actually each has replicated lengthwise and the members of each resulting pair are separating. Each member goes to a different pole of the cell, so that two equal complements of chromosomes are established (*3*). Nuclear membranes surround each complement and the cell divides into two identical daughter cells (*4*).

of motion to the egg—a view that, as we have seen, Miescher accepted.

The biologists who at this time succeeded in discovering what actually happens in fertilization, and thereupon recognized the role of Miescher's nuclein, were (with several notable exceptions) not interested in physical science and in modes of motion. They were the founders of cell biology. The essential step taken by these biologists was the minute observation of what actually happens when a sperm fertilizes an egg. Although salmon sperm was an ideal material for Miescher's experiments on the chemistry of the sperm nucleus, it was of little value for observing the act of fertilization. For this purpose Oskar Hertwig of the University of Berlin and the Swiss biologist Hermann Fol chose the sea urchin and the starfish. In their classic experiments in the late 1870's they observed that the sperm cell penetrates the egg and that the sperm nucleus then fuses with the egg nucleus. At the same time these experiments were being done Flemming, in another series of observations, described the changes that occur in the nucleus during cell division [*see illustration, page 266*]. He was able to show that chromosomes provide the continuing elements from one generation of cells to the next and that they do so by replicating during cell division.

The observations on fertilization by Hertwig and Fol and those on cell division by Flemming were brought together by Edouard van Beneden of the University of Liège in a wonderful series of observations on fertilization in the threadworm *Ascaris*, a parasite of horses. *Ascaris*, unlike most other animals, has nuclei that break down before the egg and sperm nuclei fuse, so that its chromosomes can be seen with unusual clarity. Chromosome behavior can be followed readily because there are only two chromosomes in each nucleus. The male and female chromosomes, indistinguishable from each other, are brought together but do not fuse. Each chromosome replicates and soon there are two cells, formed in the way Flemming described for cell division. Van Beneden saw that in fertilization, as in cell division, continuity depends on chromosomes: the sperm's contribution to fertilization is a set of chromosomes homologous with those present in the egg [*see illustration on next two pages*]. (In the course of this work, which was published in 1883, van Beneden discovered meiosis, the halving of chromosome number that precedes the bringing together of egg and sperm chromosomes in fertilization.)

Two years earlier Miescher's nuclein had been brought into the picture, not by Miescher but by an obscure young botanist named E. Zacharias. He showed that the characteristic material of chromosomes either was nuclein or was intimately associated with it. In his work on cell division Flemming had relied extensively on stains to make the nucleus and chromosomes readily visible. In the nucleus of cells that were not in the process of division there was a somewhat formless structure that took up certain stains; the same stains were taken up by the rod-shaped chromosomes as they emerged from the nucleus during mitosis. The material that took the stain was called chromatin, and Zacharias identified it as nuclein by following the same procedure that had led Miescher to the discovery of nuclein. He found that when a cell was digested with pepsin–hydrochloric acid its nucleus remained and retained its ability to be stained. If, however, the digested cell was extracted with dilute alkali (which, as Miescher had shown, removes nuclein), then no stainable material remained. Zacharias carried out these tests on an exceedingly wide range of cells, both plant and animal, with essentially uniform results. He tested cells in the process of division and found that chromatin could be stained after pepsin-digestion but not after extraction in alkali. Moreover, the spindle that forms during mitosis did not stain, and it was removed by pepsin–hydrochloric acid digestion. All of this pointed to the coupling of nuclein and chromatin. Zacharias' conclusions were quickly accepted by Flemming and many others.

In 1884 and 1885 four biologists published papers that summarized and interpreted the work of the preceding decade, which had been so crowded with discoveries that those who participated felt the swift movement of events in much the same way that biologists do nowadays. Of the four summarizing accounts, three were by zoologists (Hertwig, Albrecht Kölliker and August Weismann) and one by a botanist (Eduard Strasburger). It was clear that an understanding of fertilization was at the same time an understanding of heredity. Continuity from one generation of an organism to the next was accomplished by the chromosomes in the nuclei of egg and sperm; continuity from one cell generation to the next was also accomplished by chromosomes in mitosis. At this point (in 1884) we suddenly come very close to what is one of the cornerstones of current biology: "I believe that I have at least made it highly probable," Hertwig said, "that nuclein is the substance that is responsible not only for fertilization but also for the transmission of hereditary characteristics.... Furthermore, nuclein is in an organized state before, during and after fertilization, so that fertilization is at the same time both a morphological and a physicochemical event."

Even before Hertwig recognized nuclein as the genetic material, the botanist Julius von Sachs had (in 1882) not only suggested this role for nuclein but also gone a step further in pointing out that the nucleins of egg and sperm could hardly be identical—that the nuclein brought into the egg by the sperm must be different from the nuclein already there. And so by 1885, only a decade after Miescher's paper on salmon sperm and 14 years after his first publication on nuclein, a number of biologists had reached a point of view that is at the heart of our present conception of DNA.

What was the attitude of the discoverer of nuclein? As far as we can tell from Miescher's letters (and those to His freely express his views on many problems of biology), he did not accept the new ideas. He doubted the association of nuclein with chromatin, which was an essential element in the ideas expressed by Hertwig and the others. Miescher had a rather low estimate of the value of staining. He wrote in a letter of 1890, "Here once again I must defend my skin against the guild of dyers who suppose there is nothing else [in the sperm head] but chromatin," and in a paper published posthumously he referred disparagingly to Zacharias' work on chromatin. Miescher's attitude was also conditioned by his belief (expressed in several letters in 1892 and 1893) that he had discovered a new substance in the sperm head, which he proposed calling "karyogen." It was this phosphorus-free substance (the nature of which is a mystery today), and not nuclein, that in his opinion was responsible for the special chromatin stain.

In a letter of 1893 Miescher wrote: "The speculations of Weismann and others are afflicted with half-chemical concepts, which are partly unclear and partly derived from an outmoded kind of chemistry. When, as is quite possible, a protein molecule has 40 asymmetric carbon atoms so that there can be a billion isomers,... my [stereochemical] theory is better suited than any other to account for the unimaginable diversity required by our knowledge of heredity." Weismann's speculations were based on the most advanced biology of the time; Miescher preferred to base his speculations on recent advances in chemistry and paid relatively little attention to advances in biology. Time—and the rediscovery of Mendel's "units" of heredity and their identification with genes by Thomas Hunt Morgan—vindicated the idea, forcefully expressed by Weismann, that chromosomes transmit heredity.

Time did not, however, deal so consistently with the idea that nuclein is the material in chromosomes that transmits heredity. That idea appeared in the 1880's, as we have seen, and was widely held in the 1890's. As late as 1895 the American cytologist E. B. Wilson wrote: "Now, chromatin is known to be closely similar to, if not identical with, a substance known as nuclein.... And thus we reach the remarkable conclusion that inheritance may, perhaps, be effected by the physical transmission of a particular chemical compound from parent to offspring." In the next few years doubts arose concerning this conclusion because the amount of chromatin in the nucleus seemed too unstable to provide continuity; the amount varied considerably with the cycle of cell division and with changes in the physiological state of the cell. In the second edition of his book *The Cell in Development and Heredity*, published in 1900, Wilson described this fluctuation in staining but was able to explain it: "We may infer that the original chromosomes contain a high percentage of nucleinic acid; that their growth and loss of staining power is due to a combination with a large amount of albuminous substance...; that their final diminution in size and resumption of staining power is caused by a giving up of the albuminous constituent."

This analysis corresponds exactly with our present understanding, but it was soon to succumb to the apparent evidence of staining. When the large "lampbrush" chromosomes present in certain egg-cell precursors came under intensive study in the 1890's, there seemed to be no chromatin in them at all, and surely if chromatin (or nuclein) was the hereditary material, it must be present in an unbroken line from one cell generation to the next. In 1909 Strasburger wrote: "Chromatin cannot itself be the hereditary substance [because] the amount of it is subject to considerable variation in the nucleus, according to its stage of development." Strasburger was an eminent authority. So was Wilson, and the third edition of his book, which was published in 1925 and influenced a generation of biologists, took the Strasburger view:

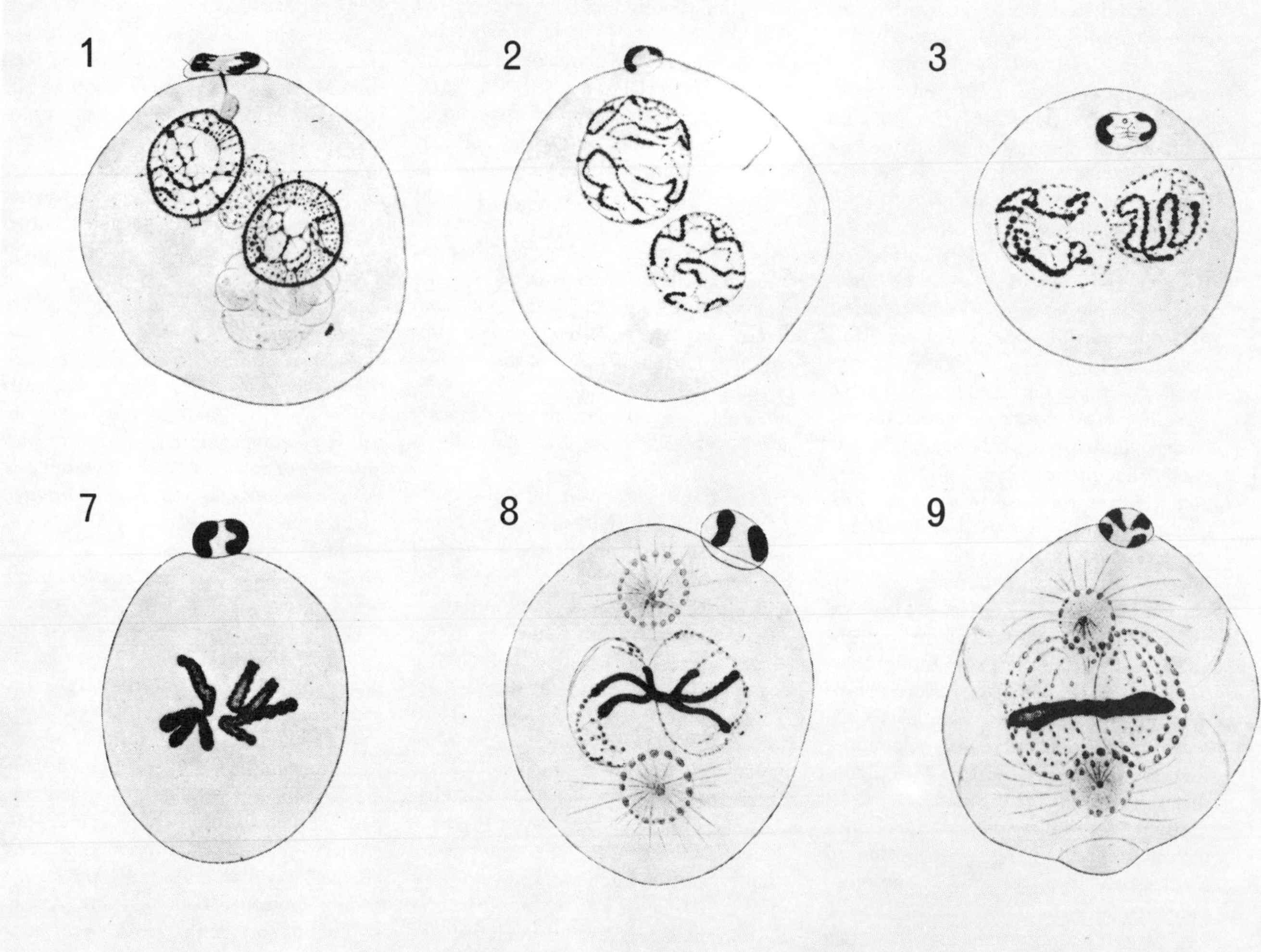

IN FERTILIZATION TOO chromosomes provide the continuity, as shown in drawings of the fertilized egg of *Ascaris*, a parasitic worm, published in 1883 by Edouard van Beneden. Sperm and egg nuclei approach each other (*1*). The nuclear membrane breaks down and the chromosomes become clearly visible; there are two chromosomes in each nucleus, half the normal "diploid" number in

that the loss of staining in the enlarged chromosomes indicates "a progressive accumulation of protein components and a giving up, or even a complete loss, of nuclein." Wilson emphasized, using italics, "These facts afford conclusive proof that *the individuality and genetic continuity of chromosomes does not depend upon a persistence of chromatin.*" Biologists maintained this position for a generation.

Then, in 1948 and 1949, groups at the Rockefeller Institute and in France independently measured the quantity of DNA in cell nuclei. They found that the amount of DNA per set of chromosomes is in general constant in the different cell types of any organism, even when there are striking differences in the intensity of staining, and that the amount of protein associated with a fixed quantity of DNA may vary considerably and so account for variations in staining capacity. Moreover, the DNA content per chromosome set is a characteristic of each particular species. The doubts and questions concerning staining that had been raised by Miescher and many others were essentially resolved by these measurements. Now the biological role of DNA, proposed in 1884 and 1885 when the chromosome theory of heredity was formulated, received solid support: It was demonstrated that the chromosome complements of egg and sperm carry identical amounts of DNA, which are combined at fertilization and then carried by successive replications to all cells of the organism. The continuity of chromosomes at fertilization and cell division has always been an essential element in the chromosome theory of heredity; the associated continuity of DNA (Miescher's nuclein) points to it as providing the molecular basis for heredity in the chromosomes.

Several years before this point was finally established investigators at the Rockefeller Institute—following a line of investigation with a different historical background—found that hereditary traits could be transmitted from one strain of bacteria to another by the transfer of DNA. Nucleic acid was thus shown to be the genetic material. It should be pointed out that if this substance had been discovered in bacteria, it would never have been called nucleic acid, because a bacterial cell does not have a formed nucleus. The use of such words as "nucleic acid" and "chromosome" in work on bacteria is a constant reminder that the contributions of Friedrich Miescher and his contemporaries form the background for the study of heredity in all living cells.

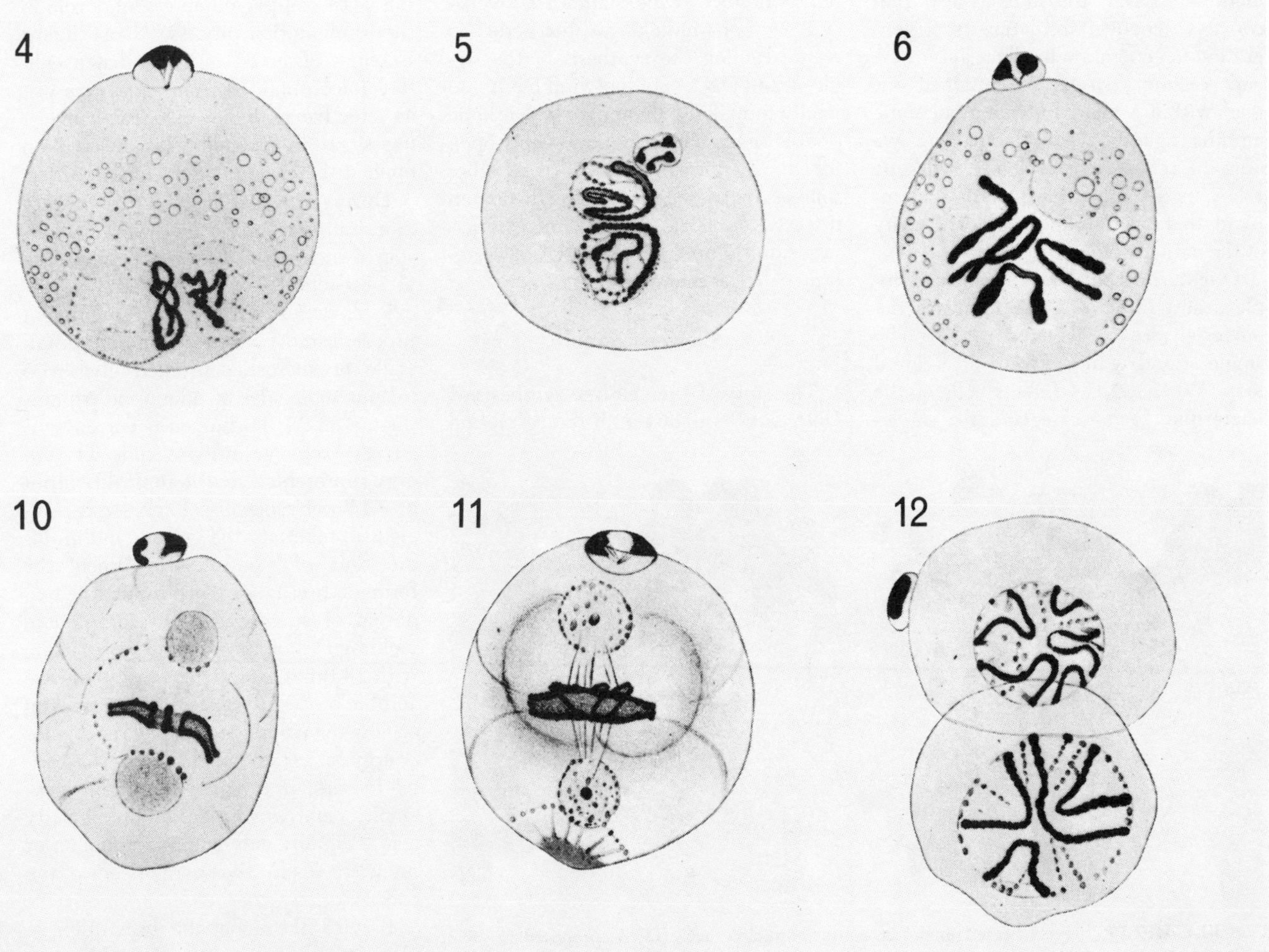

an *Ascaris* cell (*2, 3, 4*). The four chromosomes come together (*5, 6*). Each of the chromosomes has previously doubled; the replicated pairs separate slightly (*7*). Two "centrioles" appear and a spindle forms between them, and cell division begins (*8, 9, 10, 11*). Two cells are finally formed, each with a nucleus containing four chromosomes, two derived from the sperm and two from egg (*12*).

21 The Synthesis of DNA

by Arthur Kornberg
October 1968

Test-tube synthesis of the double helix that controls heredity climaxes a half-century of effort by biochemists to re-create biologically active giant molecules outside the living cell

My colleagues and I first undertook to synthesize nucleic acids outside the living cell, with the help of cellular enzymes, in 1954. A year earlier James Watson and Francis Crick had proposed their double-helix model of DNA, the nucleic acid that conveys genetic information from generation to generation in all organisms except certain viruses. We attained our goal within a year, but not until some months ago—14 years later—were we able to report a completely synthetic DNA, made with natural DNA as a template, that has the full biological activity of the native material.

Our starting point was an unusual single-strand form of DNA found in the bacterial virus designated ϕX174. The single strand is in the form of a closed loop. When ϕX174 infects cells of the bacterium *Escherichia coli,* the single-strand loop of DNA serves as the template that directs enzymes in the synthesis of a second loop of DNA. The two loops form a ring-shaped double helix similar to the DNA helixes found in bacterial cells and higher organisms. In our laboratory at the Stanford University School of Medicine we succeeded in reconstructing the synthesis of the single-strand DNA copies of viral DNA and finally in making a completely synthetic double helix. The way now seems open for the synthesis of DNA from other sources: viruses associated with human disease, bacteria, multicellular organisms and ultimately the DNA of vertebrates such as mammals.

An Earlier Beginning

The story of the cell-free synthesis of DNA does not start with the revelation of the structure of DNA in 1953. It begins around 1900 with the biochemical understanding of how the fermentation of fruit juices yields alcohol. Some 40 years earlier Louis Pasteur had convinced his contemporaries that the living yeast cell played an essential role in the fermentation process. Then Eduard Buchner observed in 1897 that a cell-free juice obtained from yeast was just as effective as intact cells for converting sugar to alcohol. This observation opened the era of modern biochemistry.

During the first half of this century biochemists resolved the overall conversion of sucrose to alcohol into a sequence of 14 reactions, each catalyzed by a specific enzyme. When this fermentation proceeds in the absence of air, each molecule of sucrose consumed gives rise to four molecules of adenosine triphosphate (ATP), the universal currency of energy exchange in living cells. The energy represented by the fourfold output of ATP per molecule of sucrose is sufficient to maintain the growth and multiplication of yeast cells. When the fermentation takes place in air, the oxidation of sucrose goes to completion, yielding carbon dioxide and water along with 18 times as much energy as the anaerobic process does. This understanding of how the combustion of sugar provides energy for cell metabolism was succeeded by similar explanations of how enzymes catalyze the oxidation of fatty acids, amino acids and the subunits of nucleic acids for the energy needs of the cell.

By 1950 the enzymatic dismantling of large molecules was well understood. Little thought or effort had yet been invested, however, in exploring how the cell makes large molecules out of small ones. In fact, many biochemists doubted that biosynthetic pathways could be suc-

DOUBLE HELIX, the celebrated model of deoxyribonucleic acid (DNA) proposed in 1953 by James D. Watson and F. H. C. Crick, consists of two strands held together by crossties (*color*) that spell out a genetic message, unique for each organism. The Watson-Crick model explained for the first time how each crosstie consists of two subunits, called bases, that form obligatory pairs (*see illustrations on page 272*). Thus each strand of the double helix and its associated sequence of bases is complementary to the other strand and its bases. Consequently each strand can serve as a template for the reconstruction of the other strand.

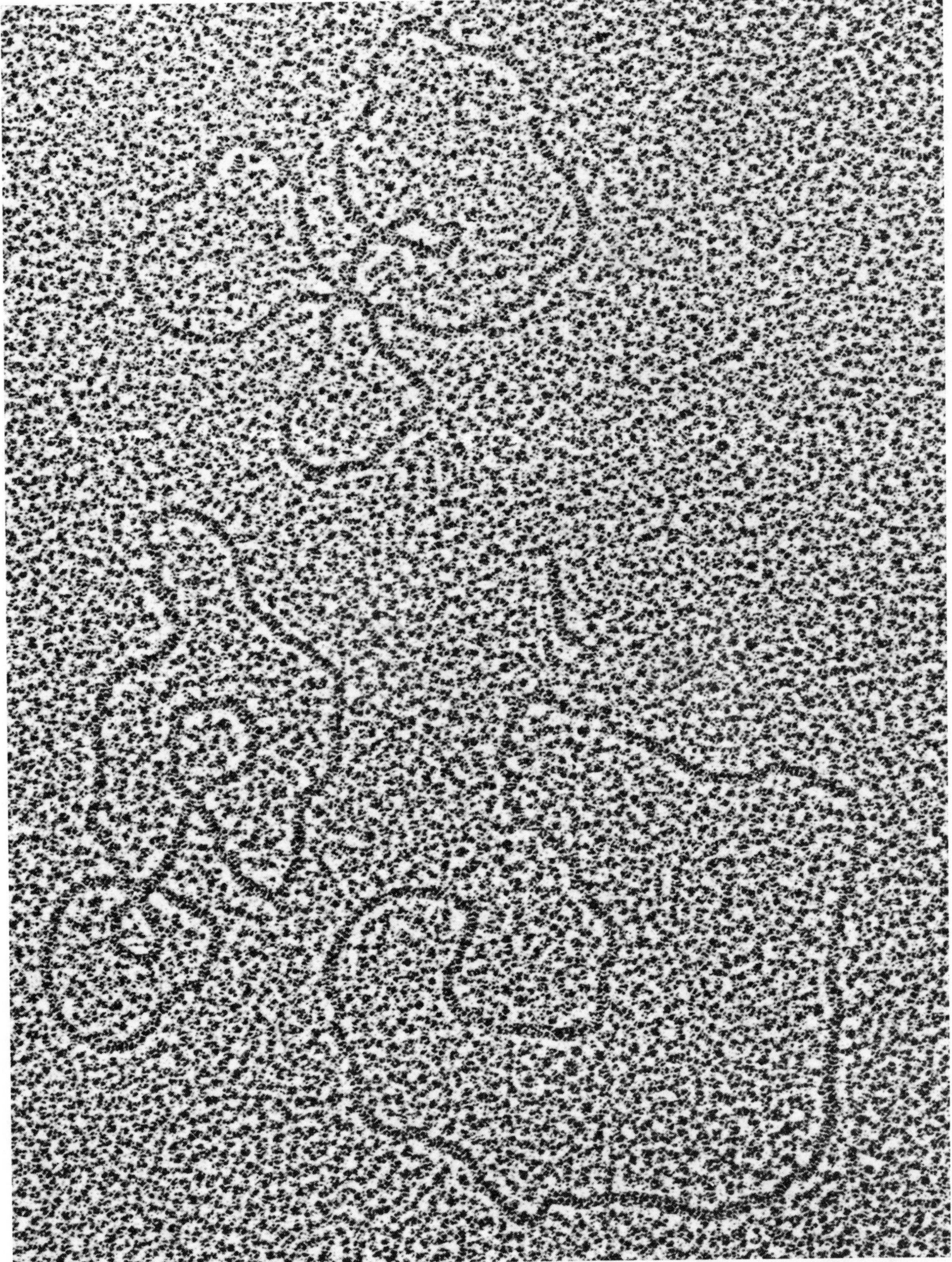

THREE CLOSED LOOPS OF DNA, each a complete double helix, are shown in this electron micrograph made in the author's laboratory at the Stanford University School of Medicine. One strand of each loop is the natural single-strand DNA of the bacterial virus ϕX174, which served as a template for the test-tube synthesis, carried out by enzymes, of a synthetic complementary strand. The hybrid molecules are biologically active. The enlargement is about 200,000 diameters. Each loop contains some 5,500 pairs of bases. If enlarged to the scale of the model on the opposite page, each loop of DNA would form a circle roughly 150 feet in circumference.

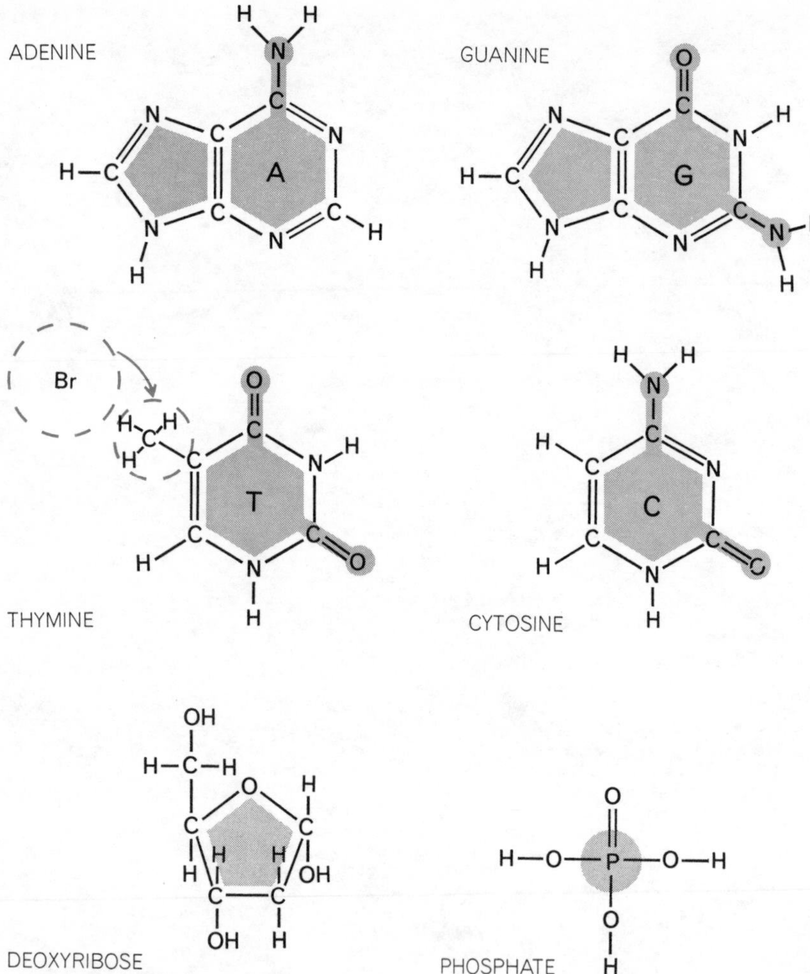

DNA CONSTITUENTS are bases of four kinds, deoxyribose (a sugar) and a simple phosphate. The bases are adenine (*A*) and thymine (*T*), which form one obligatory pair, and guanine (*G*) and cytosine (*C*), which form another. Deoxyribose and phosphate form the backbone of each strand of the DNA molecule. The bases provide the code letters of the genetic message. For purposes of tagging synthetic DNA, thymine can be replaced by 5′-bromouracil, which contains a bromine atom where thymine contains a lighter CH_3 group.

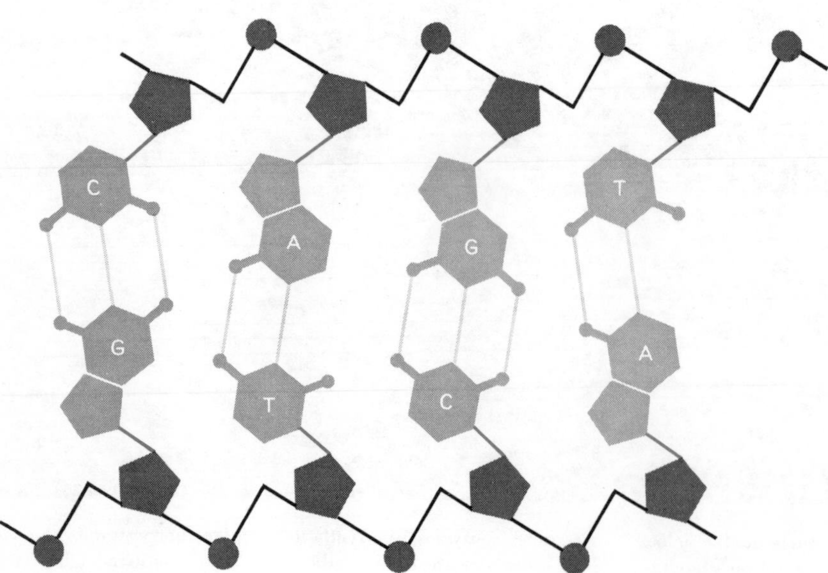

DNA STRUCTURE resembles a ladder in which the side pieces consist of alternating units of deoxyribose and phosphate. The rungs are formed by the bases paired in a special way, A with T and G with C, and held together respectively by two and three hydrogen bonds.

cessfully reconstructed in cell-free systems. Since then nearly two decades of intensive study have been devoted to the cell-free biosynthesis of large molecules. Two things above all have been made clear.

The first is that large molecules can be assembled in cell-free systems with the aid of purified enzymes and coenzymes. The second is that the routes of biosynthesis are different from those of degradation. Some biochemists had speculated that the routes of breakdown were really two-way streets whose flow might somehow be reversed. Now we know that the molecular traffic in cells flows on distinctive and divided highways. All cells have the enzymatic machinery to manufacture most of the subunits of large molecules from simple nutrients such as glucose, ammonia and carbon dioxide. Cells also have the capacity to salvage preformed subunits when they are available. On the basis of what has been learned the prospects are that in this century biochemists will assemble in the test tube complex viruses and major components of the cell. Perhaps the next century will bring the synthesis of a complete cell.

The Nucleotides

My co-workers and I were at Washington University in St. Louis when we made our first attempts to synthesize a nucleic acid in the test tube. By that time the constituents of nucleic acid were well known [*see illustrations at left*]. If one regards DNA as a chain made up of repeating links, the basic link is a structure known as a nucleotide [*see illustration on opposite page*]. It consists of a phosphate group attached to the five-carbon sugar deoxyribose, which is linked in turn to one of four different nitrogen-containing bases. The four bases are adenine (A), thymine (T), guanine (G) and cytosine (C). In the double helix of DNA the phosphate and deoxyribose units alternate to form the two sides of a twisted ladder. The rungs joining the sides consist of two bases: A is invariably linked to T and G is invariably linked to C. This particular pairing arrangement was the key insight of the Watson-Crick model. It means that if the two strands of the helix are separated, uncoupling the paired bases, each half can serve as a template for re-creating the missing half. Thus if the bases projecting from a single strand follow the sequence A, G, G, C, A, T. . . , one immediately knows that the complementary bases on the missing strand are T, C, C, G, T, A. . . . This base-pairing mechanism enables the cell to make accurate copies of the DNA molecule however many times the cell may divide.

When a strand of DNA is taken apart link by link (by treatment with acid or certain enzymes), the phosphate group of the nucleotide may be found attached to carbon No. 3 of the five-carbon deoxyribose sugar. Such a structure is called a 3′-nucleoside monophosphate. We judged, however, that better subunits for purposes of synthesis would be the 5′-nucleoside monophosphates, in which the phosphate linkage is to carbon No. 5 of deoxyribose.

This judgment was based on two lines of evidence. The first had just emerged from an understanding of how the cell itself made nucleotides from glucose, ammonia, carbon dioxide and amino acids. John M. Buchanan of the Massachusetts Institute of Technology had shown that nucleotides containing the bases A and G were naturally synthesized with a 5′ linkage. Our own work had shown the same thing for nucleotides containing T and C. The second line of evidence came from earlier studies my group had conducted at the National Institutes of Health. We had found that certain coenzymes, the simplest molecules formed from two nucleotides, were elaborated from 5′ nucleotide units. For the enzymatic linkage to take place the phosphate of the nucleotide had to be activated by an additional phosphate group [*see illustration on next page*]. Thus it seemed reasonable that activated 5′ nucleotides (nucleoside 5′ triphosphates) might combine with each other, under the proper enzymatic guidance, to form long chains of nucleic acid.

Our initial attempts at nucleic acid synthesis relied principally on two techniques. The first involved the use of radioactive atoms to label the nucleotide so that we could detect the incorporation of even minute amounts of it into nucleic acid. We sought the enzymatic machinery for synthesizing nucleic acids in the juices of the thymus gland, bone marrow and bacterial cells. Unfortunately such extracts also have a potent capacity for degrading nucleic acids. We added our labeled nucleotides to a pool of nucleic acids and hoped that a few synthesized molecules containing a labeled nucleotide would survive by being mixed into the pool. Even if there were net destruction of the pool of nucleic acids, the synthesis of a few molecules trapped in this pool might still be detected. The second technique exploited the fact that the nucleic acid could be precipitated by making the medium strongly acidic, whereas the nucleotide precursors remained behind in solution.

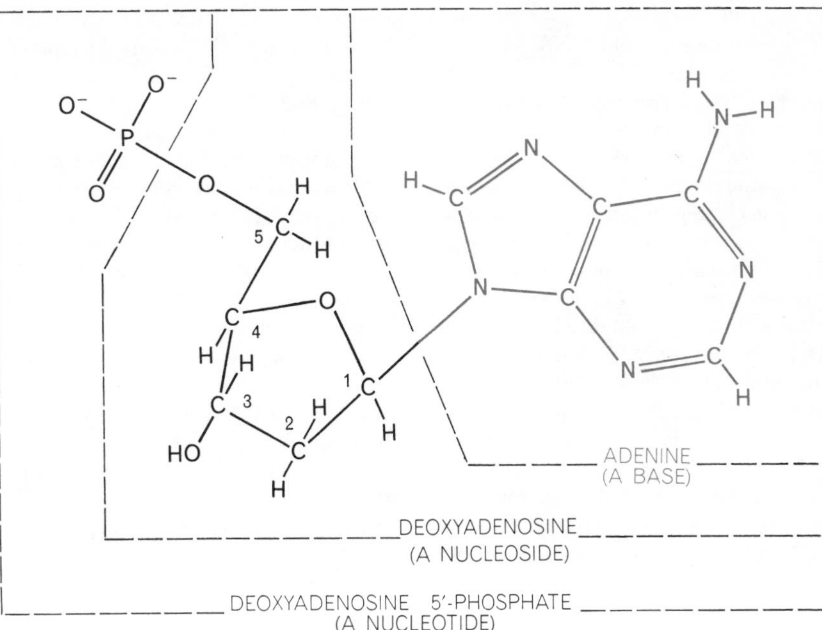

DNA BUILDING BLOCK, the monomer from which DNA polymers are constructed, is termed a nucleotide. There are four nucleotides, one for each of the four bases A, T, G and C. Deoxyadenosine 5′-phosphate, the nucleotide incorporating adenine, is shown here. If the phosphate group is replaced by a hydrogen atom, the structure is called a nucleoside.

Our first experiments with animal-cell extracts were uniformly negative. Therefore we turned to *E. coli,* which has the virtue of reproducing once every 20 minutes. Here we saw a glimmer. In samples to which we had added a quantity of labeled nucleotides whose radioactive atoms disintegrated at the rate of a million per minute we detected about 50 radioactive disintegrations per minute in the nucleic acid fraction that was precipitated by acid. Although the amount of nucleotide incorporated into nucleic acid was minuscule, it was nonetheless significantly above the level of background "noise." Through this tiny crack we tried to drive a wedge. The hammer was enzyme purification, a technique that had matured during the elucidation of alcoholic fermentation.

DNA Polymerase

In these experiments Uriel Littauer, a Fellow of the Weizmann Institute in Israel, and I observed the incorporation of adenylate (a nucleotide) from ATP into ribonucleic acid (RNA), in which the five-carbon sugar in the backbone of the chain is ribose rather than deoxyribose. Actually the first definitive demonstration of synthesis of an RNA-like molecule in a cell-free system had been achieved in the laboratory of Severo Ochoa in 1955. Working at the New York University School of Medicine, he and Marianne Grunberg-Manago were investigating an aspect of energy metabolism and made the unexpected observation that one of the reactants, adenosine diphosphate (ADP), had been polymerized by cell juices into a chain of adenylates resembling RNA.

In our first attempts to achieve DNA synthesis in a cell-free system we used the deoxyribonucleoside called deoxythymidine. To Morris E. Friedkin, who was then at Washington University, we are grateful not only for supplying the radioactively labeled compound but also for the knowledge that the compound was readily incorporated into DNA by bone marrow cells and other animal cells. We were hopeful that extracts of *E. coli* would be able to incoporate deoxythymidine into nucleic acid by converting it first into the 5′ deoxynucleotide and then activating the deoxynucleotide to the triphosphate form. I found this to be the case. In subsequent months Ernest Simms and I were able to prepare separately deoxythymidine 5′-triphosphate and the other deoxynucleoside triphosphates, using enzymes or chemical synthetic routes. (In what follows the various deoxynucleosides in their 5′ triphosphate form will be designated simply by the initial of the base followed by an asterisk. Thus deoxythymidine 5′-triphosphate will be T*.)

In November, 1955, I. Robert Lehman, who is now at Stanford, started on the purification of the enzyme system in *E. coli* extracts that is responsible for converting T* into DNA. We were joined by Maurice J. Bessman some weeks later. Those were eventful days in which the enzyme, now given the name DNA polymerase, was progressively separated from other large molecules. With each step in purification the character of this DNA synthetic reaction became clearer. By June, 1956, when we participated at a conference on the chemical basis of heredity held at Johns Hopkins University, we could report two important facts about DNA synthesis in vitro, although we still lacked the answers to many important questions.

We reported first that preformed DNA had to be present along with DNA polymerase, and that all four of the deoxynucleotides that occur in DNA (A, G, T and C) had to be furnished in the activated triphosphate form. We also reported that DNA from virtually any source—virus, bacterium or animal—could serve with the *E. coli* enzyme. What we still did not know was whether the synthetic DNA was a new molecule or an extension of a preexisting one. There were other questions. Did the synthetic DNA have the same chemical backbone and physical structure as natural DNA? Did it have a chemical composition typical of DNA, in which A equals T and G equals C, and in which, therefore, A plus G equals T plus C? Finally, and crucially: Did the chemical composition of the synthetic DNA reflect the composition of the particular natural DNA used to direct the reaction?

During the next three years these questions and related ones were resolved by the efforts of Julius Adler, Sylvy Kornberg and Steven B. Zimmerman. The synthetic DNA was shown to be a molecule with the chemical structure typical of DNA and the same ratio of A-T pairs to G-C pairs as the particular DNA used to prime, or direct, the reaction [*see illustration on page 276*]. The relative starting amounts of the four deoxynucleoside triphosphates had no influence whatever on the composition of the new DNA. The composition of the synthetic DNA was determined solely by the composition of the DNA that served as a template. An interesting illustration of this last fact justifies a slight digression.

Howard K. Schachman of the University of California at Berkeley spent his sabbatical year of 1957–1958 with us at Washington University examining the physical properties of the synthetic DNA. It had the high viscosity, the comparatively slow rate of sedimentation and other physical properties typical of natural DNA. The new DNA, like the natural one, was therefore a long, fibrous polymer molecule. Moreover, the longer the mixture of active ingredients was allowed to incubate, the greater the viscosity of the product was; this was direct evidence that the synthetic DNA was continuing to grow in length and in amount. However, we were startled to find one day that viscosity developed in a control test tube that lacked one of the essential triphosphates, G*. To be sure, no reaction was observed during the standard incubation period of one or two hours. On prolonging the incubation for several more hours, however, a viscous substance materialized!

Analysis proved this substance to be a DNA that contained only A and T nucleotides. They were arranged in a perfect alternating sequence. The isolated polymer, named dAT, behaved like any other DNA in directing DNA synthesis: it led to the immediate synthesis of more dAT polymer. Would any G* and C* be polymerized if these nucleotides were present in equal or even far greater amounts than A* and T* in a synthesis directed by dAT polymer? We found no detectable incorporation of G or C under conditions that would have measured the inclusion of even one G for every 100,000 A or T nucleotides polymerized. Thus DNA polymerase rarely, if ever, made the mistake of matching G or C with A or T.

The DNA of a chromosome is a linear array of many genes. Each gene, in turn,

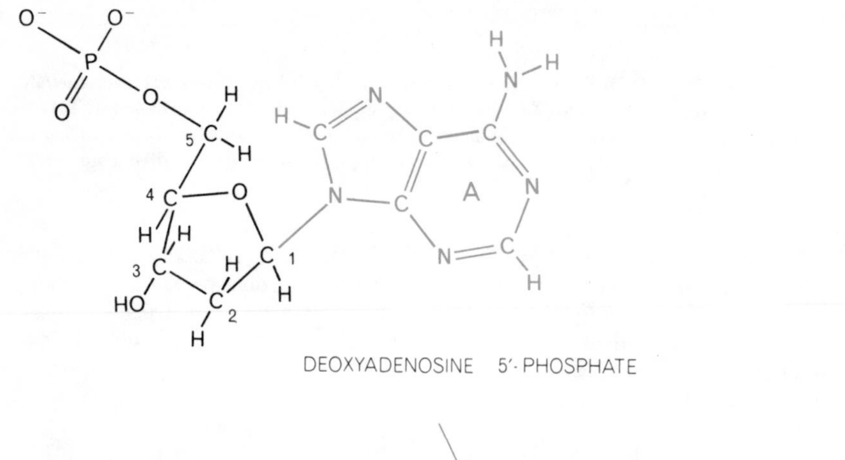

DEOXYADENOSINE 5′-PHOSPHATE

ATP

ATP

DEOXYADENOSINE 5′-TRIPHOSPHATE (A*)

ACTIVATED BUILDING BLOCK is required when synthesizing DNA on a template of natural DNA with the aid of enzymes. The activated form of the nucleotide containing adenine is deoxyadenosine 5′-triphosphate, symbolized in this article by "A*." It is made from deoxyadenosine 5′-monophosphate by two different enzymes in two steps. Each step involves the donation of a terminal phosphate group from adenosine triphosphate (ATP).

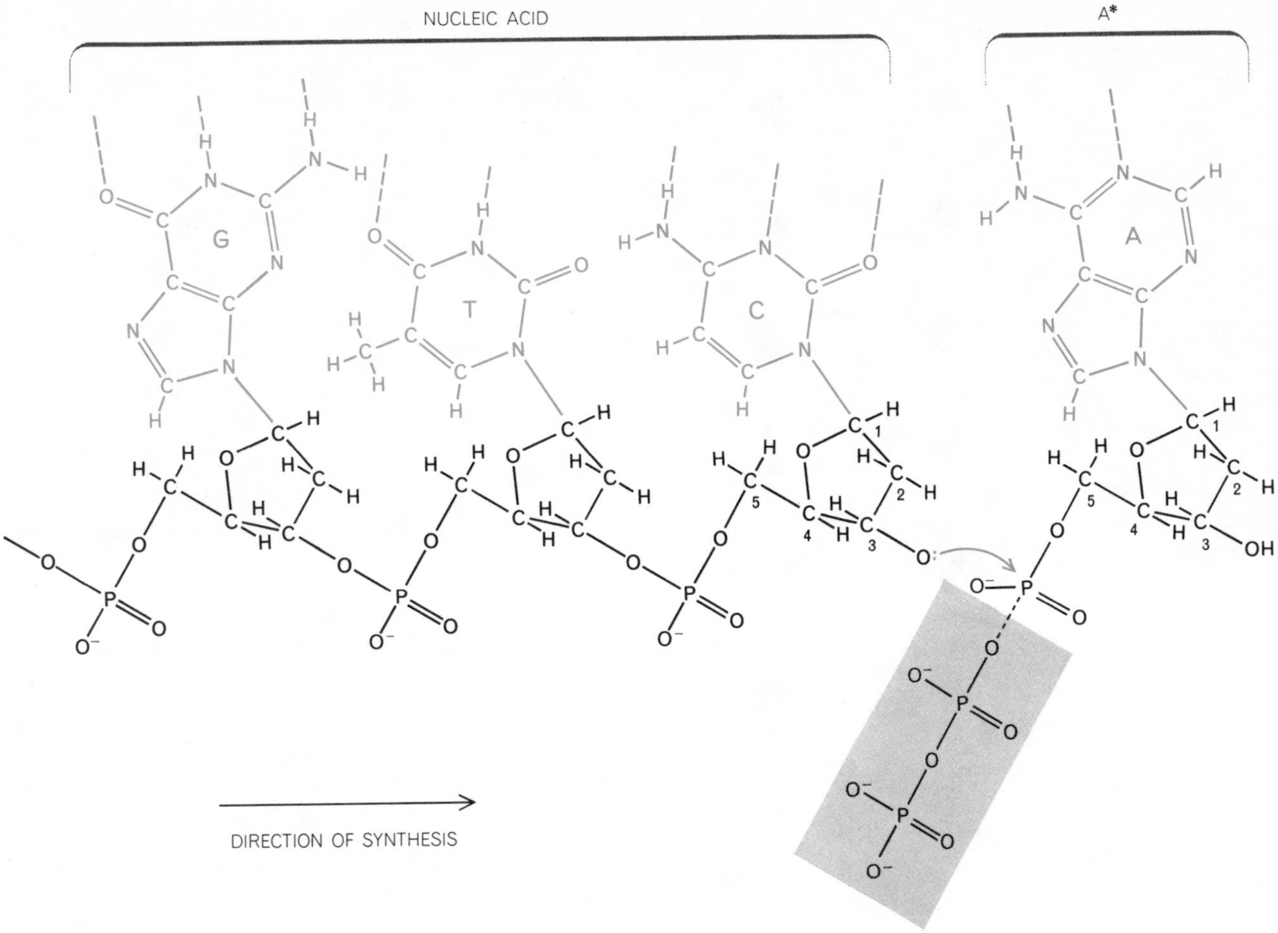

SYNTHESIS OF DNA involves the stepwise addition of activated nucleotides to the growing polymer chain. In this illustration deoxyadenosine 5′-triphosphate (A*) is being coupled through a phosphodiester bond that links the 3′ carbon in the deoxyribose portion of the last nucleotide in the growing chain to the 5′ carbon in the deoxyribose portion of the newest member of the chain.

is a chain of about 1,000 nucleotides in a precisely defined sequence, which when translated into amino acids spells out a particular protein or enzyme. Does DNA polymerase in its test-tube synthesis of DNA accurately copy the sequential arrangement of nucleotides by base-pairing (A = T, G = C) without errors of mismatching, omission, commission or transposition? Unfortunately techniques are not available for determining the precise sequence of nucleotides of even short DNA chains. Because it is impossible to spell out the base sequence of natural DNA or any copy of it, we have resorted to two other techniques to test the fidelity with which DNA polymerase copies the template DNA. One is "nearest neighbor" analysis. The other is the duplication of genes with demonstrable biological activity.

Nearest-Neighbor Analysis

The nearest-neighbor analysis devised by John Josse, A. Dale Kaiser and myself in 1959 determines the relative frequency with which two nucleotides can end up side by side in a molecule of synthetic DNA. There are 16 possible combinations in all. There are four possible nearest-neighbor sequences of A (AA, AG, AT and AC), four for G (GA, GG, GT and GC) and similarly four for T and four for C. How can the frequency of these dinucleotide sequences be determined in a synthetic DNA chain? The procedure is to use a triphosphate labeled with a radioactive phosphorus atom in conducting the synthesis and to treat the synthesized DNA with a specific enzyme that cleaves the DNA and leaves the radioactive phosphorus atom attached to its nearest neighbor. For example, DNA synthesis is carried out with A* labeled in the innermost phosphate group, the group that will be included in the DNA product. This labeled phosphate group now forms a normal linkage (10^{16} times in a typical experiment!) with the nucleotide next to it in the chain—its nearest neighbor [*see illustration on page 277*]. After the synthetic DNA is isolated it is subjected to degradation by an enzyme that cleaves every bond between the 5′ carbon of deoxyribose and the phosphate, leaving the radioactive phosphorus atom attached to the neighboring nucleotide rather than to the one (A) to which it had originally been attached. The nucleotides of the degraded DNA are readily separated by electrophoresis or paper chromatography into the four types of which DNA is composed: A, G, T and C. Radioactive assay establishes the radioactive phosphorus content in each of these nucleotides and at once indicates the frequency with which A is next to A, to G, to T and to C.

The entire experiment is repeated, this time with the radioactive label in G* instead of A*. The second experiment yields the frequency of GA, GG, GT and GC dinucleotides. Two more experiments with radioactive T* and C* complete the analysis and establish the 16 possible nearest-neighbor frequencies.

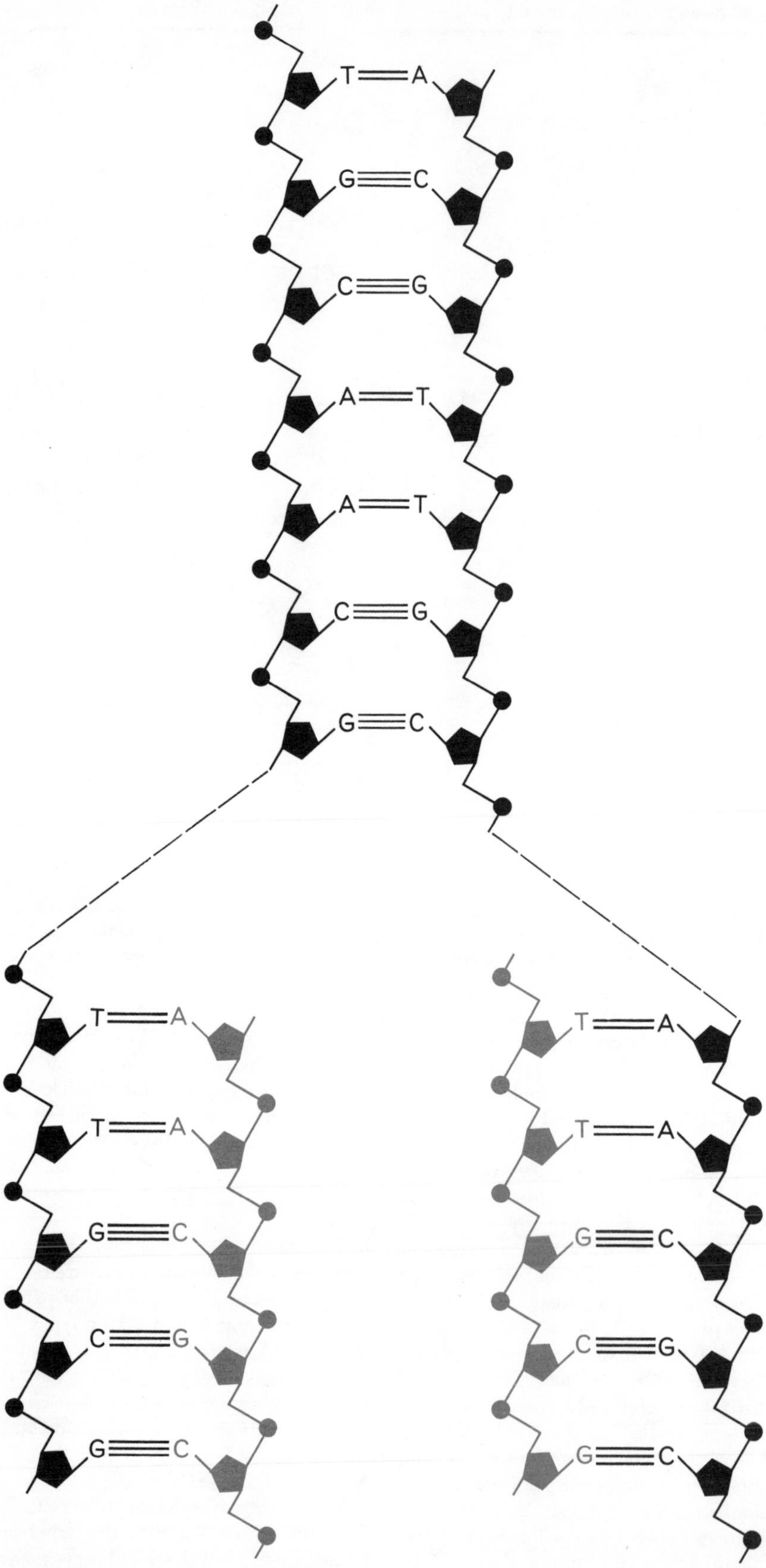

SYNTHESIS OF DUPLEX CHAIN OF DNA yields two hybrid molecules, consisting of a parental strand and a daughter strand, that are identical with each other and with the original duplex molecule. During the replicating process the parental duplex (*black*) separates into two strands, each of which then serves as the template for assembly of a daughter strand (*color*). The pairing of A with T and G with C guarantees faithful reproduction.

Many such experiments were performed with DNA templates obtained from viruses, bacteria, plants and animals. The DNA of each species guided the synthesis of DNA with what proved to be a distinctive assortment of nearest-neighbor frequencies. What is more, when a synthetic DNA was used as a template for a new round of replication, it gave rise to DNA with a nearest-neighbor frequency distribution identical with itself. Among the other insights obtained from these analyses was the recognition of a basic fact about the structure of the double helix. In replication the direction of the DNA chain being synthesized was found to run opposite to that of its template. By inference we can conclude that the chains of the double helix in natural DNA, as surmised by Watson and Crick, must also run in opposite directions.

Even with considerable care the accuracy of nearest-neighbor frequency analysis cannot be better than about 98 percent. Consequently we were still left with major uncertainties as to the precision of copying chains that contain 1,000 nucleotides or more, corresponding to the length of genes. An important question thus remained unanswered: Does DNA that is synthesized on a genetically or biologically active template duplicate the activity of that template?

One way to recognize the biological activity of bacterial DNA is to see if it can carry out "transformation," a process in which DNA from one species of bacteria alters the genetic endowment of a second species. For example, DNA from a strain of *Bacillus subtilis* resistant to streptomycin can be assimilated by a strain susceptible to the antibiotic, whereupon the recipient bacterium and all its descendants carry the trait of resistance to streptomycin. In other words, DNA molecules carrying the genes for a particular characteristic can be identified by their capacity for assimilation into the chromosome of a cell that previously lacked that trait. Yet when DNA was synthesized on a template of DNA that had transforming ability, the synthetic product invariably lacked that ability.

Part of the difficulty in synthesizing biologically active DNA lay in the persistence of trace quantities of nuclease enzymes in our DNA polymerase preparations. Nucleases are enzymes that degrade DNA. The introduction by a nuclease of one break in a long chain of DNA is enough to destroy its genetic activity. Further purification of DNA polymerase was indicated. Efforts over several years by Charles C. Richardson, Thomas Jovin, Paul T. Englund and LeRoy L. Bertsch resulted in a new procedure that was both simple and efficient. Finally, in April, 1967, with the assistance of the personnel and large-scale equipment of the New England Enzyme Center (sponsored by the National Institutes of Health at the Tufts University School of Medicine), we processed 100 kilograms of *E. coli* bacterial paste and obtained about half a gram of pure enzyme, free of the nuclease that puts random breaks in a DNA chain.

Unfortunately even this highly purified DNA polymerase has proved incapable of producing a biologically active DNA from a template of bacterial DNA. The difficulty, we believe, is that the DNA we extract from a bacterium such as *B. subtilis* provides the enzyme with a poor template. A proper template would be the natural chromosome, which is a double-strand loop about one millimeter in circumference. During its isolation from the bacterium the chromosome is broken, probably at random, into 100 or more fragments. The manner in which DNA polymerase and its related enzymes go about the replication of a DNA molecule as large and complex as the *B. subtilis* chromosome is the subject of current study in many laboratories.

The Virus φX174

It occurred to us in 1964 that the problem of synthesizing biologically active DNA might be solved by dealing with a simpler form of DNA that also has genetic activity. This is represented in viruses, such as φX174, whose DNA core is a single-strand loop. This "chromosome" not only is simpler in structure but also it is so small (about two microns in circumference) that it is fairly easy to extract without breakage. We also knew from the work of Robert L. Sinsheimer at the California Institute of Technology that when the DNA of φX174 invades *E. coli,* the first stage of infection involves the "subversion" of one of the host's enzymes to convert the single-strand loop into a double-strand helical loop. Sinsheimer called this first-stage product a "replicative form." Could the host enzyme that copies the viral DNA be the same DNA polymerase we had isolated from *E. coli?*

In undertaking the problem of copying a closed-loop DNA we could foresee some serious obstacles. Would it be possible for DNA polymerase to orient itself and start replication on a DNA template if the template had no ends? Shashanka Mitra and later Peter Reichard succeeded in finding conditions under which the enzyme, as judged by electron microscope pictures, appeared to copy the single-strand loop. We then wondered if in spite of appearances in the electron micrographs, the DNA of φX174 was really just a simple loop. Perhaps, as had been suggested by other workers, it was really more like a necklace with a clasp, the clasp consisting of substances unrelated to the nucleotides we were supplying. Finally, we were aware from Sinsheimer's work that the DNA of φX174 had to be a completely closed loop in order to be infectious. We knew that our polymerase could only catalyze the synthesis of linear DNA molecules. How could we synthesize a genuinely closed loop? We were still missing either the clasplike component to insert into our product or, if the clasp was a mistaken hypothesis, a new kind of enzyme to close the loop.

Fortunately the missing factor was provided for us by work carried on independently in five different laboratories. The discovery in 1966 of a polynucleotide-joining enzyme was made almost simultaneously by Martin F. Gellert and his co-workers at the National Institutes of Health, by Richardson and Bernard Weiss at the Harvard Medical

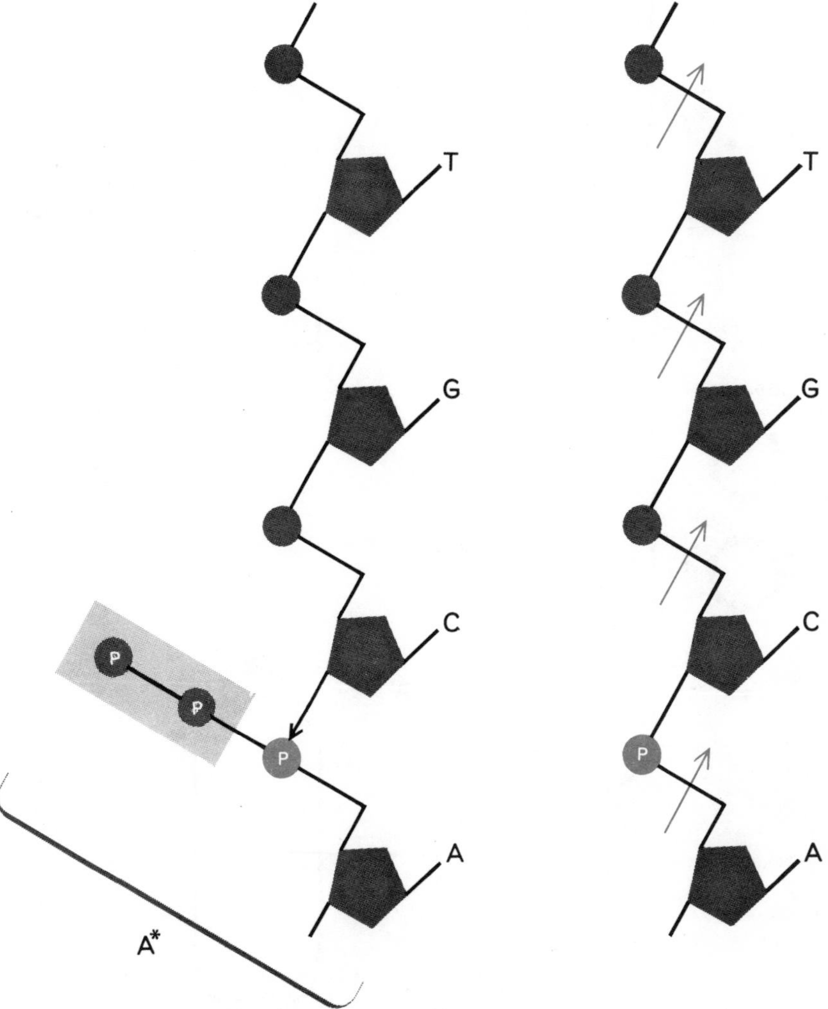

"NEAREST NEIGHBOR" ANALYSIS can reveal how often any of the four bases is located next to any other base in a single strand of synthetic DNA. Thus one can learn how often A is next to A, T, G or C, and so on. A radioactive phosphorus atom (*color*) is placed in the innermost position of one of the activated nucleotides, for example A*. The finished DNA molecule is then treated with an enzyme (*right*) that cleaves the chain between every phosphate and the 5′ carbon of the adjacent deoxyribose. Thus the phosphate is separated from the nucleotide on which it entered the chain and ends up attached to the nearest neighbor instead, C in the above example. The four kinds of nucleotide are separated by paper chromatography and the radioactivity associated with each is measured. The experiment is repeated with radioactive phosphorus linked to the other activated nucleotides.

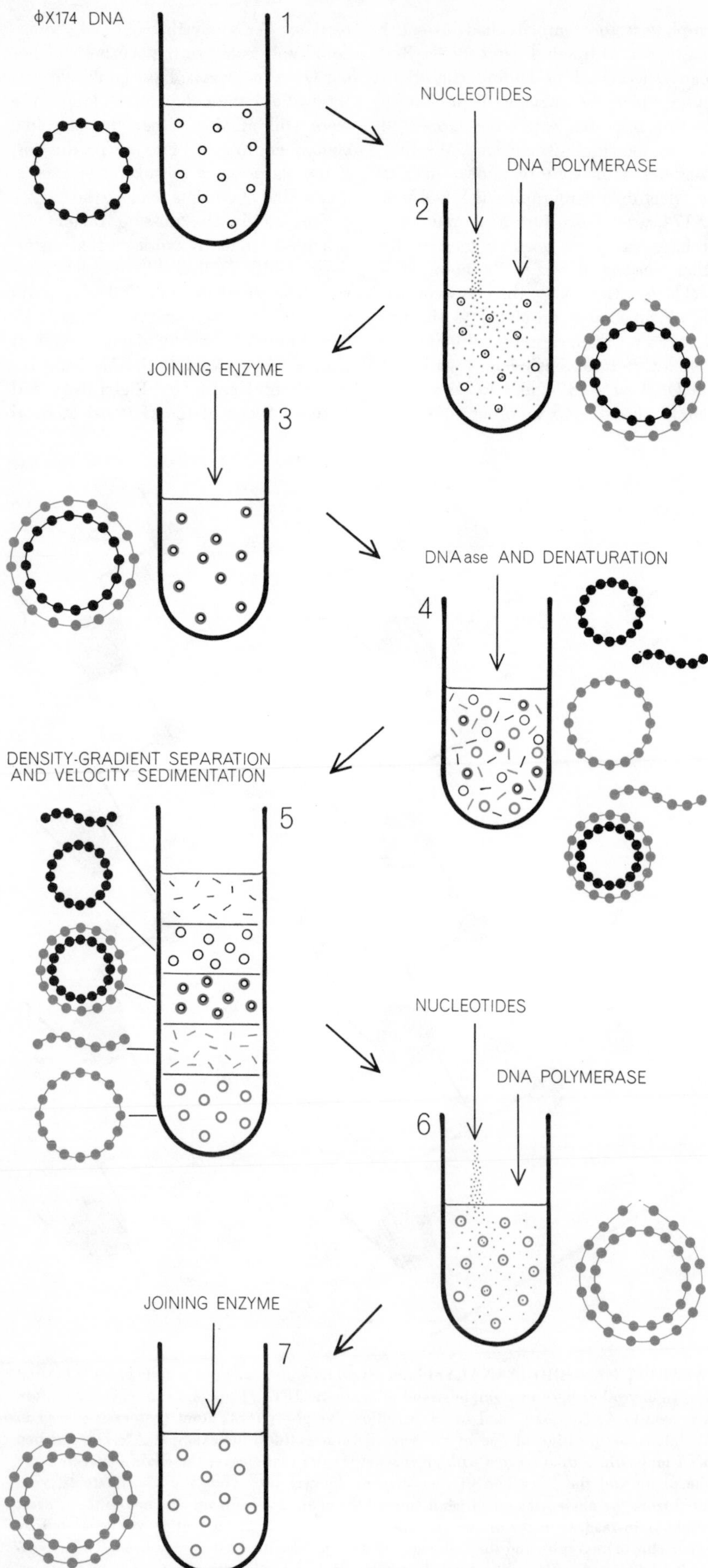

School, by Jerard Hurwitz and his colleagues at the Albert Einstein College of Medicine in New York, by Lehman and Baldomero M. Olivera at Stanford and by Nicholas R. Cozzarelli in my own group. It was the Lehman-Olivera preparation that we now employed in our experiments.

The polynucleotide-joining enzyme has the ability to repair "nicks" in the DNA strand. The nicks occur where there is a break in the sugar-phosphate backbone of one strand of the DNA molecule. The enzyme can repair a break only if all the nucleotides are intact and if what is missing is the covalent bond in the DNA backbone between a sugar and the neighboring phosphate. Provided with the joining enzyme, we were now in a position to find out whether it could work in conjunction with DNA polymerase to synthesize a completely circular and biologically active virus DNA.

By using the DNA of ϕX174 as a template we gained an important advantage over experiments based on transforming ability. Even if we were successful in synthesizing a DNA with transforming activity, this would still be of relatively limited significance. We could then say only that a restricted section of the DNA—a section as small as a part of a gene—had been assimilated by the recipient cell to replace a comparable section of its chromosome, substituting a proper sequence for a defective or incorrect one. However, Sinsheimer had demonstrated with the DNA of ϕX174 that a change in even one of its 5,500 nucleotides is sufficient to make the virus noninfective. Therefore the demonstration of infectivity in a completely synthetic

SYNTHESIS OF ϕX174 DNA was accomplished by the following steps. Circular single-strand ϕX174 DNA, tagged with tritium, served as a template (*1*). Activated nucleotides containing A, G, C and 5′-bromouracil instead of T were added to the template, together with DNA polymerase. One of the activated nucleotides was tagged with radioactive phosphorus. The DNA synthesized on the template was complete but not yet joined in a loop (*2*). The loop was closed by the joining enzyme (*3*). Enough nuclease was now added to cut one strand in about half of all the duplex loops (*4*). This left a mixture of complete duplex loops, template loops, synthetic loops, linear template strands and linear synthetic strands. Since the synthetic strands contained 5′-bromouracil, they were heavier than the template strands and could be separated by centrifugation (*5*). The synthetic loops were then isolated and used as templates for making wholly synthetic duplex loops (*6 and 7*).

virus DNA would conclusively prove that we had carried out virtually error-free synthesis of this large number of nucleotides, comprising the five or six genes that carry out the virus's biological function.

In less than a year the test-tube synthesis of ϕX174 DNA was achieved. The steps can be summarized as follows. Template DNA was obtained from ϕX174 and labeled with tritium, the radioactive isotope of hydrogen. Tritium would thereafter provide a continuing label identifying the template. To the template were added DNA polymerase, purified joining enzyme and a cofactor (diphosphopyridine nucleotide), together with A*, T*, G* and C*. One of the nucleoside triphosphates was labeled with radioactive phosphorus. The radioactive phosphorus would thus provide a label for synthetic material analogous to the tritium label for the template. The interaction of the reagents then proceeded until the number of nucleotide units polymerized was exactly equal to the number of nucleotides in the template DNA. This equality was readily determined by comparing the radioactivity from the tritium in the template with the radioactivity from the phosphorus in the nucleotides provided for synthesis.

Such comparison showed that the experiments had progressed to an extent adequate for the formation of complementary loops of synthetic DNA. Complementary loops were designated (−) to distinguish them from the template loop (+). We had to demonstrate that the synthetic (−) loops were really loops. Had the polymerase made a full turn around the template and had the two ends of the chain been united by the joining enzyme? Several physical measurements, including electron microscopy, assured us that our product was a closed loop coiled tightly around the virus-DNA template and that it was identical in size and other details with the replicative form of DNA that appears in the infected cells. We could now exclude the possibility that some clasp material different from the nucleotide-containing compounds we had employed was involved in closing the virus-DNA loop.

The critical questions remaining were whether the synthetic (−) loops had biological activity—that is, infectivity—and whether the synthetic loops could in turn act as templates for the formation of a completely synthetic "duplex" DNA analogous to the replicative forms that were produced naturally inside infected cells. In order to answer the first of these questions we had to isolate the synthetic DNA strands from the partially synthetic duplexes. For reasons that will be apparent below, we substituted bromouracil, a synthetic but biologically active analogue of thymine, for thymine [*see top illustration on page 272*]. We then introduced just enough nuclease to produce a single nick in one strand of about half the population of molecules. The duplex loops that had been nicked would release a single linear strand of DNA; these single strands could be separated from their circular companions and from unnicked duplex loops by heating. Thus we were left with a mixture that contained (+) template loops, (−) synthetic loops, (+) template linear forms, (−) synthetic linear forms—all in about equal quantities—and full duplex loops.

It was at this point that the substitution of bromouracil for thymine became useful. Because bromouracil contains a bromine atom in place of the methyl group of thymine, it is heavier than the thymine. Therefore a molecule containing bromouracil can be separated from one containing thymine by high-speed centrifugation in a heavy salt solution (the density-gradient technique perfected by Jerome R. Vinograd of Cal Tech). In this system the denser a substance is, the lower in the centrifuge tube it will settle. Thus from top to bottom of the centrifuge tube we obtained fractions containing the light single strands of thymine-containing (+) template DNA, the duplex hybrids of intermediate weight and finally the single-strand synthetic (−) DNA "weighted down" with bromouracil. The reliability of this fractionation was confirmed by three separate peaks of radioactivity corresponding to each of the fractions. We were further reassured by observations that the mean density of each fraction corresponded almost exactly to the mean density of standard samples of virus DNA containing bromouracil or thymine.

Still another physical technique involving density-gradient sedimentation was employed to separate the synthetic linear forms from the synthetic circular forms. The circular forms could then be used in tests of infectivity, by methods previously developed by Sinsheimer to demonstrate the infectivity of circular ϕX174 DNA. We tested our (−) loops by incubating them with *E. coli* cells whose walls had been removed by the action of the enzyme lysozyme. Infectivity is assayed by the ability of the virus to lyse, or dissolve, these cells when they are "plated" on a nutrient medium. Our synthetic loops showed almost exactly the same patterns of infectivity as their natural counterparts had. Their biological activity was now demonstrated.

One further set of experiments remained in which the (−) synthetic loops were employed as the template to determine if we could produce completely synthetic duplex circular forms analogous to the replicative forms found in cells infected with natural ϕX174 virus. Because the synthetic (−) loops were labeled with radioactive phosphorus, this time we added tritium to one of the nucleotide-containing subunits (C*). The remaining procedures were essentially the same as the ones described above, and we did produce fully synthetic duplex loops of ϕX174. The (+) loops were then separated and were found to be identical in all respects with the (+) loops of natural ϕX174 virus. Their infectivity could also be demonstrated. Sinsheimer had previously shown that, under these assay conditions, a change in a single nucleotide of the virus gave rise to a mutant of markedly decreased infectivity. Therefore the correspondence between the infectivity of our synthetic forms and their natural counterparts attested to the precision of the enzymatic operation.

Future Directions

The total synthesis of infective virus DNA by DNA polymerase with the four deoxynucleoside triphosphates not only demonstrates the capacity of this enzyme to copy a small chromosome (of five or six genes) without error but also shows that this chromosome, at least, is as simple and straightforward as a linear sequence of the standard four deoxynucleotide units. It is a long step to the human chromosome, some 10,000 times larger, yet we are encouraged to extrapolate our current conceptions of nucleotide composition and nucleotide linkage from the tiny ϕX174 chromosome to larger ones.

What are the major directions this research will take? I see at least three immediate and productive paths. One is the exploration of the physical and chemical nature of DNA polymerase in order to understand exactly how it performs its error-free replication of DNA. Without this knowledge of the structure of the enzyme and how it operates under defined conditions in the test tube, our understanding of the intracellular behavior of the enzyme will be incomplete.

A second direction is to clarify the

control of DNA replication in the cell and in the animal. Why is DNA synthesis arrested in a mature liver cell and what sets it in motion 24 hours after part of the liver is removed surgically? What determines the slow rate of DNA replication in adult cells compared with the rate in embryonic or cancer cells? The time is ripe for exploration of the factors that govern the initiation and rate of DNA synthesis in the intact cell and animal. Finally, there are now prospects of applying our knowledge of DNA structure and synthesis directly to human welfare. This is the realm of genetic engineering, and it is our collective responsibility to see that we exploit our great opportunities to improve the quality of human life.

An obvious area for investigation would be the synthesis of the polyoma virus, a virus known to induce a variety of malignant tumors in several species of rodents. Polyoma virus in its infective form is made up of duplex circular DNA and presumably replicates in this form on entering the cell. On the basis of our experience it would appear quite feasible to synthesize polyoma virus DNA. If this synthesis is accomplished, there would seem to be many opportunities for modifying the virus DNA and thus determining where in the chromosome its tumor-producing capacity lies. With this knowledge it might prove possible to modify the virus in order to control its tumor-producing potential.

Our speculations can extend even to large DNA molecules. For example, if a failure in the production of insulin were to be traced to a genetic deficit, then administration of the appropriate synthetic DNA might conceivably provide a cure for diabetes. Of course, a system for delivering the corrective DNA to the cells must be devised. Even this does not seem inconceivable. The extremely interesting work of Stanfield Rogers at the Oak Ridge National Laboratory suggests a possibility. Rogers has shown that the Shope papilloma virus, which is not pathogenic in man, is capable of inducing production of the enzyme arginase in rabbits at the same time that it induces tumors. Rogers found that in the blood of laboratory investigators working with the virus there is a significant reduction of the amino acid arginine, which is destroyed by arginase. This is apparently an expression of enhanced arginase activity. Might it not be possible, then, to use similar nonpathogenic viruses to carry into man pieces of DNA capable of replacing or repairing defective genes?

The Bacterial Chromosome

by John Cairns
January 1966

When bacterial DNA is labeled with radioactive atoms, it takes its own picture. Autoradiographs reveal that the bacterial chromosome is a single very long DNA molecule and show how it is duplicated

The information inherited by living things from their forebears is inscribed in their deoxyribonucleic acid (DNA). It is written there in a decipherable code in which the "letters" are the four subunits of DNA, the nucleotide bases. It is ordered in functional units—the genes—and thence translated by way of ribonucleic acid (RNA) into sequences of amino acids that determine the properties of proteins. The proteins are, in the final analysis, the executors of each organism's inheritance.

The central event in the passage of genetic information from one generation to the next is the duplication of DNA. This cannot be a casual process. The complement of DNA in a single bacterium, for example, amounts to some six million nucleotide bases; this is the bacterium's "inheritance." Clearly life's security of tenure derives in large measure from the precision with which DNA can be duplicated, and the manner of this duplication is therefore a matter of surpassing interest. This article deals with a single set of experiments on the duplication of DNA, the antecedents to them and some of the speculations they have provoked.

When James D. Watson and Francis H. C. Crick developed their two-strand model for the structure of DNA, they saw that it contained within it the seeds of a system for self-duplication. The two strands, or polynucleotide chains, were apparently related physically to each other by a strict system of *complementary* base pairing. Wherever the nucleotide base adenine occurred in one chain, thymine was present in the other; similarly, guanine was always paired with cytosine. These rules meant that the sequence of bases in each chain inexorably stipulated the sequence in the other; each chain, on its own, could generate the entire sequence of base pairs. Watson and Crick therefore suggested that accurate duplication of DNA could occur if the chains separated and each then acted as a template on which a new complementary chain was laid down. This form of duplication was later called "semiconservative" because it supposed that although the individual parental chains were conserved during duplication (in that they were not thrown away), their association ended as part of the act of duplication.

The prediction of semiconservative replication soon received precise experimental support. Matthew S. Meselson and Franklin W. Stahl, working at the California Institute of Technology, were able to show that each molecule of DNA in the bacterium *Escherichia coli* is composed of equal parts of newly synthesized DNA and of old DNA that was present in the previous generation [*see top illustration on page 283*]. They realized they had not proved that the two parts of each molecule were in fact two chains of the DNA duplex, because they had not established that the molecules they were working with consisted of only two chains. Later experiments, including some to be described in this article, showed that what they were observing was indeed the separation of the two chains during duplication.

The Meselson-Stahl experiment dealt with the end result of DNA duplication. It gave no hint about the mechanism that separates the chains and then supervises the synthesis of the new chains. Soon, however, Arthur Kornberg and his colleagues at Washington University isolated an enzyme from *E. coli* that, if all the necessary precursors were provided, could synthesize in the test tube chains that were complementary in base sequence to any DNA offered as a template. It was clear, then, that polynucleotide chains could indeed act as templates for the production of complementary chains and that this kind of reaction could be the normal process of duplication, since the enzymes for carrying it out were present in the living cell.

Such, then, was the general background of the experiments I undertook beginning in 1962 at the Australian National University. My object was simply (and literally) to look at molecules of DNA that had been caught in the act of duplication, in order to find out which of the possible forms of semiconservative replication takes place in the living cell: how the chains of parent DNA are arranged and how the new chains are laid down [*see bottom illustration on page 283*].

Various factors dictated that the experiments should be conducted with *E. coli*. For one thing, this bacterium was known from genetic studies to have only one chromosome; that is, its DNA is contained in a single functional unit in which all the genetic markers are arrayed in sequence. For another thing, the duplication of its chromosome was known to occupy virtually the entire cycle of cell division, so that one could be sure that every cell in a rapidly multiplying culture would contain replicating DNA.

Although nothing was known about the number of DNA molecules in the *E. coli* chromosome (or in any other complex chromosome, for that matter), the dispersal of the bacterium's DNA among its descendants had been shown to be semiconservative. For this and other reasons it seemed likely that the

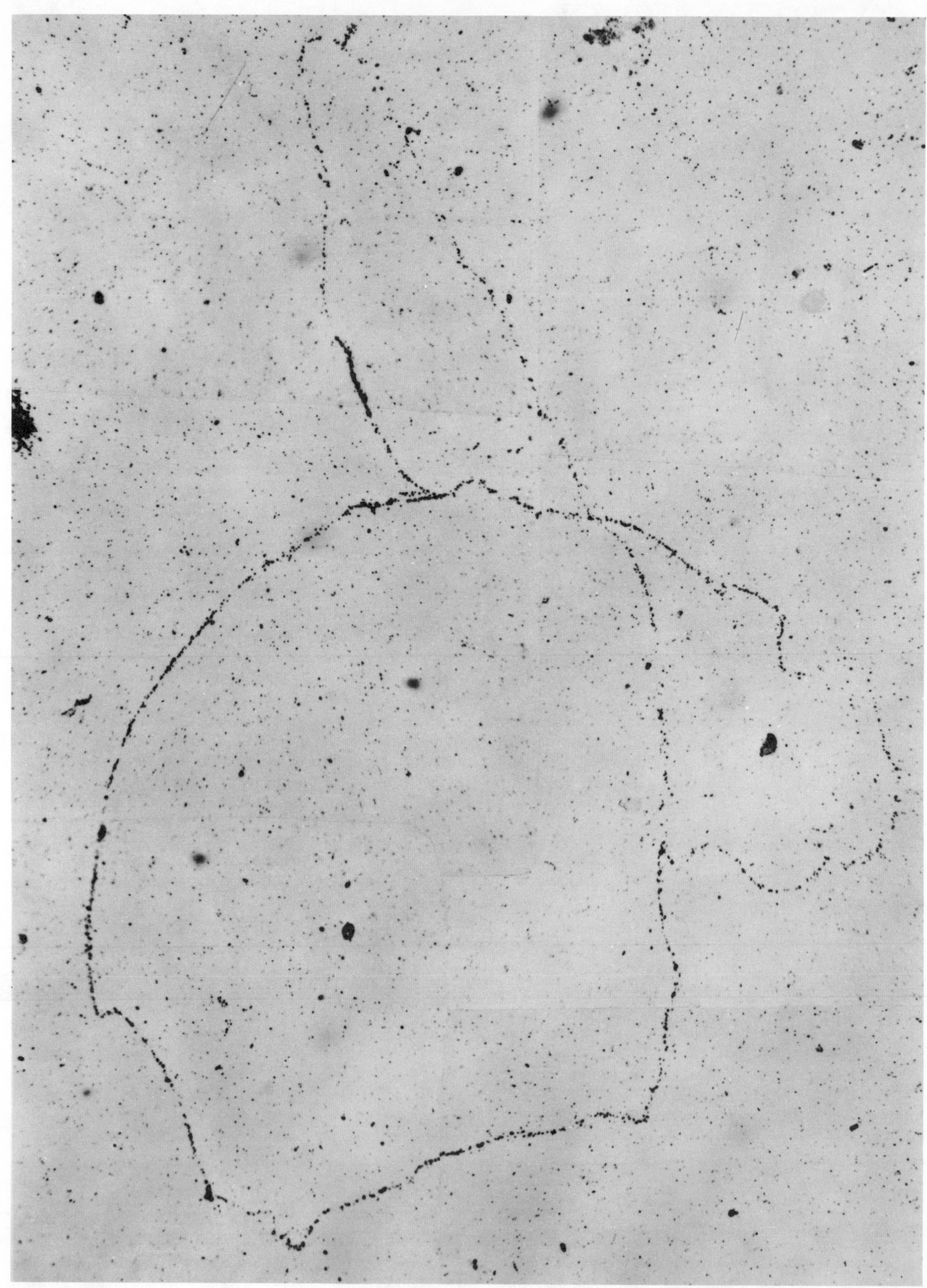

AUTORADIOGRAPH shows a duplicating chromosome from the bacterium *Escherichia coli* enlarged about 480 diameters. The DNA of the chromosome is visible because for two generations it incorporated a radioactive precursor, tritiated thymine. The thymine reveals its presence as a line of dark grains in the photographic emulsion. (Scattered grains are from background radiation.) The diagram on the opposite page shows how the picture is interpreted as demonstrating the manner of DNA duplication.

bacterial chromosome would turn out to be a single very large molecule. All the DNA previously isolated from bacteria had, to be sure, proved to be in molecules much smaller than the total chromosome, but a reason for this was suggested by studies by A. D. Hershey of the Carnegie Institution Department of Genetics at Cold Spring Harbor, N.Y. He had pointed out that the giant molecules of DNA that make up the genetic complement of certain bacterial viruses had been missed by earlier workers simply because they are so large that they are exceedingly fragile. Perhaps the same thing was true of the bacterial chromosome.

If so, the procedure for inspecting the replicating DNA of bacteria would have to be designed to cater for an exceptionally fragile molecule, since the bacterial chromosome contains some 20 times more DNA than the largest bacterial virus. It would have to be a case of looking but not touching. This was not as onerous a restriction as it may sound. The problem was, after all, a topographical one, involving delineation of strands of parent DNA and newly synthesized DNA. There was no need for manipulation, only for visualization.

Although electron microscopy is the

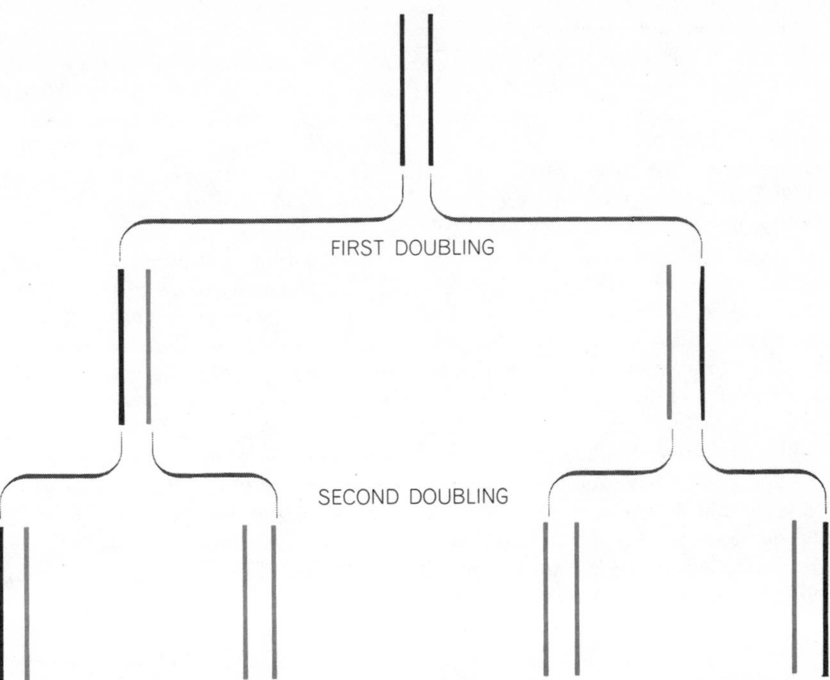

SEMICONSERVATIVE DUPLICATION was confirmed by the Meselson-Stahl experiment, which showed that each DNA molecule is composed of two parts: one that is present in the parent molecule, the other comprising new material synthesized when the parent molecule is duplicated. If radioactive labeling begins with the first doubling, the unlabeled (*black*) and labeled (*colored*) nucleotide chains of DNA form two-chain duplexes as shown here.

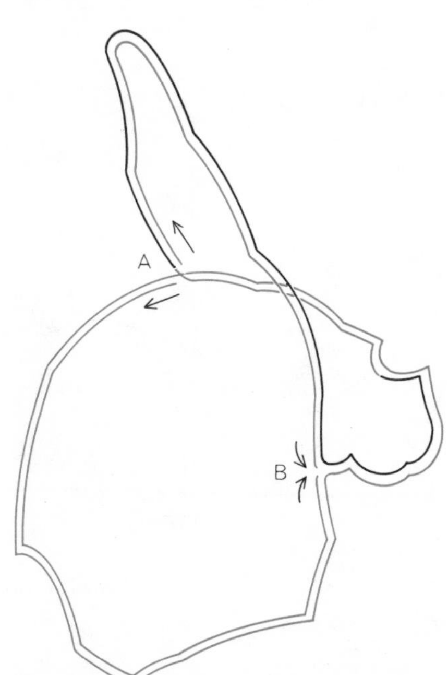

INTERPRETATION of autoradiograph on opposite page is based on the varying density of the line of grains. Excluding artifacts, dense segments represent doubly labeled DNA duplexes (*two colored lines*), faint segments singly labeled DNA (*color and black*). The parent chromosome, labeled in one strand and part of another, began to duplicate at *A*; new labeled strands have been laid down in two loops as far as *B*.

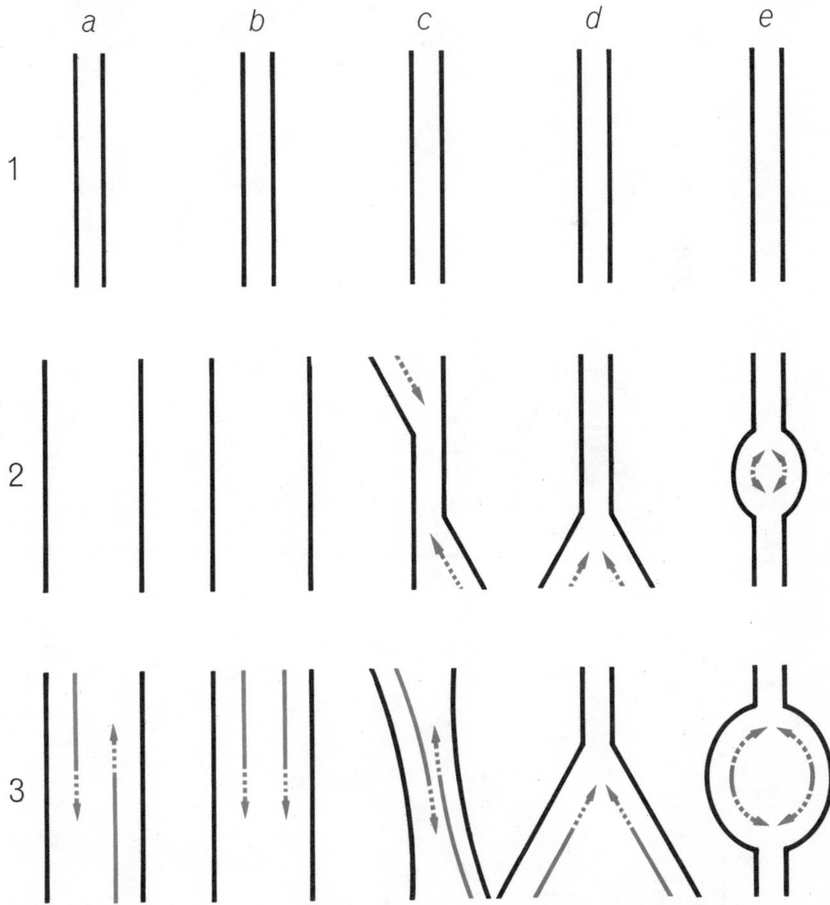

DUPLICATION could proceed in various ways (*a–e*). In these examples parental chains are shown as black lines and new chains as colored lines. The arrows show the direction of growth of the new chains, the newest parts of which are denoted by broken-line segments.

obvious way to get a look at a large molecule, I chose autoradiography in this instance because it offered certain peculiar advantages (which will become apparent) and because it had already proved to be the easier, albeit less accurate, technique for displaying large DNA molecules. Autoradiography capitalizes on the fact that electrons emitted by the decay of a radioactive isotope produce images on certain kinds of photographic emulsion. It is possible, for example, to locate the destination within a cell of a particular species of molecule by labeling such molecules with a radioactive atom, feeding them to the cell and then placing the cell in contact with an emulsion; a developed grain in the emulsion reveals the presence of a labeled molecule [see "Autobiographies of Cells," by Renato Baserga and Walter E. Kisieleski; SCIENTIFIC AMERICAN Offprint 165].

It happens that the base thymine, which is solely a precursor of DNA, is susceptible to very heavy labeling with tritium, the radioactive isotope of hydrogen. Replicating DNA incorporates the labeled thymine and thus becomes visible in autoradiographs. I had been able to extend the technique to demonstrating the form of individual DNA molecules extracted from bacterial viruses. This was possible because, in spite of the poor resolving power of autoradiography (compared with electron microscopy), molecules of DNA are so extremely long in relation to the resolving power that they appear as a linear array of grains. The method grossly exaggerates the apparent width of the DNA, but this is not a serious fault in the kind of study I was undertaking.

The general design of the experiments called for extracting labeled DNA from bacteria as gently as possible and then mounting it—without breaking the DNA molecules—for autoradiography. What I did was kill bacteria that had been fed tritiated thymine for various periods and put them, along with the enzyme lysozyme and an excess of unlabeled DNA, into a small capsule closed on one side by a semipermeable membrane. The enzyme, together with a detergent diffused into the chamber, induced the bacteria to break open and discharge their DNA. After the detergent, the enzyme and low-molecular-weight cellular debris had been diffused out of the chamber, the chamber was drained, leaving some of the DNA deposited on the membrane [*see illustration below*]. Once dry, the membrane was coated with a photographic emulsion sensitive to electrons emitted by the tritium and was left for two months. I hoped by this procedure to avoid subjecting the DNA to appreciable turbulence and so to find

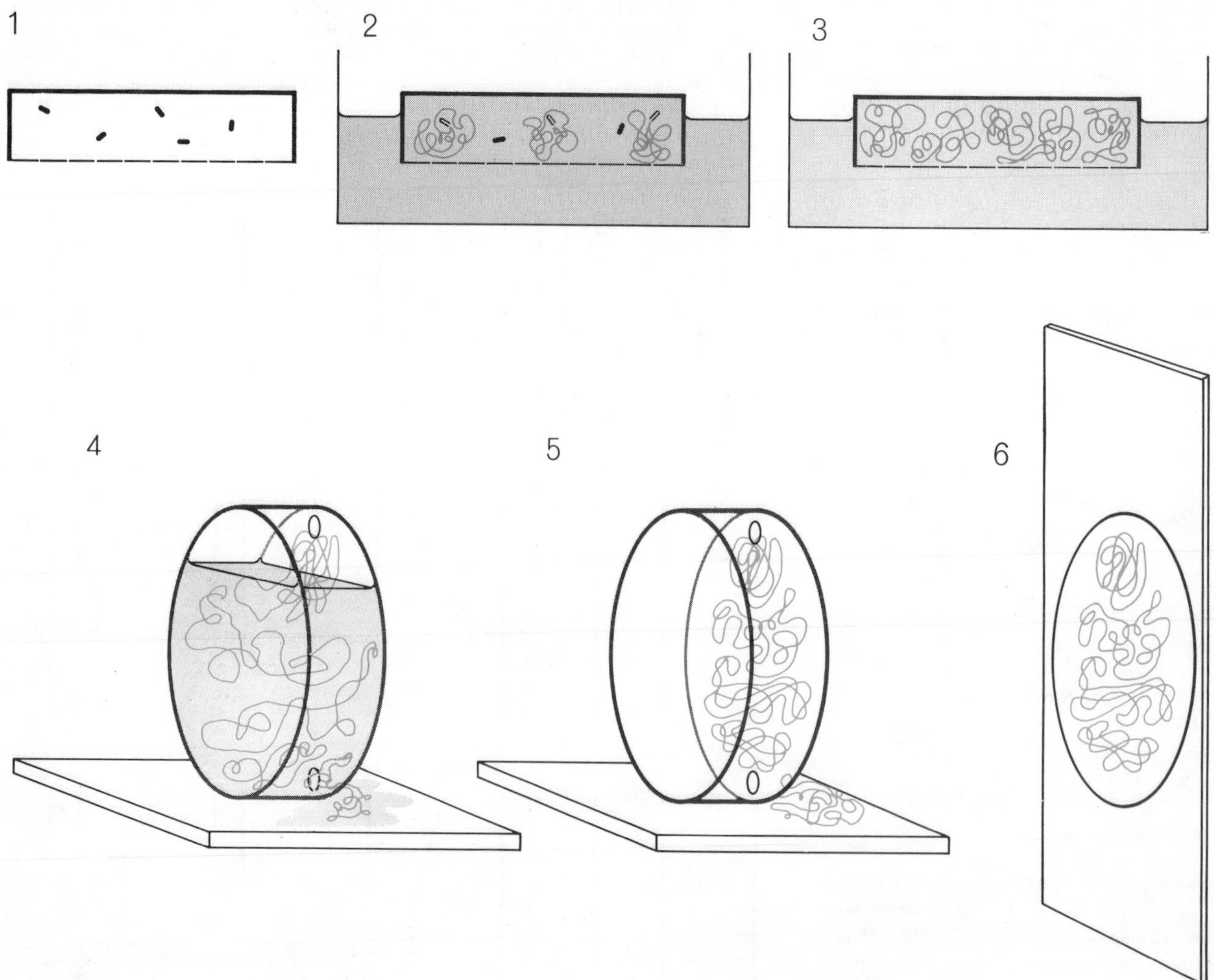

AUTORADIOGRAPHY EXPERIMENT begins with bacteria whose DNA has been labeled with radioactive thymine. The bacteria and an enzyme are placed in a small chamber closed by a semipermeable membrane (*1*). Detergent diffused into the chamber causes the bacteria to discharge their contents (*2*). The detergent and cellular debris are washed away by saline solution diffused through the chamber (*3*). The membrane is then punctured. The saline drains out slowly (*4*), leaving some unbroken DNA molecules (*color*) clinging to the membrane (*5*). The membrane, with DNA, is placed on a microscope slide and coated with emulsion (*6*).

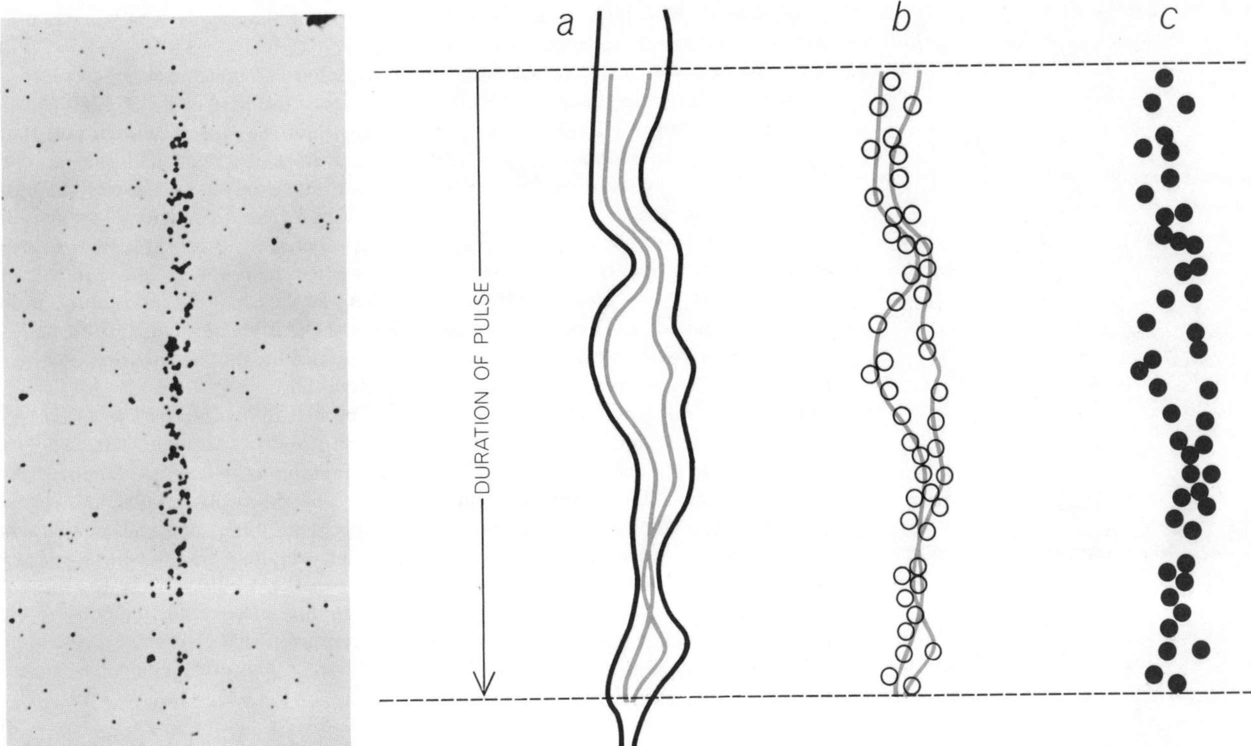

DNA synthesized in *E. coli* fed radioactive thymine for three minutes is visible in an autoradiograph, enlarged 1,200 diameters, as an array of heavy black grains (*left*). The events leading to the autoradiograph are shown at right. The region of the DNA chains synthesized during the "pulse-labeling" is radioactive and is shown in color (*a*). The radioactivity affects silver grains in the photographic emulsion (*b*). The developed grains appear in the autoradiograph (*c*), approximately delineating the new chains of DNA.

some molecules that—however big—had not been broken and see their form. Inasmuch as *E. coli* synthesizes DNA during its entire division cycle, some of the extracted DNA should be caught in the act of replication. (Since there was an excess of unlabeled DNA present, any tendency for DNA to produce artificial aggregates would not produce a spurious increase in the size of the labeled molecules or an alteration in their form.)

It is the peculiar virtue of autoradiography that one sees only what has been labeled; for this reason the technique can yield information on the history as well as the form of a labeled structure. The easiest way to determine which of the schemes of replication was correct was to look at bacterial DNA that had been allowed to duplicate for only a short time in the presence of labeled thymine. Only the most recently made DNA would be visible (corresponding to the broken-line segments in the bottom illustration on page 283), and so it should be possible to determine if the two daughter molecules were being made at the same point or in different regions of the parent molecule. A picture obtained after labeling bacteria for three minutes, or a tenth of a generation-time [*at left in illustration above*], makes it clear that two labeled structures are being made in the same place. This place is presumably a particular region of a larger (unseen) parent molecule [*see diagrams at right in illustration above*].

The autoradiograph also shows that at least 80 microns (80 thousandths of a millimeter) of the DNA has been duplicated in three minutes. Since duplication occupies the entire generation-time (which was about 30 minutes in these experiments), it follows that the process seen in the autoradiograph could traverse at least 10×80 microns, or about a millimeter, of DNA between one cell division and the next. This is roughly the total length of the DNA in the bacterial chromosome. The autoradiograph therefore suggests that the entire chromosome may be duplicated at a single locus that can move fast enough to traverse the total length of the DNA in each generation.

Finally, the autoradiograph gives evidence on the semiconservative aspect of duplication. Two structures are being synthesized. It is possible to estimate how heavily each structure is labeled (in terms of grains produced per micron of length) by counting the number of exposed grains and dividing by the length. Then the density of labeling can be compared with that of virus DNA labeled similarly but uniformly, that is, in both of its polynucleotide chains. It turns out that each of the two new structures seen in the picture must be a single polynucleotide chain. If, therefore, the picture is showing the synthesis of two daughter molecules from one parent molecule, it follows that each daughter molecule must be made up of one new (labeled) chain and one old (unlabeled) chain—just as Watson and Crick predicted.

The "pulse-labeling" experiment just described yielded information on the isolated regions of bacterial DNA actually engaged in duplication. To learn if the entire chromosome is a single molecule and how the process of duplication proceeds it was necessary to look at DNA that had been labeled with tritiated thymine for several generations. Moreover, it was necessary to find, in the jumble of chromosomes extracted from *E. coli*, autoradiographs of unbroken chromosomes that were disen-

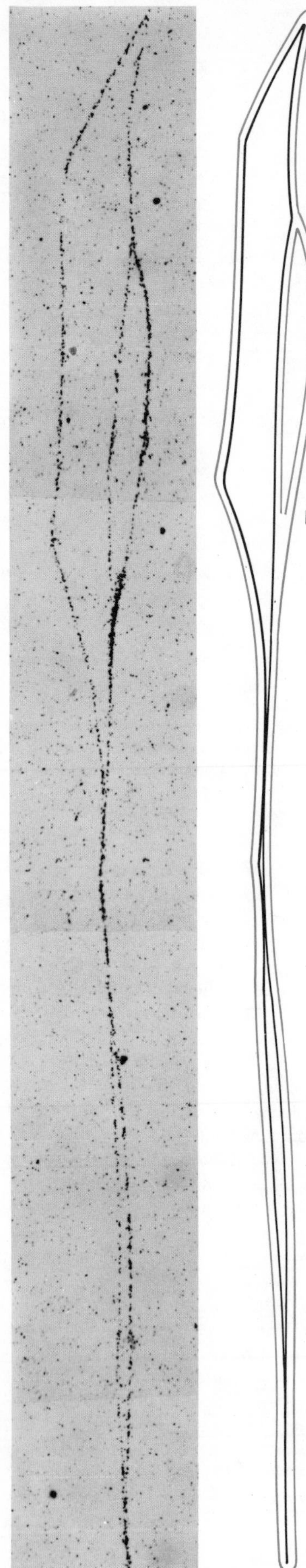

tangled enough to be seen as a whole. Rather than retrace all the steps that led, after many months, to satisfactory pictures of the entire bacterial chromosome in one piece, it is simpler to present two sample autoradiographs and explain how they can be interpreted and what they reveal.

The autoradiographs on page 282 and at the left show bacterial chromosomes in the process of duplication. All that is visible is labeled, or "hot," DNA; any unlabeled, or "cold," chain is unseen. A stretch of DNA duplex labeled in only one chain ("hot-cold") makes a faint trace of black grains. A duplex that is doubly labeled ("hot-hot") shows as a heavier trace. The autoradiographs therefore indicate, as shown in the diagrams that accompany them, the extent to which new, labeled polynucleotide chains have been laid down along labeled or unlabeled parent chains. Such data make it possible to construct a bacterial family history showing the process of duplication over several generations [*see illustration on opposite page*].

The significant conclusions are these:

1. The chromosome of *E. coli* apparently contains a single molecule of DNA roughly a millimeter in length and with a calculated molecular weight of about two billion. This is by far the largest molecule known to occur in a biological system.

2. The molecule contains two polynucleotide chains, which separate at the time of duplication.

3. The molecule is duplicated at a single locus that traverses the entire length of the molecule. At this point both new chains are being made: two chains are becoming four. This locus has come to be called the replicating "fork" because that is what it looks like.

4. Replicating chromosomes are not Y-shaped, as would be the case for a linear structure [*see "d" in bottom illustration on page 283*]. Instead the three ends of the Y are joined: the ends of the daughter molecules are joined to each other and to the far end of the parent molecule. In other words, the chromosome is circular while it is being duplicated.

COMPLETE CHROMOSOME is seen in this autoradiograph, enlarged about 370 diameters. Like the chromosome represented on pages 282 and 283, this one is circular, although it happens to have landed on the membrane in a more compressed shape and some segments are tangled. Whereas the first chromosome was more than halfway through the duplication process, this one is only about one-sixth duplicated (*from A to B*).

It is hard to conceive of the behavior of a molecule that is about 1,000 times larger than the largest protein and that exists, moreover, coiled inside a cell several hundred times shorter than itself. Apart from this general problem of comprehension, there are two special difficulties inherent in the process of DNA duplication outlined here. Both have their origin in details of the structure of DNA that I have not yet discussed.

The first difficulty arises from the opposite polarities of the two polynucleotide chains [*see illustration on page 288*]. The deoxyribose-phosphate backbone of one chain of the DNA duplex has the sequence $-O-C_3-C_4-C_5-O-P-O-C_3-C_4-C_5-O-P-\ldots$ (The C_3, C_4 and C_5 are the three carbon atoms of the deoxyribose that contribute to the backbone.) The other chain has the sequence $-P-O-C_5-C_4-C_3-O-P-O-C_5-C_4-C_3-O-\ldots$

If both chains are having their complements laid down at a single locus moving in one particular direction, it follows that one of these new chains must grow by repeated addition to the C_3 of the preceding nucleotide's deoxyribose and the other must grow by addition to a C_5. One would expect that two different enzymes should be needed for these two quite different kinds of polymerization. As yet, however, only the reaction that adds to chains ending in C_3 has been demonstrated in such experiments as Kornberg's. This fact had seemed to support a mode of replication in which the two strands grew in opposite directions [*see "a" and "c" in bottom illustration on page 283*]. If the single-locus scheme is correct, the problem of opposite polarities remains to be explained.

The second difficulty, like the first, is related to the structure of DNA. For the sake of simplicity I have been representing the DNA duplex as a pair of chains lying parallel to each other. In actuality the two chains are wound helically around a common axis, with one complete turn for every 10 base pairs, or 34 angstrom units of length (34 ten-millionths of a millimeter). It would seem, therefore, that separation of the chains at the time of duplication, like separation of the strands of an ordinary rope, must involve rotation of the parent molecule with respect to the two daughter molecules. Moreover, this rotation must be very rapid. A fast-multiplying bacterium can divide every 20 minutes;

during this time it has to duplicate—and consequently to unwind—about a millimeter of DNA, or some 300,000 turns. This implies an average unwinding rate of 15,000 revolutions per minute.

At first sight it merely adds to the difficulty to find that the chromosome is circular while all of this is going on. Obviously a firmly closed circle—whether a molecule or a rope—cannot be unwound. This complication is worth worrying about because there is increasing evidence that the chromosome of *E. coli* is not exceptional in its circularity. The DNA of numerous viruses has been shown either to be circular or to become circular just before replication begins. For all we know, circularity may therefore be the rule rather than the exception.

There are several possible explanations for this apparent impasse, only one of which strikes me as plausible.

First, one should consider the possibility that there is no impasse—that in the living cell the DNA is two-stranded but not helical, perhaps being kept that way precisely by being in the form of a circle. (If a double helix cannot be unwound when it is firmly linked into a circle, neither can relational coils ever be introduced into a pair of uncoiled circles.) This hypothesis, however, requires a most improbable structure for two-strand DNA, one that has not been observed. And it does not really avoid the unwinding problem because there would still have to be some mechanism for making nonhelical circles out of the helical rods of DNA found in certain virus particles.

Second, one could avoid the unwinding problem by postulating that at least one of the parental chains is repeatedly broken and reunited during replication, so that the two chains can be separated over short sections without rotation of the entire molecule. One rather sentimental objection to this hypothesis (which was proposed some time ago) is that it is hard to imagine such cavalier and hazardous treatment being meted out to such an important molecule, and one so conspicuous for its stability. A second objection is that it does not explain circularity.

The most satisfactory solution to the unwinding problem would be to find some reason why the ends of the chromosome actually *must* be joined together. This is the case if one postulates that there is an active mechanism for unwinding the DNA, distinct from the mechanism that copies the unwound

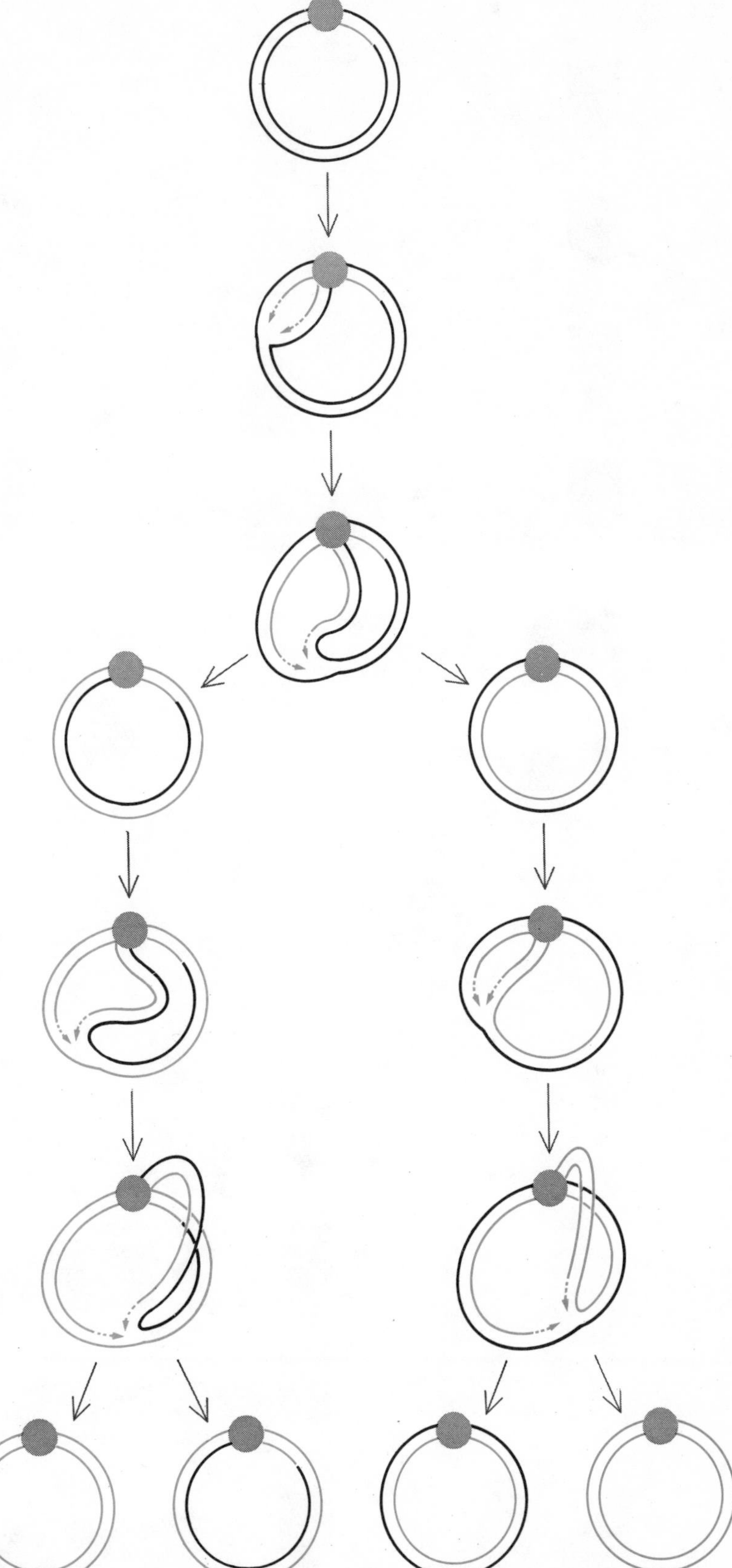

BACTERIAL DNA MOLECULE apparently replicates as in this schematic diagram. The two chains of the circular molecule are represented as concentric circles, joined at a "swivel" (*gray spot*). Labeled DNA is shown in color; part of one chain of the parent molecule is labeled, as are two generations of newly synthesized DNA. Duplication starts at the swivel and, in these drawings, proceeds counterclockwise. The arrowheads mark the replicating "fork": the point at which DNA is being synthesized in each chromosome. The drawing marked *A* is a schematic rendering of the chromosome in the autoradiograph on page 282.

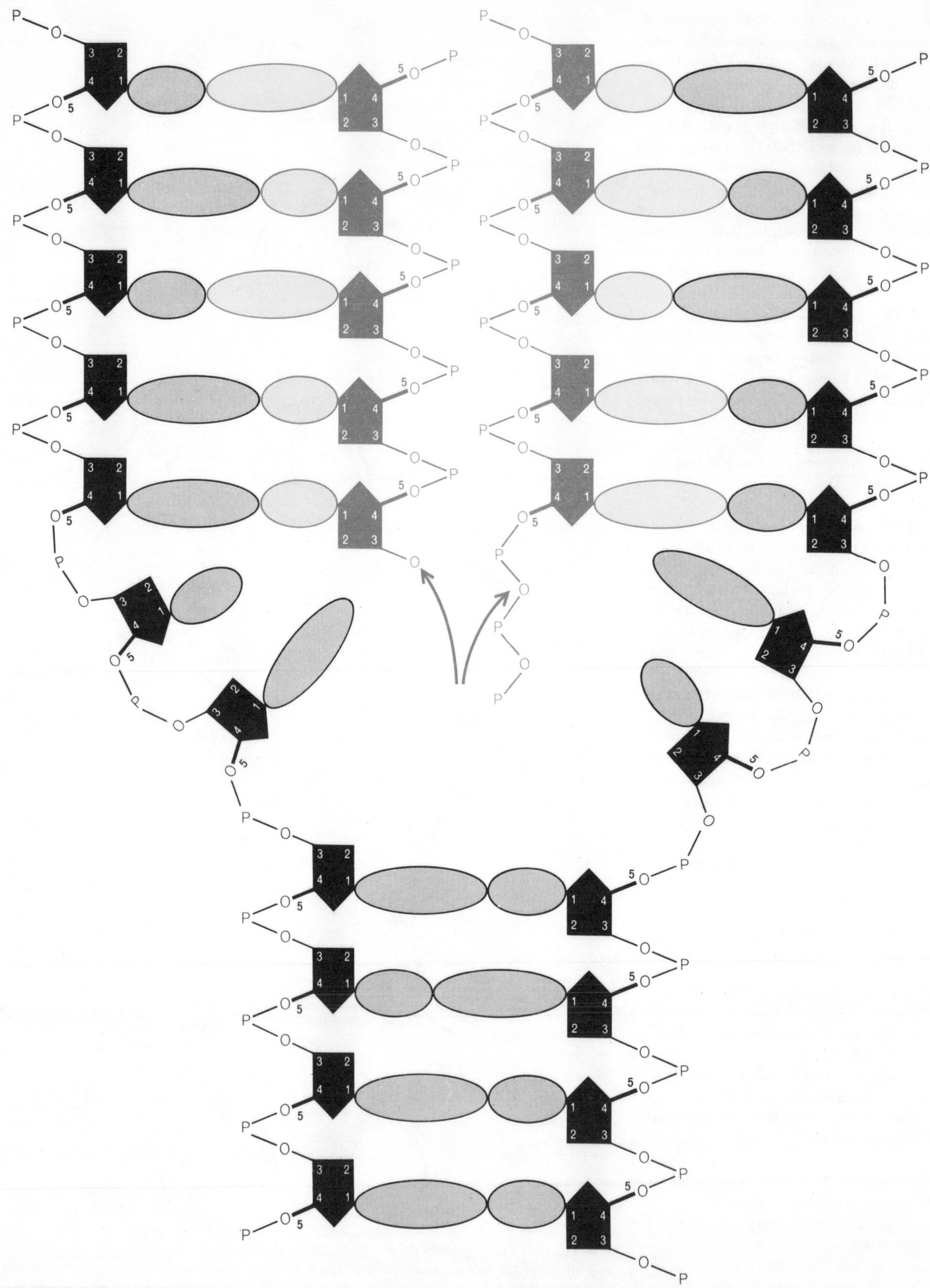

OPPOSITE POLARITIES of the two parental chains of the DNA duplex result in opposite polarities and different directions of growth in the two new chains (*color*) being laid down as complements of the old ones during duplication. Note that the numbered carbon atoms (*1 to 5*) in the deoxyribose rings (*solid black*) are in different positions in the two parental chains and therefore in the two new chains. As the replicating fork moves downward, the new chain that is complementary to the left parental chain must grow by addition to a C_3, the other new chain by addition to a C_5, as shown by the arrows. The elliptical shapes are the four bases.

chains. Now, any active unwinding mechanism must rotate the parent molecule with respect to the two new molecules—must hold the latter fast, in other words, just as the far end of a rope must be held if it is to be unwound. A little thought will show that this can be most surely accomplished by a machine attached, directly or through some common "ground," to the parent molecule and to the two daughters [*see illustration below*]. Every turn taken by such a machine would inevitably unwind the parent molecule one turn.

Although other kinds of unwinding machine can be imagined (one could be situated, for example, at the replicating fork), a practical advantage of this particular hypothesis is that it accounts for circularity. It also makes the surprising—and testable—prediction that any irreparable break in the parent molecule will instantly stop DNA synthesis, no matter how far the break is from the replicating fork. If this prediction is fulfilled, and the unwinding machine acquires the respectability that at present it lacks, we may find ourselves dealing with the first example in nature of something equivalent to a wheel.

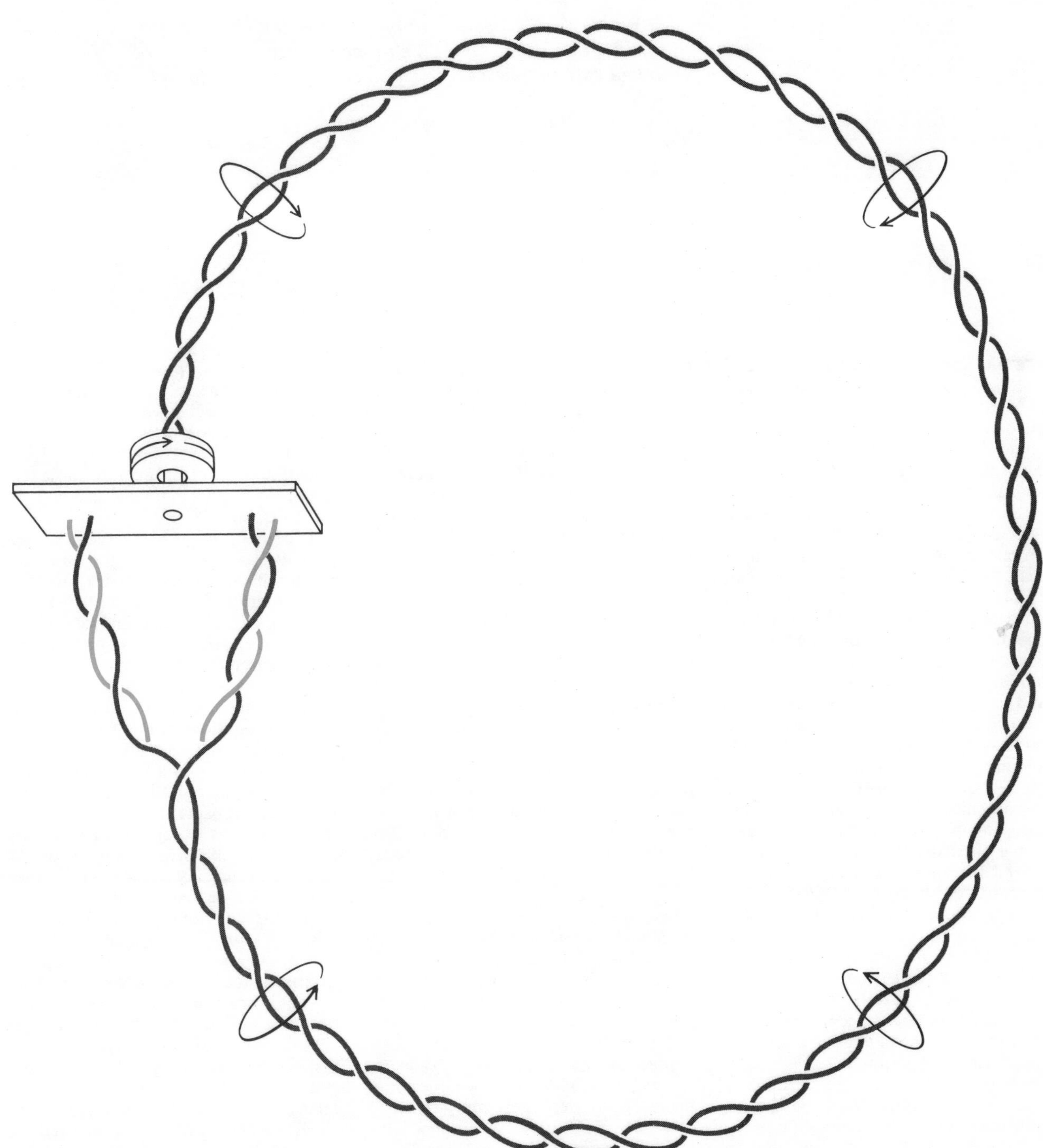

POSSIBLE MECHANISM for unwinding the DNA double helix is a swivel-like machine to which the end of the parent molecule and also the ends of the two daughter molecules are joined. The torque imparted by this machine is considered to be transmitted along the parent molecule, producing unwinding at the replicating fork. If this is correct, chromosome breakage should halt duplication.

23

Bacterial Tests for Potential Carcinogens

by Raymond Devoret
August 1979

New short-term tests can identify environmental agents that cause damage to DNA, the primary event in chemical carcinogenesis. The tests are also valuable for clarifying the mechanism of DNA damage

The fact that physical and chemical agents in the environment enhance the incidence of cancer has become of great concern. It is clear that an adequate limitation of human exposure to carcinogens would save lives. To identify potential carcinogens in the environment is therefore an urgent task. In the case of chemical carcinogens that is no easy matter. It is estimated that more than 50,000 different man-made chemicals are currently in commercial and industrial use; between 500 and 1,000 new chemicals are put on the market every year. The standard animal tests for potential carcinogenicity take a long time and cost a great deal of money.

Fortunately there is an alternative to the classical animal tests. One can take advantage of the profound unity of living matter and resort to bacteria as the test organisms. A bacterial assay for carcinogenicity takes a few hours or days rather than the two or three years required for an animal test, and it costs far less. An effective bacterial test has been developed by Bruce N. Ames of the University of California at Berkeley, based on the ability of a chemical to cause mutations in bacteria. More recently my colleagues and I at the Centre National de la Recherche Scientifique in Gif-sur-Yvette have devised a group of tests based on a chemical's ability to induce the development of a dormant virus in bacteria. The Ames test and our tests not only provide means of identifying dangerous chemicals but also are powerful new tools for learning to understand the primary events of the carcinogenic process initiated by chemicals.

Cancer is a disease of highly evolved multicellular organisms such as human beings, whereas bacteria are minute single cells at the opposite end of the evolutionary scale. It may therefore seem paradoxical that bacteria can serve to identify substances that cause cancer. Actually, however, there is no paradox.

One tends to think of a cancer as a tumor that can spread through the body to form multiple tumor colonies (metastases). That is a clinical, macroscopic view of cancer at a multicellular stage, since just one gram of malignant tumor already contains a million cancer cells. Cancer begins at the level of the single cell. A cell in an adult tissue evolves in such a way that it departs from conformity with the strict physiological rules governing the set of identical cells that constitute a tissue; it becomes a unique and distinguishable defect in an otherwise monotonous structure. The cell begins to divide and a tumor grows. Some of the daughter cells may break the tissue barrier, invading adjacent tissues and usually metastasizing to distant sites. Cancer cells have a great selective advantage, since they escape the programmed fate of most normal cells: to age and die. For cancer cells the entire body is a culture medium in which they thrive, ultimately to die with the body they kill.

Physical and chemical agents in the environment cause cancer by damaging DNA, the cell's hereditary material. DNA damage initiates a complex cellular process that in mammalian cells can eventually lead to transformation into a cancerous state. Agents that damage DNA are therefore potential carcinogens. DNA is the hereditary material of all living cells, and both DNA lesions and the cellular processes that repair them are remarkably similar in bacteria and in human cells; what is detrimental to bacterial DNA is likely to harm human DNA. That is the theoretical justification for substituting bacteria for mammalian cells in tests to detect damage to DNA.

The theoretical justification is supported by experimental and practical results: bacterial tests distinguish with more than 90 percent reliability between known carcinogens and known noncarcinogens, and they have identified as potential carcinogens new chemicals that have subsequently been shown to be carcinogenic in animal tests. Of course, the manifestations of DNA damage are very different in bacteria from the transformation to the cancerous state that is observed in mammalian cells. In compensation the bacterial tests are so much faster and less expensive as to finally make comprehensive screening for potential carcinogens a feasible objective.

The carcinogenic potential of physical agents, and notably of ionizing radiations, is much better understood than that of chemical agents. Broad popular awareness that radiations can cause cancer has come only in the era of nuclear weapons and reactors, but the danger actually became apparent soon after the discovery of X rays by Wilhelm Konrad Röntgen in 1895. Only four years later it was reported that a technician who checked newly manufactured X-ray tubes by fluoroscoping his own hand was afflicted with a skin cancer; he eventually died of it. The warning was ignored, and most of the first generation of radiation therapists died of cancer. The carcinogenic effect of the ultraviolet radiation in sunlight has also been known for some time. As long ago as 1905 a French physician named Dubreuilh observed that skin cancers of the back of the neck were particularly prevalent among workers who were exposed to the sun as they tended the vineyards and harvested grapes in the Bordeaux region.

Although X rays, gamma rays and other radiations hit all the components of cells in a random manner, it was recognized quite early that the genetic material must be the most radiation-sensitive cellular target. DNA constitutes only a minor fraction of the chemical components of a cell, but direct or indirect damage to it has great impact on the cell's future, whereas damage to proteins and other cellular components has much less effect. That is because DNA is the memory of the cell. DNA replicates to beget DNA, and so errors in DNA are transmitted from cell generation to cell generation.

The double helix of DNA consists of two chains of sugar (deoxyribose) and

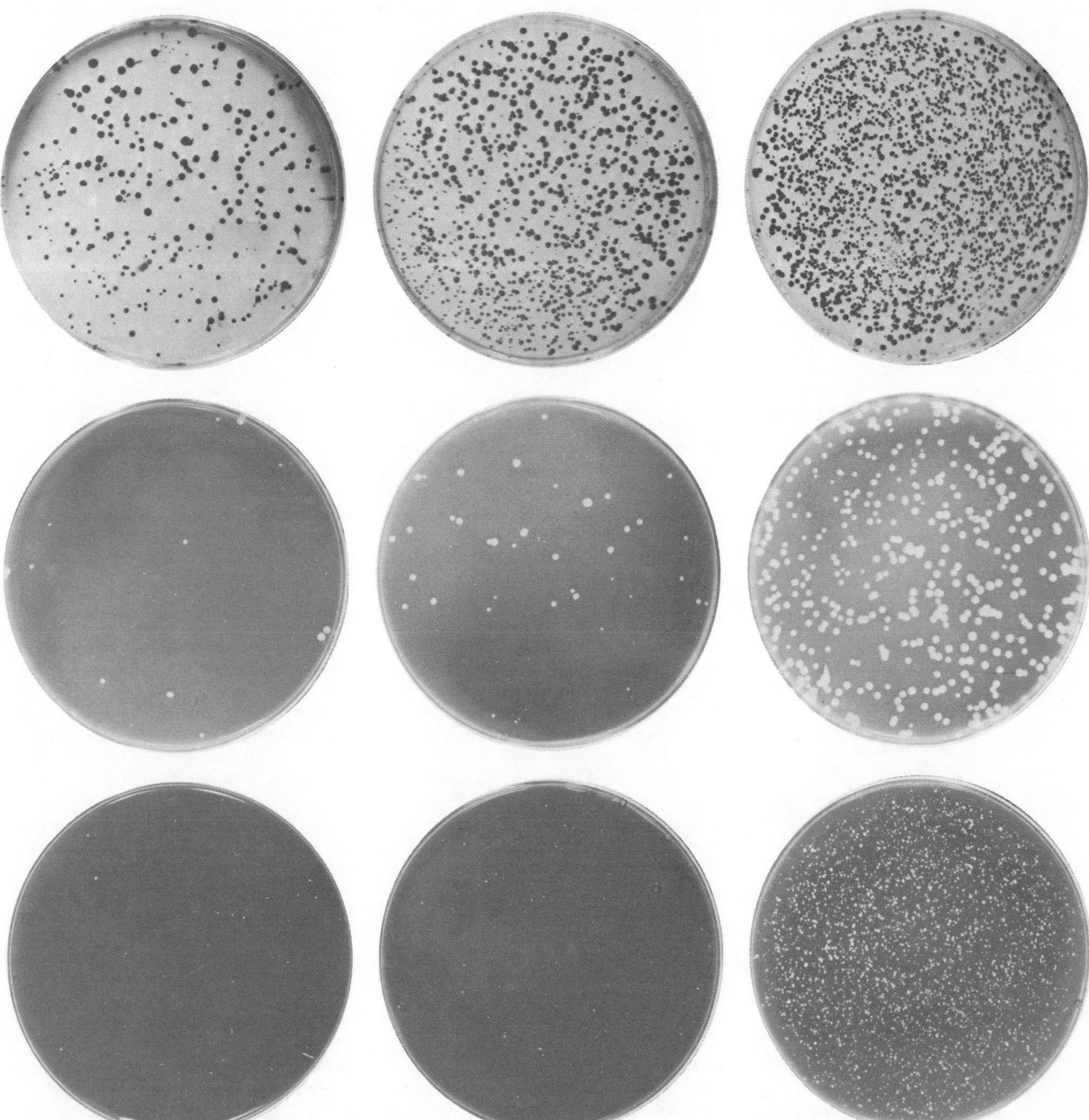

THREE BACTERIAL TESTS for potential carcinogens reveal DNA damage: the Ames test (*top*), the inductest (*middle*) and the lambda mutatest (*bottom*). In each case the culture plate at the left is an untreated control; tester bacteria on the center plate were treated with a moderate dose and those at the right with a higher dose. The Ames test shows the extent of reverse mutations in histidine-deficient *Salmonella typhimurium* that enable the revertant bacteria to proliferate. A background of spontaneous mutant colonies (*red stain*) is seen at the left. Many more colonies grow when tester bacteria are exposed to 250 nanograms (*center*) and 750 nanograms (*right*) of the potent mutagen (and carcinogen) nitrosoguanidine. In the inductest (*middle*) DNA damage is revealed by the induction of a prophage, a dormant bacterial virus integrated in the DNA of "lysogenic" *Escherichia coli;* mature phage particles burst out of the tester bacteria and create plaques on a lawn of indicator bacteria of strain *A* (*red stain*). Here the DNA damage was caused by the antitumor drug mitomycin C, 10 nanograms of it on the center plate, 200 nanograms at the right. In the mutatest (*bottom*) a modified form of the prophage makes plaques on a lawn of *E. coli* of strain *B* (on which nonmutated phage cannot form plaques) when it undergoes mutation in its "operator" regions and can no longer remain dormant. Again the treatment was 10 nanograms of mitomycin C (*center*) and 200 nanograms (*right*).

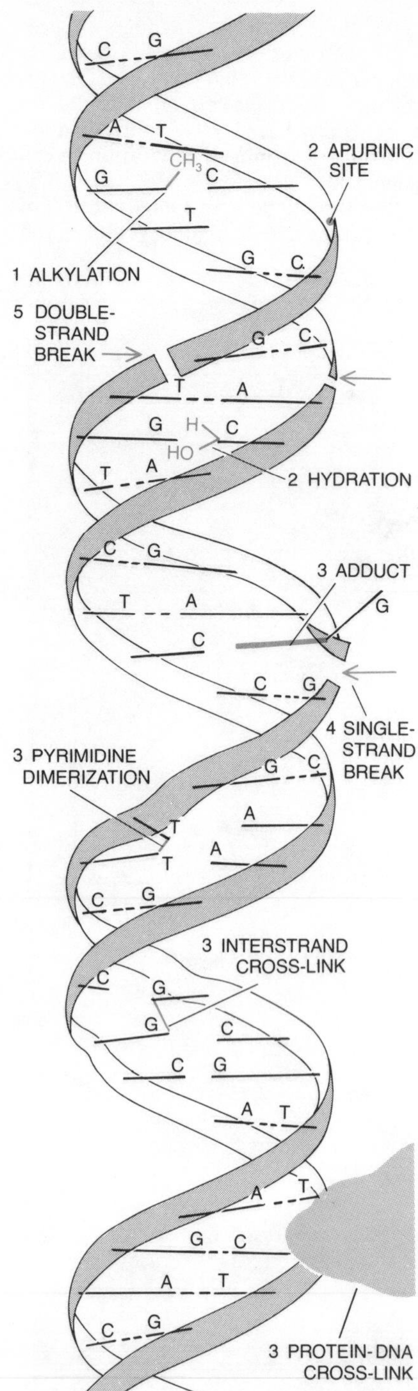

STRUCTURAL ALTERATIONS are imposed by radiations or by chemical agents on the double helix of DNA, two chains of sugar and phosphate groups (*helical bands*) linked by paired bases: either adenine (*A*) and thymine (*T*), or guanine (*G*) and cytosine (*C*). The alterations can be classified in five categories, examples of which are illustrated: negligible helix distortions (*1*), as by alkylation of one of the bases; minor distortions (*2*) caused by hydration or the absence of a base; major distortions (*3*) caused by insertion of an "adduct," linking of two bases to form a dimer, or cross-linking between the two strands or between a strand and a protein; breaks in a single strand (*4*) or in both strands (*5*). Any structural alteration affects DNA's function as a template for replication, but some negligible alterations are not sensed by a cell as damage to DNA.

phosphate groups linked, as by the rungs of a twisted ladder, by paired nitrogenous bases: two purines (adenine and guanine) and two pyrimidines (thymine and cytosine). Adenine always pairs with thymine and guanine with cytosine, but the sequence of the bases along a strand of DNA is variable and carries a particular coded message. Any alteration, even a slight one, in the structure of the double helix affects the functions of the DNA, one of which is to serve as a template for its own replication.

Not every DNA alteration is sensed in the cell as DNA damage. Defined precisely, DNA damage is an alteration that constitutes a stumbling block for the replication machinery and hence hampers the replication of DNA, endangering the survival of the cell. Once incurred, DNA damage calls for repair, which is accomplished by the interplay of at least a score of enzymes whose action is governed by as many genes. Repair is never totally efficient, and so many cells die. Some cells may survive, however, even though the lesions are not totally removed from their DNA, because a repair process has bypassed the lesions. Replication then reconstitutes an undamaged double helix, but one bearing a coded message different from the original one. The scars left on the DNA by such a process are mutations.

The correlation between radiation-induced cancer and radiation-induced DNA damage has long been apparent to radiation biologists, but it is only recently that molecular evidence has been found that DNA damage is a direct cause of cancer. The evidence comes from patients suffering from xeroderma pigmentosum, who are extremely sensitive to sunlight and, while they are still very young, develop skin cancers of which they may eventually die. James E. Cleaver of the University of California School of Medicine in San Francisco and Dirk Bootsma of Erasmus University in Rotterdam have demonstrated that xeroderma patients suffer from a well-characterized genetic defect: their cells cannot carry out a particular DNA-repairing process.

At noontime on a sunny day the flow of ultraviolet radiation that reaches the earth is strong enough to generate pyrimidine dimers in the DNA of exposed cells by linking two laterally adjacent thymines or a thymine and a cytosine. Most such DNA lesions in skin cells are repaired in normal people by an excision process strikingly similar to the "cut and patch" repair process in bacteria exposed to the same radiation. In xeroderma patients, however, the lesions go unrepaired, and their accumulation appears to bring on cell transformation: DNA damage breeds cancer.

Appreciation of the carcinogenic role of chemicals in the environment has come slowly, even though instances of cancer caused by occupational exposure have long been observed. As early as 1775 the British physician Sir Percival Pott correlated the incidence of cancer of the scrotum in men who had once been chimney sweeps with the accumulation of soot in their groin area many years before. Experimental findings on chemical carcinogens date back at least to 1918, when two Japanese investigators, K. Yamagiwa and K. Ichikawa, showed that skin cancers could be induced by repeated applications of coal tar to the ear skin of rabbits. One of the chemicals responsible for causing such cancers is benzo[*a*]pyrene, which is also present in soot, cigarette smoke and charred meat.

Progress toward understanding chemical carcinogenesis has been slow for at least three reasons. First of all, most chemical carcinogens are not biologically active in their original form, so that testing them in that form does not reveal their carcinogenic nature. Only about a decade ago did it become clear, as a result in particular of the investigations of James A. Miller and Elizabeth C. Miller of the University of Wisconsin, that normal metabolic processes, which convert food into substances the body can absorb and eliminate and convert harmful compounds into harmless ones, transform environmental chemicals into metabolites capable of inducing cancer. Those metabolites had to be characterized in order to show just how their parent substances are threats.

A second delaying factor was that the actively carcinogenic metabolites react with various cell components, including RNA and proteins as well as DNA. Since there are a lot of proteins, performing important functions, it was often assumed that damage to proteins might play the major role in the cancerous transformation initiated by chemical carcinogens. The key role of damage to DNA in carcinogenesis by chemicals was recognized only recently, and the results of the bacterial tests have provided strong (although indirect) evidence for such a mechanism.

Finally, the understanding of chemical carcinogenesis has been obscured by the fact that DNA damage, although it is the essential event in the initiation of carcinogenesis, does not usually in itself lead to cancerous transformation; additional factors are apparently required to promote the complex chain of cellular events culminating in the transformation. DNA-damaging agents in themselves are therefore only potential carcinogens.

As I pointed out at the beginning of this article, only a small fraction of the flood of chemicals reaching the market every year can be tested accurately by means of the standard animal assays. In order to obtain results with statistical significance a great many animals must be (or at least should be) exposed to each tested chemical; for that reason alone a comprehensive screening of new chemicals would be impractical (other than the few, such as food additives, drugs and cosmetics, for which testing is mandated). Even when animal tests are feasible, manufacturers need low-cost, fast tests if they are to identify DNA-damaging substances while new products are still under development and alternative ones can still be sought.

Studies of the epidemiology of human cancers have provided much information on environmental carcinogens, but such studies are of little immediate value in detecting new potential carcinogens because most human cancers appear only some 20 or 30 years after the exposure that gives rise to them. Epidemiological analysis has most frequently been successful in detecting a chemical to which an identifiable subpopulation, such as workers in a particular industry, are exposed and because of which they show a high incidence of a particular type of cancer. For some widely distributed carcinogens there are no clearly identifiable subpopulations for statistical analysis.

There is no doubt, then, that the present situation calls for simple, fast, inexpensive methods of detecting potential carcinogens, which is to say DNA-damaging agents. One possibility is to measure DNA damage directly in mammalian cells by determining biochemically the incidence of particular forms of DNA damage in cells exposed to a chemical. Such studies are being done by molecular biologists, and their results provide valuable standards of reference. For screening purposes, however, it is cheaper and faster to detect and measure the extent of DNA damage by scoring its manifestations in bacteria.

One of the great advantages of assays done with bacteria is the enormous biological amplification implicit in bacterial manipulations. It is easy to grow as many as a billion (10^9) bacteria per milliliter of culture medium. A mutational event such as a change in a single base pair in the bacterial DNA, which is impossible to detect by standard biochemical methods, will be revealed as a mutant bacterium. That single bacterium can be selected from among 10^9 cells because its daughter cells, and only they, will proliferate and form a colony visible to the unaided eye on an agar nutrient plate. Since a colony consists of about a million (10^6) bacteria, a rare single mutational event with a probability of, say, one in 100 million (a probability of 10^{-8}) would thus be amplified by a factor of 100 trillion (10^{14}).

The series of cellular events that leads to mutation, and to several other manifestations of DNA damage in bacteria, has recently been somewhat clarified. The bacterium *Escherichia coli,* which inhabits the colon of a number of mammals including human beings, is genetically programmed to divide and (under optimal conditions for growth) form two daughter bacteria in 30 minutes. When the DNA of a bacterium is damaged, the bacterium may need more than two hours to resume its cycle of division (if it is not killed by the damage). During that time there is a sequence of cellular events.

Immediately after the primary DNA lesions are incurred a first round of repair is effected: an excision repair, in which the damaged segment is cut out of the DNA and replaced by an undamaged DNA sequence. Some residual lesions will remain, however. If they are bulky or clustered, the residual lesions hamper the DNA-replication machinery, and replication stops abruptly. Unless the arrest of replication is transitory

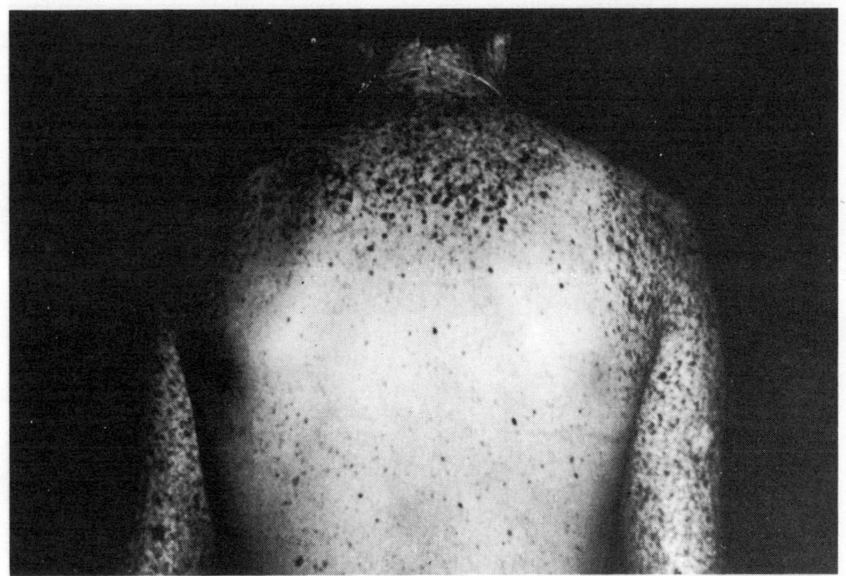

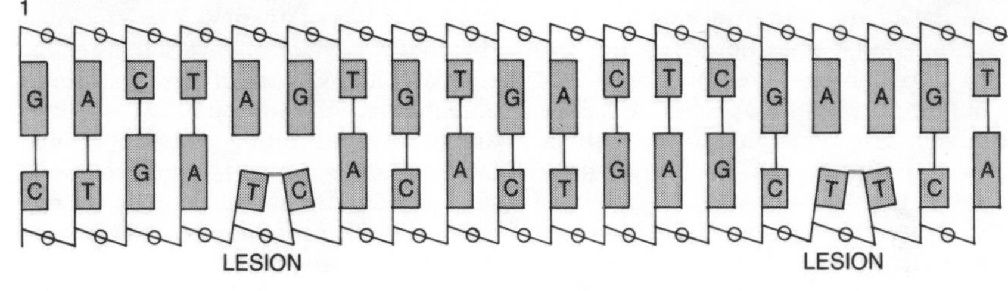

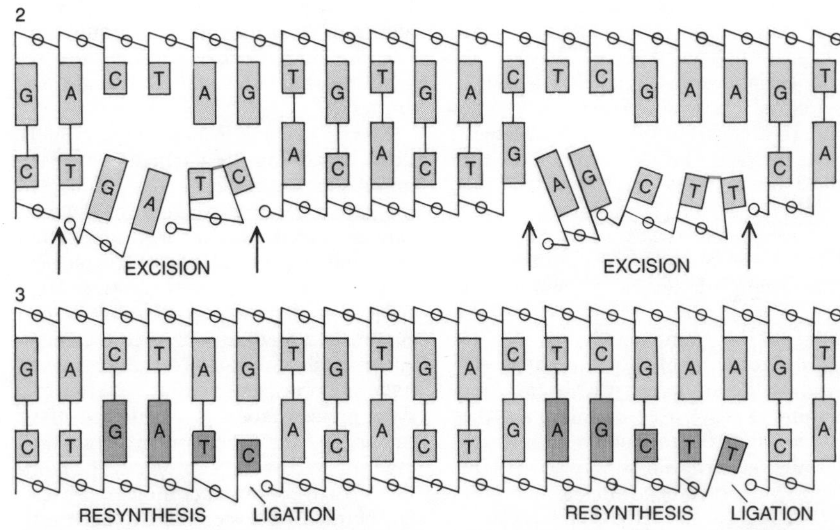

XERODERMA PIGMENTOSUM is a genetic disease whose victims develop skin cancer as a result of normal exposure to sunlight, typically on uncovered parts of the body, as is shown in the photograph at the top. These patients lack a DNA-repair process that in normal individuals removes thymine dimers, the lesions created by the ultraviolet radiation present in sunlight (*1*). In the normal repair process the bases in the lesion areas are first excised (*2*). Then the excised stretches of DNA are resynthesized and the new segments are ligated to undamaged stretches (*3*).

the cell's survival is at risk. As Miroslav Radman of the University of Brussels and Evelyn M. Witkin of Rutgers University first suggested, the following cellular adaptive mechanisms come into play to cope with the emergency:

1. A second round of repair is induced. This inducible and error-prone repair (nicknamed "SOS repair") tends to restore the structure of the DNA even though there are errors in the coded message; indeed, this adaptive response may be successful partly because it does not "bother" to follow the base-pairing rules of normal DNA replication. At any rate, the price paid for cell survival appears often to be mutagenesis.

2. Cell partition ceases. The elongation of the cell that ordinarily precedes division is protracted, and the cells may form filaments. The elongation may be adaptive in that it facilitates recombination between the two sets of damaged chromosomes in the cell, the intact segments of each combining to yield an intact chromosome.

3. If a prophage, or dormant bacterial virus, is present in the cell, it is induced to develop into a large number of mature progeny phage that burst out of the cell. This adaptive response evolved by the prophage ensures its survival when a host cell appears to be doomed to die: the rats leave the sinking ship.

Each of these adaptive responses is in part promoted by a multifunctional protein called the RecA protein (for "recombination," since defects in the protein impair recombination in general). The RecA protein appears to be required in *E. coli* not only for recombination but also for the responses listed above. The RecA protein is activated and its synthesis dramatically increased when DNA replication is blocked.

Of the various bacterial manifestations of DNA damage, Ames chose mutagenesis as the basis of his pioneering work to develop a test for potential carcinogens. His *Salmonella*–mammalian-liver assay, known generally as the Ames test, is currently the standard test and by far the most widely used. The tester organism is a strain of *Salmonella typhimurium*, another colon bacterium, bearing a mutation (*his*⁻) that renders it unable to manufacture one of the enzymes required for the synthesis of the amino acid histidine, a necessary component of proteins. As a result of the mutation the bacterium is unable to grow in a mineral nutrient medium unless the medium is supplemented with an external supply of histidine.

On very rare occasions a *his*⁻ mutation undergoes reversion: a back mutation restores the DNA's normal coding sequence for the needed enzyme and thereby restores the internal supply of histidine. The reversion can be scored because only the revertant bacteria form colonies on a medium that lacks histidine. Obviously the spontaneous rate of reversion, which is ordinarily very low, will be considerably enhanced if the *his*⁻ bacteria are exposed to a chemical that induces mutations. This is the theoretical basis of the Ames test.

Actually Ames and his colleagues had to introduce three important modifications into the original *his*⁻ strain to make it a sensitive and versatile tester bacterium. Bacteria such as *E. coli* and *S. typhimurium* have a rather impermeable envelope that reduces or even prevents the penetration of many chemicals into the cell. (This bacterial armor has evolved because the bacteria must usually survive in a hostile environment such as an intestine or a sewer.) Ames and his colleagues overcame the envelope barrier by introducing a mutation that gives rise to defects in the envelope. They went on to make the strain more sensitive to DNA-damaging agents by eliminating its capacity for excision repair, so that most of the primary lesions remain unhealed. And they introduced into the bacterium a plasmid, a foreign genetic element that makes DNA replication more error-prone. By means of these three modifications a strain was constructed in which just a few molecules of a carcinogen are able to create DNA lesions, each of which is likely to engender a mutation; of those mutations, some will be such that the internal supply of histidine is restored.

The real breakthrough, and the one that made the *Salmonella* test truly effective, was Ames's idea of mixing the tester bacteria with an extract of rat liver and thereby subjecting the tested chemical to mammalian metabolic processes. As I pointed out above, it is usually not the original form of a chemical carcinogen that is active but rather one of its metabolites. Because the liver is the body's major metabolic factory the enzymes of a rat-liver extract should convert the chemical being tested into metabolites that react with DNA—if there are any.

In practice the Ames test is usually done by adding the chemical to be examined to *his*⁻ tester bacteria immersed in a rat-liver extract and plating the mixture on a solid nutrient medium devoid of histidine. (For demonstration purposes the chemical can instead be spotted on a disk of filter paper, which is then placed on a medium on which the bacteria, mixed with the liver preparation, have previously been plated.) After two days of incubation any cells that have undergone the reversion mutation will give rise to revertant colonies. The number of such colonies per mole of the tested chemical provides a quantitative estimate of the mutagenic potency of the chemical.

The simplicity, sensitivity and accuracy of the *Salmonella* test for screening large numbers of environmental sources of potential carcinogens has resulted in its current application in more than 2,000 governmental, industrial and academic laboratories throughout the world; it is estimated that 2,600 chemicals have been subjected to the test. Ames and his collaborator Joyce McCann have themselves validated the procedure by testing more than 300 chemicals that were previously reported, on the basis of animal experiments, to be either carcinogens or noncarcinogens. About 90 percent of the reputed carcinogens turned out to be mutagenic and about the same proportion of the noncarcinogens were negative for mutagenicity. Other mutagenicity tests have since been devised with *E. coli* as the tester bacteria, and their efficiency is about as high as that of the original *Salmonella* test.

Two impressive accomplishments of mutagenicity tests can be mentioned to give an idea of their value. In Japan the chemical furyl furamide, known as AF-2, was added to a broad range of common food products for some years as an antibacterial agent. It had not shown any carcinogenic activity in standard tests on rats in 1962 or on mice in 1971. Then in 1973 T. Sugimura and his colleagues at the National Cancer Centre in Tokyo found that AF-2 was highly mutagenic in bacteria; they could easily demonstrate the mutagenic activity of the additive contained in just one slice of fish sausage! The discovery prompted a new round of more thorough animal tests for carcinogenicity, which showed that AF-2 was indeed a carcinogen. It was withdrawn from the market. If it were not for the bacterial test, AF-2—which had passed two approved animal tests and been declared negative for carcinogenicity—would presumably still be a component of fish sausage and other Japanese food products.

In 1975 Ames and his colleagues reported that 89 percent of the major class of hair dyes sold on the U.S. market contained mutagenic compounds. Since then the cosmetic industry has modified the composition of most hair dyes. It is estimated that several tens of millions of people dye their hair in the U.S. alone, which suggests that the discontinued components had presented a considerable risk.

In spite of all their advantages the mutagenicity tests have some technical and even theoretical limitations. Since mutations are revealed in the Ames assay by a restoration of enzyme activity, any mutation that does not happen to reconstruct the precise DNA sequence that codes for the histidine-making enzyme is not observed. To take one example, the antitumor drug bleomycin, which does its therapeutic work by damaging the DNA of tumor cells (as do about half of all antitumor agents), fails to induce the mutation that is scored in the *Salmonella* test. A false-negative re-

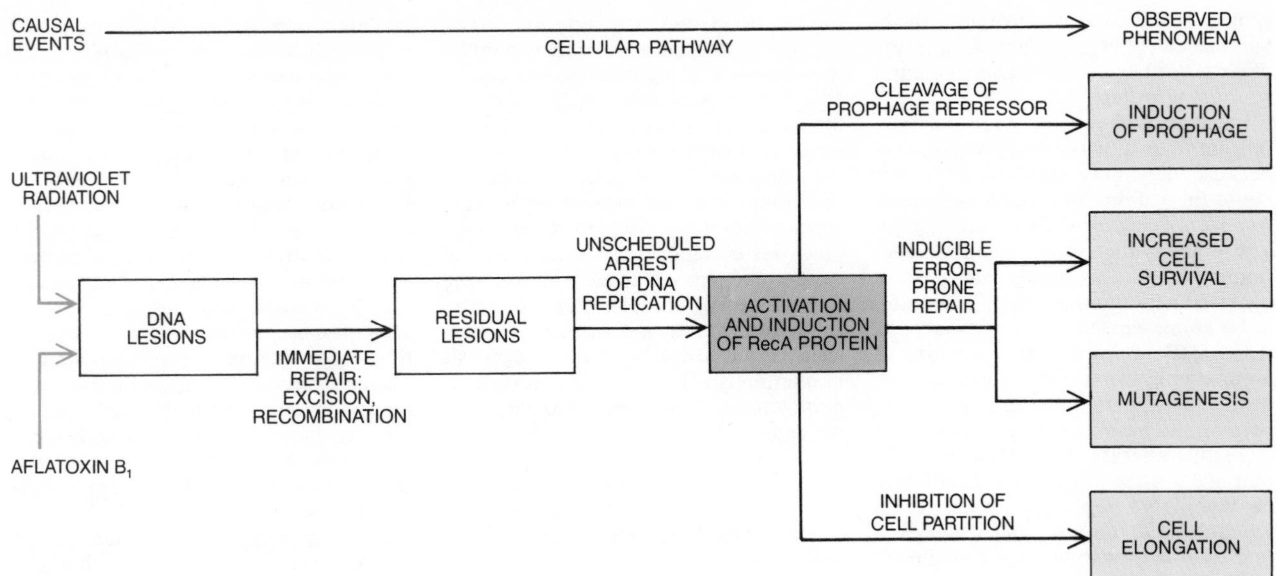

SEQUENCE OF CELLULAR EVENTS follows DNA damage imposed on *E. coli* by ultraviolet radiation or by the carcinogen aflatoxin B_1. Immediate repair processes mend most of the lesions but leave residual damage, causing an unscheduled arrest of DNA replication. This threat to cell survival activates the RecA protein to become a protein-cleaving enzyme and also induces the synthesis of more RecA. Various forms of the protein are involved in three processes that give rise to the four observed phenomena shown at right.

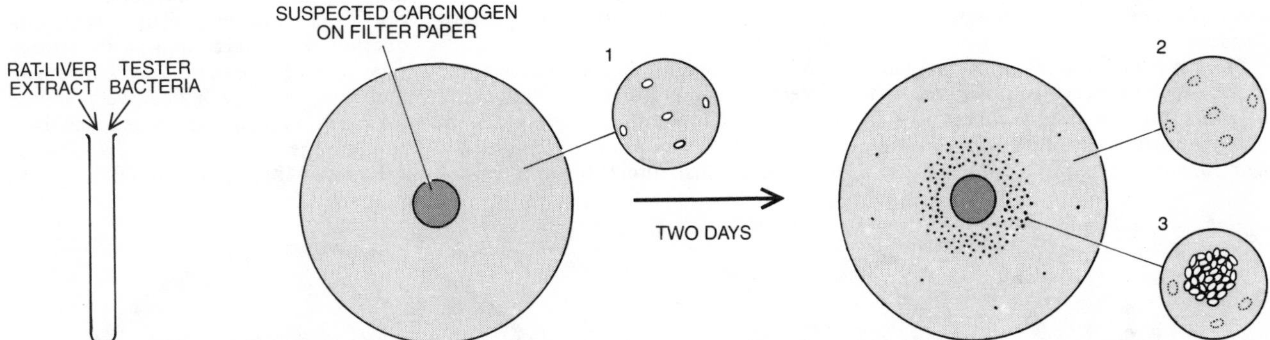

MUTAGENESIS is detected in the Ames test by mixing an extract of rat liver (which supplies mammalian metabolic functions) with tester bacteria (which cannot grow because a mutation makes them unable to manufacture histidine, a necessary nutrient) and plating the mixture on an agar medium so that a thin layer of bacteria covers the medium evenly, as is shown on a microscopic scale (*1*). In this "spot assay" a dose of the chemical to be tested is placed on a disk of filter paper on the tester bacteria. After two days most of the *his*⁻ bacteria have died for lack of histidine (*2*), but DNA damage caused by the chemical diffusing out from the disk has given rise to mutations, some of which result in reversion of the *his*⁻ mutation. The histidine-making revertant bacteria proliferate, forming visible colonies (*3*).

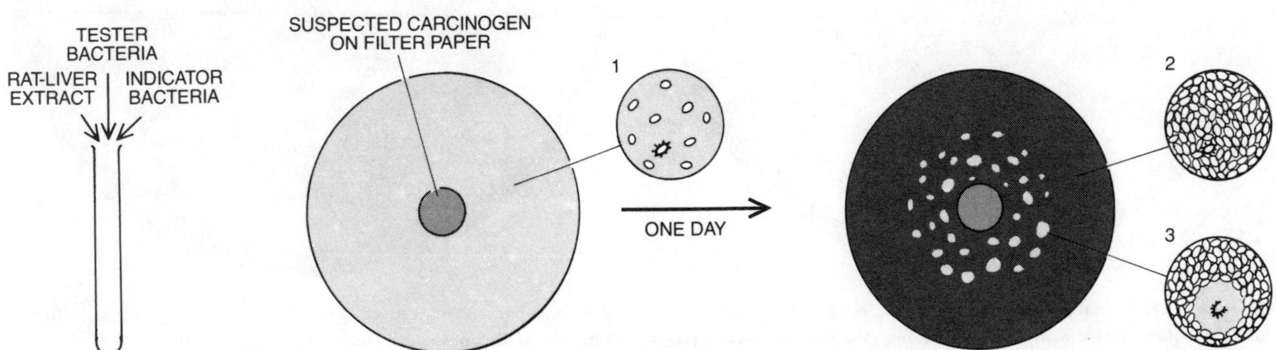

INDUCTION of a dormant bacterial virus, prophage lambda, is detected in the inductest. Lysogenic tester bacteria are mixed with rat-liver extract and then with indicator bacteria. The mixture is plated; the medium is covered with a thin layer of indicator bacteria interspersed with a few lysogenic bacteria (*1*). After a day most of the plate is covered by a thick lawn of indicator bacteria (*2*). Where the chemical that is being tested has diffused from the filter-paper disk the DNA damage it causes leads to the induction of mature lambda phage. The phage particles burst out of the lysogenic cells and kill indicator bacteria in the vicinity, making visible plaques on the lawn (*3*).

sponse of this kind is a technical problem that can be remedied by substituting other tester bacteria or a complementary short-term test.

False-positive responses are more significant from a theoretical point of view because they may raise some doubts about the validity of mutagenicity tests for identifying potential carcinogens. The point is that some chemical reactions with DNA are highly mutagenic in bacterial and mammalian cells without, as far as one can tell, being carcinogenic. Among these are the incorporation of an analogue of one of the nitrogenous bases and the methylation of certain sites on the bases. Such reactions cause negligible alterations in DNA structure that are not sensed in the cell as DNA damage; DNA replication proceeds on schedule and the new DNA carries a readable—although wrong—coded message. This process is termed direct mutagenesis, and in scoring for mutations it should be clearly distinguished from the more frequent indirect mutagenesis. In the latter process, described above, DNA damage brings about a transient arrest in replication; replication is resumed with the help of the RecA protein on a template that carries lesions, causing mutagenesis of the newly formed strand.

Since it is DNA damage that appears to initiate the cancerous transformation of a mammalian cell, a chemical is a potential carcinogen if it is a DNA-damaging agent, not simply because it causes mutations. A mutagen can have strong genetic effects on a biological population by altering the information encoded in the DNA of individuals' germ cells but nonetheless fail to cause cancer because it does no real DNA damage to the somatic cells: the cells of the rest of the body. Not all genetic toxic substances (chemicals that affect DNA) are potential carcinogens, whereas chemical carcinogens—because they damage DNA—are all indirect mutagens. In other words, mutagenic activity in bacteria is correlated with carcinogenesis in mammals primarily through indirect mutagenesis, which results from DNA damage.

In 1953, before the structure of DNA had even been established, André Lwoff of the Pasteur Institute in Paris anticipated that "inducible lysogenic bacteria might become a good test for carcinogenic and perhaps anticarcinogenic activity." A bacterium is said to be lysogenic when it carries, in a dormant state, the DNA of a "temperate" bacterial virus, which in this dormant state is called a prophage. One such temperate virus is phage lambda, which becomes a prophage when it is integrated into the DNA of certain lysogenic strains of *E. coli.*

When lysogenic *E. coli* bacteria are subjected to any treatment that halts DNA replication, the prophage is induced: its DNA loops out of the bacterial DNA and also directs the synthesis of proteins forming the virus particle, and a progeny of mature phage develops, bursting out of the host cell. The process is called lysogenic induction. Under normal conditions the dormant state of prophage lambda is maintained by a repressor, a protein that lies on the DNA's "operator" regions and blocks the operation of the lambda genes (except the gene that directs the synthesis of repressor to keep the prophage dormant). Jeffrey W. Roberts and Christine W. Roberts of Cornell University discovered that the induction of prophage lambda results from the cleavage of the lambda repressor; together with Nancy Craig they have recently shown that relatively pure repressor can be cleaved when it is mixed with an activated form of the RecA protein.

By triggering the activation of the RecA protein into the form that can cleave a viral repressor, DNA damage results in the induction of a dormant virus. The induction of prophage lambda can therefore serve as a test for DNA damage. Even before the mechanism of prophage development was understood at a molecular level, lysogenic induction was applied in the pharmaceutical industry to identify prospective antibiotics and antitumor drugs. When induction tests were done with ordinary strains of lysogenic *E. coli,* however, they did not give a positive response for such known carcinogens as benzo[*a*]pyrene.

Patrice Moreau, Adriana Bailone and

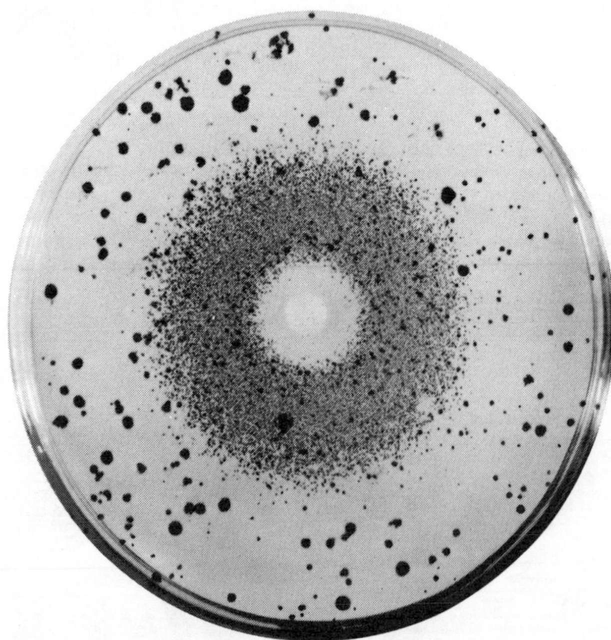

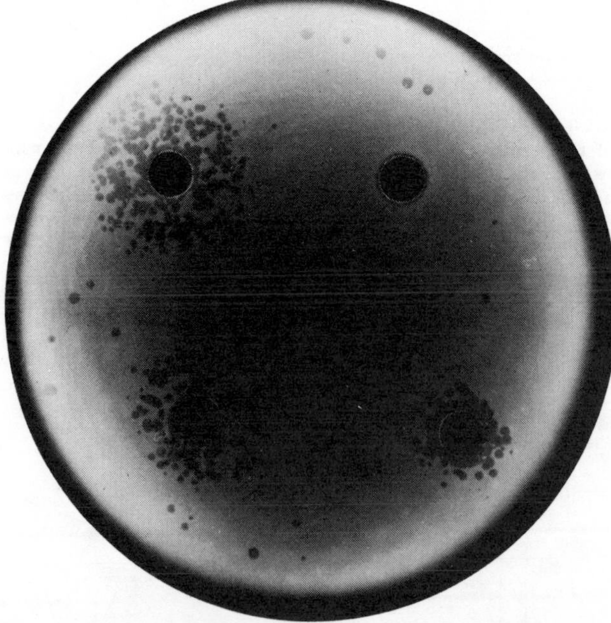

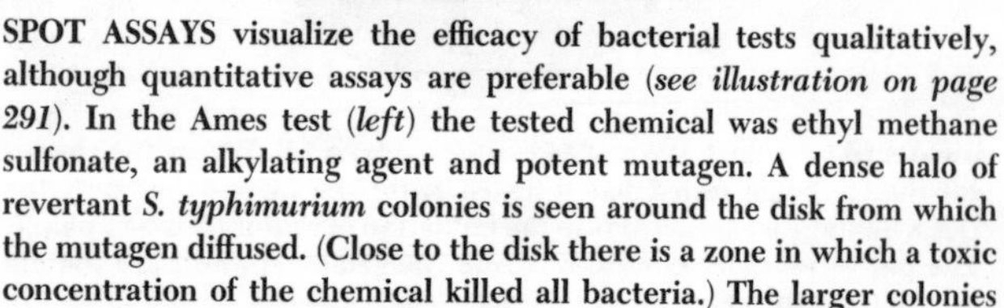

SPOT ASSAYS visualize the efficacy of bacterial tests qualitatively, although quantitative assays are preferable (*see illustration on page 291*). In the Ames test (*left*) the tested chemical was ethyl methane sulfonate, an alkylating agent and potent mutagen. A dense halo of revertant *S. typhimurium* colonies is seen around the disk from which the mutagen diffused. (Close to the disk there is a zone in which a toxic concentration of the chemical killed all bacteria.) The larger colonies are mutant bacteria that arose spontaneously without exposure to the chemical. In the inductest (*right*) the tested chemical was aflatoxin B_1, a potent carcinogen of the liver. The aflatoxin doses placed on the four test disks were (*clockwise from top right*) 0, 20, 200 and 2,000 nanograms. In the background a few of the lysogenic *E. coli* tester bacteria have spontaneously produced mature phage that have killed the nearby indicator bacteria, producing plaques at random locations.

I therefore set out five years ago to renovate the lambda-induction test to make it capable of identifying all kinds of DNA-damaging agents. We reasoned that because the molecular mechanism of lysogenic induction was understood better than that of mutagenesis, a prophage-induction test should provide more insights than mutagenicity tests could into the precise effect of chemical carcinogens on DNA and should provide a supplementary tool for screening as well.

Following Ames's lead, we constructed new tester bacteria that are permeable to chemicals and deficient in excision-repair enzymes, and we assayed potential carcinogens in the presence of a rat-liver metabolizing mixture. The prophage-induction test, or inductest, has turned out to discriminate effectively between carcinogens and noncarcinogens in our laboratory and in several others.

The inductest has certain advantages over mutagenicity tests. Mutations can be scored only in bacteria that survive exposure to a chemical treatment that is often toxic as well as mutagenic, and no more than one bacterium out of 1,000 is likely to be detected as a histidine revertant. In contrast, lysogenic induction can be observed, when it takes place, in the bulk of a cell population. Prophage induction is a mass effect, largely independent of whether or not cells would survive the chemical treatment; a cell that is induced to produce a progeny of phage would die in any case. The inductest can therefore continue to give a positive response for a highly toxic potential carcinogen at dose levels that would kill the tester bacteria in a mutagenicity test.

The fact that in the inductest most cells undergo a dramatic change led us to design a biochemical assay of lysogenic induction. Sankar Adhya, Maxwell E. Gottesman and Asis Das of the National Cancer Institute in the U.S. constructed a bacterial strain in which the gene for the enzyme galactokinase is hooked up to the lambda DNA in such a way that the lambda repressor blocks the synthesis of the enzyme. Alain Levine, Moreau, Steven Sedgwick and I demonstrated that when a DNA-damaging chemical activates the RecA protein in this strain, the repressor is cleaved and galactokinase is synthesized, and the amount of enzyme activity that can be detected reveals the extent of DNA damage. This biochemical assay may be the shortest of the short-term tests; it takes only half a day to test for a potential carcinogen.

Moreau and I have also constructed strains, containing a new form of prophage lambda, in which one can identify chemicals that induce mutagenesis without damaging DNA. This lambda mu-

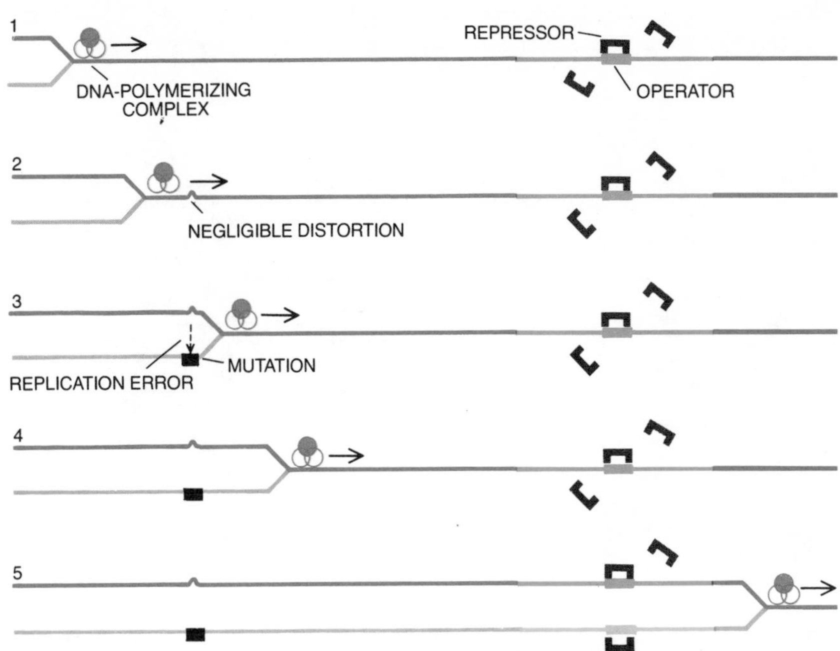

DIRECT MUTATION is a result of DNA replication on a minor distortion of a DNA strand. This diagram and the one below show the replication of only one strand of bacterial DNA (*blue*) carrying a piece of viral DNA, prophage lambda (*red*), which is kept in a dormant state by a repressor protein that blocks an operator region. The DNA strand is replicated by a polymerizing complex (*1*). A negligible distortion of the bacterial DNA (*2*) does not obstruct replication, but it does give rise to an error in replication, so that the newly formed DNA strand (*light blue*) carries a mutation (*3*). The negligible distortion is not sensed as DNA damage, RecA protein is not induced, repressor stays in place and prophage is replicated in dormant state (*4, 5*).

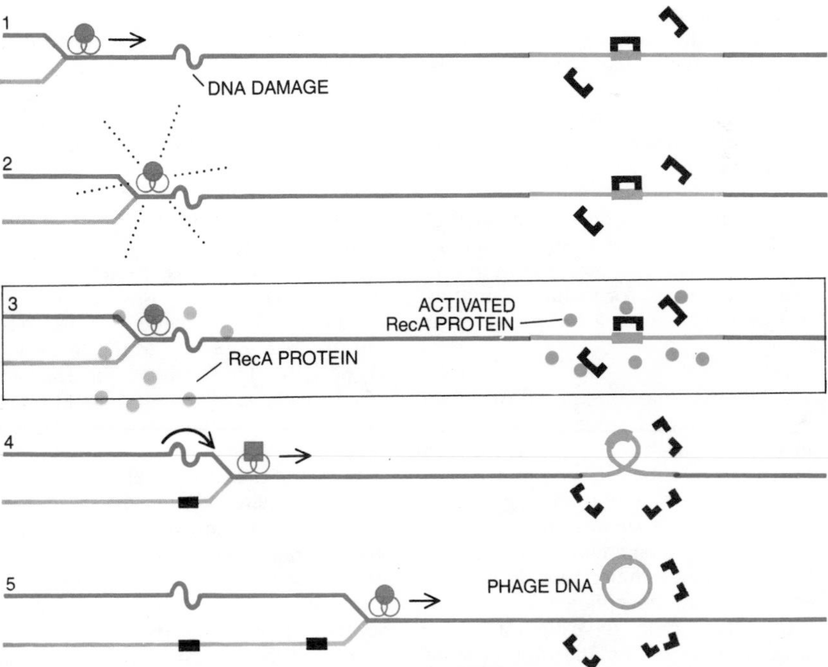

INDIRECT MUTATION results from the events that follow DNA damage (*1*), which blocks DNA replication and produces an "SOS" signal (*2*). A large amount of RecA protein (*orange*) is synthesized (*3*). The RecA somehow facilitates DNA replication on a damaged template, apparently by temporarily modifying the polymerizing complex (*4*). This replication is highly mutagenic not only at the lesion site but also elsewhere (*5*). Activated RecA cleaves the repressor (*4, right*), freeing the operator region of the prophage DNA. The prophage is thereby induced to form phage DNA (*5*), which in turn forms virus particles that burst out of bacterium.

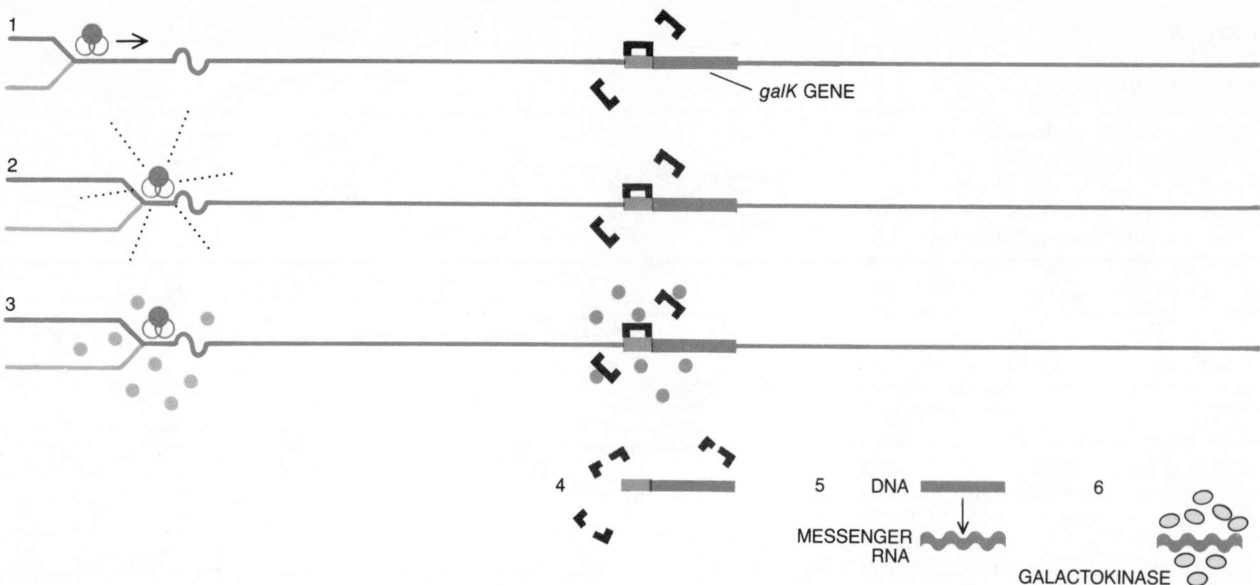

ENZYME-INDUCTION TEST is a biochemical counterpart of the inductest. The gene, *galK*, for synthesis of galactokinase is hooked up to prophage-lambda DNA in such a way that enzyme synthesis is blocked by lambda repressor. When DNA is damaged, activated RecA protein cleaves repressor. Liberated *galK* DNA (*4*) is transcribed into messenger RNA (*5*), which is translated into galactokinase (*6*).

tatest, as we call it, reveals whether a chemical gives rise to mutations directly (that is, without causing DNA damage and therefore without inducing prophage) or indirectly (by damaging the DNA and producing active virus).

Although neither the inductest nor the lambda mutatest has yet been validated as extensively as the Ames mutagenicity test, the two appear to be valuable complementary assays, in particular for determining the nature of DNA damage and for establishing a correlation between potency in producing DNA damage in bacteria and potency in causing cancer in mammals. Perhaps more important in the long run, a biological system comprising both a cell and a dormant virus should be a valuable tool for analyzing the pathway of biochemical changes, genetic or nongenetic (cleavage of repressor, for example), that are triggered when cells are exposed to carcinogens.

Given that man is exposed to a flood of chemicals, some of which surely increase the risk of cancer, there is an urgent need to identify potential carcinogens and lower human exposure to them to a level of negligible risk and keep it there. In such an effort bacterial tests can play a major role. In order to ensure broad acceptance of uniform standards, some investigators have strongly favored the adoption of a single, standardized mutagenicity test. Often, however, it is advisable to adapt a screening effort to the particular physical and chemical properties of the substance that is under study. It would therefore seem to be sensible to subject each chemical to a battery of tests. Any test has its foibles, which give rise to false-negative or false-positive responses. Several independent and complementary determinations of a chemical's DNA-damaging capacity can provide a balanced and hence more compelling assessment of carcinogenic risk.

The pronouncement that a particular chemical is a potential carcinogen has obvious social, economic and even political implications. The chemical may present a risk to a very large population, as was the case with the food additive AF-2 in Japan. That situation was dealt with rather easily: AF-2 could be replaced by an innocuous substitute. It was dispensable, and so it was simply banned from the market. Some chemical carcinogens cannot easily be dispensed with and cannot be banned; particular problems are raised by substances as different as cigarette smoke, motor-vehicle exhaust fumes and antitumor drugs. For most indispensable carcinogens it is clearly necessary to reduce human exposure to the safest levels possible. That calls for the adoption of a broadly accepted set of safety standards to be implemented by stringent regulations.

We stand today, with regard to chemical carcinogens, about where we were with regard to ionizing radiations three decades ago. The advent of nuclear power reactors made it necessary to protect workers in the industry from direct sources of radiation and to protect everyone from exposure to radioactive effluents and solid wastes. An international agency sponsored by the United Nations worked out a set of standards and regulations designed to minimize human exposure to various sources of radiation. The regulations were broadly accepted and implemented, and they have stood the test of time. They should provide a valuable reference for evolving standards for chemical carcinogens.

The experts who set the standards for radiation safety have proceeded from two basic facts. One is that mankind is exposed to a natural background of radiations whose biological effect is not perceptible (even though in principle there is no threshold for the effects of ionizing radiations). The other is that biological damage is proportional to the energy released within the living system being considered, and that this linear relation is more or less independent of the particular source of radiation. The experts assumed that for the known sources an absorbed dose only a few times higher than the dose delivered by background radiation, and less than the level that would cause a doubling of the mutation rate in a mammalian population, would constitute a negligible risk; in general such doses were adopted as being permissible for human exposure. In the case of newly devised or newly discovered radioactive materials for which there were no accumulated experimental data, permissible values were calculated on the basis of the linear relation between biological damage and the energy released within the organism.

Since chemical pollution is worldwide, an international committee of experts should propose protective regulations for adoption by individual governments. Some principles should gain broad acceptance rather easily: for example, all unnecessary potential carcinogens should be banned. A list of carcinogens that cannot be dispensed with but to which exposure should be strictly limited will also have to be drawn up. It may be difficult, however, to reach a consensus on such a list and on the permissible environmental concentrations of each chemical. One would like to begin by defining permissible concentrations of indispensable carcinogens in relation to a "natural" background. The trouble is that cultural differences, social habits and other factors have considerably distorted notions of what is a natural background level of negligible risk. (For geographical, economic and religious reasons the carcinogenic background is surely lower in Salt Lake City than it is in New York.)

If a negligible-risk level can be agreed on, the next step would be to estimate, for each indispensable carcinogen, the concentrations that should be allowed for workers in the industry and (at a much lower level) for the general population. Can bacterial tests help in setting such levels? Yes, if there is proportionality between potency in causing DNA damage in bacteria and potency in causing cancer in mammals. Matthew S. Meselson and Kenneth Russell of Harvard University have attempted to demonstrate a direct relation between mutagenic potency in the *Salmonella* test and carcinogenic potency in laboratory animals. For 10 of the 14 known chemical carcinogens they considered there was a correlation.

The many exceptions to the rule of correlation were to be expected because in a mammal the extent of DNA damage in a target cell depends on factors (which may be different in bacteria) such as the specific metabolism of carcinogens in the organ involved, the cells' permeability to particular metabolites and tendency to accumulate them and the extent of residual DNA damage after the cellular repair mechanisms have had their effect. Each of these variables must be determined for each individual carcinogen if one is to set valid levels for human exposure. Each chemical—in contrast to individual sources of radiation—requires a thorough specific study to determine its carcinogenic potency. Bacterial tests have good indicative value.

Even though establishing regulations will be much more difficult for chemical carcinogens than it was for radiations, the regulations should be established soon. It is wiser and safer to have safety regulations than not to have them. A large part of the human population contracts cancer; one person in five dies of it. It is important to determine the precise mechanism by which each DNA-damaging agent acts on DNA, not only to prevent cancer but also to develop better ways of curing it. About half of all the drugs administered in an effort to arrest cancer are DNA-damaging agents, and so a significant number of people are being regularly exposed, for therapeutic purposes, to such agents. By establishing, through studies in bacteria, the mechanism whereby antitumor agents affect DNA, investigators may be able to help design more drugs that are effective in curing cancer without also causing cancer.

Bacterial tests for potential carcinogens should help to answer two questions now being asked by industry, government and people at large. What chemicals are DNA-damaging agents? How much of each—in our air, our water and our food, in the products we manufacture and consume—constitutes a biological risk? These two questions will be answered satisfactorily, however, only if a third question is asked: What is the mechanism whereby each chemical affects DNA? Only basic knowledge can breed safety.

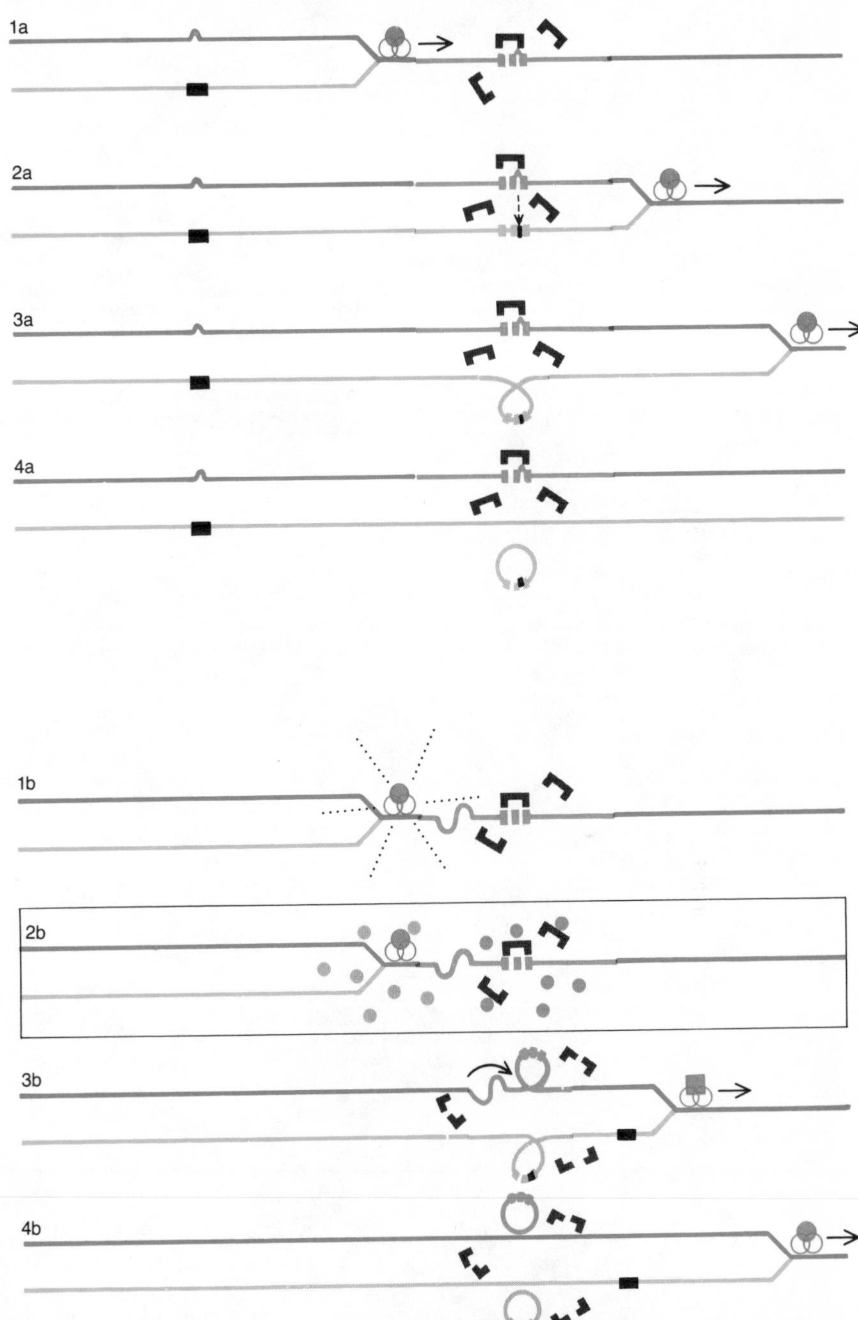

DIRECT AND INDIRECT MUTAGENESIS can be distinguished from each other by the mutatest, based on mechanisms diagrammed here. The repressor binds to a slightly modified operator (*1a, 1b*). A negligible alteration of operator DNA (*1a*) results, on replication, in a mutation (*2a*) that prevents further binding of the repressor (*3a*). The prophage develops into active phage particles (*4a*) that are selectively revealed by strain-*B* indicator bacteria, which are not sensitive to nonmutated phage; phage development independent of RecA-protein activation reveals direct mutagenesis. DNA damage (*1b*), on the other hand, leads on replication (*2b, 3b*) to RecA-dependent indirect mutagenesis, which is detected with strain-*B* indicator bacteria. The concomitant event, prophage induction (*3b, 4b*) resulting from activation of RecA, is detected with indicator bacteria of strain *A*, not with strain-*B* bacteria. Mutagenesis can be diagnosed as indirect when it occurs along with prophage induction: both are dependent on RecA.

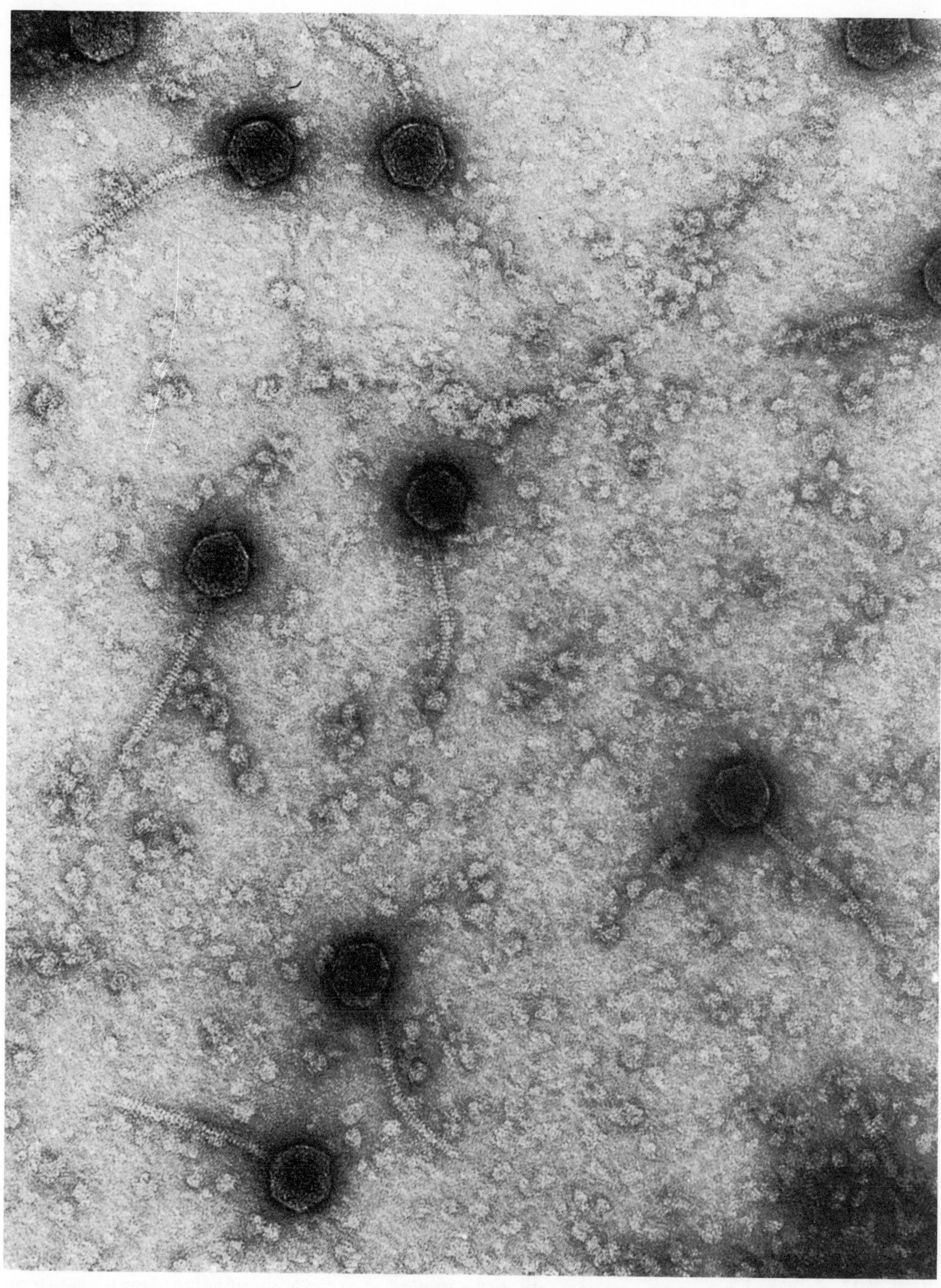

BACTERIAL VIRUSES, or phages, of the strain designated lambda are magnified 200,000 diameters in this electron micrograph made by William C. Earnshaw and Philip A. Youdarian of the Massachusetts Institute of Technology. Each virus particle has an icosahedral head containing a molecule of infectious viral DNA, and a flexible hollow tail with which the phage attaches itself to the outer membrane of its bacterial-cell host and injects its DNA into the interior of the cell. Negative staining with uranyl acetate clearly reveals the subunits that make up the virus's protein shell. The phage-lambda particles shown here were extracted from an infected strain of the colon bacillus *Escherichia coli*, together with a large number of bacterial ribosomes (*lightly stained particles scattered over background*).

How Viruses Insert Their DNA into the DNA of the Host Cell

by Allan M. Campbell
December 1976

Some viruses are able to coexist peaceably with their host cell for long periods, incorporating their genes into the host chromosome. The details of the insertion process are now fairly well understood

The imagery of medicine often seems to portray viruses as aggressive organisms bent on human destruction. In actuality their role in disease is merely a by-product of their parasitic existence. Unable to reproduce by themselves, viruses must invade living cells and redirect some of the cellular machinery to the production of new virus particles. In the process many of the host cells are destroyed, leaving the tissue of which they are a part damaged and giving rise to disease.

Although virus production and cell destruction are the most dramatic outcomes of virus infection, they are by no means the only outcomes possible. Like other parasites, many viruses find it advantageous to persist innocuously within their host cell for an indefinite period, actively multiplying only when the host weakens or stops growing. The existence of such latent viruses was first suspected in the 1920's, but it was not until the early 1950's that André Lwoff and his colleagues at the Pasteur Institute established unequivocally that a latent virus, or "provirus," can be transmitted from one cell generation to the next without external reinfection. Working with bacteriophages, the viruses that infect bacteria, they showed that the provirus exists in a "lysogenic" state. The term lysogenic refers to the fact that the provirus can come out of the dormant state and give rise to mature virus particles that lyse, or dissolve, the bacterial cells. Among the agents that can induce the provirus to resume its former mode of multiplication are ultraviolet radiation, X rays and carcinogenic chemical compounds. In the intervening 20 years the further study of bacteriophages has revealed much about the mechanism of lysogeny.

A typical bacteriophage consists of a single linear molecule of nucleic acid enclosed in a protein coat. Resembling a minute hypodermic syringe, the phage attaches itself to a bacterial cell and injects its strand of DNA into the interior. Once inside the cell, the viral DNA may begin directing the manufacture of new virus particles or become a provirus by incorporating itself into the DNA of the bacterial cell's long, threadlike chromosome.

What then switches the provirus from the lysogenic state to the active production of virus particles? A mechanism was put forward by François Jacob and Jacques Monod of the Pasteur Institute in 1961 and later demonstrated by Mark Ptashne of Harvard University. In the provirus only a few genes are expressed, and the product of one of them is a "repressor" protein that combines with the viral DNA and prevents the expression of the other viral genes, particularly those responsible for the independent replication of the viral chromosome. Inducing agents such as X rays alter the metabolism of the bacterium in such a way that a substance is produced that inactivates the repressor. If all the repressor molecules in a given lysogenic bacterium are simultaneously inactivated, the viral genes are sequentially expressed. Proteins needed for the replication of the viral DNA are made first, followed by the head and tail proteins of the virus particle, which spontaneously assemble inside the bacterium. Unit segments of viral DNA are then packaged into the phage heads. Finally, some 60 minutes after the cell was exposed to the inducing agent, the cell bursts, releasing about 100 virus particles that are capable of infecting other cells.

When a single lysogenic bacterium multiplies to form a colony, every cell of that colony is potentially capable of manufacturing virus particles and can express that potentiality when its repressor is destroyed. The viral genes have therefore been added to the bacterial genes in such a way that both sets of genetic information are inherited by the cell's descendants. In order for this to happen the viral chromosome must replicate like any normal cellular component, and it must be partitioned at cell division so that each daughter cell receives at least one copy of it.

Such a result can be achieved in at least three possible ways. One is for the viral chromosome to insert itself into the host chromosome, after which it can be passively replicated and distributed at cell division as part of the host DNA. Another way is for the viral chromosome to establish itself independently; it replicates and is distributed like the host chromosome but is separate from it. A third way is for the viral chromosome to replicate separately from the host chromosome and in multiple copies. In that case the viral chromosome might not be distributed regularly when the bacterial cell divides, but if the number of copies is large enough, the chance of a daughter cell's receiving no copies whatever is small. If we survey all known viruses, we find that the first two mechanisms exist in nature, and the third is observed in certain mutant viruses produced in the laboratory. One property all three mechanisms have in common is that once the virus has become established as a hereditary component of the cell, there is nothing particularly novel in its mode of inheritance from then on.

Bacteriophage lambda, which was discovered by Esther M. Lederberg of Stanford University as a provirus carried by the K12 strain of the colon bacillus *Escherichia coli,* is the best-understood genetically of the bacteriophages and remains one of the favorite experimental organisms of molecular biologists. Soon after Lwoff's demonstration of heritable lysogeny several investigators were able to achieve genetic crosses between strains of *E. coli* that were lysogenic for phage lambda and strains that were nonlysogenic for it. They got the surprising result that the characteristic of being lysogenic or nonlysogenic was distributed among the progeny of the cross exactly as one would expect if the trait were being determined by a gene at a specific location on the bacterial chromosome. In fact, lambda behaved in crosses as if it was closely linked to the *gal* operon, a cluster of bacterial genes concerned with the metabolism of the sugar galactose.

There is now general agreement that the lambda provirus is inserted into the *E. coli* DNA at a specific site between the *gal* operon and the *bio* operon, which is responsible for the synthesis of the vitamin biotin. The significance of the insertion process is not confined to virology. It shows one way in which two genetic units, each capable of existing and replicating independently, can coalesce into

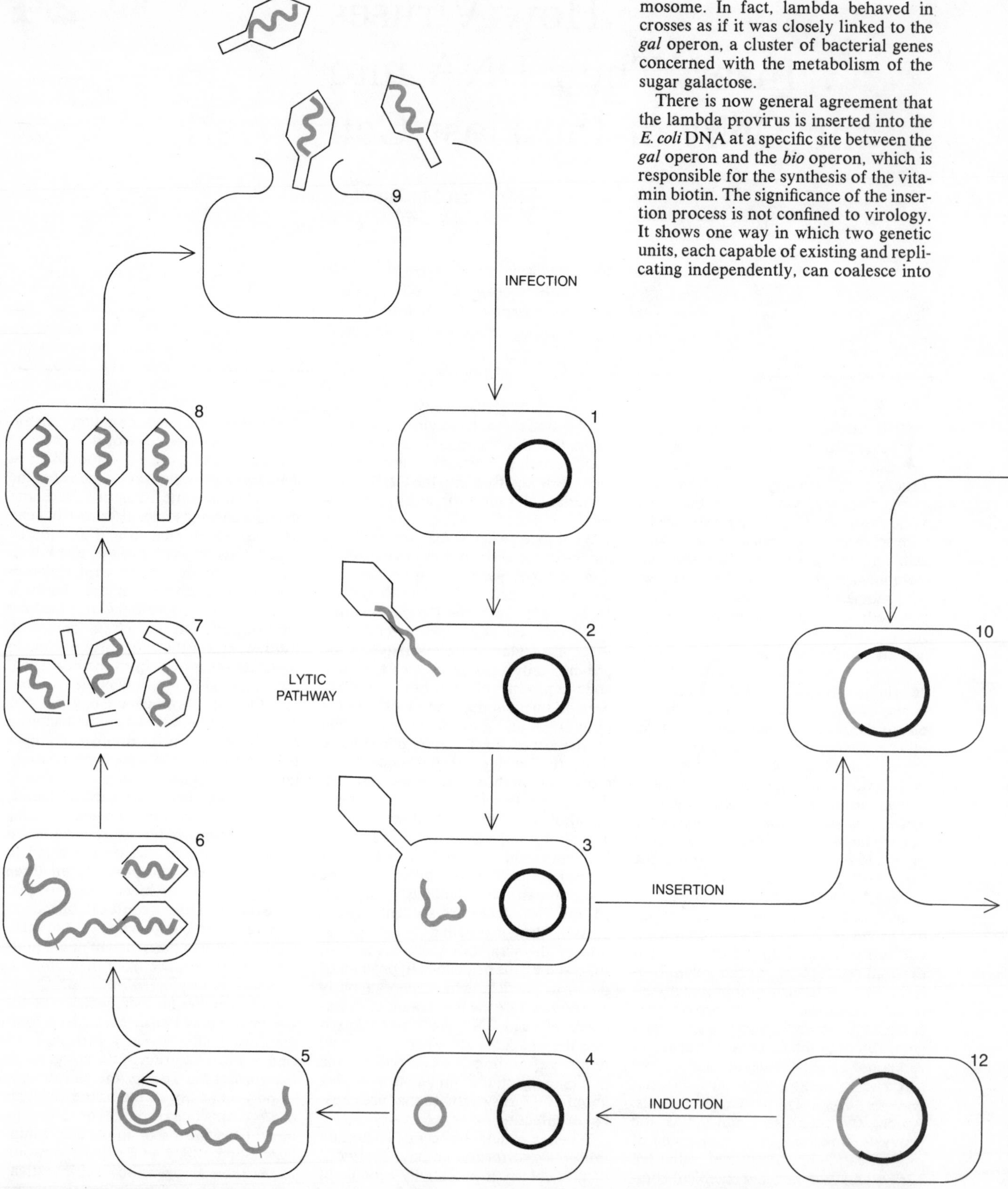

LIFE CYCLE OF PHAGE LAMBDA shows that lysis (dissolution) and death of the infected bacterial cell are not inevitable. After a molecule of viral DNA is injected into a healthy *E. coli* cell (*2*) the host may take either of two paths. In the productive, or lytic, pathway the viral DNA forms a circle (*3 and 4*) and replicates by the "rolling circle" process to give rise to a long "sausage string" of DNA containing multiple copies of the viral genes (*5*). Next the viral genes direct the synthesis and assembly of the head and tail proteins of the virus particles and the packaging of unit DNA segments into the heads (*6 and 7*). Heads and tails then assemble spontaneously inside bacterium, giving rise to mature phage particles (*8*). Finally, some 60 minutes after infection, the host cell bursts, releasing about 100 progeny virus particles that can then infect other *E. coli* cells (*9 and 1*). Alternatively, depending on the conditions of infection, the phage-lambda DNA can live quietly within the host, establishing itself as a semipermanent part of the bacterial chromosome (*10*). It is then replicated and segregated at cell division (*11 and 12*) and passed on to succeeding cell generations for an indefinite period. The latent virus,

a larger unit. Phage lambda is only one of a number of elements for which insertion is known or suspected. Most of these elements are viruses or the small pieces of bacterial DNA called plasmids, some of which can transfer resistance to antibiotics and other properties from one bacterial cell to another. In addition there are the "insertion sequences" of bacteria and the "transposable elements" of maize, which are known only for their ability to move occasionally from one chromosomal location to another and may well have no independent existence.

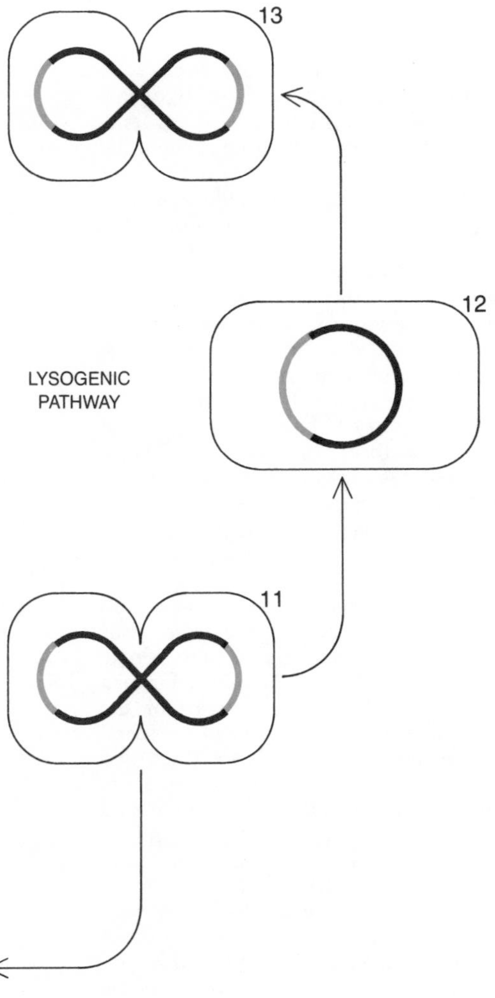

or "provirus," still retains the potential to grow lytically, however. Bacteria harboring a provirus are called lysogenic, meaning that they carry a property that can lead to viral multiplication and the fatal lysis of the host cell. Exposure of lysogenic bacterium to ultraviolet radiation, X rays or chemicals such as nitrogen mustard and organic peroxides can induce the provirus to return to lytic pathway.

Although the widespread occurrence of genetic elements that can insert themselves into preexisting chromosomes is incontestable, there is a considerable divergence of opinion about their importance in the normal life of the host or in evolution. One extreme viewpoint is that these elements are basically foreign to the cells that harbor them; they are invaders like viruses that have somehow got into cells and chromosomes and do not really belong there. The alternative view is that much of the DNA of present-day chromosomes is derived from elements that were originally foreign but that through a series of small evolutionary steps gradually became naturalized citizens of the intracellular community. A somewhat intermediate position is that the mobility of these elements plays some essential role in the normal development of multicellular organisms from the fertilized egg into the adult.

Studies of the mechanism by which new genetic elements are added to existing chromosomes cannot provide any direct information about the origin or function of such elements. The studies are, however, relevant in one respect. If it turned out that the DNA of the added elements was hooked onto the rest of the chromosome in some unusual way, by connections that were not normal features of chromosome structure, then a clear distinction between foreign DNA and indigenous DNA would be implied. It therefore seemed particularly important to firmly establish whether or not the foreign DNA of some prototypical examples such as the lambda provirus is directly inserted into the linear sequence of genes on the host cell's chromosome.

The concept of genes' being arranged in a linear sequence, and the experimental basis of that concept, have a long history. In 1913 A. H. Sturtevant of Columbia University crossed fruit flies differing from each other by several genetic traits and analyzed the frequencies of progeny exhibiting new combinations of the parental traits. He concluded that the determinants for these traits (all of which happened to be on the same chromosome) distributed themselves as though they were on some one-dimensional structure, with new combinations arising from the redistribution of connected segments of that structure. It became possible to extend this kind of formal linkage analysis in the 1950's, when Seymour Benzer of Purdue University, working with small segments of phage chromosome, proved that the linkage maps of mutational sites within genes were also one-dimensional. These purely genetic studies were followed by cytological and biochemical work showing that the one-dimensionality of linkage maps was associated with the linear arrangement of the genes along the chromosome and ultimately with the linear sequence of nucleotides along the double helix of DNA.

The question of whether all the linear DNA segments in a chromosome are joined end to end in one continuous double helix is still not completely settled, but the evidence is increasingly strong that they are. The uncertainty surrounding the question through the mid-1960's was such, however, that many bacterial geneticists found nothing particularly bizarre in the notion that the provirus might lie alongside the host chromosome rather than being inserted into it.

One way to investigate the question is to cross two lines of lysogenic bacteria, in which both the host genes and the viral genes are marked with mutations. In such bacterial crosses pieces of DNA from the donor cells are introduced into the recipient cells, where they pair with corresponding segments of the recipient cells' DNA and, in some fraction of those cells, replace them. Progeny in which a specific gene has been replaced can be recognized and selected for because they express a genetic trait characteristic of the donor, such as the ability to metabolize galactose. The recognition of the replacement of specific genes in the provirus has been made easier in recent years by the isolation of conditionally lethal mutant viruses, which fail to multiply in some condition under the control of the experimenter but which grow normally when that condition is changed. For example, temperature-sensitive mutants survive and multiply at one temperature (25 degrees Celsius) but not at another (42 degrees C.).

Such experiments throw some light on the provirus's mode of attachment. If, for example, the provirus is not linearly inserted into the host chromosome but instead projects from it sideways, one would expect that an exchange of genetic material occurring along the viral segment would not redistribute the genes lying on the main axis of the host chromosome. The fact is that redistribution of the host genes does occur, suggesting that insertion is indeed linear.

A second and more informative method of determining the topology of the lysogenic chromosome is deletion mapping, in which mutants marked by chromosomal deletions are crossed with strains that have other genetic markers. A deletion mutation involves the permanent loss from the chromosome of a string of neighboring nucleotides, numbering from one to many thousands. Since each deletion eliminates a continuous segment of chromosome, the order of the genes along the chromosome can be deduced by piecing together the information provided by the characteristics of the various deletion mutants observed. Deletion mapping provides a strict "betweenness" criterion for locating genes with respect to one another. The condition that every observed deletion should be representable as a linear segment is a highly restrictive one; a segment represented by two deletion markers must necessarily include all points that lie between them. For example, if the provirus is inserted between two identified host genes, then every deletion removing both of those genes must remove the provirus as well. This turns out to be the case in experiments, and it is again consistent with the model that the provirus is continuous with the host DNA.

Until the beginning of the 1970's genetic analysis of the type I have been describing provided the only precise information on the relation between the provirus and the chromosome; no direct physical information was available on the relevant nucleotide sequences within the DNA of the provirus. In principle the simplest approach would be to use direct methods of determining the sequence of nucleotides along the DNA chains of the virus and the lysogenic chromosome. Although such sequencing methods are improving rapidly, the identification of all 50,000 nucleotides in the lambda provirus would be a time-consuming and costly task. For many purposes adequate information can be obtained by exploiting the fact that single DNA chains with complementary nucleotide sequences can find each other and form double helixes in the test tube. Electron micrographs of DNA molecules formed this way, notably in the laboratory of Norman R. Davidson at the California Institute of Technology and that of Waclaw T. Szybalski of the University of Wisconsin, have demonstrated that the structures inferred from genetic results have a physical reality.

For these experiments double-strand DNA molecules extracted from virus particles are dissociated into single chains by heating. If the solution is then cooled slowly, double helixes of the complementary chains will re-form. When single-strand DNA's from two viruses with some nucleotide sequences in common (such as phage lambda and a deletion mutant of it) are mixed before the cooling step, new helixes can form not only between complementary chains from one virus but also between complementary chains from both viruses; the latter kind of chain is known as a heteroduplex. The nucleotides in the complementary segments of these hybrid chains pair up and form double helixes but the noncomplementary sequences do not, leaving single-strand loops that can be seen in electron micrographs.

By some ingenious manipulations Davidson and his colleagues Phillip A. Sharp and Ming-Ta Hsu were able to examine heteroduplexes between viral

INSERTED ELEMENT	APPROXIMATE LENGTH IN NUCLEOTIDES	INSERTED INTO	SPECIAL PROPERTIES
BACTERIOPHAGE LAMBDA	50,000	*E. COLI* CHROMOSOME	SPECIFIC SITES ON VIRUS AND HOST CHROMOSOMES
BACTERIOPHAGE MU-1	37,000	*E. COLI* CHROMOSOME	SPECIFIC SITE ON VIRUS, ANY SITE ON HOST CHROMOSOME
BACTERIAL SEX FACTOR *F*	100,000	*E. COLI* CHROMOSOME	MANY SITES ON HOST CHROMOSOME
DRUG-RESISTANCE PLASMID	20,000	BACTERIAL FACTOR RESEMBLING *F*	TRANSMITS RESISTANCE TO ANTIBIOTICS BETWEEN BACTERIAL STRAINS
TUMOR VIRUS SV-40	5,000	HUMAN CHROMOSOME	DERIVED ORIGINALLY FROM MONKEY CELLS
TRANSPOSABLE ELEMENTS	(UNKNOWN)	MAIZE CHROMOSOMES	NO DETECTABLE EXTRA-CHROMOSOMAL PHASE
INSERTION SEQUENCE IS2	1,400	*E. COLI* CHROMOSOME	

ADDED GENETIC ELEMENTS, small pieces of DNA that can exist as part of the main chromosome or independently, have been observed in bacterial, maize and human cells. Some are viruses; others are not. When they are inserted, they introduce into the cell instructions governing additional biochemical reactions that may be superimposed on the cell's metabolism.

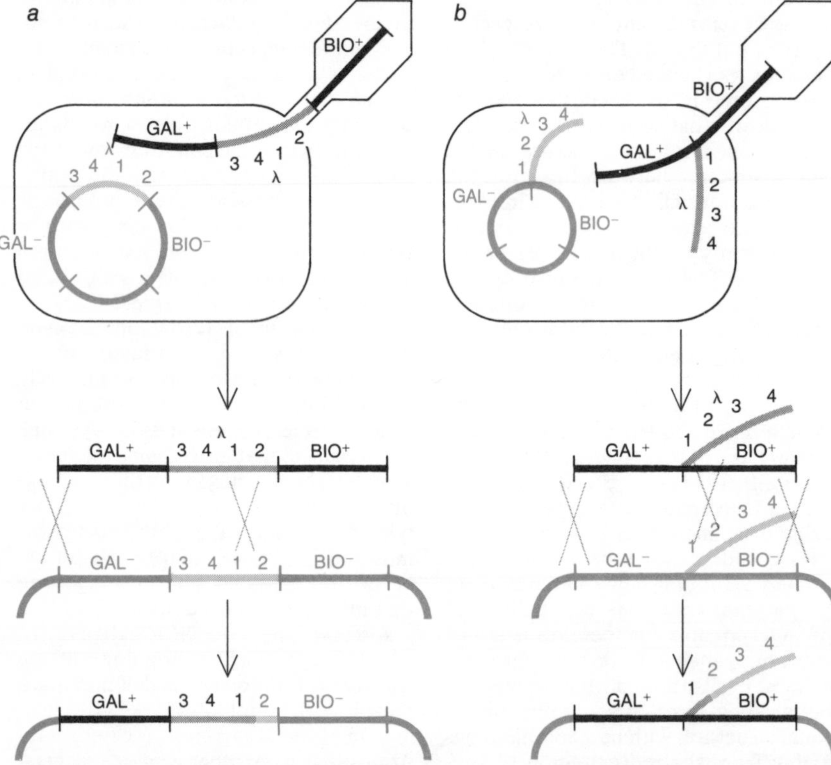

TYPES OF GENETIC EXCHANGE, or recombination, observed between two lysogenic *E. coli* chromosomes containing different genetic markers shed light on the mode of attachment of the phage-lambda genes. In the experiment depicted here a fragment of DNA from a bacterial cell capable of metabolizing galactose and synthesizing biotin (*gal*$^+$*bio*$^+$) and containing a lambda provirus with mutations in genes *1, 2, 3* and *4* is introduced by means of an infective phage coat into a recipient bacterium that is unable to utilize galactose or synthesize biotin because of genetic mutations (*gal*$^-$*bio*$^-$) and that harbors a nonmutant provirus. Type of recombination that might occur if the viral genes were linearly inserted into each chromosome is shown in *a*. The mechanism of insertion proposed by the author requires the permutation of the order of genes along the viral chromosome, so that a genetic exchange between viral genes *1* and *2* would serve to recombine the flanking bacterial markers *gal* and *bio*. The resulting recombinant bacterium would be capable of utilizing galactose but incapable of synthesizing biotin, a prediction that agrees with observed results. For comparison, *b* indicates the expectation for one kind of nonlinear topology, in which the provirus joins to the chromosome as a branch.

DNA and the DNA of an inserted provirus. Although the most straightforward approach would be to make heteroduplexes from one DNA chain of phage lambda and the complementary chain of a chromosome from a lysogenic bacterium, that experiment is not yet feasible because of the difficulties in handling a DNA molecule the size of the bacterial chromosome. The same end was achieved by letting the phage-lambda DNA insert itself not into the entire bacterial chromosome but into a smaller DNA molecule: a derivative of the bacterial sex factor that had picked up from the bacterial chromosome the specific DNA segment into which lambda inserts. Strands of the sex factor with the lambda provirus inserted into them could be readily isolated intact and used to form heteroduplexes with DNA extracted from virus particles.

The combined results of genetic and physical studies make us quite confident

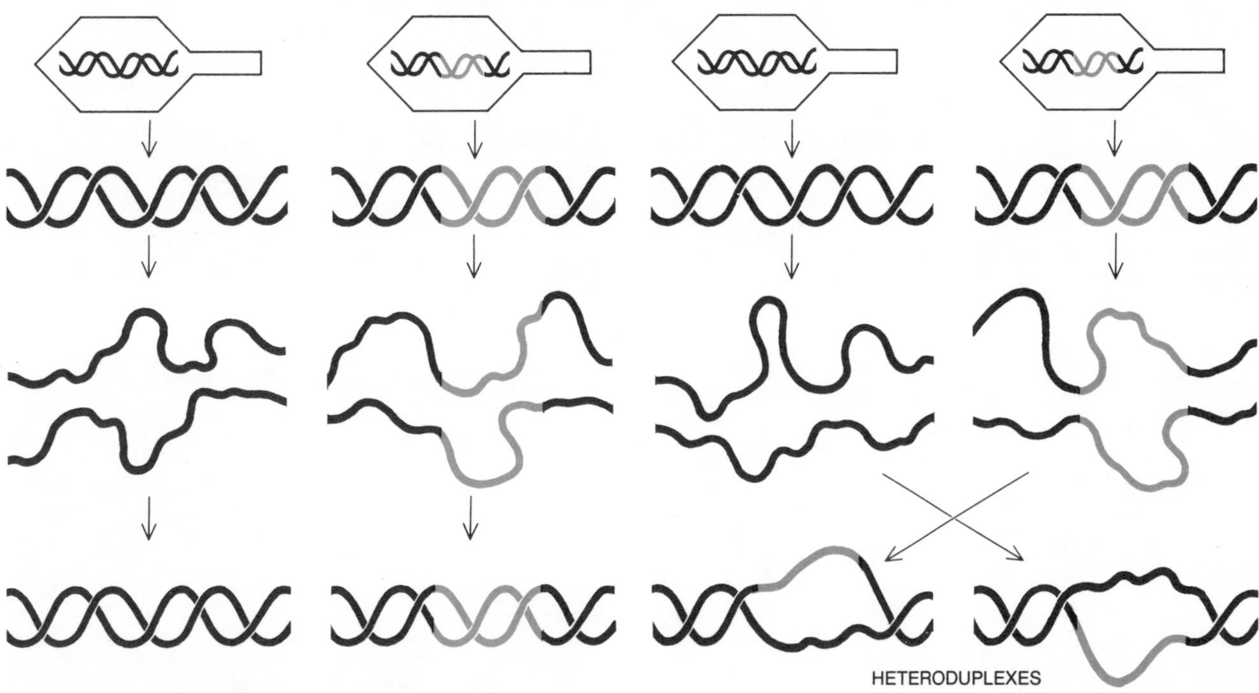

HYBRID DNA MOLECULES artificially formed in the test tube can be used to map viral genes physically. When the DNA double helix is heated, it unwinds, giving rise to two single-strand chains. If single chains having complementary and noncomplementary regions are mixed together at this stage and slowly cooled, some "duplexes" will form between the chains. When these heteroduplexes are viewed in the electron microscope, the two DNA chains will be double helixes where they have the same sequence of nucleotides and unpaired where they differ in sequence. With this method one can precisely map position of a given marker mutation along the DNA molecule.

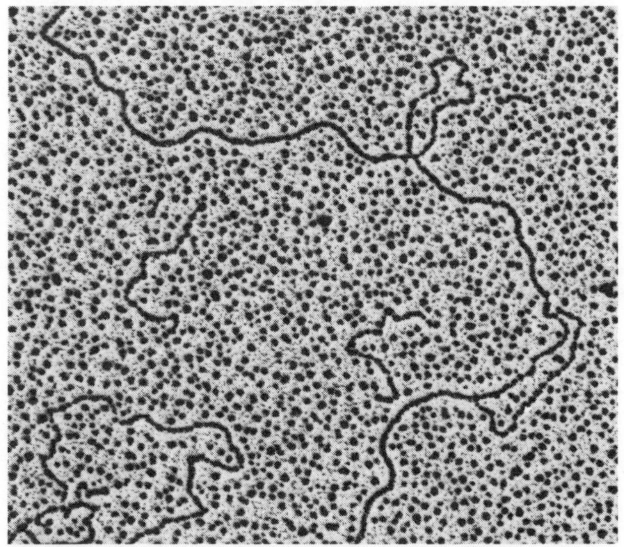

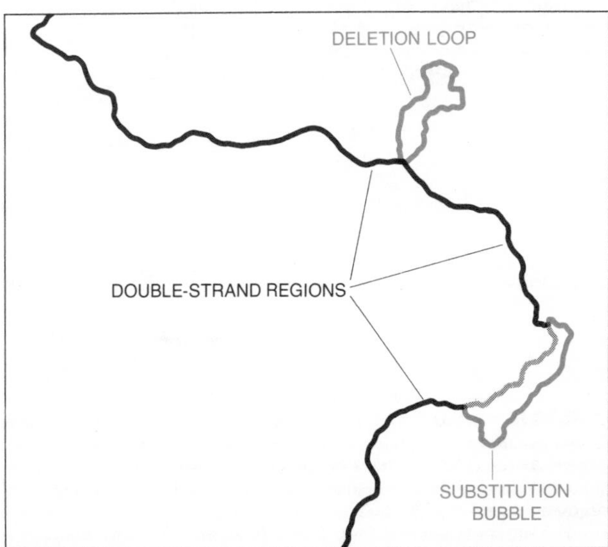

HETERODUPLEX between a strand of normal lambda DNA and a second strand incorporating two mutations is clearly visible in this electron micrograph made by Elizabeth A. Raleigh of M.I.T. (Only a small segment of the long viral DNA molecule is shown.) The loop of single-strand DNA at top right results from a deletion mutation that removed an entire segment of DNA from one strand. The "collapsed bubble" of single-strand DNA results from the substitution of several nucleotides in one strand by different ones, making that region of the two strands noncomplementary and hence unable to pair. The remaining portions of molecule shown here are double-helical.

that we now know the structure of the lysogenic chromosome. The steps by which this structure is formed and dissociated into its component parts are the subject of current research. During the life cycle of phage lambda, DNA must be cut and rejoined at the ends of the viral chromosome and at the ends of the provirus. The lambda DNA injected into the bacterial cell is in linear form, but before it is inserted into the bacterial chromosome its ends are joined so that it makes a circle. During insertion the circle is opened at a different point. As a result, although the provirus and virus chromosomes are both linear structures, the order of the genes along the two is not identical.

How the ends of the viral DNA are joined to form a circle is known, thanks largely to the work of A. D. Kaiser and

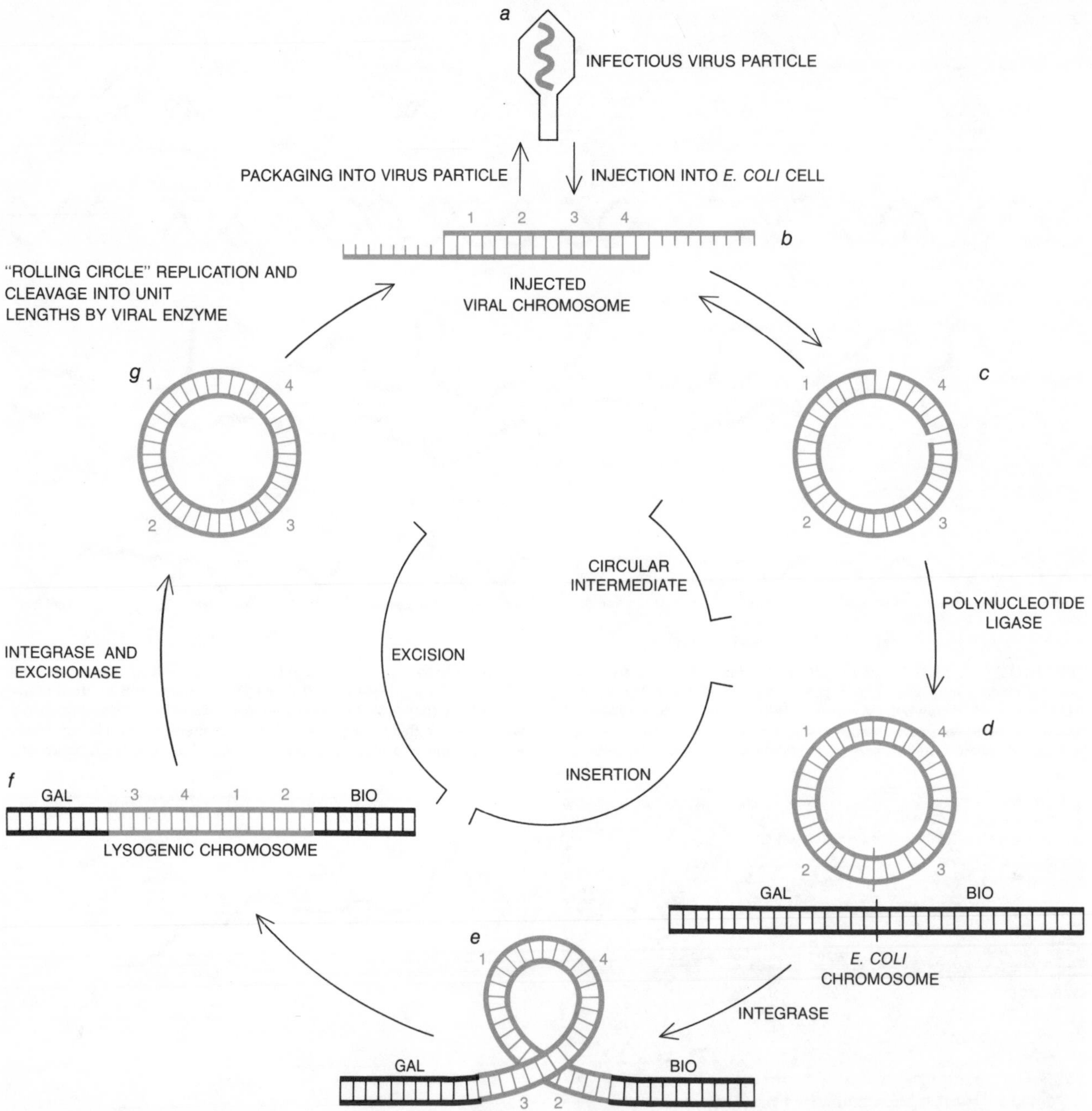

INSERTION AND EXCISION of the phage-lambda genes into the *E. coli* chromosome require the action of both bacterial and viral enzymes. As the DNA of phage lambda is packaged in the viral particle (*a*) it is linear and double-helical, except for complementary unpaired segments 12 nucleotides long at the ends of the two nucleotide chains (*b*). In solution this linear form comes to equilibrium with a circular form that has staggered "nicks" 12 nucleotides apart in the two complementary chains (*c*). When viral DNA is injected into the bacterial cell in the course of infection, the two nicks in the open circle are sealed by the bacterial enzyme polynucleotide ligase, so that both chains of the circle are now closed throughout their length (*d*). This circular intermediate then interacts with a particular segment of the *E. coli* chromosome (between the *gal* and *bio* genes). Viral and bacterial chromosomes break and rejoin at unique sites on each partner, so that viral DNA is spliced into the host DNA, a reaction catalyzed by the viral enzyme integrase (*e*). (Note that the gene order in the provirus is *3, 4, 1, 2*, a cyclic permutation of the viral gene order *1, 2, 3, 4*.) The *E. coli* chromosome is now lysogenic for phage lambda (*f*). After several cell generations radiation or chemically active compounds may induce the provirus to enter the lytic state. When this happens, the lambda repressor, which has so far blocked the expression of most of the viral genes, is inactivated, allowing the synthesis of the viral enzyme excisionase. Together with integrase, excisionase catalyzes the excision of the provirus from the host chromosome, converting it back into the circular form with the original gene order (*g*). The circle of viral DNA replicates, producing multiple copies that are then cleaved by a specific viral enzyme to give rise to the linear form with "sticky" ends (*b*). Each linear DNA segment is then packaged in a virus coat (*a*). When the host cell ruptures, the liberated phages infect healthy cells and the lysogenic cycle begins anew.

his collaborators at Stanford. Lambda DNA is a double helix throughout most of its length, but one end of each polynucleotide chain extends for 12 nucleotides beyond the double helix. These two single-strand chains are complementary to each other and are called "sticky ends." In solution the linear DNA molecules can come to equilibrium with circular molecules formed by the pairing of the two ends. When lambda infects an *E. coli* cell, the open circle formed by the viral DNA is closed by the action of polynucleotide ligase, a bacterial enzyme that seals breaks in one chain of a double helix. This step requires no viral enzymes, and it is not specific to the nucleotide sequences involved. On the other hand, the insertion of viral DNA into the bacterial chromosome requires the recognition and cutting of highly specific nucleotide sequences in both the lambda and the *E. coli* DNA.

Little is known of the biochemistry of insertion, although its genetic control has been intensively explored. At the time I proposed the circular-molecule-intermediate model for the insertion of phage-lambda DNA in 1962, the only known mechanism for breaking or rejoining two DNA molecules at corresponding points was homologous recombination, which requires that the two molecules have similar or identical base sequences in the recombining region. The chemical steps by which homologous recombination takes place are still largely conjectural, but Alvin J. Clark of the University of California at Berkeley and others have isolated bacterial mutants that are unable to carry out this process. Under conditions where homologous recombination is blocked by such mutations, however, phage lambda can still insert its DNA with the normal frequency. Hence the insertion of viral DNA seems to be accomplished not by the same bacterial enzymes that are responsible for homologous recombination but by viral enzymes that cut and join DNA molecules at highly specific sites.

Direct evidence for the existence of such viral enzymes has been provided by the genetic studies of James F. Zissler of the University of Minnesota Medical School and the biochemical investigations of Howard A. Nash of the National Institute of Mental Health, which have shown that the enzyme product of a specific viral gene (dubbed integrase) is required for the insertion of viral DNA; mutant viruses lacking this enzyme are unable to enter the lysogenic state. Similar studies of the reverse process—the excision of viral DNA from the bacterial chromosome—by Gabriel Guarneros and Harrison Echols of the University of California at Berkeley and Susan Gottesman of the National Cancer Institute—have shown that excision requires in addition to integrase the

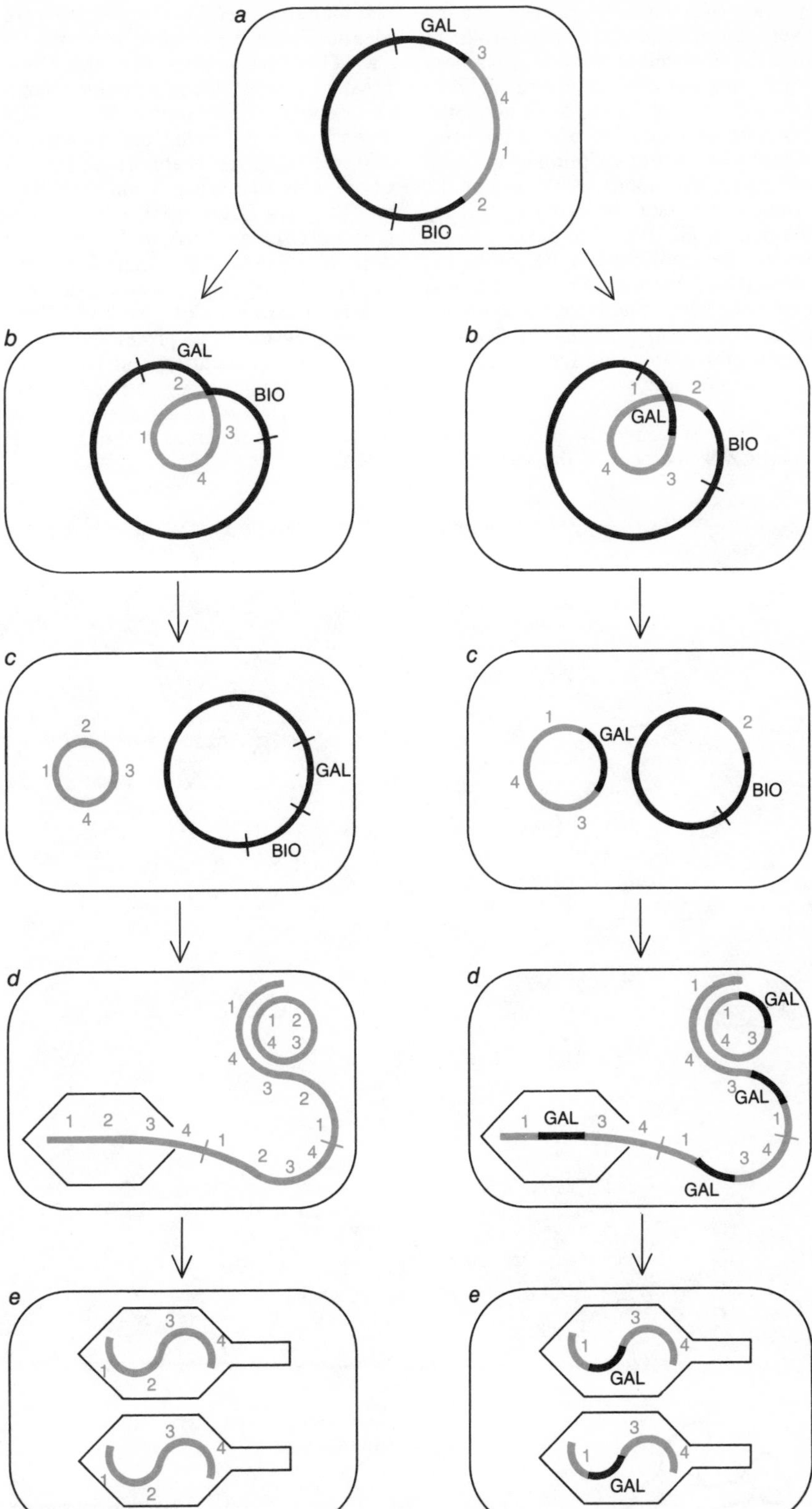

NOVEL VIRUS VARIANTS arise when certain proviruses turn into infectious viruses, killing the bacterium that harbors them and carrying away with them pieces of the DNA of the dead host. The "transduced" bacterial genes linked to the viral chromosome can subsequently replicate at the unrestricted viral pace. Transduction is believed to result from rare errors in the excision of the provirus from the host chromosome. Normal excision (*left panel*) takes place in the vast majority of cells. Viral DNA separates from the bacterial DNA (*b*) to give rise to a circular viral chromosome and a nonlysogenic bacterial chromosome (*c*). The viral chromosome replicates in several stages, ultimately as a rolling circle that generates a long sausage string of DNA in which the entire viral sequence is repeated many times (*d*). Unit DNA lengths are then packaged into infectious virus particles (*e*). In one cell out of 100,000 (*right panel*) abnormal excision generates a circular molecule including some host DNA. Infectious particles are thus formed in which a segment of bacterial DNA has replaced a segment of viral DNA.

product of a second viral gene (called excisionase). The virus thus introduces into the host cell enzymatic machinery for cutting and joining the viral and host DNA at specific sites to bring about the insertion and excision of the provirus. As long as the transcription of the gene coding for excisionase is blocked by the lambda repressor the provirus will remain inserted. When repression is released, the excisionase gene will be expressed and there will be a reciprocal exchange within the lysogenic chromosome, re-creating a circular molecule of lambda DNA and a nonlysogenic bacterial chromosome.

Excision is generally precise: more than 99 percent of the virus particles manufactured by lysogenic cells are identical with the original infecting virus. This fact implies that the DNA breaks at exactly the same point when it comes out of the chromosome as it does when it goes in. About one excision in 100,000, however, is abnormal. Instead of breaking away cleanly the host DNA and the viral DNA break and rejoin to create a circular DNA molecule incorporating some viral DNA and some host DNA. If the size and physical characteristics of the molecule allow it to be recognized by the viral proteins as being suitable for replication and packaging as a virus, it can then give rise to infectious particles in which a segment of host DNA has replaced part of the viral DNA.

M. Laurance Morse of the University of Colorado Medical Center first discovered the existence of these "transducing" virus variants when he found that some of the phage-lambda particles liberated from lysogenic bacteria contained the *gal* genes of the host. When lambda and lambda-*gal* DNA's were hybridized in the test tube, the heteroduplexes showed that nucleotides at the two ends of the molecules were complementary to one another but that in the middle of the duplex there was an unpaired region where the picked-up segment of bacterial DNA (including the *gal* operon) was not complementary to the viral DNA.

The theory that transducing phages are produced by errors in the excision of

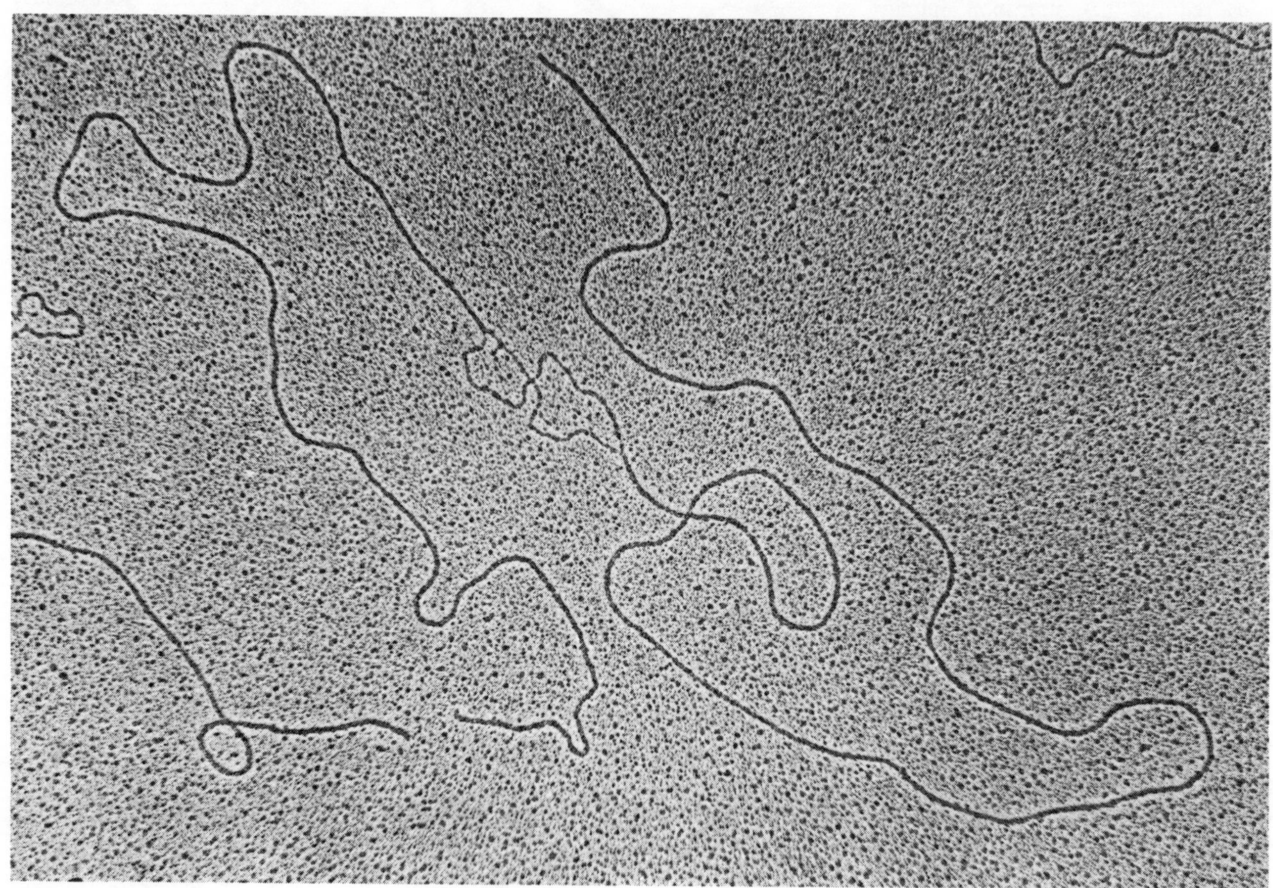

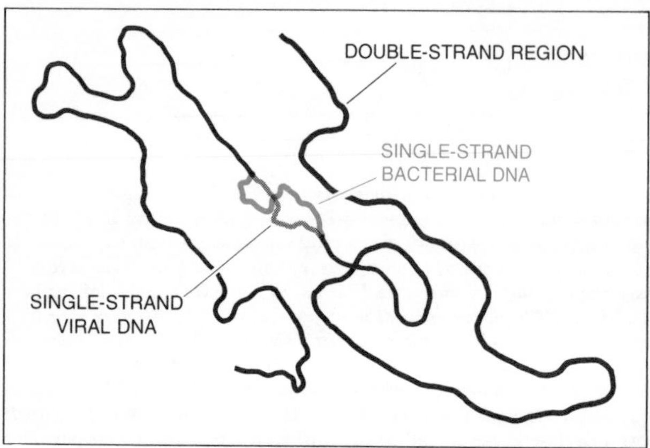

HETERODUPLEX made in the test tube between DNA from a transducing phage designated $\phi 80psu_3$ and DNA from the parental phage $\phi 80$ reveals the location of the inserted bacterial genes in the middle of the viral chromosome. The black lines on the map represent the double-helical regions where the nucleotide sequence on the DNA strands of both partners is complementary. The colored segment is the piece of *E. coli* DNA 3,000 nucleotides long carried by the transducing phage; the gray segment is the piece of viral DNA 2,000 nucleotides long that is present in the normal phage $\phi 80$ but replaced by bacterial DNA in $\phi 80psu_3$. Viral and bacterial DNA sequences on opposing strands are not complementary and cannot pair, forming a substitution bubble. Total length of duplex molecule is about 43,000 nucleotides. Electron micrograph was made by Madeline C. Wu and Norman R. Davidson of California Institute of Technology.

the provirus from the host chromosome was supported by the observation that under normal circumstances lambda can incorporate only genes, such as *gal*, that are within a few thousand nucleotides of its insertion site. This distance is a small fraction (less than 1 percent) of the total length of the host chromosome. Recently K. Shimada and his co-workers at the National Institutes of Health have studied rare bacterial lines in which lambda DNA has inserted itself into a chromosomal site other than the normal one. From such abnormal strains virus variants carrying *gal* are not obtained, but variants carrying genes close to the new attachment site are.

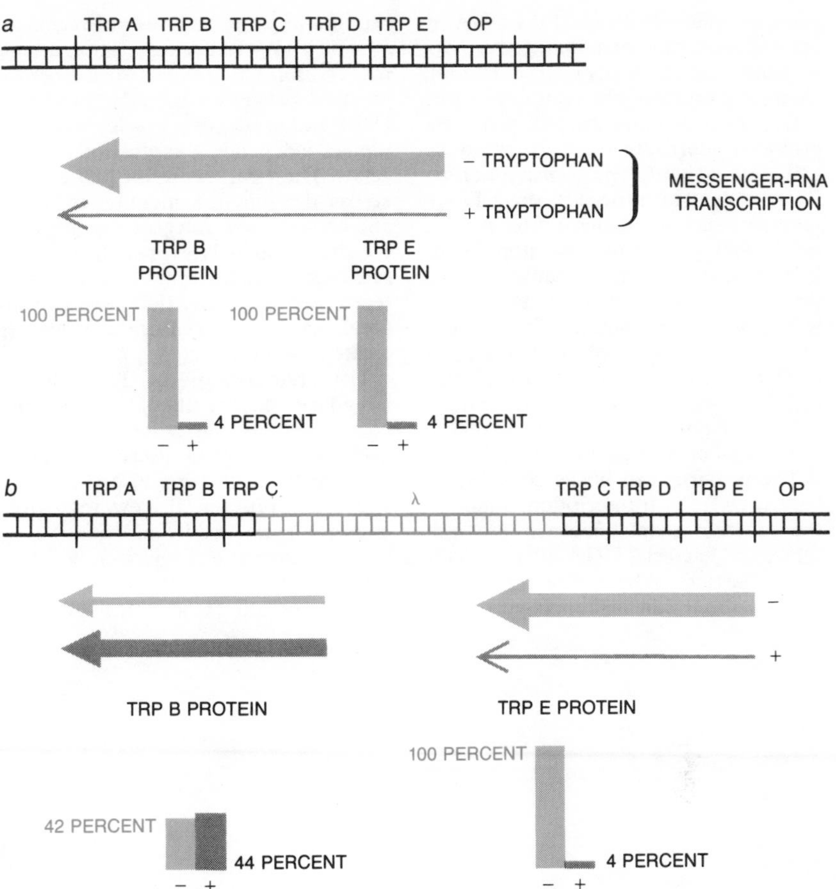

ABNORMAL INSERTION of the phage-lambda chromosome into the *trp* operon (a group of bacterial genes coding for five enzymes in the biosynthetic pathway of the amino acid tryptophan) disrupts the operon's genetic control mechanism. In the normal operon (*a*) transcription of the *trp* genes into messenger RNA is regulated by the tryptophan concentration. When tryptophan levels in the cell are high, transcription is repressed, and when tryptophan levels are low, transcription and synthesis of the *trp* enzymes is high. If phage lambda inserts its DNA into the middle of the *trpC* gene (*b*), the operon as a transcriptional unit will be disrupted. Genes "upstream" from the viral DNA (*trpD* and *trpE*) will continue to be transcribed normally, but the transcript cannot pass through the viral DNA to the genes downstream from the provirus (*trpA* and *trpB*). These genes are expressed to some extent, however, because of transcription arising within the provirus that cannot be repressed by tryptophan. (Protein levels shown are for a mutant of lambda in which the rate of this transcription is abnormally high.)

Why has the virus evolved such a complex and specific mechanism for getting its DNA into and out of chromosomes? The obvious answer is that the ability to do so at appropriate times plays an important role in the virus's survival. Little is known about the selective forces operating on viruses in nature, but one can imagine that it is to the virus's advantage for its DNA to be inserted soon after infection, for the DNA to remain stably inserted while the lysogenic bacterium is growing and for the DNA to be excised while the bacterial genes are repressed. Since insertion and excision have different enzymatic requirements, the virus can control both the direction and the extent of these activities by regulating integrase and excisionase.

The integrase reaction seen in phage lambda and similar viruses is the first case known where two DNA molecules are cut and rejoined at specific sites as part of the normal life cycle of an organism. Enzymatic cleavage and rejoining of DNA molecules in the test tube has become a common pastime of biochemists, but the bacterial restriction enzymes used for this purpose ordinarily function in DNA degradation rather than in genetic recombination. We do not yet know the actual nucleotide sequences recognized and acted on by integrase. The in vivo results require that the viral and bacterial sequences differ from each other, since the genetic requirements for insertion and excision are not the same.

The study of how the phage-lambda DNA inserts itself has provided some useful dividends, among them knowledge of the specific process of the breaking and joining of DNA molecules, which is becoming amenable to biochemical study and opens up new possibilities for the controlled translocation of DNA segments in other organisms. It has also given us the transducing virus variants, which have become workhorses of molecular biologists because they enable one to replicate specific segments of host DNA apart from the rest of the chromosome. In addition understanding the mode of viral DNA insertion has helped to define the ways in which inserted elements and chromosomes can interact functionally. The simplest examples come from instances of abnormal insertion within known genes, such as the insertion of lambda DNA into the *trpC* gene.

The work of Charles Yanofsky of Stanford and others has shown that the *trp* operon consists of five genes, each of which codes for a different enzyme catalyzing a specific step in the biosynthesis of the amino acid tryptophan; the genes are designated *trpA*, *trpB* and so on. This entire stretch of DNA is transcribed into messenger RNA as a unit starting from *trpE* and continuing to *trpA*. The messenger RNA then attaches itself to the ribosomes, the subcellular particles where the enzymes are synthesized. Near the beginning of the transcribed stretch is the specific nucleotide sequence known as an operator. In the presence of high concentrations of tryptophan a repressor protein binds to this sequence and prevents the transcription of the entire DNA segment. All five proteins are hence synthesized together when tryptophan is needed by the cell, but their synthesis is shut off by large amounts of the end product, a feedback control mechanism common to many operons.

How does the insertion of phage-lambda DNA change things? First, since the lambda DNA goes into the middle of the *trpC* gene, the complete protein product of that gene can no longer be formed. A bacterium that carries the abnormally inserted provirus is thus unable to make tryptophan, since one of the enzymes in the biosynthetic pathway is not synthesized. It is possible to recover descendants of this lysogenic bacterium that have lost the provirus. In these bacteria the two halves of the *trpC* gene are rejoined, and active enzyme is again synthesized. None of the DNA of the *trpC* gene has been damaged or permanently lost; it simply cannot code for its normal product when it is cut in two.

Besides disrupting the *trpC* gene, the provirus interrupts the *trp* operon as a functional unit. The transcription of RNA ordinarily proceeds along DNA segments such as the *trp* cluster from a fixed starting point to some stop signal. The precise nature of transcriptional stop signals is not known, but somewhere within the lambda DNA there must be one or more of them. *TrpE* and *trpD* proteins are synthesized normally in these lysogenic bacteria, but the transcription from the *trpE* end never reaches the *trpA* or *trpB* genes.

The provirus can hence constitute a barrier to RNA transcription, although the junction points between provirus DNA and bacterial DNA are not themselves barriers. Whereas bacterial transcription that can be repressed by the product tryptophan does not reach the *trpA* or *trpB* genes, transcription arising from within the provirus does cross the junction between viral DNA and host DNA and produces a low level of tryptophan-independent expression of these genes. The viral transcript that extends across the junction includes only one known gene, the integrase gene of the provirus, which is expressed at a low rate even when the other viral genes are repressed. Thus insertion can not only break up units of transcription but also create new ones.

The transcription of RNA across boundaries in the abnormal lysogenic bacterium illustrates some of the consequences of the viral insertion of DNA both for the regulation of cell function and for evolution. If new regulatory units can be created by insertion, we can be sure that natural selection will then act on them to maximize their selective value to the cell. How extensively DNA of viral origin may have become incorporated into the regulatory systems of existing chromosomes is not known.

There is an old dichotomy between those virologists who view the virus basically as a foreign invader and those who view it more as a cellular component that escapes normal regulatory controls. The argument frequently concerns matters of definition rather than of substance, but it tends to recur at different levels of sophistication as knowledge increases. Lysogenic bacteria have long constituted a prime example for the cellular-component school. At present one can say this much: The conception of the provirus as a normal cellular constituent is at least not a superficial one. The provirus not only behaves like an integral part of the host chromosome; it really is an integral part of the host chromosome. In its manner of attachment there is nothing to distinguish the DNA of the virus from the DNA of the host.

Transposable Genetic Elements

by Stanley N. Cohen and James A. Shapiro
February 1980

They bypass the rules of ordinary genetic recombination and join together segments of DNA that are unrelated, transferring groups of genes among plasmids, viruses and chromosomes in living cells

Natural selection, as Darwin recognized more than a century ago, favors individuals and populations that acquire traits conducive to survival and reproduction. The generation of biological variation, which gives rise to new and potentially advantageous combinations of genetic traits, is therefore a central requirement for the successful evolution of species in diverse and changing environments.

Hereditary information is encoded in the sequence of the building blocks, called nucleotides, that constitute a molecule of DNA, the genetic material. The basic step in the creation of genetic variation is the mutation, or alteration, of the DNA within a gene of a single individual. Mutations involve changes in nucleotide sequence, usually the replacement of one nucleotide by another. This can lead to a change in the chain of amino acids constituting the protein encoded by the gene, and the resulting change in the properties of the protein can influence the organism's biological characteristics. Spontaneous mutations are too rare, however, for genetic variation to depend on new mutations that arise in each generation. Instead variation is generated primarily by the reshuffling of large pools of mutations that have been accumulated within a population in the course of many generations.

In higher organisms this reshuffling is done in the process of sexual reproduction. The genes are arrayed on two sets of chromosomes, one set inherited from the female parent and the other set from the male parent, so that there are two copies of each gene. Sometimes the nucleotides of a genetic sequence differ slightly as a result of earlier mutation, producing alleles, or variant forms of a gene. In the formation of gametes (egg or sperm cells) the breakage of structurally similar pairs of chromosomes can result in the reciprocal exchange of alleles between the two members of a pair of chromosomes. Such genetic recombination requires that the segments of DNA undergoing exchange be homologous, that is, the sequence of nucleotides on one segment of DNA must be very similar to the sequence on the other segment, differing only at the sites where mutations have occurred.

The ability of segments of DNA on different chromosomes to recombine makes it likely that in complex plants or animals the particular collection of genes contained in each egg or sperm cell is different. An individual produces many eggs or sperms, which can potentially interact with sperms or eggs from many other individuals, so that there is a vast opportunity for the generation of genetic diversity within the population. In the absence of intentional and extended inbreeding the possibility that any two plants or animals will have an identical genetic composition is vanishingly small.

Genetic variation is also important in the evolution of lower organisms such as bacteria, and here too it arises from mutations. Bacteria have only one chromosome, however, so that different alleles of a gene are not normally present within a single cell. The reshuffling of bacterial genes therefore ordinarily requires the introduction into a bacterium of DNA carrying an allele that originated in a different cell. One mechanism accomplishing this interbacterial transfer of genes in nature is transduction: certain viruses that can infect bacterial cells pick up fragments of the bacterial DNA and carry the DNA to other cells in the course of a later infection. In another process, known as transformation, DNA released by cell death or other natural processes simply enters a new cell from the environment by penetrating the cell wall and membrane. A third mechanism, conjugation, involves certain of the self-replicating circular segments of DNA called plasmids, which can be transferred between bacterial cells that are in direct physical contact with each other.

Whether the genetic information is introduced into a bacterial cell by transduction, transformation or conjugation, it must be incorporated into the new host's hereditary apparatus if it is to be propagated as part of that apparatus when the cell divides. As in the case of higher organisms, this incorporation is ordinarily accomplished by the exchange of homologous DNA; the entering gene must have an allelic counterpart in the recipient DNA. Because homologous recombination requires overall similarity of the two DNA segments being exchanged, it can take place only between structurally and ancestrally related segments. And so, in bacteria as well as in higher organisms, the generation of genetic variability by this mechanism is limited to what can be attained by exchanges between different alleles of the same genes or between different genes that have stretches of similar nucleotide sequences. This requirement imposes severe constraints on the rate of evolution that can be attained through homologous recombination.

Until recently mutation and homologous recombination nevertheless appeared to be the only important mechanisms for generating biological diversity. They seemed to be able to account for the degree of diversity observed in most species, and the implicit constraints of homologous recombination—which prevent the exchange of genetic information between unrelated organisms lacking extensive DNA-sequence similarity—appeared to be consistent with both a modest rate of biological evolution and the persistence of distinct species that retain their basic identity generation after generation.

Within the past decade or so, however, it has become increasingly apparent that there are various "illegitimate" recombinational processes, which can join together DNA segments having little or no nucleotide-sequence homology, and that such processes play a significant role in the organization of genetic information and the regulation of its expression. Such recombination is often effected by transposable genetic elements: structurally and genetically discrete segments of DNA that have the ability to move around among the chromosomes and the extrachromosomal DNA molecules of bacteria and higher organisms. Although transposable elements have been studied largely in bacterial cells, they were originally discovered in plants and are now known to exist in animals as well. Because illegitimate recombination can join together DNA segments that have little, if any, ancestral relationship, it can affect evolution in quantum jumps as well as in small steps.

In the late 1940's Barbara McClintock of the Carnegie Institution of Washington's Department of Genetics at Cold Spring Harbor, N.Y., first reported a genetic phenomenon in the common corn plant, *Zea mays,* that would later be found to have parallels in other biological systems. While studying the inheritance of color and the distribution of pigmentation in plants that had undergone repeated cycles of chromosome breakage she found that the activity of particular genes was being turned on or off at abnormal times. Because some of these genes were associated with the development of pigments in kernels as well as in the plant itself, certain kernels were mottled, showing patches of pigmentation against an otherwise colorless background. The patterns of this variegation were reproduced in successive generations and could be analyzed like other heritable traits. After painstaking study of many generations of corn plants McClintock concluded that the variegation she observed was the result of the action of distinct genetic units, which she called controlling elements, that could apparently move from site to site on different maize chromosomes; as they did so they sometimes served as novel biological switches, turning the expression of genes on or off.

McClintock's genetic analysis showed that some patterns of variegation affected three or more genes simultaneously, suggesting that the structure of one of the plant's chromosomes had been rearranged at the site of a controlling element. Direct microscopic examination of maize chromosomes containing controlling elements confirmed that these genetic elements did in fact serve as specific sites for the breakage and resealing of DNA, thereby giving rise to either minute or gross changes in chromosome structure.

Almost 20 years after McClintock reported her earliest studies on controlling elements in the corn plant Michael Malamy, who is now at the Tufts University School of Medicine, Elke Jordan, Heinz Saedler and Peter Starlinger of the University of Cologne and one of us (Shapiro), who was then at the University of Cambridge, found a new class of mutations in genes of a laboratory strain of the common intestinal bacterium *Escherichia coli.* They were unusual in that their effects were detectable beyond the borders of the mutated genes them-

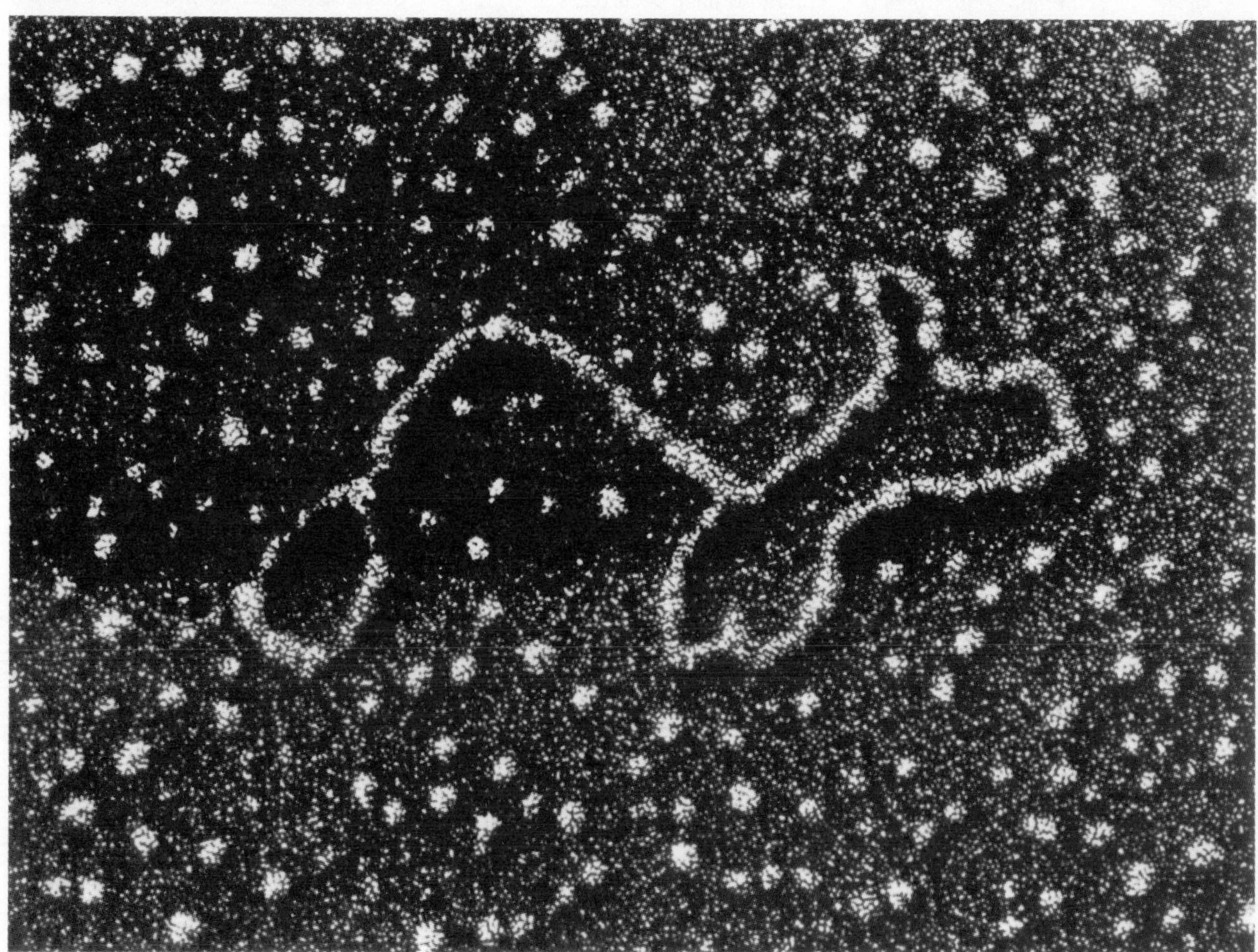

DNA OF TRANSPOSABLE GENETIC ELEMENT (a transposon) forms a characteristic stem-and-loop structure, which is seen here in an electron micrograph made by one of the authors (Cohen). The structure results from the "inverted repeat" nature of the nucleotide sequences at the two ends of the transposon DNA (*see upper illustration on page 314*). The double-strand DNA of the plasmid *pSC*105, into which the transposon had been inserted, was denatured and complementary nucleotide sequences on each strand were allowed to "reanneal." The joining of the complementary nucleotides constituting the transposon's inverted-repeat termini formed the double-strand stem. The smaller loop was formed by the segment of single-strand transposon DNA between the inverted repeats, a segment that includes a gene conferring resistance to the antibiotic kanamycin. The larger loop represents the single-strand DNA of a miniplasmid derivative of the host plasmid. DNA was spread with formamide and shadowed with platinum-palladium. Enlargement is 230,000 diameters.

selves; this property could not be explained by any known mutational mechanism.

When the DNA segments carrying these mutations were inserted into particles of a bacterial virus and the density of the virus was compared with that of viruses carrying normal genes, it became clear that the mutated DNA was longer than the normal DNA: the mutations had been caused by the insertion of sizable DNA fragments into the mutated gene. It further developed that a limited number of other kinds of distinguishable DNA segments, which were up to 2,000 nucleotides in length, could also insert themselves within many different genes, interrupting the continuity of the gene and turning off its activity. These elements were named insertion sequences, or IS elements. The observation that a small number of specific DNA segments could be inserted at a large number of different sites in the bacterial chromosome suggested that some type of nonhomologous recombination was taking place; it seemed to be unlikely that an IS element could be homologous with the nucleotide sequences at so many different insertion sites.

At about the same time that IS elements were discovered other microbiologists and geneticists made observations hinting that certain genes known to be responsible for resistance to antibiotics by bacteria were capable of transfer from one molecule of DNA to another. Results obtained by Susumu Mitsuhashi and his colleagues at the University of Tokyo in the mid-1960's suggested that a gene encoding a protein that inactivates the antibiotic chloramphenicol could move from its normal site on a plasmid-DNA molecule to the chromosome of a bacterium or to the DNA of a virus.

Similar instances of the apparent transfer of antibiotic-resistance genes between different DNA molecules in the same cell were reported from the U.S. and Britain. The first direct evidence that such transfer is by a process analogous to the insertion of IS elements was published in 1974. R. W. Hedges and A. E. Jacob of the Hammersmith Hospital in London found that the transfer from one plasmid to another of a gene conferring resistance to antibiotics such as penicillin and ampicillin was always accompanied by an increase in the size of the recipient plasmid; the recipient could donate the resistance trait to still other plasmids, which thereupon showed a similar increase in size.

Hedges and Jacob postulated that the gene for ampicillin resistance was carried by a DNA element that could be "transposed," or could move from one molecule to another, and they called such an element a transposon. Their discovery of a transposable element that carries an antibiotic-resistance gene was an important advance. In earlier studies the movement of IS elements had been tracked only indirectly by genetic techniques: by observing the effects of insertions on various genetic properties of the host organism. It now became possible to track a transposable element's intermolecular travels directly by observing the inheritance of the antibiotic-resistance trait.

DIFFERENCES IN PIGMENTATION in kernels of the corn plant *Zea mays* reflect the action of a two-element control system discovered by Barbara McClintock of the Cold Spring Harbor Laboratory. Both elements are transposable. One element is at the locus of a gene whose action it modulates to yield the faintly and homogeneously pigmented kernels. The other element acts on the first one to produce the variegated pattern that is seen in many of the kernels.

While the Hedges and Jacob experiments were being carried out Dennis J. Kopecko and one of us (Cohen), at the Stanford University School of Medicine, were studying the acquisition of a gene for resistance to ampicillin by still other plasmids. It emerged, as Hedges and Jacob had found, that the ampicillin-resistance trait present on one plasmid could be acquired by another plasmid. Surprisingly, however, it also developed that such transfer could take place in mutated bacteria lacking a particular protein, the product of a gene designated *recA*, known to be necessary for homologous recombination. Examination of the plasmid DNA with the electron microscope revealed that a 4,800-nucleotide segment carrying the ampicillin-resistance trait was being transferred as a characteristic and discrete structural unit. Moreover, the segment could become inserted at many different sites on the recipient plasmid DNA.

Electron microscopy also showed that the two ends of the transposable DNA segment had a unique feature: they consisted of nucleotide sequences that were complementary to each other but in the reverse order. This finding calls for some explanation. The four nitrogenous bases that characterize DNA nucleotides are linked in complementary pairs by hydrogen bonds to form the double helix of DNA: adenine (*A*) is linked to thymine (*T*) and guanine (*G*) to cytosine (*C*). The nucleotide sequence *AGCTT*, for example, is complementary

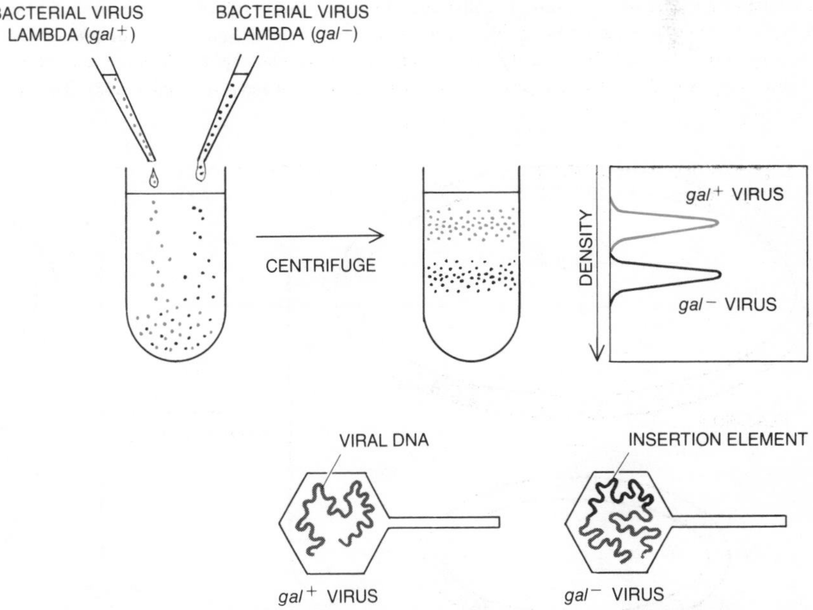

MUTATION BY INSERTION was demonstrated by one of the authors (Shapiro) with phage-lambda particles carrying the bacterial gene for galactose utilization (*gal*$^+$) and particles carrying the mutant gene *gal*$^-$. The viruses were centrifuged in a cesium chloride solution. The *gal*$^-$ particles were found to be the denser. Because the virus particles all have the same volume and their outer shells all have the same mass, increased density of *gal*$^-$ particles showed they must contain a larger DNA molecule: *gal*$^-$ mutation was caused by insertion of DNA.

to the sequence *TCGAA*. The nucleotide sequence at one end of the transposable DNA segment was complementary in reverse order to a sequence on the same strand at the other end of the element [*see upper illustration on this page*]. These "inverted repeats" were revealed when the two strands of the double-strand plasmid DNA carrying the transposon were separated in the laboratory and each of the strands was allowed to "reanneal" with itself: a characteristic stem-and-loop structure was formed by the complementary inverted repeats.

The result of the transposition process is that a segment of DNA originally present on one molecule is transferred to a different molecule that has no genetic homology with the transposable element or with the donor DNA. The fact that the process does not require a bacterial gene product known to be necessary for homologous recombination indicates that transposition is accomplished by a mechanism different from the usual recombinational processes.

Subsequent experiments done in numerous laboratories have shown that DNA segments carrying genes encoding a wide variety of antibiotic-resistance traits can be transferred between DNA molecules as discrete units. Moreover, genes encoding other traits, such as resistance to toxic mercury compounds, synthesis of bacterial toxins and the capacity to ferment sugars or metabolize hydrocarbons, have been shown to be capable of transposition. All the transposons studied so far have ends consisting of inverted-repeat sequences, which range in length from only a few nucleotides to as many as 1,400. The ends of at least two transposons actually consist of two copies of the insertion sequence IS1 (which itself has been found to have terminal inverted-repeat sequences). Recent evidence has suggested that the insertion of any gene between two trans-

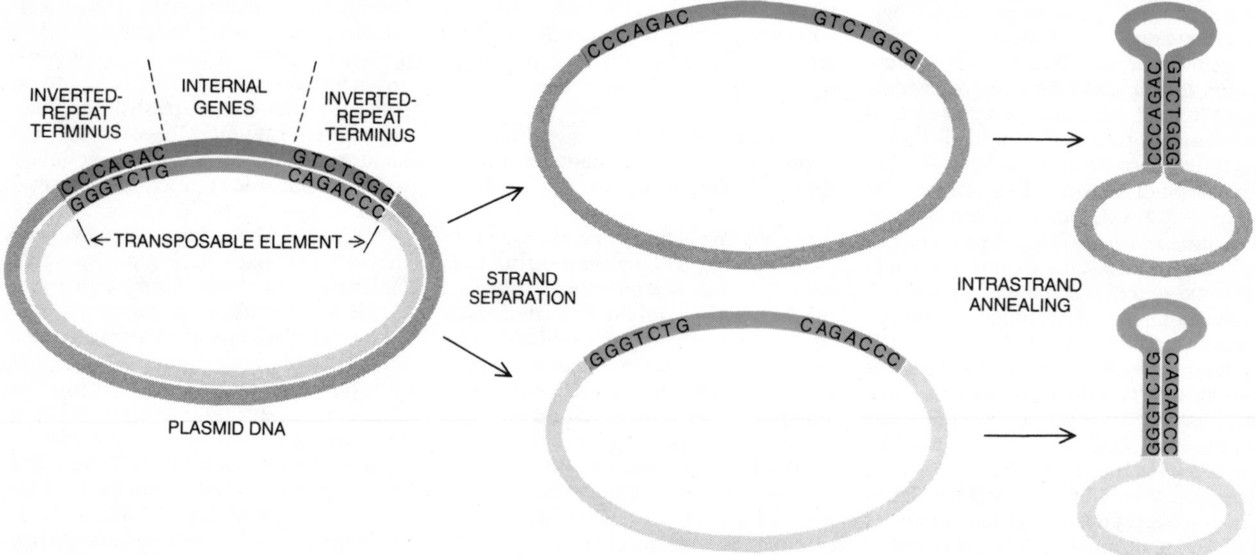

STEM-AND-LOOP STRUCTURES demonstrate the inverted-repeat nucleotide sequences of the ends of transposable elements. The four bases adenine (*A*), guanine (*G*), thymine (*T*) and cytosine (*C*) of DNA's four nucleotide building blocks are linked to form a helix (shown here schematically as a double strand); *A* always pairs with *T* and *G* pairs with *C*. The termini of a transposable element have sequences (seven nucleotides long here) that are bidirectionally and rotationally symmetrical. When the two strands of a plasmid containing an element are separated and each strand is allowed to self-anneal, the complementary nucleotides at the termini pair with each other, forming a double-strand stem (*right and in electron micrograph on page 312*). The remainder of the DNA is seen as single-strand loops.

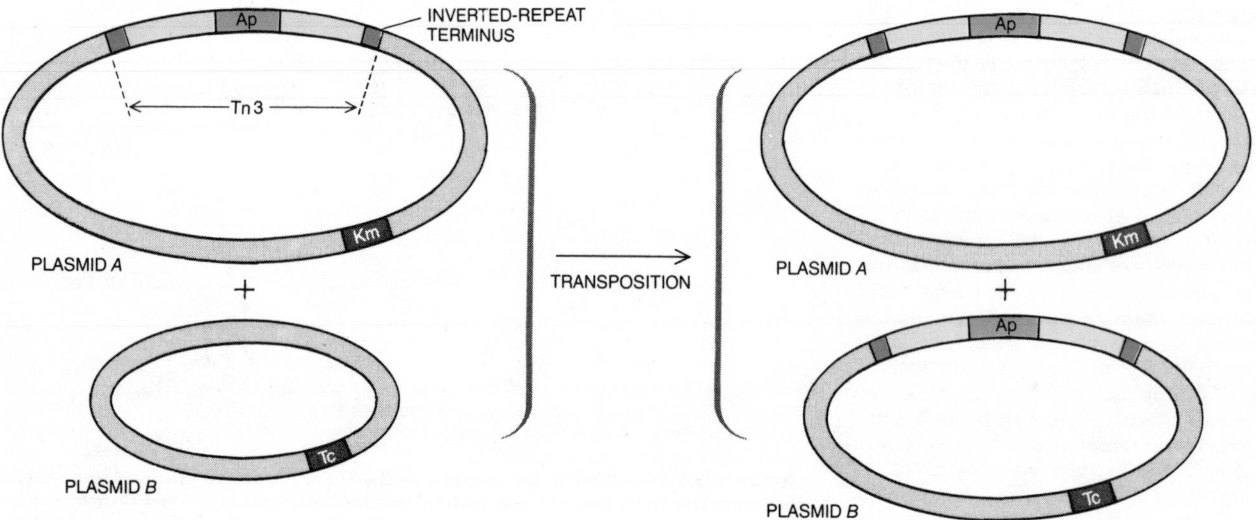

TRANSPOSITION of the transposon Tn3, which carries a gene conferring resistance to the antibiotic ampicillin (*Ap*), is diagrammed. It is shown as originally being part of plasmid *A*, which also includes a gene for resistance to kanamycin (*Km*). A plasmid *B*, which confers resistance to tetracycline (*Tc*), acquires a copy of the transposon. The new plasmid *B* confers resistance to ampicillin and tetracycline.

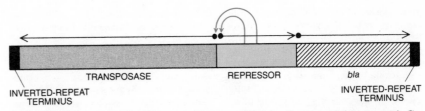

FUNCTIONAL COMPONENTS of the transposon Tn3 are diagrammed (not to scale). Genetic analysis shows there are at least four kinds of regions: the inverted-repeat termini; a gene for the enzyme beta-lactamase (*bla*), which confers resistance to ampicillin and related antibiotics; a gene encoding an enzyme required for transposition (a transposase), and a gene for a repressor protein that controls the transcription of the genes for transposase and for the repressor itself. The arrows indicate the direction in which DNA of various regions is transcribed.

posable elements makes possible the transfer of the gene to a structurally unrelated DNA molecule by nonhomologous recombination.

Since transposable elements are transferred as discrete and characteristic genetic units, there must be some highly specific enzymatic mechanism capable of recognizing their inverted-repeat ends and cleaving DNA precisely at these locations. The first evidence that genes carried by the transposable elements themselves can encode such enzymes came from a series of experiments carried out by Frederick L. Heffron, Craig Rubens and Stanley Falkow at the University of Washington and continued by Heffron and his colleagues at the University of California at San Francisco. When these investigators introduced mutations that interrupted the continuity of genes at various locations within the ampicillin-resistance transposon designated Tn3, they found alterations in the ability of the element to function as a transposon. Mutation in the inverted-repeat ends or in a particular region of the DNA segment between the ends prevented transposition. On the other hand, mutations within another region of Tn3 actually increased the frequency of movement of Tn3 between the different plasmids, suggesting that this region might contain a gene modulating the ability of Tn3 to undergo transposition.

Recently published work by Joany Chou, Peggy G. Lemaux and Malcolm J. Casadaban in the laboratory of one of us (Cohen) and by Ronald Gill in Falkow's laboratory has shown that the Tn3 transposon does in fact encode both a "transposase"—an enzyme required for transposition—and a repressor substance that regulates both the transcription into RNA of the transposase gene and the repressor's own synthesis. Analogous experiments at the University of Chicago, the University of Wisconsin and Harvard University have shown that other transposable elements also encode proteins needed for their own transposition.

Even though transposons can insert themselves at multiple sites within a recipient DNA molecule, their insertion is not random. It has been recognized for several years that certain regions of DNA are "hot spots" prone to multiple insertions of transposons. Experiments recently reported by David Tu and one of us (Cohen) have shown that Tn3 is inserted preferentially in the vicinity of nucleotide sequences similar to sequences within its inverted-repeat ends, even in a bacterial cell that does not make the *rec*A protein required for ordinary homologous recombination. It therefore appears that recognition of homologous DNA sequences may play some role in determining the frequency and site-specificity of transposon-associated recombination, even though the actual recombinational mechanism differs from the one commonly associated with the exchange of homologous segments of DNA.

The discovery of the process of transposition explains a puzzling phenomenon in bacterial evolution that has serious implications for public health: the rapid spread of antibiotic resistance among bacteria. Under the selective pressure of extensive administration of antibiotics in human and veterinary medicine and their use as a supplement in animal feeds, bacteria carrying resistance genes have a great natural advantage. For some time it has been known that resistance to several different antibiotics can be transmitted simultaneously to a new bacterial cell by a plasmid, but until transposition was discovered it was not known how a number of genes conferring resistance to different antibiotics were accumulated on a single plasmid-DNA molecule. The explanation seems to be that the resistance-determinant segments of drug-resistance plasmids have evolved as collections of transposons, each carrying a gene that confers resistance to one antibiotic or to several of them.

Work carried out at Stanford and by Phillip A. Sharp and others in the laboratory of Norman R. Davidson at the California Institute of Technology has made it clear that certain bacterial plasmids are constructed in a modular fashion. Plasmids isolated in different parts of the world show extensive sequence homology in certain of their DNA segments, whereas in other segments there is no structural similarity at all. In some instances plasmids can dissociate reversibly at specific sites. Transposable IS elements are found both at these sites and at sites where the plasmid interacts with chromosomal DNA to promote the transfer of chromosomes between different bacterial cells.

Identical transposons are commonly found in bacterial species that exchange genes with one another in nature. In addition antibiotic-resistance transposons

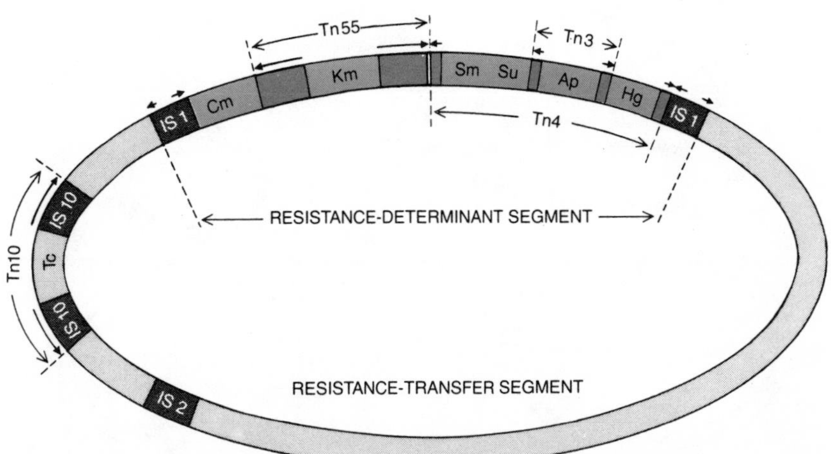

ROLE OF TRANSPOSABLE ELEMENTS in the evolution of antibiotic-resistance plasmids is illustrated by a schematic map of a plasmid carrying many resistance genes. The plasmid appears to have been formed by the joining of a resistance-determinant segment and a resistance-transfer segment; there are insertion elements (*IS1*) at the junctions, where the two segments sometimes dissociate reversibly. Genes encoding resistance to the antibiotics chloramphenicol (*Cm*), kanamycin (*Km*), streptomycin (*Sm*), sulfonamide (*Su*) and ampicillin (*Ap*) and to mercury (*Hg*) are clustered on the resistance-determinant segment, which consists of multiple transposable elements; inverted-repeat termini are designated by arrows pointing outward from the element. A transposon encoding resistance to tetracycline (*Tc*) is on the resistance-transfer segment. Transposon Tn3 is within Tn4. Each transposon can be transferred independently.

appear to be able to move among very different bacterial species that have not previously been known to exchange genes. For example, DNA sequences identical with part of Tn3 have recently been found to be responsible for penicillin resistance in two bacterial species unrelated to those commonly harboring Tn3 and in which such resistance had not previously been observed. Transposable elements seem, in other words, to accomplish in nature gene manipulations akin to the laboratory manipulations that have been called genetic engineering.

The effects of transposable genetic elements extend beyond their ability to join together unrelated DNA segments and move genes around among such segments. These elements can also promote both the rearrangement of genetic information on chromosomes and the deletion of genetic material. An awareness of these effects has emerged most clearly from studies of a peculiar phage, or bacterial virus, discovered in 1963 by Austin L. Taylor of the University of Colorado. Like other "temperate" bacterial viruses Taylor's phage could insert its DNA into a bacterial chromosome, creating a latent "prophage" that coexists with the bacterial cell and is transmitted to the bacterial progeny when the cell divides. Unlike other temperate phages, however, this one could become inserted at multiple sites within the chromosome, thereby causing many different kinds of mutation in the host bacterium. Because of this property Taylor called his phage Mu, for "mutator."

Further studies have shown that Mu is actually a transposable element that can also exist as an infectious virus. In the virus particle the Mu DNA is sandwiched between two short segments of bacterial DNA it has picked up from a bacterial chromosome. When the Mu virus infects a new cell, it sheds the old bacterial DNA and is transposed to a site in the new host chromosome. Ahmad I. Bukhari and his colleagues at the Cold Spring Harbor Laboratory have shown that Mu's ability to replicate is closely associated with its ability to be transposed; the virus has apparently evolved in such a way that its life span is dependent on transposition events.

While the structure of Mu's DNA and the details of the phage's life cycle were being unraveled Michel Faelen and Arianne Toussaint of the University of Brussels were doing genetic experiments aimed at understanding how the Mu DNA interacts with other DNA in a bacterial cell. The results of experiments carried out over a period of almost 10 years have demonstrated that Mu can catalyze a remarkable series of chromosome rearrangements. These include the fusion of two separate and independently replicating DNA molecules ("replicons"), the transposition of segments of the bacterial chromosome to plasmids, the deletion of DNA and the inversion of segments of the chromosome. Significantly, all these rearrangements seem to involve the nucleotide sequences at the ends of Mu DNA and to require the expression of a Mu gene that had been found earlier to be necessary both for transposition and for virus replication. Experiments done by Hans-Jorg Reif and Saedler at the University of Freiburg and by other groups have shown that many other transposable elements can, like Mu, promote the deletion of DNA; Nancy E. Kleckner and David Botstein and their associates at Harvard and at the Massachusetts Institute of Technology have shown that such elements can also bring about the inversion of DNA sequences. Indeed, there is evidence that some transposable elements may participate in specific re-

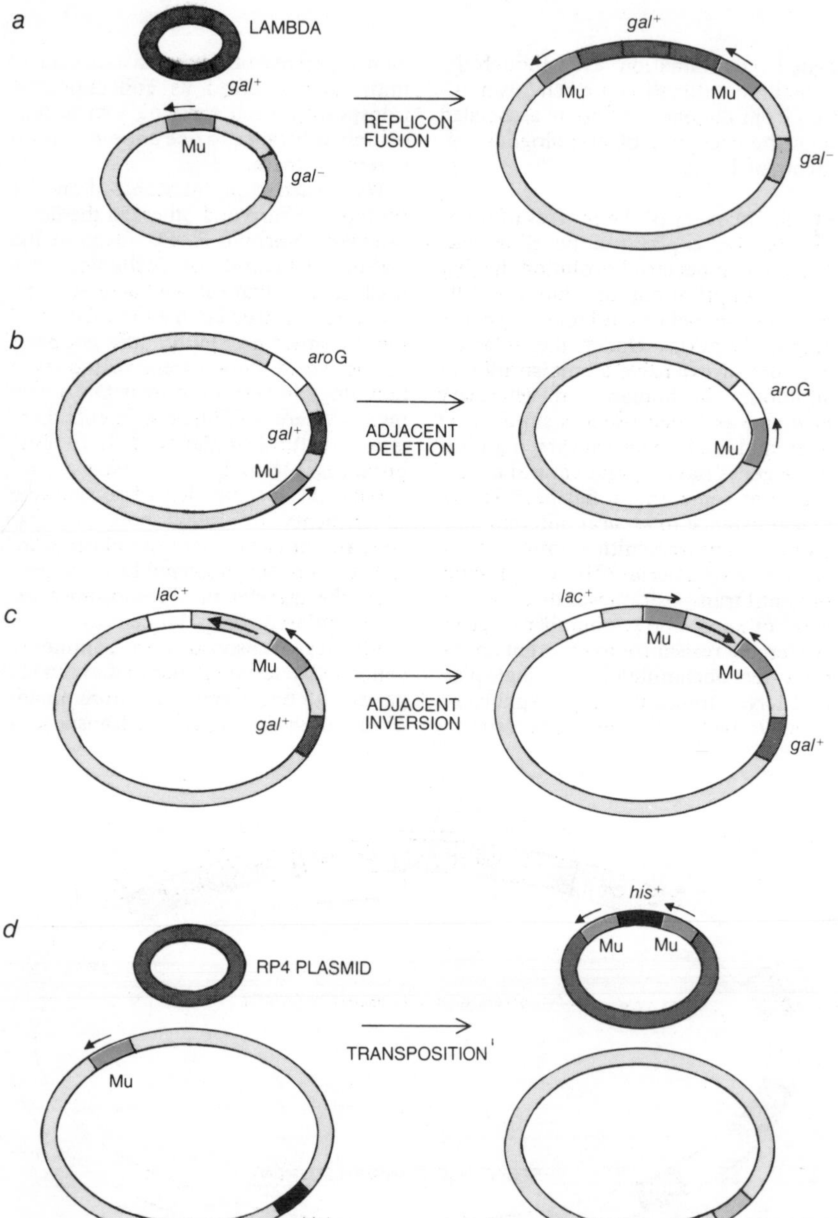

CHROMOSOME REARRANGEMENTS mediated by the bacterial virus Mu include replicon fusion, adjacent deletion, adjacent inversion and transposition of chromosome segments to a plasmid. Mu is shown in color; a small arrow gives its orientation. In a cell lysogenic for Mu (having Mu DNA integrated in its chromosome) and containing DNA of a lambda *gal*$^+$ virus (*a*) the viral DNA becomes integrated into the chromosome between two copies of Mu. In a lysogenic cell in which Mu is near integrated *gal*$^+$ genes (*b*) the *gal*$^+$ genes are deleted. In a lysogenic male bacterium (*c*) with Mu near the origin of chromosome transfer (*large arrow*) the origin becomes inverted between two oppositely oriented copies of Mu. In a lysogenic bacterium carrying a plasmid (*d*) a bacterial *his*$^+$ gene is transposed to plasmid between copies of Mu.

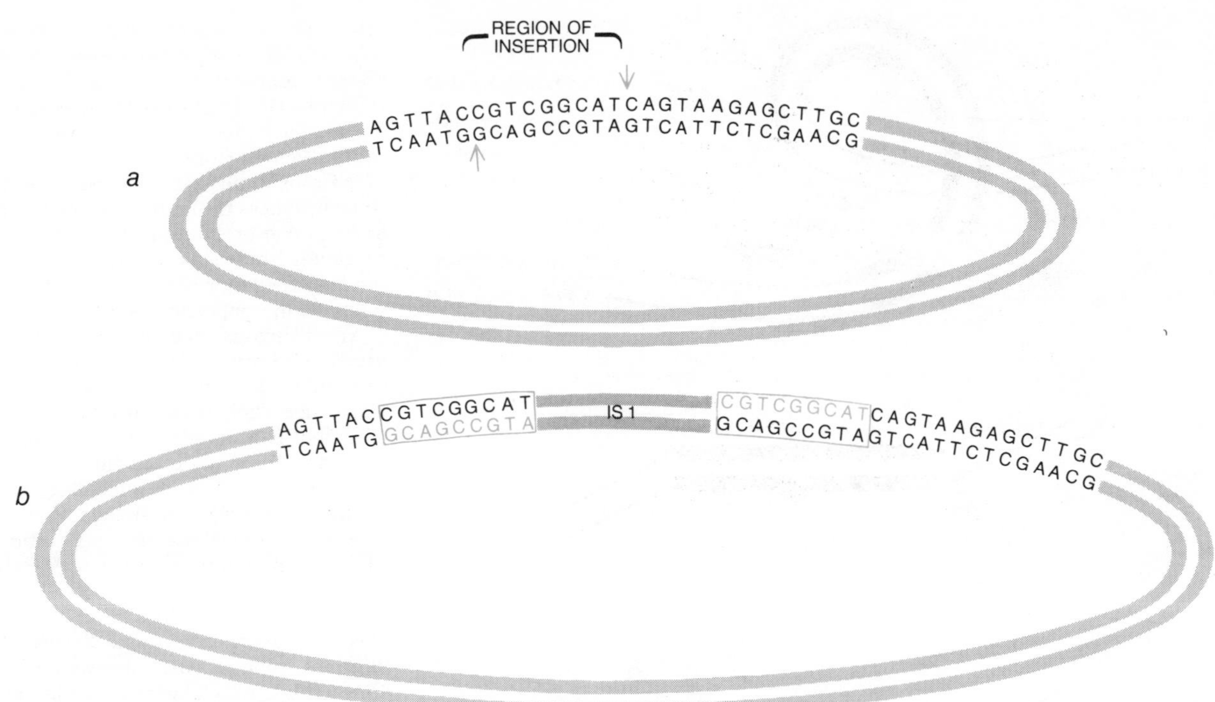

DUPLICATION of five, nine or 11 pairs of nucleotides in the recipient DNA is associated with the insertion of a transposable element; the two copies bracket the inserted element. Here the duplication that attends the insertion of IS1 is illustrated in a way that indicates how the duplication may come about. IS1 insertion causes a nine-nucleotide duplication. If the two strands of the recipient DNA are cleaved (*colored arrows*) at staggered sites that are nine nucleotides apart (*a*), then the subsequent filling in of single strands on each side of the newly inserted element (*b*) with the right complementary nucleotides (*color*) could account for the duplicated sequences (*colored boxes*).

arrangements of DNA more frequently than they do in transposition events.

New methods for determining DNA nucleotide sequences rapidly and simply have provided an important tool for elucidating the structure of transposable elements as well as the biochemical mechanisms involved in transposition and in chromosome rearrangements. The sequence of a transposable element (IS1) has been determined in its entirety by Hisako Ohtsubo and Eiichi Ohtsubo of the State University of New York at Stony Brook. An important insight into the mechanism of transposition has resulted from DNA-sequence observations initially made by Nigel Grindley of Yale University and the University of Pittsburgh and by Michele Calos and Lorraine Johnsrud of Harvard, working with Jeffrey Miller of the University of Geneva. Both groups examined the DNA sequences at the sites of several independently occurring insertions of the IS1 element. They found that the insertion of IS1 results in the duplication of a sequence of nine nucleotide pairs in the recipient DNA. The duplicated sequences bracket the insertion element and are immediately adjacent to its inverted-repeat ends. Since the sequence of the recipient DNA was different at each of the various insertion sites studied, different nucleotides were duplicated for each insertion.

Subsequent reports from many laboratories have shown that similar duplications of a short DNA sequence result from the insertion of other transposable elements. Some elements generate nine-nucleotide duplications and others generate duplications five or 11 nucleotides long. As Calos and her colleagues and Grindley have pointed out, these observations suggest that a step in the insertion process involves staggered cleavage (at positions five, nine or 11 nucleotides apart) of opposite DNA strands at the target site for transposition. The filling in of the single-strand segments following such cleavage would require the synthesis of short single-strand stretches of complementary DNA and would result in the nucleotide-sequence duplication. Faelen and Toussaint had also concluded that DNA synthesis is required in the generation of chromosome rearrangements by Mu: they had noted that the rearranged bacterial chromosome often included two copies of the prophage, the inserted form of Mu.

On the basis of these observations one of us (Shapiro) has proposed a model to explain transposition, chromosome rearrangements and the replication of transposable elements such as phage Mu as variations of a single biochemical pathway. The pathway is such that transposable elements can serve two functions in the structural reorganization of cellular DNA: they specifically duplicate themselves while remaining inserted in the bacterial chromosome, and they bring together unrelated chromosomal-DNA segments to form a variety of structural rearrangements, including fusions, deletions, inversions and transpositions.

If this model is at all close to reality, then the nonhomologous recombination events associated with transposable elements are rather different from other types of illegitimate recombination, such as the integration of the phage-lambda DNA into the bacterial chromosome, that do not involve DNA synthesis. It seems likely that bacterial cells will turn out to have several different systems for carrying out nonhomologous recombination, just as they have multiple pathways for homologous recombination.

The potential for multiple mechanisms of illegitimate recombination is important to bear in mind when comparing phenomena that appear to be similar in bacteria and higher cells. Transposition phenomena that are analogous genetically may not be similar biochemically. There is some genetic evidence indicating that the movement of controlling elements in maize from one chromosomal site to another may be brought about by a mechanism different from that of transposition in bacteria.

Genetic rearrangements can have biological importance on two time scales: on an evolutionary scale, where the effects of the rearrangement are seen

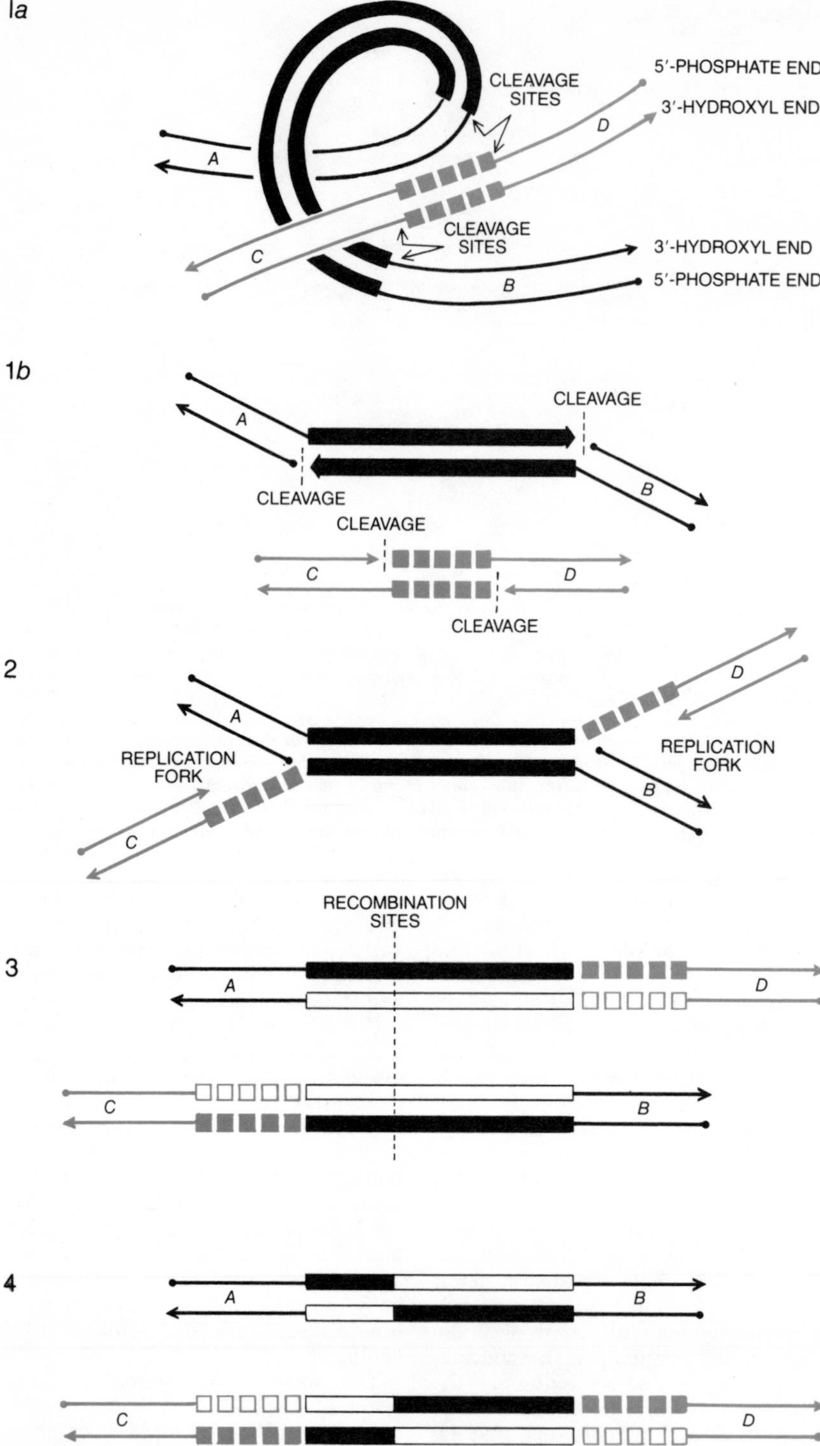

POSSIBLE MOLECULAR PATHWAY is suggested to explain transposition and chromosome rearrangements. The donor DNA, including the transposon (*thick bars*), is in black, the recipient DNA in color. Arrowheads indicate the 3′-hydroxyl ends of DNA chains, dots the 5′-phosphate ends; the letters *A, B, C* and *D* identify segments of the two DNA molecules. The pathway has four steps, beginning with single-strand cleavage (*1a*) at each end of the transposable element and at each end of the "target" nucleotide sequence (*colored squares*) that will be duplicated. The cleavages expose (*1b*) the chemical groups involved in the next step: the joining of DNA strands from donor and recipient molecules in such a way that the double-strand transposable element has a DNA-replication fork at each end (*2*). DNA synthesis (*3*) replicates the transposon (*open bars*) and the target sequence (*open squares*), accounting for the observed duplication. This step forms two new complete double-strand molecules; each copy of the transposable element joins a segment of the donor molecule and a segment of the recipient molecule. (The copies of the element serve as linkers for the recombination of two unrelated DNA molecules.) In the final step (*4*) reciprocal recombination between copies of the transposable element inserts the element at a new genetic site and regenerates the donor molecule. The mechanism of this recombination is not known; it does not require proteins needed for homologous recombination, and at least in Tn3 it is mediated by sequences within element.

after many generations, and on a developmental time scale, where the effects are apparent within a single generation. It is known that transposable genetic elements can serve as biological switches, turning genes on or off as a consequence of their insertion at specific locations. In some instances the insertion of an IS element in one orientation turns off nearby genes, whereas an unexpressed gene can be turned on when the element is inserted in the opposite orientation.

An analogous regulation of gene expression through chromosome rearrangement is "phase variation," which is seen in certain disease-producing bacteria that can invade the gastrointestinal tract. The phase, or immunological specificity, of a hairlike flagellum on these bacteria can change suddenly within a single bacterial generation. Melvin Simon of the University of California at San Diego and his colleagues have recently shown that the choice between the expression of one *Salmonella* flagellum gene and the expression of its counterpart, which specifies a different phase, is controlled by the inversion of a particular segment of the bacterial chromosome. The inversion takes place in the absence of proteins needed for homologous recombination, and so it appears to depend on recombination enzymes that recognize the ends of the invertible segment. Whether the switching mechanism responsible for phase variation operates by a molecular process similar to transposition remains to be determined, but the process clearly falls within the category of recombination events that were considered "illegitimate" a few years ago.

Although molecular studies on transposable elements have so far been carried out primarily in bacteria, there has been extensive genetic evidence for the existence of similar elements in higher organisms for years. The pioneering work of Barbara McClintock not only established the existence of transposable genetic elements in the corn plant but also showed by genetic analysis that the movement of a controlling element from one site to another in the maize chromosome depends on the action of genes on certain of the elements themselves, genes presumably analogous to those encoding the transposases of Tn elements and of phage Mu. McClintock also showed that some controlling elements (called regulators) regulate the expression of distant genes carrying insertions of other controlling elements (called receptors). Groups of genes are expressed synchronously at specific times during plant development, and McClintock suggested that the transposition of receptor elements could provide a mechanism for the rapid evolution of control mechanisms in situations in which several genes must be switched on or off at the same time, as they are in the course of development.

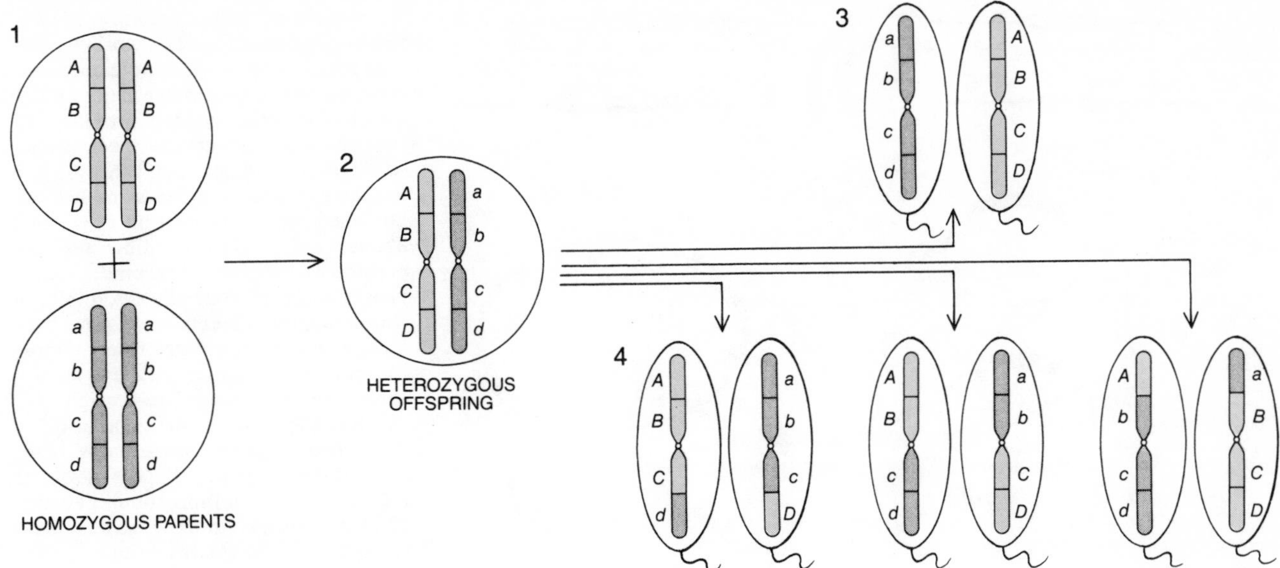

HOMOLOGOUS RECOMBINATION is accomplished in higher organisms by the "crossing over" of structurally similar chromosome segments during sexual reproduction. Here the process is shown for a hypothetical animal each of whose somatic (body) cells has a single chromosome pair carrying four genes, each of which may be present in either of two variant forms (alleles). Homozygous parents having the same set of alleles on both paired chromosomes (*1*) give rise to heterozygous offspring (*2*), which in turn can produce gametes (sperms or eggs) containing copies of the original chromosomes (*3*). As a result of crossing over and reciprocal homologous recombination, alleles can be reshuffled in various ways (*4*), producing gametes containing chromosomes that are different from either of original chromosomes.

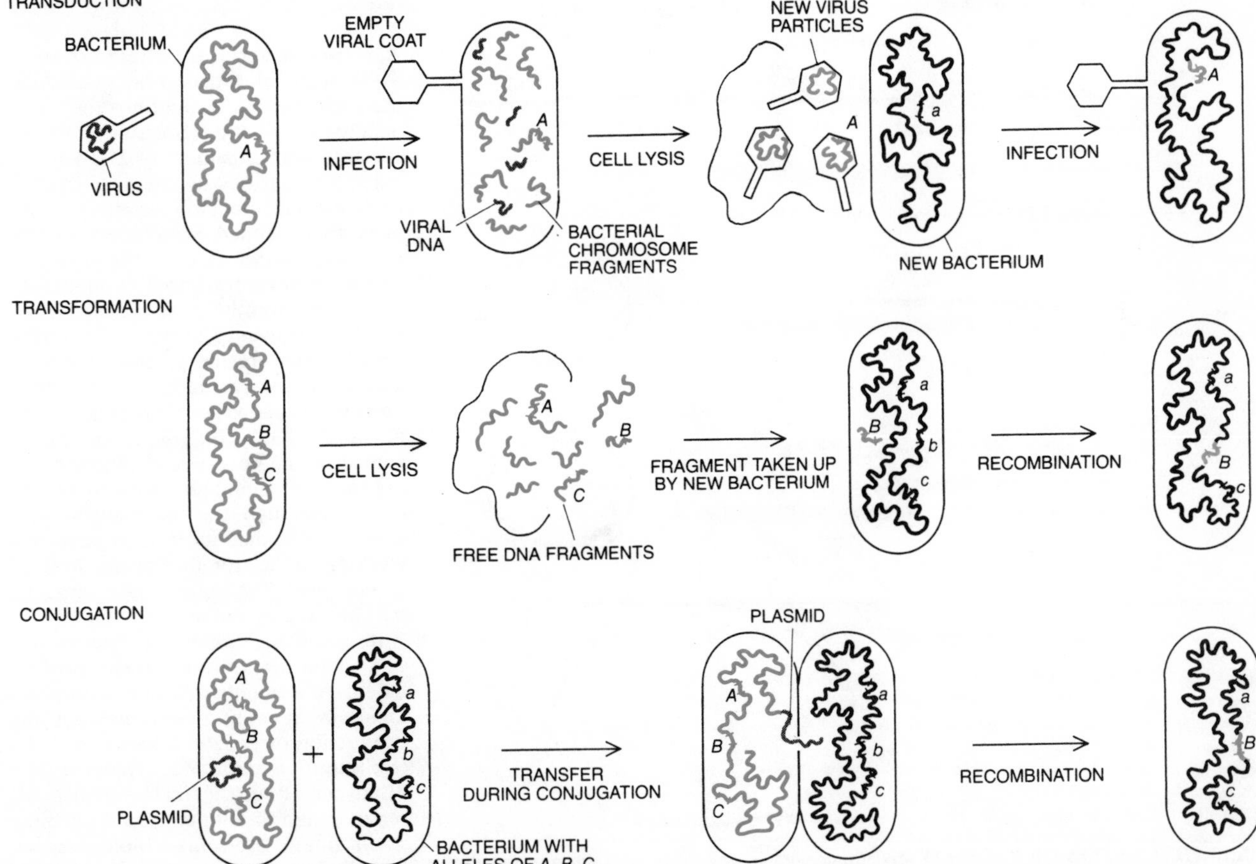

RECOMBINATION IN BACTERIA requires the introduction into a bacterial cell of an allele obtained from another cell. In transduction an infecting phage, or bacterial virus, picks up a bacterial-DNA segment carrying allele *A* and incorporates it instead of viral DNA into the virus particle. When such a particle infects another cell, the bacterial-DNA segment recombines with a homologous segment, thereby exchanging allele *A* for allele *a*. In transformation a DNA segment bearing allele *B* is taken up from the environment by a cell whose chromosome carries allele *b;* the alleles are exchanged by homologous recombination. In conjugation a plasmid inhabiting one bacterial cell can transfer the bacterium's chromosome, during cell-to-cell contact, to another cell whose chromosome carries alleles of genes on the transferred chromosome; again allele *B* is exchanged for allele *b* by recombination between homologous DNA segments.

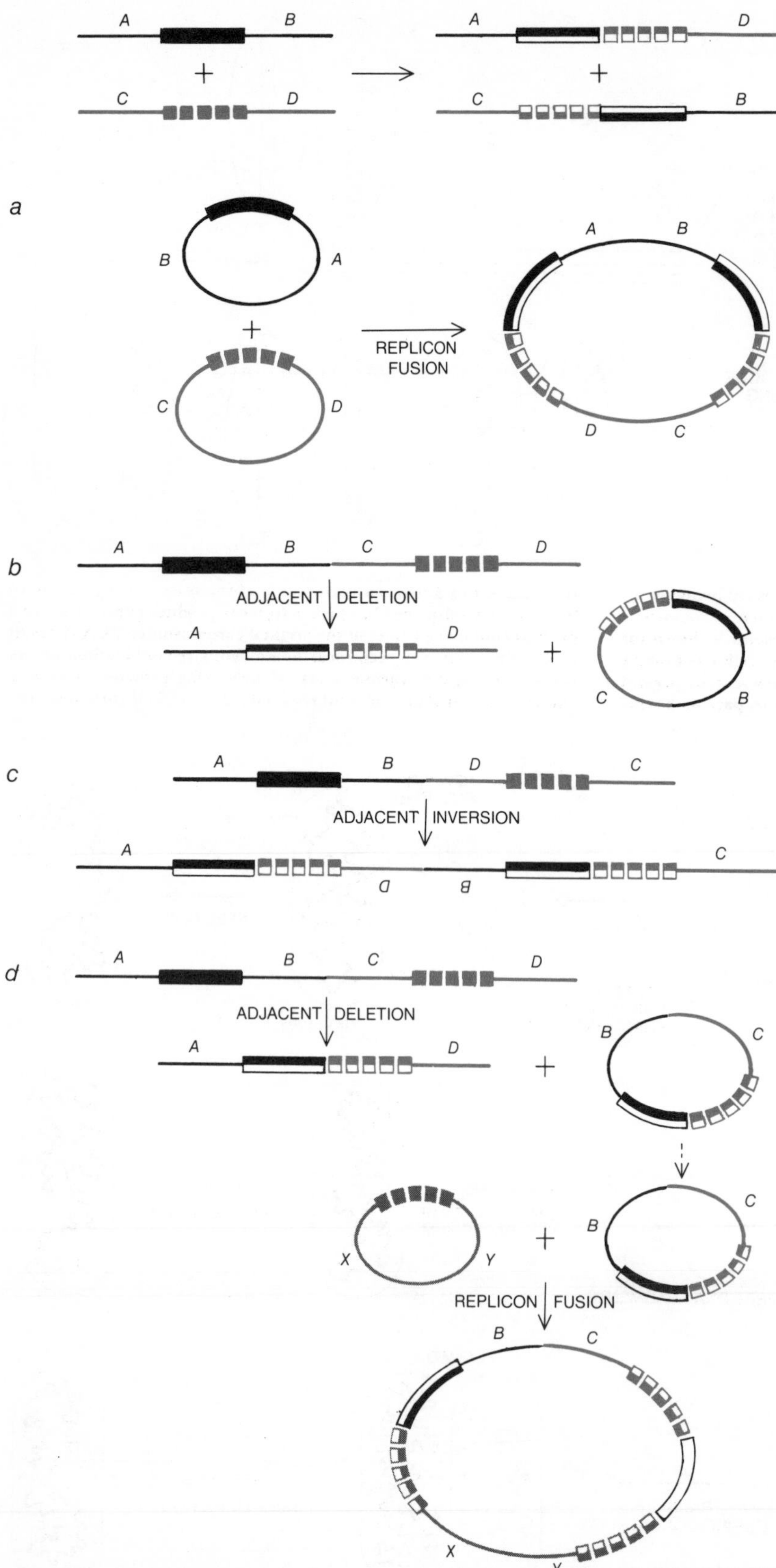

FIRST THREE STEPS OF PATHWAY are summarized schematically at the top of this illustration. These steps achieve reciprocal recombination between unrelated DNA molecules and explain all rearrangements shown in the illustration on page 316, as follows. If the donor and recipient molecules are circular, the three steps result in replicon fusion (*a*). If the donor and recipient regions are part of a single molecule, the steps generate an adjacent deletion (*b*) or an adjacent inversion (*c*), depending on the positions of regions *A, B, C* and *D*. Two successive events (deletion and then replicon fusion) can result in the transposition to a plasmid of a DNA segment adjacent to the transposable element, along with two copies of the element (*d*).

As often happens in science, the significance of McClintock's work was not entirely understood or appreciated until later studies carried out with the much simpler bacterial systems provided actual physical evidence for the existence of insertion sequences and transposons as discrete DNA segments and also established that transposition is brought about by a mechanism different from previously understood recombinational processes. Numerous other examples of transposable elements have now been recognized in higher organisms, such as the fruit fly *Drosophila* and the yeast *Saccharomyces cerevisiae.* The possible role of these elements in the generation of chromosome rearrangements is being actively investigated. Recent work on the control of immunoglobulin synthesis in mice by Susumu Tonegawa and his associates at the Basel Institute for Immunology has shown that the ability of mammalian cells to produce specific antibody molecules in response to injected foreign proteins also involves chromosome rearrangements. There is little doubt that additional instances will soon be found in which illegitimate recombination events play a major role in the expression of genes during cellular differentiation.

Even in bacteria much remains to be learned about the basic molecular mechanisms that accomplish the transposition of genetic elements and the associated rearrangement of DNA molecules. The various biochemical steps in the transposition pathway need to be more fully defined. What is the mechanism for recognition of the inverted-repeat ends of transposable elements? What proteins other than those encoded by the transposon play a role in transposition? What are the additional genetic aspects of the regulation of transposition? In a broader sense, what is the role of illegitimate recombination in the organization and expression of genes, not only in bacteria but also in higher organisms? Although the mechanisms that have been studied in bacteria provide a working model for the mechanisms of similar events in higher organisms, the parallels are probably incomplete.

It is already clear that the joining of structurally and ancestrally dissimilar DNA segments by transposable elements is of great importance for the production of genetic diversity and the evolution of biological systems. The discovery of such a fundamentally different recombinational process at a time when many molecular biologists believed virtually all the important aspects of bacterial genetics were understood in principle—with only the details of particular instances remaining to be learned—leads one to wonder whether still other fundamentally new and significant basic biological processes remain to be discovered.

The Manipulation of Genes

by Stanley N. Cohen
July 1975

Techniques for cleaving DNA and splicing it into a carrier molecule make it possible to transfer genetic information from one organism to an unrelated one. There the DNA replicates and expresses itself

Mythology is full of hybrid creatures such as the Sphinx, the Minotaur and the Chimera, but the real world is not; it is populated by organisms that have been shaped not by the union of characteristics derived from very dissimilar organisms but by evolution within species that retain their basic identity generation after generation. This is because there are natural barriers that normally prevent the exchange of genetic information between unrelated organisms. The barriers are still poorly understood, but they are of fundamental biological importance.

The basic unit of biological relatedness is the species, and in organisms that reproduce sexually species are defined by the ability of their members to breed with one another. Species are determined and defined by the genes they carry, so that in organisms that reproduce asexually the concept of species depends on nature's ability to prevent the biologically significant exchange of genetic material—the nucleic acid DNA—between unrelated groups.

The persistence of genetic uniqueness is perhaps most remarkable in simple organisms such as bacteria. Even when they occupy the same habitat most bacterial species do not exchange genetic information. Even rather similar species of bacteria do not ordinarily exchange the genes on their chromosomes, the structures that carry most of their genetic information. There are exceptions, however. There are bits of DNA, called plasmids, that exist apart from the chromosomes in some bacteria. Sometimes a plasmid can pick up a short segment of DNA from the chromosome of its own cell and transfer it to the cell of a related bacterial species, and sometimes the plasmid and the segment of chromosomal DNA can become integrated into the chromosome of the recipient cell. This transfer of genes between species by extrachromosomal elements has surely played some role in bacterial evolution, but apparently it has not been widespread in nature. Otherwise the characteristics of the common bacterial species would not have remained so largely intact over the huge number of bacterial generations that have existed during the era of modern bacteriology.

In 1973 Annie C. Y. Chang and I at the Stanford University School of Medicine and Herbert W. Boyer and Robert B. Helling at the University of California School of Medicine at San Francisco reported the construction in a test tube of biologically functional DNA molecules that combined genetic information from two different sources. We made the molecules by splicing together segments of two different plasmids found in the colon bacillus *Escherichia coli* and then inserting the composite DNA into *E. coli* cells, where it replicated itself and expressed the genetic information of both parent plasmids. Soon afterward we introduced plasmid genes from an unrelated bacterial species, *Staphylococcus aureus,* into *E. coli,* where they too expressed the biological properties they had displayed in their original host; then, applying the same procedures with John F. Morrow of Stanford and Howard M. Goodman in San Francisco, we were able to insert into *E. coli* some genes from an animal: the toad *Xenopus laevis.*

We called our composite molecules DNA chimeras because they were conceptually similar to the mythological Chimera (a creature with the head of a lion, the body of a goat and the tail of a serpent) and were the molecular counterparts of hybrid plant chimeras produced by agricultural grafting. The procedure we described has since been used and extended by workers in several laboratories. It has been called plasmid engineering, because it utilizes plasmids to introduce the foreign genes, and molecular cloning, because it provides a way to propagate a clone, or line of genetically alike organisms, all containing identical composite DNA molecules. Because of the method's potential for creating a wide variety of novel genetic combinations in microorganisms it is also known as genetic engineering and genetic manipulation. The procedure actually consists of several distinct biochemical and biological manipulations that were made possible by a series of independent discoveries made in rapid succession in the late 1960's and early 1970's. There are four essential elements: a method of breaking and joining DNA molecules derived from different sources; a suitable gene carrier that can replicate both itself and a foreign DNA segment linked to it; a means of introducing the composite DNA molecule, or chimera, into a functional bacterial cell, and a method of selecting from a large population of cells a clone of recipient cells that has acquired the molecular chimera.

In 1967 DNA ligases—enzymes that can repair breaks in DNA and under certain conditions can join together the loose ends of DNA strands—were discovered almost simultaneously in five laboratories. A DNA strand is a chain of nucleotides, each consisting of a deoxyribose sugar ring, a phosphate group and one of four organic bases: adenine, thymine, guanine and cytosine. The sugars and phosphates form the backbone of the strand, from which the bases project. The individual nucleotide building blocks are connected by phosphodiester bonds between the carbon atom at position No. 3 on one sugar and the carbon atom at position No. 5 on the adjacent sugar. Double-strand DNA, the form found in most organisms, consists of two

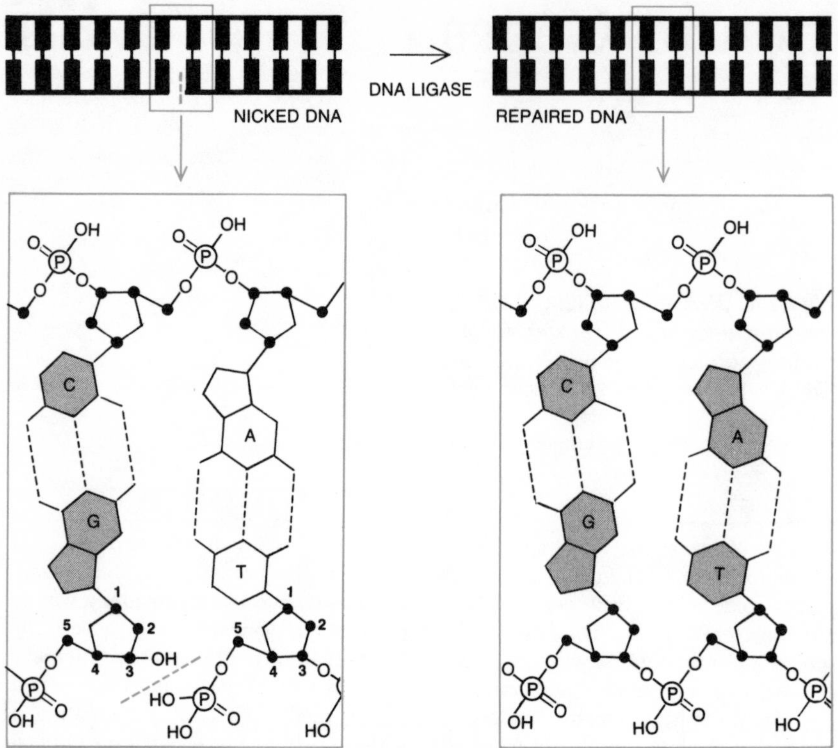

DNA LIGASE is an enzyme that repairs "nicks," or breaks in one strand of a double-strand molecule of DNA (*top*). A strand of DNA is a chain of nucleotides (*bottom*), each consisting of a deoxyribose sugar and a phosphate group and one of four organic bases: adenine (*A*), thymine (*T*), guanine (*G*) and cytosine (*C*). The sugars and phosphates constitute the backbone of the strand, and paired bases, linked by hydrogen bonds (*broken black lines*), connect two strands. The ligase catalyzes synthesis of a bond at the site of the break (*broken colored line*) between the phosphate of one nucleotide and the sugar of the next nucleotide.

chains of nucleotides linked by hydrogen bonds between their projecting bases. The bases are complementary: adenine (*A*) is always opposite thymine (*T*), and guanine (*G*) is always opposite cytosine (*C*). The function of the ligase is to repair "nicks," or breaks in single DNA strands, by synthesizing a phosphodiester bond between adjoining nucleotides [*see illustration above*].

In 1970 a group working in the laboratory of H. Gobind Khorana, who was then at the University of Wisconsin, found that the ligase produced by the bacterial virus T4 could sometimes catalyze the end-to-end linkage of completely separated double-strand DNA segments. The reaction required that the ends of two segments be able to find each other; such positioning of two DNA molecules was a matter of chance, and so the reaction was inefficient. It was clear that efficient joining of DNA molecules required a mechanism for holding the two DNA ends together so that the ligase could act.

An ingenious way of accomplishing this was developed and tested independently in two laboratories at Stanford: by Peter Lobban and A. Dale Kaiser and by David Jackson, Robert Symons and Paul Berg. Earlier work by others had shown that the ends of the DNA molecules of certain bacterial viruses can be joined by base-pairing between complementary sequences of nucleotides that are naturally present on single-strand segments projecting from the ends of those molecules: *A*'s pair with *T*'s, *G*'s pair with *C*'s and the molecules are held together by hydrogen bonds that form between the pairs. The principle of linking DNA molecules by means of the single-strand projections had been exploited in Khorana's laboratory for joining short synthetic sequences of nucleotides into longer segments of DNA.

The Stanford groups knew too that an enzyme, terminal transferase, would catalyze the stepwise addition, specifically at what are called the 3′ ends of single strands of DNA, of a series of identical nucleotides. If the enzyme worked also with double-strand DNA, then a block of identical nucleotides could be added to one population of DNA molecules and a block of the complementary nucleotides could be added to another population from another source. Molecules of the two populations could then be annealed by hydrogen bonding and sealed together by DNA ligase. The method was potentially capable of joining any two species of DNA. While Lobban and Kaiser tested the terminal-transferase procedure with the DNA of the bacterial virus P22, Jackson, Symons and Berg applied the procedure to link the DNA of the animal virus SV40 to bacterial-virus DNA.

The SV40 and bacterial-virus DNA molecules Berg's group worked with are closed loops, and the loops had first to be cleaved to provide linear molecules with free ends for further processing and linkage [*see illustration on opposite page*]. (As it happened, the particular enzyme chosen to cleave the loops was the *Eco RI* endonuclease, which was later to be used in a different procedure for making the first biologically functional gene combinations. At the time, however, the enzyme's special property of producing complementary single-strand ends all by itself had not yet been discovered.)

The cleaved linear molecules were treated with an enzyme, produced by the bacterial virus lambda, called an exonuclease because it operates by cutting off nucleotides at the end of a DNA molecule. The lambda exonuclease chewed back the 5′ ends of DNA molecules and thus left projecting single-strand ends that had 3′ termini to which the blocks of complementary nucleotides could be added. The next step was to add, with the help of terminal transferase, a block of *A*'s at the 3′ end of one of the two DNA species to be linked and a block of *T*'s at the 3′ ends of the other species. The species were mixed together. Fragments having complementary blocks at their ends could find each other, line up and become annealed by hydrogen bonding, thus forming combined molecules. To fill the gaps at the 5′ ends of the original segments the investigators supplied nucleotides and two more enzymes: exonuclease III and DNA polymerase. Finally the nicks in the molecules were sealed with DNA ligase.

The method of making cohesive termini for joining DNA molecules in the first successful genetic-manipulation experiments was conceptually and operationally different from the terminal-transferase procedure. It was also much simpler. It depended on the ability of one of a group of enzymes called restriction endonucleases to make complementary-ended fragments during the cleavage of DNA at a site within the molecule, instead of requiring the addition of new blocks of complementary nucleotides to DNA termini.

Viruses grown on certain strains of *E. coli* were known to be restricted in their ability to grow subsequently on other strains. Investigations had shown that this restriction was due to bacterial enzymes that recognize specific sites on a "foreign" viral DNA and cleave that DNA. (To protect its own DNA the bacterial cell makes a modification enzyme that adds methyl groups to nucleotides constituting the recognition sites for the restriction endonuclease, making them resistant to cleavage.) Restriction endonucleases (and modification methylases) are widespread in microorganisms; genes for making them were found on viral chromosomes and extrachromosomal plasmid DNA as well as on many bacterial chromosomes. During the early 1970's the nucleotide sequences at the cleavage sites recognized by several re-

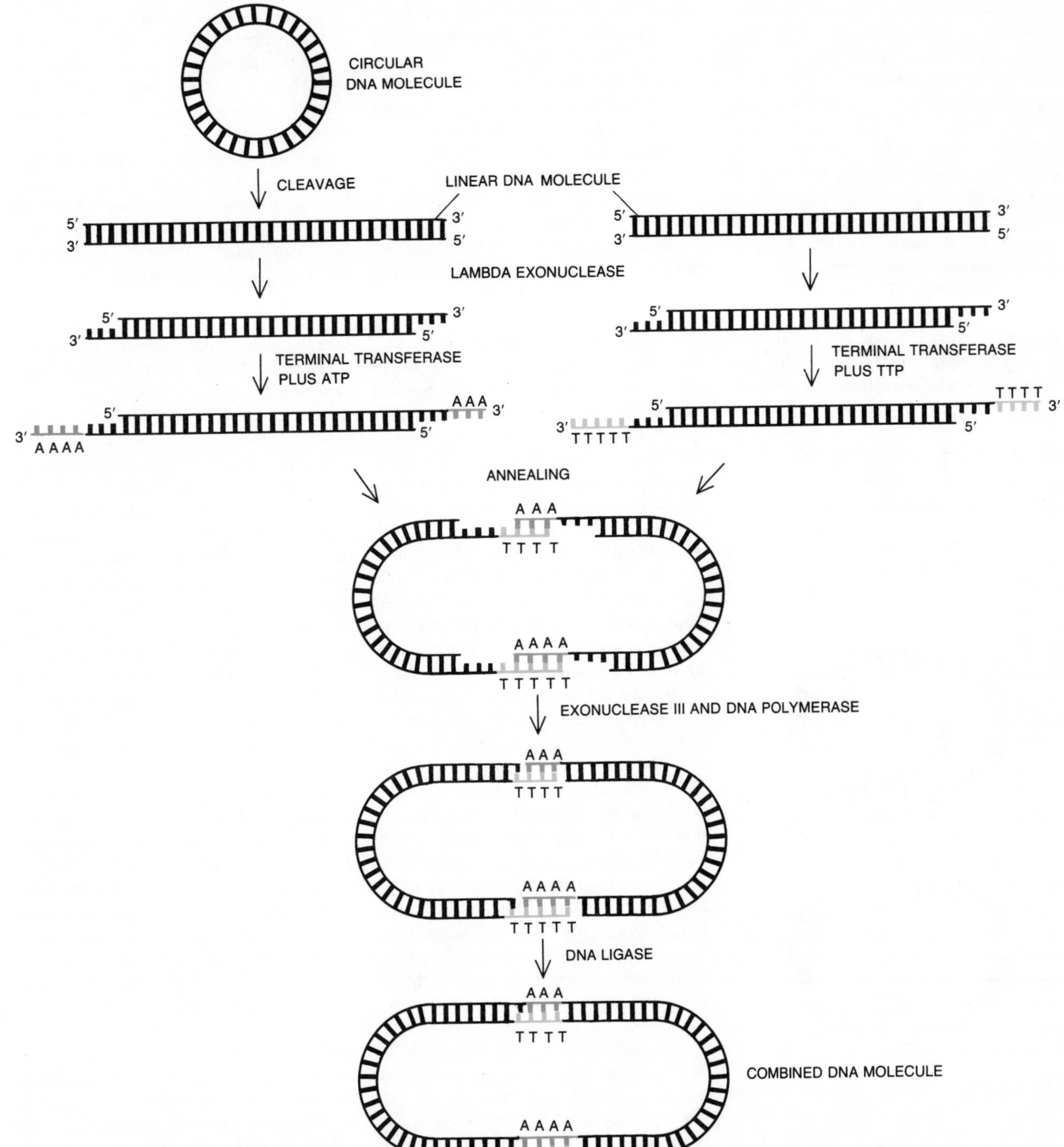

TERMINAL-TRANSFERASE procedure for joining DNA molecules involves a number of steps, each dependent on a different enzyme. If one of the molecules to be joined is a closed loop, it must first be cleaved. The linear molecules are treated with lambda exonuclease, an enzyme that cuts nucleotides off the 5′ end of DNA strands (the end with a phosphate group on the No. 5 carbon). Then specific nucleotides are added to the 3′ end (the end with an OH group on the No. 3 carbon) by the action of the enzyme terminal transferase. One DNA species is supplied with adenosine triphosphate (ATP), the other with thymidine triphosphate (TTP), so that *A* nucleotides are added to one species and complementary *T* nucleotides to the other. When the two species are mixed, the complementary bases pair up, annealing the molecules. Nucleotides and the enzymes DNA polymerase and exonuclease III are added to fill gaps and DNA ligase is added to seal the DNA backbones. The result is a double molecule composed of two separate DNA segments.

striction endonucleases were identified. In every instance, it developed, the cleavage was at or near an axis of rotational symmetry: a palindrome where the nucleotide base sequences read the same on both strands in the 5′-to-3′ direction [*see illustration below*].

In some instances the breaks in the DNA strands made by restriction enzymes were opposite each other. One particular endonuclease, however, the *Eco RI* enzyme isolated by Robert N. Yoshimori in Boyer's laboratory in San Francisco, had a property that was of special interest. Unlike the other nucleases known at the time, this enzyme introduced breaks in the two DNA strands that were separated by several nucleotides. Because of the symmetrical, palindromic arrangement of the nucleotides in the region of cleavage this separation of the cleavage points on the two strands yielded DNA termini with projecting complementary nucleotide sequences: "sticky" mortise-and-tenon termini. The *Eco RI* enzyme thus produced in one step DNA molecules that were functionally equivalent to the cohesive-end molecules produced by the complicated terminal-transferase procedure.

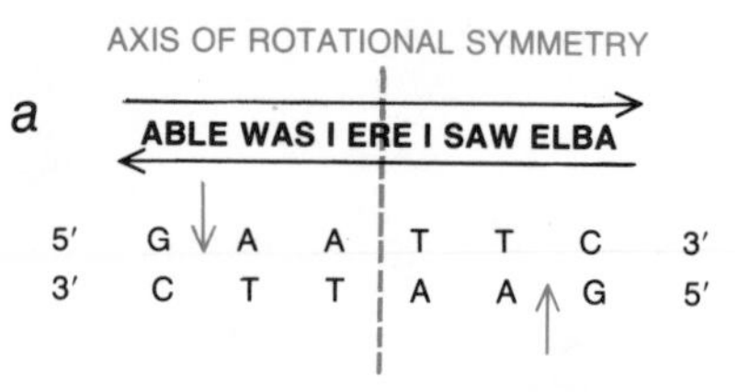

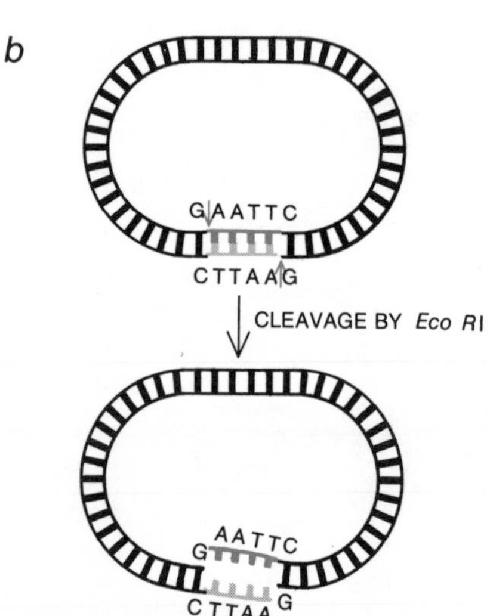

RESTRICTION ENDONUCLEASES cleave DNA at sites where complementary nucleotides are arranged in rotational symmetry: a palindrome, comparable to a word palindrome (*a*). The endonuclease *Eco RI* has the additional property of cleaving complementary strands of DNA at sites (*colored arrows*) four nucleotides apart. Such cleavage (*b*) yields DNA fragments with complementary, overlapping single-strand ends. As a result the end of any DNA fragment produced by *Eco RI* cleavage can anneal with any other fragment produced by the enzyme.

The experiments that led to the discovery of the capabilities of *Eco RI* were reported independently and simultaneously in November, 1972, by Janet Mertz and Ronald W. Davis of Stanford and by another Stanford investigator, Vittorio Sgaramella. Sgaramella found that molecules of the bacterial virus P22 could be cleaved with *Eco RI* and would then link up end to end to form DNA segments equal in length to two or more viral-DNA molecules. Mertz and Davis observed that closed-loop SV40-DNA molecules cleaved by *Eco RI* would re-form themselves into circular molecules by hydrogen bonding and could be sealed with DNA ligase; the reconstituted molecules were infectious in animal cells growing in tissue culture. Boyer and his colleagues analyzed the nucleotide sequences at the DNA termini produced by *Eco RI*, and their evidence confirmed the complementary nature of the termini, which accounted for their cohesive activity.

In late 1972, then, several methods were available by which one could join double-strand molecules of DNA. That was a major step in the development of a system for manipulating genes. More was necessary, however. Most segments of DNA do not have an inherent capacity for self-replication; in order to reproduce themselves in a biological system they need to be integrated into DNA molecules that can replicate in the particular system. Even a DNA segment that can replicate in its original host was not likely to have the specific genetic signals required for replication in a different environment. If foreign DNA was to be propagated in bacteria, as had long been proposed in speculative scenarios of genetic engineering, a suitable vehicle, or carrier, was required. A composite DNA molecule consisting of the vehicle and the desired foreign DNA would have to be introduced into a population of functional host bacteria. Finally, it would be necessary to select, or identify, those cells in the bacterial population that took up the DNA chimeras. In 1972 it still seemed possible that the genetic information on totally foreign DNA molecules might produce an aberrant situation that would prevent the propagation of hybrid molecules in a new host.

Molecular biologists had focused for many years on viruses and their relations with bacteria, and so it was natural that bacterial viruses were thought of as the most likely vehicles for genetic manipulation. For some time there had been speculation and discussion about using viruses, such as lambda, that occasionally acquire bits of the *E. coli* chromosome by natural recombination mechanisms for cloning DNA from foreign sources. It was not a virus, however, but a plasmid that first served as a vehicle for introducing foreign genes into a bacterium and that provided a mechanism for the replication and selection of the foreign DNA.

A ubiquitous group of plasmids that confer on their host bacteria the ability to resist a number of antibiotics had been studied intensively for more than a decade. Antibiotic-resistant *E. coli* isolated in many parts of the world, for example, were found to contain plasmids, designated *R* factors (for "resistance"), carrying the genetic information for products that in one way or another could interfere with the action of specific antibiotics [see "Infectious Drug Resistance," by Tsutomu Watanabe; SCIENTIFIC AMERICAN, December, 1967]. Double-strand circular molecules of *R*-factor DNA had been separated from bacterial chromosomal DNA by centrifugation in density gradients and had been characterized by biochemical and physical techniques [see "The Molecule of Infectious Drug Resistance," by Royston C. Clowes; SCIENTIFIC AMERICAN Offprint 1269].

In 1970 Morton Mandel and A. Higa of the University of Hawaii School of Medicine had discovered that treatment of *E. coli* with calcium salts enabled the bacteria to take up viral DNA. At Stanford, Chang and I, with Leslie Hsu, found that if we made the cell membranes of *E. coli* permeable by treating them with calcium chloride, purified *R*-factor DNA could be introduced into them [*see illustration on opposite page*]. The *R*-factor DNA is taken up in this transformation process by only about one bacterial cell in a million, but those few cells can be selected because they live and multiply in the presence of the antibiotics to which the *R* factor confers resistance, whereas other cells die. Each transformed cell gives rise to a clone that contains exact replicas of the parent plasmid DNA molecules, and so we reasoned that plasmids might serve as vehicles for propagating new genetic information in a line of *E. coli* cells.

In an effort to explore the genetic and molecular properties of various regions of the *R*-factor DNA we had begun to take plasmids apart by shearing their DNA mechanically and then transforming *E. coli* with the resulting fragments. Soon afterward we began to cleave the plasmids with the *Eco RI* enzyme, which had been shown to produce multiple site-specific breaks in several viruses. It might therefore be counted on to cleave all molecules of a bacterial plasmid in the same way, so that any particular species of DNA would yield a specific set of cleavage fragments, and do so reproducibly. The fragments could then be separated and identified according to the different rates at which they would migrate through a gel under the influence of an electric current.

When the DNA termini produced by *Eco RI* endonuclease were found to be cohesive, Chang and I, in collaboration with Boyer and Helling in San Francisco, proceeded to search for a plasmid that the enzyme would cleave without affecting the plasmid's ability to replicate or to confer antibiotic resistance. We hoped that if such a plasmid could be found, we could insert a segment of foreign DNA at the *Eco RI* cleavage site, and that it might be possible to propagate the foreign DNA in *E. coli.*

In our collection at Stanford there was a small plasmid, *pSC*101, that had been isolated following the mechanical shearing of a large plasmid bearing genes for multiple antibiotic resistance. It was less than a twelfth as long as the parent plasmid, but it did retain the genetic information for its replication in *E. coli* and for conferring resistance to one antibiotic, tetracycline. When we subjected *pSC*101 DNA to cleavage by *Eco RI* and analyzed the products by gel electrophoresis, we found that the enzyme had cut the plasmid molecule in only one place, producing a single linear fragment. We were able to join the ends of that fragment again by hydrogen bonding and reseal them with DNA ligase, and when we introduced the reconstituted circular DNA molecules into *E. coli* by transformation, they were biologically functional plasmids: they replicated and conferred tetracycline resistance.

The next step was to see if a fragment of foreign DNA could be inserted at the cleavage site without interfering with replication or expression of tetracycline resistance and thus destroying the plasmid's ability to serve as a cloning vehicle. We mixed the DNA of another *E. coli* plasmid, which carried resistance to the antibiotic kanamycin, with the *pSC*101 DNA. We subjected the mixed DNA to cleavage by *Eco RI* and then to ligation, transformed *E. coli* with the resulting DNA and found that some of the transformed bacteria were indeed resist-

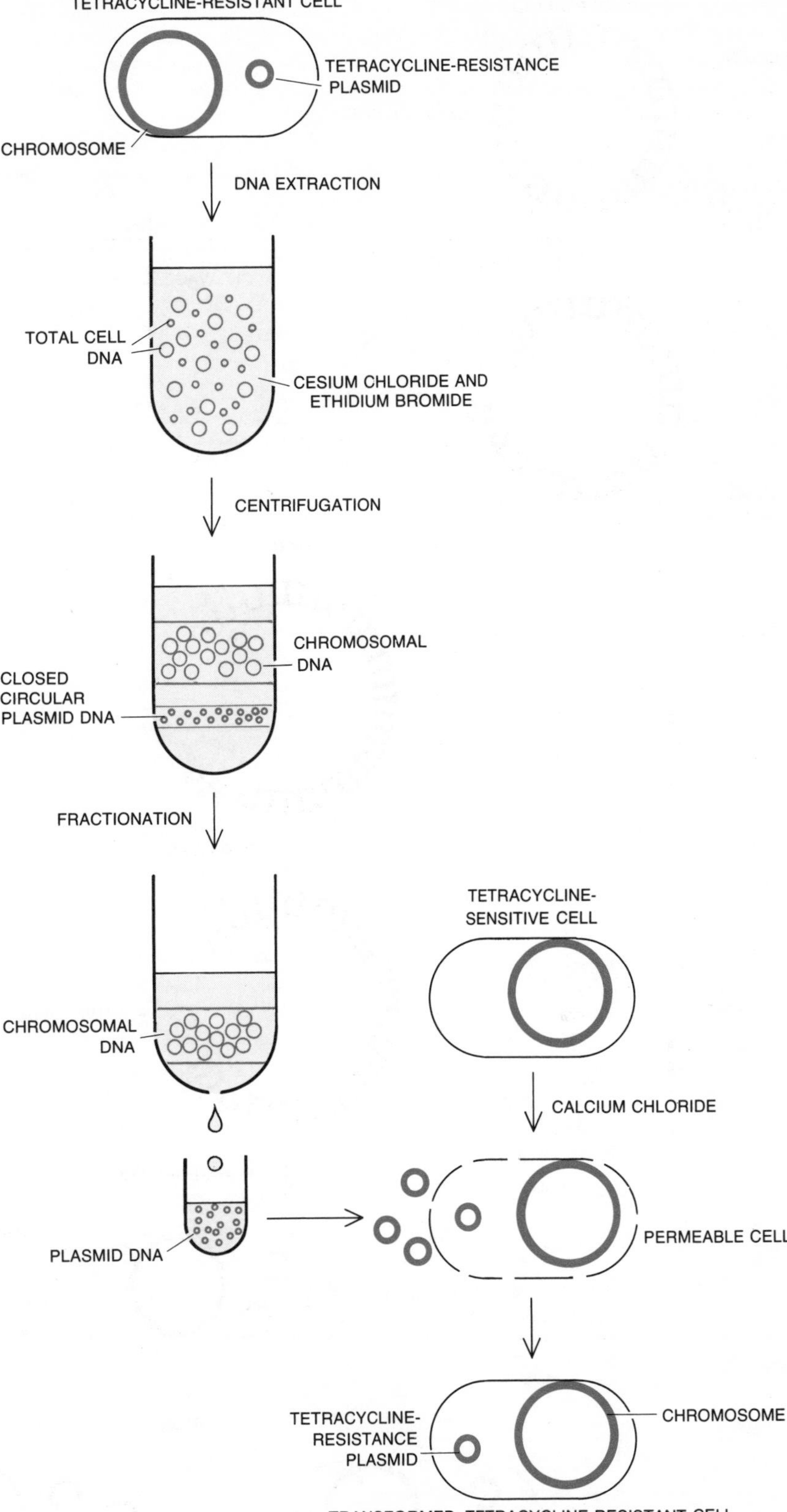

PLASMID DNA can be introduced into a bacterial cell by the procedure called transformation. Plasmids carrying genes for resistance to the antibiotic tetracycline (*top left*) are separated from bacterial chromosomal DNA. Because differential binding of ethidium bromide by the two DNA species makes the circular plasmid DNA denser than the chromosomal DNA, the plasmids form a distinct band on centrifugation in a cesium chloride gradient and can be separated (*bottom left*). The plasmid DNA is mixed with bacterial cells that are not resistant to tetracycline and that have been made permeable by treatment with a calcium salt. The DNA enters the cells, replicates there and makes the cells resistant to tetracycline.

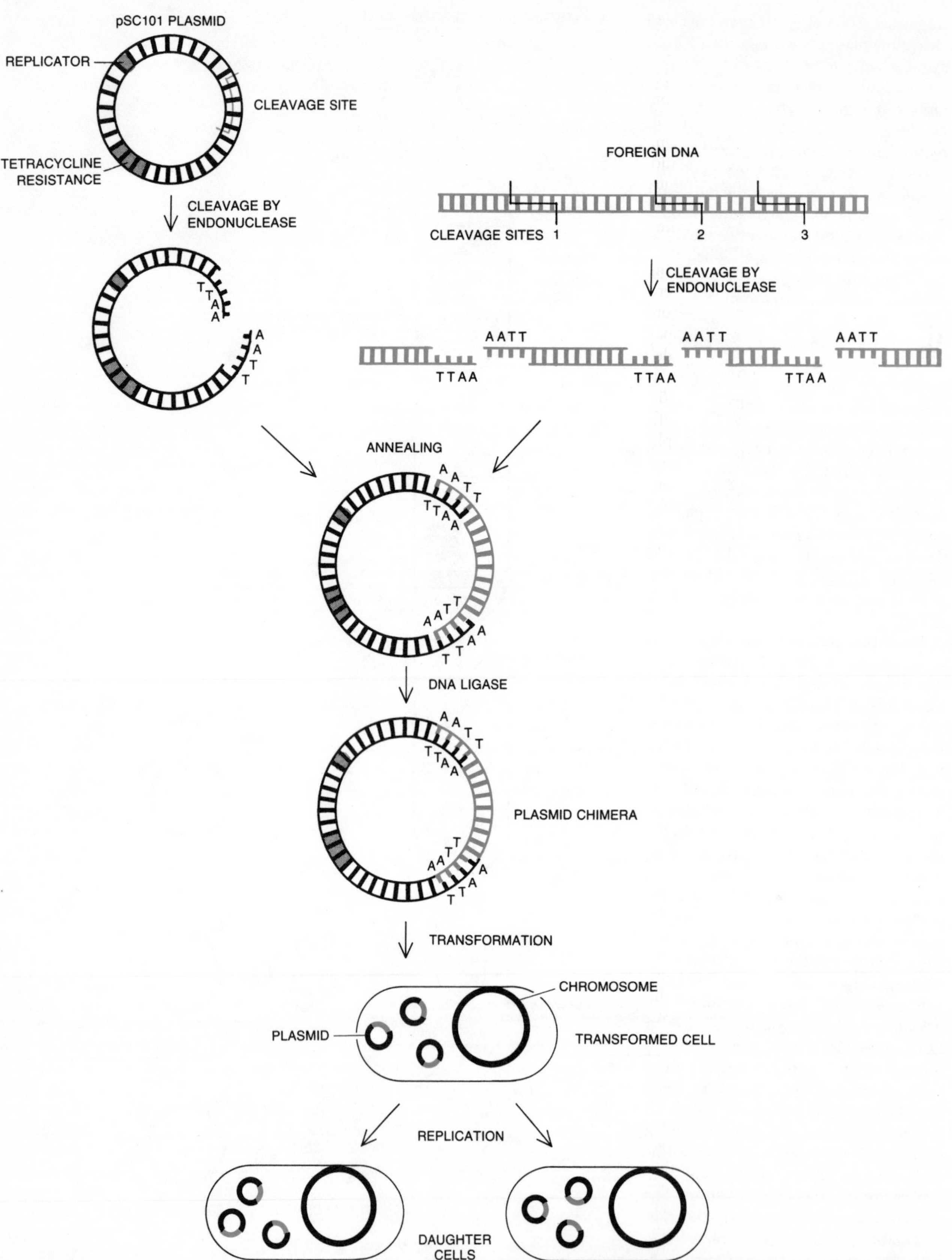

FOREIGN DNA is spliced into the *pSC*101 plasmid and introduced with the plasmid into the bacterium *Escherichia coli.* The plasmid is cleaved by the endonuclease *Eco RI* at a single site that does not interfere with the plasmid's genes for replication or for resistance to tetracycline (*top left*). The nucleotide sequence recognized by *Eco RI* is present also in other DNA, so that a foreign DNA exposed to the endonuclease is cleaved about once in every 4,000 to 16,000 nucleotide pairs on a random basis (*top right*). Fragments of cleaved foreign DNA are annealed to the plasmid DNA by hydrogen bonding of the complementary base pairs, and the new composite molecules are sealed by DNA ligase. The DNA chimeras, each consisting of the entire plasmid and a foreign DNA fragment, are introduced into *E. coli* by transformation, and the foreign DNA is replicated by virtue of the replication functions of the plasmid.

ant to both tetracycline and kanamycin. The plasmids isolated from such transformants contained the entire *pSC*101 DNA segment and also a second DNA fragment that carried the information for kanamycin resistance, although it lacked replication functions of its own. The results meant that the *pSC*101 could serve as a cloning vehicle for introducing at least a nonreplicating segment of a related DNA into *E. coli.* And the procedure was extraordinarily simple.

Could genes from other species be introduced into *E. coli* plasmids, however? There might be genetic signals on foreign DNA that would prevent its propagation or expression in *E. coli.* We decided to try to combine DNA from a plasmid of another bacterium, the *pI*258 plasmid of *Staphylococcus aureus,* with our original *E. coli* plasmid. The staphylococcal plasmid had already been studied in several laboratories; we had found that it was cleaved into four DNA fragments by *Eco RI.* Since *pI*258 was not native to *E. coli* or to related bacteria, it could not on its own propagate in an *E. coli* host. And it was known to carry a gene for resistance to still another antibiotic, penicillin, that would serve as a marker for selecting any transformed clones. (Penicillin resistance, like combined resistance to tetracycline and kanamycin, was already widespread among *E. coli* strains in nature. That was important; if genes from a bacterial species that cannot normally exchange genetic information with the colon bacillus were to be introduced into it, it was essential that they carry only antibiotic-resistance traits that were already prevalent in *E. coli.* Otherwise we would be extending the species' antibiotic-resistance capabilities.)

Chang and I repeated the experiment that had been successful with two kinds of *E. coli* plasmids, but this time we did it with a mixture of the *E. coli*'s *pSC*-101 and the staphylococcal *pI*258: we cleaved the mixed plasmids with *Eco RI* endonuclease, treated them with ligase and then transformed *E. coli.* Next we isolated transformed bacteria that expressed the penicillin resistance coded for by the *S. aureus* plasmid as well as the tetracycline resistance of the *E. coli* plasmid. These doubly resistant cells were found to contain a new DNA species that had the molecular characteristics of the staphylococcal plasmid DNA as well as the characteristics of *pSC*101.

The replication and expression in *E. coli* of genes derived from an organism ordinarily quite unable to exchange genes with *E. coli* represented a breach in the barriers that normally separate biological species. The bulk of the genetic information expressed in the transformed bacteria defined it as *E. coli,* but the transformed cells also carried replicating DNA molecules that had molecular and biological characteristics derived from an unrelated species, *S. aureus.* The fact that the foreign genes were on a plasmid meant that they would be easy to isolate and purify in large quantities for further study. Moreover, there was a possibility that one might introduce genes into the easy-to-grow *E. coli* that specify a wide variety of metabolic or synthesizing functions (such as photosynthesis or antibiotic production) and that are indigenous to other biological classes. Potentially the *pSC*101 plasmid and the molecular-cloning procedure could serve to introduce DNA molecules from complex higher organisms into bacterial hosts, making it possible to apply relatively simple bacterial genetic and biochemical techniques to the study of animal-cell genes.

Could animal-cell genes in fact be introduced into bacteria, and would they replicate there? Boyer, Chang, Helling and I, together with Morrow and Goodman, immediately undertook to find out. We picked certain genes that had been well studied and characterized and were available, purified, in quantity: the genes that code for a precursor of the ribosomes (the structure on which proteins are synthesized) in the toad *Xenopus laevis.* The genes had properties that would enable us to identify them if we succeeded in getting them to propagate in bacteria. The toad DNA was suitable for another reason: although we would be constructing a novel biological combination containing genes from both animal cells and bacteria, we and others expected that no hazard would result from transplanting the highly purified ribosomal genes of a toad.

Unlike the foreign DNA's of our earlier experiments, the toad genes did not express traits (such as antibiotic resistance) that could help us to select bacteria carrying plasmid chimeras. The tetracycline resistance conferred by *pSC*101 would make it possible to select transformed clones, however, and we could then proceed to examine the DNA isolated from such clones to see if any clones contained a foreign DNA having the molecular properties of toad ribosomal DNA. The endonuclease-generated fragments of toad ribosomal DNA have characteristic sizes and base compositions; DNA from the transformed cells could be tested for those characteristics. The genes propagated in bacteria could also be tested for nucleotide-sequence homology with DNA isolated directly from the toad.

When we did the experiment and analyzed the resulting transformed cells, we found that the animal-cell genes were indeed reproducing themselves in generation after generation of bacteria by means of the plasmid's replication functions. In addition, the nucleotide sequences of the toad DNA were being transcribed into an RNA product in the bacterial cells.

Within a very few months after the first DNA-cloning experiments the procedure was being used in a number of laboratories to clone bacterial and animal-cell DNA from a variety of sources. Soon two plasmids other than *pSC*101 were discovered that have a single *Eco RI* cleavage site at a location that does not interfere with essential genes. One of these plasmids is present in many copies in the bacterial cell, making it possible to "amplify," or multiply many times, any DNA fragments linked to it. Investigators at the University of Edinburgh and at Stanford went on to develop mutants of the virus lambda (which ordinarily infects *E. coli*) that made the virus too an effective cloning vehicle. Other restriction endonucleases were discovered that also make cohesive termini but that cleave DNA at different sites from the *Eco RI* enzymes, so that chromosomes can now be taken apart and put together in various ways.

The investigative possibilities of DNA cloning are already being explored intensively. Some workers have isolated from complex chromosomes certain regions that are implicated in particular functions such as replication. Others are making plasmids to order with specific properties that should clarify aspects of extrachromosomal-DNA biology that have been hard to study. The organization of complex chromosomes, such as those of the fruit fly *Drosophila,* is being studied by cloning the animal genes in bacteria. Within the past few months methods have been developed for selectively cloning specific genes of higher organisms through the use of radioactively labeled RNA probes: instead of purifying the genes to be studied before introducing them into bacteria, one can transform bacteria with a heterogeneous population of animal-cell DNA and then isolate those genes that produce a particular species of RNA. It is also possible to isolate groups of genes that are expressed concurrently at a particular stage in the animal's development.

The potential seems to be even broader. Gene manipulation opens the prospect of constructing bacterial cells, which can be grown easily and inexpensively, that will synthesize a variety of biologically produced substances such as antibiotics and hormones, or enzymes that can convert sunlight directly into food substances or usable energy. Perhaps it even provides an experimental basis for introducing new genetic information into plant or animal cells.

It has been clear from the beginning of experimentation in molecular cloning that the construction of some kinds of novel gene combinations may have a potential for biological hazard, and the scientific community has moved quickly to make certain that research in genetic manipulation would not endanger the public. For a time after our initial experiments the *pSC*101 plasmid was the only vehicle known to be suitable for cloning foreign DNA in *E. coli,* and our colleagues asked for supplies with which to pursue studies we knew were of major scientific and medical importance. Investigators normally facilitate the free exchange of bacteria and other experimental strains they have isolated or developed, but Chang and I were concerned that manipulation of certain genes could give rise to novel organisms whose infectious properties and ecological effects could not be predicted. In agreeing to provide the plasmid we therefore asked for assurance that our colleagues would neither introduce tumor viruses into bacteria nor create antibiotic-resistance combinations that were not already present in nature; we also asked the recipients not to send the plasmid on to other laboratories, so that we could keep track of its distribution.

When still other cloning vehicles were discovered, it became apparent that a more general mechanism for ensuring experimental safety in gene-manipulation research was advisable. The groundwork for such control had been established earlier: the National Academy of Sciences had been urged to consider the "possibility that potentially biohazardous consequences might result from widespread or injudicious use" of these techniques and had asked Paul Berg to form an advisory committee that would consider the issue. Berg too had been concerned about the potential hazards of certain kinds of experimentation for some years, and had himself decided to abandon plans to try to introduce genes from the tumor virus SV40 into bacteria because of the possible danger if the experiment were successful.

Berg brought together a number of investigators, including some who were then directly involved in molecular cloning, in the spring of 1974. In a report released in July and in a letter to leading professional journals the members of the committee expressed their "concern about the possible unfortunate consequences of indiscriminate application" of the techniques and formally asked all investigators to join them in voluntarily deferring two types of experiments (which had, as a matter of fact, been avoided by informal consensus up until that time). Experiments of Type I involved the construction of novel organisms containing combinations of toxin-producing capabilities or of antibiotic-resistance genes not found in nature. Type 2 experiments involved the introduction of DNA from tumor viruses or other animal viruses into bacteria; the committee noted that "such recombinant molecules might be more easily disseminated to bacterial populations in humans and other species, and might thus increase the incidence of cancer or other diseases."

The Academy committee was concerned largely because of our inability to assess the hazards of certain experiments accurately before the experiments were undertaken. Guidelines for safety had long been available in other areas of potentially hazardous research, such as studies involving known disease-causing bacteria and viruses, radioactive isotopes or toxic chemicals. Because of the newness of the microbial gene-manipulation methods, no such guidelines had yet been developed for work in this area, however; there was the possibility that potentially hazardous experiments might proceed before appropriate guidelines could be considered and implemented. We recognized that most work with the new methods did not and would not involve experiments of a hazardous nature but we recommended the deferral of Type I and Type II experiments until the hazards were more carefully assessed, until it was determined whether or not the work could be undertaken safely and until adequate safety precautions were available. The committee also proposed that an international meeting be held early in 1975 to consider the matter more fully.

Such a meeting was held in February at the Asilomar Conference Center near Pacific Grove, Calif. It brought together 86 American biologists and 53 investigators from 16 other countries, who spent three and a half days reviewing progress in the field of molecular cloning and formulating guidelines that would allow most types of new hereditary characteristics to be introduced into bacteria and

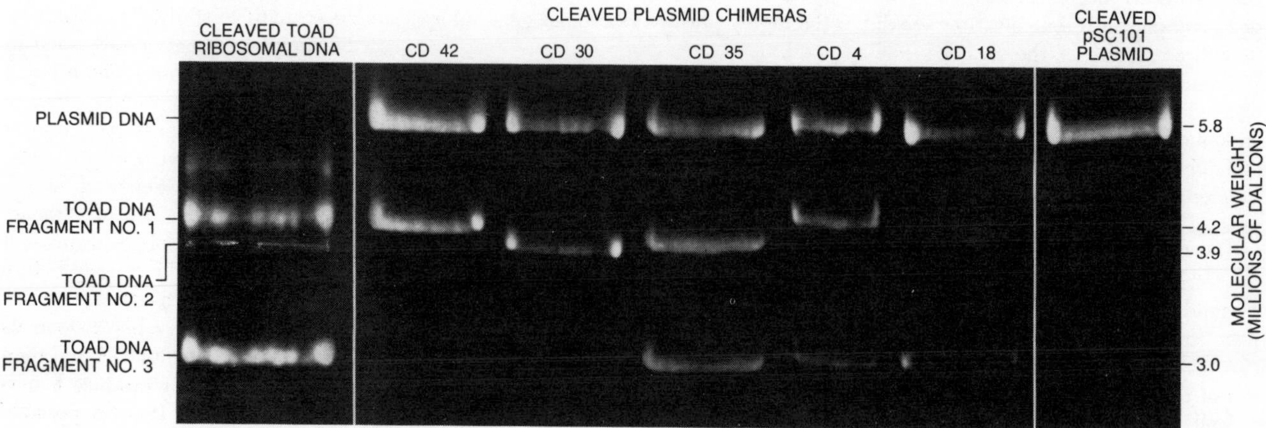

GEL ELECTROPHORESIS demonstrates the presence of toad DNA in chimeric plasmids. Fragments of DNA migrate through a gel at different rates under the influence of an electric current, depending on their size. Linear molecules of plasmid DNA (*right*) and the cleavage products of toad ribosomal DNA (*left*) therefore have characteristic sizes and migrate characteristic distances in a given time. The bands of DNA, visualized by a fluorescent dye, are photographed in ultraviolet. All five chimeric plasmids (*center*) contain a plasmid DNA molecule; in addition each chimera includes one or more fragments characteristic of original toad DNA.

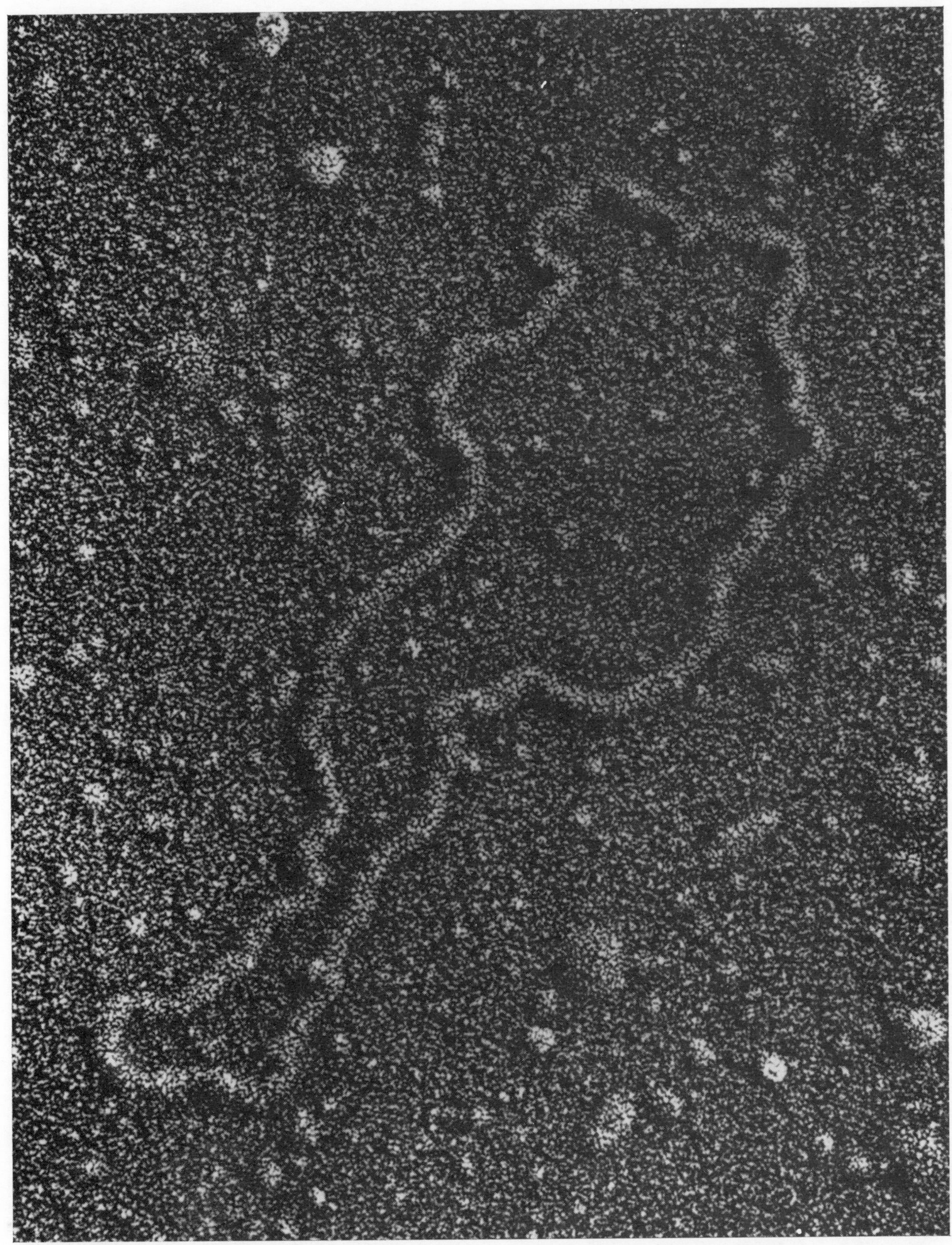

PLASMID *pSC*101 is shadowed with platinum-palladium and enlarged 230,000 diameters in an electron micrograph made by the author. A plasmid is a molecule of DNA that exists apart from the chromosome in a bacterium and replicates on its own, often carrying the genes for some supplementary activity such as resistance to antibiotics. This plasmid, a small one made by shearing a larger plasmid native to the bacterium *Escherichia coli,* is a circular, or closed-loop, molecule of DNA about three micrometers in circumference that carries the genetic information for replicating itself in *E. coli* and for conferring resistance to the antibiotic tetracycline. It was the "vehicle" for the first gene-manipulation experiments by the author and his colleagues. Foreign DNA was spliced to it and the plasmid was introduced into *E. coli,* where it replicated and expressed both its own and the foreign DNA's genetic information.

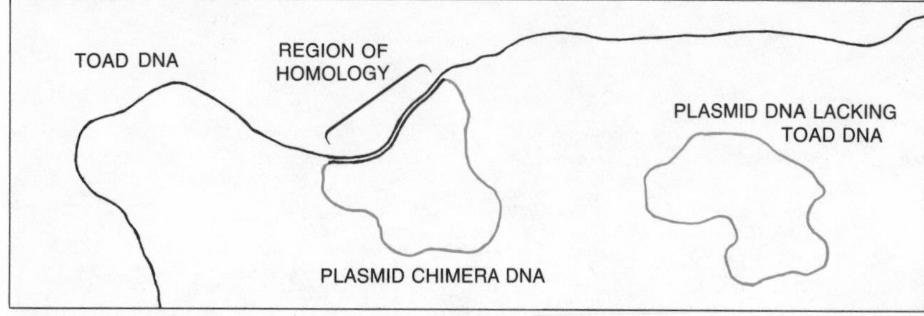

HETERODUPLEX ANALYSIS identifies regions of a toad DNA (*black*) that have been incorporated in a chimeric plasmid DNA molecule. DNA isolated from toad eggs and the DNA of the chimera are denatured, that is, each natural double-strand molecule is split into two single strands of DNA, by alkali treatment. The toad and the chimeric DNA's are mixed together, and any complementary sequences are allowed to find each other. The toad DNA incorporated in the chimeras has nucleotide sequences that are complementary to sequences in the DNA taken directly from the animal source. Those homologous sequences anneal to form heteroduplex double-strand DNA that can be identified in electron micrographs.

viruses safely. Invited nonscientists from the fields of law and ethics participated in the discussions and decisions at Asilomar, along with representatives of agencies that provide Federal funds for scientific research; the meetings were open to the press and were fully reported. The issues were complex and there were wide differences of opinion on many of them, but there was consensus on three major points. First, the newly developed cloning methods offer the prospect of dealing with a wide variety of important scientific and medical problems as well as other problems that trouble society, such as environmental pollution and food and energy shortages. Second, the accidental dissemination of certain novel biological combinations may present varying degrees of potential risk. The construction of such combinations should proceed only under a graded series of precautions, principally biological and physical barriers, adequate to prevent the escape of any hazardous organisms; the extent of the actual risk should be explored by experiments conducted under strict containment conditions. Third, some experiments are potentially too hazardous to be carried out for the present, even with the most careful containment. Future research and experience may show that many of the potential hazards considered at the meeting are less serious and less probable than we now suspect. Nevertheless, it was agreed that standards of protection should be high at the beginning and that they can be modified later if the assessment of risk changes.

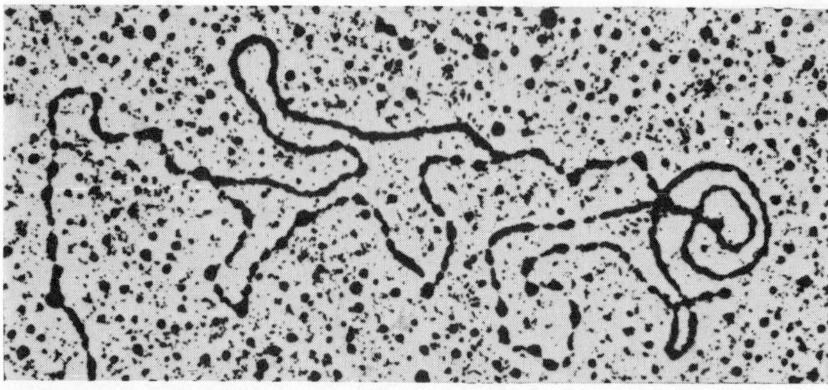

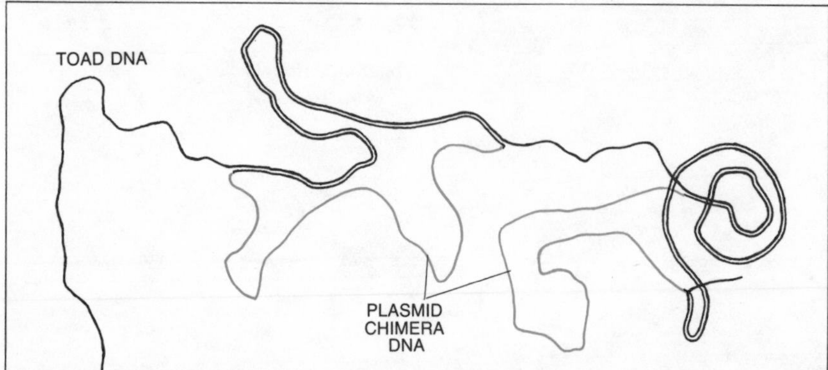

PRESENCE OF TOAD DNA in two separate chimeric plasmid molecules is demonstrated by an electron micrograph made by John F. Morrow at the Stanford University School of Medicine. As is indicated in the drawing (*bottom*), there are DNA strands from two plasmids and a strand of toad DNA. The micrograph shows thickened regions of DNA where nucleotide sequences are homologous and two single strands have been annealed. The toad DNA in the chimeras codes for ribosomes, and the space between the two heteroduplex regions is compatible with the spacing of multiple ribosomal genes in toad DNA.

Physical containment barriers have long been used in the U.S. space-exploration program to minimize the possibility of contamination of the earth by extraterrestrial microbes. Containment procedures are also employed routinely to protect laboratory workers and the public from hazards associated with radioactive isotopes and toxic chemicals and in work with disease-causing bacteria and viruses. The Asilomar meeting formulated the additional concept of biological barriers, which involve fastidious cloning vehicles that are able to propagate only in specialized hosts and equally fastidious bacterial strains that are unable to live except under stringent laboratory conditions.

In the past the scientific community has commonly policed its own actions informally, responding to ethical concerns with self-imposed restraint. Usually, but not always, society at large has also considered the public well-being in determining how knowledge obtained by basic scientific research should be applied. Extensive public scrutiny and open discussion by scientists and nonscientists of the possible risks and benefits of a particular line of basic research has been rare, however, when (as in this case) the hazards in question are only potential and, for some experiments, even hypothetical. As this article is being written it is still too early to know what the long-range outcome of the public discussions initiated by scientists working in genetic manipulation will be. One can hope that the forthright approach and the rigorous standards that have been adopted for research in the cloning of recombinant DNA molecules will promote a sharper focus on other issues relevant to public and environmental safety.

NAME INDEX

SUBJECT INDEX